Matrix Mathematics

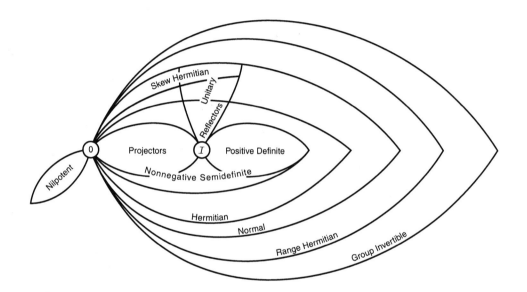

Matrix Mathematics

Theory, Facts, and Formulas with
Application to Linear Systems Theory

Dennis S. Bernstein

PRINCETON UNIVERSITY PRESS

PRINCETON AND OXFORD

Published by Princeton University Press,
41 William Street, Princeton, New Jersey 08540

In the United Kingdom: Princeton University Press,
3 Market Place, Woodstock, Oxfordshire OX20 1SY

Library of Congress Cataloging-in-Publication Data

Bernstein, Dennis S., 1954–
    Matrix mathematics: theory, facts, and formulas with
application to linear systems theory / Dennis S. Bernstein.
        p. cm.
    Includes bibliographical references and index.
    ISBN 0-691-11802-7 (acid-free paper)
    1. Matrices. 2. Linear systems. I. Title.
QA188.B475 2005
512.9'434—dc22

                        2004048903

British Library Cataloging-in-Publication Data is available

This book has been composed in Computer Modern and Helvetica.

The publisher would like to acknowledge the author of this volume for providing the
camera-ready copy from which this book was printed.

Printed on acid-free paper. ∞

pup.princeton.edu

Printed in the United States of America

10 9 8 7 6 5 4 3 2

To the memory of my parents

... vessels, unable to contain the great light flowing into them, shatter and break. ... the remains of the broken vessels fall ... into the lowest world, where they remain scattered and hidden

— D. W. Menzi and Z. Padeh, *The Tree of Life, Chayyim Vital's Introduction to the Kabbalah of Isaac Luria*, Jason Aaronson, Northvale, 1999

Thor ... placed the horn to his lips ... He drank with all his might and kept drinking as long as ever he was able; when he paused to look, he could see that the level had sunk a little, ... for the other end lay out in the ocean itself.

— P. A. Munch, *Norse Mythology*, AMS Press, New York, 1970

Contents

Special Symbols

General Notation

π	$3.14159\cdots$
e	$2.71828\cdots$
$\triangleq$	equals by definition
$\lim_{p\downarrow 0}$	limit from the right
$\binom{n}{m}$	$\frac{n!}{m!(n-m)!}$
$\lfloor a \rfloor$	largest integer less than or equal to a
δ_{ij}	1 if $i = j$, 0 if $i \neq j$ (Kronecker delta)
$\log$	logarithm with base e
$\operatorname{sign} \alpha$	1 if $\alpha > 0$, -1 if $\alpha < 0$, 0 if $\alpha = 0$
$\sinh x, \cosh x$	$\frac{1}{2}(e^x - e^{-x}), \frac{1}{2}(e^x + e^{-x})$

Chapter 1

$\{\,\}$	set (p. 2)
$\{\,\}_\mathrm{m}$	multiset (p. 2)
$\varnothing$	empty set (p. 2)
$\in$	is an element of (p. 2)
$\notin$	is not an element of (p. 2)
$\cap$	intersection (p. 2)
$\cup$	union (p. 2)
$\mathcal{Y}\backslash\mathcal{X}$	complement of $\mathcal{X}$ relative to $\mathcal{Y}$ (p. 2)
$\mathcal{X}^\sim$	complement of $\mathcal{X}$ (p. 2)
$\subseteq$	is a subset of (p. 2)
$\subset$	is a proper subset of (p. 2)
$x \overset{\mathcal{R}}{=} y$	(x, y) is an element of the equivalence relation $\mathcal{R}$ (p. 3)
$f\colon \mathcal{X} \mapsto \mathcal{Y}$	f is a function with domain $\mathcal{X}$ and codomain $\mathcal{Y}$ (p. 4)

$f^{-1}(\mathcal{S})$	inverse image of $\mathcal{S}$ (p. 4)
$f \bullet g$	composition of functions f and g (p. 4)

Chapter 2

$\mathbb{Z}$	integers (p. 13)
$\mathbb{N}$	nonnegative integers (p. 13)
$\mathbb{P}$	positive integers (p. 13)
$\mathbb{R}$	real numbers (p. 13)
$\mathbb{C}$	complex numbers (p. 13)
$\mathbb{F}$	$\mathbb{R}$ or $\mathbb{C}$ (p. 13)
$\jmath$	$\sqrt{-1}$ (p. 13)
$\bar{z}$	complex conjugate of $z \in \mathbb{C}$ (p. 13)
$\mathrm{Re}\, z$	real part of $z \in \mathbb{C}$ (p. 13)
$\mathrm{Im}\, z$	imaginary part of $z \in \mathbb{C}$ (p. 13)
$|z|$	absolute value of $z \in \mathbb{C}$ (p. 13)
CLHP	closed left half plane in $\mathbb{C}$ (p. 14)
OLHP	open left half plane in $\mathbb{C}$ (p. 14)
CRHP	closed right half plane in $\mathbb{C}$ (p. 14)
ORHP	open right half plane in $\mathbb{C}$ (p. 14)
$\jmath\mathbb{R}$	imaginary numbers (p. 14)
$\mathbb{R}^n$	$\mathbb{R}^{n \times 1}$ (real column vectors) (p. 14)
$\mathbb{C}^n$	$\mathbb{C}^{n \times 1}$ (complex column vectors) (p. 14)
$\mathbb{F}^n$	$\mathbb{R}^n$ or $\mathbb{C}^n$ (p. 14)
$x_{(i)}$	ith component of $x \in \mathbb{F}^n$ (p. 14)
$x \geq\geq y$	$x_{(i)} \geq y_{(i)}$ for all i ($x - y$ is nonnegative) (p. 15)
$x >> y$	$x_{(i)} > y_{(i)}$ for all i ($x - y$ is positive) (p. 15)
$\mathbb{R}^{n \times m}$	$n \times m$ real matrices (p. 15)
$\mathbb{C}^{n \times m}$	$n \times m$ complex matrices (p. 15)

$\mathbb{F}^{n \times m}$	$\mathbb{R}^{n \times m}$ or $\mathbb{C}^{n \times m}$ (p. 15)
$\mathrm{row}_i(A)$	ith row of A (p. 15)
$\mathrm{col}_i(A)$	ith column of A (p. 15)
$A_{(i,j)}$	(i,j) entry of A (p. 15)
$A \geq\geq B$	$A_{(i,j)} \geq B_{(i,j)}$ for all i,j $(A - B$ is nonnegative) (p. 16)
$A >> B$	$A_{(i,j)} > B_{(i,j)}$ for all i,j $(A - B$ is positive) (p. 16)
$A \overset{i}{\leftarrow} b$	matrix obtained from $A \in \mathbb{F}^{n \times m}$ by replacing $\mathrm{col}_i(A)$ with $b \in \mathbb{F}^n$ or $\mathrm{row}_i(A)$ with $b \in \mathbb{F}^{1 \times m}$ (p. 16)
$\mathrm{d}_{\max}(A) \triangleq \mathrm{d}_1(A)$	largest diagonal entry of $A \in \mathbb{F}^{n \times n}$ having real diagonal entries (p. 16)
$\mathrm{d}_i(A)$	ith largest diagonal entry of $A \in \mathbb{F}^{n \times n}$ having real diagonal entries (p. 16)
$\mathrm{d}_{\min}(A) \triangleq \mathrm{d}_n(A)$	smallest diagonal entry of $A \in \mathbb{F}^{n \times n}$ having real diagonal entries (p. 16)
$[A,B]$	commutator $AB - BA$ (p. 18)
$\mathrm{ad}_A(X)$	adjoint operator $[A,X]$ (p. 18)
$x \times y$	cross product of vectors $x,y \in \mathbb{R}^3$ (p. 18)
$K(x)$	cross-product matrix for $x \in \mathbb{R}^3$ (p. 18)
$0_{n \times m}, 0$	$n \times m$ zero matrix (p. 19)
I_n, I	$n \times n$ identity matrix (p. 19)
$e_{i,n}, e_i$	$\mathrm{col}_i(I_n)$ (p. 20)
$E_{i,j,n \times m}, E_{i,j}$	$e_{i,n} e_{j,m}^{\mathrm{T}}$ (p. 20)
$\hat{I}_n, \hat{I}$	$n \times n$ reverse identity matrix (p. 20) $\begin{bmatrix} & & 1 \\ & \cdot^{\cdot^{\cdot}} & \\ 1 & & 0 \end{bmatrix}$
$1_{n \times m}, 1$	$n \times m$ ones matrix (p. 21)
A^{T}	transpose of A (p. 22)
$\mathrm{tr}\, A$	trace of A (p. 22)
$\overline{C}$	complex conjugate of $C \in \mathbb{C}^{n \times m}$ (p. 23)
A^*	$\overline{A}^{\mathrm{T}}$ conjugate transpose of A (p. 23)
$\mathrm{Re}\, A$	real part of $A \in \mathbb{F}^{n \times m}$ (p. 23)

$\mathrm{Im}\, A$	imaginary part of $A \in \mathbb{F}^{n \times m}$ (p. 23)
$\overline{\mathcal{S}}$	$\{\overline{Z}:\ Z \in \mathcal{S}\}$ or $\{\overline{Z}:\ Z \in \mathcal{S}\}_{\mathrm{m}}$ (p. 23)
$A^{\hat{\mathrm{T}}}$	$\hat{I} A^{\mathrm{T}} \hat{I}$ reverse transpose of A (p. 24)
$A^{\hat{*}}$	$\hat{I} A^{*} \hat{I}$ reverse complex conjugate transpose of A (p. 24)
$\mathrm{co}\, \mathcal{S}$	convex hull of $\mathcal{S}$ (p. 25)
$\mathrm{cone}\, \mathcal{S}$	conical hull of $\mathcal{S}$ (p. 25)
$\mathrm{coco}\, \mathcal{S}$	convex conical hull of $\mathcal{S}$ (p. 25)
$\mathrm{span}\, \mathcal{S}$	span of $\mathcal{S}$ (p. 25)
$\mathrm{aff}\, \mathcal{S}$	affine hull of $\mathcal{S}$ (p. 25)
$\mathrm{dim}\, \mathcal{S}$	dimension of $\mathcal{S}$ (p. 26)
$\mathcal{S}^{\perp}$	orthogonal complement of $\mathcal{S}$ (p. 26)
$\mathrm{dcone}\, \mathcal{S}$	dual cone of $\mathcal{S}$ (p. 27)
$\mathcal{R}(A)$	range of A (p. 29)
$\mathcal{N}(A)$	null space of A (p. 30)
$\mathrm{rank}\, A$	rank of A (p. 32)
$\mathrm{def}\, A$	defect of A (p. 32)
A^{L}	left inverse of A (p. 34)
A^{R}	right inverse of A (p. 34)
A^{-1}	inverse of A (p. 37)
$A^{-\mathrm{T}}$	$\left(A^{\mathrm{T}}\right)^{-1}$ (p. 38)
A^{-*}	$\left(A^{*}\right)^{-1}$ (p. 38)
$\mathrm{det}\, A$	determinant of A (p. 39)
$A_{[i,j]}$	submatrix of A obtained by deleting $\mathrm{row}_i(A)$ and $\mathrm{col}_j(A)$ (p. 41)
A^{A}	adjugate of A (p. 42)

Chapter 3

N_n, N $n \times n$ standard nilpotent matrix (p. 82)

$\mathrm{diag}(a_1, \ldots, a_n)$ $\begin{bmatrix} a_1 & & 0 \\ & \ddots & \\ 0 & & a_n \end{bmatrix}$ (p. 83)

$\mathrm{revdiag}(a_1, \ldots, a_n)$ $\begin{bmatrix} 0 & & a_1 \\ & \reflectbox{$\ddots$} & \\ a_n & & 0 \end{bmatrix}$ (p. 83)

$\mathrm{diag}(A_1, \ldots, A_k)$ block-diagonal matrix $\begin{bmatrix} A_1 & & 0 \\ & \ddots & \\ 0 & & A_k \end{bmatrix}$, where $A_i \in \mathbb{F}^{n_i \times m_i}$ (p. 83)

J_{2n}, J $\begin{bmatrix} 0 & I_n \\ -I_n & 0 \end{bmatrix}$ (p. 85)

$\mathrm{gl}_{\mathbb{F}}(n)$, $\mathrm{pl}_{\mathbb{C}}(n)$, $\mathrm{sl}_{\mathbb{F}}(n)$, $\mathrm{u}(n)$, $\mathrm{su}(n)$, $\mathrm{so}(n)$, $\mathrm{sp}(n)$, $\mathrm{aff}_{\mathbb{F}}(n)$, $\mathrm{se}_{\mathbb{F}}(n)$, $\mathrm{trans}_{\mathbb{F}}(n)$ Lie algebras (p. 87)

$\mathrm{GL}_{\mathbb{F}}(n)$, $\mathrm{PL}_{\mathbb{F}}(n)$, $\mathrm{SL}_{\mathbb{F}}(n)$, $\mathrm{U}(n)$, $\mathrm{O}(n)$, $\mathrm{U}(n,m)$, $\mathrm{O}(n,m)$, $\mathrm{SU}(n)$, $\mathrm{SO}(n)$, $\mathrm{Sp}(n)$, $\mathrm{Aff}_{\mathbb{F}}(n)$, $\mathrm{SE}_{\mathbb{F}}(n)$, $\mathrm{Trans}_{\mathbb{F}}(n)$ groups (p. 88)

Chapter 4

$\mathbb{F}[s]$ polynomials with coefficients in $\mathbb{F}$ (p. 121)

$\deg p$ degree of $p \in \mathbb{F}[s]$ (p. 121)

$\mathrm{mroots}(p)$ multiset of roots of $p \in \mathbb{F}[s]$ (p. 122)

$\mathrm{roots}(p)$ set of roots of $p \in \mathbb{F}[s]$ (p. 122)

$\mathrm{m}_p(\lambda)$ multiplicity of λ as a root of $p \in \mathbb{F}[s]$ (p. 122)

$\mathbb{F}^{n \times m}[s]$ $n \times m$ matrices with entries in $\mathbb{F}[s]$ ($n \times m$ matrix polynomials with coefficients in $\mathbb{F}$) (p. 124)

$\mathrm{rank}\, P$ rank of $P \in \mathbb{F}^{n \times m}[s]$ (p. 125)

χ_A characteristic polynomial of A (p. 130)

$\lambda_{\max}(A) \triangleq \lambda_1(A)$	largest eigenvalue of $A \in \mathbb{F}^{n \times n}$ having real eigenvalues (p. 130)
$\lambda_i(A)$	ith largest eigenvalue of $A \in \mathbb{F}^{n \times n}$ having real eigenvalues (p. 130)
$\lambda_{\min}(A) \triangleq \lambda_n(A)$	smallest eigenvalue of $A \in \mathbb{F}^{n \times n}$ having real eigenvalues (p. 130)
$\mathrm{am}_A(\lambda)$	algebraic multiplicity of $\lambda \in \mathrm{spec}(A)$ (p. 130)
$\mathrm{spec}(A)$	spectrum of A (p. 130)
$\mathrm{mspec}(A)$	multispectrum of A (p. 130)
$\mathrm{gm}_A(\lambda)$	geometric multiplicity of $\lambda \in \mathrm{spec}(A)$ (p. 135)
$\mathrm{spabs}(A)$	spectral abscissa of A (p. 137)
$\mathrm{sprad}(A)$	spectral radius of A (p. 137)
$\mathrm{In}\, A$	inertia of A (p. 137)
$\mathrm{sig}\, A$	signature of A, that is, $\nu_+(A) - \nu_-(A)$ (p. 137)
$\nu_-(A), \nu_0(A), \nu_+(A)$	number of eigenvalues of A counting algebraic multiplicity having negative, zero, and positive real part, respectively (p. 137)
μ_A	minimal polynomial of A (p. 137)
$\mathbb{F}(s)$	rational functions with coefficients in $\mathbb{F}$ (SISO rational transfer functions) (p. 139)
$\mathbb{F}_{\mathrm{prop}}(s)$	proper rational functions with coefficients in $\mathbb{F}$ (SISO proper rational transfer functions) (p. 139)
$\mathrm{reldeg}\, g$	relative degree of $g \in \mathbb{F}_{\mathrm{prop}}(s)$ (p. 139)
$\mathbb{F}^{n \times m}(s)$	$n \times m$ matrices with entries in $\mathbb{F}(s)$ (MIMO rational transfer functions) (p. 139)
$\mathbb{F}_{\mathrm{prop}}^{n \times m}(s)$	$n \times m$ matrices with entries in $\mathbb{F}_{\mathrm{prop}}(s)$ (MIMO proper rational transfer functions) (p. 139)
$\mathrm{reldeg}\, G$	relative degree of $G \in \mathbb{F}_{\mathrm{prop}}^{n \times m}(s)$ (p. 140)

$\mathrm{rank}\,G$	rank of $G \in \mathbb{F}^{n \times m}(s)$ (p. 140)
$B(p,q)$	Bezout matrix of $p, q \in \mathbb{F}[s]$ (p. 143, Fact 4.8.6)
$H(g)$	Hankel matrix of $g \in \mathbb{F}(s)$ (p. 145, Fact 4.8.8)

Chapter 5

$C(p)$	companion matrix for monic polynomial p (p. 162)
$\mathcal{H}_l(q)$	$l \times l$ or $2l \times 2l$ hypercompanion matrix (p. 166)
$\mathcal{J}_l(q)$	$l \times l$ or $2l \times 2l$ real Jordan matrix (p. 168)
$\mathrm{ind}_A(\lambda)$	index of λ with respect to A (p. 174)
$\mathrm{ind}\,A$	index of A, that is, $\mathrm{ind}_A(0)$ (p. 174)
$A_\perp$	complementary idempotent matrix or projector $I - A$ corresponding to the idempotent matrix or projector A (p. 175)
$\sigma_i(A)$	ith largest singular value of $A \in \mathbb{F}^{n \times m}$ (p. 181)
$\sigma_{\max}(A) \triangleq \sigma_1(A)$	largest singular value of $A \in \mathbb{F}^{n \times m}$ (p. 182)
$\sigma_{\min}(A) \triangleq \sigma_n(A)$	minimum singular value of $A \in \mathbb{F}^{n \times n}$ (p. 182)
$P_{A,B}$	matrix pencil of (A, B), where $A, B \in \mathbb{F}^{n \times n}$ (p. 203, Fact 5.11.11)
$\chi_{A,B}$	characteristic polynomial of (A, B), where $A, B \in \mathbb{F}^{n \times n}$ (p. 203, Fact 5.11.11)
$V(\lambda_1, \ldots, \lambda_n)$	Vandermonde matrix (p. 211, Fact 5.13.1)
$\mathrm{circ}(a_0, \ldots, a_{n-1})$	circulant matrix of $a_0, \ldots, a_{n-1} \in \mathbb{F}$ (p. 213, Fact 5.13.7)

Chapter 6

A^+	(Moore-Penrose) generalized inverse of A (p. 223)
$D\|\mathcal{A}$	Schur complement of D with respect to $\mathcal{A}$ (p. 227)
A^{D}	Drazin generalized inverse of A (p. 227)
$A^{\#}$	group generalized inverse of A (p. 228)

Chapter 7

vec A	vector formed by stacking columns of A (p. 247)
$\otimes$	Kronecker product (p. 247)
$P_{n,m}$	Kronecker permutation matrix (p. 250)
$\oplus$	Kronecker sum (p. 251)
$A \circ B$	Schur product of A and B (p. 252)
$A^{\{\alpha\}}$	Schur power of A, $\left(A^{\{\alpha\}}\right)_{(i,j)} = \left(A_{(i,j)}\right)^{\alpha}$ (p. 252)

Chapter 8

$\mathbf{H}^n$	$n \times n$ Hermitian matrices (p. 263)
$\mathbf{N}^n$	$n \times n$ positive-semidefinite matrices (p. 263)
$\mathbf{P}^n$	$n \times n$ positive-definite matrices (p. 263)
$A \geq B$	$A - B \in \mathbf{N}^n$ (p. 263)
$A > B$	$A - B \in \mathbf{P}^n$ (p. 263)
$\langle A \rangle$	$(A^*A)^{1/2}$ (p. 278)
$A\#B$	geometric mean of A and B (p. 298, Fact 8.8.23)
$A:B$	parallel sum of A and B (p. 332, Fact 8.15.10)

Chapter 9

$\lvert x \rvert$	absolute value of $x \in \mathbb{F}^n$ (p. 343)
$\lvert A \rvert$	absolute value of $A \in \mathbb{F}^{n \times m}$ (p. 343)
$\lVert x \rVert_p$	Hölder norm $\left[\sum_{i=1}^{n} \lvert x_{(i)} \rvert^p \right]^{1/p}$ (p. 344)
$\lVert A \rVert_p$	Hölder norm $\left[\sum_{i,j=1}^{n,m} \lvert A_{(i,j)} \rvert^p \right]^{1/p}$ (p. 347)
$\lVert A \rVert_{\mathrm{F}}$	Frobenius norm $\sqrt{\operatorname{tr} A^* A}$ (p. 348)
$\lVert A \rVert_{\sigma p}$	Schatten norm $\left[\sum_{i=1}^{\operatorname{rank} A} \sigma_i^p(A) \right]^{1/p}$ (p. 349)
$\lVert A \rVert_{q,p}$	Hölder-induced norm (p. 354)
$\lVert A \rVert_{\mathrm{col}}$	column norm $\lVert A \rVert_{1,1} = \max_{i \in \{1,\ldots,m\}} \lVert \operatorname{col}_i(A) \rVert_1$ (p. 356)
$\lVert A \rVert_{\mathrm{row}}$	row norm $\lVert A \rVert_{\infty,\infty} = \max_{i \in \{1,\ldots,n\}} \lVert \operatorname{row}_i(A) \rVert_1$ (p. 356)
$\ell(A)$	induced lower bound of A (p. 358)
$\ell_{q,p}(A)$	Hölder-induced lower bound of A (p. 359)
$\lVert \cdot \rVert_{\mathrm{D}}$	dual norm (p. 365, Fact 9.7.9)

Chapter 10

$\mathbb{B}_\varepsilon(x)$	open ball of radius ε centered at x (p. 401)
$\mathbb{S}_\varepsilon(x)$	sphere of radius ε centered at x (p. 401)
$\operatorname{int} \mathcal{S}$	interior of $\mathcal{S}$ (p. 401)
$\operatorname{int}_{\mathcal{S}'} \mathcal{S}$	interior of $\mathcal{S}$ relative to $\mathcal{S}'$ (p. 401)
$\operatorname{cl} \mathcal{S}$	closure of $\mathcal{S}$ (p. 401)
$\operatorname{cl}_{\mathcal{S}'} \mathcal{S}$	closure of $\mathcal{S}$ relative to $\mathcal{S}'$ (p. 402)
$\operatorname{bd} \mathcal{S}$	boundary of $\mathcal{S}$ (p. 402)

$\mathrm{bd}_{\mathcal{S}'}\,\mathcal{S}$ boundary of $\mathcal{S}$ relative to $\mathcal{S}'$ (p. 402)

$\mathrm{vcone}\,\mathcal{D}$ variational cone of $\mathcal{D}$ (p. 405)

$\mathrm{D}_+ f(x_0; \xi)$ one-sided directional derivative of f at x_0 in the direction ξ (p. 405)

$\dfrac{\partial f(x_0)}{\partial x_{(i)}}$ partial derivative of f with respect to $x_{(i)}$ at x_0 (p. 405)

$f'(x)$ Fréchet derivative of f at x (p. 406)

$\dfrac{\mathrm{d}f(x_0)}{\mathrm{d}x_{(i)}}$ $f'(x_0)$ (p. 406)

$f^{(k)}(x)$ kth Fréchet derivative of f at x (p. 407)

Chapter 11

e^A or $\exp(A)$ matrix exponential (p. 419)

$\mathcal{L}$ Laplace transform (p. 423)

$\mathcal{S}_{\mathrm{s}}(A)$ asymptotically stable subspace of A (p. 440)

$\mathcal{S}_{\mathrm{u}}(A)$ unstable subspace of A (p. 440)

OUD open unit disk in $\mathbb{C}$ (p. 446)

CUD closed unit disk in $\mathbb{C}$ (p. 446)

Chapter 12

$\mathcal{U}(A, C)$ unobservable subspace of (A, C) (p. 493)

$\mathcal{O}(A, C)$ $\begin{bmatrix} C \\ CA \\ CA^2 \\ \vdots \\ CA^{n-1} \end{bmatrix}$ (p. 493)

$\mathcal{C}(A, B)$ controllable subspace of (A, B) (p. 501)

$\mathcal{K}(A, B)$ $\begin{bmatrix} B & AB & A^2B & \cdots & A^{n-1}B \end{bmatrix}$ (p. 501)

$G \sim \left[\begin{array}{c|c} A & B \\ \hline C & D \end{array}\right]$ state space realization of $G \in \mathbb{F}^{l \times m}_{\mathrm{prop}}[s]$ (p. 513)

$G \overset{\min}{\sim} \left[\begin{array}{c|c} A & B \\ \hline C & D \end{array}\right]$ state space realization of $G \in \mathbb{F}^{l \times m}_{\mathrm{prop}}[s]$ (p. 517)

$\mathcal{H}(A, B, C)$ Markov block-Hankel matrix
$\mathcal{O}(A, C)\mathcal{K}(A, B)$ (p. 515)

$\mathcal{H}$ Hamiltonian $\begin{bmatrix} A & \Sigma \\ R_1 & -A^{\mathrm{T}} \end{bmatrix}$ (p. 539)

Conventions, Notation, and Terminology

When a word is defined, it is italicized.

The definition of a word, phrase, or symbol should always be understood as an "if and only if" statement, although for brevity "only if" is omitted. The symbol $\triangleq$ means equal by definition, where $A \triangleq B$ means that the left-hand expression A is defined to be the right-hand expression B.

Analogous statements are written in parallel using the following style: If n is (even, odd), then $n + 1$ is (odd, even).

The variables i, j, k, l, m, n always denote integers. Hence, $k \geq 0$ denotes a nonnegative integer, $k \geq 1$ denotes a positive integer, and the limit $\lim_{k \to \infty} A^k$ is taken over positive integers.

The imaginary unit $\sqrt{-1}$ is always denoted by dotless $\jmath$.

The variable s always represents a complex scalar.

The inequalities $c \leq a \leq d$ and $c \leq b \leq d$ are written as

$$c \leq \left\{ \begin{array}{c} a \\ b \end{array} \right\} \leq d.$$

The prefix "non" means "not" in the words nonconstant, nonempty, nonintegral, nonnegative, nonreal, nonsingular, nonsquare, nonunique, and nonzero. In some traditional usage, "non" may mean "not necessarily."

"Increasing" and "decreasing" indicate strict change for a change in the argument. The word "strict" is superfluous, and thus is omitted. Nonincreasing means nowhere increasing, while nondecreasing means nowhere decreasing.

Multisets can have repeated elements. Hence, $\{x\}_m$ and $\{x, x\}_m$ are different. The listed elements α, β, γ of the conventional set $\{\alpha, \beta, \gamma\}$ need not be distinct, although $\{\alpha, \beta, \alpha\} = \{\alpha, \beta\}$.

The composition of functions f and g is denoted by $f \bullet g$. The traditional notation $f \circ g$ is reserved for the Schur product.

$\mathcal{S}_1 \subset \mathcal{S}_2$ means that $\mathcal{S}_1$ is a proper subset of $\mathcal{S}_2$, whereas $\mathcal{S}_1 \subseteq \mathcal{S}_2$ means that $\mathcal{S}_1$ is either a proper subset of $\mathcal{S}_2$ or is equal to $\mathcal{S}_2$. Hence, $\mathcal{S}_1 \subset \mathcal{S}_2$ is equivalent to $\mathcal{S}_1 \subseteq \mathcal{S}_2$ and $\mathcal{S}_1 \neq \mathcal{S}_2$, while $\mathcal{S}_1 \subseteq \mathcal{S}_2$ is equivalent to either $\mathcal{S}_1 \subset \mathcal{S}_2$ or $\mathcal{S}_1 = \mathcal{S}_2$.

$1/\infty \triangleq 0$ and $0! \triangleq 1$.

$A^0 \triangleq I$ for all square matrices A. In particular, $0_{n \times n}^0 = I_n$. With this convention, it is possible to write

$$\sum_{i=0}^{\infty} \alpha^i = \frac{1}{1 - \alpha}$$

for all $-1 < \alpha < 1$. Of course, $\lim_{x \downarrow 0} 0^x = 0$, $\lim_{x \downarrow 0} x^0 = 1$, and $\lim_{x \downarrow 0} x^x = 1$.

The symbol $\mathbb{F}$ denotes either $\mathbb{R}$ or $\mathbb{C}$ consistently in every result. For example, in Theorem 5.6.3, the three appearances of "$\mathbb{F}$" can be read as either all "$\mathbb{C}$" or all "$\mathbb{R}$."

The imaginary numbers are denoted by $\jmath\mathbb{R}$. Hence, 0 is both a real number and an imaginary number.

The notation $\mathrm{Re}\, A$ and $\mathrm{Im}\, A$ represents the real and imaginary parts of A, respectively. Some books use $\mathrm{Re}\, A$ and $\mathrm{Im}\, A$ to denote the Hermitian and skew-Hermitian matrices $\frac{1}{2}(A + A^*)$ and $\frac{1}{2}(A - A^*)$.

For the scalar ordering "$\leq$," if $x \leq y$, then $x < y$ if and only if $x \neq y$. For the entrywise vector and matrix orderings, $x \leq y$ and $x \neq y$ do not imply that $x < y$.

Operations denoted by superscripts are applied before operations represented by preceding operators. For example, $\mathrm{tr}\,(A+B)^2$ means $\mathrm{tr}\left[(A + B)^2\right]$ and $\mathrm{cl}\,\mathcal{S}^\sim$ means $\mathrm{cl}(\mathcal{S}^\sim)$. This convention simplifies many formulas.

A vector in $\mathbb{F}^n$ is a column vector, which is also a matrix with one column. In mathematics, "vector" generally refers to an abstract vector not resolved in coordinates.

Sets have elements, vectors have components, and matrices have entries. This terminology has no mathematical consequence.

The notation $x_{(i)}$ represents the ith component of the vector x.

The notation $A_{(i,j)}$ represents the scalar (i, j) entry of A. $A_{i,j}$ or A_{ij} denotes a block or submatrix of A.

All matrices have nonnegative integral dimensions. If at least one of the dimensions of a matrix is zero, then the matrix is empty.

The entries of a submatrix $\hat{A}$ of a matrix A are the entries of A lying in specified rows and columns. $\hat{A}$ is a block of A if $\hat{A}$ is a submatrix of A whose entries are entries of adjacent rows and columns of A. Every matrix is both a submatrix and block of itself.

The determinant of a submatrix is a subdeterminant. Some books use "minor." The determinant of a matrix is also a subdeterminant of the matrix.

The dimension of the null space of a matrix is its defect. Some books use "nullity."

A block of a square matrix is diagonally located if the block is square and the diagonal entries of the block are also diagonal entries of the matrix; otherwise, the block is off-diagonally located. This terminology avoids confusion with a "diagonal block," which is a block that is also a square, diagonal submatrix.

For the partitioned matrix $\begin{bmatrix} A & B \\ C & D \end{bmatrix} \in \mathbb{F}^{(n+m)\times(k+l)}$, it can be inferred that $A \in \mathbb{F}^{n\times k}$ and similarly for $B, C,$ and D.

The Schur product of matrices A and B is denoted by $A \circ B$. Matrix multiplication is given priority over Schur multiplication, that is, $A \circ BC$ means $A \circ (BC)$.

The adjugate of $A \in \mathbb{F}^{n\times n}$ is denoted by A^{A}. The traditional notation is $\mathrm{adj}\, A$, while the notation A^{A} is used in [665].

If $\mathbb{F} = \mathbb{R}$, then $\overline{A}$ becomes A, A^* becomes A^{T}, "Hermitian" becomes "symmetric," "unitary" becomes "orthogonal," "unitarily" becomes "orthogonally," and "congruence" becomes "T-congruence." A square complex matrix A is symmetric if $A^{\mathrm{T}} = A$ and orthogonal if $A^{\mathrm{T}}A = I$.

The diagonal entries of a matrix $A \in \mathbb{F}^{n\times n}$ all of whose diagonal entries are real are ordered as $\mathrm{d}_{\max}(A) = \mathrm{d}_1(A) \geq \mathrm{d}_2(A) \geq \cdots \geq \mathrm{d}_n(A) = \mathrm{d}_{\min}(A)$.

The eigenvalues of a matrix $A \in \mathbb{F}^{n\times n}$ all of whose eigenvalues are real are ordered as $\lambda_{\max}(A) = \lambda_1(A) \geq \lambda_2(A) \geq \cdots \geq \lambda_n(A) = \lambda_{\min}(A)$.

For $A \in \mathbb{F}^{n\times n}$, $\mathrm{am}_A(\lambda)$ is the number of copies of λ in the multispectrum of A, $\mathrm{gm}_A(\lambda)$ is the number of Jordan blocks of A associated with λ, and $\mathrm{ind}_A(\lambda)$ is the size of the largest Jordan block of A associated with λ. The index of A, denoted by $\mathrm{ind}\, A = \mathrm{ind}_A(0)$, is the size of the largest Jordan block of A associated with the eigenvalue 0.

An $n \times m$ matrix has exactly $\min\{n, m\}$ singular values, exactly $\mathrm{rank}\, A$ of which are positive.

The $\min\{n, m\}$ singular values of a matrix $A \in \mathbb{F}^{n \times m}$ are ordered as $\sigma_{\max}(A)$ $\triangleq \sigma_1(A) \geq \sigma_2(A) \geq \cdots \geq \sigma_{\min\{n,m\}}(A)$. If $n = m$, then $\sigma_{\min}(A) \triangleq \sigma_n(A)$. The notation $\sigma_{\min}(A)$ is defined only for square matrices.

Positive-semidefinite and positive-definite matrices are Hermitian.

A square matrix with entries in $\mathbb{F}$ is diagonalizable over $\mathbb{F}$ if and only if it can be transformed into a diagonal matrix whose entries are in $\mathbb{F}$ by means of a similarity transformation whose entries are in $\mathbb{F}$. Therefore, a complex matrix is diagonalizable over $\mathbb{C}$ if and only if all of its eigenvalues are semisimple, whereas a real matrix is diagonalizable over $\mathbb{R}$ if and only if all of its eigenvalues are semisimple and real. The real matrix $\begin{bmatrix} 0 & 1 \\ -1 & 0 \end{bmatrix}$ is diagonalizable over $\mathbb{C}$, although it is not diagonalizable over $\mathbb{R}$. The Hermitian matrix $\begin{bmatrix} 1 & j \\ -j & 2 \end{bmatrix}$ is diagonalizable over $\mathbb{C}$, and also has real eigenvalues.

An idempotent matrix $A \in \mathbb{F}^{n \times n}$ satisfies $A^2 = A$, while a projector is a Hermitian, idempotent matrix. Some books use "projector" for idempotent and "orthogonal projector" for projector. A reflector is a Hermitian, involutory matrix. A projector is a normal matrix each of whose eigenvalues is 1 or 0, while a reflector is a normal matrix each of whose eigenvalues is 1 or -1.

A range-Hermitian matrix is a square matrix whose range is equal to the range of its complex conjugate transpose. These matrices are traditionally called "EP" matrices.

An elementary matrix is a nonsingular matrix formed by adding an outer-product matrix to the identity matrix. An elementary reflector is a reflector exactly one of whose eigenvalues is -1. An elementary projector is a projector exactly one of whose eigenvalues is 0. Elementary reflectors are elementary matrices. However, elementary projectors are not elementary matrices since elementary projectors are singular.

The rank of a matrix polynomial P is the maximum rank of $P(s)$ over $\mathbb{C}$. This quantity is also called the normal rank. We denote this quantity by $\operatorname{rank} P$ as distinct from $\operatorname{rank} P(s)$, which denotes the rank of the matrix $P(s)$.

The rank of a rational transfer function G is the maximum rank of $G(s)$ over $\mathbb{C}$ excluding poles of the entries of G. This quantity is also called the normal rank. We denote this quantity by $\operatorname{rank} G$ as distinct from $\operatorname{rank} G(s)$, which denotes the rank of the matrix $G(s)$.

The symbol $\oplus$ denotes the Kronecker sum. Some books use $\oplus$ to denote the direct sum of matrices or subspaces.

The notation $|A|$ represents the matrix obtained by replacing every entry of A by its absolute value.

The notation $\langle A \rangle$ represents the matrix $(A^*A)^{1/2}$. Some books use $|A|$ to denote this matrix.

The Hölder norms for vectors and matrices are denoted by $\|\cdot\|_p$. The matrix norm induced by $\|\cdot\|_q$ on the domain and $\|\cdot\|_p$ on the codomain is denoted by $\|\cdot\|_{p,q}$.

The Schatten norms for matrices are denoted by $\|\cdot\|_{\sigma p}$, and the Frobenius norm is denoted by $\|\cdot\|_F$. Hence, $\|\cdot\|_{\sigma\infty} = \|\cdot\|_{2,2} = \sigma_{\max}(\cdot)$ and $\|\cdot\|_{\sigma 2} = \|\cdot\|_F$.

Preface

The idea for this book began with the realization that at the heart of the solution to many problems in science, mathematics, and engineering often lies a "matrix fact," that is, an identity, inequality, or property of matrices that is crucial to the solution of the problem. Although there are numerous excellent books on linear algebra and matrix theory, no one book contains all or even most of the vast number of matrix facts that appear throughout the scientific, mathematical, and engineering literature. This book is an attempt to organize many of these facts into a reference source for users of matrix theory in diverse applications areas.

Viewed as an extension of scalar mathematics, matrix mathematics provides the means to manipulate and analyze multidimensional quantities. Matrix mathematics thus provides powerful tools for a broad range of problems in science and engineering. For example, the matrix-based analysis of systems of ordinary differential equations accounts for interaction among all of the state variables. The discretization of partial differential equations by means of finite differences and finite elements yields linear algebraic or differential equations whose matrix structure reflects the nature of physical solutions [674]. Multivariate probability theory and statistical analysis use matrix methods to represent probability distributions, to compute moments, and to perform linear regression for data analysis [274,313,339,502,637]. The study of linear differential equations [357,358,381] depends heavily on matrix analysis, while linear systems and control theory are matrix-intensive areas of engineering [2, 38, 76, 80, 177, 178, 204, 239, 272, 388, 445, 454, 495, 589, 622, 626, 645, 647, 654, 724, 743, 794, 818]. In addition, matrices are widely used in rigid body mechanics [15, 422, 444, 515, 551, 574, 646, 659, 732], structural dynamics [452, 528, 593], fluid dynamics [173, 258, 773], circuit theory [17], queuing and stochastic systems [334,558], game theory [127,481,670], computer graphics [36, 269], signal processing [629, 740], classical and quantum information theory [584], statistical mechanics [10, 85, 86, 745], demography [421], combinatorics, networks, and graph theory [149,171,260,487,621], optics [289, 343], dimensional analysis [333, 681], and number theory [439].

In all applications involving matrices, computational techniques are essential for obtaining numerical solutions. The development of efficient and reliable algorithms for matrix computations is therefore an important area of research that has been extensively developed [54, 172, 215, 299, 354, 378, 393, 661, 662, 664, 666, 717, 744, 774, 778, 779, 804]. To facilitate the solution of matrix problems, entire computer packages have been developed using the language of matrices. However, this book is concerned with the analytical properties of matrices rather than their computational aspects.

This book encompasses a broad range of fundamental questions in matrix theory, which, in many cases can be viewed as extensions of related questions in scalar mathematics. A few such questions follow.

What are the basic properties of matrices? How can matrices be characterized, classified, and quantified?

How can a matrix be decomposed into simpler matrices? A matrix decomposition may involve addition, multiplication, and partition. Decomposing a matrix into its fundamental components provides insight into its algebraic and geometric properties. For example, the polar decomposition states that every square matrix can be written as the product of a rotation and a dilation analogous to the polar representation of a complex number.

Given a pair of matrices having certain properties, what can be inferred about the sum, product, and concatenation of these matrices? In particular, if a matrix has a given property, to what extent does that property change or remain unchanged if the matrix is perturbed by another matrix of a certain type by means of addition, multiplication, or concatenation? For example, if a matrix is nonsingular, how large can an additive perturbation to that matrix be without the sum becoming singular?

How can properties of a matrix be determined by means of simple operations? For example, how can the location of the eigenvalues of a matrix be estimated directly in terms of the entries of the matrix?

To what extent do matrices satisfy the formal properties of the real numbers? For example, while $0 \leq a \leq b$ implies that $a^r \leq b^r$ for real numbers a, b and a positive integer r, when does $0 \leq A \leq B$ imply $A^r \leq B^r$ for positive-semidefinite matrices A and B and with the positive-semidefinite ordering?

Questions of these types have occupied matrix theorists for at least a century, with motivation from diverse applications. The existing scope and depth of knowledge are enormous. Taken together, this body of knowledge provides a powerful framework for developing and analyzing models for scientific and engineering applications.

This book is intended to be useful to at least four groups of readers. Since linear algebra is a standard course in the mathematical sciences and engineering, graduate students in these fields can use this book to expand the scope of their linear algebra text. For instructors, many of the Facts can be used as exercises to augment standard material in matrix courses. For researchers in the mathematical sciences, including statistics, physics, and engineering, this book can be used as a general reference on matrix theory. Finally, for users of matrices in the applied sciences, this book will provide access to a large body of results in matrix theory. By collecting these results in a single source, it is my hope that this book will prove to be convenient and useful for a broad range of applications. The material in this book is thus intended to complement the large number of classical and modern texts and reference works on linear algebra and matrix theory [5, 273, 285, 286, 292, 307, 364, 409, 456, 497, 507, 511, 542, 562, 566, 592, 621, 643, 674].

After a review of mathematical preliminaries in Chapter 1, fundamental properties of matrices are described in Chapter 2. Chapter 3 summarizes the major classes of matrices and various matrix transformations. In Chapter 4 we turn to polynomial and rational matrices whose basic properties are essential for understanding the structure of constant matrices. Chapter 5 is concerned with various decompositions of matrices including the Jordan, Schur, and singular value decompositions. Chapter 6 provides a brief treatment of generalized inverses, while Chapter 7 describes the Kronecker and Schur product operations. Chapter 8 is concerned with the properties of positive-semidefinite matrices. A detailed treatment of vector and matrix norms is given in Chapter 9, while formulas for matrix derivatives are given in Chapter 10. Next, Chapter 11 focuses on the matrix exponential and stability theory, which are central to the study of linear differential equations. In Chapter 12 we apply matrix theory to the analysis of linear systems, their state space realizations, and their transfer function representation. This chapter also includes a discussion of the matrix Riccati equation of control theory.

Each chapter provides a core of results with, in many cases, complete proofs. Sections at the end of each chapter provide a collection of Facts organized to correspond to the order of topics in the chapter. These Facts include corollaries and special cases of results presented in the chapter, as well as related results that go beyond the results of the chapter. In some cases the Facts include open problems, illuminating remarks, and hints regarding proofs. The Facts are intended to provide the reader with a useful reference collection of matrix results as well as a gateway to the matrix theory literature.

Acknowledgments

The writing of this book spanned more than a decade and a half, during which time numerous individuals contributed both directly and indirectly. I am grateful for the helpful comments of many people who contributed technical material and insightful suggestions, all of which greatly improved the presentation and content of the book. In addition, numerous individuals graciously agreed to read sections or chapters of the book for clarity and accuracy. I wish to thank Jasim Ahmed, Suhail Akhtar, David Bayard, Sanjay Bhat, Tony Bloch, Peter Bullen, Steve Campbell, Agostino Capponi, Ramu Chandra, Jaganath Chandrasekhar, Nalin Chaturvedi, Vijay Chellaboina, Jie Chen, David Clements, Dan Davison, Dimitris Dimogianopoulos, Jiu Ding, D. Z. Djokovic, R. Scott Erwin, R. W. Farebrother, Danny Georgiev, Joseph Grcar, Wassim Haddad, Yoram Halevi, Jesse Hoagg, Roger Horn, David Hyland, Pierre Kabamba, Vikram Kapila, Fuad Kittaneh, Seth Lacy, Thomas Laffey, Cedric Langbort, Alan Laub, Alexander Leonessa, Kai Yew Lum, Pertti Makila, Roy Mathias, N. Harris McClamroch, Boris Mordukhovich, Sergei Nersesov, Jin Oh, Harish Palanthandalum-Madapusi, Concetta Pilotto, Michael Piovoso, Leiba Rodman, Phil Roe, Carsten Scherer, Wasin So, Andy Sparks, Edward Tate, Yongge Tian, Panagiotis Tsiotras, Feng Tyan, Ravi Venugopal, Jan Willems, Hong Wong, Vera Zeidan, Xingzhi Zhan, and Fuzhen Zhang for their assistance. Nevertheless, I take full responsibility for any remaining errors, and I encourage readers to alert me to any mistakes, corrections of which will be posted on the web. Solutions to the open problems are also welcome.

Portions of the manuscript were typed by Jill Straehla and Linda Smith at Harris Corporation, and by Debbie Laird, Kathy Stolaruk, and Suzanne Gurney at the University of Michigan. John Rogosich of Techsetters, Inc., provided invaluable assistance with LATEX issues, and Jennifer Slater carefully copyedited the entire manuscript. I also thank Jin Oh and Joshua Kang for writing C code to refine the index.

I especially thank Vickie Kearn of Princeton University Press for her wise guidance and constant encouragement. Vickie managed to address all of my concerns and anxieties, and helped me improve the manuscript in many ways.

Finally, I extend my greatest appreciation for the (uncountably) infinite patience of my family, who endured the days, weeks, months, and years that this project consumed. The writing of this book began with toddlers and ended with a teenager and a twenty-year old. We can all be thankful it is finally finished.

Dennis S. Bernstein
Ann Arbor, Michigan
dsbaero@umich.edu
January 2005

Chapter One

Preliminaries

In this chapter we review some basic terminology and results concerning logic, sets, functions, and related concepts. This material is used throughout the book.

1.1 Logic and Sets

Let A and B be conditions. The *negation* of A is (not A), the *both* of A and B is (A and B), and the *either* of A and B is (A or B). (A or B) does not contradict (A and B).

Let A and B be conditions. The *implication* or *statement* "if A is satisfied, then B is satisfied" or, equivalently, "A implies B" is written as $A \implies B$, while $A \iff B$ is equivalent to $[(A \implies B)$ and $(A \impliedby B)]$. Of course, $A \impliedby B$ means $B \implies A$.

Suppose $A \iff B$. Then, A is satisfied *if and only if* B is satisfied. By convention, the implication $A \implies B$ (the "only if" part) is *necessity*, while $B \implies A$ (the "if" part) is *sufficiency*. The *converse* of $A \implies B$ is $B \implies A$. The statement $A \implies B$ is equivalent to its *contrapositive* (not B) $\implies$ (not A).

A *theorem* is a significant result, while a *proposition* is less significant. The primary role of a *lemma* is to support the proof of a theorem or proposition. Furthermore, a *corollary* is a direct consequence of a theorem or proposition. Finally, a *fact* is either a theorem, proposition, lemma, or corollary.

Suppose that $A' \implies A \implies B \implies B'$. Then, $A' \implies B'$ is a corollary of $A \implies B$.

Let A, B, and C be conditions, and assume that $A \implies B$. Then, $A \implies B$ is a *strengthening* of (A and C) $\implies B$. If, in addition, $A \implies C$, then the statement (A and C) $\implies B$ has *redundant assumptions*.

Let $\mathfrak{X} \triangleq \{x, y, z\}$ be a *set*. Then,

$$x \in \mathfrak{X} \tag{1.1.1}$$

means that x is an *element* of $\mathfrak{X}$. If w is not an element of $\mathfrak{X}$, then we write

$$w \notin \mathfrak{X}. \tag{1.1.2}$$

The set with no elements, denoted by $\varnothing$, is the *empty set*. If $\mathfrak{X} \neq \varnothing$, then $\mathfrak{X}$ is *nonempty*.

A set cannot have repeated elements. For example, $\{x, x\} = \{x\}$. However, a *multiset* is a collection of elements that allows for repetition. The multiset consisting of two copies of x is written as $\{x, x\}_{\mathrm{m}}$. However, we do not assume that the listed elements x, y of the conventional set $\{x, y\}$ are distinct.

There are two basic types of mathematical statements involving quantifiers. An *existential statement* is of the form

$$\text{there exists } x \in \mathfrak{X} \text{ such that condition } Z \text{ is satisfied,} \tag{1.1.3}$$

while a *universal statement* has the structure

$$\text{condition } Z \text{ is satisfied for all } x \in \mathfrak{X}. \tag{1.1.4}$$

Let $\mathfrak{X}$ and $\mathfrak{Y}$ be sets. The *intersection* of $\mathfrak{X}$ and $\mathfrak{Y}$ is the set of common elements of $\mathfrak{X}$ and $\mathfrak{Y}$ given by

$$\mathfrak{X} \cap \mathfrak{Y} \triangleq \{x \colon \ x \in \mathfrak{X} \text{ and } x \in \mathfrak{Y}\} = \{x \in \mathfrak{X} \colon \ x \in \mathfrak{Y}\} \tag{1.1.5}$$
$$= \{x \in \mathfrak{Y} \colon \ x \in \mathfrak{X}\} = \mathfrak{Y} \cap \mathfrak{X}, \tag{1.1.6}$$

while the set of elements in either $\mathfrak{X}$ or $\mathfrak{Y}$ (the *union* of $\mathfrak{X}$ and $\mathfrak{Y}$) is

$$\mathfrak{X} \cup \mathfrak{Y} \triangleq \{x \colon \ x \in \mathfrak{X} \text{ or } x \in \mathfrak{Y}\} = \mathfrak{Y} \cup \mathfrak{X}. \tag{1.1.7}$$

The *complement* of $\mathfrak{X}$ *relative* to $\mathfrak{Y}$ is

$$\mathfrak{Y} \backslash \mathfrak{X} \triangleq \{x \in \mathfrak{Y} \colon \ x \notin \mathfrak{X}\}. \tag{1.1.8}$$

If $\mathfrak{Y}$ is specified, then the *complement* of $\mathfrak{X}$ is

$$\mathfrak{X}^{\sim} \triangleq \mathfrak{Y} \backslash \mathfrak{X}. \tag{1.1.9}$$

If $x \in \mathfrak{X}$ implies that $x \in \mathfrak{Y}$, then $\mathfrak{X}$ is *contained* in $\mathfrak{Y}$ ($\mathfrak{X}$ is a *subset* of $\mathfrak{Y}$), which is written as

$$\mathfrak{X} \subseteq \mathfrak{Y}. \tag{1.1.10}$$

The statement $\mathfrak{X} = \mathfrak{Y}$ is equivalent to the validity of both $\mathfrak{X} \subseteq \mathfrak{Y}$ and $\mathfrak{Y} \subseteq \mathfrak{X}$. If $\mathfrak{X} \subseteq \mathfrak{Y}$ and $\mathfrak{X} \neq \mathfrak{Y}$, then $\mathfrak{X}$ is a *proper subset* of $\mathfrak{Y}$ and we write $\mathfrak{X} \subset \mathfrak{Y}$. The sets $\mathfrak{X}$ and $\mathfrak{Y}$ are *disjoint* if $\mathfrak{X} \cap \mathfrak{Y} = \varnothing$. A *partition* of $\mathfrak{X}$ is a set of pairwise disjoint subsets of $\mathfrak{X}$ whose union is equal to $\mathfrak{X}$.

The operations "∩," "∪," and "\" and the relations "⊂" and "⊆" extend directly to multisets. For example,

$$\{x, x\}_{\mathrm{m}} \cup \{x\}_{\mathrm{m}} = \{x, x, x\}_{\mathrm{m}}. \tag{1.1.11}$$

By ignoring repetitions, a multiset can be converted to a set, while a set can be viewed as a multiset with distinct elements.

1.2 Relations and Functions

The *Cartesian product* $\mathfrak{X}_1 \times \cdots \times \mathfrak{X}_n$ of sets $\mathfrak{X}_1, \ldots, \mathfrak{X}_n$ is the set consisting of *ordered elements* of the form $(x_1, \ldots, x_n)$, where $x_i \in \mathfrak{X}_i$ for all $i = 1, \ldots, n$. A *relation* $\mathcal{R}$ on a set $\mathfrak{X}$ is a subset of $\mathfrak{X} \times \mathfrak{X}$. For convenience, $(x_1, x_2) \in \mathcal{R}$ is denoted by $x_1 \leq x_2$, whereas $x_1 \not\leq x_2$ denotes $(x_1, x_2) \notin \mathcal{R}$.

Definition 1.2.1. Let $\mathcal{R}$ be a relation on $\mathfrak{X}$. Then, the following terminology is defined:

i) $\mathcal{R}$ is *reflexive* if $x \leq x$ for all $x \in \mathfrak{X}$.

ii) $\mathcal{R}$ is *antisymmetric* if $x_1 \leq x_2$ and $x_2 \leq x_1$ imply that $x_1 = x_2$.

iii) $\mathcal{R}$ is *symmetric* if $x_1 \leq x_2$ implies that $x_2 \leq x_1$.

iv) $\mathcal{R}$ is *transitive* if $x_1 \leq x_2$ and $x_2 \leq x_3$ imply that $x_1 \leq x_3$.

v) $\mathcal{R}$ is *pairwise connected* if $x_1, x_2 \in \mathfrak{X}$ implies that either $x_1 \leq x_2$ or $x_2 \leq x_1$.

vi) $\mathcal{R}$ is a *partial ordering* if $\mathcal{R}$ is reflexive, antisymmetric, and transitive.

vii) $\mathcal{R}$ is a *total ordering* if $\mathcal{R}$ is a pairwise-connected partial ordering.

viii) $\mathcal{R}$ is an *equivalence relation* if $\mathcal{R}$ is reflexive, symmetric, and transitive.

For an equivalence relation $\mathcal{R}$, $x_1 \leq x_2$ is denoted by $x_1 \stackrel{\mathcal{R}}{=} x_2$. If $\mathcal{R}$ is an equivalence relation and $x \in \mathfrak{X}$, then the subset $\{y \in \mathfrak{X}: \ y \stackrel{\mathcal{R}}{=} x\}$ of $\mathfrak{X}$ is the *equivalence class of x induced by $\mathcal{R}$.*

Theorem 1.2.2. Let $\mathcal{R}$ be an equivalence relation on a set $\mathfrak{X}$. Then, the set of distinct equivalence classes of $\mathfrak{X}$ induced by $\mathcal{R}$ is a partition of $\mathfrak{X}$.

Proof. For $x \in \mathfrak{X}$, let $\mathcal{S}_x$ denote the equivalence class of x induced by $\mathcal{R}$. Clearly, $\mathfrak{X} = \bigcup_{x \in \mathfrak{X}} \mathcal{S}_x$. It suffices to show that if $x, y \in \mathfrak{X}$, then either $\mathcal{S}_x = \mathcal{S}_y$ or $\mathcal{S}_x \cap \mathcal{S}_y = \varnothing$. Hence, let $x, y \in \mathfrak{X}$, and suppose that $\mathcal{S}_x$ and $\mathcal{S}_y$ are not disjoint so that there exists $z \in \mathcal{S}_x \cap \mathcal{S}_y$. Thus, $(x, z) \in \mathcal{R}$ and $(z, y) \in \mathcal{R}$. Now, let $w \in \mathcal{S}_x$. Then, $(w, x) \in \mathcal{R}$, $(x, z) \in \mathcal{R}$, and $(z, y) \in \mathcal{R}$ imply that $(w, y) \in \mathcal{R}$. Hence, $w \in \mathcal{S}_y$, which implies that $\mathcal{S}_x \subseteq \mathcal{S}_y$. By a similar argument, $\mathcal{S}_y \subseteq \mathcal{S}_x$. Consequently, $\mathcal{S}_x = \mathcal{S}_y$. $\square$

Theorem 1.2.3. Let $\mathcal{X}$ be a set, consider a partition of $\mathcal{X}$, and define the relation $\mathcal{R}$ on $\mathcal{X}$ by $(x, y) \in \mathcal{R}$ if and only if x and y belong to the same partition subset of $\mathcal{X}$. Then, $\mathcal{R}$ is an equivalence relation on $\mathcal{X}$.

Let $\mathcal{X}$ and $\mathcal{Y}$ be sets. Then, a *function* f that maps $\mathcal{X}$ into $\mathcal{Y}$ is a rule $f \colon \mathcal{X} \mapsto \mathcal{Y}$ that assigns a unique element $f(x)$ (the *image* of x) of $\mathcal{Y}$ to every element x in $\mathcal{X}$. Equivalently, a function $f \colon \mathcal{X} \mapsto \mathcal{Y}$ can be viewed as a subset $\mathcal{F}$ of $\mathcal{X} \times \mathcal{Y}$ such that, for all $x \in \mathcal{X}$, there exists $y \in \mathcal{Y}$ such that $(x, y) \in \mathcal{F}$ and, if $(x_1, y_1) \in \mathcal{F}$, $(x_2, y_2) \in \mathcal{F}$, and $x_1 = x_2$, then $y_1 = y_2$. In this case, $\mathcal{F} = \mathrm{graph}(f) \triangleq \{(x, f(x)) \colon x \in \mathcal{X}\}$. The set $\mathcal{X}$ is the *domain* of f, while the set $\mathcal{Y}$ is the *codomain* of f. For $\mathcal{X}_1 \subseteq \mathcal{X}$, it is convenient to define $f(\mathcal{X}_1) \triangleq \{f(x) \colon x \in \mathcal{X}_1\}$. The set $f(\mathcal{X})$, which is denoted by $\mathcal{R}(f)$, is the *range* of f. If, in addition, $\mathcal{Z}$ is a set and $g \colon \mathcal{Y} \mapsto \mathcal{Z}$, then $g \bullet f \colon \mathcal{X} \mapsto \mathcal{Z}$ (the *composition* of g and f) is the function $(g \bullet f)(x) \triangleq g[f(x)]$. If $x_1, x_2 \in \mathcal{X}$ and $f(x_1) = f(x_2)$ implies that $x_1 = x_2$, then f is *one-to-one*; if $\mathcal{R}(f) = \mathcal{Y}$, then f is *onto*. The function $I_{\mathcal{X}} \colon \mathcal{X} \mapsto \mathcal{X}$ defined by $I_{\mathcal{X}}(x) \triangleq x$ for all $x \in \mathcal{X}$ is the *identity* on $\mathcal{X}$.

Let $f \colon \mathcal{X} \mapsto \mathcal{Y}$. Then, f is *left invertible* if there exists a function $g \colon \mathcal{Y} \mapsto \mathcal{X}$ (a *left inverse* of f) such that $g \bullet f = I_{\mathcal{X}}$, whereas f is *right invertible* if there exists a function $h \colon \mathcal{Y} \mapsto \mathcal{X}$ (a *right inverse* of f) such that $f \bullet h = I_{\mathcal{Y}}$. In addition, the function $f \colon \mathcal{X} \mapsto \mathcal{Y}$ is *invertible* if there exists a function $f^{-1} \colon \mathcal{Y} \mapsto \mathcal{X}$ (the *inverse* of f) such that $f^{-1} \bullet f = I_{\mathcal{X}}$ and $f \bullet f^{-1} = I_{\mathcal{Y}}$. The *inverse image* $f^{-1}(\mathcal{S})$ of $\mathcal{S} \subseteq \mathcal{Y}$ is defined by

$$f^{-1}(\mathcal{S}) \triangleq \{x \in \mathcal{X} \colon f(x) \in \mathcal{S}\}. \tag{1.2.1}$$

Theorem 1.2.4. Let $\mathcal{X}$ and $\mathcal{Y}$ be sets, and let $f \colon \mathcal{X} \mapsto \mathcal{Y}$. Then, the following statements hold:

i) f is left invertible if and only if f is one-to-one.

ii) f is right invertible if and only if f is onto.

Furthermore, the following statements are equivalent:

iii) f is invertible.

iv) f has a unique inverse.

v) f is one-to-one and onto.

vi) f is left invertible and right invertible.

vii) f has a unique left inverse.

viii) f has a unique right inverse.

Proof. To prove *i)*, suppose that f is left invertible with left inverse $g \colon \mathcal{Y} \mapsto \mathcal{X}$. Furthermore, suppose that $x_1, x_2 \in \mathcal{X}$ satisfy $f(x_1) = f(x_2)$.

Then, $x_1 = g[f(x_1)] = g[f(x_2)] = x_2$, which shows that f is one-to-one. Conversely, suppose that f is one-to-one so that, for all $y \in \mathcal{R}(f)$, there exists a unique $x \in \mathcal{X}$ such that $f(x) = y$. Hence, define the function $g \colon \mathcal{Y} \mapsto \mathcal{X}$ by $g(y) \triangleq x$ for all $y = f(x) \in \mathcal{R}(f)$ and by $g(y)$ arbitrary for all $y \in \mathcal{Y} \backslash \mathcal{R}(f)$. Consequently, $g[f(x)] = x$ for all $x \in \mathcal{X}$, which shows that g is a left inverse of f.

To prove $ii)$, suppose that f is right invertible with right inverse $g \colon \mathcal{Y} \mapsto \mathcal{X}$. Then, for all $y \in \mathcal{Y}$, it follows that $f[g(y)] = y$, which shows that f is onto. Conversely, suppose that f is onto so that, for all $y \in \mathcal{Y}$, there exists at least one $x \in \mathcal{X}$ such that $f(x) = y$. Selecting one such x arbitrarily, define $g \colon \mathcal{Y} \mapsto \mathcal{X}$ by $g(y) \triangleq x$. Consequently, $f[g(y)] = y$ for all $y \in \mathcal{Y}$, which shows that g is a right inverse of f. $\square$

1.3 Facts on Logic, Sets, and Functions

Fact 1.3.1. Let A and B be conditions. Then, the following statements hold:

  $i)$ $(A \text{ or } B) \iff [(\text{not } A) \implies B]$.

  $ii)$ $\text{not}(A \text{ or } B) \iff [(\text{not } A) \text{ and } (\text{not } B)]$.

  $iii)$ $\text{not}(A \text{ and } B) \iff (\text{not } A) \text{ or } (\text{not } B)$.

  $iv)$ $(A \implies B) \iff [(\text{not } A) \text{ or } B]$.

  $v)$ $[\text{not}(A \implies B)] \iff [A \text{ and } (\text{not } B)]$.

Fact 1.3.2. The following statements are equivalent:

$i)$ $A \implies (B \text{ or } C)$.

$ii)$ $[A \text{ and } (\text{not } B)] \implies C$.

Fact 1.3.3. The following statements are equivalent:

$i)$ $A \iff B$.

$ii)$ $[A \text{ or } (\text{not } B)]$ and $(\text{not } [A \text{ and } (\text{not } B)])$.

Fact 1.3.4. Let $\mathcal{A}, \mathcal{B}, \mathcal{C}$ be subsets of a set $\mathcal{X}$. Then, the following identities hold:

  $i)$ $\mathcal{A} \cap \mathcal{A} = \mathcal{A} \cup \mathcal{A} = \mathcal{A}$.

  $ii)$ $(\mathcal{A} \cup \mathcal{B})^{\sim} = \mathcal{A}^{\sim} \cap \mathcal{B}^{\sim}$.

  $iii)$ $\mathcal{A}^{\sim} \cup \mathcal{B}^{\sim} = (\mathcal{A} \cap \mathcal{B})^{\sim}$.

  $iv)$ $[\mathcal{A} \backslash (\mathcal{A} \cap \mathcal{B})] \cup \mathcal{B} = \mathcal{A} \cup \mathcal{B}$.

$v)$ $(\mathcal{A} \cup \mathcal{B}) \backslash (\mathcal{A} \cap \mathcal{B}) = (\mathcal{A} \cap \mathcal{B}^{\sim}) \cup (\mathcal{A}^{\sim} \cap \mathcal{B})$.

$vi)$ $\mathcal{A} \cap (\mathcal{B} \cup \mathcal{C}) = (\mathcal{A} \cap \mathcal{B}) \cup (\mathcal{A} \cap \mathcal{C})$.

$vii)$ $\mathcal{A} \cup (\mathcal{B} \cap \mathcal{C}) = (\mathcal{A} \cup \mathcal{B}) \cap (\mathcal{A} \cup \mathcal{C})$.

$viii)$ $(\mathcal{A} \cap \mathcal{B}) \backslash \mathcal{C} = (\mathcal{A} \backslash \mathcal{C}) \cap (\mathcal{B} \backslash \mathcal{C})$.

$ix)$ $(\mathcal{A} \cup \mathcal{B}) \backslash \mathcal{C} = (\mathcal{A} \backslash \mathcal{C}) \cup (\mathcal{B} \backslash \mathcal{C})$.

$x)$ $(\mathcal{A} \cup \mathcal{B}) \cap (\mathcal{A} \cup \mathcal{B}^{\sim}) = \mathcal{A}$.

$xi)$ $(\mathcal{A} \cup \mathcal{B}) \cap (\mathcal{A}^{\sim} \cup \mathcal{B}) \cap (\mathcal{A} \cup \mathcal{B}^{\sim}) = \mathcal{A} \cap \mathcal{B}$.

Fact 1.3.5. Let $(x_1, y_1), (x_2, y_2) \in \mathbb{R} \times \mathbb{R}$. Then, the relation $(x_1, y_1) \leq (x_2, y_2)$ defined by $x_1 \leq x_2$ and $y_1 \leq y_2$ is a partial ordering.

Fact 1.3.6. Let $f \colon \mathcal{X} \mapsto \mathcal{Y}$, and assume that f is invertible. Then,

$$(f^{-1})^{-1} = f.$$

Fact 1.3.7. Let $f \colon \mathcal{X} \mapsto \mathcal{Y}$ and $g \colon \mathcal{Y} \mapsto \mathcal{Z}$, and assume that f and g are invertible. Then, $g \bullet f$ is invertible and

$$(g \bullet f)^{-1} = f^{-1} \bullet g^{-1}.$$

Fact 1.3.8. Let $\mathcal{X}$ be a set, and let $\mathfrak{X}$ denote the class of subsets of $\mathcal{X}$. Then, "$\subset$" and "$\subseteq$" are transitive relations on $\mathfrak{X}$, and "$\subseteq$" is a partial ordering on $\mathfrak{X}$.

1.4 Facts on Scalar Inequalities

Fact 1.4.1. Let x be a positive number, and let α be a real number. If $\alpha \in [0, 1]$, then

$$x^{\alpha} \leq \alpha x + 1 - \alpha,$$

whereas, if either $\alpha \leq 0$ or $\alpha \geq 1$, then

$$\alpha x + 1 - \alpha \leq x^{\alpha}.$$

(Remark: A matrix version of this result is given by Fact 8.7.41.)

Fact 1.4.2. Let x be a positive number. Then,

$$1 - x^{-1} \leq \log x \leq x - 1.$$

Furthermore, equality holds if and only if $x = 1$.

Fact 1.4.3. Let x be a real number. Then,

$$2 - e^{-x} \leq x + 1 \leq e^{x}.$$

Furthermore, equality holds if and only if $x = 0$.

Fact 1.4.4. Let $x \geq -1$ and $p \geq 1$. Then,

$$1 + px \leq (1 + x)^p.$$

(Remark: This inequality is a generalization of *Bernoulli's inequality*.)

Fact 1.4.5. Let x and y be nonnegative numbers, and let $\alpha \in [0, 1]$. Then,

$$x^\alpha y^{1-\alpha} \leq \alpha x + (1 - \alpha)y.$$

(Remark: See Fact 8.10.14 and Fact 8.10.15.)

Fact 1.4.6. Let x and y be real numbers, and let $\alpha \in [0, 1]$. Then,

$$e^{\alpha x + (1-\alpha)y} \leq \alpha e^x + (1 - \alpha)e^y.$$

(Proof: Replace x and y by e^x and e^y, respectively, in Fact 1.4.5.) (Remark: This inequality is a convexity condition. See Definition 8.5.13 for the convexity of matrix-valued functions.)

Fact 1.4.7. Let x and y be nonnegative numbers, and let $p, q \in [1, \infty)$ satisfy $1/p + 1/q = 1$. Then,

$$xy \leq \frac{x^p}{p} + \frac{y^q}{q}.$$

(Remark: This result is *Young's inequality*. A matrix version is given by Fact 9.12.23.)

Fact 1.4.8. Let x and y be positive numbers, and let $0 \leq p \leq q$. Then,

$$\frac{x^p + y^p}{(xy)^{p/2}} \leq \frac{x^q + y^q}{(xy)^{q/2}}.$$

(Remark: This inequality is a monotonicity property. See Fact 8.7.31.)

Fact 1.4.9. Let x and y be distinct positive numbers, and let p and q be real numbers such that $p < q$. Then,

$$\left(\frac{x^p + y^p}{2}\right)^{1/p} < \left(\frac{x^q + y^q}{2}\right)^{1/q}.$$

(Proof: See [486].) (Remark: This result is a *power mean inequality*. Letting $q = 1$ and $p \to 0$ yields the arithmetic-mean-geometric-mean inequality $\sqrt{xy} \leq \frac{1}{2}(x + y)$.)

Fact 1.4.10. Let x and y be distinct positive numbers, and let $1/3 \leq p < 1 < q$. Then,

$$\sqrt{xy} < \frac{y - x}{\log y - \log x} < \left(\frac{x^p + y^p}{2}\right)^{1/p} < \frac{x + y}{2} < \left(\frac{x^q + y^q}{2}\right)^{1/q}.$$

(Proof: See [486].) (Remark: These inequalities are a refinement of the arithmetic-mean-geometric-mean inequality. Additional inequalities in n variables and related references are given in [798].)

Fact 1.4.11. Let $x_1, \ldots, x_n$ be nonnegative numbers. Then,

$$\left(\prod_{i=1}^{n} x_i \right)^{1/n} \leq \frac{1}{n} \sum_{i=1}^{n} x_i.$$

Furthermore, equality holds if and only if $x_1 = x_2 = \cdots = x_n$. (Remark: This result is the *arithmetic-mean-geometric-mean inequality*. Several proofs are given in [153]. Bounds for the difference between these quantities are given in [16, 167, 726].)

Fact 1.4.12. Let $x_1, \ldots, x_n$ be nonnegative numbers. Then,

$$\left(\sum_{i=1}^{n} x_i \right)^{2} \leq n \sum_{i=1}^{n} x_i^2.$$

Furthermore, equality holds if and only if $x_1, \ldots, x_n$ are equal. (Remark: The result is equivalent to i) of Fact 9.8.10 with $m = 1$.)

Fact 1.4.13. Let $x_1, \ldots, x_n$ be positive real numbers, let p be a real number, and define

$$M_p \triangleq \begin{cases} \left(\prod_{i=1}^{n} x_i \right)^{1/n}, & p = 0, \\ \left(\frac{1}{n} \sum_{i=1}^{n} x_i^p \right)^{1/p}, & p \neq 0. \end{cases}$$

Now, let p, q be real numbers such that $p \leq q$. Then,

$$M_p \leq M_q.$$

Furthermore, $p < q$ and at least two of the numbers $x_1, \ldots, x_n$ are distinct if and only if

$$M_p < M_q.$$

(Proof: See [151, p. 210] and [511, p. 105].) If p and q are nonzero and $p \leq q$, then

$$\left(\sum_{i=1}^{n} x_i^p \right)^{1/p} \leq \left(\tfrac{1}{n} \right)^{1/q - 1/p} \left(\sum_{i=1}^{n} x_i^q \right)^{1/q},$$

which is a reverse form of Fact 1.4.16. (Remark: This result is a *power mean inequality*. $M_0 \leq M_1$ is the arithmetic-mean-geometric-mean inequality given by Fact 1.4.11.)

Fact 1.4.14. Let $x_1, \ldots, x_n$ be nonnegative numbers, and let $\alpha_1, \ldots, \alpha_n$ be nonnegative numbers such that $\sum_{i=1}^{n} \alpha_i = 1$. Then,

$$\prod_{i=1}^{n} x_i^{\alpha_i} \le \sum_{i=1}^{n} \alpha_i x_i.$$

Furthermore, equality holds if and only if $x_1 = x_2 = \cdots = x_n$. (Remark: This result is the *weighted arithmetic-mean-geometric-mean* inequality.) (Proof: Since $f(x) = -\log x$ is convex, it follows that

$$\log \prod_{i=1}^{n} x_i^{\alpha_i} = \sum_{i=1}^{n} \alpha_i \log x_i \le \log \sum_{i=1}^{n} \alpha_i x_i.$$

To prove the second statement, define $f\colon [0, \infty)^n \mapsto [0, \infty)$ by $f(\mu_1, \ldots, \mu_n) \triangleq \sum_{i=1}^{n} \alpha_i \mu_i - \prod_{i=1}^{n} \mu_i^{\alpha_i}$. Note that $f(\mu, \ldots, \mu) = 0$ for all $\mu \ge 0$. If $x_1, \ldots, x_n$ minimizes f, then $\partial f / \partial \mu_i(x_1, \ldots, x_n) = 0$ for all $i = 1, \ldots, n$, which implies that $x_1 = x_2 = \cdots = x_n$.)

Fact 1.4.15. Let $x_1, \ldots, x_n$ be nonnegative numbers. Then,

$$1 + \left(\prod_{i=1}^{n} x_i \right)^{1/n} \le \left[\prod_{i=1}^{n} (1 + x_i) \right]^{1/n}.$$

Furthermore, equality holds if and only if $x_1 = x_2 = \cdots = x_n$. (Proof: Use Fact 1.4.11.) (Remark: This inequality is used to prove Corollary 8.4.15.)

Fact 1.4.16. Let $x_1, \ldots, x_n$ be nonnegative real numbers, and let p, q be positive real numbers such that $p \le q$. Then,

$$\left(\sum_{i=1}^{n} x_i^q \right)^{1/q} \le \left(\sum_{i=1}^{n} x_i^p \right)^{1/p}.$$

Furthermore, the inequality is strict if and only if $p < q$ and at least two of the numbers $x_1, \ldots, x_n$ are nonzero. (Proof: See Proposition 9.1.5.) (Remark: This result is a *power-sum inequality* or *Jensen's inequality*. See [151, p. 213]. The result implies that the Hölder norm is a monotonic function of the exponent.)

Fact 1.4.17. Let a and b be positive real numbers, and let $p \in [1, \infty)$. Then,

$$a^p + b^p \le (a + b)^p.$$

(Proof: Set $p = 1$ and $n = 2$ in Fact 1.4.16.)

Fact 1.4.18. Let $0 < x_1 < \cdots < x_n$, and let $\alpha_1, \ldots, \alpha_n \geq 0$ satisfy $\sum_{i=1}^{n} \alpha_i = 1$. Then,

$$\left(\sum_{i=1}^{n} \alpha_i x_i \right) \left(\sum_{i=1}^{n} \frac{\alpha_i}{x_i} \right) \leq \frac{(x_1 + x_n)^2}{4 x_1 x_n}.$$

(Remark: This result is the *Kantorovich inequality*. See Fact 8.12.5 and [491].)

Fact 1.4.19. Let $x_1, \ldots, x_n$ and $y_1, \ldots, y_n$ be nonnegative numbers. Then,

$$\sum_{i=1}^{n} x_i y_i \leq \left(\sum_{i=1}^{n} x_i^2 \right)^{1/2} \left(\sum_{i=1}^{n} y_i^2 \right)^{1/2}.$$

Furthermore, equality holds if and only if $\begin{bmatrix} x_1 & \cdots & x_n \end{bmatrix}^{\mathrm{T}}$ and $\begin{bmatrix} y_1 & \cdots & y_n \end{bmatrix}^{\mathrm{T}}$ are linearly dependent. (Remark: This is the *Cauchy-Schwarz inequality*.)

Fact 1.4.20. Let $x_1, \ldots, x_n$ and $y_1, \ldots, y_n$ be nonnegative numbers, and let $\alpha \in [0, 1]$. Then,

$$\sum_{i=1}^{n} x_i^\alpha y_i^{1-\alpha} \leq \left(\sum_{i=1}^{n} x_i \right)^{\alpha} \left(\sum_{i=1}^{n} y_i \right)^{1-\alpha}.$$

Now, let $p, q \in [1, \infty]$ satisfy $1/p + 1/q = 1$. Then, equivalently,

$$\sum_{i=1}^{n} x_i y_i \leq \left(\sum_{i=1}^{n} x_i^p \right)^{1/p} \left(\sum_{i=1}^{n} y_i^q \right)^{1/q}.$$

Furthermore, equality holds if and only if $\begin{bmatrix} x_1^p & \cdots & x_n^p \end{bmatrix}^{\mathrm{T}}$ and $\begin{bmatrix} y_1^q & \cdots & y_n^q \end{bmatrix}^{\mathrm{T}}$ are linearly dependent. (Remark: This result is *Hölder's inequality*.) (Remark: Note the relationship between the *conjugate parameters* p, q and the *barycentric coordinates* $\alpha, 1 - \alpha$. See Fact 8.16.37.)

Fact 1.4.21. Let $x_1, \ldots, x_n$ and $y_1, \ldots, y_n$ be nonnegative numbers. If $p \in (0, 1]$, then

$$\left[\sum_{i=1}^{n} (x_i + y_i)^p \right]^{1/p} \geq \left(\sum_{i=1}^{n} x_i^p \right)^{1/p} + \left(\sum_{i=1}^{n} y_i^p \right)^{1/p}.$$

If $p \geq 1$, then

$$\left[\sum_{i=1}^{n} (x_i + y_i)^p \right]^{1/p} \leq \left(\sum_{i=1}^{n} x_i^p \right)^{1/p} + \left(\sum_{i=1}^{n} y_i^p \right)^{1/p}.$$

Furthermore, equality holds if and only if either $p = 1$ or $\begin{bmatrix} x_1 & \cdots & x_n \end{bmatrix}^{\mathrm{T}}$ and $\begin{bmatrix} y_1 & \cdots & y_n \end{bmatrix}^{\mathrm{T}}$ are linearly dependent. (Remark: This result is *Minkowski's inequality*.)

Fact 1.4.22. Let x and y be real numbers. If $k \in \mathbb{P}$, then

$$|x - y|^{2k+1} \leq 2^{2n}|x^{2k+1} - y^{2k+1}|.$$

Now, assume that x and y are nonnegative. If $r \geq 1$, then

$$|x - y|^r \leq |x^r - y^r|.$$

(Proof: See [360].) (Remark: Matrix versions of these results are given in [360].)

Fact 1.4.23. Let z be a complex scalar with complex conjugate $\bar{z}$, real part $\operatorname{Re} z$, and imaginary part $\operatorname{Im} z$. Then, the following statements hold:

i) $|\operatorname{Re} z| \leq |z|$.

ii) If $z \neq 0$, then $z^{-1} = \bar{z}/|z|^2$.

iii) If $z \neq 0$, then $\operatorname{Re} z^{-1} = (\operatorname{Re} z)/|z|^2$.

iv) If $|z| = 1$, then $z^{-1} = \bar{z}$.

v) If $\operatorname{Re} z \neq 0$ and $\operatorname{Re} z^{-1} \neq 0$, then $|z| = \sqrt{(\operatorname{Re} z)/(\operatorname{Re} z^{-1})}$.

vi) $|z^2| = |z|^2 = z\bar{z}$.

vii) $z^2 + \bar{z}^2 + 4(\operatorname{Im} z)^2 = 2|z|^2$.

viii) $z^2 + \bar{z}^2 + 2|z|^2 = 4(\operatorname{Re} z)^2$.

ix) $|z^2 + \bar{z}^2| \leq 2|z|^2$.

x) $|e^z| \leq e^{|z|}$.

Now, let z_1 and z_2 be complex scalars. Then, the following statements hold:

xi) $|z_1 z_2| = |z_1||z_2|$.

xii) $|z_1 + z_2| \leq |z_1| + |z_2|$.

xiii) $|z_1 + z_2| = |z_1| + |z_2|$ if and only if there exists $\alpha \geq 0$ such that either $z_1 = \alpha z_2$ or $z_2 = \alpha z_1$, that is, if and only if z_1 and z_2 have the same phase angle.

xiv) If $p \geq 2$, then

$$2(|z_1|^p + |z_2|^p) \leq |z_1 + z_2|^p + |z_1 - z_2|^p.$$

Finally, if θ_1 and θ_2 are real numbers, then

xv) $|e^{j\theta_1} - e^{j\theta_2}| \leq |\theta_1 - \theta_2|$.

(Remark: Matrix versions of some of these results are given in [697]. Result *xiv)* is due to Clarkson. See [360]. A matrix version of *xv)* is given by Fact 11.14.8.)

1.5 Notes

Most of the preliminary material in this chapter can be found in [555]. A related treatment of mathematical preliminaries is given in [613]. Reference works on inequalities include [87, 151–153, 188, 511, 516, 543, 658]. See also [282, pp. 194, 195]. In the mathematical literature, alternative terminology for "one-to-one" and "onto" is *injective* and *surjective*, respectively, while a function that is injective and surjective is *bijective*. Recommended texts on complex variables include [556, 572].

Chapter Two

Basic Matrix Properties

In this chapter we provide a detailed treatment of the basic properties of matrices such as range, null space, rank, and invertibility. We also consider properties of convex sets, cones, and subspaces.

2.1 Matrix Algebra

The symbols $\mathbb{Z}$, $\mathbb{N}$, and $\mathbb{P}$ denote the sets of integers, nonnegative integers, and positive integers, respectively. The symbols $\mathbb{R}$ and $\mathbb{C}$ denote the real and complex number fields, respectively, whose elements are *scalars*. Since $\mathbb{R}$ is a proper subset of $\mathbb{C}$, we state many results for $\mathbb{C}$. In other cases, we treat $\mathbb{R}$ and $\mathbb{C}$ separately. To do this efficiently, we use the symbol $\mathbb{F}$ to consistently denote either $\mathbb{R}$ or $\mathbb{C}$.

Let $x \in \mathbb{C}$. Then, $x = y + \jmath z$, where $y, z \in \mathbb{R}$ and $\jmath \triangleq \sqrt{-1}$. Define the *complex conjugate* $\overline{x}$ of x by

$$\overline{x} \triangleq y - \jmath z \tag{2.1.1}$$

and the real part $\operatorname{Re} x$ of x and the imaginary part $\operatorname{Im} x$ of x by

$$\operatorname{Re} x \triangleq \tfrac{1}{2}(x + \overline{x}) = y \tag{2.1.2}$$

and

$$\operatorname{Im} x \triangleq \tfrac{1}{2\jmath}(x - \overline{x}) = z. \tag{2.1.3}$$

Furthermore, the *absolute value* $|x|$ of x is defined by

$$|x| \triangleq \sqrt{y^2 + z^2}. \tag{2.1.4}$$

The *closed left half plane* (CLHP), *open left half plane* (OLHP), *closed right half plane* (CRHP), and *open right half plane* (ORHP) are the subsets of $\mathbb{C}$ defined by

$$\mathrm{CLHP} \triangleq \{s \in \mathbb{C}: \ \mathrm{Re}\, s \leq 0\}, \tag{2.1.5}$$

$$\mathrm{OLHP} \triangleq \{s \in \mathbb{C}: \ \mathrm{Re}\, s < 0\}, \tag{2.1.6}$$

$$\mathrm{CRHP} \triangleq \{s \in \mathbb{C}: \ \mathrm{Re}\, s \geq 0\}, \tag{2.1.7}$$

$$\mathrm{ORHP} \triangleq \{s \in \mathbb{C}: \ \mathrm{Re}\, s > 0\}. \tag{2.1.8}$$

The imaginary numbers are represented by $\jmath\mathbb{R}$. Note that 0 is both a real number and an imaginary number.

The set $\mathbb{F}^n$ consists of *vectors* x of the form

$$x = \begin{bmatrix} x_{(1)} \\ \vdots \\ x_{(n)} \end{bmatrix}, \tag{2.1.9}$$

where $x_{(1)}, \ldots, x_{(n)} \in \mathbb{F}$ are the *components* of x. Hence, the elements of $\mathbb{F}^n$ are *column vectors*. Since $\mathbb{F}^1 = \mathbb{F}$, it follows that every scalar is also a vector. If $x \in \mathbb{R}^n$ and every component of x is nonnegative, then x is *nonnegative*, while, if every component of x is positive, then x is *positive*.

Definition 2.1.1. Let $x, y \in \mathbb{R}^n$, and assume that $x_{(1)} \geq \cdots \geq x_{(n)}$ and $y_{(1)} \geq \cdots \geq y_{(n)}$. Then, the following terminology is defined:

i) y *weakly majorizes* x if, for all $k = 1, \ldots, n$,

$$\sum_{i=1}^{k} x_{(i)} \leq \sum_{i=1}^{k} y_{(i)}. \tag{2.1.10}$$

ii) y *strongly majorizes* x if y weakly majorizes x and

$$\sum_{i=1}^{n} x_{(i)} = \sum_{i=1}^{n} y_{(i)}. \tag{2.1.11}$$

Now, assume that x and y are nonnegative. Then, the following terminology is defined:

iii) y *weakly log majorizes* x if, for all $k = 1, \ldots, n$,

$$\prod_{i=1}^{k} x_{(i)} \leq \prod_{i=1}^{k} y_{(i)}. \tag{2.1.12}$$

iv) y *strongly log majorizes* x if y weakly log majorizes x and

$$\prod_{i=1}^{n} x_{(i)} = \prod_{i=1}^{n} y_{(i)}. \tag{2.1.13}$$

If $\alpha \in \mathbb{F}$ and $x \in \mathbb{F}^n$, then $\alpha x \in \mathbb{F}^n$ is given by

$$\alpha x = \begin{bmatrix} \alpha x_{(1)} \\ \vdots \\ \alpha x_{(n)} \end{bmatrix}. \tag{2.1.14}$$

If $x, y \in \mathbb{F}^n$, then x and y are *linearly dependent* if there exists $\alpha \in \mathbb{F}$ such that either $x = \alpha y$ or $y = \alpha x$. Linear dependence for a set of two or more vectors is defined in Section 2.3. Furthermore, vectors add component by component, that is, if $x, y \in \mathbb{F}^n$, then

$$x + y = \begin{bmatrix} x_{(1)} + y_{(1)} \\ \vdots \\ x_{(n)} + y_{(n)} \end{bmatrix}. \tag{2.1.15}$$

Thus, if $\alpha, \beta \in \mathbb{F}$, then the *linear combination* $\alpha x + \beta y$ is given by

$$\alpha x + \beta y = \begin{bmatrix} \alpha x_{(1)} + \beta y_{(1)} \\ \vdots \\ \alpha x_{(n)} + \beta y_{(n)} \end{bmatrix}. \tag{2.1.16}$$

If $x \in \mathbb{R}^n$ and x is nonnegative, then we write $x \geq\geq 0$, while, if x is positive, then we write $x >> 0$. If $x, y \in \mathbb{R}^n$, then $x \geq\geq y$ means that $x - y \geq\geq 0$, while $x >> y$ means that $x - y >> 0$.

The vectors $x_1, \ldots, x_m \in \mathbb{F}^n$ placed side by side form the *matrix*

$$A \triangleq \begin{bmatrix} x_1 & \cdots & x_m \end{bmatrix}, \tag{2.1.17}$$

which has n *rows* and m *columns*. The components of the vectors $x_1, \ldots, x_m$ are the *entries* of A. We write $A \in \mathbb{F}^{n \times m}$ and say that A has *size* $n \times m$. Since $\mathbb{F}^n = \mathbb{F}^{n \times 1}$, it follows that every vector is also a matrix. Note that $\mathbb{F}^{1 \times 1} = \mathbb{F}^1 = \mathbb{F}$. If $n = m$, then n is the *order* of A, and A is *square*. The ith row of A and the jth column of A are denoted by $\mathrm{row}_i(A)$ and $\mathrm{col}_j(A)$, respectively. Hence,

$$A = \begin{bmatrix} \mathrm{row}_1(A) \\ \vdots \\ \mathrm{row}_n(A) \end{bmatrix} = \begin{bmatrix} \mathrm{col}_1(A) & \cdots & \mathrm{col}_m(A) \end{bmatrix}. \tag{2.1.18}$$

The entry $x_{j(i)}$ of A in both the ith row of A and the jth column of A is denoted by $A_{(i,j)}$. Therefore, $x \in \mathbb{F}^n$ can be written as

$$x = \begin{bmatrix} x_{(1)} \\ \vdots \\ x_{(n)} \end{bmatrix} = \begin{bmatrix} x_{(1,1)} \\ \vdots \\ x_{(n,1)} \end{bmatrix}. \tag{2.1.19}$$

Let $A \in \mathbb{F}^{n \times m}$. For $b \in \mathbb{F}^n$, the matrix obtained from A by replacing $\text{col}_i(A)$ with b is denoted by

$$A \overset{i}{\leftarrow} b. \qquad (2.1.20)$$

Likewise, for $b \in \mathbb{F}^{1 \times m}$, the matrix obtained from A by replacing $\text{row}_i(A)$ with b is denoted by (2.1.20).

Let $A \in \mathbb{F}^{n \times m}$, and let $l \triangleq \min\{n, m\}$. Then, the entries $A_{(i,i)}$ for all $i = 1, \ldots, l$ and $A_{(i,j)}$ for all $i \neq j$ are the *diagonal entries* and *off-diagonal entries* of A, respectively. Moreover, for all $i = 1, \ldots, l - 1$, the entries $A_{(i,i+1)}$ and $A_{(i+1,i)}$ are the *superdiagonal entries* and *subdiagonal entries* of A, respectively. In addition, the entries $A_{(i,l+1-i)}$ for all $i = 1, \ldots, l$ are the *reverse-diagonal entries* of A. If the diagonal entries $A_{(1,1)}, \ldots, A_{(l,l)}$ of A are real, then the diagonal entries of A are labeled from largest to smallest as

$$d_1(A) \geq \cdots \geq d_l(A), \qquad (2.1.21)$$

and we define

$$d_{\max}(A) \triangleq d_1(A), \quad d_{\min}(A) \triangleq d_l(A). \qquad (2.1.22)$$

Partitioned matrices are of the form

$$\begin{bmatrix} A_{11} & \cdots & A_{1l} \\ \vdots & \ddots & \vdots \\ A_{k1} & \cdots & A_{kl} \end{bmatrix}, \qquad (2.1.23)$$

where, for all $i = 1, \ldots, k$ and $j = 1, \ldots, l$, the *block* A_{ij} of A is a matrix of size $n_i \times m_j$. If $n_i = m_j$ and the diagonal entries of A_{ij} lie on the diagonal of A, then the square matrix A_{ij} is a *diagonally located block*; otherwise, A_{ij} is an *off-diagonally located block*.

Matrices of the same size add entry by entry, that is, if $A, B \in \mathbb{F}^{n \times m}$, then, for all $i = 1, \ldots, n$ and $j = 1, \ldots, m$, $(A + B)_{(i,j)} = A_{(i,j)} + B_{(i,j)}$. Furthermore, for all $i = 1, \ldots, n$ and $j = 1, \ldots, m$, $(\alpha A)_{(i,j)} = \alpha A_{(i,j)}$ for all $\alpha \in \mathbb{F}$ so that $(\alpha A + \beta B)_{(i,j)} = \alpha A_{(i,j)} + \beta B_{(i,j)}$ for all $\alpha, \beta \in \mathbb{F}$. If $A, B \in \mathbb{F}^{n \times m}$, then A and B are *linearly dependent* if there exists $\alpha \in \mathbb{F}$ such that either $A = \alpha B$ or $B = \alpha A$.

Let $A \in \mathbb{R}^{n \times m}$. If every entry of A is nonnegative, then A is *nonnegative*, which is written as $A \geq\geq 0$. If every entry of A is positive, then A is *positive*, which is written as $A >> 0$. If $A, B \in \mathbb{R}^{n \times m}$, then $A \geq\geq B$ means that $A - B \geq\geq 0$, while $A >> B$ means that $A - B >> 0$.

Let $z \in \mathbb{F}^{1 \times n}$ and $y \in \mathbb{F}^n = \mathbb{F}^{n \times 1}$. Then, the scalar $zy \in \mathbb{F}$ is defined by

$$zy \triangleq \sum_{i=1}^{n} z_{(1,i)} y_{(i)}. \tag{2.1.24}$$

Now, let $A \in \mathbb{F}^{n \times m}$ and $x \in \mathbb{F}^m$. Then, the matrix-vector product Ax is defined by

$$Ax \triangleq \begin{bmatrix} \mathrm{row}_1(A)x \\ \vdots \\ \mathrm{row}_n(A)x \end{bmatrix}. \tag{2.1.25}$$

It can be seen that Ax is a linear combination of the columns of A, that is,

$$Ax = \sum_{i=1}^{m} x_{(i)} \mathrm{col}_i(A). \tag{2.1.26}$$

The matrix A can be associated with the function $f\colon \mathbb{F}^m \mapsto \mathbb{F}^n$ defined by $f(x) \triangleq Ax$ for all $x \in \mathbb{F}^m$. The function $f\colon \mathbb{F}^m \mapsto \mathbb{F}^n$ is *linear* since, for all $\alpha, \beta \in \mathbb{F}$ and $x, y \in \mathbb{F}^m$, it follows that

$$f(\alpha x + \beta y) = \alpha Ax + \beta Ay. \tag{2.1.27}$$

The function $f\colon \mathbb{F}^m \mapsto \mathbb{F}^n$ defined by

$$f(x) \triangleq Ax + z, \tag{2.1.28}$$

where $z \in \mathbb{F}^n$, is *affine*.

Theorem 2.1.2. Let $A \in \mathbb{F}^{n \times m}$ and $B \in \mathbb{F}^{m \times l}$, and define $f\colon \mathbb{F}^m \mapsto \mathbb{F}^n$ and $g\colon \mathbb{F}^l \mapsto \mathbb{F}^m$ by $f(x) \triangleq Ax$ and $g(y) \triangleq By$. Furthermore, define the composition $h \triangleq f \bullet g\colon \mathbb{F}^l \mapsto \mathbb{F}^n$. Then, for all $y \in \mathbb{R}^l$,

$$h(y) = (AB)y, \tag{2.1.29}$$

where, for all $i = 1, \ldots, n$ and $j = 1, \ldots, l$, $AB \in \mathbb{F}^{n \times l}$ is defined by

$$(AB)_{(i,j)} \triangleq \sum_{k=1}^{m} A_{(i,k)} B_{(k,j)}. \tag{2.1.30}$$

Let $A \in \mathbb{F}^{n \times m}$ and $B \in \mathbb{F}^{m \times l}$. Then, $AB \in \mathbb{F}^{n \times l}$ is the *product* of A and B. The matrices A and B are *conformable*, and the product (2.1.30) defines *matrix multiplication*.

Let $A \in \mathbb{F}^{n \times m}$ and $B \in \mathbb{F}^{m \times l}$. Then, AB can be written as

$$AB = \begin{bmatrix} A\mathrm{col}_1(B) & \cdots & A\mathrm{col}_l(B) \end{bmatrix} = \begin{bmatrix} \mathrm{row}_1(A)B \\ \vdots \\ \mathrm{row}_n(A)B \end{bmatrix}. \tag{2.1.31}$$

Thus, for all $i = 1, \ldots, n$ and $j = 1, \ldots, l$,

$$(AB)_{(i,j)} = \text{row}_i(A)\text{col}_j(B), \tag{2.1.32}$$

$$\text{col}_j(AB) = A\text{col}_j(B), \tag{2.1.33}$$

$$\text{row}_i(AB) = \text{row}_i(A)B. \tag{2.1.34}$$

The expression (2.1.29) for the composite function $h \triangleq f \bullet g$ in Theorem 2.1.2 implies that $h(y) = f[g(y)] = A(By) = (AB)y$. More generally, for conformable matrices A, B, C, the associative and distributive identities

$$(AB)C = A(BC), \tag{2.1.35}$$

$$A(B + C) = AB + AC, \tag{2.1.36}$$

$$(A + B)C = AC + BC \tag{2.1.37}$$

are valid. Hence, we write ABC for $(AB)C$ and $A(BC)$.

Let $A, B \in \mathbb{F}^{n \times n}$. Then, the *commutator* $[A, B] \in \mathbb{F}^{n \times n}$ of A and B is the matrix

$$[A, B] \triangleq AB - BA. \tag{2.1.38}$$

The *adjoint operator* $\text{ad}_A \colon \mathbb{F}^{n \times n} \mapsto \mathbb{F}^{n \times n}$ is defined by

$$\text{ad}_A(X) \triangleq [A, X]. \tag{2.1.39}$$

Let $x, y \in \mathbb{R}^3$. Then, the *cross product* $x \times y \in \mathbb{R}^3$ of x and y is defined by

$$x \times y \triangleq \begin{bmatrix} x_{(2)}y_{(3)} - x_{(3)}y_{(2)} \\ x_{(3)}y_{(1)} - x_{(1)}y_{(3)} \\ x_{(1)}y_{(2)} - x_{(2)}y_{(1)} \end{bmatrix}. \tag{2.1.40}$$

Furthermore, the 3×3 *cross-product matrix* is defined by

$$K(x) \triangleq \begin{bmatrix} 0 & -x_{(3)} & x_{(2)} \\ x_{(3)} & 0 & -x_{(1)} \\ -x_{(2)} & x_{(1)} & 0 \end{bmatrix}. \tag{2.1.41}$$

Note that

$$x \times y = K(x)y. \tag{2.1.42}$$

Multiplication of partitioned matrices is analogous to matrix multiplication with scalar entries. For example, for matrices with conformable blocks,

$$\begin{bmatrix} A & B \end{bmatrix} \begin{bmatrix} C \\ D \end{bmatrix} = AC + BD, \tag{2.1.43}$$

$$\left[\begin{array}{c} A \\ B \end{array}\right] C = \left[\begin{array}{c} AC \\ BC \end{array}\right], \tag{2.1.44}$$

$$\left[\begin{array}{c} A \\ B \end{array}\right]\left[\begin{array}{cc} C & D \end{array}\right] = \left[\begin{array}{cc} AC & AD \\ BC & BD \end{array}\right], \tag{2.1.45}$$

$$\left[\begin{array}{cc} A & B \\ C & D \end{array}\right]\left[\begin{array}{cc} E & F \\ G & H \end{array}\right] = \left[\begin{array}{cc} AE+BG & AF+BH \\ CE+DG & CF+DH \end{array}\right]. \tag{2.1.46}$$

The $n \times m$ *zero matrix*, all of whose entries are zero, is written as $0_{n \times m}$. If the dimensions are unambiguous, then we write just 0. Let $x \in \mathbb{F}^m$ and $A \in \mathbb{F}^{n \times m}$. Then, the zero matrix satisfies

$$0_{k \times m} x = 0_k, \tag{2.1.47}$$
$$A 0_{m \times l} = 0_{n \times l}, \tag{2.1.48}$$
$$0_{k \times n} A = 0_{k \times m}. \tag{2.1.49}$$

Another special matrix is the *empty matrix*. For $n \in \mathbb{N}$, the $0 \times n$ empty matrix, which is written as $0_{0 \times n}$, has zero rows and n columns, while the $n \times 0$ empty matrix, which is written as $0_{n \times 0}$, has n rows and zero columns. For $A \in \mathbb{F}^{n \times m}$, where $n, m \in \mathbb{N}$, the empty matrix satisfies the multiplication rules

$$0_{0 \times n} A = 0_{0 \times m} \tag{2.1.50}$$

and

$$A 0_{m \times 0} = 0_{n \times 0}. \tag{2.1.51}$$

Although empty matrices have no entries, it is useful to define the product

$$0_{n \times 0} 0_{0 \times m} \triangleq 0_{n \times m}. \tag{2.1.52}$$

Also, we define

$$I_0 \triangleq \hat{I}_0 \triangleq 0_{0 \times 0}. \tag{2.1.53}$$

For $n, m \in \mathbb{N}$, we define $\mathbb{F}^{0 \times m} \triangleq \{0_{0 \times m}\}$, $\mathbb{F}^{n \times 0} \triangleq \{0_{n \times 0}\}$, and $\mathbb{F}^0 \triangleq \mathbb{F}^{0 \times 1}$. The empty matrix can be viewed as a useful device for matrices just as 0 is for real numbers and $\varnothing$ is for sets.

The $n \times n$ *identity matrix*, which has ones on the diagonal and zeros elsewhere, is denoted by I_n or just I. Let $x \in \mathbb{F}^n$ and $A \in \mathbb{F}^{n \times m}$. Then, the identity matrix satisfies

$$I_n x = x \tag{2.1.54}$$

and

$$A I_m = I_n A = A. \tag{2.1.55}$$

Let $A \in \mathbb{F}^{n \times n}$. Then, $A^2 \triangleq AA$ and, for all $k \in \mathbb{P}$, $A^k \triangleq A A^{k-1}$. We use the convention $A^0 \triangleq I$ even if A is the zero matrix.

The $n \times n$ *reverse identity matrix*, which has ones on the reverse diagonal and zeros elsewhere, is denoted by $\hat{I}_n$ or just $\hat{I}$. Left multiplication of $A \in \mathbb{F}^{n \times m}$ by $\hat{I}_n$ reverses the rows of A, while right multiplication of A by $\hat{I}_m$ reverses the columns of A. Note that

$$\hat{I}_n^2 = I_n. \tag{2.1.56}$$

2.2 Transpose and Inner Product

A fundamental vector and matrix operation is the transpose. If $x \in \mathbb{F}^n$, then the *transpose* x^{T} of x is defined to be the row vector

$$x^{\mathrm{T}} \triangleq \left[\begin{array}{ccc} x_{(1)} & \cdots & x_{(n)} \end{array} \right] \in \mathbb{F}^{1 \times n}. \tag{2.2.1}$$

Similarly, if $x = \left[\begin{array}{ccc} x_{(1,1)} & \cdots & x_{(1,n)} \end{array} \right] \in \mathbb{F}^{1 \times n}$, then

$$x^{\mathrm{T}} = \left[\begin{array}{c} x_{(1,1)} \\ \vdots \\ x_{(1,n)} \end{array} \right] \in \mathbb{F}^{n \times 1}. \tag{2.2.2}$$

Let $x, y \in \mathbb{F}^n$. Then, $x^{\mathrm{T}} y \in \mathbb{F}$ is a scalar, and

$$x^{\mathrm{T}} y = y^{\mathrm{T}} x = \sum_{i=1}^{n} x_{(i)} y_{(i)}. \tag{2.2.3}$$

Note that

$$x^{\mathrm{T}} x = \sum_{i=1}^{n} x_{(i)}^2. \tag{2.2.4}$$

The vector $e_{i,n} \in \mathbb{R}^n$, or just e_i, has 1 as its ith component and zeros elsewhere. Thus,

$$e_{i,n} = \mathrm{col}_i(I_n). \tag{2.2.5}$$

Let $A \in \mathbb{F}^{n \times m}$. Then, $e_i^{\mathrm{T}} A = \mathrm{row}_i(A)$ and $A e_i = \mathrm{col}_i(A)$. Furthermore, the (i,j) entry of A can be written as

$$A_{(i,j)} = e_i^{\mathrm{T}} A e_j. \tag{2.2.6}$$

The $n \times m$ matrix $E_{i,j,n \times m} \in \mathbb{R}^{n \times m}$, or just $E_{i,j}$, has 1 as its (i,j) entry and zeros elsewhere. Thus,

$$E_{i,j,n \times m} = e_{i,n} e_{j,m}^{\mathrm{T}}. \tag{2.2.7}$$

Note that $E_{i,1,n \times 1} = e_{i,n}$ and

$$I_n = E_{1,1} + \cdots + E_{n,n} = \sum_{i=1}^{n} e_i e_i^{\mathrm{T}}. \tag{2.2.8}$$

Finally, the $n \times m$ *ones matrix*, all of whose entries are 1, is written as $1_{n \times m}$ or just 1. Thus,

$$1_{n \times m} = \sum_{i,j=1}^{n,m} E_{i,j,n \times m}. \tag{2.2.9}$$

Note that

$$1_{n \times 1} = \sum_{i=1}^{n} e_{i,n} = \begin{bmatrix} 1 \\ \vdots \\ 1 \end{bmatrix} \tag{2.2.10}$$

and

$$1_{n \times m} = 1_{n \times 1} 1_{1 \times m}. \tag{2.2.11}$$

Lemma 2.2.1. Let $x \in \mathbb{R}$. Then, $x^{\mathrm{T}} x = 0$ if and only if $x = 0$.

Let $x, y \in \mathbb{R}^n$. Then, $x^{\mathrm{T}} y \in \mathbb{R}$ is the *inner product* of x and y. Furthermore, x is *orthogonal* to y if $x^{\mathrm{T}} y = 0$.

Let $x \in \mathbb{C}^n$. Then, $x = y + \jmath z$, where $y, z \in \mathbb{R}^n$. Therefore, the transpose x^{T} of x is given by

$$x^{\mathrm{T}} = y^{\mathrm{T}} + \jmath z^{\mathrm{T}}. \tag{2.2.12}$$

The *complex conjugate* $\bar{x}$ of x is defined by

$$\bar{x} \triangleq y - \jmath z, \tag{2.2.13}$$

while the *complex conjugate transpose* x^* of x is defined by

$$x^* \triangleq \bar{x}^{\mathrm{T}} = y^{\mathrm{T}} - \jmath z^{\mathrm{T}}. \tag{2.2.14}$$

The vectors y and z are the *real* and *imaginary* parts $\operatorname{Re} x$ and $\operatorname{Im} x$ of x, respectively, which are defined by

$$\operatorname{Re} x \triangleq \tfrac{1}{2}(x + \bar{x}) = y \tag{2.2.15}$$

and

$$\operatorname{Im} x \triangleq \tfrac{1}{2\jmath}(x - \bar{x}) = z. \tag{2.2.16}$$

Note that

$$x^* x = \sum_{i=1}^{n} \bar{x}_{(i)} x_{(i)} = \sum_{i=1}^{n} |x_{(i)}|^2 = \sum_{i=1}^{n} \left[y_{(i)}^2 + z_{(i)}^2 \right]. \tag{2.2.17}$$

If $w, x \in \mathbb{C}^n$, then $w^{\mathrm{T}} x = x^{\mathrm{T}} w$.

Lemma 2.2.2. Let $x \in \mathbb{C}^n$. Then, $x^* x = 0$ if and only if $x = 0$.

Let $x, y \in \mathbb{C}^n$. Then, $x^* y \in \mathbb{C}$ is the *inner product* of x and y, which is given by

$$x^* y = \sum_{i=1}^{n} \bar{x}_{(i)} y_{(i)}. \tag{2.2.18}$$

Furthermore, x is *orthogonal* to y if $x^*y = 0$.

Let $A \in \mathbb{F}^{n \times m}$. Then, the *transpose* $A^{\mathrm{T}} \in \mathbb{F}^{m \times n}$ of A is defined by

$$A^{\mathrm{T}} \triangleq \left[\begin{array}{ccc} [\mathrm{row}_1(A)]^{\mathrm{T}} & \cdots & [\mathrm{row}_n(A)]^{\mathrm{T}} \end{array} \right] = \left[\begin{array}{c} [\mathrm{col}_1(A)]^{\mathrm{T}} \\ \vdots \\ [\mathrm{col}_m(A)]^{\mathrm{T}} \end{array} \right], \qquad (2.2.19)$$

that is, $\mathrm{col}_i\left(A^{\mathrm{T}}\right) = [\mathrm{row}_i(A)]^{\mathrm{T}}$ for all $i = 1, \ldots, n$ and $\mathrm{row}_i\left(A^{\mathrm{T}}\right) = [\mathrm{col}_i(A)]^{\mathrm{T}}$ for all $i = 1, \ldots, m$. Hence, $\left(A^{\mathrm{T}}\right)_{(i,j)} = A_{(j,i)}$ and $\left(A^{\mathrm{T}}\right)^{\mathrm{T}} = A$. If $B \in \mathbb{F}^{m \times l}$, then

$$(AB)^{\mathrm{T}} = B^{\mathrm{T}}A^{\mathrm{T}}. \qquad (2.2.20)$$

In particular, if $x \in \mathbb{F}^m$, then

$$(Ax)^{\mathrm{T}} = x^{\mathrm{T}}A^{\mathrm{T}}, \qquad (2.2.21)$$

while, if, in addition, $y \in \mathbb{F}^n$, then $y^{\mathrm{T}}Ax$ is a scalar and

$$y^{\mathrm{T}}Ax = \left(y^{\mathrm{T}}Ax\right)^{\mathrm{T}} = x^{\mathrm{T}}A^{\mathrm{T}}y. \qquad (2.2.22)$$

If $B \in \mathbb{F}^{n \times m}$, then, for all $\alpha, \beta \in \mathbb{F}$,

$$(\alpha A + \beta B)^{\mathrm{T}} = \alpha A^{\mathrm{T}} + \beta B^{\mathrm{T}}. \qquad (2.2.23)$$

Let $x \in \mathbb{F}^n$ and $y \in \mathbb{F}^m$. Then, the matrix $xy^{\mathrm{T}} \in \mathbb{F}^{n \times m}$ is the *outer product* of x and y. The outer product xy^{T} is nonzero if and only if both x and y are nonzero.

The *trace* of a square matrix $A \in \mathbb{F}^{n \times n}$, denoted by $\mathrm{tr}\, A$, is defined to be the sum of its diagonal entries, that is,

$$\mathrm{tr}\, A \triangleq \sum_{i=1}^{n} A_{(i,i)}. \qquad (2.2.24)$$

Note that

$$\mathrm{tr}\, A = \mathrm{tr}\, A^{\mathrm{T}}. \qquad (2.2.25)$$

Let $A \in \mathbb{F}^{n \times m}$ and $B \in \mathbb{F}^{m \times n}$. Then, AB and BA are square,

$$\mathrm{tr}\, AB = \mathrm{tr}\, BA = \mathrm{tr}\, A^{\mathrm{T}}B^{\mathrm{T}} = \mathrm{tr}\, B^{\mathrm{T}}A^{\mathrm{T}} = \sum_{i,j=1}^{n,m} A_{(i,j)}B_{(j,i)}, \qquad (2.2.26)$$

and

$$\mathrm{tr}\, AA^{\mathrm{T}} = \mathrm{tr}\, A^{\mathrm{T}}A = \sum_{i,j=1}^{n,m} A_{(i,j)}^2. \qquad (2.2.27)$$

Furthermore, if $n = m$, then, for all $\alpha, \beta \in \mathbb{F}$,

$$\operatorname{tr}(\alpha A + \beta B) = \alpha \operatorname{tr} A + \beta \operatorname{tr} B. \tag{2.2.28}$$

Lemma 2.2.3. Let $A \in \mathbb{R}^{n \times m}$. Then, $\operatorname{tr} A^{\mathrm{T}} A = 0$ if and only if $A = 0$.

Let $A, B \in \mathbb{R}^{n \times m}$. Then, the *inner product* of A and B is $\operatorname{tr} A^{\mathrm{T}} B$. Furthermore, A is *orthogonal* to B if $\operatorname{tr} A^{\mathrm{T}} B = 0$.

Let $C \in \mathbb{C}^{n \times m}$. Then, $C = A + \jmath B$, where $A, B \in \mathbb{R}^{n \times m}$. Therefore, the transpose C^{T} of C is given by

$$C^{\mathrm{T}} = A^{\mathrm{T}} + \jmath B^{\mathrm{T}}. \tag{2.2.29}$$

The *complex conjugate* $\overline{C}$ of C is

$$\overline{C} \triangleq A - \jmath B, \tag{2.2.30}$$

while the *complex conjugate transpose* C^* of C is

$$C^* \triangleq \overline{C}^{\mathrm{T}} = A^{\mathrm{T}} - \jmath B^{\mathrm{T}}. \tag{2.2.31}$$

Note that $\overline{C} = C$ if and only if $B = 0$, and that

$$\left(C^{\mathrm{T}}\right)^{\mathrm{T}} = \overline{\overline{C}} = (C^*)^* = C. \tag{2.2.32}$$

The matrices A and B are the real and imaginary parts $\operatorname{Re} C$ and $\operatorname{Im} C$ of C, respectively, which are denoted by

$$\operatorname{Re} C \triangleq \tfrac{1}{2}(C + \overline{C}) = A, \tag{2.2.33}$$

and

$$\operatorname{Im} C \triangleq \tfrac{1}{2\jmath}(C - \overline{C}) = B. \tag{2.2.34}$$

If C is square, then

$$\operatorname{tr} C = \operatorname{tr} A + \jmath \operatorname{tr} B \tag{2.2.35}$$

and

$$\operatorname{tr} C = \operatorname{tr} C^{\mathrm{T}} = \overline{\operatorname{tr} \overline{C}} = \overline{\operatorname{tr} C^*}. \tag{2.2.36}$$

If $\mathcal{S} \subseteq \mathbb{C}^{n \times m}$, then

$$\overline{\mathcal{S}} \triangleq \{\overline{A} : A \in \mathcal{S}\}. \tag{2.2.37}$$

If $\mathcal{S}$ is a multiset with elements in $\mathbb{C}^{n \times m}$, then

$$\overline{\mathcal{S}} = \{\overline{A} : A \in \mathcal{S}\}_{\mathrm{m}}. \tag{2.2.38}$$

Let $A \in \mathbb{F}^{n \times n}$. Then, for all $k \in \mathbb{N}$,

$$A^{k\mathrm{T}} \triangleq \left(A^k\right)^{\mathrm{T}} = (A^{\mathrm{T}})^k \tag{2.2.39}$$

and

$$A^{k*} \triangleq \left(A^k\right)^* = (A^*)^k. \tag{2.2.40}$$

Lemma 2.2.4. Let $A \in \mathbb{C}^{n \times m}$. Then, $\operatorname{tr} A^*A = 0$ if and only if $A = 0$.

Let $A, B \in \mathbb{C}^{n \times m}$. Then, the *inner product* of A and B is $\operatorname{tr} A^*B$. Furthermore, A is *orthogonal* to B if $\operatorname{tr} A^*B = 0$.

If $A, B \in \mathbb{C}^{n \times m}$, then, for all $\alpha, \beta \in \mathbb{C}$,

$$(\alpha A + \beta B)^* = \bar{\alpha} A^* + \bar{\beta} B^*, \tag{2.2.41}$$

while, if $A \in \mathbb{C}^{n \times m}$ and $B \in \mathbb{C}^{m \times l}$, then

$$\overline{AB} = \bar{A}\,\bar{B} \tag{2.2.42}$$

and

$$(AB)^* = B^*A^*. \tag{2.2.43}$$

In particular, if $A \in \mathbb{C}^{n \times m}$ and $x \in \mathbb{C}^m$, then

$$(Ax)^* = x^*A^*, \tag{2.2.44}$$

while, if, in addition, $y \in \mathbb{C}^n$, then

$$y^*Ax = (y^*Ax)^{\mathrm{T}} = x^{\mathrm{T}}A^{\mathrm{T}}\bar{y} \tag{2.2.45}$$

and

$$(y^*Ax)^* = \left(\overline{y^*Ax}\right)^{\mathrm{T}} = \left(y^{\mathrm{T}}\bar{A}\bar{x}\right)^{\mathrm{T}} = x^*A^*y. \tag{2.2.46}$$

For $A \in \mathbb{F}^{n \times m}$ define the *reverse transpose* of A by

$$A^{\hat{\mathrm{T}}} \triangleq \hat{I}_m A^{\mathrm{T}} \hat{I}_n \tag{2.2.47}$$

and the *reverse complex conjugate transpose* of A by

$$A^{\hat{*}} \triangleq \hat{I}_m A^* \hat{I}_n. \tag{2.2.48}$$

For example,

$$\begin{bmatrix} 1 & 2 & 3 \\ 4 & 5 & 6 \end{bmatrix}^{\hat{\mathrm{T}}} = \begin{bmatrix} 6 & 3 \\ 5 & 2 \\ 4 & 1 \end{bmatrix}. \tag{2.2.49}$$

In general,

$$\left(A^*\right)^{\hat{*}} = \left(A^{\hat{*}}\right)^* = \left(A^{\mathrm{T}}\right)^{\hat{\mathrm{T}}} = \left(A^{\hat{\mathrm{T}}}\right)^{\mathrm{T}} = \hat{I}_n A \hat{I}_m \tag{2.2.50}$$

and

$$\left(A^{\hat{*}}\right)^{\hat{*}} = \left(A^{\hat{\mathrm{T}}}\right)^{\hat{\mathrm{T}}} = A. \tag{2.2.51}$$

Note that, if $B \in \mathbb{F}^{m \times l}$, then

$$(AB)^{\hat{*}} = B^{\hat{*}}A^{\hat{*}} \tag{2.2.52}$$

and

$$(AB)^{\hat{\mathrm{T}}} = B^{\hat{\mathrm{T}}}A^{\hat{\mathrm{T}}}. \tag{2.2.53}$$

2.3 Convex Sets, Cones, and Subspaces

Let $\mathcal{S} \subseteq \mathbb{F}^n$. If $\alpha \in \mathbb{F}$, then $\alpha \mathcal{S} \triangleq \{\alpha x\colon\ x \in \mathcal{S}\}$ and, if $y \in \mathbb{F}^n$, then $y + \mathcal{S} = \mathcal{S} + y \triangleq \{y + x\colon\ x \in \mathcal{S}\}$. We write $-\mathcal{S}$ for $(-1)\mathcal{S}$. The set $\mathcal{S}$ is *symmetric* if $\mathcal{S} = -\mathcal{S}$, that is, $x \in \mathcal{S}$ if and only if $-x \in \mathcal{S}$. For $\mathcal{S}_1, \mathcal{S}_2 \subseteq \mathbb{F}^n$ define $\mathcal{S}_1 + \mathcal{S}_2 \triangleq \{x + y\colon\ x \in \mathcal{S}_1 \text{ and } y \in \mathcal{S}_2\}$.

If $x, y \in \mathbb{F}^n$ and $\alpha \in [0, 1]$, then $\alpha x + (1 - \alpha)y$ is a *convex combination* of x and y with *barycentric coordinates* α and $1 - \alpha$. The set $\mathcal{S} \subseteq \mathbb{F}^n$ is *convex* if, for all $x, y \in \mathcal{S}$, every convex combination of x and y is an element of $\mathcal{S}$.

Let $\mathcal{S} \subseteq \mathbb{F}^n$. Then, $\mathcal{S}$ is a *cone* if, for all $x \in \mathcal{S}$ and all $\alpha > 0$, the vector αx is an element of $\mathcal{S}$. Now, assume that $\mathcal{S}$ is a cone. Then, $\mathcal{S}$ is *pointed* if $0 \in \mathcal{S}$, while $\mathcal{S}$ is *one-sided* if $x, -x \in \mathcal{S}$ implies that $x = 0$. Hence, $\mathcal{S}$ is one-sided if and only if $\mathcal{S} \cap -\mathcal{S} \subseteq \{0\}$. Finally, $\mathcal{S}$ is a *convex cone* if it is convex.

Let $\mathcal{S} \subseteq \mathbb{F}^n$ be nonempty. Then, $\mathcal{S}$ is a *subspace* if, for all $x, y \in \mathcal{S}$ and $\alpha, \beta \in \mathbb{F}$, the vector $\alpha x + \beta y$ is an element of $\mathcal{S}$. Note that, if $\{x_1, \ldots, x_r\} \subset \mathbb{F}^n$, then the set $\{\sum_{i=1}^{r} \alpha_i x_i\colon\ \alpha_1, \ldots, \alpha_r \in \mathbb{F}\}$ is a subspace. In addition, $\mathcal{S}$ is an *affine subspace* if there exists a vector $z \in \mathbb{F}^n$ such that $\mathcal{S} + z$ is a subspace. Affine subspaces $\mathcal{S}_1, \mathcal{S}_2 \subseteq \mathbb{F}^n$ are *parallel* if there exists a vector $z \in \mathbb{F}^n$ such that $\mathcal{S}_1 + z = \mathcal{S}_2$. If $\mathcal{S}$ is an affine subspace, then there exists a unique subspace parallel to $\mathcal{S}$. Trivially, the empty set is a convex cone, although it is neither a subspace nor an affine subspace. All of these definitions also apply to subsets of $\mathbb{F}^{n \times m}$.

Let $\mathcal{S} \subseteq \mathbb{F}^n$. The *convex hull* of $\mathcal{S}$, denoted by $\mathrm{co}\,\mathcal{S}$, is the smallest convex set containing $\mathcal{S}$. Hence, $\mathrm{co}\,\mathcal{S}$ is the intersection of all convex subsets of $\mathbb{F}^n$ that contain $\mathcal{S}$. The *conical hull* of $\mathcal{S}$, denoted by $\mathrm{cone}\,\mathcal{S}$, is the smallest cone in $\mathbb{F}^n$ containing $\mathcal{S}$, while the *convex conical hull* of $\mathcal{S}$, denoted by $\mathrm{coco}\,\mathcal{S}$, is the smallest convex cone in $\mathbb{F}^n$ containing $\mathcal{S}$. If $\mathcal{S}$ has a finite number of elements, then $\mathrm{co}\,\mathcal{S}$ is a *polytope* and $\mathrm{coco}\,\mathcal{S}$ is a *polyhedral convex cone*. The *span* of $\mathcal{S}$, denoted by $\mathrm{span}\,\mathcal{S}$, is the smallest subspace in $\mathbb{F}^n$ containing $\mathcal{S}$, while, if $\mathcal{S}$ is nonempty, then the *affine hull* of $\mathcal{S}$, denoted by $\mathrm{aff}\,\mathcal{S}$, is the smallest affine subspace in $\mathbb{F}^n$ containing $\mathcal{S}$. Note that $\mathcal{S}$ is convex if and only if $\mathcal{S} = \mathrm{co}\,\mathcal{S}$, while similar statements hold for $\mathrm{cone}\,\mathcal{S}$, $\mathrm{coco}\,\mathcal{S}$, $\mathrm{span}\,\mathcal{S}$, and $\mathrm{aff}\,\mathcal{S}$. Trivially, $\mathrm{co}\,\varnothing = \mathrm{cone}\,\varnothing = \mathrm{coco}\,\varnothing = \varnothing$, whereas, viewing $\varnothing \subset \mathbb{F}^n$, it follows that $\mathrm{span}\,\varnothing = \{0_{n \times 1}\}$. We define $\mathrm{aff}\,\varnothing \triangleq \{0_{n \times 1}\}$. All of these definitions also apply to subsets of $\mathbb{F}^{n \times m}$.

Let $x_1, \ldots, x_r \in \mathbb{F}^n$. Then, $x_1, \ldots, x_r$ are *linearly independent* if $\alpha_1, \ldots, \alpha_r \in \mathbb{F}$ and

$$\sum_{i=1}^{r} \alpha_i x_i = 0 \qquad (2.3.1)$$

imply that $\alpha_1 = \alpha_2 = \cdots = \alpha_r = 0$. Clearly, $x_1, \ldots, x_r$ is linearly independent if and only if $\overline{x_1}, \ldots, \overline{x_r}$ are linearly independent. If $x_1, \ldots, x_r$ are not linearly independent, then $x_1, \ldots, x_r$ are *linearly dependent*. Note that $0_{n \times 1}$ is linearly dependent.

Let $\mathcal{S} \subseteq \mathbb{F}^n$. If $\mathcal{S}$ is a subspace not equal to $\{0_{n \times 1}\}$, then there exist vectors $x_1, \ldots, x_r \in \mathbb{F}^n$ such that $x_1, \ldots, x_r$ are linearly independent over $\mathbb{F}$ and such that $\operatorname{span}\{x_1, \ldots, x_r\} = \mathcal{S}$. The set of vectors $\{x_1, \ldots, x_r\}$ is a *basis* for $\mathcal{S}$. The positive integer r, which is the *dimension* of the subspace $\mathcal{S}$, is uniquely defined. The dimension of $\mathcal{S} = \{0_{n \times 1}\}$ is defined to be zero since $\operatorname{span} \varnothing = \{0_{n \times 1}\}$. The *dimension* of an arbitrary set $\mathcal{S} \subseteq \mathbb{F}^n$, denoted by $\dim \mathcal{S}$, is the dimension of the subspace parallel to aff $\mathcal{S}$. Hence, $\dim \mathcal{S} = \dim \operatorname{aff} \mathcal{S}$ for all $\mathcal{S} \subseteq \mathbb{F}^n$. We define $\dim \varnothing \triangleq -\infty$.

The following result is the *dimension theorem*.

Theorem 2.3.1. Let $\mathcal{S}_1, \mathcal{S}_2 \subseteq \mathbb{F}^n$ be subspaces. Then,

$$\dim(\mathcal{S}_1 + \mathcal{S}_2) + \dim(\mathcal{S}_1 \cap \mathcal{S}_2) = \dim \mathcal{S}_1 + \dim \mathcal{S}_2. \qquad (2.3.2)$$

Proof. See [329, p. 227]. $\qquad \square$

Let $\mathcal{S}_1, \mathcal{S}_2 \subseteq \mathbb{F}^n$ be subspaces. Then, $\mathcal{S}_1$ and $\mathcal{S}_2$ are *complementary* if $\mathcal{S}_1 \cap \mathcal{S}_2 = \{0\}$ and $\mathcal{S}_1 + \mathcal{S}_2 = \mathbb{F}^n$. In this case, we say that $\mathcal{S}_1$ is complementary to $\mathcal{S}_2$, or vice versa.

Corollary 2.3.2. Let $\mathcal{S}_1, \mathcal{S}_2 \subseteq \mathbb{F}^n$ be subspaces. Then, $\mathcal{S}_1, \mathcal{S}_2$ are complementary if and only if $\mathcal{S}_1 \cap \mathcal{S}_2 = \{0\}$ and

$$\dim \mathcal{S}_1 + \dim \mathcal{S}_2 = n. \qquad (2.3.3)$$

Let $\mathcal{S} \subseteq \mathbb{F}^n$ be nonempty. Then, the *orthogonal complement* $\mathcal{S}^\perp$ of $\mathcal{S}$ is defined by

$$\mathcal{S}^\perp \triangleq \{x \in \mathbb{F}^n: \ x^* y = 0 \text{ for all } y \in \mathcal{S}\}. \qquad (2.3.4)$$

The orthogonal complement $\mathcal{S}^\perp$ of $\mathcal{S}$ is a subspace even if $\mathcal{S}$ is not.

Let $y \in \mathbb{F}^n$ be nonzero. Then, the subspace $\{y\}^\perp$, whose dimension is $n-1$, is a *hyperplane*. Furthermore, $\mathcal{S}$ is an *affine hyperplane* if there exists a vector $z \in \mathbb{F}^n$ such that $\mathcal{S}+z$ is a hyperplane. The set $\{x \in \mathbb{F}^n: \operatorname{Re} x^* y \leq 0\}$ is a *closed half space*, while the set $\{x \in \mathbb{F}^n: \operatorname{Re} x^* y < 0\}$ is an *open half space*. Finally, $\mathcal{S}$ is an *affine (closed, open) half space* if there exists a vector $z \in \mathbb{F}^n$ such that $\mathcal{S} + z$ is a (closed, open) half space.

Let $\mathcal{S} \subseteq \mathbb{F}^n$. Then,

$$\operatorname{dcone} \mathcal{S} \triangleq \{x \in \mathbb{F}^n: \operatorname{Re} x^* y \leq 0 \text{ for all } y \in \mathcal{S}\} \tag{2.3.5}$$

is the *dual cone* of $\mathcal{S}$. Note that $\operatorname{dcone} \mathcal{S}$ is a pointed convex cone and that

$$\operatorname{dcone} \mathcal{S} = \operatorname{dcone} \operatorname{cone} \mathcal{S} = \operatorname{dcone} \operatorname{coco} \mathcal{S}. \tag{2.3.6}$$

Let $\mathcal{S}_1, \mathcal{S}_2 \subseteq \mathbb{F}^n$ be subspaces. Then, $\mathcal{S}_1$ and $\mathcal{S}_2$ are *orthogonally complementary* if $\mathcal{S}_1$ and $\mathcal{S}_2$ are complementary and $x^* y = 0$ for all $x \in \mathcal{S}_1$ and $y \in \mathcal{S}_2$.

Proposition 2.3.3. Let $\mathcal{S}_1, \mathcal{S}_2 \subseteq \mathbb{F}^n$ be subspaces. Then, $\mathcal{S}_1$ and $\mathcal{S}_2$ are orthogonally complementary if and only if $\mathcal{S}_1 = \mathcal{S}_2^\perp$.

For the next result, note that "$\subset$" indicates proper inclusion.

Lemma 2.3.4. Let $\mathcal{S}_1, \mathcal{S}_2 \subseteq \mathbb{F}^n$ be subspaces such that $\mathcal{S}_1 \subseteq \mathcal{S}_2$. Then, $\mathcal{S}_1 \subset \mathcal{S}_2$ if and only if $\dim \mathcal{S}_1 < \dim \mathcal{S}_2$. Equivalently, $\mathcal{S}_1 = \mathcal{S}_2$ if and only if $\dim \mathcal{S}_1 = \dim \mathcal{S}_2$.

The following result provides constructive characterizations of $\operatorname{co} \mathcal{S}$, $\operatorname{cone} \mathcal{S}$, $\operatorname{coco} \mathcal{S}$, $\operatorname{span} \mathcal{S}$, and $\operatorname{aff} \mathcal{S}$.

Theorem 2.3.5. Let $\mathcal{S} \subseteq \mathbb{R}^n$ be nonempty. Then,

$$\operatorname{co} \mathcal{S} = \bigcup_{k \in \mathbb{P}} \left\{ \sum_{i=1}^{k} \alpha_i x_i: \; \alpha_i \geq 0, \; x_i \in \mathcal{S}, \text{ and } \sum_{i=1}^{k} \alpha_i = 1 \right\} \tag{2.3.7}$$

$$= \left\{ \sum_{i=1}^{n+1} \alpha_i x_i: \; \alpha_i \geq 0, \; x_i \in \mathcal{S}, \text{ and } \sum_{i=1}^{n+1} \alpha_i = 1 \right\}, \tag{2.3.8}$$

$$\operatorname{cone} \mathcal{S} = \{\alpha x: \; x \in \mathcal{S} \text{ and } \alpha > 0\}, \tag{2.3.9}$$

$$\operatorname{coco} \mathcal{S} = \bigcup_{k \in \mathbb{P}} \left\{ \sum_{i=1}^{k} \alpha_i x_i: \; \alpha_i \geq 0, \; x_i \in \mathcal{S}, \text{ and } \sum_{i=1}^{k} \alpha_i > 0 \right\} \tag{2.3.10}$$

$$= \left\{ \sum_{i=1}^{n+1} \alpha_i x_i: \; \alpha_i \geq 0, \; x_i \in \mathcal{S}, \text{ and } \sum_{i=1}^{n} \alpha_i > 0 \right\}, \tag{2.3.11}$$

$$\text{span } S = \bigcup_{k \in \mathbb{P}} \left\{ \sum_{i=1}^{k} \alpha_i x_i \colon \ \alpha_i \in \mathbb{R} \text{ and } x_i \in S \right\} \tag{2.3.12}$$

$$= \left\{ \sum_{i=1}^{n} \alpha_i x_i \colon \ \alpha_i \in \mathbb{R} \text{ and } x_i \in S \right\}, \tag{2.3.13}$$

$$\text{aff } S = \bigcup_{k \in \mathbb{P}} \left\{ \sum_{i=1}^{k} \alpha_i x_i \colon \ \alpha_i \in \mathbb{R}, \ x_i \in S, \text{ and } \sum_{i=1}^{k} \alpha_i = 1 \right\} \tag{2.3.14}$$

$$= \left\{ \sum_{i=1}^{n+1} \alpha_i x_i \colon \ \alpha_i \in \mathbb{R}, \ x_i \in S, \text{ and } \sum_{i=1}^{n+1} \alpha_i = 1 \right\}. \tag{2.3.15}$$

Now, let $S \subseteq \mathbb{C}^n$. Then,

$$\text{co } S = \bigcup_{k \in \mathbb{P}} \left\{ \sum_{i=1}^{k} \alpha_i x_i \colon \ \alpha_i \geq 0, \ x_i \in S, \text{ and } \sum_{i=1}^{k} \alpha_i = 1 \right\} \tag{2.3.16}$$

$$= \left\{ \sum_{i=1}^{2n+1} \alpha_i x_i \colon \ \alpha_i \geq 0, \ x_i \in S, \text{ and } \sum_{i=1}^{2n+1} \alpha_i = 1 \right\}, \tag{2.3.17}$$

$$\text{cone } S = \{ \alpha x \colon \ x \in S \text{ and } \alpha > 0 \}, \tag{2.3.18}$$

$$\text{coco } S = \bigcup_{k \in \mathbb{P}} \left\{ \sum_{i=1}^{k} \alpha_i x_i \colon \ \alpha_i \geq 0, \ x_i \in S, \text{ and } \sum_{i=1}^{k} \alpha_i > 0 \right\} \tag{2.3.19}$$

$$= \left\{ \sum_{i=1}^{2n+1} \alpha_i x_i \colon \ \alpha_i \geq 0, \ x_i \in S, \text{ and } \sum_{i=1}^{2n} \alpha_i > 0 \right\}, \tag{2.3.20}$$

$$\text{span } S = \bigcup_{k \in \mathbb{P}} \left\{ \sum_{i=1}^{k} \alpha_i x_i \colon \ \alpha_i \in \mathbb{C} \text{ and } x_i \in S \right\} \tag{2.3.21}$$

$$= \left\{ \sum_{i=1}^{n} \alpha_i x_i \colon \ \alpha_i \in \mathbb{C} \text{ and } x_i \in S \right\}, \tag{2.3.22}$$

$$\text{aff } S = \bigcup_{k \in \mathbb{P}} \left\{ \sum_{i=1}^{k} \alpha_i x_i \colon \ \alpha_i \in \mathbb{C}, \ x_i \in S, \text{ and } \sum_{i=1}^{k} \alpha_i = 1 \right\} \tag{2.3.23}$$

$$= \left\{ \sum_{i=1}^{n+1} \alpha_i x_i \colon \ \alpha_i \in \mathbb{C}, \ x_i \in S, \text{ and } \sum_{i=1}^{n+1} \alpha_i = 1 \right\}. \tag{2.3.24}$$

Proof. Result (2.3.7) is immediate, while (2.3.8) is proved in [462, p. 17]. Furthermore, (2.3.9) is immediate. Next, note that, since $\text{coco } S =$

co cone S, it follows that (2.3.7) and (2.3.9) imply (2.3.11) with n replaced by $n + 1$. However, every element of coco S lies in the convex hull of $n + 1$ points one of which is the origin. It thus follows that we can set $x_{n+1} = 0$, which yields (2.3.11). Similar arguments yield (2.3.13). Finally, note that all vectors of the form $x_1 + \beta(x_2 - x_1)$, where $x_1, x_2 \in S$ and $\beta \in \mathbb{R}$, are elements of aff S. Forming the convex hull of these vectors yields (2.3.15). $\square$

The following result shows that cones can be used to induce relations on $\mathbb{F}^n$.

Proposition 2.3.6. Let $S \subseteq \mathbb{F}^n$ be a cone and, for $x, y \in \mathbb{F}^n$, let $x \le y$ denote the relation $y - x \in S$. Then, the following statements hold:

i) "$\le$" is reflexive if and only if S is a pointed cone.

ii) "$\le$" is antisymmetric if and only if S is a one-sided cone.

iii) "$\le$" is symmetric if and only if S is a symmetric cone.

iv) "$\le$" is transitive if and only if S is a convex cone.

Proof. The proofs of *i)*, *ii)*, and *iii)* are immediate. To prove *iv)*, suppose that "$\le$" is transitive, and let $x, y \in S$ so that $0 \le \alpha x \le \alpha x + (1 - \alpha)y$ for all $\alpha \in (0, 1]$. Hence, $\alpha x + (1 - \alpha)y \in S$ for all $\alpha \in (0, 1]$, and thus S is convex. Conversely, suppose that S is a convex cone, and assume that $x \le y$ and $y \le z$. Then, $y - x \in S$ and $z - y \in S$ imply that $z - x = 2\left[\frac{1}{2}(y - x) + \frac{1}{2}(z - y)\right] \in S$. Hence, $x \le z$, and thus "$\le$" is transitive. $\square$

2.4 Range and Null Space

Two important features of a matrix $A \in \mathbb{F}^{n \times m}$ are its range and null space, denoted by $\mathcal{R}(A)$ and $\mathcal{N}(A)$, respectively. The *range* of A is defined by

$$\mathcal{R}(A) \triangleq \{Ax \colon x \in \mathbb{F}^m\}. \tag{2.4.1}$$

Note that $\mathcal{R}(0_{n \times 0}) = \{0_{n \times 1}\}$ and $\mathcal{R}(0_{0 \times m}) = \{0_{0 \times 1}\}$. Letting α_i denote $x_{(i)}$, it can be seen that

$$\mathcal{R}(A) = \left\{\sum_{i=1}^{m} \alpha_i \mathrm{col}_i(A) \colon \alpha_1, \ldots, \alpha_m \in \mathbb{F}\right\}, \tag{2.4.2}$$

which shows that $\mathcal{R}(A)$ is a subspace of $\mathbb{F}^n$. It thus follows from Theorem 2.3.5 that

$$\mathcal{R}(A) = \mathrm{span}\,\{\mathrm{col}_1(A), \ldots, \mathrm{col}_m(A)\}. \tag{2.4.3}$$

By viewing A as a function from $\mathbb{F}^m$ into $\mathbb{F}^n$, we can also write $\mathcal{R}(A) = A\mathbb{F}^m$.

The *null space* of $A \in \mathbb{F}^{n \times m}$ is defined by

$$\mathcal{N}(A) \triangleq \{x \in \mathbb{F}^m \colon \ Ax = 0\}. \tag{2.4.4}$$

Note that $\mathcal{N}(0_{n \times 0}) = \mathbb{F}^0 = \{0_{0 \times 1}\}$ and $\mathcal{N}(0_{0 \times m}) = \mathbb{F}^m$. Equivalently,

$$\mathcal{N}(A) = \left\{x \in \mathbb{F}^m \colon \ x^{\mathrm{T}}[\mathrm{row}_i(A)]^{\mathrm{T}} = 0 \text{ for all } i = 1, \ldots, n\right\} \tag{2.4.5}$$

$$= \left\{[\mathrm{row}_1(A)]^{\mathrm{T}}, \ldots, [\mathrm{row}_n(A)]^{\mathrm{T}}\right\}^{\perp}, \tag{2.4.6}$$

which shows that $\mathcal{N}(A)$ is a subspace of $\mathbb{F}^m$. Note that if $\alpha \in \mathbb{F}$ is nonzero, then $\mathcal{R}(\alpha A) = \mathcal{R}(A)$ and $\mathcal{N}(\alpha A) = \mathcal{N}(A)$. Finally, if $\mathbb{F} = \mathbb{C}$, then $\mathcal{R}(A)$ and $\mathcal{R}(\overline{A})$ are not necessarily identical. For example, let $A \triangleq \left[\begin{smallmatrix} \jmath \\ 1 \end{smallmatrix}\right]$.

Let $A \in \mathbb{F}^{n \times n}$, and let $\mathcal{S} \subseteq \mathbb{F}^n$ be a subspace. Then, $\mathcal{S}$ is an *invariant subspace* of A if $A\mathcal{S} \subseteq \mathcal{S}$. Note that $A\mathcal{R}(A) \subseteq A\mathbb{F}^n = \mathcal{R}(A)$ and $A\mathcal{N}(A) = \{0_n\} \subseteq \mathcal{N}(A)$. Hence, $\mathcal{R}(A)$ and $\mathcal{N}(A)$ are invariant subspaces of A.

If $A \in \mathbb{F}^{n \times m}$ and $B \in \mathbb{F}^{m \times l}$, then it is easy to see that

$$\mathcal{R}(AB) = A\mathcal{R}(B). \tag{2.4.7}$$

Hence, the following result is not surprising.

Lemma 2.4.1. Let $A \in \mathbb{F}^{n \times m}$, $B \in \mathbb{F}^{m \times l}$, and $C \in \mathbb{F}^{k \times n}$. Then,

$$\mathcal{R}(AB) \subseteq \mathcal{R}(A) \tag{2.4.8}$$

and

$$\mathcal{N}(A) \subseteq \mathcal{N}(CA). \tag{2.4.9}$$

Proof. Since $\mathcal{R}(B) \subseteq \mathbb{F}^m$, it follows that $\mathcal{R}(AB) = A\mathcal{R}(B) \subseteq A\mathbb{F}^m = \mathcal{R}(A)$. Furthermore, $y \in \mathcal{N}(A)$ implies that $Ay = 0$, and thus $CAy = 0$. $\square$

Corollary 2.4.2. Let $A \in \mathbb{F}^{n \times n}$, and let $k \in \mathbb{P}$. Then,

$$\mathcal{R}\left(A^k\right) \subseteq \mathcal{R}(A) \tag{2.4.10}$$

and

$$\mathcal{N}(A) \subseteq \mathcal{N}\left(A^k\right). \tag{2.4.11}$$

Although $\mathcal{R}(AB) \subseteq \mathcal{R}(A)$ for arbitrary conformable matrices A, B, we now show that equality holds in the special case $B = A^*$. This result, along with others, is the subject of the following basic theorem.

Theorem 2.4.3. Let $A \in \mathbb{F}^{n \times m}$. Then, the following identities hold:

i) $\mathcal{R}(A)^{\perp} = \mathcal{N}(A^*)$.

ii) $\mathcal{R}(A) = \mathcal{R}(AA^*)$.

iii) $\mathcal{N}(A) = \mathcal{N}(A^*\!A)$.

Proof. To prove *i*), we first show that $\mathcal{R}(A)^\perp \subseteq \mathcal{N}(A^*)$. Let $x \in \mathcal{R}(A)^\perp$. Then, $x^*z = 0$ for all $z \in \mathcal{R}(A)$. Hence, $x^*\!Ay = 0$ for all $y \in \mathbb{R}^m$. Equivalently, $y^*\!A^*x = 0$ for all $y \in \mathbb{R}^m$. Letting $y = A^*x$, it follows that $x^*\!AA^*x = 0$. Now, Lemma 2.2.2 implies that $A^*x = 0$. Thus, $x \in \mathcal{N}(A^*)$. Conversely, let us show that $\mathcal{N}(A^*) \subseteq \mathcal{R}(A)^\perp$. Letting $x \in \mathcal{N}(A^*)$, it follows that $A^*x = 0$, and, hence, $y^*\!A^*x = 0$ for all $y \in \mathbb{R}^m$. Equivalently, $x^*\!Ay = 0$ for all $y \in \mathbb{R}^m$. Hence, $x^*z = 0$ for all $z \in \mathcal{R}(A)$. Thus, $x \in \mathcal{R}(A)^\perp$, which proves *i*).

To prove *ii*), note that Lemma 2.4.1 with $B = A^*$ implies that $\mathcal{R}(AA^*)$ $\subseteq \mathcal{R}(A)$. To show that $\mathcal{R}(A) \subseteq \mathcal{R}(AA^*)$, let $x \in \mathcal{R}(A)$, and suppose that $x \notin \mathcal{R}(AA^*)$. Then, it follows from Proposition 2.3.3 that $x = x_1 + x_2$, where $x_1 \in \mathcal{R}(AA^*)$ and $x_2 \in \mathcal{R}(AA^*)^\perp$ with $x_2 \neq 0$. Thus, $x_2^*\!AA^*y = 0$ for all $y \in \mathbb{R}^n$, and setting $y = x_2$ yields $x_2^*\!AA^*x_2 = 0$. Hence, Lemma 2.2.2 implies that $A^*x_2 = 0$, so that, by *i*), $x_2 \in \mathcal{N}(A^*) = \mathcal{R}(A)^\perp$. Since $x \in \mathcal{R}(A)$, it follows that $0 = x_2^*x = x_2^*x_1 + x_2^*x_2$. However, $x_2^*x_1 = 0$ so that $x_2^*x_2 = 0$ and $x_2 = 0$, which is a contradiction. This proves *ii*).

To prove *iii*), note that *ii*) with A replaced by A^* implies that $\mathcal{R}(A^*\!A)^\perp$ $= \mathcal{R}(A^*)^\perp$. Furthermore, replacing A by A^* in *i*) yields $\mathcal{R}(A^*)^\perp = \mathcal{N}(A)$. Hence, $\mathcal{N}(A) = \mathcal{R}(A^*\!A)^\perp$. Now, *i*) with A replaced by $A^*\!A$ implies that $\mathcal{R}(A^*\!A)^\perp = \mathcal{N}(A^*\!A)$. Hence, $\mathcal{N}(A) = \mathcal{N}(A^*\!A)$, which proves *iii*). $\square$

Result *i*) of Theorem 2.4.3 can be written equivalently as

$$\mathcal{N}(A)^\perp = \mathcal{R}(A^*), \tag{2.4.12}$$

$$\mathcal{N}(A) = \mathcal{R}(A^*)^\perp, \tag{2.4.13}$$

$$\mathcal{N}(A^*)^\perp = \mathcal{R}(A), \tag{2.4.14}$$

while replacing A by A^* in *ii*) and *iii*) of Theorem 2.4.3 yields

$$\mathcal{R}(A^*) = \mathcal{R}(A^*\!A), \tag{2.4.15}$$

$$\mathcal{N}(A^*) = \mathcal{N}(AA^*). \tag{2.4.16}$$

Using *ii*) of Theorem 2.4.3 and (2.4.15), it follows that

$$\mathcal{R}(AA^*\!A) = A\mathcal{R}(A^*\!A) = A\mathcal{R}(A^*) = \mathcal{R}(AA^*) = \mathcal{R}(A). \tag{2.4.17}$$

Letting $A \triangleq \begin{bmatrix} 1 & \jmath \end{bmatrix}$ shows that $\mathcal{R}(A)$ and $\mathcal{R}(AA^\mathrm{T})$ are generally different.

2.5 Rank and Defect

The *rank* of $A \in \mathbb{F}^{n \times m}$ is defined by

$$\text{rank}\, A \triangleq \dim \mathcal{R}(A). \tag{2.5.1}$$

It can be seen that the rank of A is equal to the number of linearly independent columns of A over $\mathbb{F}$. For example, if $\mathbb{F} = \mathbb{C}$, then $\text{rank} \begin{bmatrix} 1 & j \end{bmatrix} = 1$, while, if either $\mathbb{F} = \mathbb{C}$ or $\mathbb{F} = \mathbb{C}$, then $\text{rank} \begin{bmatrix} 1 & 1 \end{bmatrix} = 1$ Furthermore, $\text{rank}\, A = \text{rank}\, \overline{A}$, $\text{rank}\, A^{\mathrm{T}} = \text{rank}\, A^*$, $\text{rank}\, A \leq m$, and $\text{rank}\, A^{\mathrm{T}} \leq n$. If $\text{rank}\, A = m$, then A has *full column rank*, while, if $\text{rank}\, A^{\mathrm{T}} = n$, then A has *full row rank*. If A has either full column rank or full row rank, then A has *full rank*. Finally, the *defect* of A is

$$\text{def}\, A \triangleq \dim \mathcal{N}(A). \tag{2.5.2}$$

The following result follows from Theorem 2.4.3.

Corollary 2.5.1. Let $A \in \mathbb{F}^{n \times m}$. Then, the following identities hold:

i) $\text{rank}\, A^* + \text{def}\, A = m$.

ii) $\text{rank}\, A = \text{rank}\, AA^*$.

iii) $\text{def}\, A = \text{def}\, A^*A$.

Proof. It follows from (2.4.12) and Proposition 2.3.2 that $\text{rank}\, A^* = \dim \mathcal{R}(A^*) = \dim \mathcal{N}(A)^{\perp} = m - \dim \mathcal{N}(A) = m - \text{def}\, A$, which proves *i)*. Results *ii)* and *iii)* follow from *ii)* and *iii)* of Theorem 2.4.3. $\qquad \square$

Replacing A by A^* in Corollary 2.5.1 yields

$$\text{rank}\, A + \text{def}\, A^* = n, \tag{2.5.3}$$

$$\text{rank}\, A^* = \text{rank}\, A^*A, \tag{2.5.4}$$

$$\text{def}\, A^* = \text{def}\, AA^*. \tag{2.5.5}$$

Furthermore, note that

$$\text{def}\, A = \text{def}\, \overline{A} \tag{2.5.6}$$

and

$$\text{def}\, A^{\mathrm{T}} = \text{def}\, A^*. \tag{2.5.7}$$

Lemma 2.5.2. Let $A \in \mathbb{F}^{n \times m}$ and $B \in \mathbb{F}^{m \times l}$. Then,

$$\text{rank}\, AB \leq \min\{\text{rank}\, A, \text{rank}\, B\}. \tag{2.5.8}$$

Proof. Since, by Lemma 2.4.1, $\mathcal{R}(AB) \subseteq \mathcal{R}(A)$, it follows that $\text{rank}\, AB \leq \text{rank}\, A$. Next, suppose that $\text{rank}\, B < \text{rank}\, AB$. Let $\{y_1, \ldots, y_r\} \subset \mathbb{F}^n$ be a basis for $\mathcal{R}(AB)$, where $r \triangleq \text{rank}\, AB$, and, since $y_i \in A\mathcal{R}(B)$ for all $i = 1, \ldots, r$, let $x_i \in \mathcal{R}(B)$ be such that $y_i = Ax_i$ for all $i = 1, \ldots, r$. Since $\text{rank}\, B < r$, it follows that $x_1, \ldots, x_r$ are linearly dependent. Hence, there exist $\alpha_1, \ldots, \alpha_r \in \mathbb{F}$, not all zero, such that $\sum_{i=1}^{r} \alpha_i x_i = 0$, which implies that $\sum_{i=1}^{r} \alpha_i A x_i = \sum_{i=1}^{r} \alpha_i y_i = 0$. Thus, $y_1, \ldots, y_r$ are linearly dependent, which is a contradiction. $\square$

Corollary 2.5.3. Let $A \in \mathbb{F}^{n \times m}$. Then,

$$\operatorname{rank} A = \operatorname{rank} A^* \tag{2.5.9}$$

and

$$\operatorname{def} A = \operatorname{def} A^* + m - n. \tag{2.5.10}$$

If, in addition, $n = m$, then

$$\operatorname{def} A = \operatorname{def} A^*. \tag{2.5.11}$$

Proof. It follows from (2.5.8) with $B = A^*$ that $\operatorname{rank} AA^* \le \operatorname{rank} A^*$. Furthermore, $ii)$ of Corollary 2.5.1 implies that $\operatorname{rank} A = \operatorname{rank} AA^*$. Hence, $\operatorname{rank} A \le \operatorname{rank} A^*$. Interchanging A and A^* and repeating this argument yields $\operatorname{rank} A^* \le \operatorname{rank} A$. Hence, $\operatorname{rank} A = \operatorname{rank} A^*$. Next, using $i)$ of Corollary 2.5.1, (2.5.9), and (2.5.3) it follows that $\operatorname{def} A = m - \operatorname{rank} A^* = m - \operatorname{rank} A = m - (n - \operatorname{def} A^*)$, which proves (2.5.10). $\square$

Corollary 2.5.4. Let $A \in \mathbb{F}^{n \times m}$. Then,

$$\operatorname{rank} A \le \min\{m, n\}. \tag{2.5.12}$$

Proof. By definition, $\operatorname{rank} A \le m$, while it follows from (2.5.9) that $\operatorname{rank} A = \operatorname{rank} A^* \le n$. $\square$

The *fundamental theorem of linear algebra* is given by (2.5.13) in the following result.

Corollary 2.5.5. Let $A \in \mathbb{F}^{n \times m}$. Then,

$$\operatorname{rank} A + \operatorname{def} A = m \tag{2.5.13}$$

and

$$\operatorname{rank} A = \operatorname{rank} A^*A. \tag{2.5.14}$$

Proof. The result (2.5.13) follows from $i)$ of Corollary 2.5.1 and (2.5.9), while (2.5.14) follows from (2.5.4) and (2.5.9). $\square$

Corollary 2.5.6. Let $A \in \mathbb{F}^{n \times n}$ and $k \in \mathbb{P}$. Then,

$$\operatorname{rank} A^k \le \operatorname{rank} A \tag{2.5.15}$$

and

$$\operatorname{def} A \le \operatorname{def} A^k. \tag{2.5.16}$$

Proposition 2.5.7. Let $A \in \mathbb{F}^{n \times n}$. If rank $A^2 =$ rank A, then rank A^k = rank A for all $k \in \mathbb{P}$. Equivalently, if def $A^2 =$ def A, then def $A^k =$ def A for all $k \in \mathbb{P}$.

Proof. Since rank $A^2 =$ rank A and $\mathcal{R}(A^2) \subseteq \mathcal{R}(A)$, it follows from Lemma 2.3.4 that $\mathcal{R}(A^2) = \mathcal{R}(A)$. Hence, $\mathcal{R}(A^3) = A\mathcal{R}(A^2) = A\mathcal{R}(A) = \mathcal{R}(A^2)$. Thus, rank $A^3 =$ rank A. Similar arguments yield rank $A^k =$ rank A for all $k \in \mathbb{P}$. $\qquad\square$

We now prove *Sylvester's inequality*, which provides a lower bound for the rank of the product of two matrices.

Proposition 2.5.8. Let $A \in \mathbb{F}^{n \times m}$ and $B \in \mathbb{F}^{m \times l}$. Then,

$$\text{rank } A + \text{rank } B \leq m + \text{rank } AB. \qquad (2.5.17)$$

Proof. Using (2.5.8) to obtain the second inequality below, it follows that

$$\begin{aligned}
\text{rank } A + \text{rank } B &= \text{rank} \begin{bmatrix} 0 & A \\ B & 0 \end{bmatrix} \\
&\leq \text{rank} \begin{bmatrix} 0 & A \\ B & I \end{bmatrix} \\
&= \text{rank} \begin{bmatrix} I & A \\ 0 & I \end{bmatrix} \begin{bmatrix} -AB & 0 \\ B & I \end{bmatrix} \\
&\leq \text{rank} \begin{bmatrix} -AB & 0 \\ B & I \end{bmatrix} \\
&\leq \text{rank} \begin{bmatrix} -AB & 0 \end{bmatrix} + \text{rank} \begin{bmatrix} B & I \end{bmatrix} \\
&= \text{rank } AB + m. \qquad\square
\end{aligned}$$

Combining (2.5.8) with (2.5.17) yields the following result.

Corollary 2.5.9. Let $A \in \mathbb{F}^{n \times m}$ and $B \in \mathbb{F}^{m \times l}$. Then,

$$\text{rank } A + \text{rank } B - m \leq \text{rank } AB \leq \min\{\text{rank } A, \text{rank } B\}. \qquad (2.5.18)$$

2.6 Invertibility

Let $A \in \mathbb{F}^{n \times m}$. Then, A is *left invertible* if there exists a matrix $A^{\mathrm{L}} \in \mathbb{F}^{m \times n}$ such that $A^{\mathrm{L}}A = I_m$, while A is *right invertible* if there exists a matrix $A^{\mathrm{R}} \in \mathbb{F}^{m \times n}$ such that $AA^{\mathrm{R}} = I_n$. These definitions are consistent with the definitions of left and right invertibility given in Chapter 1 applied to the function $f \colon \mathbb{F}^m \mapsto \mathbb{F}^n$ given by $f(x) = Ax$.

Theorem 2.6.1. Let $A \in \mathbb{F}^{n \times m}$. Then, the following statements are equivalent:

i) A is left invertible.

ii) A is one-to-one.

iii) def $A = 0$.

iv) rank $A = m$.

v) A has full column rank.

The following statements are also equivalent:

vi) A is right invertible.

vii) A is onto.

viii) def $A = m - n$.

ix) rank $A = n$.

x) A has full row rank.

Note that A is left invertible if and only if A^* is right invertible.

The following result shows that the rank and defect of a matrix are not affected by either left multiplication by a left invertible matrix or right multiplication by a right invertible matrix.

Proposition 2.6.2. Let $A \in \mathbb{F}^{n \times m}$, and let $C \in \mathbb{F}^{k \times n}$ be left invertible and $B \in \mathbb{F}^{m \times l}$ be right invertible. Then,

$$\text{rank } A = \text{rank } CA = \text{rank } AB \qquad (2.6.1)$$

and

$$\text{def } A = \text{def } CA = \text{def } AB + m - l. \qquad (2.6.2)$$

Proof. Let C^{L} be a left inverse of C. Using both inequalities in (2.5.18) and the fact that rank $A \leq n$, it follows that

$$\text{rank } A = \text{rank } A + \text{rank } C^{\mathrm{L}}C - n \leq \text{rank } C^{\mathrm{L}}CA \leq \text{rank } CA \leq \text{rank } A,$$

which implies that rank $A = \text{rank } CA$. A similar argument implies that rank $A = \text{rank } AB$. Next, (2.5.13) and (2.6.1) imply that $m - \text{def } A = m - \text{def } CA = l - \text{def } AB$, which yields (2.6.2). $\qquad \square$

In general, left and right inverses are not unique. For example, the matrix $A = \begin{bmatrix} 0 \\ 1 \end{bmatrix}$ is left invertible and has left inverses $\begin{bmatrix} 0 & 1 \end{bmatrix}$ and $\begin{bmatrix} 1 & 1 \end{bmatrix}$. In spite of this nonuniqueness, however, left inverses are useful for solving equations of the form $Ax = b$, where $A \in \mathbb{F}^{n \times m}$, $x \in \mathbb{F}^m$, and $b \in \mathbb{F}^n$. If A is left invertible, then one can formally (although not rigorously) solve $Ax = b$ by noting that $x = A^{\mathrm{L}}Ax = A^{\mathrm{L}}b$, where $A^{\mathrm{L}} \in \mathbb{R}^{m \times n}$ is a left inverse of A. However, it is necessary to determine beforehand whether or not there actually exists a vector x satisfying $Ax = b$. For example, if $A = \begin{bmatrix} 0 \\ 1 \end{bmatrix}$ and $b = \begin{bmatrix} 1 \\ 0 \end{bmatrix}$, then A is left invertible although there does not exist a vector x satisfying $Ax = b$. The following result addresses the various possibilities that can arise. One interesting feature of this result is that, if there exists a solution of $Ax = b$ and A is left invertible, then the solution is unique even if A does not have a unique left inverse. For this result, $\begin{bmatrix} A & b \end{bmatrix}$ denotes the $n \times (m+1)$ partitioned matrix formed from A and b. Note that $\operatorname{rank} A \leq \operatorname{rank} \begin{bmatrix} A & b \end{bmatrix} \leq m+1$, while $\operatorname{rank} A = \operatorname{rank} \begin{bmatrix} A & b \end{bmatrix}$ is equivalent to $b \in \mathcal{R}(A)$.

Theorem 2.6.3. Let $A \in \mathbb{F}^{n \times m}$ and $b \in \mathbb{F}^n$. Then, the following statements hold:

i)   There does not exist a vector $x \in \mathbb{F}^m$ satisfying $Ax = b$ if and only if $\operatorname{rank} A < \operatorname{rank} \begin{bmatrix} A & b \end{bmatrix}$.

ii) There exists a unique vector $x \in \mathbb{F}^m$ satisfying $Ax = b$ if and only if $\operatorname{rank} A = \operatorname{rank} \begin{bmatrix} A & b \end{bmatrix} = m$. In this case, if $A^{\mathrm{L}} \in \mathbb{F}^{m \times n}$ is a left inverse of A, then the solution is given by $x = A^{\mathrm{L}}b$.

iii) There exist infinitely many $x \in \mathbb{F}^m$ satisfying $Ax = b$ if and only if $\operatorname{rank} A = \operatorname{rank} \begin{bmatrix} A & b \end{bmatrix} < m$. In this case, let $\hat{x} \in \mathbb{F}^m$ satisfy $A\hat{x} = b$. Then, the set of solutions of $Ax = b$ is given by $\hat{x} + \mathcal{N}(A)$.

iv) Assume that $\operatorname{rank} A = n$. Then, there exists at least one vector $x \in \mathbb{F}^m$ satisfying $Ax = b$. Furthermore, if $A^{\mathrm{R}} \in \mathbb{F}^{m \times n}$ is a right inverse of A, then $x = A^{\mathrm{R}}b$ satisfies $Ax = b$. If $n = m$, then $x = A^{\mathrm{R}}b$ is the unique solution of $Ax = b$. If $n < m$ and $\hat{x} \in \mathbb{F}^n$ satisfies $A\hat{x} = b$, then the set of solutions of $Ax = b$ is given by $\hat{x} + \mathcal{N}(A)$.

Proof. To prove i), note that $\operatorname{rank} A < \operatorname{rank} \begin{bmatrix} A & b \end{bmatrix}$ is equivalent to the fact that b cannot be represented as a linear combination of columns of A, that is, $Ax = b$ does not have a solution $x \in \mathbb{F}^m$. To prove ii), suppose that $\operatorname{rank} A = \operatorname{rank} \begin{bmatrix} A & b \end{bmatrix} = m$ so that, by i), $Ax = b$ has a solution $x \in \mathbb{F}^m$. If $\hat{x} \in \mathbb{F}^m$ satisfies $A\hat{x} = b$, then $A(x - \hat{x}) = 0$. Since $\operatorname{rank} A = m$, it follows from Theorem 2.6.1 that A has a left inverse $A^{\mathrm{L}} \in \mathbb{F}^{m \times n}$. Hence, $x - \hat{x} = A^{\mathrm{L}}A(x-\hat{x}) = 0$, which proves that $Ax = b$ has a unique solution. Conversely, suppose that $\operatorname{rank} A = \operatorname{rank} \begin{bmatrix} A & b \end{bmatrix} = m$ and there exist vectors $x, \hat{x} \in \mathbb{F}^m$, where $x \neq \hat{x}$, such that $Ax = b$ and $A\hat{x} = b$. Then, $A(x - \hat{x}) = 0$, which implies that $\operatorname{def} A \geq 1$. Therefore, $\operatorname{rank} A = m - \operatorname{def} A \leq m - 1$, which is a contradiction. This proves the first statement of ii). Assuming $Ax = b$ has a unique solution $x \in \mathbb{F}^m$, multiplying by A^{L} yields $x = A^{\mathrm{L}}b$. To prove iii), note that it follows from i) that $Ax = b$ has at least one solution $\hat{x} \in \mathbb{F}^m$. Hence, $x \in \mathbb{F}^m$ is a solution of $Ax = b$ if and only if $A(x - \hat{x}) = 0$, or, equivalently, $x \in \hat{x} + \mathcal{N}(A)$. To prove iv), note that, since $\operatorname{rank} A = n$, it follows that $\operatorname{rank} A = \operatorname{rank} \begin{bmatrix} A & b \end{bmatrix}$, and thus either ii) or iii) applies. $\square$

The set of solutions $x \in \mathbb{F}^m$ to $Ax = b$ is explicitly characterized by Proposition 6.1.7.

Let $A \in \mathbb{F}^{n \times m}$. Then, A is *nonsingular* if there exists a matrix $B \in \mathbb{F}^{m \times n}$, the *inverse* of A, such that $AB = I_n$ and $BA = I_m$, that is, B is both a right and left inverse for A. If A is nonsingular, then, since B is both right and left invertible, it follows from (2.6.1) that $\text{rank}\, A = \text{rank}\, AB = m$ and $\text{rank}\, A = \text{rank}\, BA = n$, and thus $m = n$. Hence, only square matrices can be nonsingular. Furthermore, the inverse $B \in \mathbb{F}^{n \times n}$ of $A \in \mathbb{F}^{n \times n}$ is unique since, if $C \in \mathbb{F}^{n \times n}$ is a left inverse of A, then $C = CI_n = CAB = I_nB = B$, while, if $D \in \mathbb{F}^{n \times n}$ is a right inverse of A, then $D = I_nD = BAD = BI_n = B$. The following result follows from similar arguments and Theorem 2.6.1. This result can be viewed as a specialization of Theorem 1.2.4 to the function $f \colon \mathbb{F}^n \mapsto \mathbb{F}^n$, where $f(x) = Ax$.

Corollary 2.6.4. Let $A \in \mathbb{F}^{n \times n}$. Then, the following statements are equivalent:

  i) A is nonsingular.

 ii) A has a unique inverse.

iii) A is one-to-one.

 iv) A is onto.

  v) A is left invertible.

 vi) A is right invertible.

vii) A has a unique left inverse.

viii) A has a unique right inverse.

 ix) $\text{rank}\, A = n$.

  x) $\text{def}\, A = 0$.

Let $A \in \mathbb{F}^{n \times n}$ be nonsingular. Then, the inverse of A, denoted by A^{-1}, is a unique $n \times n$ matrix with entries in $\mathbb{F}$. If A is not nonsingular, then A is *singular*.

The following result is a specialization of Theorem 2.6.3 to the case $n = m$.

Corollary 2.6.5. Let $A \in \mathbb{F}^{n \times n}$ and $b \in \mathbb{F}^n$. Then, the following statements hold:

  i) A is nonsingular if and only if there exists a unique vector $x \in \mathbb{F}^n$ satisfying $Ax = b$. In this case, $x = A^{-1}b$.

 ii) A is singular and $\text{rank}\, A = \text{rank} \begin{bmatrix} A & b \end{bmatrix}$ if and only if there exist infinitely many $x \in \mathbb{R}^n$ satisfying $Ax = b$. In this case, let $\hat{x} \in \mathbb{F}^m$ satisfy $A\hat{x} = b$. Then, the set of solutions of $Ax = b$ is given by $\hat{x} + \mathcal{N}(A)$.

Proposition 2.6.6. Let $A \in \mathbb{F}^{n \times n}$. Then, the following statements are equivalent:

i) A is nonsingular.

ii) $\overline{A}$ is nonsingular.

iii) A^{T} is nonsingular.

iv) A^* is nonsingular.

In this case,

$$(\overline{A})^{-1} = \overline{A^{-1}}, \tag{2.6.3}$$

$$(A^{\mathrm{T}})^{-1} = (A^{-1})^{\mathrm{T}}, \tag{2.6.4}$$

$$(A^*)^{-1} = (A^{-1})^*. \tag{2.6.5}$$

Proof. Since $AA^{-1} = I$, it follows that $(A^{-1})^*A^* = I$. Hence, $(A^{-1})^* = (A^*)^{-1}$. $\qquad\square$

We thus use $A^{-\mathrm{T}}$ to denote $(A^{\mathrm{T}})^{-1}$ or $(A^{-1})^{\mathrm{T}}$ and A^{-*} to denote $(A^*)^{-1}$ or $(A^{-1})^*$.

Proposition 2.6.7. Let $A, B \in \mathbb{F}^{n \times n}$ be nonsingular. Then,

$$(AB)^{-1} = B^{-1}A^{-1}, \tag{2.6.6}$$

$$(AB)^{-\mathrm{T}} = A^{-\mathrm{T}}B^{-\mathrm{T}}, \tag{2.6.7}$$

$$(AB)^{-*} = A^{-*}B^{-*}. \tag{2.6.8}$$

Proof. Note that $ABB^{-1}A^{-1} = AIA^{-1} = I$, which shows that $B^{-1}A^{-1}$ is the inverse of AB. Similarly, $(AB)^*A^{-*}B^{-*} = B^*A^*A^{-*}B^{-*} = B^*IB^{-*} = I$, which shows that $A^{-*}B^{-*}$ is the inverse of $(AB)^*$. $\qquad\square$

For a nonsingular matrix $A \in \mathbb{F}^{n \times n}$ and $r \in \mathbb{Z}$ we write

$$A^{-r} \triangleq (A^r)^{-1} = (A^{-1})^r, \tag{2.6.9}$$

$$A^{-r\mathrm{T}} \triangleq (A^r)^{-\mathrm{T}} = (A^{-\mathrm{T}})^r = (A^{-r})^{\mathrm{T}} = (A^{\mathrm{T}})^{-r}, \tag{2.6.10}$$

$$A^{-r*} \triangleq (A^r)^{-*} = (A^{-*})^r = (A^{-r})^* = (A^*)^{-r}. \tag{2.6.11}$$

For example, $A^{-2*} = (A^{-*})^2$.

2.7 Determinants

One of the most important quantities associated with a square matrix is its determinant. In this section we develop some basic results pertaining to the determinant of a matrix.

The *determinant* of $A \in \mathbb{F}^{n \times n}$ is defined by

$$\det A \triangleq \sum_{\sigma} (-1)^{N_\sigma} \prod_{i=1}^{n} A_{(i,\sigma(i))}, \qquad (2.7.1)$$

where the sum is taken over all $n!$ permutations $\sigma = (\sigma(1), \ldots, \sigma(n))$ of the column indices $1, \ldots, n$, and where N_σ is the minimal number of pairwise transpositions needed to transform $\sigma(1), \ldots, \sigma(n)$ to $1, \ldots, n$. The following result is an immediate consequence of this definition.

Proposition 2.7.1. Let $A \in \mathbb{F}^{n \times n}$. Then,

$$\det A^{\mathrm{T}} = \det A, \qquad (2.7.2)$$

$$\det \overline{A} = \overline{\det A}, \qquad (2.7.3)$$

$$\det A^* = \overline{\det A}, \qquad (2.7.4)$$

and, for all $\alpha \in \mathbb{F}$,

$$\det \alpha A = \alpha^n \det A. \qquad (2.7.5)$$

If, in addition, $B \in \mathbb{F}^{m \times n}$ and $C \in \mathbb{F}^{m \times m}$, then

$$\det \begin{bmatrix} A & 0 \\ B & C \end{bmatrix} = (\det A)(\det C). \qquad (2.7.6)$$

The following observations are immediate consequences of the definition of the determinant.

Proposition 2.7.2. Let $A, B \in \mathbb{F}^{n \times n}$. Then, the following statements hold:

i) If every off-diagonal entry of A is zero, then

$$\det A = \prod_{i=1}^{n} A_{(i,i)}. \qquad (2.7.7)$$

In particular, $\det I_n = 1$.

ii) If A has a row or column consisting entirely of zeros, then $\det A = 0$.

iii) If A has two identical rows or two identical columns, then $\det A = 0$.

iv) If $x \in \mathbb{F}^n$ and $i \in \{1, \ldots, n\}$, then

$$\det\left(A + x e_i^{\mathrm{T}}\right) = \det A + \det\left(A \overset{i}{\leftarrow} x\right). \qquad (2.7.8)$$

v) If $x \in \mathbb{F}^{1 \times n}$ and $i \in \{1, \ldots, n\}$, then

$$\det(A + e_i x) = \det A + \det\left(A \overset{i}{\leftarrow} x\right). \qquad (2.7.9)$$

vi) If B is identical to A except that, for some $i \in \{1, \ldots, n\}$ and $\alpha \in \mathbb{F}$, either $\mathrm{col}_i(B) = \alpha \mathrm{col}_i(A)$ or $\mathrm{row}_i(B) = \alpha \mathrm{row}_i(A)$, then $\det B = \alpha \det A$.

vii) If B is formed from A by interchanging two rows or two columns of A, then $\det B = -\det A$.

$viii$) If B is formed from A by adding a multiple of a (row, column) of A to another (row, column) of A, then $\det B = \det A$.

Statements vi)–$viii$) correspond, respectively, to multiplying the matrix A on the left or right by matrices of the form

$$I_n + (\alpha - 1)E_{i,i} = \begin{bmatrix} I_{i-1} & 0 & 0 \\ 0 & \alpha & 0 \\ 0 & 0 & I_{n-i} \end{bmatrix}, \qquad (2.7.10)$$

$$I_n + E_{i,j} + E_{j,i} - E_{i,i} - E_{j,j} = \begin{bmatrix} I_{i-1} & 0 & 0 & 0 & 0 \\ 0 & 0 & 0 & 1 & 0 \\ 0 & 0 & I_{j-i-1} & 0 & 0 \\ 0 & 1 & 0 & 0 & 0 \\ 0 & 0 & 0 & 0 & I_{n-j} \end{bmatrix}, \qquad (2.7.11)$$

where $i \neq j$, and

$$I_n + \beta E_{i,j} = \begin{bmatrix} I_{i-1} & 0 & 0 & 0 & 0 \\ 0 & 1 & 0 & \beta & 0 \\ 0 & 0 & I_{j-i-1} & 0 & 0 \\ 0 & 0 & 0 & 1 & 0 \\ 0 & 0 & 0 & 0 & I_{n-j} \end{bmatrix}, \qquad (2.7.12)$$

where $\beta \in \mathbb{F}$ and $i \neq j$. The matrices in (2.7.11) and (2.7.12) illustrate the case $i < j$. Since $I + (\alpha - 1)E_{i,i} = I + (\alpha - 1)e_i e_i^{\mathrm{T}}$, $I + E_{i,j} + E_{j,i} - E_{i,i} - E_{j,j} = I - (e_i - e_j)(e_i - e_j)^{\mathrm{T}}$, and $I + \beta E_{i,j} = I + \beta e_i e_j^{\mathrm{T}}$, it follows that all of these matrices are of the form $I - xy^{\mathrm{T}}$. In terms of Definition 3.1.1, (2.7.10) is an elementary matrix if and only if $\alpha \neq 0$, (2.7.11) is an elementary matrix, and (2.7.12) is an elementary matrix if and only if either $i \neq j$ or $\beta \neq -1$.

Proposition 2.7.3. Let $A, B \in \mathbb{F}^{n \times n}$. Then,

$$\det AB = \det BA = (\det A)(\det B). \qquad (2.7.13)$$

Proof. First note the identity

$$\begin{bmatrix} A & 0 \\ I & B \end{bmatrix} = \begin{bmatrix} I & A \\ 0 & I \end{bmatrix} \begin{bmatrix} -AB & 0 \\ 0 & I \end{bmatrix} \begin{bmatrix} I & 0 \\ B & I \end{bmatrix} \begin{bmatrix} 0 & I \\ I & 0 \end{bmatrix}.$$

The first and third matrices on the right-hand side of this identity add multiples of rows and columns of $\left[\begin{smallmatrix} -AB & 0 \\ 0 & I \end{smallmatrix}\right]$ to other rows and columns of $\left[\begin{smallmatrix} -AB & 0 \\ 0 & I \end{smallmatrix}\right]$. As already noted, these operations do not affect the determinant of $\left[\begin{smallmatrix} -AB & 0 \\ 0 & I \end{smallmatrix}\right]$. In addition, the fourth matrix on the right-hand side of this identity interchanges n pairs of columns of $\left[\begin{smallmatrix} 0 & A \\ B & I \end{smallmatrix}\right]$. Using (2.7.5), (2.7.6), and the fact that every interchange of a pair of columns of $\left[\begin{smallmatrix} 0 & A \\ B & I \end{smallmatrix}\right]$ entails a factor of -1, it thus follows that $(\det A)(\det B) = \det\left[\begin{smallmatrix} A & 0 \\ I & B \end{smallmatrix}\right] = (-1)^n\det\left[\begin{smallmatrix} -AB & 0 \\ 0 & I \end{smallmatrix}\right] = (-1)^n\det(-AB) = \det AB$. $\qquad\square$

Corollary 2.7.4. Let $A \in \mathbb{F}^{n\times n}$ be nonsingular. Then, $\det A \neq 0$ and

$$\det A^{-1} = (\det A)^{-1}. \qquad (2.7.14)$$

Proof. Since $AA^{-1} = I_n$, it follows that $\det AA^{-1} = (\det A)(\det A^{-1}) = 1$. Hence, $\det A \neq 0$. In addition, $\det A^{-1} = 1/\det A$. $\qquad\square$

Let $A \in \mathbb{F}^{n\times m}$. Then, a *submatrix* of A is formed by deleting rows and columns of A. By convention, A is a submatrix of A. If A is a partitioned matrix, then every block of A is a submatrix of A. A block is thus a submatrix whose entries are entries of adjacent rows and adjacent columns. The determinant of a square submatrix of A is a *subdeterminant* of A. By convention, the determinant of A is a subdeterminant of A.

Let $A \in \mathbb{F}^{n\times n}$. If like-numbered rows and columns of A are deleted, then the resulting square submatrix of A is a *principal submatrix* of A. If, in particular, rows and columns $j+1,\ldots,n$ of A are deleted, then the resulting $j \times j$ submatrix of A is a *leading principal submatrix* of A. Every diagonally located block is a principal submatrix. Finally, the determinant of a $j \times j$ (principal, leading principal) submatrix of A is a $j \times j$ (*principal, leading principal*) *subdeterminant* of A.

Let $A \in \mathbb{F}^{n\times n}$. Then, the *cofactor* of $A_{(i,j)}$, denoted by $A_{[i,j]}$, is the $(n-1)\times(n-1)$ submatrix of A obtained by deleting the ith row and jth column of A. The following result provides a cofactor expansion of $\det A$.

Proposition 2.7.5. Let $A \in \mathbb{F}^{n\times n}$. Then, for all $i = 1,\ldots,n$,

$$\det A = \sum_{k=1}^n (-1)^{i+k}A_{(i,k)}\det A_{[i,k]}. \qquad (2.7.15)$$

Furthermore, for all $i,j = 1,\ldots,n$ such that $j \neq i$,

$$0 = \sum_{k=1}^n (-1)^{i+k}A_{(j,k)}\det A_{[i,k]}. \qquad (2.7.16)$$

Proof. Identity (2.7.15) is an equivalent recursive form of the definition $\det A$, while the right-hand side of (2.7.16) is equal to $\det B$, where B is obtained from A by replacing $\text{row}_i(A)$ by $\text{row}_j(A)$. As already noted, $\det B = 0$. $\qquad\square$

Let $A \in \mathbb{F}^{n\times n}$. To simplify (2.7.15) and (2.7.16) it is useful to define the *adjugate* of A, denoted by $A^{\mathrm{A}} \in \mathbb{F}^{n\times n}$, where, for all $i, j = 1, \ldots, n$,

$$\left(A^{\mathrm{A}}\right)_{(i,j)} \triangleq (-1)^{i+j}\det A_{[j,i]}. \tag{2.7.17}$$

Then, (2.7.15) and (2.7.16) imply that, for all $i = 1, \ldots, n$,

$$\left(AA^{\mathrm{A}}\right)_{(i,i)} = \left(A^{\mathrm{A}}A\right)_{(i,i)} = \det A \tag{2.7.18}$$

and, for all $i, j = 1, \ldots, n$ such that $j \neq i$,

$$\left(AA^{\mathrm{A}}\right)_{(i,j)} = \left(A^{\mathrm{A}}A\right)_{(i,j)} = 0. \tag{2.7.19}$$

Thus,

$$AA^{\mathrm{A}} = A^{\mathrm{A}}A = (\det A)I. \tag{2.7.20}$$

Consequently, if $\det A \neq 0$, then

$$A^{-1} = (\det A)^{-1}A^{\mathrm{A}}. \tag{2.7.21}$$

The following result provides the converse of Corollary 2.7.4 by using (2.7.21) to construct A^{-1} in terms of $(n-1) \times (n-1)$ subdeterminants of A.

Corollary 2.7.6. Let $A \in \mathbb{F}^{n\times n}$. Then, A is nonsingular if and only if $\det A \neq 0$. In this case, for all $i, j = 1, \ldots, n$, the (i,j) entry of A^{-1} is given by

$$\left(A^{-1}\right)_{(i,j)} = (-1)^{i+j}\frac{\det A_{[j,i]}}{\det A}. \tag{2.7.22}$$

Finally, the following result uses the nonsingularity of submatrices to characterize the rank of a matrix.

Proposition 2.7.7. Let $A \in \mathbb{F}^{n\times m}$. Then, $\text{rank}\, A$ is the largest order of all nonsingular submatrices of A.

2.8 Properties of Partitioned Matrices

Partitioned matrices were used to state or prove several results in this chapter including Proposition 2.5.8, Theorem 2.6.3, Proposition 2.7.1, and Proposition 2.7.3. In this section we give several useful identities involving partitioned matrices.

Proposition 2.8.1. Let $A_{ij} \in \mathbb{F}^{n_i \times m_j}$ for all $i = 1, \ldots, k$ and $j = 1, \ldots, l$. Then,

$$\begin{bmatrix} A_{11} & \cdots & A_{1l} \\ \vdots & \ddots & \vdots \\ A_{k1} & \cdots & A_{kl} \end{bmatrix}^{\mathrm{T}} = \begin{bmatrix} A_{11}^{\mathrm{T}} & \cdots & A_{k1}^{\mathrm{T}} \\ \vdots & \ddots & \vdots \\ A_{1l}^{\mathrm{T}} & \cdots & A_{kl}^{\mathrm{T}} \end{bmatrix} \qquad (2.8.1)$$

and

$$\begin{bmatrix} A_{11} & \cdots & A_{1l} \\ \vdots & \ddots & \vdots \\ A_{k1} & \cdots & A_{kl} \end{bmatrix}^{*} = \begin{bmatrix} A_{11}^{*} & \cdots & A_{k1}^{*} \\ \vdots & \ddots & \vdots \\ A_{1l}^{*} & \cdots & A_{kl}^{*} \end{bmatrix}. \qquad (2.8.2)$$

If, in addition, $k = l$ and $n_i = m_i$ for all $i = 1, \ldots, m$, then

$$\mathrm{tr} \begin{bmatrix} A_{11} & \cdots & A_{1k} \\ \vdots & \ddots & \vdots \\ A_{k1} & \cdots & A_{kk} \end{bmatrix} = \sum_{i=1}^{k} \mathrm{tr}\, A_{ii} \qquad (2.8.3)$$

and

$$\det \begin{bmatrix} A_{11} & A_{12} & \cdots & A_{1k} \\ 0 & A_{22} & \cdots & A_{2k} \\ \vdots & \ddots & \ddots & \vdots \\ 0 & 0 & \cdots & A_{kk} \end{bmatrix} = \prod_{i=1}^{k} \det A_{ii}. \qquad (2.8.4)$$

Lemma 2.8.2. Let $B \in \mathbb{F}^{n \times m}$ and $C \in \mathbb{F}^{m \times n}$. Then,

$$\begin{bmatrix} I & B \\ 0 & I \end{bmatrix}^{-1} = \begin{bmatrix} I & -B \\ 0 & I \end{bmatrix} \qquad (2.8.5)$$

and

$$\begin{bmatrix} I & 0 \\ C & I \end{bmatrix}^{-1} = \begin{bmatrix} I & 0 \\ -C & I \end{bmatrix}. \qquad (2.8.6)$$

Let $A \in \mathbb{F}^{n \times n}$ and $D \in \mathbb{F}^{m \times m}$ be nonsingular. Then,

$$\begin{bmatrix} A & 0 \\ 0 & D \end{bmatrix}^{-1} = \begin{bmatrix} A^{-1} & 0 \\ 0 & D^{-1} \end{bmatrix}. \qquad (2.8.7)$$

Proposition 2.8.3. Let $A \in \mathbb{F}^{n \times n}$, $B \in \mathbb{F}^{n \times m}$, $C \in \mathbb{F}^{l \times n}$, and $D \in \mathbb{F}^{l \times m}$, and assume that A is nonsingular. Then,

$$\begin{bmatrix} A & B \\ C & D \end{bmatrix} = \begin{bmatrix} I & 0 \\ CA^{-1} & I \end{bmatrix} \begin{bmatrix} A & 0 \\ 0 & D - CA^{-1}B \end{bmatrix} \begin{bmatrix} I & A^{-1}B \\ 0 & I \end{bmatrix} \qquad (2.8.8)$$

and

$$\mathrm{rank} \begin{bmatrix} A & B \\ C & D \end{bmatrix} = n + \mathrm{rank}(D - CA^{-1}B). \qquad (2.8.9)$$

If, furthermore, $l = m$, then

$$\det \begin{bmatrix} A & B \\ C & D \end{bmatrix} = (\det A) \det(D - CA^{-1}B). \qquad (2.8.10)$$

Proposition 2.8.4. Let $A \in \mathbb{F}^{n \times m}$, $B \in \mathbb{F}^{n \times l}$, $C \in \mathbb{F}^{l \times m}$, and $D \in \mathbb{F}^{l \times l}$, and assume that D is nonsingular. Then,

$$\begin{bmatrix} A & B \\ C & D \end{bmatrix} = \begin{bmatrix} I & BD^{-1} \\ 0 & I \end{bmatrix} \begin{bmatrix} A - BD^{-1}C & 0 \\ 0 & D \end{bmatrix} \begin{bmatrix} I & 0 \\ D^{-1}C & I \end{bmatrix} \qquad (2.8.11)$$

and

$$\operatorname{rank} \begin{bmatrix} A & B \\ C & D \end{bmatrix} = l + \operatorname{rank}(A - BD^{-1}C). \qquad (2.8.12)$$

If, furthermore, $n = m$, then

$$\det \begin{bmatrix} A & B \\ C & D \end{bmatrix} = (\det D) \det(A - BD^{-1}C). \qquad (2.8.13)$$

Corollary 2.8.5. Let $A \in \mathbb{F}^{n \times m}$ and $B \in \mathbb{F}^{m \times n}$. Then,

$$\det \begin{bmatrix} I_n & A \\ -B & I_m \end{bmatrix} = \det(I_n + AB) = \det(I_m + BA). \qquad (2.8.14)$$

Hence, $I_n + AB$ is nonsingular if and only if $I_m + BA$ is nonsingular.

Lemma 2.8.6. Let $A \in \mathbb{F}^{n \times n}$, $B \in \mathbb{F}^{n \times m}$, $C \in \mathbb{F}^{m \times n}$, and $D \in \mathbb{F}^{m \times m}$. If A and D are nonsingular, then

$$(\det A) \det(D - CA^{-1}B) = (\det D) \det(A - BD^{-1}C), \qquad (2.8.15)$$

and thus $D - CA^{-1}B$ is nonsingular if and only if $A - BD^{-1}C$ is nonsingular.

Proposition 2.8.7. Let $A \in \mathbb{F}^{n \times n}$, $B \in \mathbb{F}^{n \times m}$, $C \in \mathbb{F}^{m \times n}$, and $D \in \mathbb{F}^{m \times m}$. If A and $D - CA^{-1}B$ are nonsingular, then

$$\begin{bmatrix} A & B \\ C & D \end{bmatrix}^{-1}$$

$$= \begin{bmatrix} A^{-1} + A^{-1}B(D - CA^{-1}B)^{-1}CA^{-1} & -A^{-1}B(D - CA^{-1}B)^{-1} \\ -(D - CA^{-1}B)^{-1}CA^{-1} & (D - CA^{-1}B)^{-1} \end{bmatrix}.$$

$$(2.8.16)$$

If D and $A - BD^{-1}C$ are nonsingular, then

$$
\begin{bmatrix} A & B \\ C & D \end{bmatrix}^{-1}
$$

$$
= \begin{bmatrix} \left(A - BD^{-1}C\right)^{-1} & -\left(A - BD^{-1}C\right)^{-1}BD^{-1} \\ -D^{-1}C\left(A - BD^{-1}C\right)^{-1} & D^{-1} + D^{-1}C\left(A - BD^{-1}C\right)^{-1}BD^{-1} \end{bmatrix}.
$$
(2.8.17)

If A, D, and $D - CA^{-1}B$ are nonsingular, then $A - BD^{-1}C$ is nonsingular, and

$$
\begin{bmatrix} A & B \\ C & D \end{bmatrix}^{-1}
$$

$$
= \begin{bmatrix} \left(A - BD^{-1}C\right)^{-1} & -\left(A - BD^{-1}C\right)^{-1}BD^{-1} \\ -\left(D - CA^{-1}B\right)^{-1}CA^{-1} & \left(D - CA^{-1}B\right)^{-1} \end{bmatrix}.
$$
(2.8.18)

The following result is the *matrix inversion lemma*.

Corollary 2.8.8. Let $A \in \mathbb{F}^{n \times n}$, $B \in \mathbb{F}^{n \times m}$, $C \in \mathbb{F}^{m \times n}$, and $D \in \mathbb{F}^{m \times m}$. If A, $D - CA^{-1}B$, and D are nonsingular, then $A - BD^{-1}C$ is nonsingular, and

$$
\left(A - BD^{-1}C\right)^{-1} = A^{-1} + A^{-1}B\left(D - CA^{-1}B\right)^{-1}CA^{-1}.
$$
(2.8.19)

If A and $I - CA^{-1}B$ are nonsingular, then $A - BC$ is nonsingular, and

$$
(A - BC)^{-1} = A^{-1} + A^{-1}B\left(I - CA^{-1}B\right)^{-1}CA^{-1}.
$$
(2.8.20)

If $D - CB$, and D are nonsingular, then $I_n - BD^{-1}C$ is nonsingular, and

$$
\left(I_n - BD^{-1}C\right)^{-1} = I_n + B(D - CB)^{-1}C.
$$
(2.8.21)

If $I - CB$ is nonsingular, then $I - BC$ is nonsingular, and

$$
(I - BC)^{-1} = I + B(I - CB)^{-1}C.
$$
(2.8.22)

Corollary 2.8.9. Let $A, B, C, D \in \mathbb{F}^{n \times n}$. If A, B, $C - DB^{-1}A$, and $D - CA^{-1}B$ are nonsingular, then

$$
\begin{bmatrix} A & B \\ C & D \end{bmatrix}^{-1} = \begin{bmatrix} A^{-1} - \left(C - DB^{-1}A\right)^{-1}CA^{-1} & \left(C - DB^{-1}A\right)^{-1} \\ -\left(D - CA^{-1}B\right)^{-1}CA^{-1} & \left(D - CA^{-1}B\right)^{-1} \end{bmatrix}.
$$
(2.8.23)

If A, C, $B - AC^{-1}D$, and $D - CA^{-1}B$ are nonsingular, then

$$\begin{bmatrix} A & B \\ C & D \end{bmatrix}^{-1} = \begin{bmatrix} A^{-1} - A^{-1}B(B - AC^{-1}D)^{-1} & -A^{-1}B(D - CA^{-1}B)^{-1} \\ (B - AC^{-1}D)^{-1} & (D - CA^{-1}B)^{-1} \end{bmatrix}.$$
$$(2.8.24)$$

If A, B, C, $B - AC^{-1}D$, and $D - CA^{-1}B$ are nonsingular, then $C - DB^{-1}A$ is nonsingular, and

$$\begin{bmatrix} A & B \\ C & D \end{bmatrix}^{-1} = \begin{bmatrix} A^{-1} - A^{-1}B(B - AC^{-1}D)^{-1} & (C - DB^{-1}A)^{-1} \\ (B - AC^{-1}D)^{-1} & (D - CA^{-1}B)^{-1} \end{bmatrix}.$$
$$(2.8.25)$$

If B, D, $A - BD^{-1}C$, and $C - DB^{-1}A$ are nonsingular, then

$$\begin{bmatrix} A & B \\ C & D \end{bmatrix}^{-1} = \begin{bmatrix} (A - BD^{-1}C)^{-1} & (C - DB^{-1}A)^{-1} \\ -D^{-1}C(A - BD^{-1}C)^{-1} & D^{-1} - D^{-1}C(C - DB^{-1}A)^{-1} \end{bmatrix}.$$
$$(2.8.26)$$

If C, D, $A - BD^{-1}C$, and $B - AC^{-1}D$ are nonsingular, then

$$\begin{bmatrix} A & B \\ C & D \end{bmatrix}^{-1} = \begin{bmatrix} (A - BD^{-1}C)^{-1} & -(A - BD^{-1}C)^{-1}BD^{-1} \\ (B - AC^{-1}D)^{-1} & D^{-1} - (B - AC^{-1}D)^{-1}BD^{-1} \end{bmatrix}.$$
$$(2.8.27)$$

If B, C, D, $A - BD^{-1}C$, and $C - DB^{-1}A$ are nonsingular, then $B - AC^{-1}D$ is nonsingular, and

$$\begin{bmatrix} A & B \\ C & D \end{bmatrix}^{-1} = \begin{bmatrix} (A - BD^{-1}C)^{-1} & (C - DB^{-1}A) \\ (B - AC^{-1}D)^{-1} & D^{-1} - D^{-1}C(C - DB^{-1}A)^{-1} \end{bmatrix}.$$
$$(2.8.28)$$

Finally, if A, B, C, D, $A - BD^{-1}C$, and $B - AC^{-1}D$, are nonsingular, then $C - DB^{-1}A$ and $D - CA^{-1}B$ are nonsingular, and

$$\begin{bmatrix} A & B \\ C & D \end{bmatrix}^{-1} = \begin{bmatrix} (A - BD^{-1}C)^{-1} & (C - DB^{-1}A)^{-1} \\ (B - AC^{-1}D)^{-1} & (D - CA^{-1}B)^{-1} \end{bmatrix}.$$
$$(2.8.29)$$

Corollary 2.8.10. Let $A, B \in \mathbb{F}^{n \times n}$, and assume that A and $I - A^{-1}B$ are nonsingular. Then, $A - B$ is nonsingular, and

$$(A - B)^{-1} = A^{-1} + A^{-1}B(I - A^{-1}B)^{-1}A^{-1}.$$
$$(2.8.30)$$

If, in addition, B is nonsingular, then

$$(A - B)^{-1} = A^{-1} + A^{-1}(B^{-1} - A^{-1})^{-1}A^{-1}. \qquad (2.8.31)$$

2.9 Facts on Cones, Convex Hulls, and Subspaces

Fact 2.9.1. Let $S \subseteq \mathbb{F}^n$. Then, the following statements hold:

i) $\operatorname{coco} S = \operatorname{co} \operatorname{cone} S = \operatorname{cone} \operatorname{co} S$.

ii) $S^{\perp\perp} = \operatorname{span} S = \operatorname{coco}(S \cup -S)$.

iii) $S \subseteq \operatorname{co} S \subseteq (\operatorname{aff} S \cap \operatorname{coco} S) \subseteq \left\{ {\operatorname{aff} S \atop \operatorname{coco} S} \right\} \subseteq \operatorname{span} S$.

iv) $S \subseteq (\operatorname{co} S \cap \operatorname{cone} S) \subseteq \left\{ {\operatorname{co} S \atop \operatorname{cone} S} \right\} \subseteq \operatorname{coco} S \subseteq \operatorname{span} S$.

v) $\operatorname{dcone} \operatorname{dcone} S = \operatorname{coco} S$.

(Proof: See [97, p. 52] for a proof of *v*). Note that "pointed" in [97] means one-sided.)

Fact 2.9.2. Let $S \subseteq \mathbb{F}^n$, and assume that S is convex. Then, $S + S = 2S$. (Proof: "$\supseteq$" is immediate. To prove "$\subseteq$," let $x, y \in S$. Then, $x + y = 2\left(\frac{1}{2}x + \frac{1}{2}y\right)$.)

Fact 2.9.3. Let $S \subset \mathbb{F}^n$. Then, S is an affine hyperplane if and only if there exist a nonzero vector $y \in \mathbb{F}^n$ and $\alpha \in \mathbb{R}$ such that $S = \{x: \operatorname{Re} x^*y = \alpha\}$. Furthermore, S is an affine closed half space if and only if there exist a nonzero vector $y \in \mathbb{F}^n$ and $\alpha \in \mathbb{R}$ such that $S = \{x \in \mathbb{F}^n: \operatorname{Re} x^*y \leq \alpha\}$. Finally, S is an affine open half space if and only if there exist a nonzero vector $y \in \mathbb{F}^n$ and $\alpha \in \mathbb{R}$ such that $S = \{x \in \mathbb{F}^n: \operatorname{Re} x^*y \leq \alpha\}$. (Proof: Let $z \in \mathbb{F}^n$ satisfy $z^*y = \alpha$. Then, $\{x: x^*y = \alpha\} = \{y\}^\perp + z$.)

Fact 2.9.4. Let $S_1, S_2 \subseteq \mathbb{F}^n$ be (cones, convex sets, convex cones, subspaces). Then, so are $S_1 \cap S_2$ and $S_1 + S_2$.

Fact 2.9.5. Let $S_1, S_2 \subseteq \mathbb{F}^n$ be pointed convex cones. Then,

$$\operatorname{co}(S_1 \cup S_2) = S_1 + S_2.$$

Fact 2.9.6. Let $S_1, S_2 \subseteq \mathbb{F}^n$ be subspaces. Then, $S_1 \cup S_2$ is a subspace if and only if either $S_1 \subseteq S_2$ or $S_2 \subseteq S_1$.

Fact 2.9.7. Let $S_1, S_2 \subseteq \mathbb{F}^n$ be subspaces. Then,

$$\operatorname{span}(S_1 \cup S_2) = S_1 + S_2.$$

Fact 2.9.8. Let $\mathcal{S}_1, \mathcal{S}_2 \subseteq \mathbb{F}^n$ be subspaces. Then, $\mathcal{S}_1 \subseteq \mathcal{S}_2$ if and only if $\mathcal{S}_2^\perp \subseteq \mathcal{S}_1^\perp$. Furthermore, $\mathcal{S}_1 \subset \mathcal{S}_2$ if and only if $\mathcal{S}_2^\perp \subset \mathcal{S}_1^\perp$.

Fact 2.9.9. Let $\mathcal{S}_1, \mathcal{S}_2 \subseteq \mathbb{F}^n$. Then,

$$\mathcal{S}_1^\perp \cap \mathcal{S}_2^\perp \subseteq (\mathcal{S}_1 + \mathcal{S}_2)^\perp.$$

(Problem: Determine necessary and sufficient conditions under which equality holds.)

Fact 2.9.10. Let $\mathcal{S}_1, \mathcal{S}_2 \subseteq \mathbb{F}^n$ be subspaces. Then,

$$(\mathcal{S}_1 \cap \mathcal{S}_2)^\perp = \mathcal{S}_1^\perp + \mathcal{S}_2^\perp$$

and

$$(\mathcal{S}_1 + \mathcal{S}_2)^\perp = \mathcal{S}_1^\perp \cap \mathcal{S}_2^\perp.$$

Fact 2.9.11. Let $\mathcal{S}_1, \mathcal{S}_2, \mathcal{S}_3 \subseteq \mathbb{F}^n$ be subspaces. Then,

$$\mathcal{S}_1 + (\mathcal{S}_2 \cap \mathcal{S}_3) \subseteq (\mathcal{S}_1 + \mathcal{S}_2) \cap (\mathcal{S}_1 + \mathcal{S}_3)$$

and

$$\mathcal{S}_1 \cap (\mathcal{S}_2 + \mathcal{S}_3) \supseteq (\mathcal{S}_1 \cap \mathcal{S}_2) + (\mathcal{S}_1 \cap \mathcal{S}_3).$$

Fact 2.9.12. Let $\mathcal{S}_1, \mathcal{S}_2 \subseteq \mathbb{F}^n$ be subspaces. Then, $\mathcal{S}_1, \mathcal{S}_2$ are complementary if and only if $\mathcal{S}_1^\perp, \mathcal{S}_2^\perp$ are complementary. (Remark: See Fact 3.8.1.)

Fact 2.9.13. Let $\mathcal{S}_1, \ldots, \mathcal{S}_k \subseteq \mathbb{F}^n$ be subspaces having the same dimension. Then, there exists a subspace $\hat{\mathcal{S}} \subseteq \mathbb{F}^n$ such that, for all $i = 1, \ldots, k$, $\hat{\mathcal{S}}$ and $\mathcal{S}_i$ are complementary. (Proof: See [328, pp. 78, 79, 259, 260].)

Fact 2.9.14. Let $\mathcal{S}_1, \mathcal{S}_2 \subseteq \mathbb{F}^n$ be subspaces. Then,

$$\dim(\mathcal{S}_1 \cap \mathcal{S}_2) \leq \min\{\dim \mathcal{S}_1, \dim \mathcal{S}_2\}$$
$$\leq \left\{ \begin{matrix} \dim \mathcal{S}_1 \\ \dim \mathcal{S}_2 \end{matrix} \right\}$$
$$\leq \max\{\dim \mathcal{S}_1, \dim \mathcal{S}_2\}$$
$$\leq \dim(\mathcal{S}_1 + \mathcal{S}_2)$$
$$\leq \min\{\dim \mathcal{S}_1 + \dim \mathcal{S}_2, n\}.$$

Fact 2.9.15. Let $\mathcal{S} \subseteq \mathbb{F}^n$ be a subspace. Then, for all $m \geq \dim \mathcal{S}$, there exists a matrix $A \in \mathbb{F}^{n \times m}$ such that $\mathcal{S} = \mathcal{R}(A)$.

Fact 2.9.16. Let $A \in \mathbb{F}^{n \times n}$, let $\mathcal{S} \subseteq \mathbb{F}^n$, assume that $\mathcal{S}$ is a subspace, let $k \triangleq \dim \mathcal{S}$, let $S \in \mathbb{F}^{n \times k}$, and assume that $\mathcal{R}(S) = \mathcal{S}$. Then, $\mathcal{S}$ is an invariant subspace of A if and only if there exists a matrix $M \in \mathbb{F}^{k \times k}$ such that $AS = SM$. (Proof: See [456, p. 99].) (Remark: See Fact 5.11.11.)

Fact 2.9.17. Let $S \subseteq \mathbb{F}^m$ and $A \in \mathbb{F}^{n \times m}$. If S is convex, then AS is convex. Conversely, if A is left invertible and AS is convex, then S is convex.

Fact 2.9.18. Let $S \subseteq \mathbb{F}^m$, assume that S is a cone, and let $A \in \mathbb{F}^{n \times m}$. Then, AS is a cone. If, in addition, S is a pointed cone, then AS is a pointed cone. Finally, if S is a (subspace, affine subspace), then so is AS.

Fact 2.9.19. Let $S_1, S_2 \subseteq \mathbb{F}^m$ be subspaces, and let $A \in \mathbb{F}^{n \times m}$. Then,

$$A(S_1 \cap S_2) \subseteq AS_1 \cap AS_2$$

and

$$A(S_1 + S_2) = AS_1 + AS_2.$$

(Proof: See [177, p. 12].)

Fact 2.9.20. Let $A \in \mathbb{F}^{n \times m}$, and let $S_1 \subseteq \mathbb{R}^m$ and $S_2 \subseteq \mathbb{F}^n$ be subspaces. Then, the following statements are equivalent:

i) $AS_1 \subseteq S_2$.

ii) $A^* S_2^{\perp} \subseteq S_1^{\perp}$.

(Proof: See [177, p. 12].)

Fact 2.9.21. Let $S, S_1, S_2 \subseteq \mathbb{F}^n$ be subspaces, let $A \in \mathbb{F}^{n \times m}$, and define $f: \mathbb{F}^m \mapsto \mathbb{F}^n$ by $f(x) \triangleq Ax$. Then,

$$[f^{-1}(S)]^{\perp} = A^* S^{\perp},$$

$$f^{-1}(S_1 \cap S_2) = f^{-1}(S_1) \cap f^{-1}(S_2)$$

and

$$f^{-1}(S_1 + S_2) \supseteq f^{-1}(S_1) + f^{-1}(S_2).$$

(Proof: See [177, p. 12].) (Problem: For a subspace $S \subseteq \mathbb{F}^n$, $A \in \mathbb{F}^{n \times m}$, and $f(x) \triangleq Ax$, determine $B \in \mathbb{F}^{m \times n}$ such that $f^{-1}(S) = BS$, that is, $ABS \subseteq S$ and BS is maximal.)

2.10 Facts on Range, Null Space, Rank, and Defect

Fact 2.10.1. Let $n, m, k \in \mathbb{P}$. Then, $\operatorname{rank} 1_{n \times m} = 1$ and $1_{n \times n}^k = n^{k-1} 1_{n \times n}$.

Fact 2.10.2. Let $A \in \mathbb{F}^{n \times m}$. Then, $\operatorname{rank} A = 1$ if and only if there exist vectors $x \in \mathbb{F}^n$ and $y \in \mathbb{F}^m$ such that $x \neq 0$, $y \neq 0$, and $A = xy^{\mathrm{T}}$. In this case, $\operatorname{tr} A = y^{\mathrm{T}} x$.

Fact 2.10.3. Let $A \in \mathbb{F}^{n \times n}$, $k \in \mathbb{P}$, and $l \in \mathbb{N}$. Then, the following identities hold:

i) $\mathcal{R}\left[(AA^*)^k\right] = \mathcal{R}\left[(AA^*)^l A\right]$.

ii) $\mathcal{N}\left[(A^*A)^k\right] = \mathcal{N}\left[A(A^*A)^l\right]$.

iii) rank $(AA^*)^k = $ rank $(AA^*)^l A$.

iv) def $(A^*A)^k = $ def $A(A^*A)^l$.

Fact 2.10.4. Let $A \in \mathbb{F}^{n \times n}$. Then,

$$\text{rank } A + \text{rank}(A - A^3) = \text{rank}(A + A^2) + \text{rank}(A - A^2).$$

Consequently,

$$\text{rank } A \leq \text{rank}(A + A^2) + \text{rank}(A - A^2),$$

and A is tripotent if and only if

$$\text{rank } A = \text{rank}(A + A^2) + \text{rank}(A - A^2).$$

(Proof: See [714].) (Remark: This result is due to Anderson and Styan.)

Fact 2.10.5. Let $A, B \in \mathbb{F}^{n \times n}$, and assume there exists $\alpha \in \mathbb{F}$ such that $\alpha A + B$ is nonsingular. Then, $\mathcal{N}(A) \cap \mathcal{N}(B) = \{0\}$. (Remark: The converse is not true. Let $A \triangleq \left[\begin{smallmatrix} 1 & 0 \\ 2 & 0 \end{smallmatrix}\right]$ and $B \triangleq \left[\begin{smallmatrix} 0 & 1 \\ 0 & 2 \end{smallmatrix}\right]$.)

Fact 2.10.6. Let $A, B \in \mathbb{F}^{n \times m}$. Then,

$$\mathcal{N}(A) \cap \mathcal{N}(B) = \mathcal{N}(A) \cap \mathcal{N}(A + B) = \mathcal{N}(A + B) \cap \mathcal{N}(B).$$

Fact 2.10.7. Let $A, B \in \mathbb{F}^{n \times m}$. Then,

$$|\text{rank } A - \text{rank } B| \leq \text{rank}(A + B) \leq \text{rank } A + \text{rank } B.$$

If, in addition, rank $B \leq k$, then

$$(\text{rank } A) - k \leq \text{rank}(A + B) \leq (\text{rank } A) + k.$$

Fact 2.10.8. Let $A, B \in \mathbb{F}^{n \times m}$, and assume that $A^*B = 0$ and $BA^* = 0$. Then,

$$\text{rank}(A + B) = \text{rank } A + \text{rank } B.$$

(Proof: Use Fact 2.10.9 and Proposition 6.1.6. See [187].)

Fact 2.10.9. Let $A, B \in \mathbb{F}^{n \times m}$. Then,

$$\text{rank}(A + B) = \text{rank } A + \text{rank } B$$

if and only if there exists a matrix $C \in \mathbb{F}^{m \times n}$ such that $ACA = A$, $CB = 0$, and $BC = 0$. (Proof: See [187].)

Fact 2.10.10. Let $x, y \in \mathbb{F}^n$. Then,

$$\operatorname{rank}(xy^{\mathrm{T}} + yx^{\mathrm{T}}) \leq 2.$$

Furthermore, $\operatorname{rank}(xy^{\mathrm{T}} + yx^{\mathrm{T}}) = 1$ if and only if there exists $\alpha \in \mathbb{F}$ such that $x = \alpha y \neq 0$.

Fact 2.10.11. Let $A \in \mathbb{F}^{n \times m}$, $x \in \mathbb{F}^n$, and $y \in \mathbb{F}^m$. Then,

$$(\operatorname{rank} A) - 1 \leq \operatorname{rank}(A + xy^{\mathrm{T}}) \leq (\operatorname{rank} A) + 1.$$

In addition, the following statements hold:

i) $\operatorname{rank}(A + xy^{\mathrm{T}}) = (\operatorname{rank} A) - 1$ if and only if there exist vectors $\hat{x} \in \mathbb{F}^m$ and $\hat{y} \in \mathbb{F}^n$ such that $\hat{y}^{\mathrm{T}} A \hat{x} \neq 0$, $x = -(\hat{y}^{\mathrm{T}} A \hat{x})^{-1} A \hat{x}$, and $y = A^{\mathrm{T}} \hat{y}$.

ii) If there exists a vector $\hat{x} \in \mathbb{F}^m$ such that $x = A\hat{x}$ and $\hat{x}^{\mathrm{T}} y \neq -1$, then $\operatorname{rank}(A + xy^{\mathrm{T}}) = \operatorname{rank} A$.

iii) If $xy^{\mathrm{T}} \neq 0$, $A^* x = 0$, and $A\overline{y} = 0$, then $\operatorname{rank}(A + xy^{\mathrm{T}}) = (\operatorname{rank} A) + 1$.

(Proof: To prove *ii)*, note that $A + xy^{\mathrm{T}} = A(I + xy^{\mathrm{T}})$ and $I + xy^{\mathrm{T}}$ is nonsingular. See [378, p. 33] and [182].)

Fact 2.10.12. Let $A \triangleq \left[\begin{smallmatrix} 1 & 0 \\ 0 & 0 \end{smallmatrix}\right]$ and $B \triangleq \left[\begin{smallmatrix} 0 & 1 \\ 0 & 0 \end{smallmatrix}\right]$. Then, $\operatorname{rank} AB = 1$ and $\operatorname{rank} BA = 0$.

Fact 2.10.13. Let $A \in \mathbb{F}^{n \times m}$ and $B \in \mathbb{F}^{m \times l}$. Then, the following statements hold:

i) $\operatorname{rank} AB + \operatorname{def} A = \dim[\mathcal{N}(A) + \mathcal{R}(B)]$.

ii) $\operatorname{rank} AB + \dim[\mathcal{N}(A) \cap \mathcal{R}(B)] = \operatorname{rank} B$.

iii) $\operatorname{rank} AB + \dim[\mathcal{N}(A^*) \cap \mathcal{R}(B^*)] = \operatorname{rank} A$.

iv) $\operatorname{def} AB + \operatorname{rank} A + \dim[\mathcal{N}(A) + \mathcal{R}(B)] = l + m$.

v) $\operatorname{def} AB = \operatorname{def} B + \dim[\mathcal{N}(A) \cap \mathcal{R}(B)]$.

vi) $\operatorname{def} AB + m = \operatorname{def} A + \dim[\mathcal{N}(A^*) \cap \mathcal{R}(B^*)] + l$.

(Remark: $\operatorname{rank} B - \operatorname{rank} AB = \dim[\mathcal{N}(A) \cap \mathcal{R}(B)] \leq \dim \mathcal{N}(A) = m - \operatorname{rank} A$ yields (2.5.17).)

Fact 2.10.14. Let $A \in \mathbb{F}^{n \times m}$ and $B \in \mathbb{F}^{m \times l}$. Then,

$$\max\{\operatorname{def} A + l - m, \operatorname{def} B\} \leq \operatorname{def} AB \leq \operatorname{def} A + \operatorname{def} B.$$

If, in addition, $m = l$, then

$$\max\{\operatorname{def} A, \operatorname{def} B\} \leq \operatorname{def} AB.$$

(Remark: The first inequality is *Sylvester's law of nullity*.)

Fact 2.10.15. Let $S \subseteq \mathbb{F}^m$, and let $A \in \mathbb{F}^{n \times m}$. Then, the following statements hold:

   i) $\operatorname{rank} A + \dim S - m \leq \dim AS \leq \min\{\operatorname{rank} A, \dim S\}$.

   ii) $\dim(AS) + \dim[\mathcal{N}(A) \cap S] = \dim S$.

   iii) $\operatorname{aff} AS \subseteq A\operatorname{aff} S$.

   iv) $\dim AS = \dim \operatorname{aff} AS \leq \dim A\operatorname{aff} S \leq \dim S$.

   v) If A is left invertible, then $\dim AS = \dim S$ and $\operatorname{aff} AS = A\operatorname{aff} S$.

(Proof: For *ii)*, see [613, p. 413].) (Remark: See Fact 10.7.15.)

Fact 2.10.16. Let $A \in \mathbb{F}^{n \times m}$ and $B \in \mathbb{F}^{1 \times m}$. Then, $\mathcal{N}(A) \subseteq \mathcal{N}(B)$ if and only if there exists a vector $\lambda \in \mathbb{F}^n$ such that $B = \lambda^* A$.

Fact 2.10.17. Let $A \in \mathbb{F}^{n \times m}$ and $b \in \mathbb{F}^n$. Then, there exists a vector $x \in \mathbb{F}^n$ satisfying $Ax = b$ if and only if $b^* \lambda = 0$ for all $\lambda \in \mathcal{N}(A^*)$. (Proof: Assume that $A^* \lambda = 0$ implies that $b^* \lambda = 0$. Then, $\mathcal{N}(A^*) \subseteq \mathcal{N}(b^*)$. Hence, $b \in \mathcal{R}(b) \subseteq \mathcal{R}(A)$.)

Fact 2.10.18. Let $A \in \mathbb{F}^{n \times m}$ and $B \in \mathbb{F}^{l \times m}$. Then, $\mathcal{N}(B) \subseteq \mathcal{N}(A)$ if and only if there exists a matrix $C \in \mathbb{F}^{n \times l}$ such that $A = CB$. Now, let $A \in \mathbb{F}^{n \times m}$ and $B \in \mathbb{F}^{n \times l}$. Then, $\mathcal{R}(A) \subseteq \mathcal{R}(B)$ if and only if there exists a matrix $C \in \mathbb{F}^{l \times m}$ such that $A = BC$.

Fact 2.10.19. Let $A, B \in \mathbb{F}^{n \times m}$, and let $C \in \mathbb{F}^{m \times l}$ be right invertible. If $\mathcal{R}(A) \subseteq \mathcal{R}(B)$, then $\mathcal{R}(AC) \subseteq \mathcal{R}(BC)$. Furthermore, $\mathcal{R}(A) = \mathcal{R}(B)$ if and only if $\mathcal{R}(AC) = \mathcal{R}(BC)$.

Fact 2.10.20. Let $A \in \mathbb{F}^{n \times m}$ and $B \in \mathbb{F}^{m \times l}$. Then, $\operatorname{rank} AB = \operatorname{rank} A$ if and only if $\mathcal{R}(AB) = \mathcal{R}(A)$. (Proof: If $\mathcal{R}(AB) \subset \mathcal{R}(A)$ (note proper inclusion), then $\operatorname{rank} AB < \operatorname{rank} A$.)

Fact 2.10.21. Let $A \in \mathbb{F}^{n \times m}$, $B \in \mathbb{F}^{m \times l}$, and $C \in \mathbb{F}^{l \times k}$. If $\operatorname{rank} AB = \operatorname{rank} B$, then $\operatorname{rank} ABC = \operatorname{rank} BC$. (Proof: $\operatorname{rank} B^{\mathrm{T}} A^{\mathrm{T}} = \operatorname{rank} B^{\mathrm{T}}$ implies that $\mathcal{R}(C^{\mathrm{T}} B^{\mathrm{T}} A^{\mathrm{T}}) = \mathcal{R}(C^{\mathrm{T}} B^{\mathrm{T}})$.)

Fact 2.10.22. Let $A \in \mathbb{F}^{n \times m}$ and $B \in \mathbb{F}^{n \times l}$. Then,

$$\mathcal{R}(\begin{bmatrix} A & B \end{bmatrix}) = \mathcal{R}(A) + \mathcal{R}(B).$$

Fact 2.10.23. Let $A \in \mathbb{F}^{n \times m}$ and $B \in \mathbb{F}^{n \times l}$. Then,

$$\mathcal{R}(A) = \mathcal{R}(B)$$

if and only if

$$\mathrm{rank}\, A = \mathrm{rank}\, B = \mathrm{rank}\, \begin{bmatrix} A & B \end{bmatrix}.$$

Fact 2.10.24. Let $A \in \mathbb{F}^{n \times m}$ and $B \in \mathbb{F}^{n \times l}$. Then,

$$\max\{\mathrm{rank}\, A, \mathrm{rank}\, B\} \leq \mathrm{rank}\, \begin{bmatrix} A & B \end{bmatrix}$$
$$= \mathrm{rank}\, A + \mathrm{rank}\, B - \dim[\mathcal{R}(A) \cap \mathcal{R}(B)]$$
$$\leq \mathrm{rank}\, A + \mathrm{rank}\, B$$

and

$$\mathrm{def}\, A + \mathrm{def}\, B \leq \mathrm{def}\, A + \mathrm{def}\, B + \dim[\mathcal{R}(A) \cap \mathcal{R}(B)]$$
$$= \mathrm{def}\, \begin{bmatrix} A & B \end{bmatrix}$$
$$\leq \min\{l + \mathrm{def}\, A, m + \mathrm{def}\, B\}.$$

If, in addition, $A^*B = 0$, then

$$\mathrm{rank}\, \begin{bmatrix} A & B \end{bmatrix} = \mathrm{rank}\, A + \mathrm{rank}\, B$$

and

$$\mathrm{def}\, \begin{bmatrix} A & B \end{bmatrix} = \mathrm{def}\, A + \mathrm{def}\, B.$$

(Proof: Use Fact 2.9.14. Assume $A^*B = 0$. Then,

$$\mathrm{rank}\, \begin{bmatrix} A & B \end{bmatrix} = \mathrm{rank}\, \begin{bmatrix} A^* \\ B^* \end{bmatrix} \begin{bmatrix} A & B \end{bmatrix} = \begin{bmatrix} A^*A & 0 \\ 0 & B^*B \end{bmatrix}$$
$$= \mathrm{rank}\, A^*A + \mathrm{rank}\, B^*B = \mathrm{rank}\, A + \mathrm{rank}\, B.)$$

(Remark: See Fact 6.4.20.)

Fact 2.10.25. Let $A \in \mathbb{F}^{n \times m}$ and $B \in \mathbb{F}^{l \times m}$. Then,

$$\max\{\mathrm{rank}\, A, \mathrm{rank}\, B\} \leq \mathrm{rank}\, \begin{bmatrix} A \\ B \end{bmatrix}$$
$$= \mathrm{rank}\, A + \mathrm{rank}\, B - \dim[\mathcal{R}(A^*) \cap \mathcal{R}(B^*)]$$
$$\leq \mathrm{rank}\, A + \mathrm{rank}\, B$$

and

$$\mathrm{def}\, A - \mathrm{rank}\, B \leq \mathrm{def}\, A - \mathrm{rank}\, B + \dim[\mathcal{R}(A^*) \cap \mathcal{R}(B^*)]$$
$$= \mathrm{def}\, \begin{bmatrix} A \\ B \end{bmatrix}$$
$$\leq \min\{\mathrm{def}\, A, \mathrm{def}\, B\}.$$

If, in addition, $AB^* = 0$, then

$$\mathrm{rank}\, \begin{bmatrix} A \\ B \end{bmatrix} = \mathrm{rank}\, A + \mathrm{rank}\, B$$

and

$$\det \begin{bmatrix} A \\ B \end{bmatrix} = \det A - \operatorname{rank} B.$$

(Proof: Use Fact 2.10.24 and Fact 2.9.14.) (Remark: See Fact 6.4.20.)

Fact 2.10.26. Let $A, B \in \mathbb{F}^{n \times m}$. Then,

$$\left. \begin{matrix} \max\{\operatorname{rank} A, \operatorname{rank} B\} \\ \\ \operatorname{rank}(A + B) \end{matrix} \right\} \leq \left\{ \begin{matrix} \operatorname{rank} \begin{bmatrix} A & B \end{bmatrix} \\ \\ \operatorname{rank} \begin{bmatrix} A \\ B \end{bmatrix} \end{matrix} \right\} \leq \operatorname{rank} A + \operatorname{rank} B$$

and

$$\det A - \operatorname{rank} B \leq \left\{ \begin{matrix} \det \begin{bmatrix} A & B \end{bmatrix} - m \\ \\ \det \begin{bmatrix} A \\ B \end{bmatrix} \end{matrix} \right\} \leq \left\{ \begin{matrix} \min\{\det A, \det B\} \\ \\ \det(A + B) \end{matrix} \right\}$$

(Proof: $\operatorname{rank}(A + B) = \operatorname{rank} \begin{bmatrix} A & B \end{bmatrix} \begin{bmatrix} I \\ I \end{bmatrix} \leq \operatorname{rank} \begin{bmatrix} A & B \end{bmatrix}$, and $\operatorname{rank}(A + B) = \operatorname{rank} \begin{bmatrix} I & I \end{bmatrix} \begin{bmatrix} A \\ B \end{bmatrix} \leq \operatorname{rank} \begin{bmatrix} A \\ B \end{bmatrix}$.)

Fact 2.10.27. Let $A \in \mathbb{F}^{n \times m}$, $B \in \mathbb{F}^{l \times k}$, and $C \in \mathbb{F}^{l \times m}$. Then,

$$\operatorname{rank} A + \operatorname{rank} B = \operatorname{rank} \begin{bmatrix} A & 0 \\ 0 & B \end{bmatrix} \leq \operatorname{rank} \begin{bmatrix} A & 0 \\ C & B \end{bmatrix}$$

and

$$\operatorname{rank} A + \operatorname{rank} B = \operatorname{rank} \begin{bmatrix} 0 & A \\ B & 0 \end{bmatrix} \leq \operatorname{rank} \begin{bmatrix} 0 & A \\ B & C \end{bmatrix}.$$

Fact 2.10.28. Let $A \in \mathbb{F}^{n \times m}$, $B \in \mathbb{F}^{m \times l}$, and $C \in \mathbb{F}^{l \times k}$. Then,

$$\operatorname{rank} AB + \operatorname{rank} BC \leq \operatorname{rank} \begin{bmatrix} 0 & AB \\ BC & B \end{bmatrix} = \operatorname{rank} B + \operatorname{rank} ABC.$$

Consequently,

$$\operatorname{rank} AB + \operatorname{rank} BC - \operatorname{rank} B \leq \operatorname{rank} ABC.$$

(Remark: This result is *Frobenius' inequality*.) (Proof: Use Fact 2.10.27 and $\begin{bmatrix} 0 & AB \\ BC & B \end{bmatrix} = \begin{bmatrix} I & A \\ 0 & I \end{bmatrix} \begin{bmatrix} -ABC & 0 \\ 0 & B \end{bmatrix} \begin{bmatrix} I & 0 \\ C & I \end{bmatrix}$.) (Remark: See Fact 6.4.25 for the case of equality.)

Fact 2.10.29. Let $A, B \in \mathbb{F}^{n \times m}$. Then,

$$\operatorname{rank} \begin{bmatrix} A & B \end{bmatrix} + \operatorname{rank} \begin{bmatrix} A \\ B \end{bmatrix} \leq \operatorname{rank} \begin{bmatrix} 0 & A & B \\ A & A & 0 \\ B & 0 & B \end{bmatrix}$$

$$= \operatorname{rank} A + \operatorname{rank} B + \operatorname{rank}(A + B).$$

(Proof: Use Frobenius' inequality with $A \triangleq C^{\mathrm{T}} \triangleq \begin{bmatrix} I & I \end{bmatrix}$ and with B replaced by $\begin{bmatrix} A & 0 \\ 0 & B \end{bmatrix}$.)

Fact 2.10.30. Let $A \in \mathbb{F}^{n \times m}$ and $B \in \mathbb{F}^{k \times l}$, and assume that B is a submatrix of A. Then,

$$k + l - \operatorname{rank} B \leq n + m - \operatorname{rank} A.$$

(Proof: See [70].)

2.11 Facts on Identities

Fact 2.11.1. Let $A \in \mathbb{F}^{2 \times 2}$, assume that A is positive semidefinite and nonzero, and define $B \in \mathbb{F}^{2 \times 2}$ by

$$B \triangleq \left(\operatorname{tr} A + 2\sqrt{\det A} \right)^{-1/2} \left(A + \sqrt{\det A} I \right).$$

Then, $B^2 = A$. (Problem: Consider more general matrices A.) (Proof: See [328, pp. 84, 266, 267].)

Fact 2.11.2. $\begin{bmatrix} -\frac{1}{2} & \frac{\sqrt{3}}{2} \\ \frac{-\sqrt{3}}{2} & -\frac{1}{2} \end{bmatrix}^3 = \begin{bmatrix} -1 & -1 \\ 1 & 0 \end{bmatrix}^3 = I_2.$

Fact 2.11.3. Let $A \in \mathbb{F}^{n \times m}$ and $B \in \mathbb{F}^{l \times k}$. Then, $A E_{i,j,m \times l} B = \operatorname{col}_i(A) \operatorname{row}_j(B)$.

Fact 2.11.4. Let $A \in \mathbb{F}^{n \times m}$, $B \in \mathbb{F}^{m \times l}$, and $C \in \mathbb{F}^{l \times n}$. Then,

$$\operatorname{tr} ABC = \sum_{i=1}^{n} \operatorname{row}_i(A) B \operatorname{col}_i(C).$$

Fact 2.11.5. Let $A \in \mathbb{F}^{n \times m}$. Then, $Ax = 0$ for all $x \in \mathbb{F}^m$ if and only if $A = 0$.

Fact 2.11.6. Let $x, y \in \mathbb{F}^n$. Then, $x^*x = y^*y$ and $\operatorname{Im} x^*y = 0$ if and only if $x - y$ is orthogonal to $x + y$.

Fact 2.11.7. Let $x, y \in \mathbb{R}^n$. Then, $xx^{\mathrm{T}} = yy^{\mathrm{T}}$ if and only if either $x = y$ or $x = -y$.

Fact 2.11.8. Let $x, y \in \mathbb{R}^n$. Then, $xy^{\mathrm{T}} = yx^{\mathrm{T}}$ if and only if x and y are linearly dependent.

Fact 2.11.9. Let $x, y \in \mathbb{R}^n$. Then, $xy^{\mathrm{T}} = -yx^{\mathrm{T}}$ if and only if either $x = 0$ or $y = 0$. (Proof: If $x_{(i)} \neq 0$ and $y_{(j)} \neq 0$, then $x_{(j)} = y_{(i)} = 0$ and

$0 \neq x_{(i)}y_{(j)} \neq x_{(j)}y_{(i)} = 0$.)

Fact 2.11.10. Let $x, y \in \mathbb{R}^n$. Then, $yx^{\mathrm{T}} + xy^{\mathrm{T}} = y^{\mathrm{T}}yxx^{\mathrm{T}}$ if and only if either $x = 0$ or $y = \frac{1}{2}y^{\mathrm{T}}yx$.

Fact 2.11.11. Let $x, y \in \mathbb{F}^n$. Then,

$$(xy^*)^r = (y^*x)^{r-1}xy^*.$$

Fact 2.11.12. Let $y \in \mathbb{F}^n$ and $x \in \mathbb{F}^m$. Then, there exists a matrix $A \in \mathbb{F}^{n \times m}$ such that $y = Ax$ if and only if either $y = 0$ or $x \neq 0$. If $y = 0$, then one such matrix is $A = 0$. If $x \neq 0$, then one such matrix is

$$A = (x^*x)^{-1}yx^*.$$

(Remark: See Fact 3.5.28.)

Fact 2.11.13. Let $A \in \mathbb{F}^{n \times m}$. Then, $A = 0$ if and only if $\operatorname{tr} AA^* = 0$.

Fact 2.11.14. Let $A, B \in \mathbb{F}^{n \times n}$, and define $\mathcal{A} \triangleq \left[\begin{smallmatrix} A & A \\ A & A \end{smallmatrix}\right]$ and $\mathcal{B} \triangleq \left[\begin{smallmatrix} B & -B \\ -B & B \end{smallmatrix}\right]$. Then,

$$\mathcal{A}\mathcal{B} = \mathcal{B}\mathcal{A} = 0.$$

Fact 2.11.15. Let $A \in \mathbb{F}^{n \times n}$ and $k \in \mathbb{P}$. Then,

$$\operatorname{Re} \operatorname{tr} A^{2k} \leq \operatorname{tr} A^k A^{k*} \leq \operatorname{tr} (AA^*)^k.$$

(Remark: To prove the left-hand inequality, consider $\operatorname{tr} (A^k - A^{k*})(A^{k*} - A^k)$. For the right-hand inequality when $k = 2$, consider $\operatorname{tr} (AA^* - A^*A)^2$.)

Fact 2.11.16. Let $A \in \mathbb{F}^{n \times n}$. Then, $\operatorname{tr} A^k = 0$ for all $k = 1, \ldots, n$ if and only if $A^n = 0$. (Proof: For sufficiency, Fact 4.10.2 implies that $\operatorname{spec}(A) = \{0\}$, and thus the Jordan form of A is a block-diagonal matrix each of whose diagonally located blocks is a standard nilpotent matrix. For necessity, see [811, p. 112].)

Fact 2.11.17. Let $A \in \mathbb{F}^{n \times n}$, and assume that $\operatorname{tr} A = 0$. If $A^2 = A$, then $A = 0$. If $A^k = A$, where $k \geq 4$ and $2 \leq n < p$, where p is the smallest prime divisor of $k - 1$, then $A = 0$. (Proof: See [191].)

Fact 2.11.18. Let $A, B \in \mathbb{F}^{n \times n}$, and assume that $AB = 0$. Then, for all $k \in \mathbb{P}$,

$$\operatorname{tr} (A + B)^k = \operatorname{tr} A^k + \operatorname{tr} B^k.$$

Fact 2.11.19. Let $A, B \in \mathbb{F}^{n \times n}$. Then, the following statements hold:

i) $AB + BA = \frac{1}{2}[(A + B)^2 - (A - B)^2]$.

ii) $(A + B)(A - B) = A^2 - B^2 - [A, B]$.

iii) $(A - B)(A + B) = A^2 - B^2 + [A, B]$.

iv) $A^2 - B^2 = \frac{1}{2}[(A + B)(A - B) + (A - B)(A + B)]$.

Fact 2.11.20. Let $A, B \in \mathbb{F}^{n \times n}$ and $k \in \mathbb{P}$. Then,

$$A^k - B^k = \sum_{i=0}^{k-1} A^i(A - B)B^{k-1-i}.$$

Fact 2.11.21. Let $\alpha \in \mathbb{R}$ and $A \in \mathbb{R}^{n \times n}$. Then, the matrix equation $\alpha A + A^{\mathrm{T}} = 0$ has a nonzero solution A if and only if $\alpha = 1$ or $\alpha = -1$.

2.12 Facts on Determinants

Fact 2.12.1. $\det \hat{I}_n = (-1)^{\lfloor n/2 \rfloor} = (-1)^{n(n-1)/2}$.

Fact 2.12.2. $\det(I_n + \alpha 1_{n \times n}) = 1 + \alpha n$.

Fact 2.12.3. Let $A \in \mathbb{F}^{n \times m}$, $B \in \mathbb{F}^{m \times n}$, and $m < n$. Then, $\det AB = 0$.

Fact 2.12.4. Let $A \in \mathbb{F}^{n \times m}$, $B \in \mathbb{F}^{m \times n}$, and $n \le m$. Then, $\det AB$ is equal to the sum of all $\binom{n}{m}$ products of pairs of subdeterminants of A and B formed by choosing n columns of A and the corresponding n rows of B. (Remark: This identity is the *Binet-Cauchy formula*, which yields Proposition 2.7.1 in the case $n = m$.) (Remark: Determinantal identities are given in [150].)

Fact 2.12.5. Let $A \in \mathbb{F}^{n \times n}$, assume that A is nonsingular, and let $b \in \mathbb{F}^n$. Then, the solution $x \in \mathbb{F}^n$ of $Ax = b$ is given by

$$x = \begin{bmatrix} \dfrac{\det\left(A \overset{1}{\leftarrow} b\right)}{\det A} \\ \vdots \\ \dfrac{\det\left(A \overset{n}{\leftarrow} b\right)}{\det A} \end{bmatrix}.$$

(Proof: Note that $A\left(I \overset{i}{\leftarrow} x\right) = A \overset{i}{\leftarrow} b$. Since $\det\left(I \overset{i}{\leftarrow} x\right) = x_{(i)}$, it follows that $(\det A)x_{(i)} = \det\left(A \overset{i}{\leftarrow} b\right)$.) (Remark: This identity is *Cramer's rule*.)

Fact 2.12.6. Let $A \in \mathbb{F}^{n \times m}$ be right invertible, and let $b \in \mathbb{F}^n$. Then, a solution $x \in \mathbb{F}^m$ of $Ax = b$ is given by

$$x_{(i)} = \frac{\det\left[\left(A \overset{i}{\leftarrow} b\right)A^*\right] - \det\left[\left(A \overset{i}{\leftarrow} 0\right)A^*\right]}{\det(AA^*)},$$

for all $i = 1, \ldots, m$. (Proof: See [451].)

Fact 2.12.7. Let $A \in \mathbb{F}^{n \times n}$, and assume that either $A_{(i,j)} = 0$ for all i, j such that $i + j < n + 1$ or $A_{(i,j)} = 0$ for all i, j such that $i + j > n + 1$. Then,

$$\det A = (-1)^{\lfloor n/2 \rfloor} \prod_{i=1}^{n} A_{(i,n+1-i)}.$$

(Remark: A is *lower reverse triangular*.)

Fact 2.12.8. Define $A \in \mathbb{R}^{n \times n}$ by

$$A \triangleq \begin{bmatrix} 0 & 1 & 0 & \cdots & 0 & 0 \\ 0 & 0 & 1 & \cdots & 0 & 0 \\ 0 & 0 & 0 & \ddots & 0 & 0 \\ \vdots & \vdots & \vdots & \ddots & \ddots & \vdots \\ 0 & 0 & 0 & \cdots & 0 & 1 \\ 1 & 0 & 0 & \cdots & 0 & 0 \end{bmatrix}.$$

Then,

$$\det A = (-1)^{n+1}.$$

Fact 2.12.9. Let $a_1, \ldots, a_n \in \mathbb{F}$. Then,

$$\det \begin{bmatrix} 1 + a_1 & a_2 & \cdots & a_n \\ a_1 & 1 + a_2 & \cdots & a_n \\ \vdots & \vdots & \ddots & \vdots \\ a_1 & a_2 & \cdots & 1 + a_n \end{bmatrix} = 1 + \sum_{i=1}^{n} a_i.$$

Fact 2.12.10. Let $a_1, \ldots, a_n \in \mathbb{F}$ be nonzero. Then,

$$\det \begin{bmatrix} \frac{1+a_1}{a_1} & 1 & \cdots & 1 \\ 1 & \frac{1+a_2}{a_2} & \cdots & 1 \\ \vdots & \vdots & \ddots & \vdots \\ 1 & 1 & \cdots & \frac{1+a_n}{a_n} \end{bmatrix} = \frac{1 + \sum_{i=1}^{n} a_i}{\prod_{i=1}^{n} a_i}.$$

Fact 2.12.11. Let $a, b, c_1, \ldots, c_n \in \mathbb{F}$, define $A \in \mathbb{F}^{n \times n}$ by

$$A \triangleq \begin{bmatrix} c_1 & a & a & \cdots & a \\ b & c_2 & a & \cdots & a \\ b & b & c_3 & \ddots & a \\ \vdots & \vdots & \ddots & \ddots & \vdots \\ b & b & b & \cdots & c_n \end{bmatrix},$$

and let $p(x) = (c_1 - x)(c_2 - x) \cdots (c_n - x)$ and $p_i(x) = p(x)/(c_i - x)$ for all $i = 1, \ldots, n$. Then,

$$\det A = \begin{cases} \dfrac{bp(a) - ap(b)}{b - a}, & b \neq a, \\ a \displaystyle\sum_{i=1}^{n-1} p_i(a) + c_n p_n(a), & b = a. \end{cases}$$

In particular,

$$\det \begin{bmatrix} a & b & b & \cdots & b \\ b & a & b & \cdots & b \\ b & b & a & \ddots & b \\ \vdots & \vdots & \ddots & \ddots & \vdots \\ b & b & b & \cdots & a \end{bmatrix} = (a - b)^{n-1}[a + (n-1)b]$$

and

$$\det(aI_n + b1_{n \times n}) = a^{n-1}(a + bn).$$

(Remark: See Fact 4.10.12.) (Remark: The matrix $aI_n + b1_{n \times n}$ arises in combinatorics. See [147, 149].)

Fact 2.12.12. Define the tridiagonal matrix $A \in \mathbb{F}^{n \times n}$ by

$$A \triangleq \begin{bmatrix} a+b & ab & 0 & \cdots & 0 & 0 \\ 1 & a+b & ab & \cdots & 0 & 0 \\ 0 & 1 & a+b & \ddots & 0 & 0 \\ \vdots & \vdots & \ddots & \ddots & \ddots & \vdots \\ 0 & 0 & 0 & \ddots & a+b & ab \\ 0 & 0 & 0 & \cdots & 1 & a+b \end{bmatrix}.$$

Then,

$$\det A = \begin{cases} (n+1)a^n, & a = b, \\ \dfrac{a^{n+1} - b^{n+1}}{a - b}, & a \neq b. \end{cases}$$

(Proof: See [439, pp. 401, 621].)

2.13 Facts on Determinants of Partitioned Matrices

Fact 2.13.1. Let $A \in \mathbb{F}^{n \times n}$, $x, y \in \mathbb{F}^n$, and $a \in \mathbb{F}$. Then,

$$\begin{bmatrix} A & x \\ y^{\mathrm{T}} & a \end{bmatrix} = \begin{cases} \begin{bmatrix} I & 0 \\ y^{\mathrm{T}}A^{-1} & 1 \end{bmatrix}\begin{bmatrix} A & 0 \\ 0 & a - y^{\mathrm{T}}A^{-1}x \end{bmatrix}\begin{bmatrix} I & A^{-1}x \\ 0 & 1 \end{bmatrix}, & \det A \neq 0, \\ \begin{bmatrix} I & a^{-1}x \\ 0 & 1 \end{bmatrix}\begin{bmatrix} A - a^{-1}xy^{\mathrm{T}} & 0 \\ 0 & a \end{bmatrix}\begin{bmatrix} I & 0 \\ a^{-1}y^{\mathrm{T}} & 1 \end{bmatrix}, & a \neq 0. \end{cases}$$

(Remark: See Fact 6.4.38.)

Fact 2.13.2. Let $A \in \mathbb{F}^{n \times n}$, $x, y \in \mathbb{F}^n$, and $a \in \mathbb{F}$. Then,

$$\det \begin{bmatrix} A & x \\ y^{\mathrm{T}} & a \end{bmatrix} = a(\det A) - y^{\mathrm{T}}A^{\mathrm{A}}x.$$

Hence,

$$\det \begin{bmatrix} A & x \\ y^{\mathrm{T}} & a \end{bmatrix} = \begin{cases} (\det A)(a - y^{\mathrm{T}}A^{-1}x), & \det A \neq 0, \\ a\det(A - a^{-1}xy^{\mathrm{T}}), & a \neq 0, \\ -y^{\mathrm{T}}A^{\mathrm{A}}x, & a = 0. \end{cases}$$

In particular,

$$\det \begin{bmatrix} A & Ax \\ y^{\mathrm{T}}A & y^{\mathrm{T}}Ax \end{bmatrix} = 0.$$

Finally,

$$\det(A + xy^{\mathrm{T}}) = \det A + y^{\mathrm{T}}A^{\mathrm{A}}x = -\det \begin{bmatrix} A & x \\ y^{\mathrm{T}} & -1 \end{bmatrix}.$$

(Remark: See Fact 2.13.4 and Fact 2.14.3.)

Fact 2.13.3. $\det \begin{bmatrix} 0 & I_n \\ I_m & 0 \end{bmatrix} = (-1)^{nm}$.

Fact 2.13.4. Let $A \in \mathbb{R}^{n \times n}$, $b \in \mathbb{R}^n$, and $a \in \mathbb{R}$. Then,

$$\det \begin{bmatrix} A & b \\ b^{\mathrm{T}} & a \end{bmatrix} = a(\det A) - b^{\mathrm{T}}A^{\mathrm{A}}b.$$

In particular,

$$\det \begin{bmatrix} A & b \\ b^{\mathrm{T}} & a \end{bmatrix} = \begin{cases} (\det A)(a - b^{\mathrm{T}}A^{-1}b), & \det A \neq 0, \\ a\det(A - a^{-1}bb^{\mathrm{T}}), & a \neq 0, \\ -b^{\mathrm{T}}A^{\mathrm{A}}b, & a = 0. \end{cases}$$

(Remark: This identity is a specialization of Fact 2.13.2.)

Fact 2.13.5. Let $A \in \mathbb{F}^{n \times n}$. Then,

$$\operatorname{rank} \begin{bmatrix} A & A \\ A & A \end{bmatrix} = \operatorname{rank} \begin{bmatrix} A & -A \\ -A & A \end{bmatrix} = \operatorname{rank} A,$$

$$\operatorname{rank} \begin{bmatrix} A & A \\ -A & A \end{bmatrix} = 2\operatorname{rank} A,$$

$$\det \begin{bmatrix} A & A \\ A & A \end{bmatrix} = \det \begin{bmatrix} A & -A \\ -A & A \end{bmatrix} = 0,$$

$$\det \begin{bmatrix} A & A \\ -A & A \end{bmatrix} = 2^n (\det A)^2.$$

(Remark: See Fact 2.13.6.)

Fact 2.13.6. Let $a, b, c, d \in \mathbb{F}$, let $A \in \mathbb{F}^{n \times n}$, and define $\mathcal{A} \triangleq \begin{bmatrix} aA & bA \\ cA & dA \end{bmatrix}$. Then,

$$\operatorname{rank} \mathcal{A} = \left(\operatorname{rank} \begin{bmatrix} a & b \\ c & d \end{bmatrix} \right) \operatorname{rank} A$$

and

$$\det \mathcal{A} = (ad - bc)^n (\det A)^2.$$

(Remark: See Fact 2.13.5.) (Proof: See Proposition 7.1.11 and Fact 7.4.20.)

Fact 2.13.7. Let A, B, C, D be conformable matrices with entries in $\mathbb{F}$. Then,

$$\begin{bmatrix} A & AB \\ C & D \end{bmatrix} = \begin{bmatrix} I & 0 \\ C & I \end{bmatrix} \begin{bmatrix} A & 0 \\ C - CA & D - CB \end{bmatrix} \begin{bmatrix} I & B \\ 0 & I \end{bmatrix},$$

$$\det \begin{bmatrix} A & AB \\ C & D \end{bmatrix} = (\det A)\det(D - CB),$$

$$\begin{bmatrix} A & B \\ CA & D \end{bmatrix} = \begin{bmatrix} I & 0 \\ C & I \end{bmatrix} \begin{bmatrix} A & B - AB \\ 0 & D - CB \end{bmatrix} \begin{bmatrix} I & B \\ 0 & I \end{bmatrix},$$

$$\det \begin{bmatrix} A & B \\ CA & D \end{bmatrix} = (\det A)\det(D - CB),$$

$$\begin{bmatrix} A & BD \\ C & D \end{bmatrix} = \begin{bmatrix} I & B \\ 0 & I \end{bmatrix} \begin{bmatrix} A - BC & 0 \\ C - DC & D \end{bmatrix} \begin{bmatrix} I & 0 \\ C & I \end{bmatrix},$$

$$\det \begin{bmatrix} A & BD \\ C & D \end{bmatrix} = \det(A - BC)\det D,$$

$$\begin{bmatrix} A & B \\ DC & D \end{bmatrix} = \begin{bmatrix} I & B \\ 0 & I \end{bmatrix} \begin{bmatrix} A - BC & B - BD \\ 0 & D \end{bmatrix} \begin{bmatrix} I & 0 \\ C & I \end{bmatrix},$$

$$\det \begin{bmatrix} A & B \\ DC & D \end{bmatrix} = \det(A - BC)\det D.$$

(Remark: See Fact 6.4.38.)

Fact 2.13.8. Let $A_1, A_2, B_1, B_2 \in \mathbb{F}^{n \times m}$, and define $\mathcal{A} \triangleq \begin{bmatrix} A_1 & A_2 \\ A_2 & A_1 \end{bmatrix}$ and $\mathcal{B} \triangleq \begin{bmatrix} B_1 & B_2 \\ B_2 & B_1 \end{bmatrix}$. Then,

$$\operatorname{rank} \begin{bmatrix} \mathcal{A} & \mathcal{B} \\ \mathcal{B} & \mathcal{A} \end{bmatrix} = \sum_{i=1}^{4} \operatorname{rank} C_i,$$

where $C_1 \triangleq A_1 + A_2 + B_1 + B_2$, $C_2 \triangleq A_1 + A_2 - B_1 - B_2$, $C_3 \triangleq A_1 - A_2 + B_1 - B_2$, and $C_4 \triangleq A_1 - A_2 - B_1 + B_2$. If, in addition, $n = m$, then

$$\det \begin{bmatrix} \mathcal{A} & \mathcal{B} \\ \mathcal{B} & \mathcal{A} \end{bmatrix} = \prod_{i=1}^{4} \det C_i.$$

(Proof: See [711].) (Remark: See Fact 3.14.3.)

Fact 2.13.9. Let $A, B, C, D \in \mathbb{F}^{n \times n}$, and assume that rank $\begin{bmatrix} A & B \\ C & D \end{bmatrix} = n$. Then,

$$\det \begin{bmatrix} \det A & \det B \\ \det C & \det D \end{bmatrix} = 0.$$

Fact 2.13.10. Let $A, B, C, D \in \mathbb{F}^{n \times n}$. Then,

$$\det \begin{bmatrix} A & B \\ C & D \end{bmatrix} = \begin{cases} \det(DA - CB), & AB = BA, \\ \det(AD - CB), & AC = CA, \\ \det(AD - BC), & DC = CD, \\ \det(DA - BC), & DB = BD. \end{cases}$$

(Remark: These identities are *Schur's formulas*. See [80, p. 11].) (Proof: If

A is nonsingular, then

$$\det\begin{bmatrix} A & B \\ C & D \end{bmatrix} = (\det A)\det(D - CA^{-1}B) = \det(DA - CA^{-1}BA)$$

$$= \det(DA - CB).$$

Alternatively, note the identity

$$\begin{bmatrix} A & B \\ C & D \end{bmatrix} = \begin{bmatrix} A & 0 \\ C & DA - CB \end{bmatrix}\begin{bmatrix} I & BA^{-1} \\ 0 & A^{-1} \end{bmatrix}.$$

If A is singular, then replace A by $A + \varepsilon I$ and use continuity.) (Problem: Find a direct proof for the case in which A is singular.)

Fact 2.13.11. Let $A, B, C, D \in \mathbb{F}^{n \times n}$. Then,

$$\det\begin{bmatrix} A & B \\ C & D \end{bmatrix} = \begin{cases} \det(AD^{\mathrm{T}} - B^{\mathrm{T}}C^{\mathrm{T}}), & AB = BA^{\mathrm{T}}, \\ \det(AD^{\mathrm{T}} - BC), & DC = CD^{\mathrm{T}}, \\ \det(A^{\mathrm{T}}D - CB), & A^{\mathrm{T}}C = CA, \\ \det(A^{\mathrm{T}}D - C^{\mathrm{T}}B^{\mathrm{T}}), & D^{\mathrm{T}}B = BD. \end{cases}$$

(Proof: Define the nonsingular matrix $A_\varepsilon \triangleq A + \varepsilon I$, which satisfies $A_\varepsilon B = BA_\varepsilon^{\mathrm{T}}$. Then,

$$\det\begin{bmatrix} A_\varepsilon & B \\ C & D \end{bmatrix} = (\det A_\varepsilon)\det(D - CA_\varepsilon^{-1}B)$$

$$= \det(DA_\varepsilon^{\mathrm{T}} - CA_\varepsilon^{-1}BA_\varepsilon^{\mathrm{T}}) = \det(DA_\varepsilon^{\mathrm{T}} - CB).)$$

Fact 2.13.12. Let $A, B, C, D \in \mathbb{F}^{n \times n}$. Then,

$$\det\begin{bmatrix} A & B \\ C & D \end{bmatrix} = \begin{cases} (-1)^{\mathrm{rank}\,C}\det(A^{\mathrm{T}}D + C^{\mathrm{T}}B), & A^{\mathrm{T}}C = -C^{\mathrm{T}}A, \\ (-1)^{n+\mathrm{rank}\,A}\det(A^{\mathrm{T}}D + C^{\mathrm{T}}B), & A^{\mathrm{T}}C = -C^{\mathrm{T}}A, \\ (-1)^{\mathrm{rank}\,B}\det(A^{\mathrm{T}}D + C^{\mathrm{T}}B), & B^{\mathrm{T}}D = -D^{\mathrm{T}}B, \\ (-1)^{n+\mathrm{rank}\,D}\det(A^{\mathrm{T}}D + C^{\mathrm{T}}B), & B^{\mathrm{T}}D = -D^{\mathrm{T}}B, \\ (-1)^{\mathrm{rank}\,B}\det(AD^{\mathrm{T}} + BC^{\mathrm{T}}), & AB^{\mathrm{T}} = -BA^{\mathrm{T}}, \\ (-1)^{n+\mathrm{rank}\,A}\det(AD^{\mathrm{T}} + BC^{\mathrm{T}}), & AB^{\mathrm{T}} = -BA^{\mathrm{T}}, \\ (-1)^{\mathrm{rank}\,C}\det(AD^{\mathrm{T}} + BC^{\mathrm{T}}), & CD^{\mathrm{T}} = -DC^{\mathrm{T}}, \\ (-1)^{n+\mathrm{rank}\,D}\det(AD^{\mathrm{T}} + BC^{\mathrm{T}}), & CD^{\mathrm{T}} = -DC^{\mathrm{T}}. \end{cases}$$

(Proof: See [509, 761].) (Remark: This result is due to Callan. See [761].) (Remark: If $A^{\mathrm{T}}C = -C^{\mathrm{T}}A$ and $\mathrm{rank}\,A + \mathrm{rank}\,C + n$ is odd, then $\begin{bmatrix} A & B \\ C & D \end{bmatrix}$ is singular.)

Fact 2.13.13. Let $A, B, C, D \in \mathbb{F}^{n \times n}$. Then,

$$\det \begin{bmatrix} A & B \\ C & D \end{bmatrix} = \begin{cases} \det(AD^{\mathrm{T}} - BC^{\mathrm{T}}), & AB^{\mathrm{T}} = BA^{\mathrm{T}}, \\ \det(AD^{\mathrm{T}} - BC^{\mathrm{T}}), & DC^{\mathrm{T}} = CD^{\mathrm{T}}, \\ \det(A^{\mathrm{T}}D - C^{\mathrm{T}}B), & A^{\mathrm{T}}C = C^{\mathrm{T}}A, \\ \det(A^{\mathrm{T}}D - C^{\mathrm{T}}B), & D^{\mathrm{T}}B = B^{\mathrm{T}}D. \end{cases}$$

(Proof: See [509].)

Fact 2.13.14. Let $A \in \mathbb{F}^{n \times m}$, $B \in \mathbb{F}^{n \times l}$, $C \in \mathbb{F}^{k \times m}$, and $D \in \mathbb{F}^{k \times l}$, and assume that $n + k = m + l$. If $AC^{\mathrm{T}} + BD^{\mathrm{T}} = 0$, then

$$\det \begin{bmatrix} A & B \\ C & D \end{bmatrix}^2 = \det(AA^{\mathrm{T}} + BB^{\mathrm{T}})\det(CC^{\mathrm{T}} + DD^{\mathrm{T}}).$$

Alternatively, if $A^{\mathrm{T}}B + C^{\mathrm{T}}D = 0$, then

$$\det \begin{bmatrix} A & B \\ C & D \end{bmatrix}^2 = \det(A^{\mathrm{T}}A + C^{\mathrm{T}}C)\det(B^{\mathrm{T}}B + D^{\mathrm{T}}D).$$

(Proof: Form $\begin{bmatrix} A & B \\ C & D \end{bmatrix}\begin{bmatrix} A & B \\ C & D \end{bmatrix}^{\mathrm{T}}$ and $\begin{bmatrix} A & B \\ C & D \end{bmatrix}^{\mathrm{T}}\begin{bmatrix} A & B \\ C & D \end{bmatrix}$.)

Fact 2.13.15. Let $A \in \mathbb{F}^{n \times m}$, $B \in \mathbb{F}^{n \times m}$, $C \in \mathbb{F}^{k \times m}$, and $D \in \mathbb{F}^{k \times m}$, and assume that $n + k = 2m$. If $AD^{\mathrm{T}} + BC^{\mathrm{T}} = 0$, then

$$\det \begin{bmatrix} A & B \\ C & D \end{bmatrix}^2 = (-1)^m \det(AB^{\mathrm{T}} + BA^{\mathrm{T}})\det(CD^{\mathrm{T}} + DC^{\mathrm{T}}).$$

Alternatively, if $AB^{\mathrm{T}} + BA^{\mathrm{T}} = 0$ or $CD^{\mathrm{T}} + DC^{\mathrm{T}} = 0$, then

$$\det \begin{bmatrix} A & B \\ C & D \end{bmatrix}^2 = (-1)^{m^2 + nk}\det(AD^{\mathrm{T}} + BC^{\mathrm{T}})^2.$$

(Proof: Form $\begin{bmatrix} A & B \\ C & D \end{bmatrix}\begin{bmatrix} B^{\mathrm{T}} & D^{\mathrm{T}} \\ A^{\mathrm{T}} & C^{\mathrm{T}} \end{bmatrix}$ and $\begin{bmatrix} A & B \\ C & D \end{bmatrix}\begin{bmatrix} D^{\mathrm{T}} & B^{\mathrm{T}} \\ C^{\mathrm{T}} & A^{\mathrm{T}} \end{bmatrix}$. See [761].)

Fact 2.13.16. Let $A \in \mathbb{F}^{n \times m}$, $B \in \mathbb{F}^{n \times l}$, $C \in \mathbb{F}^{n \times m}$, and $D \in \mathbb{F}^{n \times l}$, and assume that $m + l = 2n$. If $A^{\mathrm{T}}D + C^{\mathrm{T}}B = 0$, then

$$\det \begin{bmatrix} A & B \\ C & D \end{bmatrix}^2 = (-1)^n \det(C^{\mathrm{T}}A + A^{\mathrm{T}}C)\det(D^{\mathrm{T}}B + B^{\mathrm{T}}D).$$

Alternatively, if $B^{\mathrm{T}}D + D^{\mathrm{T}}B = 0$ or $A^{\mathrm{T}}C + C^{\mathrm{T}}A = 0$, then

$$\det \begin{bmatrix} A & B \\ C & D \end{bmatrix}^2 = (-1)^{n^2 + ml}\det(A^{\mathrm{T}}D + C^{\mathrm{T}}B)^2.$$

(Proof: Form $\begin{bmatrix} C^{\mathrm{T}} & A^{\mathrm{T}} \\ D^{\mathrm{T}} & B^{\mathrm{T}} \end{bmatrix}\begin{bmatrix} A & B \\ C & D \end{bmatrix}$ and $\begin{bmatrix} D^{\mathrm{T}} & B^{\mathrm{T}} \\ C^{\mathrm{T}} & A^{\mathrm{T}} \end{bmatrix}\begin{bmatrix} A & B \\ C & D \end{bmatrix}$.)

Fact 2.13.17. Let $A \in \mathbb{F}^{n \times n}$, $B \in \mathbb{F}^{n \times k}$, $C \in \mathbb{F}^{k \times n}$, and $D \in \mathbb{F}^{k \times k}$. If $AB + BD = 0$ or $CA + DC = 0$, then

$$\det \begin{bmatrix} A & B \\ C & D \end{bmatrix}^2 = \det(A^2 + BC)\det(CB + D^2).$$

Alternatively, if $A^2 + BC = 0$ or $CB + D^2 = 0$, then

$$\det \begin{bmatrix} A & B \\ C & D \end{bmatrix}^2 = (-1)^{nk}\det(AB + BD)\det(CA + DC).$$

(Proof: Form $\begin{bmatrix} A & B \\ C & D \end{bmatrix}^2$ and $\begin{bmatrix} A & B \\ C & D \end{bmatrix}\begin{bmatrix} B & A \\ D & C \end{bmatrix}$.)

Fact 2.13.18. Let $A \in \mathbb{F}^{n \times m}$, $B \in \mathbb{F}^{n \times n}$, $C \in \mathbb{F}^{m \times m}$, and $D \in \mathbb{F}^{m \times n}$. If $AD + B^2 = 0$ or $C^2 + DA = 0$, then

$$\det \begin{bmatrix} A & B \\ C & D \end{bmatrix}^2 = (-1)^{nm}\det(AC + BA)\det(CD + DB).$$

Alternatively, if $AC + BA = 0$ or $CD + DB = 0$, then

$$\det \begin{bmatrix} A & B \\ C & D \end{bmatrix}^2 = \det(AD + B^2)\det(C^2 + DA).$$

(Proof: Form $\begin{bmatrix} A & B \\ C & D \end{bmatrix}\begin{bmatrix} C & D \\ A & B \end{bmatrix}$ and $\begin{bmatrix} A & B \\ C & D \end{bmatrix}\begin{bmatrix} D & C \\ B & A \end{bmatrix}$.)

Fact 2.13.19. Let $A \in \mathbb{F}^{n \times m}$, $B \in \mathbb{F}^{n \times l}$, $C \in \mathbb{F}^{k \times m}$, and $D \in \mathbb{F}^{k \times l}$, and assume that $n + k = m + l$. If $AC^* + BD^* = 0$, then

$$\left| \det \begin{bmatrix} A & B \\ C & D \end{bmatrix} \right|^2 = \det(AA^* + BB^*)\det(CC^* + DD^*).$$

Alternatively, if $A^*B + C^*D = 0$, then

$$\left| \det \begin{bmatrix} A & B \\ C & D \end{bmatrix} \right|^2 = \det(A^*A + C^*C)\det(B^*B + D^*D).$$

(Proof: Form $\begin{bmatrix} A & B \\ C & D \end{bmatrix}\begin{bmatrix} A & B \\ C & D \end{bmatrix}^*$ and $\begin{bmatrix} A & B \\ C & D \end{bmatrix}^*\begin{bmatrix} A & B \\ C & D \end{bmatrix}$.) (Remark: See Fact 8.11.20.)

Fact 2.13.20. Let $A \in \mathbb{F}^{n \times m}$, $B \in \mathbb{F}^{n \times m}$, $C \in \mathbb{F}^{k \times m}$, and $D \in \mathbb{F}^{k \times m}$, and assume that $n + k = 2m$. If $AD^* + BC^* = 0$, then

$$\left| \det \begin{bmatrix} A & B \\ C & D \end{bmatrix} \right|^2 = (-1)^m\det(AB^* + BA^*)\det(CD^* + DC^*).$$

Alternatively, if $AB^* + BA^* = 0$ or $CD^* + DC^* = 0$, then

$$\left| \det \begin{bmatrix} A & B \\ C & D \end{bmatrix} \right|^2 = (-1)^{m^2+nk}|\det(AD^* + BC^*)|^2.$$

(Proof: Form $\begin{bmatrix} A & B \\ C & D \end{bmatrix}\begin{bmatrix} B^* & D^* \\ A^* & C^* \end{bmatrix}$ and $\begin{bmatrix} A & B \\ C & D \end{bmatrix}\begin{bmatrix} D^* & B^* \\ C^* & A^* \end{bmatrix}$.) (Remark: If $m^2 + nk$ is odd, then $\begin{bmatrix} A & B \\ C & D \end{bmatrix}$ is singular.)

Fact 2.13.21. Let $A \in \mathbb{F}^{n \times m}$, $B \in \mathbb{F}^{n \times l}$, $C \in \mathbb{F}^{n \times m}$, and $D \in \mathbb{F}^{n \times l}$, and assume that $m + l = 2n$. If $A^*D + C^*B = 0$, then

$$\left| \det \begin{bmatrix} A & B \\ C & D \end{bmatrix} \right|^2 = (-1)^m \det(C^*A + A^*C) \det(D^*B + B^*D).$$

Alternatively, if $D^*B + B^*D = 0$ or $C^*A + A^*C = 0$, then

$$\left| \det \begin{bmatrix} A & B \\ C & D \end{bmatrix} \right|^2 = (-1)^{n^2 + ml} |\det(A^*D + C^*B)|^2.$$

(Proof: Form $\begin{bmatrix} C^* & A^* \\ D^* & B^* \end{bmatrix} \begin{bmatrix} A & B \\ C & D \end{bmatrix}$ and $\begin{bmatrix} D^* & B^* \\ C^* & A^* \end{bmatrix} \begin{bmatrix} A & B \\ C & D \end{bmatrix}$.) (Remark: If $n^2 + ml$ is odd, then $\begin{bmatrix} A & B \\ C & D \end{bmatrix}$ is singular.)

Fact 2.13.22. Let $A \in \mathbb{F}^{n \times m}$ and $B \in \mathbb{F}^{n \times l}$. Then,

$$\det \begin{bmatrix} A^*A & A^*B \\ B^*A & B^*B \end{bmatrix} = \begin{cases} \det(A^*A) \det[B^*B - B^*A(A^*A)^{-1}A^*B], & \text{rank } A = m, \\ \det(B^*B) \det[A^*A - A^*B(B^*B)^{-1}B^*A], & \text{rank } B = l, \\ 0, & n < m + l. \end{cases}$$

Fact 2.13.23. Let $A, B \in \mathbb{F}^{n \times n}$, and define $\mathcal{A} \in \mathbb{F}^{kn \times kn}$ by

$$\mathcal{A} \triangleq \begin{bmatrix} A & B & B & \cdots & B \\ B & A & B & \cdots & B \\ B & B & A & \ddots & B \\ \vdots & \vdots & \ddots & \ddots & \vdots \\ B & B & B & \cdots & A \end{bmatrix}.$$

Then,

$$\det \mathcal{A} = [\det(A - B)]^{k-1} \det[A + (k-1)B].$$

If $k = 2$, then

$$\det \begin{bmatrix} A & B \\ B & A \end{bmatrix} = \det[(A + B)(A - B)] = \det(A^2 - B^2 - [A, B]).$$

(Proof: See [301].)

2.14 Facts on Adjugates and Inverses

Fact 2.14.1. Let $x, y \in \mathbb{F}^n$. Then,

$$(I + xy^{\mathrm{T}})^{\mathrm{A}} = (1 + y^{\mathrm{T}}x)I - xy^{\mathrm{T}}$$

and

$$\det(I + xy^{\mathrm{T}}) = \det(I + yx^{\mathrm{T}}) = 1 + x^{\mathrm{T}}y = 1 + y^{\mathrm{T}}x.$$

If, in addition, $x^{\mathrm{T}}y \neq -1$, then

$$\left(I + xy^{\mathrm{T}}\right)^{-1} = I - \left(1 + x^{\mathrm{T}}y\right)^{-1}xy^{\mathrm{T}}.$$

Fact 2.14.2. Let $A \in \mathbb{F}^{n \times n}$, assume that A is nonsingular, and let $x, y \in \mathbb{F}^n$. Then,

$$\det\left(A + xy^{\mathrm{T}}\right) = \left(1 + y^{\mathrm{T}}A^{-1}x\right)\det A$$

and

$$\left(A + xy^{\mathrm{T}}\right)^{\mathrm{A}} = \left(1 + y^{\mathrm{T}}A^{-1}x\right)(\det A)I - A^{\mathrm{A}}xy^{\mathrm{T}}.$$

Furthermore, $\det\left(A + xy^{\mathrm{T}}\right) \neq 0$ if and only if $y^{\mathrm{T}}A^{-1}x \neq -1$. In this case,

$$\left(A + xy^{\mathrm{T}}\right)^{-1} = A^{-1} - \left(1 + y^{\mathrm{T}}A^{-1}x\right)^{-1}A^{-1}xy^{\mathrm{T}}A^{-1}.$$

(Remark: This identity is the *Sherman-Morrison-Woodbury formula*.)

Fact 2.14.3. Let $A \in \mathbb{F}^{n \times n}$, assume that A is nonsingular, let $x, y \in \mathbb{F}^n$, let $a \in \mathbb{F}$, and assume that $y^{\mathrm{T}}A^{-1}x \neq a$. Then,

$$\begin{bmatrix} A & x \\ y^{\mathrm{T}} & a \end{bmatrix}^{-1} = \frac{1}{a - y^{\mathrm{T}}A^{-1}x} \begin{bmatrix} (a - y^{\mathrm{T}}A^{-1}x)A^{-1} + A^{-1}xy^{\mathrm{T}}A^{-1} & -A^{-1}x \\ -y^{\mathrm{T}}A^{-1} & 1 \end{bmatrix}$$

$$= \frac{1}{a\det A - y^{\mathrm{T}}A^{\mathrm{A}}x} \begin{bmatrix} [(a - y^{\mathrm{T}}A^{-1}x)I + A^{-1}xy^{\mathrm{T}}]A^{\mathrm{A}} & -A^{\mathrm{A}}x \\ -y^{\mathrm{T}}A^{\mathrm{A}} & 1 \end{bmatrix}.$$

(Problem: Find an expression for $\begin{bmatrix} A & x \\ y^{\mathrm{T}} & a \end{bmatrix}^{-1}$ in the case $\det A = 0$ and $y^{\mathrm{T}}A^{\mathrm{A}}x \neq 0$. See Fact 2.13.2.)

Fact 2.14.4. Let $A \in \mathbb{F}^{n \times n}$. Then, the following statements hold:

i) $\left(\overline{A}\right)^{\mathrm{A}} = \overline{A^{\mathrm{A}}}$.

ii) $\left(A^{\mathrm{T}}\right)^{\mathrm{A}} = \left(A^{\mathrm{A}}\right)^{\mathrm{T}}$.

iii) $\left(A^*\right)^{\mathrm{A}} = \left(A^{\mathrm{A}}\right)^*$.

iv) If $\alpha \in \mathbb{F}$, then $(\alpha A)^{\mathrm{A}} = \alpha^{n-1}A^{\mathrm{A}}$.

v) $\det A^{\mathrm{A}} = (\det A)^{n-1}$.

vi) $\left(A^{\mathrm{A}}\right)^{\mathrm{A}} = (\det A)^{n-2}A$.

vii) $\det\left(A^{\mathrm{A}}\right)^{\mathrm{A}} = (\det A)^{(n-1)^2}$.

Fact 2.14.5. Let $A \in \mathbb{F}^{n \times n}$. Then,

$$\det(A + 1_{n \times n}) - \det A = 1_{1 \times n}^{\mathrm{T}}A^{\mathrm{A}}1 = \sum_{i=1}^n \det\left(A \overset{i}{\leftarrow} 1_{n \times 1}\right).$$

(Proof: See [126].) (Remark: See Fact 2.13.2, Fact 2.14.8, and Fact 10.8.13.)

Fact 2.14.6. Let $A \in \mathbb{F}^{n \times n}$, and assume that A is singular. Then,

$$\mathcal{R}(A) \subseteq \mathcal{N}(A^{\mathrm{A}}).$$

Hence,

$$\operatorname{rank} A \leq \operatorname{def} A^{\mathrm{A}}$$

and

$$\operatorname{rank} A + \operatorname{rank} A^{\mathrm{A}} \leq n.$$

Furthermore, if $n \geq 2$, then $\mathcal{R}(A) = \mathcal{N}(A^{\mathrm{A}})$ if and only if $\operatorname{rank} A = n - 1$.

Fact 2.14.7. Let $A \in \mathbb{F}^{n \times n}$ and $n \geq 2$. Then, the following statements hold:

  i) $\operatorname{rank} A^{\mathrm{A}} = n$ if and only if $\operatorname{rank} A = n$.

  ii) $\operatorname{rank} A^{\mathrm{A}} = 1$ if and only if $\operatorname{rank} A = n - 1$.

  iii) $A^{\mathrm{A}} = 0$ if and only if $\operatorname{rank} A < n - 1$.

(Proof: See [592, p. 12].) (Remark: See Fact 4.10.3.)

Fact 2.14.8. Let $A, B \in \mathbb{F}^{n \times n}$. Then,

$$\left(A^{\mathrm{A}}B\right)_{(i,j)} = \det\left[A \overset{i}{\leftarrow} \operatorname{col}_j(B)\right].$$

(Remark: See Fact 10.8.13.)

Fact 2.14.9. Let $A, B \in \mathbb{F}^{n \times n}$. Then, the following statements hold:

  i) $(AB)^{\mathrm{A}} = B^{\mathrm{A}}A^{\mathrm{A}}$.

  ii) If B is nonsingular, then $\left(BAB^{-1}\right)^{\mathrm{A}} = BA^{\mathrm{A}}B^{-1}$.

  iii) If $AB = BA$, then $A^{\mathrm{A}}B = BA^{\mathrm{A}}$, $AB^{\mathrm{A}} = B^{\mathrm{A}}A$, and $A^{\mathrm{A}}B^{\mathrm{A}} = B^{\mathrm{A}}A^{\mathrm{A}}$.

Fact 2.14.10. Let $A, B, C, D \in \mathbb{F}^{n \times n}$ and $ABCD = I$. Then, $ABCD = DABC = CDAB = BCDA$.

Fact 2.14.11. Let $A = \begin{bmatrix} a & b \\ c & d \end{bmatrix} \in \mathbb{F}^{2 \times 2}$, where $ad - bc \neq 0$. Then,

$$A^{-1} = (ad - bc)^{-1} \begin{bmatrix} d & -b \\ -c & a \end{bmatrix}.$$

Furthermore, if $A = \begin{bmatrix} a & b & c \\ d & e & f \\ g & h & i \end{bmatrix} \in \mathbb{F}^{3 \times 3}$ and $\beta = a(ei - fh) - b(di - fg) + c(dh - eg) \neq 0$, then

$$A^{-1} = \beta^{-1} \begin{bmatrix} ei - fh & -(bi - ch) & bf - ce \\ -(di - fg) & ai - cg & -(af - cd) \\ dh - eg & -(ah - bg) & ae - bd \end{bmatrix}.$$

Fact 2.14.12. Let $A, B \in \mathbb{F}^{n \times n}$, and assume that $A + B$ is nonsingular. Then,

$$A(A + B)^{-1}B = B(A + B)^{-1}A = A - A(A + B)^{-1}A = B - B(A + B)^{-1}B.$$

Fact 2.14.13. Let $A, B \in \mathbb{F}^{n \times n}$, and assume that A and B are nonsingular. Then,

$$A^{-1} + B^{-1} = A^{-1}(A + B)B^{-1}.$$

Furthermore, $A^{-1} + B^{-1}$ is nonsingular if and only if $A + B$ is nonsingular. In this case,

$$\begin{aligned}
\left(A^{-1} + B^{-1}\right)^{-1} &= A(A + B)^{-1}B \\
&= B(A + B)^{-1}A \\
&= A - A(A + B)^{-1}A \\
&= B - B(A + B)^{-1}B.
\end{aligned}$$

Fact 2.14.14. Let $A, B \in \mathbb{F}^{n \times n}$, assume that A and B are nonsingular, and assume that $A - B$ is nonsingular. Then,

$$\left(A^{-1} - B^{-1}\right)^{-1} = A - A(A - B)^{-1}A.$$

Fact 2.14.15. Let $A \in \mathbb{F}^{n \times m}$ and $B \in \mathbb{F}^{m \times n}$, and assume that $I + AB$ is nonsingular. Then, $I + BA$ is nonsingular and

$$(I_n + AB)^{-1}A = A(I_m + BA)^{-1}.$$

(Remark: This result is the *push-through identity*.) Furthermore,

$$(I + AB)^{-1} = I - (I + AB)^{-1}AB.$$

Fact 2.14.16. Let $A, B \in \mathbb{F}^{n \times n}$, and assume that $I + BA$ is nonsingular. Then,

$$(I + AB)^{-1} = I - A(I + BA)^{-1}B.$$

Fact 2.14.17. Let $A \in \mathbb{F}^{n \times n}$, and assume that A and $A + I$ are nonsingular. Then,

$$(A + I)^{-1} + \left(A^{-1} + I\right)^{-1} = (A + I)^{-1} + (A + I)^{-1}A = I.$$

Fact 2.14.18. Let $A \in \mathbb{F}^{n \times m}$. Then,

$$(I + AA^*)^{-1} = I - A(I + A^*A)^{-1}A^*.$$

Fact 2.14.19. Let $A \in \mathbb{F}^{n \times n}$, assume that A is nonsingular, let $B \in \mathbb{F}^{n \times m}$, let $C \in \mathbb{F}^{m \times n}$, and assume that $A + BC$ and $I + CA^{-1}B$ are nonsingular. Then,

$$(A + BC)^{-1}B = A^{-1}B\left(I + CA^{-1}B\right)^{-1}.$$

Fact 2.14.20. Let $A, B \in \mathbb{F}^{n \times n}$, and assume that B is nonsingular. Then,

$$A = B[I + B^{-1}(A - B)].$$

Fact 2.14.21. Let $A, B \in \mathbb{F}^{n \times n}$, and assume that A and $A + B$ are nonsingular. Then, for all $k \in \mathbb{N}$,

$$(A + B)^{-1} = \sum_{i=0}^{k} A^{-1}(-BA^{-1})^i + (-A^{-1}B)^{k+1}(A + B)^{-1}$$

$$= \sum_{i=0}^{k} A^{-1}(-BA^{-1})^i + A^{-1}(-BA^{-1})^{k+1}(I + BA^{-1})^{-1}.$$

Fact 2.14.22. Let $A, B \in \mathbb{F}^{n \times n}$ and $\alpha \in \mathbb{F}$, and assume that A, B, $\alpha A^{-1} + (1 - \alpha)B^{-1}$, and $\alpha B + (1 - \alpha)A$ are nonsingular. Then,

$$\alpha A + (1 - \alpha)B - [\alpha A^{-1} + (1 - \alpha)B^{-1}]^{-1}$$
$$= \alpha(1 - \alpha)(A - B)[\alpha B + (1 - \alpha)A]^{-1}(A - B).$$

Fact 2.14.23. Let $A \in \mathbb{F}^{n \times m}$. If $\operatorname{rank} A = m$, then $(A^*A)^{-1}A^*$ is a left inverse of A. If $\operatorname{rank} A = n$, then $A^*(AA^*)^{-1}$ is a right inverse of A. (Remark: See Fact 3.5.20, Fact 3.5.21, and Fact 3.9.4.) (Problem: If $\operatorname{rank} A = n$ and $b \in \mathbb{R}^n$, then, for every solution $x \in \mathbb{R}^m$ of $Ax = b$, does there exist a right inverse A^{R} of A such that $x = A^{\mathrm{R}}b$?)

Fact 2.14.24. Let $A \in \mathbb{F}^{n \times m}$, and assume that $\operatorname{rank} A = m$. Then, $A^{\mathrm{L}} \in \mathbb{F}^{m \times n}$ is a left inverse of A if and only if there exists a matrix $B \in \mathbb{F}^{m \times n}$ such that BA is nonsingular and

$$A^{\mathrm{L}} = (BA)^{-1}B.$$

(Proof: For necessity, let $B = A^{\mathrm{L}}$.)

Fact 2.14.25. Let $A \in \mathbb{F}^{n \times m}$ and $B \in \mathbb{F}^{m \times l}$, and assume that A and B are right invertible. Then, AB is right invertible. If, in addition, A^{R} is a right inverse of A and B^{R} is a right inverse of B, then $B^{\mathrm{R}}A^{\mathrm{R}}$ is a right inverse of AB.

Fact 2.14.26. Let $A \in \mathbb{F}^{n \times m}$ and $B \in \mathbb{F}^{m \times l}$, and assume that A and B are left invertible. Then, AB is left invertible. If, in addition, A^{L} is a left inverse of A and B^{L} is a left inverse of B, then $B^{\mathrm{L}}A^{\mathrm{L}}$ is a left inverse of AB.

Fact 2.14.27. Let $A \in \mathbb{F}^{n \times n}$, assume that A is nonsingular, and define $A_0 \triangleq I_n$. Furthermore, for all $k = 1, \ldots, n$, let

$$\alpha_k = \tfrac{1}{k} \operatorname{tr} AA_{k-1},$$

and, for all $k = 1, \ldots, n - 1$, let

$$A_k = AA_{k-1} - \alpha_k I.$$

Then,

$$A^{-1} = \tfrac{1}{\alpha_n} A_{n-1}.$$

(Remark: This result is due to Frame. See [92, p. 99].)

Fact 2.14.28. Let $A \in \mathbb{F}^{n \times n}$, assume that A is nonsingular, and define $\{B_i\}_{i=1}^{\infty}$ by

$$B_{i+1} \triangleq 2B_i - B_i A B_i,$$

where $B_0 \in \mathbb{F}^{n \times n}$ satisfies $\mathrm{sprad}(I - B_0 A) < 1$. Then,

$$B_i \to A^{-1}$$

as $i \to \infty$. (Proof: See [78, p. 167].) (Remark: This sequence is given by a Newton-Raphson algorithm.) (Remark: See Fact 6.3.27 for the case in which A is singular or not square.)

Fact 2.14.29. Let $A \in \mathbb{F}^{n \times n}$, and assume that A is nonsingular. Then, $A + A^{-*}$ is nonsingular. (Proof: Note that $AA^* + I$ is positive definite.)

2.15 Facts on Inverses of Partitioned Matrices

Fact 2.15.1. Let $A \in \mathbb{F}^{n \times n}$, $B \in \mathbb{F}^{n \times m}$, $C \in \mathbb{F}^{m \times n}$, and $D \in \mathbb{F}^{m \times m}$, and assume that A and D are nonsingular. Then,

$$\begin{bmatrix} A & B \\ 0 & D \end{bmatrix}^{-1} = \begin{bmatrix} A^{-1} & -A^{-1}BD^{-1} \\ 0 & D^{-1} \end{bmatrix}$$

and

$$\begin{bmatrix} A & 0 \\ C & D \end{bmatrix}^{-1} = \begin{bmatrix} A^{-1} & 0 \\ -D^{-1}CA^{-1} & D^{-1} \end{bmatrix}.$$

Fact 2.15.2. Let $A \in \mathbb{F}^{n \times n}$, $B \in \mathbb{F}^{m \times m}$, and $C \in \mathbb{F}^{m \times n}$. Then,

$$\det \begin{bmatrix} 0 & A \\ B & C \end{bmatrix} = \det \begin{bmatrix} C & B \\ A & 0 \end{bmatrix} = (-1)^{nm}(\det A)(\det B).$$

If, in addition, A and B are nonsingular, then

$$\begin{bmatrix} 0 & A \\ B & C \end{bmatrix}^{-1} = \begin{bmatrix} -B^{-1}CA^{-1} & B^{-1} \\ A^{-1} & 0 \end{bmatrix}$$

and

$$\begin{bmatrix} C & B \\ A & 0 \end{bmatrix}^{-1} = \begin{bmatrix} 0 & A^{-1} \\ B^{-1} & -B^{-1}CA^{-1} \end{bmatrix}.$$

Fact 2.15.3. Let $A \in \mathbb{F}^{n \times n}$, $B \in \mathbb{F}^{n \times m}$, and $C \in \mathbb{F}^{m \times m}$, and assume that C is nonsingular. Then,

$$\begin{bmatrix} A & B \\ B^\mathrm{T} & C \end{bmatrix} = \begin{bmatrix} A - BC^{-1}B^\mathrm{T} & B \\ 0 & C \end{bmatrix} \begin{bmatrix} I & 0 \\ C^{-1}B^\mathrm{T} & I \end{bmatrix}.$$

If, in addition, $A - BC^{-1}B^\mathrm{T}$ is nonsingular, then $\begin{bmatrix} A & B \\ B^\mathrm{T} & C \end{bmatrix}$ is nonsingular and

$$\begin{bmatrix} A & B \\ B^\mathrm{T} & C \end{bmatrix}^{-1}$$

$$= \begin{bmatrix} (A - BC^{-1}B^\mathrm{T})^{-1} & -(A - BC^{-1}B^\mathrm{T})^{-1}BC^{-1} \\ -C^{-1}B^\mathrm{T}(A - BC^{-1}B^\mathrm{T})^{-1} & C^{-1}B^\mathrm{T}(A - BC^{-1}B^\mathrm{T})^{-1}BC^{-1} + C^{-1} \end{bmatrix}.$$

Fact 2.15.4. Let $A, B \in \mathbb{F}^{n \times n}$. Then,

$$\det \begin{bmatrix} I & A \\ B & I \end{bmatrix} = \det(I - AB) = \det(I - BA).$$

If $\det(I - BA) \neq 0$, then

$$\begin{bmatrix} I & A \\ B & I \end{bmatrix}^{-1} = \begin{bmatrix} I + A(I - BA)^{-1}B & -A(I - BA)^{-1} \\ -(I - BA)^{-1}B & (I - BA)^{-1} \end{bmatrix}$$

$$= \begin{bmatrix} (I - AB)^{-1} & -(I - AB)^{-1}A \\ -B(I - AB)^{-1} & I + B(I - AB)^{-1}A \end{bmatrix}.$$

Fact 2.15.5. Let $A, B \in \mathbb{F}^{n \times m}$. Then,

$$\begin{bmatrix} A & B \\ B & A \end{bmatrix} = \tfrac{1}{2} \begin{bmatrix} I & I \\ I & -I \end{bmatrix} \begin{bmatrix} A + B & 0 \\ 0 & A - B \end{bmatrix} \begin{bmatrix} I & I \\ I & -I \end{bmatrix}.$$

Therefore,

$$\mathrm{rank} \begin{bmatrix} A & B \\ B & A \end{bmatrix} = \mathrm{rank}(A + B) + \mathrm{rank}(A - B).$$

Now, assume that $n = m$. Then,

$$\det \begin{bmatrix} A & B \\ B & A \end{bmatrix} = \det[(A + B)(A - B)] = \det(A^2 - B^2 - [A, B]).$$

If, in addition, $A + B$ and $A - B$ are nonsingular, then

$$\begin{bmatrix} A & B \\ B & A \end{bmatrix}^{-1} = \tfrac{1}{2} \begin{bmatrix} (A + B)^{-1} + (A - B)^{-1} & (A + B)^{-1} - (A - B)^{-1} \\ (A + B)^{-1} - (A - B)^{-1} & (A + B)^{-1} + (A - B)^{-1} \end{bmatrix}.$$

Fact 2.15.6. Let $\mathcal{A} \triangleq \left[\begin{smallmatrix} A & B \\ 0_{m \times m} & C \end{smallmatrix} \right]$, where $A \in \mathbb{F}^{n \times m}$, $B \in \mathbb{F}^{n \times n}$, and $C \in \mathbb{F}^{m \times n}$, and assume that CA is nonsingular. Furthermore, define $P \triangleq A(CA)^{-1}C$ and $P_{\perp} \triangleq I - P$. Then, $\mathcal{A}$ is nonsingular if and only if $P + P_{\perp}BP_{\perp}$ is nonsingular. In this case,

$$\mathcal{A}^{-1} = \left[\begin{array}{cc} (CA)^{-1}(C - CBD) & -(CA)^{-1}CB(A - DBA)(CA)^{-1} \\ D & (A - DBA)(CA)^{-1} \end{array} \right],$$

where $D \triangleq (P + P_{\perp}BP_{\perp})^{-1}P_{\perp}$. (Proof: See [331].)

Fact 2.15.7. Let $A \in \mathbb{F}^{n \times m}$ and $B \in \mathbb{F}^{n \times (n-m)}$, and assume that $\left[\begin{array}{cc} A & B \end{array} \right]$ is nonsingular and $A^*B = 0$. Then,

$$\left[\begin{array}{cc} A & B \end{array} \right]^{-1} = \left[\begin{array}{c} (A^*A)^{-1}A^* \\ (B^*B)^{-1}B^* \end{array} \right].$$

(Remark: See Fact 6.4.27.) (Problem: Find an expression for $\left[\begin{array}{cc} A & B \end{array} \right]^{-1}$ without assuming $A^*B = 0$.)

Fact 2.15.8. Let $A \in \mathbb{F}^{n \times n}$, $B \in \mathbb{F}^{n \times m}$, $C \in \mathbb{F}^{m \times n}$, and $D \in \mathbb{F}^{m \times m}$, define $M \triangleq \left[\begin{smallmatrix} A & B \\ C & D \end{smallmatrix} \right] \in \mathbb{F}^{(n+m) \times (n+m)}$, and assume that M is nonsingular. Furthermore, let $\left[\begin{smallmatrix} A' & B' \\ C' & D' \end{smallmatrix} \right] \triangleq M^{-1}$, where $A' \in \mathbb{F}^{n \times n}$ and $D' \in \mathbb{F}^{m \times m}$. Then,

$$\det D' = \frac{\det A}{\det M}$$

and

$$\det A' = \frac{\det D}{\det M}.$$

Consequently, A is nonsingular if and only if D' is nonsingular, and D is nonsingular if and only if A' is nonsingular. (Proof: Use $M \left[\begin{smallmatrix} I & B' \\ 0 & D' \end{smallmatrix} \right] = \left[\begin{smallmatrix} A & 0 \\ C & I \end{smallmatrix} \right]$. See [640].) (Remark: This identity is a special case of *Jacobi's identity*. See [367, p. 21].) (Remark: See Fact 3.7.6.)

Fact 2.15.9. Let $A \in \mathbb{F}^{n \times m}$, $B \in \mathbb{F}^{n \times l}$, and $C \in \mathbb{F}^{m \times l}$. Then,

$$\left[\begin{array}{ccc} I_n & A & B \\ 0 & I_m & C \\ 0 & 0 & I_l \end{array} \right]^{-1} = \left[\begin{array}{ccc} I_n & -A & AC - B \\ 0 & I_m & -C \\ 0 & 0 & I_l \end{array} \right].$$

Fact 2.15.10. Let $A \in \mathbb{F}^{n \times n}$, and assume that A is nonsingular. Then, $X = A^{-1}$ is the unique matrix satisfying

$$\text{rank} \left[\begin{array}{cc} A & I \\ I & X \end{array} \right] = \text{rank} \, A.$$

(Remark: See Fact 6.3.22 and Fact 6.5.5.) (Proof: See [261].)

2.16 Facts on Commutators

Fact 2.16.1. Let $A, B \in \mathbb{F}^{2 \times 2}$. Then,

$$[A, B]^2 = \tfrac{1}{2}(\mathrm{tr}\,[A, B]^2)I_2.$$

(Remark: See [270, 271].)

Fact 2.16.2. Let $A, B \in \mathbb{F}^{n \times n}$, assume that $[A, B] = 0$, and let $k, l \in \mathbb{N}$. Then, $[A^k, B^l] = 0$.

Fact 2.16.3. Let $A, B, C \in \mathbb{F}^{n \times n}$. Then, the following identities hold:

i) $[A, A] = 0$.

ii) $[A, B] = [-A, -B] = -[B, A]$.

iii) $[A, B + C] = [A, B] + [A, C]$.

iv) $[\alpha A, B] = [A, \alpha B] = \alpha[A, B]$ for all $\alpha \in \mathbb{F}$.

v) $[A, [B, C]] + [B, [C, A]] + [C, [A, B]] = 0$.

vi) $[A, B]^{\mathrm{T}} = [B^{\mathrm{T}}, A^{\mathrm{T}}] = -[A^{\mathrm{T}}, B^{\mathrm{T}}]$.

vii) $\mathrm{tr}\,[A, B] = 0$.

viii) $\mathrm{tr}\, A^k[A, B] = \mathrm{tr}\, B^k[A, B] = 0$ for all $k \in \mathbb{P}$.

ix) $[[A, B], B - A] = [[B, A], A - B]$.

x) $[A, [A, B]] = -[A, [B, A]]$.

(Remark: *v)* is the *Jacobi identity*.)

Fact 2.16.4. Let $A, B \in \mathbb{F}^{n \times n}$. Then, for all $X \in \mathbb{F}^{n \times n}$,

$$\mathrm{ad}_{[A,B]} = [\mathrm{ad}_A, \mathrm{ad}_B],$$

that is,

$$\mathrm{ad}_{[A,B]}(X) = \mathrm{ad}_A[\mathrm{ad}_B(X)] - \mathrm{ad}_B[\mathrm{ad}_A(X)]$$

or

$$[[A, B], X] = [A, [B, X]] - [B, [A, X]].$$

Fact 2.16.5. Let $A \in \mathbb{F}^{n \times n}$ and, for all $X \in \mathbb{F}^{n \times n}$, define

$$\mathrm{ad}_A^k(X) \triangleq \begin{cases} \mathrm{ad}_A(X), & k = 1, \\ \mathrm{ad}_A^{k-1}[\mathrm{ad}_A(X)], & k \geq 2. \end{cases}$$

Then, for all $X \in \mathbb{F}^{n \times n}$ and for all $k \geq 1$,

$$\mathrm{ad}_A^2(X) = [A, [A, X]] - [[A, X], A]$$

and

$$\mathrm{ad}_A^k(X) = \sum_{i=0}^{k} (-1)^{k-i} \binom{k}{i} A^i X A^{k-i}.$$

(Remark: The proof of Proposition 11.4.9 is based on $g\left(e^{t\,\mathrm{ad}_A} e^{t\,\mathrm{ad}_B}\right)$, where $g(z) \triangleq (\log z)/(z-1)$. See [628, p. 35].) (Remark: See Fact 11.12.4.) (Proof: For the last identity, see [592, pp. 176, 207].)

Fact 2.16.6. Let $A, B \in \mathbb{F}^{n \times n}$, and assume that $[A, B] = A$. Then, A is singular. (Proof: If A is nonsingular, then $\mathrm{tr}\, B = \mathrm{tr}\, ABA^{-1} = \mathrm{tr}\, B + n$.)

Fact 2.16.7. Let $A, B \in \mathbb{R}^{n \times n}$ be such that $AB = BA$. Then, there exists a matrix $C \in \mathbb{R}^{n \times n}$ such that $A^2 + B^2 = C^2$. (Proof: See [228].) (Remark: This result applies to real matrices only.)

2.17 Facts on Complex Matrices

Fact 2.17.1. Let $a, b \in \mathbb{R}$. Then, $\left[\begin{smallmatrix} a & b \\ -b & a \end{smallmatrix}\right]$ is a representation of the complex number $a + \jmath b$ that preserves addition, multiplication and inversion of complex numbers. In particular, if $a^2 + b^2 \neq 0$, then

$$\begin{bmatrix} a & b \\ -b & a \end{bmatrix}^{-1} = \begin{bmatrix} \frac{a}{a^2+b^2} & \frac{-b}{a^2+b^2} \\ \frac{b}{a^2+b^2} & \frac{a}{a^2+b^2} \end{bmatrix}$$

and

$$(a + \jmath b)^{-1} = \frac{a}{a^2 + b^2} - \jmath \frac{b}{a^2 + b^2}.$$

(Remark: $\left[\begin{smallmatrix} a & b \\ -b & a \end{smallmatrix}\right]$ is a *rotation-dilation*. See Fact 3.14.1.)

Fact 2.17.2. Let $\nu, \omega \in \mathbb{R}$. Then,

$$\begin{bmatrix} \nu & \omega \\ -\omega & \nu \end{bmatrix} = \frac{1}{\sqrt{2}}\begin{bmatrix} 1 & 1 \\ \jmath & -\jmath \end{bmatrix}\begin{bmatrix} \nu + \jmath\omega & 0 \\ 0 & \nu - \jmath\omega \end{bmatrix}\frac{1}{\sqrt{2}}\begin{bmatrix} 1 & 1 \\ \jmath & -\jmath \end{bmatrix}^*$$

and

$$\begin{bmatrix} \nu & \omega \\ -\omega & \nu \end{bmatrix}^{-1} = \frac{1}{\nu^2 + \omega^2}\begin{bmatrix} \nu & -\omega \\ \omega & \nu \end{bmatrix}.$$

(Remark: See Fact 2.17.1.)

Fact 2.17.3. Let $A, B \in \mathbb{R}^{n \times m}$. Then,

$$\begin{bmatrix} A & B \\ -B & A \end{bmatrix} = \frac{1}{2} \begin{bmatrix} I & I \\ \jmath I & -\jmath I \end{bmatrix} \begin{bmatrix} A + \jmath B & 0 \\ 0 & A - \jmath B \end{bmatrix} \begin{bmatrix} I & -\jmath I \\ I & \jmath I \end{bmatrix}$$

$$= \frac{1}{2} \begin{bmatrix} I & \jmath I \\ -\jmath I & -I \end{bmatrix} \begin{bmatrix} A - \jmath B & 0 \\ 0 & A + \jmath B \end{bmatrix} \begin{bmatrix} I & \jmath I \\ -\jmath I & -I \end{bmatrix}$$

$$= \begin{bmatrix} I & 0 \\ \jmath I & I \end{bmatrix} \begin{bmatrix} A + \jmath B & B \\ 0 & A - \jmath B \end{bmatrix} \begin{bmatrix} I & 0 \\ -\jmath I & I \end{bmatrix}$$

and

$$\mathrm{rank}(A + \jmath B) = \mathrm{rank}(A - \jmath B) = \tfrac{1}{2}\mathrm{rank} \begin{bmatrix} A & B \\ -B & A \end{bmatrix}.$$

Now, assume that $n = m$. Then,

$$\det \begin{bmatrix} A & B \\ -B & A \end{bmatrix} = \det(A + \jmath B)\det(A - \jmath B)$$

$$= |\det(A + \jmath B)|^2$$

$$= \det\left[A^2 + B^2 + \jmath(AB - BA)\right]$$

$$\geq 0$$

and

$$\mathrm{mspec}\left(\begin{bmatrix} A & B \\ -B & A \end{bmatrix}\right) = \mathrm{mspec}(A + \jmath B) \cup \mathrm{mspec}(A - \jmath B).$$

If A is nonsingular, then

$$\det \begin{bmatrix} A & B \\ -B & A \end{bmatrix} = \det\left(A^2 + ABA^{-1}B\right).$$

If $AB = BA$, then

$$\det \begin{bmatrix} A & B \\ -B & A \end{bmatrix} = \det\left(A^2 + B^2\right).$$

(Proof: If A is nonsingular, then use

$$\begin{bmatrix} A & B \\ -B & A \end{bmatrix} = \begin{bmatrix} A & 0 \\ 0 & A \end{bmatrix} \begin{bmatrix} I & A^{-1}B \\ -A^{-1}B & I \end{bmatrix}$$

and

$$\det \begin{bmatrix} I & A^{-1}B \\ -A^{-1}B & I \end{bmatrix} = \det\left[I + \left(A^{-1}B\right)^2\right].)$$

(Remark: See Fact 4.10.19 and [45, 704].)

Fact 2.17.4. Let $A, B \in \mathbb{R}^{n \times m}$ and $C, D \in \mathbb{R}^{m \times l}$. Then, $\begin{bmatrix} A & B \\ -B & A \end{bmatrix}$, $\begin{bmatrix} C & D \\ -D & C \end{bmatrix}$. and $\begin{bmatrix} A+C & B+D \\ -(B+D) & A+C \end{bmatrix}$ are addition-preserving representations of the complex matrices $A + \jmath B$, $C + \jmath D$, and $A + \jmath B + C + \jmath D$.

Fact 2.17.5. Let $A, B \in \mathbb{R}^{n \times m}$ and $C, D \in \mathbb{R}^{m \times l}$. Then, $\left[\begin{smallmatrix} A & B \\ -B & A \end{smallmatrix}\right]$, $\left[\begin{smallmatrix} C & D \\ -D & C \end{smallmatrix}\right]$, and $\left[\begin{smallmatrix} AC-BD & AD+BC \\ -(AD+BC) & AC-BD \end{smallmatrix}\right]$ are product-preserving representations of the complex matrices $A + jB$, $C + jD$, and $(A + jB)(C + jD)$.

Fact 2.17.6. Let $A, B \in \mathbb{R}^{n \times n}$. Then, $\left[\begin{smallmatrix} A & B \\ -B & A \end{smallmatrix}\right]$ is a representation of the complex matrix $A + jB$ that preserves addition, multiplication, and inversion of complex matrices. In particular, $A + jB$ is nonsingular if and only if $\left[\begin{smallmatrix} A & B \\ -B & A \end{smallmatrix}\right]$ is nonsingular. Furthermore, if A is nonsingular, then $A + jB$ is nonsingular if and only if $A + BA^{-1}B$ is nonsingular. In this case,

$$\left[\begin{array}{cc} A & B \\ -B & A \end{array}\right]^{-1} = \left[\begin{array}{cc} \left(A + BA^{-1}B\right)^{-1} & -A^{-1}B\left(A + BA^{-1}B\right)^{-1} \\ A^{-1}B\left(A + BA^{-1}B\right)^{-1} & \left(A + BA^{-1}B\right)^{-1} \end{array}\right]$$

and

$$(A + jB)^{-1} = \left(A + BA^{-1}B\right)^{-1} - jA^{-1}B\left(A + BA^{-1}B\right)^{-1}.$$

Finally, assume that B is nonsingular. Then, $A + jB$ is nonsingular if and only if $B + AB^{-1}A$ is nonsingular. In this case,

$$\left[\begin{array}{cc} A & B \\ -B & A \end{array}\right]^{-1} = \left[\begin{array}{cc} B^{-1}A\left(B + AB^{-1}A\right)^{-1} & -\left(B + AB^{-1}A\right)^{-1} \\ \left(B + AB^{-1}A\right)^{-1} & B^{-1}A\left(B + AB^{-1}A\right)^{-1} \end{array}\right]$$

and

$$(A + jB)^{-1} = B^{-1}A\left(B + AB^{-1}A\right)^{-1} - j\left(B + AB^{-1}A\right)^{-1}.$$

(Problem: Consider the case in which A and B are singular.)

Fact 2.17.7. Let $A \in \mathbb{F}^{n \times n}$. Then,

$$\det\left(I + A\overline{A}\right) \geq 0.$$

(Proof: See [229].)

Fact 2.17.8. Let $A, B \in \mathbb{F}^{n \times n}$. Then,

$$\det\left[\begin{array}{cc} A & B \\ -\overline{B} & \overline{A} \end{array}\right] \geq 0.$$

If, in addition, A is nonsingular, then

$$\det\left[\begin{array}{cc} A & B \\ -\overline{B} & \overline{A} \end{array}\right] = |\det A|^2 \det\left(I + \overline{A^{-1}B}A^{-1}B\right).$$

(Proof: See [810].) (Remark: Fact 2.17.7 implies that $\det\left(I + \overline{A^{-1}B}A^{-1}B\right) \geq 0$.)

Fact 2.17.9. Let $A, B \in \mathbb{R}^{n \times n}$, and define $C \in \mathbb{R}^{2n \times 2n}$ by $C \triangleq$
$\begin{bmatrix} C_{11} & C_{12} & \cdots \\ C_{21} & \cdots \\ \vdots \end{bmatrix}$, where $C_{ij} \triangleq \begin{bmatrix} A_{(i,j)} & B_{(i,j)} \\ -B_{(i,j)} & A_{(i,j)} \end{bmatrix}$ for all $i, j = 1, \ldots, n$. Then,

$$\det C = |\det(A + \jmath B)|^2.$$

(Proof: Note that

$$C = A \otimes I_2 + B \otimes J_2 = P_{2,n}(I_2 \otimes A + J_2 \otimes B)P_{2,n} = P_{2,n} \begin{bmatrix} A & B \\ -B & A \end{bmatrix} P_{2,n}.$$

See [142].)

2.18 Facts on Geometry

Fact 2.18.1. The points $x, y, z \in \mathbb{R}^2$ lie on one line if and only if

$$\det \begin{bmatrix} x & y & z \\ 1 & 1 & 1 \end{bmatrix} = 0.$$

Furthermore, the points $x, y, z \in \mathbb{R}^3$ lie on one plane if and only if

$$\det \begin{bmatrix} x & y & z \end{bmatrix} = 0.$$

Fact 2.18.2. Let $\mathcal{S} \subset \mathbb{R}^2$ denote the triangle with vertices $\begin{bmatrix} 0 \\ 0 \end{bmatrix}, \begin{bmatrix} x_1 \\ y_1 \end{bmatrix}$, $\begin{bmatrix} x_2 \\ y_2 \end{bmatrix} \in \mathbb{R}^2$. Then,
$$\text{area}(\mathcal{S}) = \tfrac{1}{2} \left| \det \begin{bmatrix} x_1 & x_2 \\ y_1 & y_2 \end{bmatrix} \right|.$$

Fact 2.18.3. Let $\mathcal{S} \subset \mathbb{R}^2$ denote the polygon with vertices $\begin{bmatrix} x_1 \\ y_1 \end{bmatrix}, \ldots,$ $\begin{bmatrix} x_n \\ y_n \end{bmatrix} \in \mathbb{R}^2$ arranged in counterclockwise order, and assume that the interior of the polygon is either empty or simply connected. Then,

$$\text{area}(\mathcal{S}) = \tfrac{1}{2}\det \begin{bmatrix} x_1 & x_2 \\ y_1 & y_2 \end{bmatrix} + \tfrac{1}{2}\det \begin{bmatrix} x_2 & x_3 \\ y_2 & y_3 \end{bmatrix} + \cdots$$

$$+ \tfrac{1}{2}\det \begin{bmatrix} x_{n-1} & x_n \\ y_{n-1} & y_n \end{bmatrix} + \tfrac{1}{2}\det \begin{bmatrix} x_n & x_1 \\ y_n & y_1 \end{bmatrix}.$$

(Remark: The polygon need not be convex, while "counterclockwise" is determined with respect to a point in the interior of the polygon. "Simply connected" means that the polygon has no holes. See [673].)

Fact 2.18.4. Let $\mathcal{S} \subset \mathbb{R}^3$ denote the triangle with vertices $x, y, z \in \mathbb{R}^3$. Then,
$$\text{area}(\mathcal{S}) = \tfrac{1}{2}\sqrt{[(y-x) \times (z-x)]^{\mathrm{T}}[(y-x) \times (z-x)]}.$$

Fact 2.18.5. Let $\mathcal{S} \subset \mathbb{R}^3$ denote the tetrahedron with vertices $x, y, z, w \in \mathbb{R}^3$. Then,

$$\text{volume}(\mathcal{S}) = \tfrac{1}{6}\left|(x - w)^\mathrm{T}[(y - w) \times (z - w)]\right|.$$

(Proof: The volume of the unit simplex $\mathcal{S} \subset \mathbb{R}^3$ with vertices $(0, 0, 0), (1, 0, 0)$, $(0, 1, 0), (0, 0, 1)$ is $1/6$. Now, Fact 2.18.8 implies that the volume of $A\mathcal{S}$ is $(1/6)|\det A|$.) (Remark: The connection between the *signed volume* of a simplex and the determinant is discussed in [461, pp. 32, 33].)

Fact 2.18.6. Let $\mathcal{S} \subset \mathbb{R}^3$ denote the parallelepiped with vertices x, y, z, $w, y + z - x, y + w - x, z + w - x, y + z + w - 2x \in \mathbb{R}^3$. Then,

$$\text{volume}(\mathcal{S}) = \left|(w - x)^\mathrm{T}[(y - x) \times (z - x)]\right|.$$

Fact 2.18.7. Let $A \in \mathbb{R}^{n \times m}$, assume that rank $A = m$, and let $\mathcal{S} \subset \mathbb{R}^n$ denote the parallelepiped in $\mathbb{R}^n$ generated by the columns of A. Then,

$$\text{volume}(\mathcal{S}) = \left[\det\left(A^\mathrm{T}A\right)\right]^{1/2}.$$

If, in addition, $m = n$, then

$$\text{volume}(\mathcal{S}) = |\det A|.$$

Fact 2.18.8. Let $\mathcal{S} \subset \mathbb{R}^n$ and $A \in \mathbb{R}^{n \times n}$. Then,

$$\text{volume}(A\mathcal{S}) = |\det A|\text{volume}(\mathcal{S}).$$

(Remark: See [535, p. 468].)

2.19 Notes

The theory of determinants is discussed in [549, 728, 745]. The empty matrix is discussed in [209], [557], [613, pp. 462–464], and [672, p. 3]. Convexity is the subject of [98, 135, 141, 240, 462, 614, 672, 735, 765]. Convex optimization theory is developed in [97, 141]. Our development of rank properties is based on [514]. Theorem 2.6.3 is based on [562]. The term "subdeterminant" is used in [581] and is equivalent to *minor*. The notation A^A for adjugate is used in [665]. Numerous papers on basic topics in matrix theory and linear algebra are collected in [164, 165]. A geometric interpretation of $\mathcal{N}(A), \mathcal{R}(A), \mathcal{N}(A^*)$, and $\mathcal{R}(A^\mathrm{T})$ is given in [675]. Some reflections on matrix theory are given in [684, 698]. Applications of the matrix inversion lemma are discussed in [322].

Chapter Three

Matrix Classes and Transformations

This chapter presents definitions of various types of matrices as well as transformations for analyzing matrices.

3.1 Matrix Classes

In this section we categorize various types of matrices based on their algebraic and structural properties.

The following definition introduces various types of square matrices.

Definition 3.1.1. For $A \in \mathbb{F}^{n \times n}$ define the following types of matrices:

i) A is *group invertible* if $\mathcal{R}(A) = \mathcal{R}(A^2)$.

ii) A is *involutory* if $A^2 = I$.

iii) A is *skew involutory* if $A^2 = -I$.

iv) A is *idempotent* if $A^2 = A$.

v) A is *tripotent* if $A^3 = A$.

vi) A is *nilpotent* if there exists $k \in \mathbb{P}$ such that $A^k = 0$.

vii) A is *unipotent* if $A - I$ is nilpotent.

viii) A is *range Hermitian* if $\mathcal{R}(A) = \mathcal{R}(A^*)$.

ix) A is *range symmetric* if $\mathcal{R}(A) = \mathcal{R}(A^{\mathrm{T}})$.

x) A is *Hermitian* if $A = A^*$.

xi) A is *symmetric* if $A = A^{\mathrm{T}}$.

xii) A is *skew Hermitian* if $A = -A^*$.

xiii) A is *skew symmetric* if $A = -A^{\mathrm{T}}$.

xiv) A is *normal* if $AA^* = A^*A$.

   xv) A is *positive semidefinite* ($A \geq 0$) if A is Hermitian and $x^*Ax \geq 0$ for all $x \in \mathbb{F}^n$.

  xvi) A is *negative semidefinite* ($A \leq 0$) if $-A$ is positive semidefinite.

 xvii) A is *positive definite* ($A > 0$) if A is Hermitian and $x^*Ax > 0$ for all $x \in \mathbb{F}^n$ such that $x \neq 0$.

xviii) A is *negative definite* ($A < 0$) if $-A$ is positive definite.

  xix) A is *semidissipative* if $A + A^*$ is negative semidefinite.

   xx) A is *dissipative* if $A + A^*$ is negative definite.

  xxi) A is *unitary* if $A^*A = I$.

 xxii) A is *orthogonal* if $A^{\mathrm{T}}A = I$.

xxiii) A is a *projector* if A is Hermitian and idempotent.

xxiv) A is a *reflector* if A is Hermitian and unitary.

 xxv) A is an *elementary projector* if there exists a nonzero vector $x \in \mathbb{F}^n$ such that $A = I - (x^*x)^{-1}xx^*$.

xxvi) A is an *elementary reflector* if there exists a nonzero vector $x \in \mathbb{F}^n$ such that $A = I - 2(x^*x)^{-1}xx^*$.

xxvii) A is an *elementary matrix* if there exist vectors $x, y \in \mathbb{F}^n$ such that $A = I - xy^{\mathrm{T}}$ and $x^{\mathrm{T}}y \neq 1$.

xxviii) A is *reverse Hermitian* if $A = A^{\hat{*}}$.

xxix) A is *reverse symmetric* if $A = A^{\hat{\mathrm{T}}}$.

 xxx) A is a *permutation matrix* if each row of A and each column of A possesses one 1 and zeros otherwise.

Let $A \in \mathbb{F}^{n \times n}$ be Hermitian. Then, the function $f \colon \mathbb{F}^n \mapsto \mathbb{R}$ defined by

$$f(x) \triangleq x^*Ax \tag{3.1.1}$$

is a *quadratic form*.

The $n \times n$ *standard nilpotent matrix*, which has ones on the superdiagonal and zeros elsewhere, is denoted by N_n or just N. We define $N_1 \triangleq 0$ and $N_0 \triangleq 0_{0 \times 0}$.

The following definition considers matrices that are not necessarily square.

Definition 3.1.2. For $A \in \mathbb{F}^{n \times m}$ define the following types of matrices:

   i) A is *semicontractive* if $I_n - AA^*$ is positive semidefinite.

ii) A is *contractive* if $I_n - AA^*$ is positive definite.

iii) A is *left inner* if $A^*A = I_m$.

iv) A is *right inner* if $AA^* = I_n$.

v) A is *centrohermitian* if $A = \hat{I}_n \overline{A} \hat{I}_m$.

vi) A is *centrosymmetric* if $A = \hat{I}_n A \hat{I}_m$.

vii) A is an *outer-product matrix* if there exist $x \in \mathbb{F}^n$ and $y \in \mathbb{F}^m$ such that $A = xy^{\mathrm{T}}$.

The following definition introduces various types of structured matrices.

Definition 3.1.3. For $A \in \mathbb{F}^{n \times m}$ with $l \triangleq \min\{n, m\}$ define the following types of matrices:

i) A is *diagonal* if $A_{(i,j)} = 0$ for all $i \neq j$. If $n = m$, then

$$A = \operatorname{diag}\big(A_{(1,1)}, \ldots, A_{(n,n)}\big).$$

ii) A is *tridiagonal* if $A_{(i,j)} = 0$ for all $|i - j| > 1$.

iii) A is *reverse diagonal* if $A_{(i,j)} = 0$ for all $i + j \neq l + 1$. If $n = m$, then

$$A = \operatorname{revdiag}\big(A_{(1,n)}, \ldots, A_{(n,1)}\big).$$

iv) A is (*upper triangular, strictly upper triangular*) if $A_{(i,j)} = 0$ for all $(i \geq j, i > j)$.

v) A is (*lower triangular, strictly lower triangular*) if $A_{(i,j)} = 0$ for all $(i \leq j, i < j)$.

vi) A is (*upper Hessenberg, lower Hessenberg*) if $A_{(i,j)} = 0$ for all $(i > j + 1, i < j + 1)$.

vii) A is *Toeplitz* if $A_{(i,j)} = A_{(k,l)}$ for all $k - i = l - j$, that is,

$$A = \begin{bmatrix} a & b & c & \cdots \\ d & a & b & \ddots \\ e & d & a & \ddots \\ \vdots & \ddots & \ddots & \ddots \end{bmatrix}.$$

viii) A is *Hankel* if $A_{(i,j)} = A_{(k,l)}$ for all $i + j = k + l$, that is,

$$A = \begin{bmatrix} a & b & c & \cdots \\ b & c & d & \iddots \\ c & d & e & \iddots \\ \vdots & \iddots & \iddots & \iddots \end{bmatrix}.$$

$ix)$ A is *block diagonal* if

$$A = \begin{bmatrix} A_1 & & 0 \\ & \ddots & \\ 0 & & A_k \end{bmatrix} = \text{diag}(A_1, \ldots, A_k),$$

where $A_i \in \mathbb{F}^{n_i \times m_i}$ for all $i = 1, \ldots, k$.

$x)$ A is *upper block triangular* if

$$A = \begin{bmatrix} A_{11} & A_{12} & \cdots & A_{1k} \\ 0 & A_{22} & \cdots & A_{2k} \\ \vdots & \ddots & \ddots & \vdots \\ 0 & 0 & \cdots & A_{kk} \end{bmatrix},$$

where $A_{ij} \in \mathbb{F}^{n_i \times n_j}$ for all $i, j = 1, \ldots, k$.

$xi)$ A is *lower block triangular* if

$$A = \begin{bmatrix} A_{11} & 0 & \cdots & 0 \\ A_{21} & A_{22} & \ddots & 0 \\ \vdots & \vdots & \ddots & \vdots \\ A_{k1} & A_{k2} & \cdots & A_{kk} \end{bmatrix},$$

where $A_{ij} \in \mathbb{F}^{n_i \times n_j}$ for all $i, j = 1, \ldots, k$.

$xii)$ A is *block Toeplitz* if $A_{(i,j)} = A_{(k,l)}$ for all $k - i = l - j$, that is,

$$A = \begin{bmatrix} A_1 & A_2 & A_3 & \cdots \\ A_4 & A_1 & A_2 & \ddots \\ A_5 & A_4 & A_1 & \ddots \\ \vdots & \ddots & \ddots & \ddots \end{bmatrix},$$

where $A_i \in \mathbb{F}^{n_i \times m_i}$.

$xiii)$ A is *block Hankel* if $A_{(i,j)} = A_{(k,l)}$ for all $i + j = k + l$, that is,

$$A = \begin{bmatrix} A_1 & A_2 & A_3 & \cdots \\ A_2 & A_3 & A_4 & \cdot^{\cdot^\cdot} \\ A_3 & A_4 & A_5 & \cdot^{\cdot^\cdot} \\ \vdots & \cdot^{\cdot^\cdot} & \cdot^{\cdot^\cdot} & \cdot^{\cdot^\cdot} \end{bmatrix},$$

where $A_i \in \mathbb{F}^{n_i \times m_i}$.

Definition 3.1.4. For $A \in \mathbb{R}^{n \times m}$ define the following types of matrices:

$i)$ A is *nonnegative* ($A \geq\geq 0$) if $A_{(i,j)} \geq 0$ for all $i = 1, \ldots, n$ and $j = 1, \ldots, m$.

ii) A is *positive* ($A >> 0$) if $A_{(i,j)} > 0$ for all $i = 1, \ldots, n$ and $j = 1, \ldots, m$.

Define the matrix $J_{2n} \in \mathbb{R}^{2n \times 2n}$ (or just J) by

$$J_{2n} \triangleq \begin{bmatrix} 0 & I_n \\ -I_n & 0 \end{bmatrix}. \tag{3.1.2}$$

In particular,

$$J_2 = \begin{bmatrix} 0 & 1 \\ -1 & 0 \end{bmatrix}. \tag{3.1.3}$$

The following definition introduces structured matrices of even order. Note that

$$J_{2n}^2 = -I_{2n}. \tag{3.1.4}$$

Definition 3.1.5. For $A \in \mathbb{F}^{2n \times 2n}$ define the following types of matrices:

i) A is *Hamiltonian* if $J^{-1}A^{\mathrm{T}}J = -A$.

ii) A is *symplectic* if A is nonsingular and $J^{-1}A^{\mathrm{T}}J = A^{-1}$.

Proposition 3.1.6. Let $A \in \mathbb{F}^{n \times n}$. Then, the following statements hold:

i) If A is Hermitian or skew Hermitian, then A is normal.

ii) If A is nonsingular or normal, then A is range Hermitian.

iii) If A is range Hermitian, idempotent, or tripotent, then A is group invertible.

iv) If A is a reflector, then A is tripotent.

v) If A is a permutation matrix, then A is orthogonal.

Proof. *i)* is immediate. To prove *ii)*, note that, if A is nonsingular, then $\mathcal{R}(A) = \mathcal{R}(A^*) = \mathbb{F}^n$, and thus A is range Hermitian. If A is normal, then it follows from Theorem 2.4.3 that $\mathcal{R}(A) = \mathcal{R}(AA^*) = \mathcal{R}(A^*A) = \mathcal{R}(A^*)$, which proves that A is range Hermitian. To prove *iii)*, note that, if A is range Hermitian, then $\mathcal{R}(A) = \mathcal{R}(AA^*) = A\mathcal{R}(A^*) = A\mathcal{R}(A) = \mathcal{R}(A^2)$, while, if A is idempotent, then $\mathcal{R}(A) = \mathcal{R}(A^2)$. If A is tripotent, then $\mathcal{R}(A) = \mathcal{R}(A^3) = A^2\mathcal{R}(A) \subseteq \mathcal{R}(A^2) = A\mathcal{R}(A) \subseteq \mathcal{R}(A)$. Hence, $\mathcal{R}(A) = \mathcal{R}(A^2)$. $\square$

Proposition 3.1.7. Let $\mathcal{A} \in \mathbb{F}^{2n \times 2n}$. Then, $\mathcal{A}$ is Hamiltonian if and only if there exist matrices $A, B, C \in \mathbb{F}^{n \times n}$ such that B and C are symmetric and

$$\mathcal{A} = \begin{bmatrix} A & B \\ C & -A^{\mathrm{T}} \end{bmatrix}. \tag{3.1.5}$$

3.2 Matrix Transformations

A variety of transformations can be employed for analyzing matrices.

Definition 3.2.1. Let $A, B \in \mathbb{F}^{n \times m}$. Then, the following terminology is defined:

i) A and B are *left equivalent* if there exists a nonsingular matrix $S_1 \in \mathbb{F}^{n \times n}$ such that $A = S_1 B$.

ii) A and B are *right equivalent* if there exists a nonsingular matrix $S_2 \in \mathbb{F}^{m \times m}$ such that $A = B S_2$.

iii) A and B are *biequivalent* if there exist nonsingular matrices $S_1 \in \mathbb{F}^{n \times n}$ and $S_2 \in \mathbb{F}^{m \times m}$ such that $A = S_1 B S_2$.

iv) A and B are *unitarily left equivalent* if there exists a unitary matrix $S_1 \in \mathbb{F}^{n \times n}$ such that $A = S_1 B$.

v) A and B are *unitarily right equivalent* if there exists a unitary matrix $S_2 \in \mathbb{F}^{m \times m}$ such that $A = B S_2$.

vi) A and B are *unitarily biequivalent* if there exist unitary matrices $S_1 \in \mathbb{F}^{n \times n}$ and $S_2 \in \mathbb{F}^{m \times m}$ such that $A = S_1 B S_2$.

Definition 3.2.2. Let $A, B \in \mathbb{F}^{n \times n}$. Then, the following terminology is defined:

i) A and B are *similar* if there exists a nonsingular matrix $S \in \mathbb{F}^{n \times n}$ such that $A = SBS^{-1}$.

ii) A and B are *congruent* if there exists a nonsingular matrix $S \in \mathbb{F}^{n \times n}$ such that $A = SBS^*$.

iii) A and B are *T-congruent* if there exists a nonsingular matrix $S \in \mathbb{F}^{n \times n}$ such that $A = SBS^{\mathrm{T}}$.

iv) A and B are *unitarily similar* if there exists a unitary matrix $S \in \mathbb{F}^{n \times n}$ such that $A = SBS^* = SBS^{-1}$.

The following results summarize some matrix properties that are preserved under left equivalence, right equivalence, biequivalence, similarity, congruence, and unitary similarity.

Proposition 3.2.3. Let $A, B \in \mathbb{F}^{n \times n}$. If A and B are similar, then the following statements hold:

i) A and B are biequivalent.

ii) $\operatorname{tr} A = \operatorname{tr} B$.

iii) $\det A = \det B$.

iv) A^k and B^k are similar for all $k \in \mathbb{P}$.

v) A^{k*} and B^{k*} are similar for all $k \in \mathbb{P}$.

vi) A is nonsingular if and only if B is; in this case, A^{-k} and B^{-k} are similar for all $k \in \mathbb{P}$.

vii) A is (group invertible, involutory, skew involutory, idempotent, tripotent, nilpotent) if and only if B is.

If A and B are congruent, then the following statements hold:

viii) A and B are biequivalent.

ix) A^* and B^* are congruent.

x) A is nonsingular if and only if B is; in this case, A^{-1} and B^{-1} are congruent.

xi) A is (range Hermitian, Hermitian, skew Hermitian, positive semidefinite, positive definite) if and only if B is.

If A and B are unitarily similar, then the following statements hold:

xii) A and B are similar.

xiii) A and B are congruent.

xiv) A is (range Hermitian, group invertible, normal, Hermitian, skew Hermitian, positive semidefinite, positive definite, unitary, involutory, skew involutory, idempotent, tripotent, nilpotent) if and only if B is.

3.3 Lie Algebras and Groups

In this section we introduce Lie algebras and groups. Lie groups are discussed in Section 11.5.

Definition 3.3.1. Let $\mathcal{S} \subseteq \mathbb{F}^{n \times n}$. Then, $\mathcal{S}$ is a *Lie algebra* if the following conditions are satisfied:

i) If $A, B \in \mathcal{S}$ and $\alpha, \beta \in \mathbb{R}$, then $\alpha A + \beta B \in \mathcal{S}$.

ii) If $A, B \in \mathcal{S}$, then $[A, B] \in \mathcal{S}$.

Proposition 3.3.2. The following sets are Lie algebras:

i) $\mathrm{gl}_{\mathbb{F}}(n) \triangleq \mathbb{F}^{n \times n}$.

ii) $\mathrm{pl}_{\mathbb{C}}(n) \triangleq \{A \in \mathbb{C}^{n \times n} : \ \mathrm{tr}\, A \in \mathbb{R}\}$.

iii) $\mathrm{sl}_{\mathbb{F}}(n) \triangleq \{A \in \mathbb{F}^{n \times n} : \ \mathrm{tr}\, A = 0\}$.

iv) $\mathrm{u}(n) \triangleq \{A \in \mathbb{C}^{n \times n}: A \text{ is skew Hermitian}\}$.

v) $\mathrm{su}(n) \triangleq \{A \in \mathbb{C}^{n \times n}: A \text{ is skew Hermitian and } \operatorname{tr} A = 0\}$.

vi) $\mathrm{so}(n) \triangleq \{A \in \mathbb{R}^{n \times n}: A \text{ is skew symmetric}\}$.

vii) $\mathrm{su}(n, m) \triangleq \{A \in \mathbb{C}^{(n+m) \times (n+m)}: \operatorname{diag}(I_n, -I_m) A^* \operatorname{diag}(I_n, -I_m) = -A$
and $\operatorname{tr} A = 0\}$.

viii) $\mathrm{so}(n, m) \triangleq \{A \in \mathbb{R}^{(n+m) \times (n+m)}: \operatorname{diag}(I_n, -I_m) A^{\mathrm{T}} \operatorname{diag}(I_n, -I_m) = -A\}$.

ix) $\mathrm{sp}_{\mathbb{F}}(n) \triangleq \{A \in \mathbb{F}^{2n \times 2n}: A \text{ is Hamiltonian}\}$.

x) $\mathrm{aff}_{\mathbb{F}}(n) \triangleq \left\{ \begin{bmatrix} A & b \\ 0 & 0 \end{bmatrix}: A \in \mathrm{gl}_{\mathbb{F}}(n), \ b \in \mathbb{F}^n \right\}$.

xi) $\mathrm{se}_{\mathbb{C}}(n) \triangleq \left\{ \begin{bmatrix} A & b \\ 0 & 0 \end{bmatrix}: A \in \mathrm{su}(n), \ b \in \mathbb{C}^n \right\}$.

xii) $\mathrm{se}_{\mathbb{R}}(n) \triangleq \left\{ \begin{bmatrix} A & b \\ 0 & 0 \end{bmatrix}: A \in \mathrm{so}(n), \ b \in \mathbb{R}^n \right\}$.

xiii) $\mathrm{trans}_{\mathbb{F}}(n) \triangleq \left\{ \begin{bmatrix} 0 & b \\ 0 & 0 \end{bmatrix}: b \in \mathbb{F}^n \right\}$.

Definition 3.3.3. Let $\mathcal{S} \subset \mathbb{F}^{n \times n}$. Then, $\mathcal{S}$ is a *group* if the following conditions are satisfied:

i) If $A \in \mathcal{S}$, then A is nonsingular.

ii) If $A \in \mathcal{S}$, then $A^{-1} \in \mathcal{S}$.

iii) If $A, B \in \mathcal{S}$, then $AB \in \mathcal{S}$.

Note that, if $\mathcal{S} \subset \mathbb{F}^{n \times n}$ is a group, then $I_n \in \mathcal{S}$.

The following result lists several classical groups that are of importance in physics and engineering. In particular, $\mathrm{O}(1, 3)$ is the *Lorentz group*; see, for example, [628, p. 16] or [639, p. 126].

Proposition 3.3.4. The following sets are groups:

i) $\mathrm{GL}_{\mathbb{F}}(n) \triangleq \{A \in \mathbb{F}^{n \times n}: \det A \neq 0\}$.

ii) $\mathrm{PL}_{\mathbb{F}}(n) \triangleq \{A \in \mathbb{F}^{n \times n}: \det A > 0\}$.

iii) $\mathrm{SL}_{\mathbb{F}}(n) \triangleq \{A \in \mathbb{F}^{n \times n}: \det A = 1\}$.

iv) $\mathrm{U}(n) \triangleq \{A \in \mathbb{C}^{n \times n}: A \text{ is unitary}\}$.

v) $\mathrm{O}(n) \triangleq \{A \in \mathbb{R}^{n \times n}: A \text{ is orthogonal}\}$.

vi) $\mathrm{SU}(n) \triangleq \{A \in \mathrm{U}(n): \det A = 1\}$.

vii) $\mathrm{SO}(n) \triangleq \{A \in \mathrm{O}(n): \ \det A = 1\}$.

viii) $\mathrm{U}(n,m) \triangleq \{A \in \mathbb{C}^{(n+m)\times(n+m)} : A^* \mathrm{diag}(I_n, -I_m)A = \mathrm{diag}(I_n, -I_m)\}$.

ix) $\mathrm{O}(n,m) \triangleq \{A \in \mathbb{R}^{(n+m)\times(n+m)} : A^{\mathrm{T}} \mathrm{diag}(I_n, -I_m)A = \mathrm{diag}(I_n, -I_m)\}$.

x) $\mathrm{SU}(n,m) \triangleq \{A \in \mathrm{U}(n,m): \ \det A = 1\}$.

xi) $\mathrm{SO}(n,m) \triangleq \{A \in \mathrm{O}(n,m): \ \det A - 1\}$.

xii) $\mathrm{Sp}_{\mathbb{F}}(n) \triangleq \{A \in \mathbb{F}^{2n\times 2n} : \ A \text{ is symplectic}\}$.

xiii) $\mathrm{Aff}_{\mathbb{F}}(n) \triangleq \left\{ \begin{bmatrix} A & b \\ 0 & 1 \end{bmatrix} : \ A \in \mathrm{GL}_{\mathbb{F}}(n), \ b \in \mathbb{F}^n \right\}$.

xiv) $\mathrm{SE}_{\mathbb{C}}(n) \triangleq \left\{ \begin{bmatrix} A & b \\ 0 & 1 \end{bmatrix} : \ A \in \mathrm{SU}(n), \ b \in \mathbb{C}^n \right\}$.

xv) $\mathrm{SE}_{\mathbb{R}}(n) \triangleq \left\{ \begin{bmatrix} A & b \\ 0 & 1 \end{bmatrix} : \ A \in \mathrm{SO}(n), \ b \in \mathbb{R}^n \right\}$.

xvi) $\mathrm{Trans}_{\mathbb{F}}(n) \triangleq \left\{ \begin{bmatrix} I & b \\ 0 & 1 \end{bmatrix} : \ b \in \mathbb{F}^n \right\}$.

The following result shows that groups can be used to define equivalence relations on $\mathbb{F}^{n\times m}$.

Proposition 3.3.5. Let $\mathcal{S}_1 \subset \mathbb{R}^{n\times n}$ and $\mathcal{S}_2 \subset \mathbb{R}^{m\times m}$ be groups. Then, the relation $\mathcal{R}$ defined on $\mathbb{F}^{n\times m}$ by

$(A, B) \in \mathcal{R} \iff$ there exist $S_1 \in \mathcal{S}_1$ and $S_2 \in \mathcal{S}_2$ such that $A = S_1 B S_2$

is an equivalence relation.

3.4 Facts on Group-Invertible and Range-Hermitian Matrices

Fact 3.4.1. Let $A \in \mathbb{F}^{n\times n}$. Then, the following statements are equivalent:

i) A is group invertible.

ii) A^* is group invertible.

iii) $\mathcal{N}(A) = \mathcal{N}(A^2)$.

iv) $\mathcal{N}(A) \cap \mathcal{R}(A) = \{0\}$.

v) $\mathcal{N}(A) + \mathcal{R}(A) = \mathbb{F}^n$.

vi) A and A^2 are left equivalent.

vii) A and A^2 are right equivalent.

viii) rank $A = $ rank A^2.

  ix) def $A = $ def A^2.

   x) def $A = \text{am}_A(0)$.

(Remark: See Corollary 5.5.7, Proposition 5.5.9, and Corollary 5.5.15.)

Fact 3.4.2. Let $A \in \mathbb{F}^{n \times n}$, and assume that A is range Hermitian. Then, A is group invertible.

Fact 3.4.3. Let $A \in \mathbb{F}^{n \times n}$. Then, A is range Hermitian if and only if $\mathcal{N}(A) = \mathcal{N}(A^*)$. (Proof: See [699].)

Fact 3.4.4. Let $A \in \mathbb{F}^{n \times n}$. Then, A is range Hermitian if and only if A and A^* are right equivalent.

Fact 3.4.5. Let $A \in \mathbb{F}^{n \times n}$. Then, A is range Hermitian if and only if $\text{rank}\, A = \text{rank} \begin{bmatrix} A & A^* \end{bmatrix}$. (Proof: See [180, 699].)

Fact 3.4.6. Let $A, B \in \mathbb{F}^{n \times n}$, and assume that A and B are range Hermitian. Then,
$$\text{rank}\, AB = \text{rank}\, BA.$$
(Proof: See [65].)

Fact 3.4.7. Let $A, B \in \mathbb{F}^{n \times n}$, and assume that A is dissipative and B is range Hermitian. Then,
$$\text{ind}\, B = \text{ind}\, AB.$$
(Proof: See [105].)

3.5 Facts on Normal, Hermitian, and Skew-Hermitian Matrices

Fact 3.5.1. Let $A \in \mathbb{F}^{n \times m}$. Then, $AA^{\mathrm{T}} \in \mathbb{F}^{n \times n}$ and $A^{\mathrm{T}}A \in \mathbb{F}^{m \times m}$ are symmetric.

Fact 3.5.2. Let $A \in \mathbb{F}^{n \times n}$, assume that A is Hermitian, and let $k \in \mathbb{P}$. Then, $\mathcal{R}(A) = \mathcal{R}(A^k)$ and $\mathcal{N}(A) = \mathcal{N}(A^k)$.

Fact 3.5.3. Let $A \in \mathbb{R}^{n \times n}$. Then, the following statements hold:

  i) $x^{\mathrm{T}}Ax = 0$ for all $x \in \mathbb{R}^n$ if and only if A is skew symmetric.

  ii) A is symmetric and $x^{\mathrm{T}}Ax = 0$ for all $x \in \mathbb{R}^n$ if and only if $A = 0$.

Fact 3.5.4. Let $A \in \mathbb{C}^{n \times n}$. Then, the following statements hold:

i) x^*Ax is real for all $x \in \mathbb{C}^n$ if and only if A is Hermitian.

ii) x^*Ax is imaginary for all $x \in \mathbb{C}^n$ if and only if A is skew Hermitian.

iii) $x^*Ax = 0$ for all $x \in \mathbb{C}^n$ if and only if $A = 0$.

Fact 3.5.5. Let $A \in \mathbb{C}^{n \times n}$. Then, the following statements hold:

i) A is skew Hermitian if and only if $\jmath A$ is Hermitian.

ii) A is Hermitian if and only if $\jmath A$ is skew Hermitian.

iii) A is Hermitian if and only if $\mathrm{Re}\, A$ is symmetric and $\mathrm{Im}\, A$ is skew symmetric.

iv) A is skew Hermitian if and only if $\mathrm{Re}\, A$ is skew symmetric and $\mathrm{Im}\, A$ is symmetric.

v) A is positive semidefinite if and only if $\mathrm{Re}\, A$ is positive semidefinite.

vi) A is positive definite if and only if $\mathrm{Re}\, A$ is positive definite.

Fact 3.5.6. Let $A \in \mathbb{F}^{n \times n}$. Then, the following statements hold:

i) If A is (Hermitian, positive semidefinite, positive definite), then so is A^{A}.

ii) If A is skew Hermitian and n is odd, then A^{A} is Hermitian.

iii) If A is skew Hermitian and n is even, then A^{A} is skew Hermitian.

iv) If A is normal, then so is A^{A}.

v) If A is diagonal, then so is A^{A}, and, for all $i = 1, \ldots, n$,

$$\left(A^{\mathrm{A}}\right)_{(i,i)} = \prod_{\substack{j=1 \\ j \neq i}}^{n} A_{(j,j)}.$$

(Proof: Use Fact 2.14.9.) (Remark: See Fact 5.12.3.)

Fact 3.5.7. Let $A \in \mathbb{F}^{n \times n}$, assume that n is even, let $x \in \mathbb{F}^n$, and let $\alpha \in \mathbb{F}$. Then,
$$\det(A + \alpha x x^*) = \det A.$$
(Proof: Use Fact 2.14.2 and Fact 3.5.6.)

Fact 3.5.8. Let $A \in \mathbb{F}^{n \times n}$. Then, the following statements are equivalent:

i) A is normal.

ii) $\mathrm{tr}\, (AA^*)^2 = \mathrm{tr}\, A^2 A^{2*}$.

iii) There exists $k \in \mathbb{P}$ such that

$$\text{tr } (AA^*)^k = \text{tr } A^k A^{k*}.$$

iv) There exist $k, l \in \mathbb{P}$ such that

$$\text{tr } (AA^*)^{kl} = \text{tr } \left(A^k A^{k*} \right)^l.$$

v) A is range Hermitian, and $AA^*A^2 = A^2 A^*A$.

vi) $AA^* - A^*A$ is positive semidefinite.

vii) $[A, A^*A] = 0$.

viii) $[A, [A, A^*]] = 0$.

(Proof: See [180, 241, 243, 309, 650].) (Remark: See Fact 3.7.10, Fact 5.12.11, Fact 5.14.2, Fact 6.3.8, Fact 6.5.10, Fact 8.7.21, Fact 11.13.3, and Fact 11.14.10.)

Fact 3.5.9. Let $A \in \mathbb{F}^{n \times n}$. Then, the following statements are equivalent:

i) A is Hermitian.

ii) $A^2 = A^*A$.

iii) $\text{tr } A^2 = \text{tr } A^*A$.

(Proof: Use the Schur decomposition Theorem 5.4.1. See [447].)

Fact 3.5.10. Let $A \in \mathbb{R}^{n \times n}$, assume that A is skew symmetric, and let $\alpha > 0$. Then, $-A^2$ is positive semidefinite, $\det A \geq 0$, and $\det(\alpha I + A) > 0$. If, in addition, n is odd, then $\det A = 0$.

Fact 3.5.11. Let $A \in \mathbb{F}^{n \times n}$, and assume that A is skew Hermitian. If n is even, then $\det A \geq 0$. If n is odd, then $\det A$ is imaginary. (Proof: The first statement follows from Proposition 5.5.25.)

Fact 3.5.12. Let $x, y \in \mathbb{F}^n$, and define

$$A \triangleq \begin{bmatrix} x & y \end{bmatrix}.$$

Then,

$$xy^* - yx^* = A J_2 A^*.$$

Furthermore, $xy^* - yx^*$ is skew Hermitian and has rank 0 or 2.

Fact 3.5.13. Let $x, y \in \mathbb{F}^n$. Then, the following statements hold:

i) xy^T is idempotent if and only if either $xy^\mathrm{T} = 0$ or $x^\mathrm{T}y = 1$.

ii) xy^T is Hermitian if and only if there exists $\alpha \in \mathbb{R}$ such that either $y = \alpha \overline{x}$ or $x = \alpha \overline{y}$.

Fact 3.5.14. Let $x, y \in \mathbb{F}^n$, and define $A \triangleq I - xy^{\mathrm{T}}$. Then, the following statements hold:

i) $\det A = 1 - x^{\mathrm{T}}y$.

ii) A is nonsingular if and only if $x^{\mathrm{T}}y \neq 1$.

iii) A is nonsingular if and only if A is elementary.

iv) $\operatorname{rank} A = n - 1$ if and only if $x^{\mathrm{T}}y = 1$.

v) A is Hermitian if and only if there exists $\alpha \in \mathbb{R}$ such that either $y = \alpha\bar{x}$ or $x = \alpha\bar{y}$.

vi) A is positive semidefinite if and only if A is Hermitian and $x^{\mathrm{T}}y \leq 1$.

vii) A is positive definite if and only if A is Hermitian and $x^{\mathrm{T}}y < 1$.

viii) A is idempotent if and only if either $xy^{\mathrm{T}} = 0$ or $x^{\mathrm{T}}y = 1$.

ix) A is orthogonal if and only if either $x = 0$ or $y = \frac{1}{2}y^{\mathrm{T}}yx$.

x) A is involutory if and only if $x^{\mathrm{T}}y = 2$.

xi) A is a projector if and only if either $y = 0$ or $x = x^*xy$.

xii) A is a reflector if and only if either $y = 0$ or $2x = x^*xy$.

xiii) A is an elementary projector if and only if $x \neq 0$ and $y = (x^*x)^{-1}x$.

xiv) A is an elementary reflector if and only if $x \neq 0$ and $y = 2(x^*x)^{-1}x$.

(Remark: See Fact 3.9.7.)

Fact 3.5.15. Let $x, y \in \mathbb{F}^n$ satisfy $x^{\mathrm{T}}y \neq 1$. Then, $I - xy^{\mathrm{T}}$ is nonsingular and

$$\left(I - xy^{\mathrm{T}}\right)^{-1} = I - \frac{1}{x^{\mathrm{T}}y - 1}xy^{\mathrm{T}}.$$

(Remark: The inverse of an elementary matrix is an elementary matrix.)

Fact 3.5.16. Let $A \in \mathbb{F}^{n \times n}$, and assume that A is Hermitian. Then, $\det A$ is real.

Fact 3.5.17. Let $A \in \mathbb{F}^{n \times n}$, and assume that A is Hermitian. Then,

$$(\operatorname{tr} A)^2 \leq (\operatorname{rank} A)\operatorname{tr} A^2.$$

Furthermore, equality holds if and only if there exists $\alpha \in \mathbb{R}$ such that $A^2 = \alpha A$. (Remark: See Fact 5.10.27.)

Fact 3.5.18. Let $A \in \mathbb{R}^{n \times n}$, and assume that A is skew symmetric. Then, $\operatorname{tr} A = 0$. If, in addition, $B \in \mathbb{R}^{n \times n}$ is symmetric, then $\operatorname{tr} AB = 0$.

Fact 3.5.19. Let $A \in \mathbb{F}^{n \times n}$, and assume that A is skew Hermitian. Then, $\operatorname{Re} \operatorname{tr} A = 0$. If, in addition, $B \in \mathbb{F}^{n \times n}$ is Hermitian, then $\operatorname{Re} \operatorname{tr} AB = 0$.

Fact 3.5.20. Let $A \in \mathbb{F}^{n \times m}$. Then, A^*A is positive semidefinite. Furthermore, A^*A is positive definite if and only if A is left invertible. In this case, $A^{\mathrm{L}} \in \mathbb{F}^{m \times n}$ defined by

$$A^{\mathrm{L}} \triangleq (A^*A)^{-1}A^*$$

is a left inverse of A. (Remark: See Fact 2.14.23, Fact 3.5.21, and Fact 3.9.4.)

Fact 3.5.21. Let $A \in \mathbb{F}^{n \times m}$. Then, AA^* is positive semidefinite. Furthermore, AA^* is positive definite if and only if A is right invertible. In this case, $A^{\mathrm{R}} \in \mathbb{F}^{m \times n}$ defined by

$$A^{\mathrm{R}} \triangleq A^*(AA^*)^{-1}$$

is a right inverse of A. (Remark: See Fact 2.14.23, Fact 3.9.4, and Fact 3.5.20.)

Fact 3.5.22. Let $A \in \mathbb{F}^{n \times m}$. Then, A^*A, AA^*, $A + A^*$, and $\begin{bmatrix} 0 & A^* \\ A & 0 \end{bmatrix}$ are Hermitian, and $\begin{bmatrix} 0 & A^* \\ -A & 0 \end{bmatrix}$ and $A - A^*$ are skew Hermitian.

Fact 3.5.23. Let $A \in \mathbb{F}^{n \times n}$. Then, there exist a unique Hermitian matrix $B \in \mathbb{F}^{n \times n}$ and a unique skew-Hermitian matrix $C \in \mathbb{F}^{n \times n}$ such that $A = B + C$. Specifically, if $A = \hat{B} + \jmath\hat{C}$, where $\hat{B}, \hat{C} \in \mathbb{R}^{n \times n}$, then $\hat{B}$ and $\hat{C}$ are given by

$$B = \tfrac{1}{2}(A + A^*) = \tfrac{1}{2}\left(\hat{B} + \hat{B}^{\mathrm{T}}\right) + \jmath\tfrac{1}{2}\left(\hat{C} - \hat{C}^{\mathrm{T}}\right)$$

and

$$C = \tfrac{1}{2}(A - A^*) = \tfrac{1}{2}\left(\hat{B} - \hat{B}^{\mathrm{T}}\right) + \jmath\tfrac{1}{2}\left(\hat{C} + \hat{C}^{\mathrm{T}}\right).$$

Furthermore, A is normal if and only if $BC = CB$. (Remark: See Fact 11.11.8.)

Fact 3.5.24. Let $A \in \mathbb{F}^{n \times n}$. Then, there exist unique Hermitian matrices $B, C \in \mathbb{C}^{n \times n}$ such that $A = B + \jmath C$. Specifically, if $A = \hat{B} + \jmath\hat{C}$, where $\hat{B}, \hat{C} \in \mathbb{R}^{n \times n}$, then $\hat{B}$ and $\hat{C}$ are given by

$$B = \tfrac{1}{2}(A + A^*) = \tfrac{1}{2}\left(\hat{B} + \hat{B}^{\mathrm{T}}\right) + \jmath\tfrac{1}{2}\left(\hat{C} - \hat{C}^{\mathrm{T}}\right)$$

and

$$C = \tfrac{1}{2\jmath}(A - A^*) = \tfrac{1}{2}\left(\hat{C} + \hat{C}^{\mathrm{T}}\right) - \jmath\tfrac{1}{2}\left(\hat{B} - \hat{B}^{\mathrm{T}}\right).$$

Furthermore, A is normal if and only if $BC = CB$. (Remark: This result is the *Cartesian decomposition*.)

Fact 3.5.25. Let $x, y, z, w \in \mathbb{R}^3$. Then, the following statements hold for the cross-product matrix defined in (2.1.41):

i) $x \times x = K(x)x = 0.$

ii) $K^2(x) = xx^\mathrm{T} - (x^\mathrm{T}x)I.$

iii) $K^3(x) = -x^\mathrm{T}xK(x).$

iv) $[I - K(x)]^{-1} = I + (1 + x^\mathrm{T}x)^{-1}[K(x) + K^2(x)].$

v) $[I + \frac{1}{2}K(x)][I - \frac{1}{2}K(x)]^{-1} = I + \frac{4}{4+x^\mathrm{T}x}[K(x) + \frac{1}{2}K^2(x)].$

vi) $x \times y = -(y \times x) = K(x)y = -K(y)x.$

vii) If $x \times y \neq 0$, then $\mathcal{N}[(x \times y)^\mathrm{T}] = \{x \times y\}^\perp = \mathcal{R}([\; x \quad y \;]).$

viii) $K(x \times y) = K[K(x)y] = [K(x), K(y)]$

ix) $K(x \times y) = yx^\mathrm{T} - xy^\mathrm{T} = [\; x \quad y \;]\begin{bmatrix} -y^\mathrm{T} \\ x^\mathrm{T} \end{bmatrix}.$

x) $(x \times y) \times x = (x^\mathrm{T}xI - xx^\mathrm{T})y.$

xi) $K[(x \times y) \times x] = x^\mathrm{T}xK(y) - x^\mathrm{T}yK(x).$

xii) $(x \times y)^\mathrm{T}(x \times y) = \det [\; x \quad y \quad x \times y \;].$

xiii) $(x \times y)^\mathrm{T}z = x^\mathrm{T}(y \times z) = \det [\; x \quad y \quad z \;].$

xiv) $(x \times y) \times z = (x^\mathrm{T}z)y - (y^\mathrm{T}z)x.$

xv) $x \times (y \times z) = (x^\mathrm{T}z)y - (x^\mathrm{T}y)z.$

xvi) $K[(x \times y) \times z] = (x^\mathrm{T}z)K(y) - (y^\mathrm{T}z)K(x).$

xvii) $K[x \times (y \times z)] = (x^\mathrm{T}z)K(y) - (x^\mathrm{T}y)K(z).$

xviii) $(x \times y)^\mathrm{T}(x \times y) = x^\mathrm{T}xy^\mathrm{T}y - (x^\mathrm{T}y)^2.$

xix) $K(x)K(y)K(x) = -(x^\mathrm{T}y)K(x).$

xx) $K^2(x)K(y) + K(y)K^2(x) = -(x^\mathrm{T}x)K(y) - (x^\mathrm{T}y)K(x).$

xxi) $K^2(x)K^2(y) - K^2(y)K^2(x) = -(x^\mathrm{T}y)K(x \times y).$

xxii) $\sqrt{(x \times y)^\mathrm{T}(x \times y)} = \sqrt{x^\mathrm{T}xy^\mathrm{T}y}\,\sin\theta$, where θ is the angle between x and y.

xxiii) $(x \times y)^\mathrm{T}(z \times w) = x^\mathrm{T}zy^\mathrm{T}w - x^\mathrm{T}wy^\mathrm{T}z = \det \begin{bmatrix} x^\mathrm{T}z & x^\mathrm{T}w \\ y^\mathrm{T}z & y^\mathrm{T}w \end{bmatrix}.$

xxiv) $(x \times y) \times (z \times w) = x^\mathrm{T}(y \times w)z - x^\mathrm{T}(y \times z)w = x^\mathrm{T}(z \times w)y - y^\mathrm{T}(z \times w)x.$

xxv) $x \times [y \times (z \times w)] = (y^\mathrm{T}w)(x \times z) - (y^\mathrm{T}z)(x \times w).$

xxvi) $x \times [y \times (y \times x)] = y \times [x \times (y \times x)] = (y^\mathrm{T}x)(x \times y).$

xxvii) If $A \in \mathbb{R}^{3 \times 3}$, then $A^{\mathrm{T}}(Ax \times Ay) = (\det A)(x \times y)$.

xxviii) If $A \in \mathbb{R}^{3 \times 3}$ is orthogonal and $\det A = 1$, then $A(x \times y) = Ax \times Ay$.

(Proof: Using *xiii*), $e_i^{\mathrm{T}} A^{\mathrm{T}}(Ax \times Ay) = \det \begin{bmatrix} Ax & Ay & Ae_i \end{bmatrix} = (\det A) e_i^{\mathrm{T}}(x \times y)$ for all $i = 1, 2, 3$, which proves *xxvii*). Results *iv*), *v*), and *xix*)–*xxi*) are given in [390, p. 363].) (Remark: See [224, 255, 390, 570, 642, 687].)

Fact 3.5.26. Let $A, B \in \mathbb{R}^3$, and assume that A and B are skew symmetric. Then,

$$\mathrm{tr}\, AB^3 = \tfrac{1}{2}(\mathrm{tr}\, AB)(\mathrm{tr}\, B^2)$$

and

$$\mathrm{tr}\, A^3 B^3 = \tfrac{1}{4}(\mathrm{tr}\, A^2)(\mathrm{tr}\, AB)(\mathrm{tr}\, B^2) + \tfrac{1}{3}(\mathrm{tr}\, A^3)(\mathrm{tr}\, B^3).$$

(Proof: See [45].)

Fact 3.5.27. Let $A \in \mathbb{F}^{n \times n}$ and $k \in \mathbb{P}$. If A is (normal, Hermitian, unitary, involutory, positive semidefinite, positive definite, idempotent, nilpotent), then so is A^k. If A is (skew Hermitian, skew involutory), then so is A^{2k+1}. If A is Hermitian, then A^{2k} is positive semidefinite. If A is tripotent, then so is A^{3k}.

Fact 3.5.28. Let $x, y \in \mathbb{F}^n$, and assume that $x \neq 0$. Then, there exists a Hermitian matrix $A \in \mathbb{F}^{n \times n}$ such that $y = Ax$ if and only if $x^* y$ is real. One such matrix is

$$A = (x^* x)^{-1}[yx^* + xy^* - x^* y I].$$

(Remark: See Fact 2.11.12.)

Fact 3.5.29. Let $x, y \in \mathbb{F}^n$, and assume that $x \neq 0$. Then, there exists a positive-definite matrix $A \in \mathbb{F}^{n \times n}$ such that $y = Ax$ if and only if $x^* y$ is real and positive. One such matrix is

$$A = I + (x^* y)^{-1} yy^* - (x^* x)^{-1} xx^*.$$

(Proof: To show that A is positive definite, note that the elementary projector $I - (x^* x)^{-1} xx^*$ is positive semidefinite and $\mathrm{rank}[I - (x^* x)^{-1} xx^*] = n - 1$. Since $(x^* y)^{-1} yy^*$ is positive semidefinite, it follows that $\mathcal{N}(A) \subseteq \mathcal{N}[I - (x^* x)^{-1} xx^*]$. Next, since $x^* y > 0$, it follows that $y^* x \neq 0$ and $y \neq 0$, and thus $x \notin \mathcal{N}(A)$. Consequently, $\mathcal{N}(A) \subset \mathcal{N}[I - (x^* x)^{-1} xx^*]$ (note proper inclusion), and thus $\mathrm{def}\, A < 1$. Hence, A is nonsingular.)

Fact 3.5.30. Let $x, y \in \mathbb{F}^n$. Then, there exists a skew-Hermitian matrix $A \in \mathbb{F}^{n \times n}$ such that $y = Ax$ if and only if either $y = 0$ or $x \neq 0$ and $x^* y = 0$. If $x \neq 0$ and $x^* y = 0$, then one such matrix is

$$A = (x^* x)^{-1}(yx^* - xy^*).$$

(Proof: See [489].)

Fact 3.5.31. Let $A \in \mathbb{R}^{n \times n}$, and assume that A is positive definite. Then,

$$\mathcal{E} \triangleq \{x \in \mathbb{R}^n : x^{\mathrm{T}}Ax \leq 1\}$$

is a hyperellipsoid. Furthermore, the volume V of $\mathcal{E}$ is given by

$$V = \frac{\alpha(n)}{\sqrt{\det A}},$$

where

$$\alpha(n) \triangleq \begin{cases} \pi^{n/2}/(n/2)!, & n \text{ even}, \\ 2^n \pi^{(n-1)/2}[(n-1)/2]!/n!, & n \text{ odd}. \end{cases}$$

(Remark: $\alpha(n)$ is the volume of the unit *n-dimensional hypersphere*.) (Remark: See [417, p. 36].)

3.6 Facts on the Commutator

Fact 3.6.1. Let $A, B \in \mathbb{F}^{n \times n}$. If either A and B are Hermitian or A and B are skew Hermitian, then $[A, B]$ is skew Hermitian. Furthermore, if A is Hermitian and B is skew Hermitian, or vice versa, then $[A, B]$ is Hermitian.

Fact 3.6.2. Let $A \in \mathbb{F}^{n \times n}$. Then, the following statements are equivalent:

i) $\operatorname{tr} A = 0$.

ii) There exist matrices $B, C \in \mathbb{F}^{n \times n}$ such that B is Hermitian, $\operatorname{tr} C = 0$, and $A = [B, C]$.

(Proof: See [284] and Fact 5.8.14. If every diagonal entry of A is zero, then let $B \triangleq \operatorname{diag}(1, \ldots, n)$, $C_{(i,i)} \triangleq 0$, and, for $i \neq j$, $C_{(i,j)} \triangleq A_{(i,j)}/(i - j)$. See [808, p. 110]. See also [592, p. 172].)

Fact 3.6.3. Let $A \in \mathbb{F}^{n \times n}$. Then, the following statements are equivalent:

i) A is Hermitian, and $\operatorname{tr} A = 0$.

ii) There exists a nonsingular matrix $B \in \mathbb{F}^{n \times n}$ such that $A = [B, B^*]$.

iii) There exist a Hermitian matrix $B \in \mathbb{F}^{n \times n}$ and a skew-Hermitian matrix $C \in \mathbb{F}^{n \times n}$ such that $A = [B, C]$.

iv) There exist a skew-Hermitian matrix $B \in \mathbb{F}^{n \times n}$ and a Hermitian matrix $C \in \mathbb{F}^{n \times n}$ such that $A = [B, C]$.

(Proof: See [690] and [284].)

Fact 3.6.4. Let $A \in \mathbb{F}^{n \times n}$. Then, the following statements are equivalent:

   $i)$ A is skew Hermitian, and $\operatorname{tr} A = 0$.

   $ii)$ There exists a nonsingular matrix $B \in \mathbb{F}^{n \times n}$ such that $A = [\jmath B, B^*]$.

   $iii)$ If $A \in \mathbb{C}^{n \times n}$ is skew Hermitian, then there exist Hermitian matrices $B, C \in \mathbb{F}^{n \times n}$ such that $A = [B, C]$.

(Proof: See [284] or use Fact 3.6.3.)

Fact 3.6.5. Let $A \in \mathbb{F}^{n \times n}$, and assume that A is skew symmetric. Then, there exist symmetric matrices $B, C \in \mathbb{F}^{n \times n}$ such that $A = [B, C]$. (Proof: Use Fact 5.14.22. See [592, pp. 83, 89].) (Remark: "Symmetric" is correct for $\mathbb{F} = \mathbb{C}$.)

Fact 3.6.6. Let $A \in \mathbb{F}^{n \times n}$, and assume that $[A, [A, A^*]] = 0$. Then, A is normal. (Remark: See [808, p. 32].)

Fact 3.6.7. Let $A \in \mathbb{F}^{n \times n}$. Then, there exist $B, C \in \mathbb{F}^{n \times n}$ such that B is normal, C is Hermitian, and

$$A = B + [C, B].$$

(Remark: See [237].)

3.7 Facts on Unitary Matrices

Fact 3.7.1. Let $A \in \mathbb{F}^{n \times n}$, and assume that A is unitary. Then, the following statements hold:

   $i)$ $A = A^{-*}$.

   $ii)$ $A^{\mathrm{T}} = \overline{A}^{-1} = \overline{A}^*$.

   $iii)$ $\overline{A} = A^{-\mathrm{T}} = \overline{A}^{-*}$.

   $iv)$ $A^* = A^{-1}$.

Fact 3.7.2. Let $A \in \mathbb{F}^{n \times n}$, and assume that A is unitary. Then,

$$-n \leq \operatorname{Re} \operatorname{tr} A \leq n,$$
$$-n \leq \operatorname{Im} \operatorname{tr} A \leq n,$$

and

$$|\operatorname{tr} A| \leq n.$$

Fact 3.7.3. Let $x, y \in \mathbb{F}^n$, let $A \in \mathbb{F}^{n \times n}$, and assume that A is unitary. Then, $x^* y = 0$ if and only if $(Ax)^* Ay = 0$.

Fact 3.7.4. Let $A \in \mathbb{F}^{n \times m}$. If A is (left inner, right inner), then A is (left invertible, right invertible) and A^* is a (left inverse, right inverse) of A.

Fact 3.7.5. Let $A \in \mathbb{C}^{n \times n}$, and assume that A is unitary. Then, $|\det A| = 1$.

Fact 3.7.6. Let $M \triangleq \begin{bmatrix} A & B \\ C & D \end{bmatrix} \in \mathbb{F}^{(n+m) \times (n+m)}$, and assume that M is unitary. Then,

$$\det A = (\det M)\overline{\det D}.$$

(Proof: Let $\begin{bmatrix} \hat{A} & \hat{B} \\ \hat{C} & \hat{D} \end{bmatrix} \triangleq A^{-1}$, and take the determinant of $A \begin{bmatrix} I & \hat{B} \\ 0 & \hat{D} \end{bmatrix} = \begin{bmatrix} A & 0 \\ C & I \end{bmatrix}$. See [6] or [640].) (Remark: See Fact 2.15.8.)

Fact 3.7.7. Let $A \in \mathbb{F}^{n \times n}$, and assume that A is Hermitian, skew Hermitian, or unitary. Then, A is normal.

Fact 3.7.8. Let $A \in \mathbb{F}^{n \times n}$, assume that A is nonsingular, and assume that A is (normal, Hermitian, skew Hermitian, unitary). Then, so is A^{-1}.

Fact 3.7.9. Let $A \in \mathbb{F}^{n \times n}$, and assume that A is block diagonal. Then, A is (normal, Hermitian, unitary) if and only if every diagonally located block has the same property.

Fact 3.7.10. Let $A \in \mathbb{F}^{n \times n}$, and assume that A is nonsingular. Then, the following statements are equivalent:

   i) A is normal

  ii) $A^{-1}A^*$ is unitary.

 iii) $[A, A^*] = 0$.

 iv) $[A, A^{-*}] = 0$.

  v) $[A^{-1}, A^{-*}] = 0$.

(Proof: See [309].) (Remark: See Fact 3.5.8, Fact 5.14.2, Fact 6.3.8, and Fact 6.5.10.)

Fact 3.7.11. Let $A, B \in \mathbb{R}^{n \times n}$. Then, $A + \jmath B$ is (Hermitian, skew Hermitian, unitary) if and only if $\begin{bmatrix} A & B \\ -B & A \end{bmatrix}$ is (symmetric, skew symmetric, orthogonal).

Fact 3.7.12. Let $A \in \mathbb{F}^{n \times n}$ be semicontractive. Then, $B \in \mathbb{F}^{2n \times 2n}$ defined by

$$B \triangleq \begin{bmatrix} A & (I - AA^*)^{1/2} \\ (I - A^*A)^{1/2} & -A^* \end{bmatrix}$$

is unitary. (Remark: See [275, p. 180].)

Fact 3.7.13. Let $\theta \in \mathbb{R}$, and define the orthogonal matrix

$$A(\theta) \triangleq \left[\begin{array}{cc} \cos\theta & \sin\theta \\ -\sin\theta & \cos\theta \end{array} \right].$$

Now, let $\theta_1, \theta_2 \in \mathbb{R}$. Then,

$$A(\theta_1)A(\theta_2) = A(\theta_1 + \theta_2).$$

Consequently,

$$\cos(\theta_1 + \theta_2) = (\cos\theta_1)\cos\theta_2 - (\sin\theta_1)\sin\theta_2,$$
$$\sin(\theta_1 + \theta_2) = (\cos\theta_1)\sin\theta_2 + (\sin\theta_1)\cos\theta_2.$$

Furthermore,

$$\mathrm{SO}(2) = \{A(\theta) \colon \ \theta \in \mathbb{R}\}$$

and

$$\mathrm{O}(2)\backslash\mathrm{SO}(2) = \left\{ \left[\begin{array}{cc} \cos\theta & \sin\theta \\ \sin\theta & -\cos\theta \end{array} \right] \colon \ \theta \in \mathbb{R} \right\}.$$

(Remark: See Proposition 3.3.4 and Fact 11.10.3.)

Fact 3.7.14. Let $x, y, z \in \mathbb{R}^2$. If x is rotated according to the right hand rule through an angle $\theta \in \mathbb{R}$ about y, then the resulting vector $\hat{x} \in \mathbb{R}^2$ is given by

$$\hat{x} = \left[\begin{array}{cc} \cos\theta & -\sin\theta \\ \sin\theta & \cos\theta \end{array} \right] x + \left[\begin{array}{c} y_{(1)}(1 - \cos\theta) + y_{(2)}\sin\theta \\ y_{(2)}(1 - \cos\theta) + y_{(1)}\sin\theta \end{array} \right].$$

If x is reflected across the line passing through 0 and z and parallel to the line passing through 0 and y, then the resulting vector $\hat{x} \in \mathbb{R}^2$ is given by

$$\hat{x} = \left[\begin{array}{cc} y_{(1)}^2 - y_{(2)}^2 & 2y_{(1)}y_{(2)} \\ 2y_{(1)}y_{(2)} & y_{(2)}^2 - y_{(1)}^2 \end{array} \right] x + \left[\begin{array}{c} -z_{(1)}\left(y_{(1)}^2 - y_{(2)}^2 - 1\right) - 2z_{(2)}y_{(1)}y_{(2)} \\ -z_{(2)}\left(y_{(1)}^2 - y_{(2)}^2 - 1\right) - 2z_{(1)}y_{(1)}y_{(2)} \end{array} \right].$$

(Remark: These *affine planar transformations* are used in computer graphics. See [36, 269, 590].)

Fact 3.7.15. Let $x, y \in \mathbb{R}^3$, and assume that $y^{\mathrm{T}}y = 1$. If x is rotated according to the right-hand rule through an angle $\theta \in \mathbb{R}$ about the line passing through 0 and y, then the resulting vector $\hat{x} \in \mathbb{R}^3$ is given by

$$\hat{x} = x + (\sin\theta)(y \times x) + (1 - \cos\theta)[y \times (y \times x)].$$

(Proof: See [14].)

Fact 3.7.16. Let $x, y \in \mathbb{R}^n$. Then, there exists an orthogonal matrix $A \in \mathbb{R}^{n \times n}$ such that $Ax = y$ if and only if $x^{\mathrm{T}}x = y^{\mathrm{T}}y$. (Remark: One such matrix is given by a product of n plane rotations given by Fact 5.14.14. Another matrix is given by the product of elementary reflectors given by

Fact 5.14.13. See Fact 11.10.9 and Fact 3.10.4.) (Problem: Extend this result to $\mathbb{C}$.)

Fact 3.7.17. Let $A \in \mathbb{F}^{n \times n}$, assume that A is unitary, and let $x \in \mathbb{F}^n$ be such that $x^*x = 1$ and $Ax = -x$. Then, the following statements hold:

i) $\det(A + I) = 0$.

ii) $A + 2xx^*$ is unitary.

iii) $A = (A + 2xx^*)(I_n - 2xx^*) = (I_n - 2xx^*)(A + 2xx^*)$.

iv) $\det(A + 2xx^*) = -\det A$.

Fact 3.7.18. Let $A \in \mathbb{R}^{3 \times 3}$. Then, A is orthogonal if and only if there exist real numbers a, b, c, d, not all zero, such that

$$A = \frac{\pm 1}{\alpha} \begin{bmatrix} a^2 + b^2 - c^2 - d^2 & 2(bc + da) & 2(bd - ca) \\ 2(bc - da) & a^2 - b^2 + c^2 - d^2 & 2(cd + ba) \\ 2(bd + ca) & 2(cd - ba) & a^2 - b^2 - c^2 + d^2 \end{bmatrix},$$

where $\alpha \triangleq a^2 + b^2 + c^2 + d^2$. (Remark: This result is due to Rodrigues.)

Fact 3.7.19. Let $A \in \mathbb{R}^{n \times n}$, and assume that A is orthogonal. Then, either $\det A = 1$ or $\det A = -1$.

Fact 3.7.20. Let $A \in \mathbb{F}^{n \times n}$, and assume that A is involutory. Then, either $\det A = 1$ or $\det A = -1$.

Fact 3.7.21. Let $A \in \mathbb{F}^{n \times n}$, and assume that A is unitary. Then, $\frac{1}{\sqrt{2}} \begin{bmatrix} A & -A \\ A & A \end{bmatrix}$ is also unitary.

Fact 3.7.22. If $A \in \mathbb{F}^{n \times n}$ is Hermitian, then $I + \jmath A$ is nonsingular and $B \triangleq (A - \jmath I)(A + \jmath I)^{-1}$ is unitary and $B - I$ is nonsingular. Conversely, if $B \in \mathbb{F}^{n \times n}$ is unitary and $B - I$ is nonsingular, then $A \triangleq \jmath(I + B)(I - B)^{-1}$ is Hermitian. (Proof: See [275, pp. 168, 169].) (Remark: $(A - \jmath I)(A + \jmath I)^{-1}$ is the *Cayley transform* of A. See Fact 3.7.23, Fact 3.7.24, Fact 3.12.9, Fact 8.7.23, and Fact 11.18.10.) (Remark: The linear fractional transformation $f(s) \triangleq (s - \jmath)(s + \jmath)$ maps the upper half plane of $\mathbb{C}$ onto the unit disk in $\mathbb{C}$, and the real line onto the unit circle in $\mathbb{C}$.)

Fact 3.7.23. If $A \in \mathbb{F}^{n \times n}$ is skew Hermitian, then $I + A$ is nonsingular, $B \triangleq (I - A)(I + A)^{-1} = (I + A)^{-1}(I - A)$ is unitary, and $|\det B| = 1$. Conversely, if $B \in \mathbb{F}^{n \times n}$ is unitary and $I + B$ is nonsingular, then $A \triangleq (I + B)^{-1}(I - B)$ is skew Hermitian. Furthermore, if B is unitary, then there exist $\lambda \in \mathbb{C}$ and a skew-Hermitian matrix $A \in \mathbb{F}^{n \times n}$ such that $|\lambda| = 1$ and $B \triangleq (I - A)(I + A)^{-1}$. (Proof: See [369, p. 440] and [275, p. 184].)

Fact 3.7.24. If $A \in \mathbb{R}^{n \times n}$ is skew symmetric, then $I + A$ is nonsingular, $B \triangleq (I - A)(I + A)^{-1} = (I + A)^{-1}(I - A)$ is orthogonal, and $I + B$ is nonsingular. Equivalently, if $A \in \mathbb{R}^{n \times n}$ is skew symmetric, then there exists an orthogonal matrix $B \in \mathbb{R}^{n \times n}$ such that $I + B$ is nonsingular and $A = (I + B)^{-1}(I - B)$. Conversely, if $B \in \mathbb{R}^{n \times n}$ is orthogonal and $I + B$ is nonsingular, then $\det B = 1$ and $A \triangleq (I + B)^{-1}(I - B)$ is skew symmetric. Equivalently, if $B \in \mathbb{R}^{n \times n}$ is orthogonal and $I + B$ is nonsingular, then there exists a skew-symmetric matrix $A \in \mathbb{R}^{n \times n}$ such that $B = (I - A)(I + A)^{-1}$.

Fact 3.7.25. Let $x \in \mathbb{R}^3$, assume that $x_{(1)}^2 + x_{(2)}^2 + x_{(3)}^2 = 1$, let $\theta \in [0, 2\pi)$, assume that $\theta \neq \pi$, and define the skew-symmetric matrix $A \in \mathbb{R}^{3 \times 3}$ by

$$A \triangleq \begin{bmatrix} 0 & x_{(3)}\tan(\theta/2) & -x_{(2)}\tan(\theta/2) \\ -x_{(3)}\tan(\theta/2) & 0 & x_{(1)}\tan(\theta/2) \\ x_{(2)}\tan(\theta/2) & -x_{(1)}\tan(\theta/2) & 0 \end{bmatrix}.$$

Then, the matrix $B \in \mathbb{R}^{3 \times 3}$ defined by

$$B \triangleq (I - A)(I + A)^{-1}$$

is an orthogonal matrix that rotates vectors about x through an angle equal to θ according to the right-hand rule. (Proof: See [542, pp. 243–244].) (Remark: Every 3×3 skew-symmetric matrix has a representation of the form given by A.) (Remark: See Fact 11.10.9.)

Fact 3.7.26. Let $A \in \mathbb{R}^{n \times n}$, and assume that A is orthogonal. Then, there exist a skew-symmetric matrix $B \in \mathbb{R}^{n \times n}$ and a diagonal matrix $C \in \mathbb{R}^{n \times n}$, each of whose diagonal entries is either 1 or -1, such that

$$A = C(I - B)(I + B)^{-1}.$$

(Proof: See [592, p. 101].) (Remark: This result is due to Hsu.)

Fact 3.7.27. Let $x, y, z \in \mathbb{F}^n$, and assume that $x^*x = y^*y = z^*z = 1$. Then,
$$\sqrt{1 - |x^*y|^2} \leq \sqrt{1 - |x^*z|^2} + \sqrt{1 - |y^*z|^2}.$$
Furthermore, if $A, B \in \mathbb{F}^{n \times n}$ are unitary, then

$$\sqrt{1 - \left|\tfrac{1}{n}\operatorname{tr} AB\right|^2} \leq \sqrt{1 - \left|\tfrac{1}{n}\operatorname{tr} A\right|^2} + \sqrt{1 - \left|\tfrac{1}{n}\operatorname{tr} B\right|^2}.$$

(Proof: See [753].)

3.8 Facts on Idempotent Matrices

Fact 3.8.1. Let $\mathcal{S}_1, \mathcal{S}_2 \subseteq \mathbb{F}^n$ be complementary subspaces, and let $A \in \mathbb{F}^{n \times n}$ be the idempotent matrix associated with $\mathcal{S}_1, \mathcal{S}_2$. Then, A^{T} is the idempotent matrix associated with $\mathcal{S}_2^\perp, \mathcal{S}_1^\perp$. (Remark: See Fact 2.9.12.)

Fact 3.8.2. Let $A \in \mathbb{F}^{n \times n}$. Then,

$$\mathcal{N}(A) \subseteq \mathcal{R}(I - A)$$

and

$$\mathcal{R}(A) \supseteq \mathcal{N}(I - A).$$

Furthermore, the following statements are equivalent:

i) A is idempotent.

ii) $\mathcal{N}(A) = \mathcal{R}(I - A)$.

iii) $\mathcal{R}(A) = \mathcal{N}(I - A)$.

(Proof: See [339, p. 146].)

Fact 3.8.3. Let $A \in \mathbb{F}^{n \times n}$, and assume that A is idempotent. Then,

$$\mathcal{R}(I - AA^*) = \mathcal{R}(2I - A - A^*).$$

(Proof: See [707].)

Fact 3.8.4. Let $A \in \mathbb{F}^{n \times n}$. Then, A is idempotent and rank $A = 1$ if and only if there exist vectors $x, y \in \mathbb{F}^n$ such that $y^{\mathrm{T}} x = 1$ and $A = xy^{\mathrm{T}}$.

Fact 3.8.5. Let $A \in \mathbb{F}^{n \times n}$, and assume that A is idempotent. Then, $A^{\mathrm{T}}, \overline{A}$, and A^* are idempotent.

Fact 3.8.6. Let $A \in \mathbb{F}^{n \times n}$. Then, A is idempotent if and only if rank $A + \mathrm{rank}(I - A) = n$.

Fact 3.8.7. Let $A \in \mathbb{F}^{n \times m}$. If $A^{\mathrm{L}} \in \mathbb{F}^{m \times n}$ is a left inverse of A, then AA^{L} is idempotent and rank $A^{\mathrm{L}} = \mathrm{rank}\, A$. Furthermore, if $A^{\mathrm{R}} \in \mathbb{F}^{m \times n}$ is a right inverse of A, then $A^{\mathrm{R}}A$ is idempotent and rank $A^{\mathrm{R}} = \mathrm{rank}\, A$.

Fact 3.8.8. Let $A \in \mathbb{F}^{n \times n}$, and assume that A is nonsingular and idempotent. Then, $A = I_n$.

Fact 3.8.9. Let $A \in \mathbb{F}^{n \times n}$ be idempotent. Then, so is $A_\perp \triangleq I - A$, and, furthermore, $AA_\perp = A_\perp A = 0$.

Fact 3.8.10. Let $A \in \mathbb{F}^{n \times n}$ be idempotent. Then,

$$\det(I + A) = 2^{\mathrm{tr}\, A}$$

and

$$(I + A)^{-1} = I - \tfrac{1}{2}A.$$

Fact 3.8.11. Let $A \in \mathbb{F}^{n \times n}$ and $\alpha \in \mathbb{F}$, where $\alpha \neq 0$. Then, the matrices

$$\begin{bmatrix} A & A^* \\ A^* & A \end{bmatrix}, \quad \begin{bmatrix} A & \alpha^{-1}A \\ \alpha(I - A) & I - A \end{bmatrix}, \quad \begin{bmatrix} A & \alpha^{-1}A \\ -\alpha A & -A \end{bmatrix}$$

are, respectively, normal, idempotent, and nilpotent.

Fact 3.8.12. Let $A \in \mathbb{F}^{n \times n}$, and assume that $2\mathrm{rank}\, A - 2 \leq \mathrm{tr}\, A \leq 2n$. Then, there exist idempotent matrices $B, C, D, E \in \mathbb{F}^{n \times n}$ such that $A = B + C + D + E$. (Proof: See [458].) (Remark: See Fact 3.9.13.)

Fact 3.8.13. Let $A \in \mathbb{F}^{n \times n}$. If $n = 2$ or $n = 3$, then there exist $b, c \in \mathbb{F}$ and idempotent matrices $B, C \in \mathbb{F}^{n \times n}$ such that $A = bB + cC$. Furthermore, if $n \geq 4$, then there exist $b, c, d \in \mathbb{F}$ and idempotent matrices $B, C, D \in \mathbb{F}^{n \times n}$ such that $A = bB + cC + dD$. (Proof: See [599].)

Fact 3.8.14. Let $A, B \in \mathbb{F}^{n \times n}$ be idempotent, and define $A_\perp \triangleq I - A$ and $B_\perp \triangleq I - B$. Then, the following identities hold:

 i) $(A - B)^2 + (A_\perp - B)^2 = I$.

 ii) $[A, B] = [B, A_\perp] = [B_\perp, A] = [A_\perp, B_\perp]$.

 iii) $A - B = AB_\perp - A_\perp B$.

 iv) $AB_\perp + BA_\perp = AB_\perp A + A_\perp BA_\perp$.

 v) $A[A, B] = [A, B]A_\perp$.

 vi) $B[A, B] = [A, B]B_\perp$.

(Proof: See [561].)

Fact 3.8.15. Let $A, B \in \mathbb{R}^{n \times n}$. Then, the following statements hold:

 i) Assume that $A^3 = -A$ and $B = I + A + A^2$. Then, $B^4 = I$, $B^{-1} = I - A + A^2$, $B^3 - B^2 + B - I = 0$, $A = \frac{1}{2}(B - B^3)$, and $I + A^2$ is idempotent.

 ii) Assume that $B^3 - B^2 + B - I = 0$ and $A = \frac{1}{2}(B - B^3)$. Then, $A^3 = -A$ and $B = I + A + A^2$.

 iii) Assume that $B^4 = I$ and $A = \frac{1}{2}(B - B^{-1})$. Then, $A^3 = -A$, and $\frac{1}{4}(I + B + B^2 + B^3)$ is idempotent.

(Remark: The geometric interpretation of these results is discussed in [255, pp. 153, 212–214, 242].)

Fact 3.8.16. Let $A \in \mathbb{F}^{n \times n}$, $B \in \mathbb{F}^{n \times m}$, and $C \in \mathbb{F}^{l \times n}$, and assume that A is idempotent, $\mathrm{rank}\begin{bmatrix} C^* & B \end{bmatrix} = n$, and $CB = 0$. Then,

$$\mathrm{rank}\, CAB = \mathrm{rank}\, CA + \mathrm{rank}\, AB - \mathrm{rank}\, A.$$

(Proof: See [713].) (Remark: See Fact 3.8.17.)

Fact 3.8.17. $A \triangleq \begin{bmatrix} A_{11} & A_{12} \\ A_{12}^* & A_{22} \end{bmatrix} \in \mathbb{F}^{(n+m)\times(n+m)}$, and assume that A is idempotent. Then,

$$\operatorname{rank} A = \operatorname{rank} \begin{bmatrix} A_{12} \\ A_{22} \end{bmatrix} + \operatorname{rank} \begin{bmatrix} A_{11} & A_{12} \end{bmatrix} - \operatorname{rank} A_{12}$$

$$= \operatorname{rank} \begin{bmatrix} A_{11} \\ A_{21} \end{bmatrix} + \operatorname{rank} \begin{bmatrix} A_{21} & A_{22} \end{bmatrix} - \operatorname{rank} A_{21}.$$

(Proof: See [713] and Fact 3.8.16.) (Remark: See Fact 3.9.10 and Fact 6.4.23.)

Fact 3.8.18. Let $A \in \mathbb{F}^{n\times m}$ and $B \in \mathbb{F}^{m\times n}$, and assume that AB is nonsingular. Then, $B(AB)^{-1}A$ is idempotent.

Fact 3.8.19. Let $A, B \in \mathbb{F}^{n\times n}$ be idempotent. Then, $A + B$ is idempotent if and only if $AB = BA = 0$. (Proof: $AB + BA = 0$ implies $AB + ABA = ABA + BA = 0$, which implies that $AB - BA = 0$, and hence $AB = 0$. See [329, p. 250] and [339, p. 435].)

Fact 3.8.20. Let $A, B \in \mathbb{F}^{n\times n}$, assume that A and B are idempotent, and let $C \in \mathbb{F}^{n\times m}$. Then,

$$\operatorname{rank}(AC - CB) = \operatorname{rank}(AC - ACB) + \operatorname{rank}(ACB - CB)$$

$$= \operatorname{rank} \begin{bmatrix} AC \\ B \end{bmatrix} + \operatorname{rank} \begin{bmatrix} CB & A \end{bmatrix} - \operatorname{rank} A - \operatorname{rank} B.$$

(Proof: See [704].)

Fact 3.8.21. Let $A, B \in \mathbb{F}^{n\times n}$, and assume that A and B are idempotent. Then,

$$\operatorname{rank}(A - B) = \operatorname{rank} \begin{bmatrix} A \\ B \end{bmatrix} + \operatorname{rank} \begin{bmatrix} A & B \end{bmatrix} - \operatorname{rank} A - \operatorname{rank} B$$

$$= \operatorname{rank}(A - AB) + \operatorname{rank}(AB - B),$$

$$\operatorname{rank}(A + B) = \operatorname{rank} \begin{bmatrix} A & B \\ B & 0 \end{bmatrix} - \operatorname{rank} B$$

$$= \operatorname{rank}(A - AB - BA + BAB) + \operatorname{rank} B$$

$$= \operatorname{rank}(A + B - AB),$$

$$\text{rank } [A, B] = \text{rank}(A - B) + \text{rank}(I - A - B) - n$$
$$= \text{rank}(A - B) + \text{rank } AB + \text{rank } BA - \text{rank } A - \text{rank } B,$$

and

$$\text{rank}(A - B) \le \text{rank}(A + B) \le \text{rank } A + \text{rank } B.$$

Furthermore, the following statements hold:

i) If $AB = 0$ or $BA = 0$, then

$$\text{rank}(A - B) = \text{rank}(A + B) = \text{rank } A + \text{rank } B.$$

ii) $\text{rank}(A - B) = \text{rank } A - \text{rank } B$ if and only if $ABA = B$.

iii) $A - B$ is nonsingular if and only if

$$\text{rank}\begin{bmatrix} A \\ B \end{bmatrix} = \text{rank}\begin{bmatrix} A & B \end{bmatrix} = \text{rank } A + \text{rank } B = n.$$

iv) $\text{rank}(I - AB) = \text{rank}(A_\perp + B_\perp)$.

v) If $AB = 0$, then $\text{rank}(A - BA) + \text{rank}(BA - B) = \text{rank } A + \text{rank } B$.

vi) If $AB = 0$, then

$$\text{rank}(A + B) = \text{rank}(A - BA) + \text{rank } B.$$

vii) If $BA = 0$, then

$$\text{rank}(A + B) = \text{rank}(A - AB) + \text{rank } B.$$

viii) If $AB = BA$, then

$$\text{rank}(A + B) = \text{rank}(A - AB) + \text{rank } B.$$

ix) $AB = BA$ if and only if

$$\text{rank}(A - B) + \text{rank}(I - A - B) = n.$$

x) $[A, B]$ is nonsingular if and only if $A - B$ and $I - A - B$ are non-singular.

xi) $AB + BA$ is nonsingular if and only if $A + B$ and $I - A - B$ are nonsingular.

(Proof: See [311, 437, 712].)

Fact 3.8.22. Let $A, B \in \mathbb{F}^{n \times n}$, and assume that A and B are idempotent. Then, the following statements are equivalent:

i) $A - B$ is idempotent.

ii) $\text{rank}(I - A + B) + \text{rank}(A - B) = n$.

iii) $ABA = B$.

iv) $\text{rank}(A - B) = \text{rank } A - \text{rank } B$.

v) $\mathcal{R}(B) \subseteq \mathcal{R}(A)$ and $\mathcal{R}(B^*) \subseteq \mathcal{R}(A^*)$.

(Proof: See [714].) (Remark: This result is due to Hartwig and Styan.)

Fact 3.8.23. Let $A, B \in \mathbb{F}^{n \times n}$, and assume that A and B are idempotent. Then, the following statements are equivalent:

i) $A + B$ is nonsingular.

ii) There exist $\alpha, \beta \in \mathbb{F}$ such that $\alpha + \beta \neq 0$ and $\alpha A + \beta B$ is nonsingular.

iii) For all $\alpha, \beta \in \mathbb{F}$ such that $\alpha + \beta \neq 0$, $\alpha A + \beta B$ is nonsingular.

(Proof: See [57].)

Fact 3.8.24. Let $A, B \in \mathbb{F}^{n \times n}$, and assume that A and B are idempotent. Then, the following statements are equivalent:

i) $A - B$ is nonsingular.

ii) $I - AB$ is nonsingular, and there exist $\alpha, \beta \in \mathbb{F}$ such that $\alpha + \beta \neq 0$ and $\alpha A + \beta B$ is nonsingular.

iii) $I - AB$ is nonsingular, and $\alpha A + \beta B$ is nonsingular for all $\alpha, \beta \in \mathbb{F}$ such that $\alpha + \beta \neq 0$.

iv) $I - AB$ and $A + B - AB$ are nonsingular.

v) $I - AB$ and $A + B$ are nonsingular.

vi) $\mathcal{R}(A) + \mathcal{R}(B) = \mathbb{F}^n$ and $\mathcal{R}(A^*) + \mathcal{R}(B^*) = \mathbb{F}^n$.

vii) $\mathcal{R}(A) + \mathcal{R}(B) = \mathbb{F}^n$ and $\mathcal{N}(A) + \mathcal{N}(B) = \mathbb{F}^n$.

$viii$) $\mathcal{R}(A) \cap \mathcal{R}(B) = \{0\}$ and $\mathcal{N}(A) \cap \mathcal{N}(B) = \{0\}$.

(Proof: See [57, 311, 437].)

Fact 3.8.25. Let $A, B \in \mathbb{F}^{n \times n}$, assume that A and B are idempotent, and assume that $A - B$ is nonsingular. Then, $A + B$ is nonsingular. Now, define $F, G \in \mathbb{F}^{n \times n}$ by

$$F \triangleq A(A - B)^{-1} = (A - B)^{-1}(I - B)$$

and

$$G \triangleq (A - B)^{-1}A = (I - A)(A - B)^{-1}.$$

Then, F and G are idempotent. Furthermore,

$$(A - B)^{-1} = F - G_\perp,$$

$$(A - B)^{-1} = (A + B)^{-1}(A - B)(A + B)^{-1},$$

$$(A + B)^{-1} = I - G_\perp F - GF_\perp,$$

$$(A + B)^{-1} = (A - B)^{-1}(A + B)(A - B)^{-1}.$$

(Proof: See [437].) (Remark: See [437] for an explicit expression for $(A + B)^{-1}$ in the case $A - B$ is nonsingular.)

Fact 3.8.26. If $A, B \in \mathbb{F}^{n \times n}$ are idempotent and $AB = 0$, then $A + B - BA$ is idempotent and $C \triangleq A - B$ is tripotent. Conversely, if $C \in \mathbb{F}^{n \times n}$ is tripotent, then $A \triangleq \frac{1}{2}(C^2 + C)$ and $B \triangleq \frac{1}{2}(C^2 - C)$ are idempotent and satisfy $C = A - B$ and $AB = BA = 0$. (Proof: See [525, p. 114].)

Fact 3.8.27. If $B \in \mathbb{F}^{n \times n}$ is unitary and skew Hermitian, then $A \triangleq \frac{1}{2}(B + I)$ satisfies

$$A + A^* = 2AA^*.$$

Conversely, if $A \in \mathbb{F}^{n \times n}$ satisfies this equation, then $B \triangleq 2A - I$ is unitary. (Remark: See Fact 3.9.12.) (Remark: This equation has normal solutions such that $B \triangleq 2A - I$ is not skew Hermitian, for example, $A = 1/3 + \jmath\sqrt{2}/3$.) (Problem: Characterize all solutions that are normal and all solutions that are not normal.)

Fact 3.8.28. If $A \in \mathbb{F}^{n \times n}$ is idempotent, then $B \triangleq 2A - I$ is involutory, while, if $B \in \mathbb{F}^{n \times n}$ is involutory, then $A \triangleq \frac{1}{2}(B + I)$ is idempotent. (Remark: See Fact 3.10.1.)

3.9 Facts on Projectors

Fact 3.9.1. Let $A \in \mathbb{F}^{n \times n}$. Then, the following statements are equivalent:

$i)$　A is a projector.

$ii)$　$A = AA^*$.

$iii)$　$A = A^*A$.

Fact 3.9.2. Let $A \in \mathbb{F}^{n \times n}$, and assume that A is a projector. Then, A is positive semidefinite.

Fact 3.9.3. Let $A \in \mathbb{F}^{n \times n}$, assume that A is a projector, and let $x \in \mathbb{F}^n$. Then, $x \in \mathcal{R}(A)$ if and only if $x = Ax$.

Fact 3.9.4. Let $A \in \mathbb{F}^{n \times m}$. If $\operatorname{rank} A = m$, then $B \triangleq A(A^*A)^{-1}A^*$ is a projector and $\operatorname{rank} B = m$. If $\operatorname{rank} A = n$, then $B \triangleq A^*(AA^*)^{-1}A$ is a projector and $\operatorname{rank} B = n$. (Remark: See Fact 2.14.23, Fact 3.5.20, and Fact 3.5.21.)

Fact 3.9.5. Let $x \in \mathbb{F}^n$ be nonzero, and define the elementary projector $A \triangleq I - (x^*x)^{-1}xx^*$. Then, the following statements hold:

i) rank $A = n - 1$.

ii) $\mathcal{N}(A) = \operatorname{span}\{x\}$.

iii) $\mathcal{R}(A) = \{x\}^{\perp}$.

iv) $2A - I$ is the elementary reflector $I - 2(x^*x)^{-1}xx^*$.

(Remark: If $y \in \mathbb{F}^n$, then Ay is the *projection* of y on $\{x\}^{\perp}$.)

Fact 3.9.6. Let $n > 1$, let $\mathcal{S} \subset \mathbb{F}^n$, and assume that $\mathcal{S}$ is a hyperplane. Then, there exists a unique elementary projector $A \in \mathbb{F}^{n \times n}$ such that $\mathcal{R}(A) = \mathcal{S}$ and $\mathcal{N}(A) = \mathcal{S}^{\perp}$. Furthermore, if $x \in \mathbb{F}^n$ is nonzero and $\mathcal{S} \triangleq \{x\}^{\perp}$, then $A = I - (x^*x)^{-1}xx^*$. (Remark: See Proposition 5.5.4.)

Fact 3.9.7. Let $A \in \mathbb{F}^{n \times n}$. Then, A is a projector and rank $A = n - 1$ if and only if there exists a nonzero vector $x \in \mathcal{N}(A)$ such that

$$A = I - (x^*x)^{-1}xx^*.$$

In this case, it follows that, for all $y \in \mathbb{F}^n$,

$$y^*y - y^*Ay = \frac{|y^*x|^2}{x^*x}.$$

Furthermore, for $y \in \mathbb{F}^n$, the following statements are equivalent:

i) $y^*Ay = y^*y$.

ii) $y^*x = 0$.

iii) $Ay = y$.

(Remark: See Fact 3.5.14.)

Fact 3.9.8. Let $A \in \mathbb{F}^{n \times n}$, assume that A is a projector, and let $x \in \mathbb{F}^n$. Then,
$$x^*Ax \leq x^*x.$$

Furthermore, the following statements are equivalent:

i) $x^*Ax = x^*x$.

ii) $Ax = x$.

iii) $x \in \mathcal{R}(A)$.

Fact 3.9.9. Let $A \in \mathbb{F}^{n \times n}$, and assume that A is idempotent. Then, A is a projector if and only if, for all $x \in \mathbb{F}^n$, $x^*Ax \leq x^*x$. (Proof: See [592, p. 105].)

Fact 3.9.10. $A \triangleq \begin{bmatrix} A_{11} & A_{12} \\ A_{12}^* & A_{22} \end{bmatrix} \in \mathbb{F}^{(n+m) \times (n+m)}$, and assume that A is a projector. Then,
$$\operatorname{rank} A = \operatorname{rank} A_{11} + \operatorname{rank} A_{22} - \operatorname{rank} A_{12}.$$

(Proof: See [714] and Fact 3.8.17.) (Remark: See Fact 3.8.17 and Fact 6.4.23.)

Fact 3.9.11. Let $A \in \mathbb{F}^{n \times n}$, and assume that A is idempotent. Then, the following statements are equivalent:

i) A is a projector.

ii) $AA^*A = A$.

iii) A is range Hermitian.

(Proof: See [723].)

Fact 3.9.12. Let $A \in \mathbb{F}^{n \times n}$, and assume that A satisfies two out of the three properties (Hermitian, idempotent, $A + A^* = 2AA^*$). Then, A satisfies the remaining property. (Proof: If A is idempotent and $2AA^* = A + A^*$, then $(2A - I)^{-1} = 2A - I = (2A^* - I)^{-1}$. Hence, A is Hermitian.) (Remark: These matrices are the projectors.) (Remark: The condition $A + A^* = 2AA^*$ is considered in Fact 3.8.27.) (Remark: See Fact 3.10.2 and Fact 3.10.6.)

Fact 3.9.13. Let $A \in \mathbb{C}^{n \times n}$, and assume that A is Hermitian. If $n = 2$ or $n = 3$, then there exist $b, c \in \mathbb{C}$ and projectors $B, C \in \mathbb{C}^{n \times n}$ such that $A = bB + cC$. Furthermore, if $4 \le n \le 7$, then there exist $b, c, d \in \mathbb{F}$ and projectors $B, C, D \in \mathbb{F}^{n \times n}$ such that $A = bB + cC + dD$. If $n \ge 8$, then there exist $b, c, d, e \in \mathbb{C}$ and projectors $B, C, D, E \in \mathbb{C}^{n \times n}$ such that $A = bB + cC + dD + eE$. (Proof: See [554].) (Remark: See Fact 3.8.12.)

Fact 3.9.14. Let $A, B \in \mathbb{F}^{n \times n}$, and assume that A and B are projectors. Then, $\mathcal{R}(A) = \mathcal{R}(B)$ if and only if $A = B$.

Fact 3.9.15. Let $A, B \in \mathbb{F}^{n \times n}$, and assume that A and B are projectors. Then, the following statements are equivalent:

i) $AB = A$.

ii) $BA = A$.

iii) $\mathcal{R}(A) \subseteq \mathcal{R}(B)$.

Fact 3.9.16. Let $A, B \in \mathbb{F}^{n \times n}$, and assume that A and B are projectors. Then, the following statements are equivalent:

i) AB is a projector.

ii) $AB = BA$.

In this case, the following statements hold:

i) $\mathcal{R}(AB) = \mathcal{R}(A) \cap \mathcal{R}(B)$.

ii) $A + B - AB$ is a projector.

iii) $\mathcal{R}(A + B - AB) = \mathcal{R}(A) + \mathcal{R}(B)$.

(Proof: See [282, pp. 42–44].)

Fact 3.9.17. Let $A, B \in \mathbb{F}^{n \times n}$, and assume that A and B are projectors. Then, the following statements are equivalent:

i) $AB = 0$.

ii) $BA = 0$.

iii) $\mathcal{R}(A) = \mathcal{R}(B)^{\perp}$.

iv) $A + B$ is a projector.

In this case, $\mathcal{R}(A + B) = \mathcal{R}(A) + \mathcal{R}(B)$. (Proof: See [282, pp. 42–44].)

Fact 3.9.18. Let $A, B \in \mathbb{F}^{n \times n}$, and assume that A and B are projectors. Then, the following statements are equivalent:

i) $A - B$ is nonsingular.

ii) $\operatorname{rank} \begin{bmatrix} A & B \end{bmatrix} = \operatorname{rank} A + \operatorname{rank} B = n$.

iii) $\mathcal{R}(A)$ and $\mathcal{R}(B)$ are complementary subspaces.

In this case, there exists a unique idempotent matrix $M \in \mathbb{F}^{n \times n}$ such that $\mathcal{R}(M) = \mathcal{R}(A)$ and $\mathcal{N}(M) = \mathcal{R}(B)$. In fact,

$$M + M^* = (A - B)^{-1} + I.$$

(Proof: See [603]. The existence of M follows from Proposition 5.5.8, and the last statement follows from Fact 9.11.14.) (Remark: See Fact 6.3.15.) (Problem: Express M in terms of A and B.)

Fact 3.9.19. Let $A, B \in \mathbb{F}^{n \times n}$, and assume that A and B are projectors. Then,

$$\operatorname{rank} [A, B] = 2 \big(\operatorname{rank} \begin{bmatrix} A & B \end{bmatrix} + \operatorname{rank} AB - \operatorname{rank} A - \operatorname{rank} B \big).$$

(Proof: See [712].)

Fact 3.9.20. Let $A, B \in \mathbb{F}^{n \times n}$, and assume that A and B are projectors. Then, the following statements are equivalent:

i) AB is a projector.

ii) AB is Hermitian.

iii) AB is normal.

iv) AB is range Hermitian.

(Proof: See [722].) (Remark: See Fact 6.4.14.)

Fact 3.9.21. Let $A, B \in \mathbb{F}^{n \times n}$, and assume that A and B are projectors. Then, AB is group invertible. (Proof: $\mathcal{N}(BA) \subseteq \mathcal{N}(BABA) \subseteq \mathcal{N}(ABABA) = \mathcal{N}(ABAABA) = \mathcal{N}(ABA) = \mathcal{N}(ABBA) = \mathcal{N}(BA)$.) (Remark: See [772].)

3.10 Facts on Reflectors

Fact 3.10.1. If $A \in \mathbb{F}^{n \times n}$ is a projector, then $B \triangleq 2A - I$ is a reflector, whereas, if $B \in \mathbb{F}^{n \times n}$ is a reflector, then $A \triangleq \frac{1}{2}(B + I)$ is a projector. (Remark: See Fact 3.8.28.)

Fact 3.10.2. Let $A \in \mathbb{F}^{n \times n}$, and assume that A satisfies two out of the three properties (Hermitian, unitary, involutory). Then, A also satisfies the remaining property. (Remark: These matrices are the reflectors.) (Remark: See Fact 3.9.12 and Fact 3.10.6.)

Fact 3.10.3. Let $x \in \mathbb{F}^n$ be nonzero, and define the elementary reflector $A \triangleq I - 2(x^*x)^{-1}xx^*$. Then, the following statements hold:

i) $\det A = -1$.

ii) If $y \in \mathbb{F}^n$, then Ay is the reflection of y across $\{x\}^{\perp}$.

iii) $Ax = -x$.

iv) $\frac{1}{2}(A + I)$ is the elementary projector $I - (x^*x)^{-1}xx^*$.

Fact 3.10.4. Let $x, y \in \mathbb{F}^n$. Then, there exists a unique elementary reflector $A \in \mathbb{F}^{n \times n}$ such that $Ax = y$ if and only if x^*y is real and $x^*x = y^*y$. If, in addition, $x \neq y$, then A is given by

$$A = I - 2[(x - y)^*(x - y)]^{-1}(x - y)(x - y)^*.$$

(Remark: This result is the *reflection theorem*. See [292, pp. 16–18] and [613, p. 357]. See Fact 3.7.16 and Fact 11.10.9.)

Fact 3.10.5. Let $n > 1$, let $\mathcal{S} \subset \mathbb{F}^n$, and assume that $\mathcal{S}$ is a hyperplane. Then, there exists a unique elementary reflector $A \in \mathbb{F}^{n \times n}$ such that, for all $y = y_1 + y_2 \in \mathbb{F}^n$, where $y_1 \in \mathcal{S}$ and $y_2 = \mathcal{S}^{\perp}$, it follows that $Ay = y_1 - y_2$. Furthermore, if $\mathcal{S} = \{x\}^{\perp}$, then $A = I - 2(x^*x)^{-1}xx^*$.

Fact 3.10.6. Let $A \in \mathbb{F}^{n \times n}$, and assume that A satisfies two out of the three properties (skew Hermitian, unitary, skew involutory). Then, A also satisfies the remaining property. In particular, J_n satisfies all three properties. In addition, A^2 is a reflector. (Problem: Does every reflector have a skew-Hermitian, unitary square root?) (Remark: See Fact 3.9.12 and Fact 3.10.2.)

Fact 3.10.7. Let $A \in \mathbb{F}^{n \times n}$. Then, A is a reflector if and only if $A = AA^* + A^* - I$. (Proof: This condition is equivalent to $A = \frac{1}{2}(A+I)(A^*+I) - I$.)

3.11 Facts on Nilpotent Matrices

Fact 3.11.1. Let $A, B \in \mathbb{F}^{n \times n}$, and assume that A and B are upper triangular. Then,

$$[A, B]^n = 0.$$

Hence, $[A, B]$ is nilpotent. (Remark: See [270, 271].)

Fact 3.11.2. Let $A, B \in \mathbb{F}^{n \times n}$, and assume that $[A, [A, B]] = 0$. Then, $[A, B]$ is nilpotent. (Remark: This result is due to Jacobson. See [266] or [367, p. 98].)

Fact 3.11.3. Let $A, B \in \mathbb{F}^{n \times n}$, and assume that $\left[A, B^2\right] = B$. Then, B is nilpotent. (Proof: See [622].)

Fact 3.11.4. Let $A \in \mathbb{R}^{n \times n}$. Then, rank A^k is a nonincreasing function of $k \in \mathbb{P}$. Furthermore, if there exists $k \in \{1, \ldots, n\}$ such that rank $A^{k+1} = $ rank A^k, then rank $A^l = $ rank A^k for all $l \geq k$. Finally, if A is nilpotent and $A^l \neq 0$, then rank $A^{k+1} < $ rank A^k for all $k = 1, \ldots, l$.

Fact 3.11.5. Let $n \in \mathbb{P}$ and $k \in \{0, \ldots, n\}$. Then, rank $N_n^k = n - k$.

Fact 3.11.6. Let $A \in \mathbb{F}^{n \times n}$. Then, A is nilpotent and rank $A = 1$ if and only if there exist nonzero vectors $x, y \in \mathbb{F}^n$ such that $y^{\mathrm{T}}x = 0$ and $A = xy^{\mathrm{T}}$.

Fact 3.11.7. Let $A \in \mathbb{R}^{n \times n}$, assume that A is nilpotent, and let $k \in \mathbb{P}$ be such that $A^k = 0$. Then,

$$\det(I - A) = 1$$

and

$$(I - A)^{-1} = \sum_{i=0}^{k-1} A^i.$$

Fact 3.11.8. Let $\lambda \in \mathbb{F}$ and $n, k \in \mathbb{P}$. Then,

$$(\lambda I_n + N_n)^k = \begin{cases} \lambda^k I_n + \binom{k}{1}\lambda^{k-1}N_n + \cdots + \binom{k}{k}N_n^k, & k < n-1, \\ \lambda^k I_n + \binom{k}{1}\lambda^{k-1}N_n + \cdots + \binom{k}{n-1}\lambda^{k-n+1}N_n^{n-1}, & k \geq n-1, \end{cases}$$

that is, for $k \geq n-1$,

$$
\begin{bmatrix}
\lambda & 1 & \cdots & 0 & 0 \\
0 & \lambda & \ddots & 0 & 0 \\
\vdots & \ddots & \ddots & \ddots & \vdots \\
0 & 0 & \ddots & \lambda & 1 \\
0 & 0 & \cdots & 0 & \lambda
\end{bmatrix}^k
=
\begin{bmatrix}
\lambda^k & \binom{k}{1}\lambda^{k-1} & \cdots & \binom{k}{n-2}\lambda^{k-n+1} & \binom{k}{n-1}\lambda^{k-n+1} \\
0 & \lambda^k & \ddots & \binom{k}{n-3}\lambda^{k-n+2} & \binom{k}{n-2}\lambda^{k-n+2} \\
\vdots & & \ddots & \ddots & \vdots \\
0 & 0 & \ddots & \lambda^k & \binom{k}{1}\lambda^{k-1} \\
0 & 0 & \cdots & 0 & \lambda^k
\end{bmatrix}.
$$

Fact 3.11.9. Let $A, B \in \mathbb{F}^{n \times n}$, and assume that A is nilpotent and $AB = BA$. Then, $\det(A + B) = \det B$. (Proof: Use Fact 5.9.15.)

Fact 3.11.10. Let $A, B \in \mathbb{R}^{n \times n}$, assume that A and B are nilpotent, and assume that $AB = BA$. Then, $A + B$ is nilpotent. (Proof: If $A^k = B^l = 0$, then $(A + B)^{k+l} = 0$.)

Fact 3.11.11. Let $A \in \mathbb{F}^{n \times n}$. Then, A is nilpotent if and only if, for all $k = 1, \ldots, n$, $\operatorname{tr} A^k = 0$. (Proof: See [592, p. 103].)

3.12 Facts on Hamiltonian and Symplectic Matrices

Fact 3.12.1. J_n is skew symmetric, skew involutory, and Hamiltonian, I_n is symplectic, and $\hat{I}_n$ is a symmetric permutation matrix.

Fact 3.12.2. Let $A \in \mathbb{F}^{2n \times 2n}$, and assume that A is symplectic. Then, $\det A = 1$. (Proof: See [56, p. 27], [639, p. 128], or [324, p. 8].)

Fact 3.12.3. Let $A \in \mathbb{F}^{2 \times 2}$. Then, A is symplectic if and only if $\det A = 1$. Hence, $\mathrm{SL}_{\mathbb{F}}(2) = \mathrm{Sp}_{\mathbb{F}}(1)$.

Fact 3.12.4. Let $A \in \mathbb{F}^{2n \times 2n}$, and assume that A is Hamiltonian and nonsingular. Then, A^{-1} is Hamiltonian.

Fact 3.12.5. Let $A \in \mathbb{F}^{2n \times 2n}$. Then, A is Hamiltonian if and only if $JA = (JA)^{\mathrm{T}}$. Furthermore, A is symplectic if and only if $A^{\mathrm{T}}JA = J$.

Fact 3.12.6. Let $A \in \mathbb{F}^{2n \times 2n}$, assume that A is Hamiltonian, and let $S \in \mathbb{F}^{2n \times 2n}$ be symplectic. Then, SAS^{-1} is Hamiltonian.

Fact 3.12.7. Let $\mathcal{A} \in \mathbb{F}^{2n \times 2n}$. Then, $\mathcal{A}$ is skew symmetric and Hamiltonian if and only if there exist a skew-symmetric matrix $A \in \mathbb{F}^{n \times n}$ and a symmetric matrix $B \in \mathbb{F}^{n \times n}$ such that $\mathcal{A} = \begin{bmatrix} A & B \\ -B & A \end{bmatrix}$.

Fact 3.12.8. Let $A \in \mathbb{R}^{2n \times 2n}$, and assume that A is skew symmetric. Then, there exists a nonsingular matrix $S \in \mathbb{R}^{2n \times 2n}$ such that $S^{\mathrm{T}}AS = J_n$. (Proof: See [56, p. 231].)

Fact 3.12.9. If $A \in \mathbb{F}^{2n \times 2n}$ is Hamiltonian and $A + I$ is nonsingular, then $B \triangleq (A-I)(A+I)^{-1}$ is symplectic and $I - B$ is nonsingular. Conversely, if $B \in \mathbb{F}^{2n \times 2n}$ is symplectic and $I - B$ is nonsingular, then $A = (I + B)(I - B)^{-1}$ is Hamiltonian. (Remark: See Fact 3.7.22, Fact 3.7.23, and Fact 3.7.24.)

3.13 Facts on Groups

Fact 3.13.1. The following subsets of $\mathbb{R}$ are groups:

i) $\{x \in \mathbb{R}: \ x \neq 0\}$.

ii) $\{x \in \mathbb{R}: \ x > 0\}$.

iii) $\{x \in \mathbb{R}: \ x \neq 0 \text{ and } x \text{ is rational}\}$.

iv) $\{x \in \mathbb{R}: \ x > 0 \text{ and } x \text{ is rational}\}$.

v) $\{-1, 1\}$.

vi) $\{1\}$.

Fact 3.13.2. The following subsets of $\mathbb{F}^{n \times n}$ are Lie algebras:

i) $\mathrm{ut}(n) \triangleq \{A \in \mathrm{gl}_{\mathbb{F}}(n): \ A \text{ is upper triangular}\}$.

ii) $\mathrm{sut}(n) \triangleq \{A \in \mathrm{gl}_{\mathbb{F}}(n): \ A \text{ is strictly upper triangular}\}$.

iii) $\{0_{n \times n}\}$.

Fact 3.13.3. The following subsets of $\mathbb{F}^{n \times n}$ are groups:

i) $\mathrm{UT}(n) \triangleq \{A \in \mathrm{GL}_{\mathbb{F}}(n): \ A \text{ is upper triangular}\}$.

ii) $\mathrm{UT}_+(n) \triangleq \{A \in \mathrm{UT}(n): \ A_{(i,i)} > 0 \text{ for all } i = 1, \ldots, n\}$.

iii) $\mathrm{UT}_{\pm 1}(n) \triangleq \{A \in \mathrm{UT}(n): \ A_{(i,i)} = \pm 1 \text{ for all } i = 1, \ldots, n\}$.

iv) $\mathrm{SUT}(n) \triangleq \{A \in \mathrm{UT}(n): \ A_{(i,i)} = 1 \text{ for all } i = 1, \ldots, n\}$.

v) $\{I_n\}$.

(Remark: The matrices in $\mathrm{SUT}(n)$ are unipotent. See Fact 5.14.7.) (Remark: $\mathrm{SUT}(3)$ for $\mathbb{F} = \mathbb{R}$ is the *Heisenberg group*.)

Fact 3.13.4. Let $\mathcal{S} \subset \mathbb{F}^{n \times n}$, and assume that $\mathcal{S}$ is a group. Then, $\{A^{\mathrm{T}}: \ A \in \mathcal{S}\}$ and $\{\overline{A}: \ A \in \mathcal{S}\}$ are groups.

3.14 Facts on Quaternions

Fact 3.14.1. Define $Q_0, Q_1, Q_2, Q_3 \in \mathbb{C}^{2\times 2}$ by

$$Q_0 \triangleq I_2, \quad Q_1 \triangleq \begin{bmatrix} 0 & 1 \\ -1 & 0 \end{bmatrix}, \quad Q_2 \triangleq \begin{bmatrix} \jmath & 0 \\ 0 & -\jmath \end{bmatrix}, \quad Q_3 \triangleq \begin{bmatrix} 0 & -\jmath \\ -\jmath & 0 \end{bmatrix}.$$

Then, the following statements hold:

i) $Q_0^* = Q_0$ and $Q_i^* = -Q_i$ for all $i = 1, 2, 3$.

ii) $Q_0^2 = Q_0$ and $Q_i^2 = -Q_0$ for all $i = 1, 2, 3$.

iii) $Q_i Q_j = -Q_j Q_i$ for all $1 \le i < j \le 3$.

iv) $Q_1 Q_2 = Q_3$, $Q_2 Q_3 = Q_1$, and $Q_3 Q_1 = Q_2$.

v) $\{\pm Q_0, \pm Q_1, \pm Q_2, \pm Q_3\}$ is a group.

For $\beta \triangleq \begin{bmatrix} \beta_0 & \beta_1 & \beta_2 & \beta_3 \end{bmatrix}^{\mathrm{T}} \in \mathbb{R}^4$ define

$$Q(\beta) \triangleq \sum_{i=0}^{3} \beta_i Q_i.$$

Then,

$$Q(\beta)Q^*(\beta) = \beta^{\mathrm{T}}\beta I_2$$

and

$$\det Q(\beta) = \beta^{\mathrm{T}}\beta.$$

Hence, if $\beta^{\mathrm{T}}\beta = 1$, then $Q(\beta)$ is unitary. Furthermore, the complex matrices Q_0, Q_1, Q_2, Q_3, and $Q(\beta)$ have the real representations

$$\mathcal{Q}_0 = I_4, \qquad \mathcal{Q}_1 = \begin{bmatrix} J_2 & 0 \\ 0 & J_2 \end{bmatrix},$$

$$\mathcal{Q}_2 = \begin{bmatrix} 0 & 0 & 1 & 0 \\ 0 & 0 & 0 & -1 \\ -1 & 0 & 0 & 0 \\ 0 & 1 & 0 & 0 \end{bmatrix}, \qquad \mathcal{Q}_3 = \begin{bmatrix} 0 & 0 & 0 & -1 \\ 0 & 0 & -1 & 0 \\ 0 & 1 & 0 & 0 \\ 1 & 0 & 0 & 0 \end{bmatrix},$$

$$\mathcal{Q}(\beta) = \begin{bmatrix} \beta_0 & \beta_1 & \beta_2 & -\beta_3 \\ -\beta_1 & \beta_0 & -\beta_3 & -\beta_2 \\ -\beta_2 & \beta_3 & \beta_0 & \beta_1 \\ \beta_3 & \beta_2 & -\beta_1 & \beta_0 \end{bmatrix}.$$

Hence,

$$\mathcal{Q}(\beta)\mathcal{Q}^{\mathrm{T}}(\beta) = \beta^{\mathrm{T}}\beta I_4$$

and

$$\det \mathcal{Q}(\beta) = \left(\beta^{\mathrm{T}}\beta\right)^2.$$

(Remark: Q_0, Q_1, Q_2, Q_3 represent the *quaternions* $1, \imath, \jmath, k$. See Fact 3.14.3. The quaternion group v) is isomorphic to SU(2).) (Remark: Matrices with quaternion entries and 4×4 matrix representations are considered in [46, 142,

252, 312, 809]. For applications of quaternions, see [15, 314, 444].) (Remark: $\mathcal{Q}(\beta)$ has the form $\begin{bmatrix} A & B \\ -B & A \end{bmatrix}$, where A and $\hat{I}B$ are rotation-dilations. See Fact 2.17.1.)

Fact 3.14.2. Let $A \in \mathbb{C}^{2 \times 2}$. Then, A is unitary if and only if there exist $\theta \in \mathbb{R}$ and $\beta \in \mathbb{R}^4$ such that $A = e^{J\theta}Q(\beta)$, where $Q(\beta)$ is defined in Fact 3.14.1. (Proof: See [613, p. 228].)

Fact 3.14.3. Let $A_0, A_1, A_2, A_3 \in \mathbb{R}^{n \times n}$, let $\imath, \jmath, k$ satisfy

$$\imath^2 = \jmath^2 = k^2 = -1,$$

$$\imath\jmath = k = -\jmath\imath,$$

$$\jmath k = \imath = -k\jmath,$$

$$k\imath = \jmath = -\imath k,$$

and let $A \triangleq A_0 + \imath A_1 + \jmath A_2 + kA_3$. Then,

$$\begin{bmatrix} A_0 & -A_1 & -A_2 & -A_3 \\ A_1 & A_0 & -A_3 & A_2 \\ A_2 & A_3 & A_0 & -A_1 \\ A_3 & -A_2 & A_1 & A_0 \end{bmatrix} = U\mathrm{diag}(A, A, A, A)U,$$

where

$$U \triangleq \frac{1}{2}\begin{bmatrix} I & \imath I & \jmath I & kI \\ -\imath I & I & kI & -\jmath I \\ -\jmath I & -kI & I & \imath I \\ -kI & \jmath I & -\imath I & I \end{bmatrix}.$$

(Proof: See [704].) (Remark: k is not an integer here. $\imath, \jmath, k$ are the unit quaternions. This identity uses a similarity transformation to construct a real representation of quaternions. See Fact 2.13.8.) (Remark: The geometric significance of the quaternions is discussed in [15, 192, 444].) (Remark: The *Clifford algebras* include the *quaternion algebra* $\mathbb{H}$ and the *octonion algebra* $\mathbb{O}$. See [55, 192, 199, 225, 314, 348, 355, 592, 702, 702].)

3.15 Facts on Miscellaneous Types of Matrices

Fact 3.15.1. Let $A \in \mathbb{F}^{n \times m}$. Then, A is centrosymmetric if and only if $A^{\mathrm{T}} = A^{\hat{\mathrm{T}}}$. Furthermore, A is centrohermitian if and only if $A^* = A^{\hat{*}}$.

Fact 3.15.2. Let $A \in \mathbb{F}^{n \times m}$ and $B \in \mathbb{F}^{m \times l}$. If A and B are both (centrohermitian, centrosymmetric), then so is AB.

Fact 3.15.3. Let $A \in \mathbb{F}^{n \times m}$. Then, A is (semicontractive, contractive) if and only if A^* is.

Fact 3.15.4. Let $A \in \mathbb{F}^{n \times n}$, and assume that A is dissipative. Then, A is nonsingular. (Proof: Suppose that A is singular, and let $x \in \mathcal{N}(A)$. Then, $x^*(A + A^*)x = 0$.) (Remark: If $A + A^*$ is nonsingular, then A is not necessarily nonsingular. Let $A = \begin{bmatrix} 0 & 1 \\ 0 & 0 \end{bmatrix}$.)

Fact 3.15.5. Let $A \in \mathbb{R}^{n \times n}$, assume that A is tridiagonal with positive diagonal entries, and assume that, for all $i = 2, \ldots, n$,

$$A_{(i,i-1)} A_{(i-1,i)} < \tfrac{1}{4}(\cos \tfrac{\pi}{n+1})^{-2} A_{(i,i)} A_{(i-1,i-1)}.$$

Then, $\det A > 0$. (Proof: See [396].)

Fact 3.15.6. Let $A \in \mathbb{F}^{n \times n}$, and assume that A is Toeplitz. Then, A is reverse symmetric.

Fact 3.15.7. Let $A \in \mathbb{F}^{n \times n}$. Then, A is Toeplitz if and only if there exist $a_0, \ldots, a_n \in \mathbb{F}$ and $b_1, \ldots, b_n \in \mathbb{F}$ such that

$$A = \sum_{i=1}^{n} b_i N_n^{i\mathrm{T}} + \sum_{i=0}^{n} a_i N_n^{i}.$$

Fact 3.15.8. Let $A \in \mathbb{F}^{n \times n}$, let $k \in \mathbb{P}$, and assume that A is (lower triangular, strictly lower triangular, upper triangular, strictly upper triangular). Then, so is A^k. If, in addition, A is Toeplitz, then so is A^k. (Remark: If A is Toeplitz, then A^2 is not necessarily Toeplitz.) (Remark: See Fact 11.11.1.)

Fact 3.15.9. Let $A \in \mathbb{F}^{n \times m}$. Then, the following statements hold:

i) If A is Toeplitz, then $\hat{I}A$ and $A\hat{I}$ are Hankel.

ii) If A is Hankel, then $\hat{I}A$ and $A\hat{I}$ are Toeplitz.

iii) A is Toeplitz if and only if $\hat{I}A\hat{I}$ is Toeplitz.

iv) A is Hankel if and only if $\hat{I}A\hat{I}$ is Hankel.

Fact 3.15.10. Let $A \in \mathbb{F}^{n \times n}$, assume that A is Hankel, and consider the following conditions:

i) A is Hermitian.

ii) A is real.

iii) A is symmetric.

Then, *i*) $\implies$ *ii*) $\implies$ *iii*).

Fact 3.15.11. Let $A \in \mathbb{F}^{n \times n}$, and assume that A is a partitioned matrix, each of whose blocks is a $k \times k$ (circulant, Hankel, Toeplitz) matrix. Then, A is similar to a block-(circulant, Hankel, Toeplitz) matrix. (Proof:

See [74].)

Fact 3.15.12. For all $i, j = 1, \ldots, n$, define $A \in \mathbb{R}^{n \times n}$ by $A_{(i,j)} \triangleq 1/(i + j - 1)$. Then, A is Hankel and

$$\det A = \frac{[1!2! \cdots (n-1)!]^4}{1!2! \cdots (2n-1)!}.$$

Furthermore, for all $i, j = 1, \ldots, n$, A^{-1} has integer entries given by

$$\left(A^{-1}\right)_{(i,j)} = (-1)^{i+j}(i+j-1)\binom{n+i-1}{n-j}\binom{n+j-1}{n-1}\binom{i+j-2}{i-1}^2.$$

Finally, for large n,

$$\det A \approx 2^{-2n^2}.$$

(Remark: A is the *Hilbert matrix*, which is a Cauchy matrix. See [354, p. 513], Fact 1.4.10, and Fact 3.15.13.)

Fact 3.15.13. Let $a_1, \ldots, a_n, b_1, \ldots, b_n \in \mathbb{R}$, assume that $a_i + b_j \neq 0$ for all $i, j = 1, \ldots, n$, and, for all $i, j = 1, \ldots, n$, define $A \in \mathbb{R}^{n \times n}$ by $A_{(i,j)} \triangleq 1/(a_i + b_j)$. Then,

$$\det A = \frac{\displaystyle\prod_{1 \leq i < j \leq n} (a_j - a_i)(b_j - b_i)}{\displaystyle\prod_{1 \leq i,j \leq n} (a_i + b_j)}.$$

Now, assume that $a_1, \ldots, a_n$ are distinct and $b_1, \ldots, b_n$ are distinct. Then, A is nonsingular and

$$\left(A^{-1}\right)_{(i,j)} = \frac{\displaystyle\prod_{1 \leq k \leq n} (a_j + b_k)(a_k + b_i)}{(a_j + b_i) \displaystyle\prod_{\substack{1 \leq k \leq n \\ k \neq j}} (a_j - a_k) \displaystyle\prod_{\substack{1 \leq k \leq n \\ k \neq i}} (b_i - b_k)}.$$

Furthermore,

$$1_{1 \times n} A^{-1} 1_{n \times 1} = \sum_{i=1}^{n} (a_i + b_i).$$

(Remark: A is a *Cauchy matrix*. See [354, p. 515], Fact 8.7.27, and Fact 1.4.10.)

Fact 3.15.14. Let $A \in \mathbb{R}^{n \times n}$ be tripotent. Then,

$$\operatorname{rank} A = \operatorname{rank} A^2 = \operatorname{tr} A^2.$$

Fact 3.15.15. Let $A \in \mathbb{F}^{n \times n}$. Then, A is nonsingular and tripotent if and only if A is involutory.

Fact 3.15.16. Let $A \in \mathbb{F}^{n \times n}$. Then, A is involutory if and only if $(A + I)(A - I) = 0$.

Fact 3.15.17. Let $A \in \mathbb{R}^{n \times n}$, and assume that A is skew involutory. Then, n is even.

Fact 3.15.18. Let $x, y \in \mathbb{R}^n$, and assume that $x_{(1)} \geq \cdots \geq x_{(n)}$ and $y_{(1)} \geq \cdots \geq y_{(n)}$. Then, there exists a doubly stochastic matrix $A \in \mathbb{R}^{n \times n}$ such that $y = Ax$ if and only if y strongly majorizes x. (Remark: The matrix A is *doubly stochastic* if it is nonnegative, $1_{1 \times n} A = 1_{1 \times n}$, and $A 1_{n \times 1} = 1_{n \times 1}$. This result is the *Hardy-Littlewood-Polya theorem*. See [111, p. 33], [367, p. 197], and [516, p. 22].)

3.16 Notes

In the literature on generalized inverses, range-Hermitian matrices are traditionally called *EP matrices*. Elementary reflectors are traditionally called *Householder matrices* or *Householder reflections*.

Left equivalence, right equivalence, and biequivalence are treated in [613]. Each of the groups defined in Proposition 3.3.4 is a *Lie group*; see Definition 11.5.1. Elementary treatments of Lie algebras and Lie groups are given in [43, 56, 198, 254, 290, 380, 580, 623], while an advanced treatment appears in [742]. Some additional groups of structured matrices are given in [501].

The terminology "idempotent" and "projector" is not standardized in the literature. Some writers use "projector" or "oblique projector" for idempotent, and "orthogonal projector" for projector. Centrosymmetric and centrohermitian matrices are discussed in [465, 764].

Chapter Four

Matrix Polynomials and Rational Transfer Functions

In this chapter we consider matrices whose entries are polynomials or rational functions. The decomposition of polynomial matrices in terms of the Smith form provides the foundation for developing canonical forms in Chapter 5. In this chapter we also present some basic properties of eigenvalues and eigenvectors as well as the minimal and characteristic polynomials of a square matrix. Finally, we consider the extension of the Smith form to the Smith-McMillan form for rational transfer functions.

4.1 Polynomials

A function $p \colon \mathbb{C} \mapsto \mathbb{C}$ of the form

$$p(s) = \beta_k s^k + \beta_{k-1} s^{k-1} + \cdots + \beta_1 s + \beta_0, \qquad (4.1.1)$$

where $k \in \mathbb{N}$ and $\beta_0, \ldots, \beta_k \in \mathbb{F}$, is a *polynomial*. The set of polynomials is denoted by $\mathbb{F}[s]$. If the coefficient $\beta_k \in \mathbb{F}$ is nonzero, then the *degree* of p, denoted by $\deg p$, is k. If, in addition, $\beta_k = 1$, then p is *monic*. If $k = 0$, then p is *constant*. The degree of a nonzero constant polynomial is zero, while the degree of the zero polynomial is defined to be $-\infty$.

Let p_1 and p_2 be polynomials. Then,

$$\deg p_1 p_2 = \deg p_1 + \deg p_2. \qquad (4.1.2)$$

If $p_1 = 0$ or $p_2 = 0$, then $\deg p_1 p_2 = \deg p_1 + \deg p_2 = -\infty$. If p_2 is a nonzero constant, then $\deg p_2 = 0$, and thus $\deg p_1 p_2 = \deg p_1$. Furthermore,

$$\deg(p_1 + p_2) \leq \max\{\deg p_1, \deg p_2\}. \qquad (4.1.3)$$

Therefore, $\deg(p_1 + p_2) = \max\{\deg p_1, \deg p_2\}$ if and only if either *i)* $\deg p_1 \neq \deg p_2$ or *ii)* $p_1 = p_2 = 0$ or *iii)* $r \triangleq \deg p_1 = \deg p_2 \neq -\infty$ and the sum of the coefficients of s^r in p_1 and p_2 is not zero. Equivalently, $\deg(p_1 + p_2) < \max\{\deg p_1, \deg p_2\}$ if and only if $r \triangleq \deg p_1 = \deg p_2 \neq -\infty$ and the sum of the coefficients of s^r in p_1 and p_2 is zero.

Let $p \in \mathbb{F}[s]$ be a polynomial of degree $k \geq 1$. Then, it follows from the *fundamental theorem of algebra* that p has k possibly repeated complex roots $\lambda_1, \ldots, \lambda_k$ so that p can be factored as

$$p(s) = \beta \prod_{i=1}^{k} (s - \lambda_i), \qquad (4.1.4)$$

where $\beta \in \mathbb{F}$. The multiplicity of a root $\lambda \in \mathbb{C}$ of p is denoted by $\mathrm{m}_p(\lambda)$. If λ is not a root of p, then $\mathrm{m}_p(\lambda) = 0$. The multiset consisting of the roots of p including multiplicity is $\mathrm{mroots}(p) = \{\lambda_1, \ldots, \lambda_k\}_{\mathrm{m}}$, while the set of roots of p ignoring multiplicity is $\mathrm{roots}(p) = \{\hat{\lambda}_1, \ldots, \hat{\lambda}_l\}$, where $\sum_{i=1}^{l} \mathrm{m}_p(\hat{\lambda}_i) = k$. If $\mathbb{F} = \mathbb{R}$, then the multiplicity of a root λ_i whose imaginary part is nonzero is equal to the multiplicity of its complex conjugate $\overline{\lambda}_i$. Hence, $\mathrm{mroots}(p)$ is *self conjugate*, that is, $\mathrm{mroots}(p) = \overline{\mathrm{mroots}(p)}$.

Let $p \in \mathbb{F}[s]$. If $p(-s) = p(s)$ for all $s \in \mathbb{C}$, then p is *even*, while, if $p(-s) = -p(s)$ for all $s \in \mathbb{C}$, then p is *odd*. If p is either odd or even, then $\mathrm{mroots}(p) = -\mathrm{mroots}(p)$. If $p \in \mathbb{R}[s]$ and there exists a polynomial $q \in \mathbb{R}[s]$ such that $p(s) = q(s)q(-s)$ for all $s \in \mathbb{C}$, then p has a *spectral factorization*. If p has a spectral factorization, then p is even and $\deg p$ is an even integer.

Proposition 4.1.1. Let $p \in \mathbb{R}[s]$. Then, the following statements are equivalent:

   i) p has a spectral factorization.

   ii) p is even, and every imaginary root of p has even multiplicity.

   iii) p is even, and $p(\jmath\omega) \geq 0$ for all $\omega \in \mathbb{R}$.

Proof. The equivalence of *i)* and *ii)* is immediate. To prove *i)* $\Longrightarrow$ *iii)*, note that, for all $\omega \in \mathbb{R}$,

$$p(\jmath\omega) = q(\jmath\omega)q(-\jmath\omega) = |q(\jmath\omega)|^2 \geq 0.$$

Conversely, to prove *iii)* $\Longrightarrow$ *i)* write $p = p_1 p_2$, where every root of p_1 is imaginary and none of the roots of p_2 are imaginary. Now, let z be a root of p_2. Then, $-z$, $\overline{z}$, and $-\overline{z}$ are also roots of p_2 with the same multiplicity as z. Hence, there exists a polynomial $p_{20} \in \mathbb{R}[s]$ such that $p_2(s) = p_{20}(s)p_{20}(-s)$ for all $s \in \mathbb{C}$.

Next, assuming that p has at least one imaginary root, write $p_1(s) = \prod_{i=1}^{k} (s^2 + \omega_i^2)^{m_i}$, where $0 \leq \omega_1 < \cdots < \omega_k$ and $m_i \triangleq \mathrm{m}_p(\jmath\omega_i)$. Let ω_{i_0} denote the smallest element of the set $\{\omega_1, \ldots, \omega_k\}$ such that m_i is odd. Then, it follows that $p_1(\jmath\omega) = \prod_{i=1}^{k} (\omega_i^2 - \omega^2)^{m_i} < 0$ for all $\omega \in (\omega_{i_0}, \omega_{i_0+1})$, where

$\omega_{k+1} \triangleq \infty$. However, note that $p_1(\jmath\omega) = p(\jmath\omega)/p_2(\jmath\omega) = p(\jmath\omega)/|p_{20}(\jmath\omega)|^2 \geq 0$ for all $\omega \in \mathbb{R}$, which is a contradiction. Therefore, m_i is even for all $i = 1, \ldots, k$, and thus $p_1(s) = p_{10}(s)p_{10}(-s)$ for all $s \in \mathbb{C}$, where $p_{10}(s) \triangleq \prod_{i=1}^{k}\left(s^2 + \omega_i^2\right)^{m_i/2}$. Consequently, $p(s) = p_{10}(s)p_{20}(s)p_{10}(-s)p_{20}(-s)$ for all $s \in \mathbb{C}$. Finally, if p has no imaginary roots, then $p_1 = 1$, and $p(s) = p_{20}(s)p_{20}(-s)$ for all $s \in \mathbb{C}$. $\qquad\square$

The following division algorithm is essential to the study of polynomials.

Lemma 4.1.2. Let $p_1, p_2 \in \mathbb{F}[s]$, and assume that p_2 is not the zero polynomial. Then, there exist unique polynomials $q, r \in \mathbb{F}[s]$ such that $\deg r < \deg p_2$ and

$$p_1 = qp_2 + r. \tag{4.1.5}$$

Proof. Define $n \triangleq \deg p_1$ and $m \triangleq \deg p_2$. If $n < m$, then $q = 0$ and $r = p_1$. Hence, $\deg r = \deg p_1 = n < m = \deg p_2$.

Now, assume that $n \geq m \geq 0$, and write $p_1(s) = \beta_n s^n + \cdots + \beta_0$ and $p_2(s) = \gamma_m s^m + \cdots + \gamma_0$. If $n = 0$, then $m = 0$, $\gamma_0 \neq 0$, $q = \beta_0/\gamma_0$, and $r = 0$. Hence, $-\infty = \deg r < 0 = \deg p_2$.

If $n = 1$, then either $m = 0$ or $m = 1$. If $m = 0$, then $p_2(s) = \gamma_0 \neq 0$, and (4.1.5) is satisfied with $q(s) = p_1(s)/\gamma_0$ and $r = 0$, in which case $-\infty = \deg r < 0 = \deg p_2$. If $m = 1$, then (4.1.5) is satisfied with $q(s) = \beta_1/\gamma_1$ and $r(s) = \beta_0 - \beta_1\gamma_0/\gamma_1$. Hence, $\deg r \leq 0 < 1 = \deg p_2$.

Now, suppose that $n = 2$. Then, $\hat{p}_1(s) = p_1(s) - (\beta_2/\gamma_m)s^{2-m}p_2(s)$ has degree 1. Applying (4.1.5) with p_1 replaced by $\hat{p}_1$, it follows that there exist polynomials $q_1, r_1 \in \mathbb{F}[s]$ such that $\hat{p}_1 = q_1p_2 + r_1$ and such that $\deg r_1 < \deg p_2$. It thus follows that $p_1(s) = q_1(s)p_2(s) + r_1(s) + (\beta_2/\gamma_m)s^{2-m}p_2(s) = q(s)p_2(s) + r(s)$, where $q(s) = q_1(s) + (\beta_2/\gamma_m)s^{n-m}$ and $r = r_1$, which verifies (4.1.5). Similar arguments apply to successively larger values of n.

To prove uniqueness, suppose there exist polynomials $\hat{q}$ and $\hat{r}$ such that $\deg \hat{r} < \deg p_2$ and $p_1 = \hat{q}p_2 + \hat{r}$. Then, it follows that $(\hat{q} - q)p_2 = r - \hat{r}$. Next, note that $\deg(r - \hat{r}) < \deg p_2$. If $\hat{q} \neq q$, then $\deg p_2 \leq \deg[(\hat{q} - q)p_2]$ so that $\deg(r - \hat{r}) < \deg[(\hat{q} - q)p_2]$, which is a contradiction. Thus, $\hat{q} = q$, and, hence, $r = \hat{r}$. $\qquad\square$

In Lemma 4.1.2, q is the *quotient* of p_1 and p_2, while r is the *remainder*. If $r = 0$, then p_2 *divides* p_1, or, equivalently, p_1 is a *multiple* of p_2. Note that, if $p_2(s) = s - \alpha$, where $\alpha \in \mathbb{F}$, then r is constant and is given by $r(s) = p_1(\alpha)$.

If a polynomial $p_3 \in \mathbb{F}[s]$ divides two polynomials $p_1, p_2 \in \mathbb{F}[s]$, then p_3 is a *common divisor* of p_1 and p_2. Given polynomials $p_1, p_2 \in \mathbb{F}[s]$, there exists a unique monic polynomial $p_3 \in \mathbb{F}[s]$, the *greatest common divisor* of p_1 and p_2, such that p_3 is a common divisor of p_1 and p_2 and such that every common divisor of p_1 and p_2 divides p_3. In addition, there exist polynomials $q_1, q_2 \in \mathbb{F}[s]$ such that the greatest common divisor p_3 of p_1 and p_2 is given by $p_3 = q_1 p_1 + q_2 p_2$. See [581, p. 113] for proofs of these results. Finally, p_1 and p_2 are *coprime* if their greatest common divisor is $p_3 = 1$, while a polynomial $p \in \mathbb{F}[s]$ is *irreducible* if there do not exist nonconstant polynomials $p_1, p_2 \in \mathbb{F}[s]$ such that $p = p_1 p_2$. For example, if $\mathbb{F} = \mathbb{R}$, then $p(s) = s^2 + s + 1$ is irreducible.

If a polynomial $p_3 \in \mathbb{F}[s]$ is a multiple of two polynomials $p_1, p_2 \in \mathbb{F}[s]$, then p_3 is a *common multiple* of p_1 and p_2. Given nonzero polynomials p_1 and p_2, there exists (see [581, p. 113]) a unique monic polynomial $p_3 \in \mathbb{F}[s]$ that is a common multiple of p_1 and p_2 and that divides every common multiple of p_1 and p_2. The polynomial p_3 is the *least common multiple* of p_1 and p_2.

The polynomial $p \in \mathbb{F}[s]$ given by (4.1.1) can be evaluated with a square matrix argument $A \in \mathbb{F}^{n \times n}$ by defining

$$p(A) \triangleq \beta_k A^k + \beta_{k-1} A^{k-1} + \cdots + \beta_1 A + \beta_0 I. \qquad (4.1.6)$$

4.2 Matrix Polynomials

The set $\mathbb{F}^{n \times m}[s]$ of *matrix polynomials* consists of matrix functions $P \colon \mathbb{C} \mapsto \mathbb{C}^{n \times m}$ whose entries are elements of $\mathbb{F}[s]$. A matrix polynomial $P \in \mathbb{F}^{n \times m}[s]$ can thus be written as

$$P(s) = s^k B_k + s^{k-1} B_{k-1} + \cdots + s B_1 + B_0, \qquad (4.2.1)$$

where $B_0, \ldots, B_k \in \mathbb{F}^{n \times m}$. If B_k is nonzero, then the *degree* of P, denoted by $\deg P$, is k, whereas, if $P = 0$, then $\deg P = -\infty$. If $n = m$ and B_k is nonsingular, then P is *regular*, while, if $B_k = I$, then P is *monic*.

The following result, which generalizes Lemma 4.1.2, provides a division algorithm for matrix polynomials.

Lemma 4.2.1. Let $P_1, P_2 \in \mathbb{F}^{n \times n}[s]$, where P_2 is regular. Then, there exist unique matrix polynomials $Q, R, \hat{Q}, \hat{R} \in \mathbb{F}^{n \times n}[s]$ such that $\deg R < \deg P_2$, $\deg \hat{R} < \deg P_2$,

$$P_1 = Q P_2 + R, \qquad (4.2.2)$$

and

$$P_1 = P_2 \hat{Q} + \hat{R}. \qquad (4.2.3)$$

Proof. See [581, pp. 134–135] or [293, p. 90]. $\qquad\square$

If $R = 0$, then P_2 *right divides* P_1, while, if $\hat{R} = 0$, then P_2 *left divides* P_1.

Let the matrix polynomial $P \in \mathbb{F}^{n \times m}[s]$ be given by (4.2.1). Then, P can be evaluated with a square matrix argument in two different ways, either from the right or from the left. For $A \in \mathbb{C}^{m \times m}$ define

$$P_{\mathrm{R}}(A) \triangleq B_k A^k + B_{k-1} A^{k-1} + \cdots + B_1 A + B_0, \qquad (4.2.4)$$

while, for $A \in \mathbb{C}^{n \times n}$, define

$$P_{\mathrm{L}}(A) \triangleq A^k B_k + A^{k-1} B_{k-1} + \cdots + A B_1 + B_0. \qquad (4.2.5)$$

If $n = m$, then $P_{\mathrm{R}}(A)$ and $P_{\mathrm{L}}(A)$ can be evaluated for all $A \in \mathbb{F}^{n \times n}$, although these matrices are generally different.

The following result is useful.

Lemma 4.2.2. Let $Q, \hat{Q} \in \mathbb{F}^{n \times n}[s]$ and $A \in \mathbb{F}^{n \times n}$. Furthermore, define $P, \hat{P} \in \mathbb{F}^{n \times n}[s]$ by $P(s) \triangleq Q(s)(sI - A)$ and $\hat{P}(s) \triangleq (sI - A)\hat{Q}(s)$. Then, $P_{\mathrm{R}}(A) = 0$ and $\hat{P}_{\mathrm{L}}(A) = 0$.

Let $p \in \mathbb{F}[s]$ be given by (4.1.1), and define $P(s) \triangleq p(s)I_n = s^k \beta_k I_n + s^{k-1}\beta_{k-1}I_n + \cdots + s\beta_1 I_n + \beta_0 I_n \in \mathbb{F}^{n \times n}[s]$. For $A \in \mathbb{C}^{n \times n}$ it follows that $p(A) = P(A) = P_{\mathrm{R}}(A) = P_{\mathrm{L}}(A)$.

The following result specializes Lemma 4.2.1 to the case of matrix polynomial divisors of degree 1.

Corollary 4.2.3. Let $P \in \mathbb{F}^{n \times n}[s]$ and $A \in \mathbb{F}^{n \times n}$. Then, there exist unique matrix polynomials $Q, \hat{Q} \in \mathbb{F}^{n \times n}[s]$ and unique matrices $R, \hat{R} \in \mathbb{F}^{n \times n}$ such that

$$P(s) = Q(s)(sI - A) + R \qquad (4.2.6)$$

and

$$P(s) = (sI - A)\hat{Q}(s) + \hat{R}. \qquad (4.2.7)$$

Furthermore, $R = P_{\mathrm{R}}(A)$ and $\hat{R} = P_{\mathrm{L}}(A)$.

Proof. In Lemma 4.2.1 set $P_1 = P$ and $P_2(s) = sI - A$. Since $\deg P_2 = 1$, it follows that $\deg R = \deg \hat{R} = 0$, and thus R and $\hat{R}$ are constant. Finally, the last statement follows from Lemma 4.2.2. $\qquad\square$

Definition 4.2.4. Let $P \in \mathbb{F}^{n \times m}[s]$. Then, $\operatorname{rank} P$ is defined by

$$\operatorname{rank} P \triangleq \max_{s \in \mathbb{C}} \operatorname{rank} P(s). \qquad (4.2.8)$$

Let $P \in \mathbb{F}^{n \times n}[s]$. Then, $P(s) \in \mathbb{C}^{n \times n}$ for all $s \in \mathbb{C}$. Furthermore, $\det P$ is a polynomial in s, that is, $\det P \in \mathbb{F}[s]$.

Definition 4.2.5. Let $P \in \mathbb{F}^{n \times n}[s]$. Then, P is *nonsingular* if $\det P$ is not the zero polynomial; otherwise, P is *singular*.

Proposition 4.2.6. Let $P \in \mathbb{F}^{n \times n}[s]$, and assume that P is regular. Then, P is nonsingular.

Let $P \in \mathbb{F}^{n \times n}[s]$. If P is nonsingular, then the *inverse* P^{-1} of P can be constructed according to (2.7.21). In general, the entries of P^{-1} are rational functions of s (see Definition 4.7.1). For example, if $P(s) = \begin{bmatrix} s+2 & s+1 \\ s-2 & s-1 \end{bmatrix}$, then $P^{-1}(s) = \frac{1}{2s} \begin{bmatrix} s-1 & -s-1 \\ -s+2 & s+2 \end{bmatrix}$. In certain cases, P^{-1} is also a matrix polynomial. For example, if $P(s) = \begin{bmatrix} s & 1 \\ s^2+s-1 & s+1 \end{bmatrix}$, then $P^{-1}(s) = \begin{bmatrix} s+1 & -1 \\ -s^2-s+1 & s \end{bmatrix}$.

The following result is an extension of Proposition 2.7.7 from constant matrices to matrix polynomials.

Proposition 4.2.7. Let $P \in \mathbb{F}^{n \times m}[s]$. Then, $\operatorname{rank} P$ is the order of the largest nonsingular matrix polynomial that is a submatrix of P.

Proof. For all $s \in \mathbb{C}$ it follows from Proposition 2.7.7 that $\operatorname{rank} P(s)$ is the order of the largest nonsingular submatrix of $P(s)$. Now, let $s_0 \in \mathbb{C}$ be such that $\operatorname{rank} P(s_0) = \operatorname{rank} P$. Then, $P(s_0)$ has a nonsingular submatrix of maximal order $\operatorname{rank} P$. Therefore, P has a nonsingular submatrix polynomial of maximal order $\operatorname{rank} P$. $\qquad\square$

A matrix polynomial can be transformed by performing elementary row and column operations of the following types:

i) Multiply a row or a column by a nonzero constant.

ii) Interchange two rows or two columns.

iii) Add a polynomial multiple of one (row, column) to another (row, column).

These operations correspond respectively to left multiplication or right multiplication by the elementary matrices

$$I_n + (\alpha - 1)E_{i,i} = \begin{bmatrix} I_{i-1} & 0 & 0 \\ 0 & \alpha & 0 \\ 0 & 0 & I_{n-i} \end{bmatrix}, \qquad (4.2.9)$$

where $\alpha \in \mathbb{F}$ is nonzero,

$$I_n + E_{i,j} + E_{j,i} - E_{i,i} - E_{j,j} = \begin{bmatrix} I_{i-1} & 0 & 0 & 0 & 0 \\ 0 & 0 & 0 & 1 & 0 \\ 0 & 0 & I_{j-i-1} & 0 & 0 \\ 0 & 1 & 0 & 0 & 0 \\ 0 & 0 & 0 & 0 & I_{n-j} \end{bmatrix}, \quad (4.2.10)$$

where $i \neq j$, and the *elementary matrix polynomial*

$$I_n + pE_{i,j} = \begin{bmatrix} I_{i-1} & 0 & 0 & 0 & 0 \\ 0 & 1 & 0 & p & 0 \\ 0 & 0 & I_{j-i-1} & 0 & 0 \\ 0 & 0 & 0 & 1 & 0 \\ 0 & 0 & 0 & 0 & I_{n-j} \end{bmatrix}, \quad (4.2.11)$$

where $i \neq j$ and $p \in \mathbb{F}[s]$. The matrices shown in (4.2.10) and (4.2.11) illustrate the case $i < j$. Applying these operations sequentially corresponds to forming products of elementary matrices and elementary matrix polynomials. Note that the elementary matrix polynomial $I + pE_{i,j}$ is nonsingular, and that $(I + pE_{i,j})^{-1} = I - pE_{i,j}$ so that the inverse of an elementary matrix polynomial is an elementary matrix polynomial.

4.3 The Smith Decomposition and Similarity Invariants

Definition 4.3.1. Let $P \in \mathbb{F}^{n \times n}[s]$. Then, P is *unimodular* if P is the product of elementary matrices and elementary matrix polynomials.

The following result provides a canonical form, known as the *Smith form*, for matrix polynomials under unimodular transformation.

Theorem 4.3.2. Let $P \in \mathbb{F}^{n \times m}[s]$, and let $r \triangleq \operatorname{rank} P$. Then, there exist unimodular matrices $S_1 \in \mathbb{F}^{n \times n}[s]$ and $S_2 \in \mathbb{F}^{m \times m}[s]$ and monic polynomials $p_1, \ldots, p_r \in \mathbb{F}[s]$ such that p_i divides p_{i+1} for all $i = 1, \ldots, r-1$ and such that

$$P = S_1 \begin{bmatrix} p_1 & & & & 0 \\ & \ddots & & & \\ & & p_r & & \\ 0 & & & 0_{(n-r) \times (m-r)} \end{bmatrix} S_2. \quad (4.3.1)$$

Furthermore, for all $i = 1, \ldots, r$, p_i is uniquely determined by

$$\Delta_i = p_1 \cdots p_i, \quad (4.3.2)$$

where Δ_i is the greatest common divisor of all $i \times i$ subdeterminants of P.

Proof. The result is obtained by sequentially applying elementary row and column operations to P. For details, see [407, pp. 390–392] or [581, pp. 125–128]. □

Corollary 4.3.3. Let $P \in \mathbb{R}^{n \times n}[s]$ be unimodular. Then, the Smith form of P is the identity.

Definition 4.3.4. The monic polynomials $p_1, \ldots, p_r \in \mathbb{F}[s]$ of the Smith form of $P \in \mathbb{F}^{n \times n}[s]$ are the *invariant polynomials* of P.

Proposition 4.3.5. Let $P \in \mathbb{F}^{n \times n}[s]$. Then, P is unimodular if and only if $\det P$ is a nonzero constant.

Proof. Necessity is immediate since every elementary matrix and every elementary matrix polynomial has a constant nonzero determinant. To prove sufficiency, note that, since $\det P$ is a nonzero constant, it follows from Theorem 4.3.2 that every invariant polynomial of P is also a nonzero constant. Consequently, P is a product of elementary matrices and elementary matrix polynomials, and thus is unimodular. □

Proposition 4.3.6. Let $P \in \mathbb{F}^{n \times n}[s]$. Then, the following statements are equivalent:

i) P is unimodular.

ii) P is nonsingular, and P^{-1} is a matrix polynomial.

iii) P is nonsingular, and P^{-1} is unimodular.

Proof. To prove *i)* $\implies$ *ii)*, suppose that P is unimodular. Then, it follows from Proposition 4.3.5 that $\det P$ is a nonzero constant. Therefore, P is nonsingular. Furthermore, since P^{A} is a matrix polynomial, it follows that $P^{-1} = (\det P)^{-1} P^{\mathrm{A}}$ is a matrix polynomial.

To prove *ii)* $\implies$ *iii)*, suppose that P is nonsingular and P^{-1} is a matrix polynomial so that $\det P^{-1}$ is a polynomial. Since $\det P$ is a nonzero constant and $\det P^{-1} = 1/\det P$, it follows that $\det P^{-1}$ is also a nonzero constant. Thus, Proposition 4.3.5 implies that P^{-1} is unimodular.

Finally, to prove *iii)* $\implies$ *i)*, suppose that P is nonsingular and P^{-1} is unimodular. Then, since $\det P^{-1}$ is a nonzero constant, it follows that $\det P = 1/\det P^{-1}$ is a nonzero constant. Proposition 4.3.5 thus implies that P is unimodular. □

Proposition 4.3.7. Let $A_1, B_1, A_2, B_2 \in \mathbb{F}^{n \times n}$, where A_2 is nonsingular, and define the matrix polynomials $P_1, P_2 \in \mathbb{F}^{n \times n}[s]$ by $P_1(s) \triangleq sA_1 + B_1$ and $P_2(s) \triangleq sA_2 + B_2$. Then, P_1 and P_2 have the same invariant polynomials if and only if there exist nonsingular matrices $S_1, S_2 \in \mathbb{F}^{n \times n}$ such that $P_2 = S_1 P_1 S_2$.

Proof. The sufficiency result is immediate. To prove necessity, note that it follows from Theorem 4.3.2 that there exist unimodular matrices $T_1, T_2 \in \mathbb{F}^{n \times n}[s]$ such that $P_2 = T_2 P_1 T_1$. Now, since P_2 is regular, it follows from Lemma 4.2.1 that there exist matrix polynomials $Q, \hat{Q} \in \mathbb{F}^{n \times n}[s]$ and constant matrices $R, \hat{R} \in \mathbb{F}^{n \times n}$ such that $T_1 = Q P_2 + R$ and $T_2 = P_2 \hat{Q} + \hat{R}$. Next, we have

$$
\begin{aligned}
P_2 &= T_2 P_1 T_1 \\
&= (P_2 \hat{Q} + \hat{R}) P_1 T_1 \\
&= \hat{R} P_1 T_1 + P_2 \hat{Q} T_2^{-1} P_2 \\
&= \hat{R} P_1 (Q P_2 + R) + P_2 \hat{Q} T_2^{-1} P_2 \\
&= \hat{R} P_1 R + (T_2 - P_2 \hat{Q}) P_1 Q P_2 + P_2 \hat{Q} T_2^{-1} P_2 \\
&= \hat{R} P_1 R + T_2 P_1 Q P_2 + P_2 \left(-\hat{Q} P_1 Q + \hat{Q} T_2^{-1} \right) P_2 \\
&= \hat{R} P_1 R + P_2 \left(T_1^{-1} Q - \hat{Q} P_1 Q + \hat{Q} T_2^{-1} \right) P_2.
\end{aligned}
$$

Since P_2 is regular and has degree 1, it follows that, if $T_1^{-1} Q - \hat{Q} P_1 Q + \hat{Q} T_2^{-1}$ is not zero, then $\deg P_2 \left(T_1^{-1} Q - \hat{Q} P_1 Q + \hat{Q} T_2^{-1} \right) P_2 \geq 2$. However, since P_2 and $\hat{R} P_1 R$ have degree less than 2, it follows that $T_1^{-1} Q - \hat{Q} P_1 Q + \hat{Q} T_2^{-1} = 0$. Hence, $P_2 = \hat{R} P_1 R$.

Next, to show that $\hat{R}$ and R are nonsingular, note that, for all $s \in \mathbb{C}$,

$$
P_2(s) = \hat{R} P_1(s) R = s \hat{R} A_1 R + \hat{R} B_1 R,
$$

which implies that $A_2 = S_1 A_1 S_2$, where $S_1 = \hat{R}$ and $S_2 = R$. Since A_2 is nonsingular, it follows that S_1 and S_2 are nonsingular. $\qquad \square$

Definition 4.3.8. Let $A \in \mathbb{F}^{n \times n}$. Then, the invariant polynomials of $sI - A$ are the *similarity invariants* of A.

The following result provides necessary and sufficient conditions for two matrices to be similar.

Theorem 4.3.9. Let $A, B \in \mathbb{F}^{n \times n}$. Then, A and B are similar if and only if they have the same similarity invariants.

Proof. To prove necessity, assume that A and B are similar. Then, the matrices $sI - A$ and $sI - B$ have the same Smith form and thus the same similarity invariants. To prove sufficiency, it follows from Proposition 4.3.7 that there exist nonsingular matrices $S_1, S_2 \in \mathbb{F}^{n \times n}$ such that $sI - A =$

$S_1(sI - B)S_2$. Thus, $S_1 = S_2^{-1}$, and, hence, $A = S_1 B S_1^{-1}$.   □

Corollary 4.3.10. Let $A \in \mathbb{F}^{n \times n}$. Then, A and A^{T} are similar.

An improved form of Corollary 4.3.10 is given by Corollary 5.3.8.

4.4 Eigenvalues

Let $A \in \mathbb{F}^{n \times n}$. Then, the matrix polynomial $sI - A \in \mathbb{F}^{n \times n}[s]$ is monic and has degree 1.

Definition 4.4.1. Let $A \in \mathbb{F}^{n \times n}$. Then, the *characteristic polynomial* of A is the polynomial $\chi_A \in \mathbb{F}[s]$ given by

$$\chi_A(s) \triangleq \det(sI - A). \tag{4.4.1}$$

Proposition 4.4.2. Let $A \in \mathbb{F}^{n \times n}$. Then, χ_A is monic and $\deg \chi_A = n$.

Let $A \in \mathbb{F}^{n \times n}$, and write the characteristic polynomial of A as

$$\chi_A(s) = s^n + \beta_{n-1} s^{n-1} + \cdots + \beta_1 s + \beta_0, \tag{4.4.2}$$

where $\beta_0, \ldots, \beta_{n-1} \in \mathbb{F}$. The *eigenvalues* of A are the n possibly repeated roots $\lambda_1, \ldots, \lambda_n \in \mathbb{C}$ of χ_A, that is, the solutions of the *characteristic equation*

$$\chi_A(s) = 0. \tag{4.4.3}$$

It is often convenient to denote the eigenvalues of A by $\lambda_1(A), \ldots, \lambda_n(A)$ or just $\lambda_1, \ldots, \lambda_n$. This notation may be ambiguous, however, since it does not uniquely specify which eigenvalue is denoted by λ_i. If, however, every eigenvalue of A is real, then we employ the notational convention

$$\lambda_1 \geq \cdots \geq \lambda_n, \tag{4.4.4}$$

and we define

$$\lambda_{\max}(A) \triangleq \lambda_1, \quad \lambda_{\min}(A) \triangleq \lambda_n. \tag{4.4.5}$$

Definition 4.4.3. Let $A \in \mathbb{F}^{n \times n}$. The *algebraic multiplicity* of an eigenvalue λ of A, denoted by $\mathrm{am}_A(\lambda)$, is the algebraic multiplicity of λ as a root of χ_A, that is,

$$\mathrm{am}_A(\lambda) \triangleq \mathrm{m}_{\chi_A}(\lambda). \tag{4.4.6}$$

The multiset consisting of the eigenvalues of A including their algebraic multiplicity, denoted by $\mathrm{mspec}(A)$, is the *multispectrum* of A, that is,

$$\mathrm{mspec}(A) \triangleq \mathrm{mroots}(\chi_A). \tag{4.4.7}$$

Ignoring algebraic multiplicity, $\mathrm{spec}(A)$ denotes the *spectrum* of A, that is,

$$\mathrm{spec}(A) \triangleq \mathrm{roots}(\chi_A). \tag{4.4.8}$$

If $\lambda \notin \mathrm{spec}(A)$, then $\lambda \notin \mathrm{roots}(\chi_A)$, and thus $\mathrm{am}_A(\lambda) = \mathrm{m}_{\chi_A}(\lambda) = 0$.

Let $A \in \mathbb{F}^{n \times n}$ and $\mathrm{mroots}(\chi_A) = \{\lambda_1, \ldots, \lambda_n\}_{\mathrm{m}}$. Then,

$$\chi_A(s) = \prod_{i=1}^{n}(s - \lambda_i). \tag{4.4.9}$$

If $\mathbb{F} = \mathbb{R}$, then $\chi_A(s)$ has real coefficients, and thus the eigenvalues of A occur in complex conjugate pairs, that is, $\overline{\mathrm{mroots}(\chi_A)} = \mathrm{mroots}(\chi_A)$. Now, let $\mathrm{spec}(A) = \{\lambda_1, \ldots, \lambda_r\}$, and, for all $i = 1, \ldots, r$, let n_i denote the algebraic multiplicity of λ_i. Then,

$$\chi_A(s) = \prod_{i=1}^{r}(s - \lambda_i)^{n_i}. \tag{4.4.10}$$

The following result gives some basic properties of the spectrum of a matrix.

Proposition 4.4.4. Let $A, B \in \mathbb{F}^{n \times n}$. Then, the following statements hold:

i) $\chi_{A^{\mathrm{T}}} = \chi_A$.

ii) For all $s \in \mathbb{C}$, $\chi_{-A}(s) = (-1)^n \chi_A(-s)$.

iii) $\mathrm{mspec}(A^{\mathrm{T}}) = \mathrm{mspec}(A)$.

iv) $\mathrm{mspec}(\overline{A}) = \overline{\mathrm{mspec}(A)}$.

v) $\mathrm{mspec}(A^*) = \overline{\mathrm{mspec}(A)}$.

vi) $0 \in \mathrm{spec}(A)$ if and only if $\det A = 0$.

vii) If $k \in \mathbb{N}$ or if A is nonsingular and $k \in \mathbb{Z}$, then

$$\mathrm{mspec}\left(A^k\right) = \left\{\lambda^k \colon \lambda \in \mathrm{mspec}(A)\right\}_{\mathrm{m}}. \tag{4.4.11}$$

viii) If $\alpha \in \mathbb{F}$, then $\mathrm{mspec}(\alpha I + A) = \alpha + \mathrm{mspec}(A)$.

ix) If $\alpha \in \mathbb{F}$, then $\mathrm{mspec}(\alpha A) = \alpha \mathrm{mspec}(A)$.

x) If A is Hermitian, then $\mathrm{spec}(A) \subset \mathbb{R}$.

xi) If A and B are similar, then $\chi_A = \chi_B$ and $\mathrm{mspec}(A) = \mathrm{mspec}(B)$.

Proof. To prove *i)*, note that

$$\det\left(sI - A^{\mathrm{T}}\right) = \det\,(sI - A)^{\mathrm{T}} = \det(sI - A).$$

To prove ii), note that

$$\chi_{-A}(s) = \det(sI + A) = (-1)^n \det(-sI - A) = (-1)^n \chi_A(-s).$$

Next, iii) follows from i). Next, iv) follows from

$$\det(sI - \overline{A}) = \det(\overline{sI - A}) = \overline{\det(\overline{s}I - A)},$$

while v) follows from iii) and iv).

Next, vi) follows from the fact that $\chi_A(0) = (-1)^n \det A$. To prove "$\supseteq$" in vii), note that, if $\lambda \in \mathrm{spec}(A)$ and $x \in \mathbb{C}^n$ is an eigenvector of A associated with λ (see Section 4.5), then $A^2 x = A(Ax) = A(\lambda x) = \lambda Ax = \lambda^2 x$. Similarly, if A is nonsingular, then $Ax = \lambda x$ implies that $A^{-1}x = \lambda^{-1}x$, and thus $A^{-2}x = \lambda^{-2}x$. Similar arguments apply to arbitrary $k \in \mathbb{Z}$. The reverse inclusion follows from the Jordan decomposition Theorem 5.3.3.

To prove $viii$), note that

$$\chi_{\alpha I + A}(s) = \det[sI - (\alpha I + A)] = \det[(s - \alpha)I - A] = \chi_A(s - \alpha).$$

Statement ix) is true if $\alpha = 0$. If $\alpha \neq 0$, then

$$\chi_{\alpha A}(s) = \det(sI - \alpha A) = \alpha^{-1} \det[(s/\alpha)I - A] = \chi_A(s/\alpha).$$

To prove x), assume that $A = A^*$, let $\lambda \in \mathrm{spec}(A)$, and let $x \in \mathbb{C}^n$ be an eigenvector of A associated with λ. Then, $\lambda = x^*Ax/x^*x$, which is real. Finally, xi) is immediate. $\square$

The following result characterizes the coefficients of χ_A in terms of the eigenvalues of A.

Proposition 4.4.5. Let $A \in \mathbb{F}^{n \times n}$, let $\mathrm{mspec}(A) = \{\lambda_1, \ldots, \lambda_n\}_{\mathrm{m}}$, and, for all $i = 1, \ldots, n$, let γ_i denote the sum of all $i \times i$ principal subdeterminants of A. Then, for all $i = 1, \ldots, n - 1$,

$$\gamma_i = \sum \lambda_{j_1} \cdots \lambda_{j_i}, \tag{4.4.12}$$

where the summation in (4.4.12) is taken over all multisubsets of $\mathrm{mspec}(A)$ having i elements. Furthermore, for all $i = 0, \ldots, n - 1$, the coefficient β_i of s^i in (4.4.2) is given by

$$\beta_i = (-1)^{n-i} \gamma_{n-i}. \tag{4.4.13}$$

In particular,

$$\beta_{n-1} = -\operatorname{tr} A = -\sum_{i=1}^{n} \lambda_i, \tag{4.4.14}$$

$$\beta_{n-2} = \tfrac{1}{2}\left[(\operatorname{tr} A)^2 - \operatorname{tr} A^2\right] = \sum \lambda_{j_1} \lambda_{j_2}, \tag{4.4.15}$$

$$\beta_1 = (-1)^{n-1} \text{tr } A^{\text{A}} = (-1)^{n-1} \sum \lambda_{j_1} \cdots \lambda_{j_{n-1}}, \qquad (4.4.16)$$

$$\beta_0 = (-1)^n \det A = (-1)^n \prod_{i=1}^{n} \lambda_i. \qquad (4.4.17)$$

Proof. The expression for γ_i given by (4.4.12) follows from the factored form of $\chi_A(s)$ given by (4.4.9), while the expression for β_i given by (4.4.13) follows by examining the cofactor expansion (2.7.15) of $\det(sI - A)$. For details, see [535, p. 495]. Equation (4.4.14) follows from (4.4.13) and the fact that the $(n-1) \times (n-1)$ principal subdeterminants of A are the diagonal entries $A_{(i,i)}$. Using

$$\sum_{i=1}^{n} \lambda_i^2 = \left(\sum_{i=1}^{n} \lambda_i \right)^2 - 2 \sum \lambda_{j_1} \lambda_{j_2}$$

and (4.4.14) yields (4.4.15). Next, if A is nonsingular, then $\chi_{A^{-1}}(s) = (-s)^n (\det A^{-1}) \chi_A(1/s)$. Using (4.4.2) with s replaced by $1/s$ and (4.4.14), it follows that $\text{tr } A^{-1} = (-1)^{n-1} (\det A^{-1}) \beta_1$, and, hence, (4.4.16) is satisfied. Using continuity for the case in which A is singular yields (4.4.16) for arbitrary A. Finally, $\beta_0 = \chi_A(0) = \det(0I - A) = (-1)^n \det A$, which verifies (4.4.17). $\qquad \square$

From the definition of the adjugate of a matrix it follows that $(sI - A)^{\text{A}} \in \mathbb{F}^{n \times n}[s]$ is a monic matrix polynomial of degree $n - 1$ of the form

$$(sI - A)^{\text{A}} = s^{n-1} I + s^{n-2} B_{n-2} + \cdots + s B_1 + B_0, \qquad (4.4.18)$$

where $B_0, B_1, \ldots, B_{n-2} \in \mathbb{F}^{n \times n}$. Since $(sI - A)^{\text{A}}$ is regular, it follows from Proposition 4.2.6 that $(sI - A)^{\text{A}}$ is a nonsingular polynomial matrix. The matrix $(sI - A)^{-1}$ is the *resolvent* of A, which is given by

$$(sI - A)^{-1} = \frac{1}{\chi_A(s)} (sI - A)^{\text{A}}. \qquad (4.4.19)$$

Therefore,

$$(sI - A)^{-1} = \frac{s^{n-1}}{\chi_A(s)} I + \frac{s^{n-2}}{\chi_A(s)} B_{n-2} + \cdots + \frac{s}{\chi_A(s)} B_1 + \frac{1}{\chi_A(s)} B_0. \qquad (4.4.20)$$

The next result is the *Cayley-Hamilton theorem*, which shows that every matrix is a "root" of its characteristic polynomial.

Theorem 4.4.6. Let $A \in \mathbb{F}^{n \times n}$. Then,

$$\chi_A(A) = 0. \qquad (4.4.21)$$

Proof. Define $P, Q \in \mathbb{F}^{n \times n}[s]$ by $P(s) \triangleq \chi_A(s) I$ and $Q(s) \triangleq (sI - A)^{\text{A}}$. Then, (4.4.19) implies that $P(s) = Q(s)(sI - A)$. It thus follows from Lemma

4.2.2 that $P_R(A) = 0$. Furthermore, $\chi_A(A) = P(A) = P_R(A)$. Hence, $\chi_A(A) = 0$. $\qquad\qquad\qquad\square$

In the notation of (4.4.10), it follows from Theorem 4.4.6 that

$$\prod_{i=1}^{r}(\lambda_i I - A)^{n_i} = 0. \qquad\qquad (4.4.22)$$

Lemma 4.4.7. Let $A \in \mathbb{F}^{n \times n}$. Then,

$$\frac{\mathrm{d}}{\mathrm{d}s}\chi_A(s) = \mathrm{tr}\big[(sI - A)^{\mathrm{A}}\big] = \sum_{i=1}^{n}\det\big(sI - A_{[i,i]}\big). \qquad (4.4.23)$$

Proof. It follows from (4.4.16) that $\frac{\mathrm{d}}{\mathrm{d}s}\chi_A(s)\big|_{s=0} = \beta_1 = (-1)^{n-1}\mathrm{tr}\,A^{\mathrm{A}}$. Hence,

$$\frac{\mathrm{d}}{\mathrm{d}s}\chi_A(s) = \frac{\mathrm{d}}{\mathrm{d}z}\det[(s+z)I - A]\bigg|_{z=0} = \frac{\mathrm{d}}{\mathrm{d}z}\det[zI - (-sI + A)]\bigg|_{z=0}$$

$$= (-1)^{n-1}\mathrm{tr}\big[(-sI + A)^{\mathrm{A}}\big] = \mathrm{tr}\big[(sI - A)^{\mathrm{A}}\big]. \qquad\square$$

The following result, known as *Leverrier's algorithm*, provides a recursive formula for the coefficients $\beta_0, \dots, \beta_{n-1}$ of χ_A and $B_0, \dots, B_{n-2}$ of $(sI - A)^{\mathrm{A}}$.

Proposition 4.4.8. Let $A \in \mathbb{F}^{n \times n}$, let χ_A be given by (4.4.2), and let $(sI - A)^{\mathrm{A}}$ be given by (4.4.18). Then, $\beta_{n-1}, \dots, \beta_0$ and $B_{n-2}, \dots, B_0$ are given by

$$\beta_k = \tfrac{1}{k-n}\mathrm{tr}\,AB_k, \quad k = n-1, \dots, 0, \qquad (4.4.24)$$

$$B_{k-1} = AB_k + \beta_k I, \quad k = n-1, \dots, 1, \qquad (4.4.25)$$

where $B_{n-1} = I$.

Proof. Since $(sI - A)(sI - A)^{\mathrm{A}} = \chi_A(s)I$, it follows that

$$s^n I + s^{n-1}(B_{n-2} - A) + s^{n-2}(B_{n-3} - AB_{n-2}) + \cdots + s(B_0 - AB_1) - AB_0$$
$$= (s^n + \beta_{n-1}s^{n-1} + \cdots + \beta_1 s + \beta_0)I.$$

Equating coefficients of powers of s yields (4.4.25) along with $-AB_0 = \beta_0 I$. Taking the trace of this last identity yields $\beta_0 = -\frac{1}{n}\mathrm{tr}\,AB_0$, which confirms (4.4.24) for $k = 0$. Next, using (4.4.23) and (4.4.18), it follows that

$$\frac{\mathrm{d}}{\mathrm{d}s}\chi_A(s) = \sum_{k=1}^{n}k\beta_k s^{k-1} = \sum_{k=1}^{n}(\mathrm{tr}\,B_{k-1})s^{k-1},$$

where $B_{n-1} \triangleq I_n$ and $\beta_n \triangleq 1$. Equating powers of s, it follows that $k\beta_k =$

$\text{tr}\, B_{k-1}$ for all $k = 1, \ldots, n$. Now, (4.4.25) implies that $k\beta_k = \text{tr}(AB_k + \beta_k I)$ for all $k = 1, \ldots, n-1$, which implies (4.4.24). $\qquad\square$

Proposition 4.4.9. Let $A \in \mathbb{F}^{n \times m}$ and $B \in \mathbb{F}^{m \times n}$, and assume that $m \leq n$. Then,

$$\chi_{AB}(s) = s^{n-m}\chi_{BA}(s). \tag{4.4.26}$$

Consequently,

$$\text{mspec}(AB) = \text{mspec}(BA) \cup \{0, \ldots, 0\}_{\text{m}}, \tag{4.4.27}$$

where the multiset $\{0, \ldots, 0\}_{\text{m}}$ contains $n - m$ zeros.

Proof. First note that

$$\begin{bmatrix} 0_{m \times m} & 0_{m \times n} \\ A & AB \end{bmatrix} = \begin{bmatrix} I_m & -B \\ 0_{n \times m} & I_n \end{bmatrix} \begin{bmatrix} BA & 0_{m \times n} \\ A & 0_{n \times n} \end{bmatrix} \begin{bmatrix} I_m & B \\ 0_{n \times m} & I_n \end{bmatrix},$$

which shows that $\begin{bmatrix} 0_{m \times m} & 0_{m \times n} \\ A & AB \end{bmatrix}$ and $\begin{bmatrix} BA & 0_{m \times n} \\ A & 0_{n \times n} \end{bmatrix}$ are similar. It thus follows from $xi)$ of Proposition 4.4.4 that $s^m\chi_{AB}(s) = s^n\chi_{BA}(s)$, which implies (4.4.26). Finally, (4.4.27) follows immediately from (4.4.26). $\qquad\square$

If $n = m$, then Proposition 4.4.9 specializes to the following result.

Corollary 4.4.10. Let $A, B \in \mathbb{F}^{n \times n}$. Then,

$$\chi_{AB} = \chi_{BA}. \tag{4.4.28}$$

Consequently,

$$\text{mspec}(AB) = \text{mspec}(BA). \tag{4.4.29}$$

4.5 Eigenvectors

Let $A \in \mathbb{F}^{n \times n}$, and let $\lambda \in \mathbb{C}$ be an eigenvalue of A. Then, $\chi_A(\lambda) = \det(\lambda I - A) = 0$, and thus $\lambda I - A \in \mathbb{C}^{n \times n}$ is singular. Furthermore, $\mathcal{N}(\lambda I - A)$ is a nontrivial subspace of $\mathbb{C}^n$, that is, $\text{def}(\lambda I - A) > 0$. If $x \in \mathcal{N}(\lambda I - A)$, that is, $Ax = \lambda x$, and $x \neq 0$, then x is an *eigenvector of A associated with λ.* By definition, all eigenvectors are nonzero. Note that, if A and λ are real, then there exists a real eigenvector associated with λ.

Definition 4.5.1. The *geometric multiplicity* of $\lambda \in \text{spec}(A)$, denoted by $\text{gm}_A(\lambda)$, is the number of linearly independent eigenvectors associated with λ, that is,

$$\text{gm}_A(\lambda) \triangleq \text{def}(\lambda I - A). \tag{4.5.1}$$

By convention, if $\lambda \notin \text{spec}(A)$, then $\text{gm}_A(\lambda) \triangleq 0$.

The spectral properties of normal matrices deserve special attention.

Lemma 4.5.2. Let $A \in \mathbb{F}^{n \times n}$ be normal, let $\lambda \in \text{spec}(A)$, and let $x \in \mathbb{C}^n$ be an eigenvector of A associated with λ. Then, x is an eigenvector of A^* associated with $\overline{\lambda} \in \text{spec}(A^*)$.

Proof. Since $\lambda \in \text{spec}(A)$, statement $v)$ of Proposition 4.4.4 implies that $\overline{\lambda} \in \text{spec}(A^*)$. Next, since x and λ satisfy $Ax = \lambda x$, $x^*A^* = \overline{\lambda}x^*$, and $AA^* = A^*A$, it follows that

$$
\begin{aligned}
(A^*x - \overline{\lambda}x)^*(A^*x - \overline{\lambda}x) &= x^*AA^*x - \overline{\lambda}x^*Ax - \lambda x^*A^*x + \lambda\overline{\lambda}x^*x \\
&= x^*A^*Ax - \lambda\overline{\lambda}x^*x - \lambda\overline{\lambda}x^*x + \lambda\overline{\lambda}x^*x \\
&= \lambda\overline{\lambda}x^*x - \lambda\overline{\lambda}x^*x = 0.
\end{aligned}
$$

Hence, $A^*x = \overline{\lambda}x$.   □

Proposition 4.5.3. Let $A \in \mathbb{F}^{n \times n}$. Then, eigenvectors associated with distinct eigenvalues of A are linearly independent. If, in addition, A is normal, then these eigenvectors are mutually orthogonal.

Proof. Let $\lambda_1, \lambda_2 \in \text{spec}(A)$ be distinct with associated eigenvectors $x_1, x_2 \in \mathbb{C}^n$. Suppose that x_1 and x_2 are linearly dependent, that is, $x_1 = \alpha x_2$, where $\alpha \in \mathbb{C}$ and $\alpha \neq 0$. Then, $Ax_1 = \lambda_1 x_1 = \lambda_1\alpha x_2$, while also $Ax_1 = A\alpha x_2 = \alpha\lambda_2 x_2$. Hence, $\alpha(\lambda_1 - \lambda_2)x_2 = 0$, which contradicts $\alpha \neq 0$. Since pairwise linear independence does not imply the linear independence of larger sets, next, let $\lambda_1, \lambda_2, \lambda_3 \in \text{spec}(A)$ be distinct with associated eigenvectors $x_1, x_2, x_3 \in \mathbb{C}^n$. Suppose that x_1, x_2, x_3 are linearly dependent. In this case, there exist $a_1, a_2, a_3 \in \mathbb{C}$, not all zero, such that $a_1 x_1 + a_2 x_2 + a_3 x_3 = 0$. If $a_1 = 0$, then $a_2 x_2 + a_3 x_3 = 0$. However, $\lambda_2 \neq \lambda_3$ implies that x_2 and x_3 are linearly independent, which in turn implies that $a_2 = 0$ and $a_3 = 0$. Since a_1, a_2, a_3 are not all zero, it follows that $a_1 \neq 0$. Therefore, $x_1 = \alpha x_2 + \beta x_3$, where $\alpha \triangleq -a_2/a_1$ and $\beta \triangleq -a_3/a_1$ are not both zero. Thus, $Ax_1 = A(\alpha x_2 + \beta x_3) = \alpha Ax_2 + \beta Ax_3 = \alpha\lambda_2 x_2 + \beta\lambda_3 x_3$. However, $Ax_1 = \lambda_1 x_1 = \lambda_1(\alpha x_2 + \beta x_3) = \alpha\lambda_1 x_2 + \beta\lambda_1 x_3$. Subtracting these relations yields $0 = \alpha(\lambda_1 - \lambda_2)x_2 + \beta(\lambda_1 - \lambda_3)x_3$. Since x_2 and x_3 are linearly independent, it follows that $\alpha(\lambda_1 - \lambda_2) = 0$ and $\beta(\lambda_1 - \lambda_3) = 0$. Since α and β are not both zero, it follows that $\lambda_1 = \lambda_2$ or $\lambda_1 = \lambda_3$, which contradicts the assumption that $\lambda_1, \lambda_2, \lambda_3$ are distinct. The same arguments apply to sets of four or more eigenvectors.

Now, suppose that A is normal, and let $\lambda_1, \lambda_2 \in \text{spec}(A)$ be distinct eigenvalues with associated eigenvectors $x_1, x_2 \in \mathbb{C}^n$. Then, by Lemma 4.5.2, $Ax_1 = \lambda_1 x_1$ implies that $A^*x_1 = \overline{\lambda}_1 x_1$. Consequently, $x_1^*A = \lambda_1 x_1^*$, which implies that $x_1^*Ax_2 = \lambda_1 x_1^*x_2$. Furthermore, $x_1^*Ax_2 = \lambda_2 x_1^*x_2$. It thus follows that $0 = (\lambda_1 - \lambda_2)x_1^*x_2$. Hence, $\lambda_1 \neq \lambda_2$ implies that $x_1^*x_2 = 0$.   □

If $A \in \mathbb{R}^{n \times n}$ is symmetric, then Lemma 4.5.2 is not needed and the proof of Proposition 4.5.3 is simpler. In this case, it follows from x) of Proposition 4.4.4 that $\lambda_1, \lambda_2 \in \mathrm{spec}(A)$ are real, and thus associated eigenvectors $x_1 \in \mathcal{N}(\lambda_1 I - A)$ and $x_2 \in \mathcal{N}(\lambda_1 I - A)$ can be chosen to be real. Hence, $Ax_1 = \lambda_1 x_1$ and $Ax_2 = \lambda_2 x_2$ imply that $x_2^{\mathrm{T}} A x_1 = \lambda_1 x_2^{\mathrm{T}} x_1$ and $x_1^{\mathrm{T}} A x_2 = \lambda_2 x_1^{\mathrm{T}} x_2$. Since $x_1^{\mathrm{T}} A x_2 = x_2^{\mathrm{T}} A^{\mathrm{T}} x_1 = x_2^{\mathrm{T}} A x_1$ and $x_1^{\mathrm{T}} x_2 = x_2^{\mathrm{T}} x_1$, it follows that $(\lambda_1 - \lambda_2) x_1^{\mathrm{T}} x_2 = 0$. Since $\lambda_1 \neq \lambda_2$, it follows that $x_1^{\mathrm{T}} x_2 = 0$.

We define the *spectral abscissa* of $A \in \mathbb{F}^{n \times n}$ by

$$\mathrm{spabs}(A) \triangleq \max\{\mathrm{Re}\,\lambda \colon\ \lambda \in \mathrm{spec}(A)\} \tag{4.5.2}$$

and the *spectral radius* of $A \in \mathbb{F}^{n \times n}$ by

$$\mathrm{sprad}(A) \triangleq \max\{|\lambda| \colon\ \lambda \in \mathrm{spec}(A)\}. \tag{4.5.3}$$

Let $A \in \mathbb{F}^{n \times n}$. Then, $\nu_-(A), \nu_0(A)$, and $\nu_+(A)$ denote the number of eigenvalues of A counting algebraic multiplicity having, respectively, negative, zero, and positive real part. Define the *inertia* of A by

$$\mathrm{In}\,A \triangleq \begin{bmatrix} \nu_-(A) \\ \nu_0(A) \\ \nu_+(A) \end{bmatrix} \tag{4.5.4}$$

and the *signature* of A by

$$\mathrm{sig}\,A \triangleq \nu_+(A) - \nu_-(A). \tag{4.5.5}$$

Note that $\mathrm{spabs}(A) < 0$ if and only if $\nu_-(A) = n$.

4.6 Minimal Polynomial

Theorem 4.4.6 showed that every square matrix $A \in \mathbb{F}^{n \times n}$ is a root of its characteristic polynomial. However, there may be polynomials of degree less than n having A as a root. In fact, the following result shows that there exists a unique monic polynomial that has A as a root and that divides all polynomials that have A as a root.

Theorem 4.6.1. Let $A \in \mathbb{F}^{n \times n}$. Then, there exists a unique monic polynomial $\mu_A \in \mathbb{F}[s]$ of minimal degree such that $\mu_A(A) = 0$. Furthermore, $\deg \mu_A \leq n$, and μ_A divides every polynomial $p \in \mathbb{F}[s]$ satisfying $p(A) = 0$.

Proof. Since $\chi_A(A) = 0$ and $\deg \chi_A = n$, it follows that there exists a minimal positive integer $n_0 \leq n$ such that there exists a monic polynomial $p_0 \in \mathbb{F}[s]$ satisfying $p_0(A) = 0$ and $\deg p_0 = n_0$. Let $p \in \mathbb{F}[s]$ satisfy $p(A) = 0$. Then, by Lemma 4.1.2, there exist polynomials $q, r \in \mathbb{F}[s]$ such that $p = qp_0 + r$ and $\deg r < \deg p_0$. However, $p(A) = p_0(A) = 0$ implies that $r(A) = 0$. If $r \neq 0$, then r can be normalized to obtain a monic polynomial of degree less than n_0, which contradicts the definition n_0. Hence, $r = 0$, which implies that p_0 divides p. This proves existence.

Now, suppose there exist two monic polynomials $p_0, \hat{p}_0 \in \mathbb{F}[s]$ of degree n_0 and such that $p_0(A) = \hat{p}_0(A) = 0$. By the previous argument, p_0 divides $\hat{p}_0$, and vice versa. Therefore, p_0 is a constant multiple of $\hat{p}_0$. Since p_0 and $\hat{p}_0$ are both monic, it follows that $p_0 = \hat{p}_0$. This proves uniqueness. Denote this polynomial by μ_A. $\square$

The monic polynomial μ_A of least order having A as a root is the *minimal polynomial* of A.

The following result relates the characteristic and minimal polynomials of $A \in \mathbb{F}^{n \times n}$ to the similarity invariants of A. Note that $\text{rank}(sI - A) = n$, so that A has n similarity invariants $p_1, \ldots, p_n \in \mathbb{F}[s]$. In this case, (4.3.1) becomes

$$sI - A = S_1(s) \begin{bmatrix} p_1(s) & & 0 \\ & \ddots & \\ 0 & & p_n(s) \end{bmatrix} S_2(s), \qquad (4.6.1)$$

where $S_1, S_2 \in \mathbb{F}^{n \times n}[s]$ are unimodular and p_i divides p_{i+1} for all $i = 1, \ldots, n - 1$.

Proposition 4.6.2. Let $A \in \mathbb{F}^{n \times n}$, and let $p_1, \ldots, p_n \in \mathbb{F}[s]$ be the similarity invariants of A, where p_i divides p_{i+1} for all $i = 1, \ldots, n - 1$. Then,

$$\chi_A = \prod_{i=1}^{n} p_i \qquad (4.6.2)$$

and

$$\mu_A = p_n. \qquad (4.6.3)$$

Proof. Using Theorem 4.3.2 and (4.6.1), it follows that

$$\chi_A(s) = \det(sI - A) = [\det S_1(s)] [\det S_2(s)] \prod_{i=1}^{n} p_i(s).$$

Since S_1 and S_2 are unimodular and χ_A and $p_1, \ldots, p_n$ are monic, it follows that $[\det S_1(s)][\det S_2(s)] = 1$, which proves (4.6.2).

To prove (4.6.3), first note that it follows from Theorem 4.3.2 that $\chi_A = \Delta_{n-1} p_n$, where $\Delta_{n-1} \in \mathbb{F}[s]$ is the greatest common divisor of all $(n - 1) \times (n - 1)$ subdeterminants of $sI - A$. Since the $(n - 1) \times (n - 1)$ subdeterminants of $sI - A$ are the entries of $\pm(sI - A)^A$, it follows that Δ_{n-1} divides every entry of $(sI - A)^A$. Hence, there exists a polynomial matrix $P \in \mathbb{F}^{n \times n}[s]$ such that $(sI - A)^A = \Delta_{n-1}(s)P(s)$. Furthermore, since $(sI - A)^A(sI - A) = \chi_A(s)I$, it follows that $\Delta_{n-1}(s)P(s)(sI - A) = \chi_A(s)I =$

$\Delta_{n-1}(s)p_n(s)I$, and thus $P(s)(sI - A) = p_n(s)I$. Lemma 4.2.2 now implies that $p_n(A) = 0$.

Since $p_n(A) = 0$, it follows from Theorem 4.6.1 that μ_A divides p_n. Hence, let $q \in \mathbb{F}[s]$ be the monic polynomial satisfying $p_n = q\mu_A$. Furthermore, since $\mu_A(A) = 0$, it follows from Corollary 4.2.3 that there exists a polynomial matrix $Q \in \mathbb{F}^{n \times n}[s]$ such that $\mu_A(s)I = Q(s)(sI - A)$. Thus, $P(s)(sI - A) = p_n(s)I = q(s)\mu_A(s)I = q(s)Q(s)(sI - A)$, which implies that $P = qQ$. Thus, q divides every entry of P. However, since P was obtained by dividing $(sI - A)^A$ by the greatest common divisor of all of its entries, it follows that the greatest common divisor of the entries of P is 1. Hence, $q = 1$, which implies that $p_n = \mu_A$, which proves (4.6.3). $\square$

Proposition 4.6.2 shows that μ_A divides χ_A, which is also a consequence of Theorem 4.4.6 and Theorem 4.6.1. Proposition 4.6.2 also shows that $\mu_A = \chi_A$ if and only if $p_1 = \cdots = p_{n-1} = 1$, that is, if and only if $p_n = \chi_A$ is the only nonconstant similarity invariant of A. Note that, in general, it follows from (4.6.2) that $\sum_{i=1}^{n} \deg p_i = n$.

Finally, note that the similarity invariants of the $n \times n$ identity matrix I_n are given by $p_i(s) = s - 1$ for all $i = 1, \ldots, n$. Thus, $\chi_{I_n}(s) = (s-1)^n$ and $\mu_{I_n}(s) = s - 1$.

Proposition 4.6.3. Let $A \in \mathbb{F}^{n \times n}$, and assume that A and B are similar. Then,

$$\mu_A = \mu_B. \tag{4.6.4}$$

4.7 Rational Transfer Functions and the Smith-McMillan Decomposition

We now turn our attention to rational functions.

Definition 4.7.1. The set $\mathbb{F}(s)$ of *rational functions* consists of functions $g\colon \mathbb{C}\backslash\mathcal{S} \mapsto \mathbb{C}$, where $g(s) = p(s)/q(s)$, $p, q \in \mathbb{F}[s]$, $q \neq 0$, and $\mathcal{S} \triangleq \text{roots}(q)$. The rational function g is *strictly proper, proper, exactly proper, improper,* respectively, if $\deg p < \deg q$, $\deg p \leq \deg q$, $\deg p = \deg q$, $\deg p > \deg q$. If p and q are coprime, then the roots of p are the *zeros* of g, while the roots of q are the *poles* of g. The set of proper rational functions is denoted by $\mathbb{F}_{\text{prop}}(s)$. The *relative degree* of $g \in \mathbb{F}_{\text{prop}}(s)$, denoted by reldeg g, is $\deg q - \deg p$.

Definition 4.7.2. The set $\mathbb{F}^{n \times m}(s)$ of *rational transfer functions* consists of matrices whose entries are elements of $\mathbb{F}(s)$. The rational transfer function $G \in \mathbb{F}^{n \times m}(s)$ is *strictly proper* if every entry of G is strictly proper, *proper* if every entry of G is proper, *exactly proper* if every entry of G is proper and at least one entry of G is exactly proper, and *improper* if at least one entry of G is improper. The set of proper rational transfer functions is denoted by $\mathbb{F}^{n \times m}_{\mathrm{prop}}(s)$.

Definition 4.7.3. Let $G \in \mathbb{F}^{n \times m}_{\mathrm{prop}}(s)$. Then, the *relative degree* of G, denoted by $\mathrm{reldeg}\, G$, is defined by

$$\mathrm{reldeg}\, G \triangleq \min_{\substack{i=1,\ldots,n \\ j=1,\ldots,m}} \mathrm{reldeg}\, G_{(i,j)}. \tag{4.7.1}$$

By writing $(sI - A)^{-1}$ as

$$(sI - A)^{-1} = \frac{1}{\chi_A(s)}(sI - A)^{\mathrm{A}}, \tag{4.7.2}$$

it follows from (4.4.18) that $(sI - A)^{-1}$ is a strictly proper rational transfer function. In fact, for all $i = 1, \ldots, n$,

$$\mathrm{reldeg}\left[(sI - A)^{-1}\right]_{(i,i)} = 1, \tag{4.7.3}$$

and thus

$$\mathrm{reldeg}\,(sI - A)^{-1} = 1. \tag{4.7.4}$$

The following result provides a canonical form, known as the *Smith-McMillan form*, for rational transfer functions under unimodular transformation. The following definition is an extension of Definition 4.2.4 for matrix polynomials.

Definition 4.7.4. Let $G \in \mathbb{F}^{n \times m}(s)$, and, for all $i = 1, \ldots, n$ and $j = 1, \ldots, m$, let $G_{(i,j)} = p_{ij}/q_{ij}$, where $p_{ij}, q_{ij} \in \mathbb{F}[s]$ and $q_{ij} \neq 0$, and define $\mathcal{S} \triangleq \cup_{i,j=1}^{n,m} \mathrm{roots}(q_{ij})$. Then, the *rank* of G is the nonnegative integer

$$\mathrm{rank}\, G \triangleq \max_{s \in \mathbb{C} \backslash \mathcal{S}} \mathrm{rank}\, G(s). \tag{4.7.5}$$

Theorem 4.7.5. Let $G \in \mathbb{F}^{n \times m}(s)$, and let $r \triangleq \mathrm{rank}\, G$. Then, there exist unimodular matrices $S_1 \in \mathbb{F}^{n \times n}[s]$ and $S_2 \in \mathbb{F}^{m \times m}[s]$ and monic polynomials $p_1, \ldots, p_r, q_1, \ldots, q_r \in \mathbb{F}[s]$ such that p_i and q_i are coprime for all $i = 1, \ldots, r$, p_i divides p_{i+1} for all $i = 1, \ldots, r-1$, q_{i+1} divides q_i for all $i = 1, \ldots, r-1$, and

$$G = S_1 \begin{bmatrix} p_1/q_1 & & & & 0 \\ & \ddots & & & \\ & & p_r/q_r & & \\ 0 & & & 0_{(n-r) \times (m-r)} \end{bmatrix} S_2. \tag{4.7.6}$$

Proof. Let n_{ij}/d_{ij} denote the (i,j) entry of G, where $n_{ij}, d_{ij} \in \mathbb{F}[s]$ are coprime, and let $d \in \mathbb{F}[s]$ denote the least common multiple of d_{ij} for all $i = 1, \ldots, n$ and $j = 1, \ldots, m$. From Theorem 4.3.2 it follows that the polynomial matrix dG has the Smith form $\mathrm{diag}(\hat{p}_1, \ldots, \hat{p}_r, 0, \ldots, 0)$, where $\hat{p}_1, \ldots, \hat{p}_r \in \mathbb{F}[s]$ and $\hat{p}_i$ divides $\hat{p}_{i+1}$ for all $i = 1, \ldots, r-1$. Now, divide this Smith form by d and express every rational function $\hat{p}_i/d$ in coprime form p_i/q_i so that p_i divides p_{i+1} for all $i = 1, \ldots, r-1$ and q_{i+1} divides q_i for all $i = 1, \ldots, r-1$. $\qquad\square$

Let $g_1, \ldots, g_r \in \mathbb{F}^n(s)$. Then, $g_1, \ldots, g_r$ are *linearly independent* if $\alpha_1, \ldots, \alpha_r \in \mathbb{F}[s]$ and $\sum_{n=1}^r \alpha_i g_i = 0$ imply that $\alpha_1 = \cdots = \alpha_r = 0$. It can be seen that this definition is unchanged if $\alpha_1, \ldots, \alpha_r \in \mathbb{F}(s)$.

Proposition 4.7.6. Let $G \in \mathbb{F}^{n \times m}(s)$. Then, $\mathrm{rank}\, G$ is equal to the number of linearly independent columns of G.

As a special case, Proposition 4.7.6 applies to polynomial matrices $G \in \mathbb{F}^{n \times m}[s]$.

Definition 4.7.7. Let $G \in \mathbb{F}^{n \times m}(s)$, let $r \triangleq \mathrm{rank}\, G$, and let $p_1, \ldots, p_r$, $q_1, \ldots, q_r \in \mathbb{F}[s]$ be given by Theorem 4.7.5. Then, the *McMillan degree* of G is $\sum_{i=1}^r \deg q_i$. Furthermore, the *poles* of G are the roots of q_1, the *transmission zeros* of G are the roots of p_r, and the *blocking zeros* of G are the roots of p_1.

4.8 Facts on Polynomials

Fact 4.8.1. Let $p \in \mathbb{R}[s]$ be monic, and define $q(s) \triangleq s^n p(1/s)$, where $n \triangleq \deg p$. If $0 \notin \mathrm{roots}(p)$, then $\deg(q) = n$ and

$$\mathrm{mroots}(q) = \{1/\lambda \colon\ \lambda \in \mathrm{mroots}(p)\}_{\mathrm{m}}.$$

If $0 \in \mathrm{roots}(p)$ with multiplicity r, then $\deg(q) = n - r$ and

$$\mathrm{mroots}(q) = \{1/\lambda \colon\ \lambda \neq 0 \text{ and } \lambda \in \mathrm{mroots}(p)\}_{\mathrm{m}}.$$

(Remark: See Fact 11.15.3 and Fact 11.15.4.)

Fact 4.8.2. Let $p \in \mathbb{F}^n$ be given by

$$p(s) = s^n + \beta_{n-1} s^{n-1} + \cdots + \beta_1 s + \beta_0,$$

let $\beta_n \triangleq 1$, let $\mathrm{mroots}(p) = \{\lambda_1, \ldots, \lambda_n\}_{\mathrm{m}}$, and define $\mu_1, \ldots, \mu_n$ by

$$\mu_i \triangleq \lambda_1^i + \cdots + \lambda_n^i.$$

Then, for all $k = 1, \ldots, n$,

$$k\beta_{n-k} + \mu_1 \beta_{n-k+1} + \mu_2 \beta_{n-k+2} + \cdots + \mu_k \beta_n = 0.$$

That is,

$$
\begin{bmatrix}
n & \mu_1 & \mu_2 & \mu_3 & \mu_4 & \cdots & \mu_n \\
0 & n-1 & \mu_1 & \mu_2 & \mu_3 & \cdots & \mu_{n-1} \\
\vdots & \ddots & \ddots & \ddots & \ddots & \ddots & \vdots \\
\vdots & \ddots & \ddots & \ddots & \ddots & \ddots & \vdots \\
0 & 0 & \cdots & 0 & 2 & \mu_1 & \mu_2 \\
0 & 0 & \cdots & 0 & 0 & 1 & \mu_1
\end{bmatrix}
\begin{bmatrix}
\beta_0 \\ \beta_1 \\ \vdots \\ \beta_{n-1} \\ \beta_n
\end{bmatrix}
= 0.
$$

Consequently, $\beta_1, \ldots, \beta_{n-1}$ are uniquely determined by $\mu_1, \ldots, \mu_n$. In particular,

$$\beta_{n-1} = -\mu_1$$

and

$$\beta_{n-2} = \tfrac{1}{2}\big[\mu_1^2 - \mu_2\big].$$

(Proof: See [367, p. 44] and [538, p. 9].) (Remark: These equations are *Newton's identities*.)

Fact 4.8.3. Let $p, q \in \mathbb{F}[s]$ be monic. Then, p and q are coprime if and only if their least common multiple is pq.

Fact 4.8.4. Let $p, q \in \mathbb{F}[s]$, where $p(s) = a_n s^n + \cdots + a_1 s + a_0$, $q(s) = b_m s^m + \cdots + b_1 s + b_0$, $\deg p = n$, and $\deg q = m$. Furthermore, define the Toeplitz matrices $[p]^{(m)} \in \mathbb{F}^{m \times (n+m)}$ and $[q]^{(n)} \in \mathbb{F}^{n \times (n+m)}$ by

$$
[p]^{(m)} \triangleq
\begin{bmatrix}
a_n & a_{n-1} & \cdots & a_1 & a_0 & 0 & 0 & \cdots & 0 \\
0 & a_n & a_{n-1} & \cdots & a_1 & a_0 & 0 & \cdots & 0 \\
\vdots & \ddots & \ddots & \ddots & \cdots & \ddots & \ddots & \ddots & \vdots
\end{bmatrix}
$$

and

$$
[q]^{(n)} \triangleq
\begin{bmatrix}
b_m & b_{m-1} & \cdots & b_1 & b_0 & 0 & 0 & \cdots & 0 \\
0 & b_m & b_{m-1} & \cdots & b_1 & b_0 & 0 & \cdots & 0 \\
\vdots & \ddots & \ddots & \ddots & \cdots & \ddots & \ddots & \ddots & \vdots
\end{bmatrix}.
$$

Then, p and q are coprime if and only if

$$
\det \begin{bmatrix} [p]^{(m)} \\ [q]^{(n)} \end{bmatrix} \neq 0.
$$

(Proof: See [260, p. 162] or [592, pp. 187–191].) (Remark: $\begin{bmatrix} A \\ B \end{bmatrix}$ is the *Sylvester matrix*, and $\det \begin{bmatrix} A \\ B \end{bmatrix}$ is the *resultant* of p and q.) (Remark: The form $\begin{bmatrix} [p]^{(m)} \\ [q]^{(n)} \end{bmatrix}$ appears in [592, pp. 187–191]. The result is given in [260, p. 162] in terms of $\begin{bmatrix} \hat{I}[p]^{(m)} \\ \hat{I}[q]^{(n)} \end{bmatrix} \hat{I}$ and in [820, p. 85] in terms of $\begin{bmatrix} [p]^{(m)} \\ \hat{I}[q]^{(n)} \end{bmatrix}$.)

Fact 4.8.5. Let $p_1, \ldots, p_n \in \mathbb{F}[s]$, and let $d \in \mathbb{F}[s]$ be the greatest common divisor of $p_1, \ldots, p_n$. Then, there exist polynomials $q_1, \ldots, q_n \in \mathbb{F}[s]$ such that

$$d = \sum_{i=1}^{n} q_i p_i.$$

In addition, $p_1, \ldots, p_n$ are coprime if and only if there exist polynomials $q_1, \ldots, q_n \in \mathbb{F}[s]$ such that

$$1 = \sum_{i=1}^{n} q_i p_i.$$

(Proof: See [275, p. 16].) (Remark: The polynomial d is given by the *Bezout equation*.)

Fact 4.8.6. Let $p, q \in \mathbb{F}[s]$, where $p(s) = a_n s^n + \cdots + a_1 s + a_0$ and $q(s) = b_n s^n + \cdots + b_1 s + b_0$, and define $[p]^{(n)}, [q]^{(n)} \in \mathbb{F}^{n \times 2n}$ as in Fact 4.8.4. Furthermore, define

$$R(p, q) \triangleq \begin{bmatrix} [p]^{(n)} \\ [q]^{(n)} \end{bmatrix} = \begin{bmatrix} A_1 & A_2 \\ B_1 & B_2 \end{bmatrix},$$

where $A_1, A_2, B_1, B_2 \in \mathbb{F}^{n \times n}$, and define $\hat{p}(s) \triangleq s^n p(-s)$ and $\hat{q}(s) \triangleq s^n q(-s)$. Then,

$$\begin{bmatrix} A_1 & A_2 \\ B_1 & B_2 \end{bmatrix} = \begin{bmatrix} \hat{p}(N_n^{\mathrm{T}}) & p(N_n) \\ \hat{q}(N_n^{\mathrm{T}}) & q(N_n) \end{bmatrix},$$

$$A_1 B_1 = B_1 A_1,$$

$$A_2 B_2 = B_2 A_2,$$

$$A_1 B_2 + A_2 B_1 = B_1 A_2 + B_2 A_1.$$

Therefore,

$$\begin{bmatrix} I & 0 \\ -B_1 & A_1 \end{bmatrix} \begin{bmatrix} A_1 & A_2 \\ B_1 & B_2 \end{bmatrix} = \begin{bmatrix} A_1 & A_2 \\ 0 & A_1 B_2 - B_1 A_2 \end{bmatrix},$$

$$\begin{bmatrix} -B_2 & A_2 \\ 0 & I \end{bmatrix} \begin{bmatrix} A_1 & A_2 \\ B_1 & B_2 \end{bmatrix} = \begin{bmatrix} A_2 B_1 - B_2 A_1 & 0 \\ B_1 & B_2 \end{bmatrix},$$

and

$$\det R(p, q) = \det(A_1 B_2 - B_1 A_2) = \det(B_2 A_1 - A_2 B_1).$$

Now, define $B(p, q) \in \mathbb{F}^{n \times n}$ by

$$B(p, q) \triangleq (A_1 B_2 - B_1 A_2) \hat{I}.$$

Then, the following statements hold:

i) For all $s, \hat{s} \in \mathbb{C}$,

$$p(s)q(\hat{s}) - q(s)p(\hat{s}) = (s - \hat{s})\begin{bmatrix} 1 \\ s \\ \vdots \\ s^{n-1} \end{bmatrix}^{\mathrm{T}} B(p,q) \begin{bmatrix} 1 \\ \hat{s} \\ \vdots \\ \hat{s}^{n-1} \end{bmatrix}.$$

ii) $B(p,q) = (B_2 A_1 - A_2 B_1)\hat{I} = \hat{I}(A_1^{\mathrm{T}} B_2^{\mathrm{T}} - B_1^{\mathrm{T}} A_2^{\mathrm{T}}) = \hat{I}(B_1^{\mathrm{T}} A_2^{\mathrm{T}} - A_1^{\mathrm{T}} B_2^{\mathrm{T}}).$

iii) $\begin{bmatrix} 0 & B(p,q) \\ -B(p,q) & 0 \end{bmatrix} = QR^{\mathrm{T}}(p,q)QR(p,q)Q$, where $Q \triangleq \begin{bmatrix} 0 & \hat{I} \\ -\hat{I} & 0 \end{bmatrix}.$

iv) $|\det B(p,q)| = |\det R(p,q)| = |\det q[C(p)]|.$

v) $B(p,q)$ and $\hat{B}(p,q)$ are symmetric.

vi) $B(p,q)$ is a linear function of (p,q).

vii) $B(p,q) = -B(q,p).$

Now, assume that $\deg q \le \deg p = n$ and p is monic. Then, the following statements hold:

viii) $\operatorname{def} B(p,q)$ is equal to the degree of the greatest common divisor of p and q.

ix) p and q are coprime if and only if $B(p,q)$ is nonsingular.

x) If $B(p,q)$ is nonsingular, then $[B(p,q)]^{-1}$ is Hankel. In fact,

$$[B(p,q)]^{-1} = H(a/p),$$

where $a, b \in \mathbb{F}[s]$ satisfy the Bezout equation $aq + bp = 1.$

xi) If $q = q_1 q_2$, where $q_1, q_2 \in \mathbb{F}[s]$, then

$$B(p,q) = B(p,q_1)q_2[C(p)] = q_1[C^{\mathrm{T}}(p)]B(p,q_2).$$

xii) $B(p,q) = B(p,q)C(p) = C^{\mathrm{T}}(p)B(p,q).$

xiii) $B(p,q) = B(p,1)q[C(p)] = q[C^{\mathrm{T}}(p)]B(p,1)$, where $B(p,1)$ is the Hankel matrix

$$B(p,1) = \begin{bmatrix} a_1 & a_2 & \cdots & a_{n-1} & 1 \\ a_2 & a_3 & \cdot^{\cdot^{\cdot}} & 1 & 0 \\ \vdots & \cdot^{\cdot^{\cdot}} & \cdot^{\cdot^{\cdot}} & \cdot^{\cdot^{\cdot}} & \vdots \\ a_{n-1} & 1 & \cdot^{\cdot^{\cdot}} & 0 & 0 \\ 1 & 0 & \cdots & 0 & 0 \end{bmatrix}.$$

In particular, for $n = 3$ and $q(s) = s$, it follows that

$$\begin{bmatrix} -a_0 & 0 & 0 \\ 0 & a_2 & 1 \\ 0 & 1 & 0 \end{bmatrix} = \begin{bmatrix} a_1 & a_2 & 1 \\ a_2 & 1 & 0 \\ 1 & 0 & 0 \end{bmatrix} \begin{bmatrix} 0 & 1 & 0 \\ 0 & 0 & 1 \\ -a_0 & -a_1 & -a_2 \end{bmatrix}.$$

$$xiv) \quad \begin{bmatrix} A_1 & A_2 \\ B_1 & B_2 \end{bmatrix} = \begin{bmatrix} 0 & I \\ A_2^{-1}\hat{I} & B_2 A_2^{-1} \end{bmatrix} \begin{bmatrix} B(p,q) & 0 \\ 0 & I \end{bmatrix} \begin{bmatrix} I & 0 \\ A_1 & A_2 \end{bmatrix}.$$

$xv)$ If p has distinct roots $\lambda_1, \ldots, \lambda_n$, then

$$V^{\mathrm{T}}(\lambda_1, \ldots, \lambda_n) B(p,q) V(\lambda_1, \ldots, \lambda_n) = \mathrm{diag}[q(\lambda_1)p'(\lambda_1), \ldots, q(\lambda_n)p'(\lambda_n)].$$

(Proof: See [260, pp. 164–167], [344], and [275, pp. 200–207]. To prove ii), note that A_1, A_2, B_1, B_2 are square and Toeplitz, and thus reverse symmetric, that is, $A_1 = A_1^{\hat{\mathrm{T}}}$. See Fact 3.15.6.) (Remark: $B(p,q)$ is the *Bezout matrix* of p and q. See [79, 379, 736, 787], [592, p. 189], and Fact 5.14.22.) (Remark: $xiii)$ is the *Barnett factorization*. See [73, 736]. The definitions of $B(p,q)$ and ii) are the *Gohberg-Semencul formulas*. See [275, p. 206].) (Remark: It follows from continuity that the determinant expressions are valid if A_1 or B_2 is singular. See Fact 2.13.10.) (Remark: The inverse of a Hankel matrix is a Bezout matrix. See [260, p. 174].)

Fact 4.8.7. Let $p, q \in \mathbb{F}[s]$, where $p(s) = \alpha_1 s + \alpha_0$ and $q(s) = s^2 + \beta_1 s + \beta_0$. Then, p and q are coprime if and only if $\alpha_0^2 + \alpha_1^2 \beta_0 \neq \alpha_0 \alpha_1 \beta_1$. (Proof: Use Fact 4.8.6.)

Fact 4.8.8. Let $p, q \in \mathbb{F}[s]$, assume that q is monic, assume that $\deg p < \deg q = n$, and define $B(p,q)$ as in Fact 4.8.6. Furthermore, define $g \in \mathbb{F}(s)$ by

$$g(s) \triangleq \frac{p(s)}{q(s)} = \sum_{i=1}^{\infty} \frac{g_i}{s^i}.$$

Finally, define the Hankel matrix

$$H(g) \triangleq \begin{bmatrix} g_1 & g_2 & \cdots & g_{n-1} & g_n \\ g_2 & g_3 & & g_n & g_{n+1} \\ \vdots & & & & \vdots \\ g_{n-1} & g_n & & g_{2n-3} & g_{2n-2} \\ g_n & g_{n+1} & \cdots & g_{2n-2} & g_{2n-1} \end{bmatrix}.$$

Then, the following statements hold:

$i)$ p and q are coprime if and only if $H(g)$ is nonsingular.

$ii)$ If p and q are coprime, then $[H(g)]^{-1} = B(q,a)$, where $a, b \in \mathbb{F}[s]$ satisfy the Bezout equation $ap + bq = 1$.

$iii)$ $B(q,p) = B(q,1)H(g)B(q,1)$.

$iv)$ $B(q,p)$ and $H(g)$ are congruent.

$v)$ $\mathrm{In}\, B(q,p) = \mathrm{In}\, H(g)$.

$vi)$ $\det H(g) = \det B(q,p)$.

(Proof: See [275, pp. 215–221].)

Fact 4.8.9. Let $q \in \mathbb{R}[s]$, define $g \in \mathbb{F}(s)$ by $g \triangleq q'/q$, and define $B(q, q')$ as in Fact 4.8.6. Then, the following statements hold:

i) The number of distinct roots of q is rank $B(q, q')$.

ii) q has n distinct roots if and only if $B(q, q')$ is nonsingular.

iii) The number of distinct real roots of q is sig $B(q, q')$.

iv) q has n distinct, real roots if and only if $B(q, q')$ is positive definite.

v) The number of distinct complex roots of q is $2\nu_-[B(q, q')]$.

vi) q has n distinct, complex roots if and only if n is even and $\nu_-[B(q,q')] = n/2$.

vii) q has n real roots if and only if $B(q, q')$ is positive semidefinite.

(Proof: See [275, p. 252].) (Remark: $q'(s) \triangleq (d/ds)q(s)$.)

Fact 4.8.10. Let $q \in \mathbb{F}[s]$, where $q(s) = \sum_{i=0}^{n} b_i s^i$, and define

$$\operatorname{coeff}(q) \triangleq \begin{bmatrix} b_n \\ \vdots \\ b_0 \end{bmatrix}.$$

Now, let $p \in \mathbb{F}[s]$, where $p(s) = \sum_{i=0}^{n} a_i s^i$. Then,

$$\operatorname{coeff}(pq) = A\operatorname{coeff}(q),$$

where $A \in \mathbb{F}^{2n \times (n+1)}$ is the Toeplitz matrix

$$A = \begin{bmatrix} a_n & 0 & 0 & \cdots & 0 \\ a_{n-1} & a_n & 0 & \cdots & 0 \\ \vdots & \ddots & \ddots & \ddots & \vdots \\ \vdots & \ddots & \ddots & \ddots & \vdots \\ a_0 & a_1 & \ddots & \ddots & a_n \\ 0 & a_0 & \ddots & \ddots & a_{n-1} \\ \vdots & \vdots & \ddots & \ddots & \vdots \\ 0 & 0 & \cdots & a_0 & a_1 \end{bmatrix}.$$

In particular, if $n = 3$, then

$$A = \begin{bmatrix} a_2 & 0 & 0 \\ a_1 & a_2 & 0 \\ a_0 & a_1 & a_2 \\ 0 & a_0 & a_1 \end{bmatrix}.$$

Fact 4.8.11. Let $\lambda_1, \ldots, \lambda_n \in \mathbb{C}$ be distinct and, for all $i = 1, \ldots, n$, define

$$p_i(s) \triangleq \prod_{\substack{j=1 \\ j \neq i}}^{n} \frac{s - \lambda_i}{\lambda_i - \lambda_j}.$$

Then, for all $i = 1, \ldots, n$,

$$p_i(\lambda_j) = \begin{cases} 1, & i = j, \\ 0, & i \neq j. \end{cases}$$

(Remark: This identity is the *Lagrange interpolation formula*.)

Fact 4.8.12. Let $A \in \mathbb{F}^{n \times n}$, and assume that $\det(I + A) \neq 0$. Then, there exists a polynomial p such that $\deg p \leq n - 1$ and $(I + A)^{-1} = p(A)$. (Remark: See Fact 4.8.12.)

Fact 4.8.13. Let $A \in \mathbb{F}^{n \times n}$, let $q \in \mathbb{F}[s]$, and assume that $q(A)$ is nonsingular. Then, there exists a polynomial p such that $\deg p \leq n - 1$ and $[q(A)]^{-1} = p(A)$. (Proof: See Fact 5.12.16.)

Fact 4.8.14. Let $p \in \mathbb{R}[s]$, where $p(s) = s^n + a_{n-1}s^{n-1} + \cdots + a_0$, and let $z \in \text{roots}(p)$. Then,

$$|z| < 1 + \max_{i=1,\ldots,n-1} |a_i|.$$

(Remark: This result is *Cauchy's estimate*. See [72, p. 184].)

Fact 4.8.15. Let $A \in \mathbb{R}^{n \times n}$, assume that A is skew symmetric, and let the components of $x_A \in \mathbb{R}^{n(n-1)/2}$ be the entries $A_{(i,j)}$ for all $i > j$. Then, there exists a polynomial function $p \colon \mathbb{R}^{n(n-1)/2} \mapsto \mathbb{R}$ such that, for all $\alpha \in \mathbb{R}$ and $x \in \mathbb{R}^{n(n-1)/2}$,

$$p(\alpha x) = \alpha^{n/2} p(x)$$

and

$$\det A = p^2(x_A).$$

In particular,

$$\det \begin{bmatrix} 0 & a \\ -a & 0 \end{bmatrix} = a^2$$

and

$$\det \begin{bmatrix} 0 & a & b & c \\ -a & 0 & d & e \\ -b & -d & 0 & f \\ -c & -e & -f & 0 \end{bmatrix} = (af - be + cd)^2.$$

(Proof: See [461, p. 224] and [592, pp. 125–127].) (Remark: The polynomial p is the *Pfaffian*, and this result is *Pfaff's theorem*.) (Remark: An extension to the product of a pair of skew-symmetric matrices is given in [236].)

Fact 4.8.16. Let $G \in \mathbb{F}^{n \times m}(s)$, and let $G_{(i,j)} = n_{ij}/d_{ij}$, where $n_{ij} \in \mathbb{F}[s]$ and $d_{ij} \in \mathbb{F}[s]$ are coprime for all $i = 1, \ldots, n$ and $j = 1, \ldots, m$. Then, q_1 given by the Smith-McMillan form is the least common multiple of $d_{11}, d_{12}, \ldots, d_{nm}$.

Fact 4.8.17. Let $G \in \mathbb{F}^{n \times m}(s)$, assume that $\operatorname{rank} G = m$, and let $\lambda \in \mathbb{C}$, where λ is not a pole of G. Then, λ is a transmission zero of G if and only if there exists a vector $u \in \mathbb{C}^m$ such that $G(\lambda)u = 0$. Furthermore, if G is square, then λ is a transmission zero of G if and only if $\det G(\lambda) = 0$.

4.9 Facts on the Characteristic and Minimal Polynomials

Fact 4.9.1. Let $A = \begin{bmatrix} a & b \\ c & d \end{bmatrix} \in \mathbb{R}^{2 \times 2}$. Then, the following identities hold:

   i) $\operatorname{mspec}(A) = \left\{ \frac{1}{2} \left[a + d \pm \sqrt{(a-d)^2 + 4bc} \right] \right\}_{\mathrm{m}}$

$$= \left\{ \frac{1}{2} \left[\operatorname{tr} A \pm \sqrt{(\operatorname{tr} A)^2 - 4 \det A} \right] \right\}_{\mathrm{m}}.$$

  ii) $\chi_A(s) = s^2 - (\operatorname{tr} A)s + \det A$.

 iii) $\det A = \frac{1}{2} \left[(\operatorname{tr} A)^2 - \operatorname{tr} A^2 \right]$.

 iv) $(sI - A)^{\mathrm{A}} = sI + A - (\operatorname{tr} A)I$.

  v) $A^{-1} = (\det A)^{-1} [(\operatorname{tr} A)I - A]$.

 vi) $A^{\mathrm{A}} = (\operatorname{tr} A)I - A$.

vii) $\operatorname{tr} A^{-1} = \operatorname{tr} A / \det A$.

Fact 4.9.2. Let $A, B \in \mathbb{F}^{2 \times 2}$. Then,

$$AB + BA - (\operatorname{tr} A)B - (\operatorname{tr} B)A + [(\operatorname{tr} A)(\operatorname{tr} B) - \operatorname{tr} AB]I = 0.$$

Furthermore,

$$\det(A + B) - \det A - \det B = (\operatorname{tr} A)(\operatorname{tr} B) - \operatorname{tr} AB.$$

(Proof: Apply the Cayley-Hamilton theorem to $A + xB$, differentiate with respect to x, and set $x = 0$. For the second identity, evaluate the Cayley-Hamilton theorem with $A + B$. See [270, 271, 472, 612] or [639, p. 37].)

Fact 4.9.3. Let $A \in \mathbb{R}^{3 \times 3}$. Then, the following identities hold:

 i) $\chi_A(s) = s^3 - (\operatorname{tr} A)s^2 + (\operatorname{tr} A^{\mathrm{A}})s - \det A$.

ii) $\operatorname{tr} A^{\mathrm{A}} = \frac{1}{2} \left[(\operatorname{tr} A)^2 - \operatorname{tr} A^2 \right]$.

$iii)$ $\det A = \frac{1}{3}\operatorname{tr} A^3 - \frac{1}{2}(\operatorname{tr} A)\operatorname{tr} A^2 + \frac{1}{6}(\operatorname{tr} A)^3.$

$iv)$ $(sI - A)^{\mathrm{A}} = s^2 I + s[A - (\operatorname{tr} A)I] + A^2 - (\operatorname{tr} A)A + \frac{1}{2}[(\operatorname{tr} A)^2 - \operatorname{tr} A^2]I.$

Fact 4.9.4. Let $A, B, C \in \mathbb{F}^{3 \times 3}$. Then,

$$\sum \left[A'B'C' - (\operatorname{tr} A')B'C' + (\operatorname{tr} A')(\operatorname{tr} B')C' - (\operatorname{tr} A'B')C' \right]$$
$$- [(\operatorname{tr} A)(\operatorname{tr} B)\operatorname{tr} C - (\operatorname{tr} A)\operatorname{tr} BC - (\operatorname{tr} B)\operatorname{tr} CA - (\operatorname{tr} C)\operatorname{tr} AB + \operatorname{tr} ABC$$
$$+ \operatorname{tr} CBA]I = 0,$$

where the sum is taken over all six permutations A', B', C' of A, B, C. (Remark: This identity is the *polarized Cayley-Hamilton theorem*. See [45, 472, 612].)

Fact 4.9.5. Let $A \in \mathbb{F}^{n \times n}$, and let $\chi_A(s) = s^n + \beta_{n-1}s^{n-1} + \cdots + \beta_0$. Then,

$$A^{\mathrm{A}} = (-1)^{n-1}\left(A^{n-1} + \beta_{n-1}A^{n-2} + \cdots + \beta_1 I\right).$$

Furthermore,

$$\operatorname{tr} A^{\mathrm{A}} = (-1)^{n-1}\chi_A'(0) = (-1)^{n-1}\beta_1.$$

(Proof: Use $A^{-1}\chi_A(A) = 0$. The second identity follows from (4.4.16) or Lemma 4.4.7.)

Fact 4.9.6. Let $A \in \mathbb{F}^{n \times n}$, assume that A is nonsingular, and let $\chi_A(s) = s^n + \beta_{n-1}s^{n-1} + \cdots + \beta_0$. Then,

$$\chi_{A^{-1}}(s) = \frac{1}{\det A}(-s)^n \chi_A(1/s)$$

$$= s^n + (\beta_1/\beta_0)s^{n-1} + \cdots + (\beta_{n-1}/\beta_0)s + 1/\beta_0.$$

(Remark: See Fact 5.13.2.)

Fact 4.9.7. Let $A \in \mathbb{F}^{n \times n}$, and assume that either A and $-A$ are similar or A^{T} and $-A$ are similar. Then,

$$\chi_A(s) = (-1)^n \chi_A(-s).$$

Furthermore, if n is even, then χ_A is even, whereas, if n is odd, then χ_A is odd. (Remark: A and A^{T} are similar. See Corollary 4.3.10 and Corollary 5.3.8.)

Fact 4.9.8. Let $A \in \mathbb{F}^{n \times n}$. Then, for all $s \in \mathbb{C}$,

$$(sI - A)^{\mathrm{A}} = \chi_A(s)(sI - A)^{-1} = \sum_{i=0}^{n-1} \chi_A^{[i]}(s)A^i,$$

where

$$\chi_A(s) = s^n + \beta_{n-1}s^{n-1} + \cdots + \beta_1 s + \beta_0$$

and, for all $i = 0, \ldots, n-1$, the polynomial $\chi_A^{[i]}$ is defined by

$$\chi_A^{[i]}(s) \triangleq s^{n-i} + \beta_{n-1}s^{n-1-i} + \cdots + \beta_{i+1}.$$

Note that

$$\chi_A^{[n-1]}(s) = s + \beta_{n-1}, \quad \chi_A^{[n]}(s) = 1,$$

and that, for all $i = 0, \ldots, n-1$ and with $\chi_A^{[0]} \triangleq \chi_A$, the polynomials $\chi_A^{[i]}$ satisfy the recursion

$$s\chi_A^{[i+1]}(s) = \chi_A^{[i]}(s) - \beta_i.$$

(Proof: See [794, p. 31].)

Fact 4.9.9. Let $A \in \mathbb{R}^{n \times n}$, and assume that A is skew symmetric. If n is even, then χ_A is even, whereas, if n is odd, then χ_A is odd.

Fact 4.9.10. Let $A \in \mathbb{F}^{n \times n}$, and assume that A is skew Hermitian. Then, for all $s \in \mathbb{C}$,

$$\chi_A(-s) = (-1)^n \overline{p(\bar{s})}.$$

Fact 4.9.11. Let $A \in \mathbb{F}^{n \times n}$. Then, $\chi_{\mathcal{A}}$ is even for the matrices $\mathcal{A}$ given by $\begin{bmatrix} 0 & A \\ A^* & 0 \end{bmatrix}$, $\begin{bmatrix} A & 0 \\ 0 & -A \end{bmatrix}$, and $\begin{bmatrix} A & 0 \\ 0 & -A^* \end{bmatrix}$.

Fact 4.9.12. Let $A, B \in \mathbb{F}^{n \times n}$, and define $\mathcal{A} \triangleq \begin{bmatrix} 0 & A \\ B & 0 \end{bmatrix}$. Then, $\chi_{\mathcal{A}}(s) = \chi_{AB}(s^2) = \chi_{BA}(s^2)$. Consequently, $\chi_{\mathcal{A}}$ is even. (Proof: Use Fact 2.13.10 and Proposition 4.4.9.)

Fact 4.9.13. Let $x, y, z, w \in \mathbb{F}^n$, and define $A \triangleq xy^{\mathrm{T}}$ and $B \triangleq xy^{\mathrm{T}} + zw^{\mathrm{T}}$. Then,

$$\chi_A(s) = s^{n-1}(s - x^{\mathrm{T}}y)$$

and

$$\chi_B(s) = s^{n-2}\big[s^2 - (x^{\mathrm{T}}y + z^{\mathrm{T}}w)s + x^{\mathrm{T}}yz^{\mathrm{T}}w - y^{\mathrm{T}}zx^{\mathrm{T}}w\big].$$

(Remark: See Fact 5.10.8.)

Fact 4.9.14. Let $x, y, z, w \in \mathbb{F}^{n-1}$, and define $A \in \mathbb{F}^{n \times n}$ by

$$A \triangleq \begin{bmatrix} 1 & x^{\mathrm{T}} \\ y & zw^{\mathrm{T}} \end{bmatrix}.$$

Then,

$$\chi_A(s) = s^{n-3}\big[s^3 - (1 + w^{\mathrm{T}}z)s^2 + (w^{\mathrm{T}}z - x^{\mathrm{T}}y)s + w^{\mathrm{T}}zx^{\mathrm{T}}y - x^{\mathrm{T}}zw^{\mathrm{T}}y\big].$$

(Proof: See [223].)

Fact 4.9.15. Let $A \in \mathbb{R}^{2n \times 2n}$, and assume that A is Hamiltonian. Then, χ_A is even.

Fact 4.9.16. Let $A, B, C \in \mathbb{R}^{n \times n}$, and define $\mathcal{A} \triangleq \begin{bmatrix} A & B \\ C & -A^T \end{bmatrix}$. If B and C are symmetric, then $\mathcal{A}$ is Hamiltonian. If B and C are skew symmetric, then $\chi_{\mathcal{A}}$ is even, although $\mathcal{A}$ is not necessarily Hamiltonian. (Proof: For the second result replace J_n by $\begin{bmatrix} 0 & I_n \\ I_n & 0 \end{bmatrix}$.)

Fact 4.9.17. Let $A \in \mathbb{R}^{n \times n}$, $R \in \mathbb{R}^{n \times n}$, and $B \in \mathbb{R}^{n \times m}$, and define $\mathcal{A} \in \mathbb{R}^{2n \times 2n}$ by

$$\mathcal{A} \triangleq \begin{bmatrix} A & BB^T \\ R & -A^T \end{bmatrix}.$$

Then, for all $s \notin \mathrm{spec}(A)$,

$$\chi_{\mathcal{A}}(s) = (-1)^n \chi_A(s) \chi_A(-s) \det\left[I + B^T(-sI - A^T)^{-1} R(sI - A)^{-1}B\right].$$

Now, assume that R is symmetric. Then, $\mathcal{A}$ is Hamiltonian, and $\chi_{\mathcal{A}}$ is even. If, in addition, R is positive semidefinite, then $(-1)^n \chi_{\mathcal{A}}$ has a spectral factorization. (Proof: Using (2.8.10) and (2.8.14), it follows that, for all $\pm s \notin \mathrm{spec}(A)$,

$$\chi_{\mathcal{A}}(s) = \det(sI - A)\det\left[sI + A^T - R(sI - A)^{-1}BB^T\right]$$
$$= (-1)^n \chi_A(s)\chi_A(-s)\det\left[I - B^T(sI + A^T)^{-1}R(sI - A)^{-1}B\right].$$

To prove the second statement, note that, for all $\omega \in \mathbb{R}$ such that $\jmath\omega \notin \mathrm{spec}(A)$, it follows that

$$\chi_{\mathcal{A}}(\jmath\omega) = (-1)^n \chi_A(\jmath\omega)\overline{\chi_A(\jmath\omega)}\det\left[I + B^T(\jmath\omega I - A)^{-*}R(\jmath\omega I - A)^{-1}B\right].$$

Thus, $(-1)^n \chi_{\mathcal{A}}(\jmath\omega) \geq 0$. By continuity, $(-1)^n \chi_{\mathcal{A}}(\jmath\omega) \geq 0$ for all $\omega \in \mathbb{R}$. Now, Proposition 4.1.1 implies that $(-1)^n \chi_{\mathcal{A}}$ has a spectral factorization.) (Remark: Not all Hamiltonian matrices have this property. Consider $\begin{bmatrix} 0 & 0 & 1 & 0 \\ 0 & 0 & 0 & 1 \\ -1 & 0 & 0 & 0 \\ 0 & -3 & 0 & 0 \end{bmatrix}$, whose spectrum is $\{\jmath, -\jmath, \sqrt{3}\jmath, -\sqrt{3}\jmath\}$.) (Remark: This result is closely related to Proposition 12.17.8.)

4.10 Facts on the Spectrum

Fact 4.10.1. Let $A \in \mathbb{F}^{n \times n}$, assume that A is nonsingular, and assume that $\mathrm{sprad}(I - A) < 1$. Then,

$$A^{-1} = \sum_{k=0}^{\infty}(I - A)^k.$$

Fact 4.10.2. Let $A \in \mathbb{F}^{n \times n}$ and $B \in \mathbb{F}^{m \times m}$. If $\mathrm{tr}\, A^k = \mathrm{tr}\, B^k$ for all $k \in \{1, \ldots, \max\{m, n\}\}$, then A and B have the same nonzero eigenvalues with the same algebraic multiplicity. Now, assume that $n = m$. Then, $\mathrm{tr}\, A^k = \mathrm{tr}\, B^k$ for all $k \in \{1, \ldots, n\}$ if and only if $\mathrm{mspec}(A) = \mathrm{mspec}(B)$. (Proof: Use *Newton's identities*. See Fact 4.8.2.) (Remark: This result yields Proposition 4.4.9 since $\operatorname{tr}(AB)^k = \operatorname{tr}(BA)^k$ for all $k \in \mathbb{P}$ and for all matrices A and B that are not square.) (Remark: Setting $B = 0_{n \times n}$ yields necessity in Fact 2.11.16.)

Fact 4.10.3. Let $A \in \mathbb{F}^{n \times n}$ and let $\operatorname{mspec}(A) = \{\lambda_1, \ldots, \lambda_n\}_{\mathrm{m}}$. Then,

$$
\operatorname{mspec}(A^{\mathrm{A}}) = \begin{cases} \left\{ \dfrac{\det A}{\lambda_1}, \ldots, \dfrac{\det A}{\lambda_n} \right\}_{\mathrm{m}}, & \operatorname{rank} A = n, \\[3mm] \left\{ \displaystyle\sum_{i=1}^{n} \det A_{[i,i]}, 0, \ldots, 0 \right\}_{\mathrm{m}}, & \operatorname{rank} A = n - 1, \\[3mm] \{0, \ldots, 0\}_{\mathrm{m}}, & \operatorname{rank} A < n - 1. \end{cases}
$$

(Remark: See Fact 2.14.7 and Fact 5.10.20.)

Fact 4.10.4. Let $a, b, c, d, \omega \in \mathbb{R}$, and define the skew-symmetric matrix $A \in \mathbb{R}^{4 \times 4}$ by

$$
A \triangleq \begin{bmatrix} 0 & \omega & a & b \\ -\omega & 0 & c & d \\ -a & -c & 0 & \omega \\ -b & -d & -\omega & 0 \end{bmatrix}.
$$

Then,

$$
\det A = \left[\omega^2 - (ad - bc) \right]^2.
$$

Furthermore, A has a repeated eigenvalue if and only if either $i)$ A is singular or $ii)$ $a = -d$ and $b = c$. In case $i)$, A has the repeated eigenvalue 0, while, in case $ii)$, A has the repeated eigenvalues $\jmath\sqrt{\omega^2 + a^2 + b^2}$ and $-\jmath\sqrt{\omega^2 + a^2 + b^2}$.

Fact 4.10.5. Let $A \in \mathbb{F}^{n \times n}$, and let $p \in \mathbb{F}[s]$. Then, μ_A divides p if and only if $\operatorname{spec}(A) \subseteq \operatorname{roots}(p)$ and, for all $\lambda \in \operatorname{spec}(A)$, $\operatorname{ind}_A(\lambda) \leq \mathrm{m}_p(\lambda)$.

Fact 4.10.6. Let $A \in \mathbb{F}^{n \times n}$, let $\operatorname{mspec}(A) = \{\lambda_1, \ldots, \lambda_n\}_{\mathrm{m}}$, and let $p \in \mathbb{F}[s]$. Then, the following statements hold:

$i)$ $\operatorname{mspec}[p(A)] = \{p(\lambda_1), \ldots, p(\lambda_n)\}_{\mathrm{m}}$.

$ii)$ $\operatorname{roots}(p) \cap \operatorname{spec}(A) = \varnothing$ if and only if $p(A)$ is nonsingular.

$iii)$ μ_A divides p if and only if $p(A) = 0$.

Fact 4.10.7. Let $A \in \mathbb{F}^{n \times n}$, $B \in \mathbb{F}^{n \times m}$, and $C \in \mathbb{F}^{m \times m}$, and let $p \in \mathbb{F}[s]$. Then,

$$
p\left(\begin{bmatrix} A & B \\ 0 & C \end{bmatrix} \right) = \begin{bmatrix} p(A) & \hat{B} \\ 0 & p(C) \end{bmatrix},
$$

where $\hat{B} \in \mathbb{F}^{n \times m}$.

Fact 4.10.8. Let $A_1 \in \mathbb{F}^{n \times n}$, $A_{12} \in \mathbb{F}^{n \times m}$, and $A_2 \in \mathbb{F}^{m \times m}$, and define $A \in \mathbb{F}^{(n+m) \times (n+m)}$ by

$$A \triangleq \begin{bmatrix} A_1 & A_{12} \\ 0 & A_2 \end{bmatrix}.$$

Then,

$$\chi_A = \chi_{A_1} \chi_{A_2}.$$

Furthermore,

$$\chi_{A_1}(A) = \begin{bmatrix} 0 & B_1 \\ 0 & \chi_{A_1}(A_2) \end{bmatrix}$$

and

$$\chi_{A_2}(A) = \begin{bmatrix} \chi_{A_2}(A_1) & B_2 \\ 0 & 0 \end{bmatrix},$$

where $B_1, B_2 \in \mathbb{F}^{n \times m}$. Therefore,

$$\mathcal{R}[\chi_{A_2}(A)] \subseteq \mathcal{R}\left(\begin{bmatrix} I_n \\ 0 \end{bmatrix} \right) \subseteq \mathcal{N}[\chi_{A_1}(A)]$$

and

$$\chi_{A_2}(A_1)B_1 + B_2 \chi_{A_1}(A_2) = 0.$$

Hence,

$$\chi_A(A) = \chi_{A_1}(A)\chi_{A_2}(A) = \chi_{A_2}(A)\chi_{A_1}(A) = 0.$$

Fact 4.10.9. Let $A_1 \in \mathbb{F}^{n \times n}$, $A_{12} \in \mathbb{F}^{n \times m}$, and $A_2 \in \mathbb{F}^{m \times m}$, assume that $\mathrm{spec}(A_1)$ and $\mathrm{spec}(A_2)$ are disjoint, and define $A \in \mathbb{F}^{(n+m) \times (n+m)}$ by

$$A \triangleq \begin{bmatrix} A_1 & A_{12} \\ 0 & A_2 \end{bmatrix}.$$

Furthermore, let $\mu_1, \mu_2 \in \mathbb{F}[s]$ be such that

$$\mu_A = \mu_1 \mu_2,$$
$$\mathrm{roots}(\mu_1) = \mathrm{spec}(A_1),$$
$$\mathrm{roots}(\mu_2) = \mathrm{spec}(A_2).$$

Then,

$$\mu_1(A) = \begin{bmatrix} 0 & B_1 \\ 0 & \mu_1(A_2) \end{bmatrix}$$

and

$$\mu_2(A) = \begin{bmatrix} \mu_2(A_1) & B_2 \\ 0 & 0 \end{bmatrix},$$

where $B_1, B_2 \in \mathbb{F}^{n \times m}$. Therefore,

$$\mathcal{R}[\mu_2(A)] \subseteq \mathcal{R}\left(\begin{bmatrix} I_n \\ 0 \end{bmatrix} \right) \subseteq \mathcal{N}[\mu_1(A)]$$

and

$$\mu_2(A_1)B_1 + B_2 \mu_1(A_2) = 0.$$

Hence,

$$\mu_A(A) = \mu_1(A)\mu_2(A) = \mu_2(A)\mu_1(A) = 0.$$

Fact 4.10.10. Let $A_1, A_2, A_3, A_4, B_1, B_2 \in \mathbb{F}^{n \times n}$, and define $A \in \mathbb{F}^{4n \times 4n}$ by

$$A \triangleq \begin{bmatrix} A_1 & B_1 & 0 & 0 \\ 0 & A_2 & 0 & 0 \\ 0 & 0 & A_3 & 0 \\ 0 & 0 & B_2 & A_4 \end{bmatrix}.$$

Then,

$$\mathrm{mspec}(A) = \bigcup_{i=1}^{4} \mathrm{mspec}(A_i).$$

Fact 4.10.11. Let $A \in \mathbb{F}^{n \times m}$ and $B \in \mathbb{F}^{m \times n}$, and assume that $m < n$. Then,

$$\mathrm{mspec}(I_n + AB) = \mathrm{mspec}(I_m + BA) \cup \{1, \dots, 1\}_\mathrm{m}.$$

Fact 4.10.12. Let $a, b \in \mathbb{F}$, and define the Toeplitz matrix $A \in \mathbb{F}^{n \times n}$ by

$$A \triangleq \begin{bmatrix} a & b & b & \cdots & b \\ b & a & b & \cdots & b \\ b & b & a & \cdots & b \\ \vdots & \vdots & \vdots & \ddots & \vdots \\ b & b & b & \cdots & a \end{bmatrix}.$$

Then,

$$\mathrm{mspec}(A) = \{a + (n-1)b, a - b, \dots, a - b\}_\mathrm{m}$$

and

$$A^2 + a_1 A + a_0 I = 0,$$

where $a_1 \triangleq -2a + (2-n)b$ and $a_0 \triangleq a^2 + (n-2)ab + (1-n)b^2$. Furthermore, if A is nonsingular, then

$$A^{-1} = \frac{1}{a-b} I_n + \frac{b}{(b-a)[a+b(n-1)]} 1_{n \times n}.$$

(Remark: See Fact 2.12.11.)

Fact 4.10.13. Let $A \in \mathbb{F}^{n \times n}$. Then,

$$\mathrm{spec}(A) \subset \bigcup_{i=1}^{n} \left\{ \lambda \in \mathbb{C} : \ |\lambda - A_{(i,i)}| \leq \sum_{j=1, j \neq i}^{n} |A_{(i,j)}| \right\}.$$

(Remark: This result is the *Gershgorin circle theorem*. See [148] for a proof and related results.)

Fact 4.10.14. Let $A \in \mathbb{F}^{n \times n}$. Then,

$$\mathrm{spec}(A) \subset \bigcup_{\substack{i,j=1 \\ i \neq j}}^{n} \left\{ \lambda \in \mathbb{C}: |\lambda - A_{(i,i)}||\lambda - A_{(j,j)}| \leq \sum_{\substack{k=1 \\ k \neq i}}^{n} |A_{(i,k)}| \sum_{\substack{k=1 \\ k \neq j}}^{n} |A_{(j,k)}| \right\}.$$

(Remark: The inclusion region is the *ovals of Cassini*. The result is due to Brauer. See [367, p. 380].)

Fact 4.10.15. Let $A \in \mathbb{F}^{n \times n}$, and assume that, for all $i = 1, \ldots, n$,

$$\sum_{j=1, j \neq i}^{n} |A_{(i,j)}| < |A_{(i,i)}|.$$

Then, A is nonsingular. (Proof: Apply the Gershgorin circle theorem.) (Remark: This result is the *diagonal dominance theorem*, and A is *diagonally dominant*. See [634] for a history of this result.) (Remark: For related results, see [244, 547, 597].) (Problem: Determine a lower bound for $|\det A|$ in terms of the difference between these quantities.)

Fact 4.10.16. Let $A \in \mathbb{F}^{n \times n}$, and, for $j = 1, \ldots, n$, define $b_j \triangleq \sum_{i=1}^{n} |A_{(i,j)}|$. Then,

$$\sum_{j=1}^{n} |A_{(j,j)}|/b_j \leq \mathrm{rank}\, A.$$

(Proof: See [592, p. 67].) (Remark: Interpret $0/0$ as 0.) (Remark: See Fact 4.10.15.)

Fact 4.10.17. Let $A_1, \ldots, A_r \in \mathbb{F}^{n \times n}$, assume that $A_1, \ldots, A_r$ are normal, and let $A \in \mathrm{co}\{A_1, \ldots, A_r\}$. Then,

$$\mathrm{spec}(A) \subseteq \mathrm{co} \bigcup_{i=1,\ldots,r} \mathrm{spec}(A_i).$$

(Proof: See [757].)

Fact 4.10.18. Let $A \in \mathbb{F}^{n \times n}$, and define the *numerical range* of A by

$$\Theta(A) \triangleq \{x^*Ax: \ x \in \mathbb{C}^n \text{ and } x^*x = 1\}.$$

Then, $\Theta(A)$ is a closed, convex subset of $\mathbb{C}$. Furthermore,

$$\mathrm{co}\,\mathrm{spec}(A) \subseteq \Theta(A) \subseteq \mathrm{co}\{\nu_1 + \jmath\mu_1, \nu_1 + \jmath\mu_n, \nu_n + \jmath\mu_1, \nu_n + \jmath\mu_n\},$$

where

$$\nu_1 \triangleq \lambda_{\max}\left[\tfrac{1}{2}(A + A^*)\right], \qquad \nu_n \triangleq \lambda_{\min}\left[\tfrac{1}{2}(A + A^*)\right],$$

$$\mu_1 \triangleq \lambda_{\max}\left[\tfrac{1}{2\jmath}(A - A^*)\right], \qquad \mu_n \triangleq \lambda_{\min}\left[\tfrac{1}{2\jmath}(A - A^*)\right].$$

If, in addition, A is normal, then

$$\Theta(A) = \text{co spec}(A).$$

Conversely, if $n \le 4$ and $\Theta(A) = \text{co spec}(A)$, then A is normal. (Proof: See [317] or [369, pp. 11, 52].) (Remark: $\Theta(A)$ is called the *field of values* in [369, p. 5].)

Fact 4.10.19. Let $A, B \in \mathbb{R}^{n \times n}$. Then,

$$\text{mspec}\left(\left[\begin{array}{cc} A & B \\ -B & A \end{array}\right]\right) = \text{mspec}(A + \jmath B) \cup \text{mspec}(A - \jmath B).$$

(Remark: See Fact 2.17.3.)

Fact 4.10.20. Let $A \in \mathbb{F}^{n \times n}$, $B \in \mathbb{F}^{m \times m}$, and $C \in \mathbb{F}^{n \times m}$, assume that A and B are Hermitian, and define $\mathcal{A}_0 \triangleq \left[\begin{smallmatrix} A & 0 \\ 0 & B \end{smallmatrix}\right]$ and $\mathcal{A} \triangleq \left[\begin{smallmatrix} A & C \\ C^* & B \end{smallmatrix}\right]$. Furthermore, define

$$\eta \triangleq \min_{\substack{i=1,\dots,n \\ j=1,\dots,m}} |\lambda_i(A) - \lambda_j(B)|.$$

Then, for all $i = 1, \dots, n + m$,

$$|\lambda_i(\mathcal{A}) - \lambda_i(\mathcal{A}_0)| \le \frac{2\sigma_{\max}^2(C)}{\eta + \sqrt{\eta^2 + 4\sigma_{\max}(C)}}.$$

(Proof: See [476].)

Fact 4.10.21. Let $A \in \mathbb{R}^{n \times n}$, let $b, c \in \mathbb{R}^n$, define $p \in \mathbb{R}[s]$ by $p(s) \triangleq c^{\mathrm{T}}(sI - A)^{\mathrm{A}}b$, assume that p and $\det(sI - A)$ are coprime, define $A_\alpha \triangleq A + \alpha bc^{\mathrm{T}}$ for all $\alpha \in [0, \infty)$, and let $\lambda \colon [0, \infty) \to \mathbb{C}$ be a continuous function such that $\lambda(\alpha) \in \text{spec}(A_\alpha)$ for all $\alpha \in [0, \infty)$. Then, either $\lim_{\alpha \to \infty} |\lambda(\alpha)| = \infty$ or $\lim_{\alpha \to \infty} \lambda(\alpha) \in \text{roots}(p)$. (Remark: This result is a consequence of *root locus* analysis from classical control theory, which determines asymptotic pole locations under high-gain feedback.)

4.11 Facts on Nonnegative Matrices

Fact 4.11.1. Let $A \in \mathbb{R}^{n \times n}$, where $n > 1$, and assume that A is nonnegative. Then, the following statements hold:

i) $\text{sprad}(A)$ is an eigenvalue of A.

ii) There exists a nonnegative vector $x \in \mathbb{R}^n$ such that $Ax = \text{sprad}(A)x$.

Furthermore, the following statements are equivalent:

iii) $(I + A)^{n-1}$ is positive.

iv) There do not exist $k \in \mathbb{P}$ and a permutation matrix $S \in \mathbb{R}^{n \times n}$ such that

$$SAS^\mathrm{T} = \begin{bmatrix} B & C \\ 0_{k \times (n-k)} & D \end{bmatrix}.$$

v) A has exactly one nonnegative eigenvector whose components sum to 1, and this eigenvector is positive.

A is *irreducible* if iii)–v) are satisfied. If A is irreducible, then the following statements hold:

vi) $\mathrm{sprad}(A) > 0$.

vii) $\mathrm{sprad}(A)$ is a simple eigenvalue of A.

$viii$) There exists a positive vector $x \in \mathbb{R}^n$ such that $Ax = \mathrm{sprad}(A)x$.

ix) A has exactly one positive eigenvector whose components sum to 1.

x) Assume that $\{\lambda_1, \ldots, \lambda_k\}_\mathrm{m} = \{\lambda \in \mathrm{mspec}(A) \colon |\lambda| = \mathrm{sprad}(A)\}_\mathrm{m}$. Then, $\lambda_1, \ldots, \lambda_k$ are distinct, and

$$\{\lambda_1, \ldots, \lambda_k\} = \{e^{2\pi \jmath i/k}\mathrm{sprad}(A) \colon i = 1, \ldots, k\}.$$

Furthermore,

$$\mathrm{mspec}(A) = e^{2\pi \jmath/k}\mathrm{mspec}(A).$$

xi) If at least one diagonal entry of A is positive, then $\mathrm{sprad}(A)$ is the only eigenvalue of A whose absolute value is $\mathrm{sprad}(A)$.

In addition, the following statements are equivalent:

xii) There exists $k \in \mathbb{P}$ such that A^k is positive.

$xiii$) A is irreducible and $|\lambda| < \mathrm{sprad}(A)$ for all $\lambda \in \mathrm{spec}(A) \backslash \{\mathrm{sprad}(A)\}$.

xiv) $A^{n^2 - 2n + 2}$ is positive.

A is *primitive* if xii)–xiv) are satisfied. (Example: $\begin{bmatrix} 0 & 1 \\ 1 & 0 \end{bmatrix}$ is irreducible but not primitive.) Finally, assume that A is irreducible and let $x \in \mathbb{R}^n$ be positive and satisfy $Ax = \mathrm{sprad}(A)x$. Then, for all positive $x_0 \in \mathbb{R}^n$, there exists a positive real number γ such that

$$\lim_{k \to \infty} \left(A^k x_0 - \gamma[\mathrm{sprad}(A)]^k x \right) = 0.$$

(Remark: For an arbitrary positive initial condition, the state of the difference equation $x_{k+1} = Ax_k$ approaches a distribution that is identical to the distribution of the eigenvector associated with the positive eigenvalue of maximum absolute value. In demography, this eigenvector is interpreted as the *stable age distribution*. See [421, pp. 47, 63].) (Proof: See [10, pp. 45–49], [99, pp. 26–28, 32, 55], [260], and [367, pp. 507–511].) (Remark: This result is the *Perron-Frobenius theorem*.) (Remark: See Fact 11.17.20.) (Remark: Statement xiv) is due to Wielandt. See [592, p. 157].)

Fact 4.11.2. Let $A \triangleq \left[\begin{smallmatrix} 1 & 1 \\ 1 & 0 \end{smallmatrix}\right]$. Then, $\chi_A(s) = s^2 - s - 1$ and $\mathrm{spec}(A) = \{\alpha, \beta\}$, where $\alpha \triangleq \frac{1}{2}(1 + \sqrt{5})$ and $\beta \triangleq \frac{1}{2}(1 - \sqrt{5})$ satisfy

$$\alpha - 1 = 1/\alpha, \qquad \beta - 1 = 1/\beta.$$

Furthermore, $\left[\begin{smallmatrix} \alpha \\ 1 \end{smallmatrix}\right]$ is an eigenvector of A associated with α. Now, for $k \geq 0$, consider the difference equation

$$x_{k+1} = Ax_k.$$

Then, for all $k \geq 0$,

$$x_k = A^k x_0$$

and

$$x_{k+2(1)} = x_{k+1(1)} + x_{k(1)}.$$

Furthermore, if x_0 is positive, then

$$\lim_{k \to \infty} \frac{x_{k(1)}}{x_{k(2)}} = \alpha.$$

In particular, if $x_0 \triangleq \left[\begin{smallmatrix} 1 \\ 1 \end{smallmatrix}\right]$, then, for all $k \geq 0$,

$$x_k = \left[\begin{array}{c} F_{k+2} \\ F_{k+1} \end{array} \right],$$

where $F_1 \triangleq F_2 \triangleq 1$ and, for all $k \geq 1$, F_k satisfies

$$F_{k+2} = F_{k+1} + F_k.$$

Furthermore,

$$A^k = \left[\begin{array}{cc} F_{k+1} & F_k \\ F_k & F_{k-1} \end{array} \right].$$

On the other hand, if $x_0 \triangleq \left[\begin{smallmatrix} 3 \\ 1 \end{smallmatrix}\right]$, then, for all $k \geq 0$,

$$x_k = \left[\begin{array}{c} L_{k+2} \\ L_{k+1} \end{array} \right],$$

where $L_1 \triangleq 1$, $L_2 \triangleq 3$, and, for all $k \geq 1$, L_k satisfies

$$L_{k+2} = L_{k+1} + L_k.$$

Furthermore,

$$\lim_{k \to \infty} \frac{F_{k+1}}{F_k} = \frac{L_{k+1}}{L_k} = \alpha.$$

(Proof: Use the last statement of Fact 4.11.1.) (Remark: F_k is the kth *Fibonacci number*, L_k is the kth *Lucas number*, and α is the *golden mean*. See [439, pp. 6–8, 239–241, 362, 363].)

Fact 4.11.3. Consider the nonnegative companion matrix $A \in \mathbb{R}^{n \times n}$ defined by

$$
A \triangleq
\begin{bmatrix}
0 & 1 & 0 & \cdots & 0 & 0 \\
0 & 0 & 1 & \ddots & 0 & 0 \\
0 & 0 & 0 & \ddots & 0 & 0 \\
\vdots & \vdots & \vdots & \ddots & \ddots & \vdots \\
0 & 0 & 0 & \cdots & 0 & 1 \\
1/n & 1/n & 1/n & \cdots & 1/n & 1/n
\end{bmatrix}.
$$

Then, A is irreducible, 1 is a simple eigenvalue of A with associated eigenvector $1_{n \times 1}$, and $|\lambda| < 1$ for all $\lambda \in \operatorname{spec}(A) \backslash \{1\}$. Furthermore, if $x \in \mathbb{R}^n$, then

$$
\lim_{k \to \infty} A^k x = \left[\frac{2}{n(n+1)} \sum_{i=1}^{n} i x_{(i-1)} \right] 1_{n \times 1}.
$$

(Proof: See [328, pp. 82, 83, 263–266].) (Remark: The result also follows from Fact 4.11.1.)

Fact 4.11.4. Let $A \in \mathbb{R}^{n \times m}$ and $b \in \mathbb{R}^m$. Then, the following statements are equivalent:

  i) If $x \in \mathbb{R}^m$ and $Ax \geq\geq 0$, then $b^\mathrm{T} x \geq 0$.

  ii) There exists a vector $y \in \mathbb{R}^n$ such that $y \geq\geq 0$ and $A^\mathrm{T} y = b$.

Equivalently, exactly one of the following two statements is satisfied:

  i) There exists a vector $x \in \mathbb{R}^m$ such that $Ax \geq\geq 0$ and $b^\mathrm{T} x < 0$.

  ii) There exists a vector $y \in \mathbb{R}^n$ such that $y \geq\geq 0$ and $A^\mathrm{T} y = b$.

(Proof: See [84, p. 47] or [135, p. 24].) (Remark: This result is *Farkas' theorem*.)

Fact 4.11.5. Let $A \in \mathbb{R}^{n \times m}$. Then, the following statements are equivalent:

  i) There exists a vector $x \in \mathbb{R}^m$ such that $Ax >> 0$.

  ii) If $y \in \mathbb{R}^n$ is nonzero and $y \geq\geq 0$, then $A^\mathrm{T} y \neq 0$.

Equivalently, exactly one of the following two statements is satisfied:

  i) There exists a vector $x \in \mathbb{R}^m$ such that $Ax >> 0$.

  ii) There exists a nonzero vector $y \in \mathbb{R}^n$ such that $y \geq\geq 0$ and $A^\mathrm{T} y = 0$.

(Proof: See [84, p. 47] or [135, p. 23].) (Remark: This result is *Gordan's theorem*.)

Fact 4.11.6. Let $A \in \mathbb{C}^{n \times n}$, and define $|A| \in \mathbb{R}^{n \times n}$ by $|A|_{(i,j)} \triangleq |A_{(i,j)}|$ for all $i, j = 1, \ldots, n$. Then,

$$\text{sprad}(A) \leq \text{sprad}(|A|).$$

(Proof: See [535, p. 619].)

Fact 4.11.7. Let $A, B \in \mathbb{R}^{n \times n}$, where $0 \leq\leq A \leq\leq B$. Then,

$$\text{sprad}(A) \leq \text{sprad}(B).$$

If, in addition, $B \neq A$ and $A + B$ is irreducible, then

$$\text{sprad}(A) < \text{sprad}(B).$$

(Proof: See [92, p. 27].)

Fact 4.11.8. Let $A \in \mathbb{R}^{n \times n}$, assume that $A >> 0$, and let $\lambda \in \text{spec}(A) \backslash \{\text{sprad}(A)\}$. Then,

$$|\lambda| \leq \frac{A_{\max} - A_{\min}}{A_{\max} + A_{\min}} \text{sprad}(A),$$

where

$$A_{\max} \triangleq \max \{A_{(i,j)} : \ i, j = 1, \ldots, n\}$$

and

$$A_{\min} \triangleq \min \{A_{(i,j)} : \ i, j = 1, \ldots, n\}.$$

(Remark: This result is *Hopf's theorem*.)

Fact 4.11.9. Let $A \in \mathbb{R}^{n \times n}$, assume that A is nonnegative and primitive, and let $x, y \in \mathbb{R}^n$, where $x > 0$ and $y > 0$ satisfy $Ax = \text{sprad}(A)x$ and $A^{\mathrm{T}}y = \text{sprad}(A)y$. Then,

$$\lim_{k \to \infty} \left[\frac{1}{\text{sprad}(A)} A\right]^k = xy^{\mathrm{T}}.$$

(Proof: See [367, p. 516].)

4.12 Notes

Much of the development in this chapter is based on [581]. Additional discussions of the Smith and Smith-McMillan forms are given in [407] and [818]. The proofs of Lemma 4.4.7 and Leverrier's algorithm Proposition 4.4.8 are based on [613, p. 432, 433], where it is called the *Souriau-Frame algorithm*. Alternative proofs of Leverrier's algorithm are given in [77, 377]. The proof of Theorem 4.6.1 is based on [367]. Polynomial-based approaches to linear algebra are given in [154, 275], while polynomial matrices and rational transfer functions are studied in [293, 743].

Chapter Five

Matrix Decompositions

In this chapter we present several matrix decompositions, namely, the Smith, multi-companion, elementary multi-companion, hypercompanion, Jordan, Schur, and singular value decompositions.

5.1 Smith Form

Our first decomposition involves rectangular matrices subject to a biequivalence transformation. This result is the specialization of the Smith decomposition given by Theorem 4.3.2 to constant matrices.

Theorem 5.1.1. Let $A \in \mathbb{F}^{n \times m}$ and $r \triangleq \operatorname{rank} A$. Then, there exist nonsingular matrices $S_1 \in \mathbb{F}^{n \times n}$ and $S_2 \in \mathbb{F}^{m \times m}$ such that

$$A = S_1 \begin{bmatrix} I_r & 0_{r \times (m-r)} \\ 0_{(n-r) \times r} & 0_{(n-r) \times (m-r)} \end{bmatrix} S_2. \tag{5.1.1}$$

Corollary 5.1.2. Let $A, B \in \mathbb{F}^{n \times m}$. Then, A and B are biequivalent if and only if A and B have the same Smith form.

Proposition 5.1.3. Let $A, B \in \mathbb{F}^{n \times m}$. Then, the following statements hold:

$i)$ A and B are left equivalent if and only if $\mathcal{N}(A) = \mathcal{N}(B)$.

$ii)$ A and B are right equivalent if and only if $\mathcal{R}(A) = \mathcal{R}(B)$.

$iii)$ A and B are biequivalent if and only if $\operatorname{rank} A = \operatorname{rank} B$.

Proof. The proof of necessity is immediate in $i)$–$iii)$. Sufficiency in $iii)$ follows from Corollary 5.1.2. For sufficiency in $i)$ and $ii)$, see [613, pp. 179–181]. $\qquad \square$

5.2 Multi-Companion Form

For the monic polynomial $p(s) = s^n + \beta_{n-1}s^{n-1} + \cdots + \beta_1 s + \beta_0 \in \mathbb{F}[s]$ of degree $n \geq 1$, the *companion matrix* $C(p) \in \mathbb{F}^{n \times n}$ associated with p is defined to be

$$C(p) \triangleq \begin{bmatrix} 0 & 1 & 0 & \cdots & 0 & 0 \\ 0 & 0 & 1 & \ddots & 0 & 0 \\ 0 & 0 & 0 & \ddots & 0 & 0 \\ \vdots & \vdots & \vdots & \ddots & \ddots & \vdots \\ 0 & 0 & 0 & \cdots & 0 & 1 \\ -\beta_0 & -\beta_1 & -\beta_2 & \cdots & -\beta_{n-2} & -\beta_{n-1} \end{bmatrix}. \qquad (5.2.1)$$

If $n = 1$, then $p(s) = s + \beta_0$ and $C(p) = -\beta_0$. Furthermore, if $n = 0$ and $p = 1$, then we define $C(p) \triangleq 0_{0 \times 0}$. Note that, if $n \geq 1$, then $\operatorname{tr} C(p) = -\beta_{n-1}$ and $\det C(p) = (-1)^n \beta_0 = (-1)^n p(0)$.

It is easy to see that the characteristic polynomial of the companion matrix $C(p)$ is p. For example, let $n = 3$ so that

$$C(p) = \begin{bmatrix} 0 & 1 & 0 \\ 0 & 0 & 1 \\ -\beta_0 & -\beta_1 & -\beta_2 \end{bmatrix}, \qquad (5.2.2)$$

and thus

$$sI - C(p) = \begin{bmatrix} s & -1 & 0 \\ 0 & s & -1 \\ \beta_0 & \beta_1 & s + \beta_2 \end{bmatrix}. \qquad (5.2.3)$$

Adding s times the second column and s^2 times the third column to the first column leaves the determinant of $sI - C(p)$ unchanged and yields

$$\begin{bmatrix} 0 & -1 & 0 \\ 0 & s & -1 \\ p(s) & \beta_1 & s + \beta_2 \end{bmatrix}. \qquad (5.2.4)$$

Hence, $\chi_{C(p)} = p$. If $n = 0$ and $p = 1$, then we define $\chi_{C(p)} \triangleq \chi_{0_{0 \times 0}} = 1$. The following result shows that companion matrices have the same characteristic and minimal polynomials.

Proposition 5.2.1. Let $p \in \mathbb{F}[s]$ be a monic polynomial having degree n. Then, there exist unimodular matrices $S_1, S_2 \in \mathbb{F}^{n \times n}[s]$ such that

$$sI - C(p) = S_1(s) \begin{bmatrix} I_{n-1} & 0_{(n-1) \times 1} \\ 0_{1 \times (n-1)} & p(s) \end{bmatrix} S_2(s). \qquad (5.2.5)$$

Furthermore,

$$\chi_{C(p)} = \mu_{C(p)} = p. \tag{5.2.6}$$

Proof. Since $\chi_{C(p)} = p$, it follows that $\mathrm{rank}[sI - C(p)] = n$. Next, since $\det\big([sI - C(p)]_{[n,1]}\big) = (-1)^{n-1}$, it follows that $\Delta_{n-1} = 1$, where Δ_{n-1} is the greatest common divisor (which is monic by definition) of all $(n-1) \times (n-1)$ subdeterminants of $sI - C(p)$. Furthermore, since Δ_{i-1} divides Δ_i for all $i = 2, \ldots, n-1$, it follows that $\Delta_1 = \cdots = \Delta_{n-2} = 1$. Consequently, $p_1 = \cdots = p_{n-1} = 1$. Since, by Proposition 4.6.2, $\chi_{C(p)} = \prod_{i=1}^{n} p_i = p_n$ and $\mu_{C(p)} = p_n$, it follows that $\chi_{C(p)} = \mu_{C(p)} = p$. $\square$

Next, we consider block-diagonal matrices whose diagonally located blocks are all companion matrices.

Lemma 5.2.2. Let $p_1, \ldots, p_n \in \mathbb{F}[s]$ be monic polynomials such that p_i divides p_{i+1} for all $i = 1, \ldots, n-1$ and $n = \sum_{i=1}^{n} \deg p_i$. Furthermore, define $C \triangleq \mathrm{diag}[C(p_1), \ldots, C(p_n)] \in \mathbb{F}^{n \times n}$. Then, there exist unimodular matrices $S_1, S_2 \in \mathbb{F}^{n \times n}[s]$ such that

$$sI - C = S_1(s) \begin{bmatrix} p_1(s) & & 0 \\ & \ddots & \\ 0 & & p_n(s) \end{bmatrix} S_2(s). \tag{5.2.7}$$

Proof. Letting $k_i = \deg p_i$, Proposition 5.2.1 implies that the Smith form of $sI_{k_i} - C(p_i)$ is $0_{0 \times 0}$ if $k_i = 0$ and $\mathrm{diag}(I_{k_i-1}, p_i)$ if $k_i \geq 1$. Note that $p_1 = \cdots = p_{n_0} = 1$, where $n_0 \triangleq \sum_{i=1}^{n} \max\{0, k_i - 1\}$. By combining these Smith forms and rearranging diagonal entries, it follows that there exist unimodular matrices $S_1, S_2 \in \mathbb{F}^{n \times n}[s]$ such that

$$sI - C = \begin{bmatrix} sI_{k_1} - C(p_1) & & \\ & \ddots & \\ & & sI_{k_n} - C(p_n) \end{bmatrix}$$

$$= S_1(s) \begin{bmatrix} p_1(s) & & 0 \\ & \ddots & \\ 0 & & p_n(s) \end{bmatrix} S_2(s).$$

Since p_i divides p_{i+1} for all $i = 1, \ldots, n-1$, it follows that this diagonal matrix is the Smith form of $sI - C$. $\square$

The following result uses Lemma 5.2.2 to construct a canonical form, known as the *multi-companion form*, for square matrices under a similarity transformation.

Theorem 5.2.3. Let $A \in \mathbb{F}^{n \times n}$, and let $p_1, \ldots, p_n \in \mathbb{F}[s]$ denote the similarity invariants of A, where p_i divides p_{i+1} for all $i = 1, \ldots, n-1$. Then, there exists a nonsingular matrix $S \in \mathbb{F}^{n \times n}$ such that

$$A = S \begin{bmatrix} C(p_1) & & 0 \\ & \ddots & \\ 0 & & C(p_n) \end{bmatrix} S^{-1}. \tag{5.2.8}$$

Proof. Lemma 5.2.2 implies that the $n \times n$ matrix $sI - C$, where $C \triangleq \mathrm{diag}[C(p_1), \ldots, C(p_n)]$, has the Smith form $\mathrm{diag}(p_1, \ldots, p_n)$. Now, since $sI - A$ has the same similarity invariants as C, it follows from Theorem 4.3.9 that A and C are similar. $\square$

Corollary 5.2.4. Let $A \in \mathbb{F}^{n \times n}$. Then, $\mu_A = \chi_A$ if and only if A is similar to $C(\chi_A)$.

Proof. Suppose that $\mu_A = \chi_A$. Then, it follows from Proposition 4.6.2 that $p_i = 1$ for all $i = 1, \ldots, n-1$ and $p_n = \chi_A$ is the only nonconstant similarity invariant of A. Thus, $C(p_i) = 0_{0 \times 0}$ for all $i = 1, \ldots, n-1$, and it follows from Theorem 5.2.3 that A is similar to $C(\chi_A)$. The converse follows from (5.2.6), xi) of Proposition 4.4.4, and Proposition 4.6.3. $\square$

Corollary 5.2.5. Let $A \in \mathbb{F}^{n \times n}$ be a companion matrix. Then, $A = C(\chi_A)$ and $\mu_A = \chi_A$.

Note that, if $A = I_n$, then the similarity invariants of A are $p_i(s) = s-1$ for all $i = 1, \ldots, n$. Thus, $C(p_i) = 1$ for all $i = 1, \ldots, n$, as expected.

Corollary 5.2.6. Let $A, B \in \mathbb{F}^{n \times n}$. Then, the following statements are equivalent:

i) A and B are similar.

ii) A and B have the same similarity invariants.

iii) A and B have the same multi-companion form.

The multi-companion form given by Theorem 5.2.3 provides a canonical form for A in terms of a block-diagonal matrix of companion matrices. As will be seen, however, the multi-companion form is only one such decomposition. The goal of the remainder of this section is to obtain an additional canonical form by applying a similarity transformation to the multi-companion form.

To begin, note that, if A_i is similar to B_i for all $i = 1, \ldots, r$, then $\mathrm{diag}(A_1, \ldots, A_r)$ is similar to $\mathrm{diag}(B_1, \ldots, B_r)$. Therefore, it follows from Corollary 5.2.6 that, if $sI - A_i$ and $sI - B_i$ have the same Smith form for all $i = 1, \ldots, r$, then $sI - \mathrm{diag}(A_1, \ldots, A_r)$ and $sI - \mathrm{diag}(B_1, \ldots, B_r)$ have the same Smith form. The following lemma is needed.

Lemma 5.2.7. Let $A = \mathrm{diag}(A_1, A_2)$, where $A_i \in \mathbb{F}^{n_i \times n_i}$ for $i = 1, 2$. Then, μ_A is the least common multiple of μ_{A_1} and μ_{A_2}. In particular, if μ_{A_1} and μ_{A_2} are coprime, then $\mu_A = \mu_{A_1}\mu_{A_2}$.

Proof. Since $0 = \mu_A(A) = \mathrm{diag}[\mu_A(A_1), \mu_A(A_2)]$, it follows that $\mu_A(A_1) = 0$ and $\mu_A(A_2) = 0$. Therefore, Theorem 4.6.1 implies that μ_{A_1} and μ_{A_2} both divide μ_A. Consequently, the least common multiple q of μ_{A_1} and μ_{A_2} also divides μ_A. Since $q(A_1) = 0$ and $q(A_2) = 0$, it follows that $q(A) = 0$. Therefore, μ_A divides q. Hence, $q = \mu_A$. If, in addition, μ_{A_1} and μ_{A_2} are coprime, then $\mu_A = \mu_{A_1}\mu_{A_2}$. $\qquad\square$

Proposition 5.2.8. Let $p \in \mathbb{F}[s]$ be a monic polynomial of positive degree n, and let $p = p_1 \cdots p_r$, where $p_1, \ldots, p_r \in \mathbb{F}[s]$ are monic and pairwise coprime polynomials. Then, the matrices $C(p)$ and $\mathrm{diag}[C(p_1), \ldots, C(p_r)]$ are similar.

Proof. Let $\hat{p}_2 = p_2 \cdots p_r$ and $\hat{C} \triangleq \mathrm{diag}[C(p_1), C(\hat{p}_2)]$. Since p_1 and $\hat{p}_2$ are coprime, it follows from Lemma 5.2.7 that $\mu_{\hat{C}} = \mu_{C(p_1)}\mu_{C(\hat{p}_2)}$. Furthermore, $\chi_{\hat{C}} = \chi_{C(p_1)}\chi_{C(\hat{p}_2)} = \mu_{\hat{C}}$. Hence, Corollary 5.2.4 implies that $\hat{C}$ is similar to $C(\chi_{\hat{C}})$. However, $\chi_{\hat{C}} = p_1 \cdots p_r = p$, so that $\hat{C}$ is similar to $C(p)$. If $r > 2$, then the same argument can be used to decompose $C(\hat{p}_2)$ to show that $C(p)$ is similar to $\mathrm{diag}[C(p_1), \ldots, C(p_r)]$. $\qquad\square$

Proposition 5.2.8 can be used to decompose every companion block of a multi-companion form into smaller companion matrices. This procedure can be carried out for every companion block whose characteristic polynomial has coprime factors. For example, suppose that $A \in \mathbb{R}^{10 \times 10}$ has the similarity invariants $p_i(s) = 1$ for all $i = 1, \ldots, 7$, $p_8(s) = (s + 1)^2$, $p_9(s) = (s + 1)^2(s + 2)$, and $p_{10}(s) = (s + 1)^2(s + 2)(s^2 + 3)$, so that, by Theorem 5.2.3, the multi-companion form of A is $\mathrm{diag}[C(p_8), C(p_9), C(p_{10})]$, where $C(p_8) \in \mathbb{R}^{2 \times 2}$, $C(p_9) \in \mathbb{R}^{3 \times 3}$, and $C(p_{10}) \in \mathbb{R}^{5 \times 5}$. According to Proposition 5.2.8, the companion matrices $C(p_9)$ and $C(p_{10})$ can be further decomposed. For example, $C(p_9)$ is similar to $\mathrm{diag}[C(p_{9,1}), C(p_{9,2})]$, where $p_{9,1}(s) = (s + 1)^2$ and $p_{9,2}(s) = s + 2$ are coprime. Furthermore, $C(p_{10})$ is similar to four different diagonal matrices, three of which have two companion blocks while the fourth has three companion blocks. Since $p_8(s) = (s + 1)^2$ does not have nonconstant coprime factors, however, it follows that the companion matrix $C(p_8)$ cannot be decomposed into smaller companion matrices.

The largest number of companion blocks achievable by similarity transformation is obtained by factoring every similarity invariant into *elementary divisors*, which are powers of irreducible polynomials that are nonconstant, monic, and pairwise coprime. In the above example, this factorization is given by $p_9(s) = p_{9,1}(s)p_{9,2}(s)$, where $p_{9,1}(s) = (s + 1)^2$ and $p_{9,2}(s) = s + 2$, and by $p_{10} = p_{10,1}p_{10,2}p_{10,3}$, where $p_{10,1}(s) = (s+1)^2$, $p_{10,2}(s) = s+2$, and $p_{10,3}(s) = s^2 + 3$. The elementary divisors of A are thus $(s+1)^2$, $(s+1)^2$, $s+2$, $(s+1)^2$, $s+2$, and s^2+3, which yields six companion blocks. Viewing $A \in \mathbb{C}^{n \times n}$ we can further factor $p_{10,3}(s) = (s+\jmath\sqrt{3})(s-\jmath\sqrt{3})$, which yields a total of seven companion blocks. From Proposition 5.2.8 and Theorem 5.2.3 we obtain the *elementary multi-companion form*, which provides another canonical form for A.

Theorem 5.2.9. Let $A \in \mathbb{F}^{n \times n}$, and let $q_1^{l_1}, \dots, q_h^{l_h} \in \mathbb{F}[s]$ be the elementary divisors of A, where $l_1, \dots, l_h \in \mathbb{P}$. Then, there exists a nonsingular matrix $S \in \mathbb{F}^{n \times n}$ such that

$$
A = S \begin{bmatrix} C\left(q_1^{l_1}\right) & & 0 \\ & \ddots & \\ 0 & & C\left(q_h^{l_h}\right) \end{bmatrix} S^{-1}. \tag{5.2.9}
$$

5.3 Hypercompanion Form and Jordan Form

In this section we present an alternative form of the companion blocks of the elementary multi-companion form (5.2.9). To do this we define the *hypercompanion matrix* $\mathcal{H}_l(q)$ associated with the elementary divisor $q^l \in \mathbb{F}[s]$, where $l \in \mathbb{P}$, as follows. For $q(s) = s - \lambda \in \mathbb{C}[s]$, define the $l \times l$ Toeplitz hypercompanion matrix

$$
\mathcal{H}_l(q) \triangleq \lambda I_l + N_l = \begin{bmatrix} \lambda & 1 & 0 & & & \\ 0 & \lambda & 1 & & 0 & \\ & & \ddots & \ddots & & \\ & & & \ddots & 1 & 0 \\ & 0 & & & \lambda & 1 \\ & & & & 0 & \lambda \end{bmatrix}, \tag{5.3.1}
$$

while, for $q(s) = s^2 - \beta_1 s - \beta_0 \in \mathbb{R}[s]$, define the $2l \times 2l$ real, tridiagonal hypercompanion matrix

$$
\mathcal{H}_l(q) \triangleq \begin{bmatrix} 0 & 1 & & & & & \\ \beta_0 & \beta_1 & 1 & & & 0 & \\ & 0 & 0 & 1 & & & \\ & & \beta_0 & \beta_1 & 1 & & \\ & & & \ddots & \ddots & \ddots & \\ & 0 & & & \ddots & 0 & 1 \\ & & & & & \beta_0 & \beta_1 \end{bmatrix}. \tag{5.3.2}
$$

The following result shows that the hypercompanion matrix $\mathcal{H}_l(q)$ is similar to the companion matrix $C(q^l)$ associated with the elementary divisor q^l of $\mathcal{H}_l(q)$.

Lemma 5.3.1. Let $l \in \mathbb{P}$, and let $q(s) = s - \lambda \in \mathbb{C}[s]$ or $q(s) = s^2 - \beta_1 s - \beta_0 \in \mathbb{R}[s]$. Then, q^l is the only elementary divisor of $\mathcal{H}_l(q)$, and $\mathcal{H}_l(q)$ is similar to $C(q^l)$.

Proof. Let k denote the order of $\mathcal{H}_l(q)$. Then, $\chi_{\mathcal{H}_l(q)} = q^l$ and $\det([sI - \mathcal{H}_l(q)]_{[k,1]}) = (-1)^{k-1}$. Hence, as in the proof of Proposition 5.2.1, it follows that $\chi_{\mathcal{H}_l(q)} = \mu_{\mathcal{H}_l(q)}$. Corollary 5.2.4 now implies that $\mathcal{H}_l(q)$ is similar to $C(q^l)$. $\qquad\qquad\square$

Proposition 5.2.8 and Lemma 5.3.1 yield the following canonical form, which is known as the *hypercompanion form*.

Theorem 5.3.2. Let $A \in \mathbb{F}^{n \times n}$, and let $q_1^{l_1}, \ldots, q_h^{l_h} \in \mathbb{F}[s]$ be the elementary divisors of A, where $l_1, \ldots, l_h \in \mathbb{P}$. Then, there exists a nonsingular matrix $S \in \mathbb{F}^{n \times n}$ such that

$$A = S \begin{bmatrix} \mathcal{H}_{l_1}(q_1) & & 0 \\ & \ddots & \\ 0 & & \mathcal{H}_{l_h}(q_h) \end{bmatrix} S^{-1}. \qquad (5.3.3)$$

Next, consider Theorem 5.3.2 with $\mathbb{F} = \mathbb{C}$. In this case, every elementary divisor $q_i^{l_i}$ is of the form $(s - \lambda_i)^{l_i}$, where $\lambda_i \in \mathbb{C}$. Furthermore, $S \in \mathbb{C}^{n \times n}$, and the hypercompanion form (5.3.3) is a block-diagonal matrix whose diagonally located blocks are of the form (5.3.1). The hypercompanion form (5.3.3) with every diagonally located block of the form (5.3.1) is the *Jordan form*, as given by the following result.

Theorem 5.3.3. Let $A \in \mathbb{C}^{n \times n}$, and let $q_1^{l_1}, \ldots, q_h^{l_h} \in \mathbb{C}[s]$ be the elementary divisors of A, where $l_1, \ldots, l_h \in \mathbb{P}$ and each of the polynomials $q_1, \ldots, q_h \in \mathbb{C}[s]$ has degree 1. Then, there exists a nonsingular matrix $S \in \mathbb{C}^{n \times n}$ such that

$$A = S \begin{bmatrix} \mathcal{H}_{l_1}(q_1) & & 0 \\ & \ddots & \\ 0 & & \mathcal{H}_{l_h}(q_h) \end{bmatrix} S^{-1}. \qquad (5.3.4)$$

Corollary 5.3.4. Let $p \in \mathbb{F}[s]$, let $\lambda_1, \ldots, \lambda_r$ denote the distinct roots of p, and, for $i = 1, \ldots, r$, let $l_i \triangleq \mathrm{m}_p(\lambda_i)$ and $p_i(s) \triangleq s - \lambda_i$. Then, $C(p)$ is similar to $\mathrm{diag}[\mathcal{H}_{l_1}(p_1), \ldots, \mathcal{H}_{l_r}(p_r)]$.

To illustrate the structure of the Jordan form, let $l_i = 3$ and $q_i(s) = s - \lambda_i$, where $\lambda_i \in \mathbb{C}$. Then, $\mathcal{H}_{l_i}(q_i)$ is the 3×3 matrix

$$\mathcal{H}_{l_i}(q_i) = \lambda_i I_3 + N_3 = \begin{bmatrix} \lambda_i & 1 & 0 \\ 0 & \lambda_i & 1 \\ 0 & 0 & \lambda_i \end{bmatrix} \tag{5.3.5}$$

so that $\text{mspec}[\mathcal{H}_{l_i}(q_i)] = \{\lambda_i, \lambda_i, \lambda_i\}_m$. If $\mathcal{H}_{l_i}(q_i)$ is the only diagonally located block of the Jordan form associated with the eigenvalue λ_i, then the algebraic multiplicity of λ_i is equal to 3, while its geometric multiplicity is equal to 1.

Now, consider Theorem 5.3.2 with $\mathbb{F} = \mathbb{R}$. In this case, every elementary divisor $q_i^{l_i}$ is either of the form $(s - \lambda_i)^{l_i}$ or of the form $(s^2 - \beta_{1i}s - \beta_{0i})^{l_i}$, where $\beta_{0i}, \beta_{1i} \in \mathbb{R}$. Furthermore, $S \in \mathbb{R}^{n \times n}$, and the hypercompanion form (5.3.3) is a block-diagonal matrix whose diagonally located blocks are real matrices of the form (5.3.1) or (5.3.2). In this case, (5.3.3) is the *real hypercompanion form*.

Applying an additional real similarity transformation to each diagonally located block of the real hypercompanion form yields the *real Jordan form*. To do this, define the *real Jordan matrix* $\mathcal{J}_l(q)$ for $l \in \mathbb{P}$ as follows. For $q(s) = s - \lambda \in \mathbb{F}[s]$ define $\mathcal{J}_l(q) \triangleq \mathcal{H}_l(q)$, while, if $q(s) = s^2 - \beta_1 s - \beta_0 \in \mathbb{F}[s]$ is irreducible with a nonreal root $\lambda = \nu + \jmath\omega$, then define the $2l \times 2l$ upper Hessenberg matrix

$$\mathcal{J}_l(q) \triangleq \begin{bmatrix} \nu & \omega & 1 & 0 & & & & \\ -\omega & \nu & 0 & 1 & \ddots & & 0 & \\ & & \nu & \omega & 1 & \ddots & & \\ & & -\omega & \nu & 0 & \ddots & \ddots & \\ & & & & \ddots & \ddots & 1 & 0 \\ & & & & & \ddots & 0 & 1 \\ & 0 & & & & & \nu & \omega \\ & & & & & & -\omega & \nu \end{bmatrix}. \tag{5.3.6}$$

Theorem 5.3.5. Let $A \in \mathbb{R}^{n \times n}$, and let $q_1^{l_1}, \ldots, q_h^{l_h} \in \mathbb{R}[s]$, where $l_1, \ldots, l_h \in \mathbb{P}$ are the elementary divisors of A. Then, there exists a nonsingular matrix $S \in \mathbb{R}^{n \times n}$ such that

$$A = S \begin{bmatrix} \mathcal{J}_{l_1}(q_1) & & 0 \\ & \ddots & \\ 0 & & \mathcal{J}_{l_h}(q_h) \end{bmatrix} S^{-1}. \tag{5.3.7}$$

Proof. For the irreducible quadratic $q(s) = s^2 - \beta_1 s - \beta_0 \in \mathbb{R}[s]$ we show that $\mathcal{J}_l(q)$ and $\mathcal{H}_l(q)$ are similar. Writing $q(s) = (s-\lambda)(s-\overline{\lambda})$, it follows from Theorem 5.3.3 that $\mathcal{H}_l(q) \in \mathbb{R}^{2l \times 2l}$ is similar to $\mathrm{diag}(\lambda I_l + N_l, \overline{\lambda} I_l + N_l)$. Next, by using a permutation similarity transformation, it follows that $\mathcal{H}_l(q)$ is similar to

$$
\begin{bmatrix}
\lambda & 0 & 1 & 0 & & & & & \\
0 & \overline{\lambda} & 0 & 1 & 0 & & & 0 & \\
 & & 0 & \lambda & 0 & 1 & 0 & & \\
 & & 0 & \overline{\lambda} & 0 & 1 & & & \\
 & & & \ddots & \ddots & \ddots & & & \\
 & & & & \ddots & \ddots & 1 & 0 & \\
 & & & & & \ddots & 0 & 1 & \\
 & 0 & & & & & \lambda & 0 & \\
 & & & & & & 0 & \overline{\lambda} &
\end{bmatrix},
$$

Finally, applying the similarity transformation $S \triangleq \mathrm{diag}(\hat{S}, \ldots, \hat{S})$ to the above matrix, where $\hat{S} \triangleq \begin{bmatrix} -\jmath & -\jmath \\ 1 & -1 \end{bmatrix}$ and $\hat{S}^{-1} = \frac{1}{2}\begin{bmatrix} \jmath & 1 \\ \jmath & -1 \end{bmatrix}$, yields $\mathcal{J}_l(q)$. $\square$

Example 5.3.6. Let $A, B \in \mathbb{R}^{4 \times 4}$ and $C \in \mathbb{C}^{4 \times 4}$ be given by

$$
A = \begin{bmatrix}
0 & 1 & 0 & 0 \\
0 & 0 & 1 & 0 \\
0 & 0 & 0 & 1 \\
-16 & 0 & -8 & 0
\end{bmatrix},
$$

$$
B = \begin{bmatrix}
0 & 1 & 0 & 0 \\
-4 & 0 & 1 & 0 \\
0 & 0 & 0 & 1 \\
0 & 0 & -4 & 0
\end{bmatrix},
$$

and

$$
C = \begin{bmatrix}
2\jmath & 1 & 0 & 0 \\
0 & 2\jmath & 0 & 0 \\
0 & 0 & -2\jmath & 1 \\
0 & 0 & 0 & -2\jmath
\end{bmatrix}.
$$

Then, A is in companion form, B is in real hypercompanion form, and C is in Jordan form. Furthermore, A, B, and C are similar.

Example 5.3.7. Let $A, B \in \mathbb{R}^{6 \times 6}$ and $C \in \mathbb{C}^{6 \times 6}$ be given by

$$A = \begin{bmatrix} 0 & 1 & 0 & 0 & 0 & 0 \\ 0 & 0 & 1 & 0 & 0 & 0 \\ 0 & 0 & 0 & 1 & 0 & 0 \\ 0 & 0 & 0 & 0 & 1 & 0 \\ 0 & 0 & 0 & 0 & 0 & 1 \\ -27 & 54 & -63 & 44 & -21 & 6 \end{bmatrix},$$

$$B = \begin{bmatrix} 0 & 1 & 0 & 0 & 0 & 0 \\ -3 & 2 & 1 & 0 & 0 & 0 \\ 0 & 0 & 0 & 1 & 0 & 0 \\ 0 & 0 & -3 & 2 & 1 & 0 \\ 0 & 0 & 0 & 0 & 0 & 1 \\ 0 & 0 & 0 & 0 & -3 & 2 \end{bmatrix},$$

and

$$C = \begin{bmatrix} 1 + j\sqrt{2} & 1 & 0 & 0 & 0 & 0 \\ 0 & 1 + j\sqrt{2} & 1 & 0 & 0 & 0 \\ 0 & 0 & 1 + j\sqrt{2} & 0 & 0 & 0 \\ 0 & 0 & 0 & 1 - j\sqrt{2} & 1 & 0 \\ 0 & 0 & 0 & 0 & 1 - j\sqrt{2} & 1 \\ 0 & 0 & 0 & 0 & 0 & 1 - j\sqrt{2} \end{bmatrix}.$$

Then, A is in companion form, B is in real hypercompanion form, and C is in Jordan form. Furthermore, A, B, and C are similar.

The next result shows that every matrix is similar to its transpose by means of a symmetric similarity transformation. This result, which improves Corollary 4.3.10, is due to Frobenius.

Corollary 5.3.8. Let $A \in \mathbb{F}^{n \times n}$. Then, there exists a symmetric non-singular matrix $S \in \mathbb{F}^{n \times n}$ such that $A = SA^{\mathrm{T}}S^{-1}$.

Proof. It follows from Theorem 5.3.3 that there exists a nonsingular matrix $\hat{S} \in \mathbb{C}^{n \times n}$ such that $A = \hat{S}B\hat{S}^{-1}$, where $B = \mathrm{diag}(B_1, \ldots, B_r)$ is the Jordan form of A, and $B_i \in \mathbb{C}^{n_i \times n_i}$ for all $i = 1, \ldots, r$. Now, define the symmetric nonsingular matrix $S \triangleq \hat{S}\tilde{I}\hat{S}^{\mathrm{T}}$, where $\tilde{I} \triangleq \mathrm{diag}\left(\hat{I}_{n_1}, \ldots, \hat{I}_{n_r}\right)$ is symmetric and involutory. Furthermore, note that $\hat{I}_{n_i} B_i \hat{I}_{n_i} = B_i^{\mathrm{T}}$ for all $i = 1, \ldots, r$ so that $\tilde{I}B\tilde{I} = B^{\mathrm{T}}$, and thus $\tilde{I}B^{\mathrm{T}}\tilde{I} = B$. Hence, it follows that

$$SA^{\mathrm{T}}S^{-1} = S\hat{S}^{-\mathrm{T}}B^{\mathrm{T}}\hat{S}^{\mathrm{T}}S^{-1} = \hat{S}\tilde{I}\hat{S}^{\mathrm{T}}\hat{S}^{-\mathrm{T}}B^{\mathrm{T}}\hat{S}^{\mathrm{T}}\hat{S}^{-\mathrm{T}}\tilde{I}\hat{S}^{-1}$$

$$= \hat{S}\tilde{I}B^{\mathrm{T}}\tilde{I}\hat{S}^{-1} = \hat{S}B\hat{S}^{-1} = A.$$

If A is real, then a similar argument based on the real Jordan form shows that S can be chosen to be real. $\square$

Corollary 5.3.9. Let $A \in \mathbb{F}^{n \times n}$. Then, there exist symmetric matrices $S_1, S_2 \in \mathbb{F}^{n \times n}$ such that S_2 is nonsingular and $A = S_1 S_2$.

Proof. From Corollary 5.3.8 it follows that there exists a symmetric, nonsingular matrix $S \in \mathbb{F}^{n \times n}$ such that $A = S A^{\mathrm{T}} S^{-1}$. Now, let $S_1 \triangleq S A^{\mathrm{T}}$ and $S_2 \triangleq S^{-1}$. Note that S_2 is symmetric and nonsingular. Furthermore, $S_1^{\mathrm{T}} = AS = SA^{\mathrm{T}} = S_1$, which shows that S_1 is symmetric.   □

Note that Corollary 5.3.8 follows from Corollary 5.3.9. If $A = S_1 S_2$, where S_1, S_2 are symmetric and S_2 is nonsingular, then $A = S_2^{-1} S_2 S_1 S_2 = S_2^{-1} A^{\mathrm{T}} S_2$.

5.4 Schur Decomposition

The *Schur decomposition* uses a unitary similarity transformation to transform an arbitrary square matrix into an upper triangular matrix.

Theorem 5.4.1. Let $A \in \mathbb{C}^{n \times n}$. Then, there exist a unitary matrix $S \in \mathbb{C}^{n \times n}$ and an upper triangular matrix $B \in \mathbb{C}^{n \times n}$ such that

$$A = SBS^*. \tag{5.4.1}$$

Proof. Let $\lambda_1 \in \mathbb{C}$ be an eigenvalue of A with associated eigenvector $x \in \mathbb{C}^n$ chosen such that $x^* x = 1$. Furthermore, let $S_1 \triangleq \begin{bmatrix} x & \hat{S}_1 \end{bmatrix} \in \mathbb{C}^{n \times n}$ be unitary, where $\hat{S}_1 \in \mathbb{C}^{n \times (n-1)}$ satisfies $\hat{S}_1^* S_1 = I_{n-1}$ and $x^* \hat{S}_1 = 0_{1 \times (n-1)}$. Then, $S_1 e_1 = x$, and

$$\mathrm{col}_1(S_1^{-1} A S_1) = S_1^{-1} A x = \lambda_1 S_1^{-1} x = \lambda_1 e_1.$$

Consequently,

$$A = S_1 \begin{bmatrix} \lambda_1 & C_1 \\ 0_{(n-1) \times 1} & A_1 \end{bmatrix} S_1^{-1},$$

where $C_1 \in \mathbb{C}^{1 \times (n-1)}$ and $A_1 \in \mathbb{C}^{(n-1) \times (n-1)}$. Next, let $S_{20} \in \mathbb{C}^{(n-1) \times (n-1)}$ be a unitary matrix such that

$$A_1 = S_{20} \begin{bmatrix} \lambda_2 & C_2 \\ 0_{(n-2) \times 1} & A_2 \end{bmatrix} S_{20}^{-1},$$

where $C_2 \in \mathbb{C}^{1 \times (n-2)}$ and $A_2 \in \mathbb{C}^{(n-2) \times (n-2)}$. Hence,

$$A = S_1 S_2 \begin{bmatrix} \lambda_1 & C_{11} & C_{12} \\ 0 & \lambda_2 & C_2 \\ 0 & 0 & A_2 \end{bmatrix} S_2^{-1} S_1,$$

where $C_1 = \begin{bmatrix} C_{11} & C_{12} \end{bmatrix}$, $C_{11} \in \mathbb{C}$, and $S_2 \triangleq \begin{bmatrix} 1 & 0 \\ 0 & S_{20} \end{bmatrix}$ is unitary. Proceeding in a similar manner yields (5.4.1) with $S \triangleq S_1 S_2 \cdots S_{n-1}$, where

$S_1, \ldots, S_{n-1} \in \mathbb{C}^{n \times n}$ are unitary. $\qquad\qquad\qquad\qquad\qquad\qquad$ $\square$

It can be seen that the diagonal entries of B are the eigenvalues of A.

The *real Schur decomposition* uses a real orthogonal similarity transformation to transform a real matrix into an upper Hessenberg matrix with real 1×1 and 2×2 diagonally located blocks.

Corollary 5.4.2. Let $A \in \mathbb{R}^{n \times n}$, and let $\mathrm{mspec}(A) = \{\lambda_1, \ldots, \lambda_r\}_\mathrm{m} \cup \{\nu_1 + \jmath\omega_1, \nu_1 - \jmath\omega_1, \ldots, \nu_l + \jmath\omega_l, \nu_l - \jmath\omega_l\}_\mathrm{m}$, where $\lambda_1, \ldots, \lambda_r \in \mathbb{R}$ and, for all $i = 1, \ldots, l$, $\nu_i, \omega_i \in \mathbb{R}$ and $\omega_i \neq 0$. Then, there exists an orthogonal matrix $S \in \mathbb{R}^{n \times n}$ such that

$$A = SBS^\mathrm{T}, \qquad\qquad (5.4.2)$$

where B is upper block triangular and the diagonally located blocks $B_1, \ldots, B_r \in \mathbb{R}$ and $\hat{B}_1, \ldots, \hat{B}_l \in \mathbb{R}^{2 \times 2}$ of B are $B_i \triangleq [\lambda_i]$ for all $i = 1, \ldots, r$ and $\hat{B}_i \triangleq \left[\begin{smallmatrix} \nu_i & \omega_i \\ -\omega_i & \nu_i \end{smallmatrix} \right]$ for all $i = 1, \ldots, l$.

Proof. The proof is analogous to the proof of Theorem 5.3.5. See also [367, p. 152]. $\qquad\qquad\qquad\qquad\qquad\qquad\qquad\qquad$ $\square$

Corollary 5.4.3. Let $A \in \mathbb{R}^{n \times n}$, and assume that the spectrum of A is real. Then, there exist an orthogonal matrix $S \in \mathbb{R}^{n \times n}$ and an upper triangular matrix $B \in \mathbb{R}^{n \times n}$ such that

$$A = SBS^\mathrm{T}. \qquad\qquad (5.4.3)$$

The Schur decomposition reveals the structure of range-Hermitian matrices and thus, as a special case, normal matrices.

Corollary 5.4.4. Let $A \in \mathbb{F}^{n \times n}$, and define $r \triangleq \mathrm{rank}\, A$. Then, A is range Hermitian if and only if there exist a unitary matrix $S \in \mathbb{F}^{n \times n}$ and a nonsingular matrix $B \in \mathbb{F}^{r \times r}$ such that

$$A = S \begin{bmatrix} B & 0 \\ 0 & 0 \end{bmatrix} S^*. \qquad\qquad (5.4.4)$$

In addition, A is normal if and only if there exist a unitary matrix $S \in \mathbb{C}^{n \times n}$ and a diagonal matrix $B \in \mathbb{C}^{r \times r}$ such that (5.4.4) is satisfied.

Proof. Suppose that A is range Hermitian, and let $A = S\hat{B}S^*$, where $\hat{B}$ is upper triangular and $S \in \mathbb{F}^{n \times n}$ is unitary. Assume that A is singular, and choose S such that $\hat{B}_{(j,j)} = \hat{B}_{(j+1,j+1)} = \cdots = \hat{B}_{(n,n)} = 0$ and such that all other diagonal entries of $\hat{B}$ are nonzero. Thus, $\mathrm{row}_n(\hat{B}) = 0$, which implies that $e_n \notin \mathcal{R}(\hat{B})$. Since A is range Hermitian, it follows that $\mathcal{R}(\hat{B}) = \mathcal{R}(\hat{B}^*)$ so that $e_n \notin \mathcal{R}(\hat{B}^*)$. Thus, $\mathrm{col}_n(\hat{B}) = \mathrm{row}_n(\hat{B}^*) = 0$. If, in addition, $\hat{B}_{(n-1,n-1)} = 0$, then $\mathrm{col}_{n-1}(\hat{B}) = 0$. Repeating this argument shows that $\hat{B}$

has the form $\begin{bmatrix} B & 0 \\ 0 & 0 \end{bmatrix}$, where $B \in \mathbb{F}^{r \times r}$ is nonsingular.

Now, suppose that A is normal, and let $A = S\hat{B}S^*$, where $\hat{B} \in \mathbb{C}^{n \times n}$ is upper triangular and $S \in \mathbb{C}^{n \times n}$ is unitary. Since A is normal, it follows that $AA^* = A^*A$, which implies that $\hat{B}\hat{B}^* = \hat{B}^*\hat{B}$. Since $\hat{B}$ is upper triangular, it follows that $(\hat{B}^*\hat{B})_{(1,1)} = \hat{B}_{(1,1)}\overline{\hat{B}_{(1,1)}}$, whereas $(\hat{B}\hat{B}^*)_{(1,1)} = \text{row}_1(\hat{B})[\text{row}_1(\hat{B})]^* = \sum_{i=1}^{n} \hat{B}_{(1,i)}\overline{\hat{B}_{(1,i)}}$. Since $(\hat{B}^*\hat{B})_{(1,1)} = (\hat{B}\hat{B}^*)_{(1,1)}$, it follows that $\hat{B}_{(1,i)} = 0$ for all $i = 2, \ldots, n$. Continuing in a similar fashion row by row, it follows that $\hat{B}$ is diagonal. $\square$

Corollary 5.4.5. Let $A \in \mathbb{F}^{n \times n}$, assume that A is Hermitian, and define $r \triangleq \text{rank } A$. Then, there exist a unitary matrix $S \in \mathbb{F}^{n \times n}$ and a diagonal matrix $B \in \mathbb{R}^{r \times r}$ such that (5.4.4) is satisfied. In addition, A is positive semidefinite if and only if the diagonal entries of B are positive, and A is positive definite if and only if A is positive semidefinite and $r = n$.

Proof. Corollary 5.4.4 and x), xi) of Proposition 4.4.4 imply that there exist a unitary matrix $S \in \mathbb{F}^{n \times n}$ and a diagonal matrix $B \in \mathbb{R}^{r \times r}$ such that (5.4.4) is satisfied. If A is positive semidefinite, then $x^*Ax \geq 0$ for all $x \in \mathbb{F}^n$. Choosing $x = Se_i$, it follows that $B_{(i,i)} = e_i^{\text{T}}S^*ASe_i \geq 0$ for all $i = 1, \ldots, r$. If A is positive definite, then $r = n$ and $B_{(i,i)} > 0$ for all $i = 1, \ldots, n$. $\square$

Proposition 5.4.6. Let $A \in \mathbb{F}^{n \times n}$ be Hermitian. Then, there exists a nonsingular matrix $S \in \mathbb{F}^{n \times n}$ such that

$$A = S \begin{bmatrix} -I_{\nu_-(A)} & 0 & 0 \\ 0 & 0_{\nu_0(A) \times \nu_0(A)} & 0 \\ 0 & 0 & I_{\nu_+(A)} \end{bmatrix} S^*. \tag{5.4.5}$$

Furthermore,

$$\text{rank } A = \nu_+(A) + \nu_-(A) \tag{5.4.6}$$

and

$$\text{def } A = \nu_0(A). \tag{5.4.7}$$

Proof. Since A is Hermitian, it follows from Corollary 5.4.5 that there exist a unitary matrix $\hat{S} \in \mathbb{F}^{n \times n}$ and a diagonal matrix $B \in \mathbb{R}^{n \times n}$ such that $A = \hat{S}B\hat{S}^*$. Choose S to order the diagonal entries of B such that $B = \text{diag}(B_1, 0, -B_2)$, where the diagonal matrices B_1, B_2 are both positive definite. Now, define $\hat{B} \triangleq \text{diag}(B_1, I, B_2)$. Then, $B = \hat{B}^{1/2}D\hat{B}^{1/2}$, where $D \triangleq \text{diag}(I_{\nu_-(A)}, 0_{\nu_0(A) \times \nu_0(A)}, -I_{\nu_+(A)})$. Hence, $A = \hat{S}\hat{B}^{1/2}D\hat{B}^{1/2}\hat{S}^*$. $\square$

Corollary 5.4.7. Let $A, B \in \mathbb{F}^{n \times n}$ be Hermitian. Then, A and B are congruent if and only if $\text{In } A = \text{In } B$.

In Proposition 4.5.3 it was shown that two or more eigenvectors associated with distinct eigenvalues of a normal matrix are mutually orthogonal. Thus, a normal matrix will have at least as many mutually orthogonal eigenvectors as it has distinct eigenvalues. The next result, which is an immediate consequence of Corollary 5.4.4, shows that every $n \times n$ normal matrix actually has n mutually orthogonal eigenvectors. In fact, the converse is also true.

Corollary 5.4.8. Let $A \in \mathbb{C}^{n \times n}$. Then, A is normal if and only if A has n mutually orthogonal eigenvectors.

The following result concerns the *real normal form*.

Corollary 5.4.9. Let $A \in \mathbb{R}^{n \times n}$ be range symmetric. Then, there exist an orthogonal matrix $S \in \mathbb{R}^{n \times n}$ and a nonsingular matrix $B \in \mathbb{R}^{r \times r}$, where $r \triangleq \operatorname{rank} A$, such that

$$A = S \begin{bmatrix} B & 0 \\ 0 & 0 \end{bmatrix} S^{\mathrm{T}}. \tag{5.4.8}$$

In addition, assume that A is normal, and let $\operatorname{mspec}(A) = \{\lambda_1, \ldots, \lambda_r\}_{\mathrm{m}} \cup \{\nu_1 + \jmath\omega_1, \nu_1 - \jmath\omega_1, \ldots, \nu_l + \jmath\omega_l, \nu_l - \jmath\omega_l\}_{\mathrm{m}}$, where $\lambda_1, \ldots, \lambda_r \in \mathbb{R}$ and, for all $i = 1, \ldots, l$, $\nu_i, \omega_i \in \mathbb{R}$ and $\omega_i \neq 0$. Then, there exists an orthogonal matrix $S \in \mathbb{R}^{n \times n}$ such that

$$A = SBS^{\mathrm{T}}, \tag{5.4.9}$$

where $B \triangleq \operatorname{diag}(B_1, \ldots, B_r, \hat{B}_1, \ldots, \hat{B}_l)$, $B_i \triangleq [\lambda_i]$ for all $i = 1, \ldots, r$, and $\hat{B}_i \triangleq \begin{bmatrix} \nu_i & \omega_i \\ -\omega_i & \nu_i \end{bmatrix}$ for all $i = 1, \ldots, l$.

5.5 Eigenstructure Properties

Definition 5.5.1. Let $A \in \mathbb{F}^{n \times n}$, and let $\lambda \in \mathbb{C}$. Then, the *index of λ with respect to A*, denoted by $\operatorname{ind}_A(\lambda)$, is the smallest nonnegative integer k such that

$$\mathcal{R}\left[(\lambda I - A)^k\right] = \mathcal{R}\left[(\lambda I - A)^{k+1}\right]. \tag{5.5.1}$$

Furthermore, the *index of A*, denoted by $\operatorname{ind} A$, is the smallest nonnegative integer k such that

$$\mathcal{R}\left(A^k\right) = \mathcal{R}\left(A^{k+1}\right), \tag{5.5.2}$$

that is, $\operatorname{ind} A = \operatorname{ind}_A(0)$.

Note that $\lambda \notin \operatorname{spec}(A)$ if and only if $\operatorname{ind}_A(\lambda) = 0$. Hence, $0 \notin \operatorname{spec}(A)$ if and only if $\operatorname{ind} A = \operatorname{ind}_A(0) = 0$. Hence, A is nonsingular if and only if $\operatorname{ind} A = 0$.

Proposition 5.5.2. Let $A \in \mathbb{F}^{n \times n}$, and let $\lambda \in \mathbb{C}$. Then, $\mathrm{ind}_A(\lambda)$ is the smallest nonnegative integer k such that

$$\mathrm{rank}\left[(\lambda I - A)^k\right] = \mathrm{rank}\left[(\lambda I - A)^{k+1}\right]. \tag{5.5.3}$$

Furthermore, $\mathrm{ind}\, A$ is the smallest nonnegative integer k such that

$$\mathrm{rank}\left(A^k\right) = \mathrm{rank}\left(A^{k+1}\right). \tag{5.5.4}$$

Proof. Corollary 2.4.2 implies that $\mathcal{R}\left[(\lambda I - A)^k\right] \subseteq \mathcal{R}\left[(\lambda I - A)^{k+1}\right]$. Now, Lemma 2.3.4 implies that $\mathcal{R}\left[(\lambda I - A)^k\right] = \mathcal{R}\left[(\lambda I - A)^{k+1}\right]$ if and only if $\mathrm{rank}\left[(\lambda I - A)^k\right] = \mathrm{rank}\left[(\lambda I - A)^{k+1}\right]$.   □

Proposition 5.5.3. Let $A \in \mathbb{F}^{n \times n}$, and let $\lambda \in \mathrm{spec}(A)$. Then, the following statements hold:

  i) The order of the largest Jordan block of A associated with λ is $\mathrm{ind}_A(\lambda)$.

 ii) The number of Jordan blocks of A associated with λ is $\mathrm{gm}_A(\lambda)$.

iii) $\mathrm{ind}_A(\lambda) \le \mathrm{am}_A(\lambda)$.

 iv) $\mathrm{gm}_A(\lambda) \le \mathrm{am}_A(\lambda)$.

  v) $\mathrm{ind}_A(\lambda) + \mathrm{gm}_A(\lambda) \le \mathrm{am}_A(\lambda) + 1$.

 vi) $\mathrm{rank}\, A = n - \mathrm{gm}_A(0)$.

Proposition 5.5.4. Let $\mathcal{S} \subseteq \mathbb{F}^n$ be a subspace. Then, there exists a unique projector $A \in \mathbb{F}^{n \times n}$ such that $\mathcal{S} = \mathcal{R}(A)$. Furthermore, $x \in \mathcal{S}$ if and only if $x = Ax$.

Proof. See [535, p. 386].   □

For a subspace $\mathcal{S} \subseteq \mathbb{F}^n$, the matrix $A \in \mathbb{F}^{n \times n}$ given by Proposition 5.5.4 is the *projector onto* $\mathcal{S}$.

Let $A \in \mathbb{F}^{n \times n}$ be an idempotent matrix. Then, the *complementary idempotent matrix* defined by

$$A_\perp \triangleq I - A \tag{5.5.5}$$

is also idempotent. If A is a projector, then $A_\perp$ is the *complementary projector*.

Proposition 5.5.5. Let $\mathcal{S} \subseteq \mathbb{F}^n$ be a subspace, and let $A \in \mathbb{F}^{n \times n}$ be the projector onto $\mathcal{S}$. Then, $A_\perp$ is the projector onto $\mathcal{S}^\perp$. Furthermore,

$$\mathcal{R}(A)^\perp = \mathcal{N}(A) = \mathcal{R}(A_\perp). \tag{5.5.6}$$

Proposition 5.5.6. Let $A \in \mathbb{F}^{n \times n}$, and let $k \in \mathbb{P}$. Then, $\operatorname{ind} A \leq k$ if and only if $\mathcal{R}(A^k)$ and $\mathcal{N}(A^k)$ are complementary subspaces.

Corollary 5.5.7. Let $A \in \mathbb{F}^{n \times n}$. Then, A is group invertible if and only if $\mathcal{R}(A)$ and $\mathcal{N}(A)$ are complementary subspaces.

Proposition 5.5.8. Let $\mathcal{S}_1, \mathcal{S}_2 \subseteq \mathbb{F}^n$ be complementary subspaces. Then, there exists a unique idempotent matrix $A \in \mathbb{F}^{n \times n}$ such that $\mathcal{R}(A) = \mathcal{S}_1$ and $\mathcal{N}(A) = \mathcal{S}_2$. Furthermore, $\mathcal{R}(A_\perp) = \mathcal{S}_2$ and $\mathcal{N}(A_\perp) = \mathcal{S}_1$.

Proof. See [100, p. 118] or [535, p. 386]. $\qquad\qquad\square$

For complementary subspaces $\mathcal{S}_1, \mathcal{S}_2 \subseteq \mathbb{F}^n$, the unique idempotent matrix $A \in \mathbb{F}^{n \times n}$ given by Proposition 5.5.8 is the *idempotent matrix onto* $\mathcal{S}_1 = \mathcal{R}(A)$ *along* $\mathcal{S}_2 = \mathcal{N}(A)$. Proposition 5.5.8 shows that $A_\perp$ is the idempotent matrix onto $\mathcal{S}_2$ along $\mathcal{S}_1$.

Proposition 5.5.9. Let $A \in \mathbb{F}^{n \times n}$, and let $r \triangleq \operatorname{rank} A$. Then, A is group invertible if and only if there exist matrices $B \in \mathbb{F}^{n \times r}$ and $C \in \mathbb{F}^{r \times n}$ such that $A = BC$ and $\operatorname{rank} B = \operatorname{rank} C = r$. In this case, the idempotent matrix $P \triangleq B(CB)^{-1}C$ is the idempotent matrix onto $\mathcal{R}(A)$ along $\mathcal{N}(A)$.

Proof. See [535, p. 634]. $\qquad\qquad\square$

An alternative expression for the idempotent matrix onto $\mathcal{R}(A)$ along $\mathcal{N}(A)$ is given by Proposition 6.2.2.

Definition 5.5.10. Let $A \in \mathbb{F}^{n \times n}$, and let $\lambda \in \operatorname{spec}(A)$. Then, the following terminology is defined:

 i) λ is *simple* if $\operatorname{am}_A(\lambda) = 1$.

 ii) A is *simple* if every eigenvalue of A is simple.

 iii) λ is *cyclic* (or *nonderogatory*) if $\operatorname{gm}_A(\lambda) = 1$.

 iv) A is *cyclic* (or *nonderogatory*) if every eigenvalue of A is cyclic.

 v) λ is *derogatory* if $\operatorname{gm}_A(\lambda) > 1$.

 vi) A is *derogatory* if A has at least one derogatory eigenvalue.

 vii) λ is *semisimple* if $\operatorname{gm}_A(\lambda) = \operatorname{am}_A(\lambda)$.

 viii) A is *semisimple* if every eigenvalue of A is semisimple.

 ix) λ is *defective* if $\operatorname{gm}_A(\lambda) < \operatorname{am}_A(\lambda)$.

 x) A is *defective* if A has at least one defective eigenvalue.

 xi) A is *diagonalizable over* $\mathbb{C}$ if A is semisimple.

xii) $A \in \mathbb{R}^{n \times n}$ is *diagonalizable over* $\mathbb{R}$ if A is semisimple and every eigenvalue of A is real.

Proposition 5.5.11. Let $A \in \mathbb{F}^{n \times n}$, and let $\lambda \in \text{spec}(A)$. Then, λ is simple if and only if λ is cyclic and semisimple.

Proposition 5.5.12. Let $A \in \mathbb{F}^{n \times n}$, and let $\lambda \in \text{spec}(A)$. Then,

$$\text{def}\left[(\lambda I - A)^{\text{ind}_A(\lambda)}\right] = \text{am}_A(\lambda). \tag{5.5.7}$$

Theorem 5.3.3 yields the following result, which shows that the subspaces $\mathcal{N}\left[(\lambda I - A)^k\right]$, where $\lambda \in \text{spec}(A)$ and $k = \text{ind}_A(\lambda)$, provide a decomposition of $\mathbb{F}^n$.

Proposition 5.5.13. Let $A \in \mathbb{F}^{n \times n}$, let $\text{spec}(A) = \{\lambda_1, \ldots, \lambda_r\}$, and, for all $i = 1, \ldots, r$, let $k_i \triangleq \text{ind}_A(\lambda_i)$. Then, the following statements hold:

i) $\mathcal{N}\left[(\lambda_i I - A)^{k_i}\right] \cap \mathcal{N}\left[(\lambda_j I - A)^{k_j}\right] = \{0\}$ for all $i, j = 1, \ldots, r$ such that $i \neq j$.

ii) $\sum_{i=1}^{r} \mathcal{N}\left[(\lambda_i I - A)^{k_i}\right] = \mathbb{F}^n$.

Proposition 5.5.14. Let $A \in \mathbb{F}^{n \times n}$, and let $\lambda \in \text{spec}(A)$. Then, the following statements are equivalent:

i) λ is semisimple.

ii) $\text{def}(\lambda I - A) = \text{def}\left[(\lambda I - A)^2\right]$.

iii) $\mathcal{N}(\lambda I - A) = \mathcal{N}\left[(\lambda I - A)^2\right]$.

iv) $\text{ind}_A(\lambda) = 1$.

Proof. To prove that *i)* implies *ii)*, suppose that λ is semisimple so that $\text{gm}_A(\lambda) = \text{am}_A(\lambda)$, and thus $\text{def}(\lambda I - A) = \text{am}_A(\lambda)$. Then, it follows from Proposition 5.5.12 that $\text{def}\left[(\lambda I - A)^k\right] = \text{am}_A(\lambda)$, where $k \triangleq \text{ind}_A(\lambda)$. Therefore, it follows from Corollary 2.5.6 that $\text{am}_A(\lambda) = \text{def}(\lambda I - A) \leq \text{def}\left[(\lambda I - A)^2\right] \leq \text{def}\left[(\lambda I - A)^k\right] = \text{am}_A(\lambda)$, which implies that $\text{def}(\lambda I - A) = \text{def}\left[(\lambda I - A)^2\right]$.

To prove that *ii)* implies *iii)*, note that it follows from Corollary 2.5.6 that $\mathcal{N}(\lambda I - A) \subseteq \mathcal{N}\left[(\lambda I - A)^2\right]$. Since, by *ii)*, these subspaces have equal dimension, it follows from Lemma 2.3.4 that these subspaces are equal. Conversely, *iii)* implies *ii)*.

Finally, *iv)* is equivalent to the fact that every Jordan block of A associated with λ has order 1, which is equivalent to the fact that the geometric multiplicity of λ is equal to the algebraic multiplicity of λ, that is, that λ is semisimple. □

Corollary 5.5.15. Let $A \in \mathbb{F}^{n \times n}$. Then, A is group invertible if and only if $\operatorname{ind} A \leq 1$.

Proposition 5.5.16. Suppose $A, B \in \mathbb{F}^{n \times n}$ are similar. Then, the following statements hold:

i) $\operatorname{mspec}(A) = \operatorname{mspec}(B)$.

ii) For all $\lambda \in \operatorname{spec}(A)$, $\operatorname{gm}_A(\lambda) = \operatorname{gm}_B(\lambda)$.

Proposition 5.5.17. Let $A \in \mathbb{F}^{n \times n}$. Then, A is semisimple if and only if A is similar to a normal matrix.

The following result is an extension of Corollary 5.3.9.

Proposition 5.5.18. Let $A \in \mathbb{F}^{n \times n}$. Then, the following statements are equivalent:

i) A is semisimple, and $\operatorname{spec}(A) \subset \mathbb{R}$.

ii) There exists a positive-definite matrix $S \in \mathbb{F}^{n \times n}$ such that $A = SA^*S^{-1}$.

iii) There exist a Hermitian matrix $S_1 \in \mathbb{F}^{n \times n}$ and a positive-definite matrix $S_2 \in \mathbb{F}^{n \times n}$ such that $A = S_1 S_2$.

Proof. To prove that *i*) implies *ii*), let $\hat{S} \in \mathbb{F}^{n \times n}$ be a nonsingular matrix such that $A = \hat{S} B \hat{S}^{-1}$, where $B \in \mathbb{R}^{n \times n}$ is diagonal. Then, $B = \hat{S}^{-1} A \hat{S} = \hat{S}^* A^* \hat{S}^{-*}$. Hence, $A = \hat{S} B \hat{S}^{-1} = \hat{S}(\hat{S}^* A^* \hat{S}^{-*})\hat{S}^{-1} = (\hat{S}\hat{S}^*) A^* (\hat{S}\hat{S}^*)^{-1} = SA^*S^{-1}$, where $S \triangleq \hat{S}\hat{S}^*$ is positive definite. To show that *ii*) implies *iii*), note that $A = SA^*S^{-1} = S_1 S_2$, where $S_1 \triangleq SA^*$ and $S_2 = S^{-1}$. Since $S_1^* = (SA^*)^* = AS^* = AS = SA^* = S_1$, it follows that S_1 is Hermitian. Furthermore, since S is positive definite, it follows that S^{-1}, and hence S_2, is also positive definite. Finally, to prove that *iii*) implies *i*), note that $A = S_1 S_2 = S_2^{-1/2}(S_2^{1/2} S_1 S_2^{1/2}) S_2^{1/2}$. Since $S_2^{1/2} S_1 S_2^{1/2}$ is Hermitian, it follows from Corollary 5.4.5 that $S_2^{1/2} S_1 S_2^{1/2}$ is unitarily similar to a real diagonal matrix. Consequently, A is semisimple and $\operatorname{spec}(A) \subset \mathbb{R}$. □

If a matrix is block triangular, then the following result shows that its eigenvalues and their algebraic multiplicity are determined by the diagonally located blocks. If, in addition, the matrix is block diagonal, then the geometric multiplicities of its eigenvalues are determined by the diagonally located blocks.

Proposition 5.5.19. Let $A \in \mathbb{F}^{n \times n}$ be either upper block triangular or lower block triangular with diagonally located blocks $A_{11}, \ldots, A_{rr}$, where $A_{ii} \in \mathbb{F}^{n_i \times n_i}$ for all $i = 1, \ldots, r$. Then,

$$\mathrm{am}_A(\lambda) = \sum_{i=1}^{r} \mathrm{am}_{A_{ii}}(\lambda). \tag{5.5.8}$$

Hence,

$$\mathrm{mspec}(A) = \bigcup_{i=1}^{r} \mathrm{mspec}(A_{ii}). \tag{5.5.9}$$

Now, assume that A is block diagonal. Then,

$$\mathrm{gm}_A(\lambda) = \sum_{i=1}^{r} \mathrm{gm}_{A_{ii}}(\lambda). \tag{5.5.10}$$

Proposition 5.5.20. Let $A \in \mathbb{F}^{n \times n}$, let $\mathrm{spec}(A) = \{\lambda_1, \ldots, \lambda_r\}$, and let $k_i \triangleq \mathrm{ind}_A(\lambda_i)$ for all $i = 1, \ldots, r$. Then,

$$\mu_A(s) = \prod_{i=1}^{r} (s - \lambda_i)^{k_i} \tag{5.5.11}$$

and

$$\deg \mu_A = \sum_{i=1}^{r} k_i. \tag{5.5.12}$$

Furthermore, the following statements are equivalent:

i) $\mu_A = \chi_A$.

ii) A is cyclic.

iii) For all $\lambda \in \mathrm{spec}(A)$, the Jordan form of A contains exactly one block associated with λ.

Proof. Let $A = SBS^{-1}$, where $B = \mathrm{diag}(B_1, \ldots, B_{n_h})$ denotes the Jordan form of A given by (5.3.4). Let $\lambda_i \in \mathrm{spec}(A)$, and let B_j be a Jordan block associated with λ_i. Then, the order of B_j is less than or equal to k_i. Consequently, $(B_j - \lambda_i I)^{k_i} = 0$.

Next, let $p(s)$ denote the right-hand side of (5.5.11). Thus,

$$p(A) = \prod_{i=1}^{r} (A - \lambda_i I)^{k_i} = S \left[\prod_{i=1}^{r} (B - \lambda_i I)^{k_i} \right] S^{-1}$$

$$= S \mathrm{diag} \left(\prod_{i=1}^{r} (B_1 - \lambda_i I)^{k_i}, \ldots, \prod_{i=1}^{r} (B_{n_h} - \lambda_i I)^{k_i} \right) S^{-1} = 0.$$

Therefore, it follows from Theorem 4.6.1 that μ_A divides p. Furthermore, note that, if k_i is replaced by $\hat{k}_i < k_i$, then $p(A) \neq 0$. Hence, p is the minimal polynomial of A. The equivalence of i) and ii) is now immediate, while the equivalence of ii) and iii) follows from Theorem 5.3.5.   □

Example 5.5.21. The matrix $\begin{bmatrix} 1 & 1 \\ -1 & 1 \end{bmatrix}$ is normal but is neither symmetric nor skew symmetric, while the matrix $\begin{bmatrix} 0 & 1 \\ -1 & 0 \end{bmatrix}$ is normal but is neither symmetric nor semisimple with real eigenvalues.

Example 5.5.22. The matrices $\begin{bmatrix} 1 & 0 \\ 2 & -1 \end{bmatrix}$ and $\begin{bmatrix} 1 & 1 \\ 0 & 2 \end{bmatrix}$ are diagonalizable over $\mathbb{R}$ but not normal, while the matrix $\begin{bmatrix} -1 & 1 \\ -2 & 1 \end{bmatrix}$ is diagonalizable but is neither normal nor diagonalizable over $\mathbb{R}$.

Example 5.5.23. The product of the Hermitian matrices $\begin{bmatrix} 1 & 2 \\ 2 & 1 \end{bmatrix}$ and $\begin{bmatrix} 2 & 1 \\ 1 & -2 \end{bmatrix}$ has no real eigenvalues.

Example 5.5.24. The matrices $\begin{bmatrix} 1 & 0 \\ 0 & 2 \end{bmatrix}$ and $\begin{bmatrix} 0 & 1 \\ -2 & 3 \end{bmatrix}$ are similar, whereas $\begin{bmatrix} 1 & 0 \\ 0 & 1 \end{bmatrix}$ and $\begin{bmatrix} 0 & 1 \\ -1 & 2 \end{bmatrix}$ have the same spectrum but are not similar.

Proposition 5.5.25. Let $A \in \mathbb{F}^{n \times n}$. Then, the following statements hold:

   i) A is singular if and only if $0 \in \mathrm{spec}(A)$.

  ii) A is group invertible if and only if either A is nonsingular or $0 \in \mathrm{spec}(A)$ is semisimple.

 iii) A is Hermitian if and only if A is normal and $\mathrm{spec}(A) \subset \mathbb{R}$.

 iv) A is skew Hermitian if and only if A is normal and $\mathrm{spec}(A) \subset \jmath\mathbb{R}$.

  v) A is positive semidefinite if and only if A is normal and $\mathrm{spec}(A) \subset [0, \infty)$.

 vi) A is positive definite if and only if A is normal and $\mathrm{spec}(A) \subset (0, \infty)$.

vii) A is unitary if and only if A is normal and $\mathrm{spec}(A) \subset \{\lambda \in \mathbb{C} : |\lambda| = 1\}$.

$viii$) A is involutory if and only if A is semisimple and $\mathrm{spec}(A) \subseteq \{-1, 1\}$.

 ix) A is skew involutory if and only if A is semisimple and $\mathrm{spec}(A) \subseteq \{-\jmath, \jmath\}$.

  x) A is idempotent if and only if A is semisimple and $\mathrm{spec}(A) \subseteq \{0, 1\}$.

 xi) A is tripotent if and only if A is semisimple and $\mathrm{spec}(A) \subseteq \{-1, 0, 1\}$.

 xii) A is nilpotent if and only if $\mathrm{spec}(A) = \{0\}$.

$xiii$) A is unipotent if and only if $\mathrm{spec}(A) = \{1\}$.

xiv) A is a projector if and only if A is normal and $\mathrm{spec}(A) \subseteq \{0, 1\}$.

$xv)$ A is a reflector if and only if A is normal and $\text{spec}(A) \subseteq \{-1, 1\}$.

$xvi)$ A is an elementary projector if and only if A is normal and $\text{mspec}(A)$ $= \{0, 1, \ldots, 1\}_\text{m}$.

$xvii)$ A is an elementary reflector if and only if A is normal and $\text{mspec}(A)$ $= \{-1, 1, \ldots, 1\}_\text{m}$.

$xviii)$ A is an elementary matrix if and only if A is normal and $\text{mspec}(A) = \{\alpha, 1, \ldots, 1\}_\text{m}$, where $\alpha \neq 0$.

If, furthermore, $A \in \mathbb{F}^{2n \times 2n}$, then the following statements hold:

$xix)$ If A is Hamiltonian, then $\text{mspec}(A) = -\text{mspec}(A)$.

$xx)$ If A is symplectic, then $\text{mspec}(A) = \{1/\lambda : \ \lambda \in \text{mspec}(A)\}_\text{m}$.

5.6 Singular Value Decomposition

The third matrix decomposition that we consider is the *singular value decomposition*. Unlike the Jordan and Schur decompositions, the singular value decomposition applies to matrices that are not necessarily square. Let $A \in \mathbb{F}^{n \times m}$, where $A \neq 0$, and consider the positive-semidefinite matrices $AA^* \in \mathbb{F}^{n \times n}$ and $A^*A \in \mathbb{F}^{m \times m}$. It follows from Proposition 4.4.9 that AA^* and A^*A have the same nonzero eigenvalues with the same algebraic multiplicities. Since AA^* and A^*A are positive semidefinite, it follows that they have the same *positive* eigenvalues with the same algebraic multiplicities. Furthermore, since AA^* is Hermitian, it follows that the number of positive eigenvalues of AA^* (or A^*A) counting algebraic multiplicity is equal to the rank of AA^* (or A^*A). Since $\text{rank } A = \text{rank } AA^* = \text{rank } A^*A$, it thus follows that AA^* and A^*A both have r positive eigenvalues, where $r \triangleq \text{rank } A$.

Definition 5.6.1. Let $A \in \mathbb{F}^{n \times m}$. Then, the *singular values* of A are the $\min\{n, m\}$ nonnegative numbers $\sigma_1(A), \ldots, \sigma_{\min\{n,m\}}(A)$, where, for all $i = 1, \ldots, \min\{n, m\}$,

$$\sigma_i(A) \triangleq [\lambda_i(AA^*)]^{1/2} = [\lambda_i(A^*A)]^{1/2}. \tag{5.6.1}$$

Let $A \in \mathbb{F}^{n \times m}$. Then,

$$\sigma_1(A) \geq \cdots \geq \sigma_{\min\{n,m\}}(A) \geq 0. \tag{5.6.2}$$

If $A \neq 0$, then

$$\sigma_1(A) \geq \cdots \geq \sigma_r(A) > \sigma_{r+1}(A) = \cdots = \sigma_{\min\{n,m\}}(A) = 0, \tag{5.6.3}$$

where $r \triangleq \text{rank } A$. For convenience, define

$$\sigma_{\max}(A) \triangleq \sigma_1(A), \tag{5.6.4}$$

and, if $n = m$,
$$\sigma_{\min}(A) \triangleq \sigma_n(A). \tag{5.6.5}$$

Note that
$$\sigma_{\max}(0_{n \times n}) = \sigma_{\min}(0_{n \times n}) = 0, \tag{5.6.6}$$

and, for all $i = 1, \ldots, \min\{n, m\}$,
$$\sigma_i(A) = \sigma_i(A^*) = \sigma_i(\overline{A}) = \sigma_i(A^{\mathrm{T}}). \tag{5.6.7}$$

Proposition 5.6.2. Let $A \in \mathbb{F}^{n \times m}$, where $A \neq 0$. Then, the following statements are equivalent:

  i) rank $A = n$.

  ii) $\sigma_n(A) > 0$.

The following statements are also equivalent:

  iii) rank $A = m$.

  iv) $\sigma_m(A) > 0$.

Now, assume that $n = m$. Then, the following statements are also equivalent:

  v) A is nonsingular.

  vi) $\sigma_{\min}(A) > 0$.

We now state the singular value decomposition.

Theorem 5.6.3. Let $A \in \mathbb{F}^{n \times m}$, assume that A is nonzero, let $r \triangleq$ rank A, and define $B \triangleq \mathrm{diag}[\sigma_1(A), \ldots, \sigma_r(A)]$. Then, there exist unitary matrices $S_1 \in \mathbb{F}^{n \times n}$ and $S_2 \in \mathbb{F}^{m \times m}$ such that

$$A = S_1 \begin{bmatrix} B & 0_{r \times (m-r)} \\ 0_{(n-r) \times r} & 0_{(n-r) \times (m-r)} \end{bmatrix} S_2. \tag{5.6.8}$$

Proof. For convenience, assume that $r < \min\{n, m\}$, since otherwise the zero matrices become empty matrices. By Corollary 5.4.5 there exists a unitary matrix $U \in \mathbb{F}^{n \times n}$ such that

$$AA^* = U \begin{bmatrix} B^2 & 0 \\ 0 & 0 \end{bmatrix} U^*.$$

Partition $U = \begin{bmatrix} U_1 & U_2 \end{bmatrix}$, where $U_1 \in \mathbb{F}^{n \times r}$ and $U_2 \in \mathbb{F}^{n \times (n-r)}$. Since $U^*U = I_n$, it follows that $U_1^*U_1 = I_r$ and $U_1^*U = \begin{bmatrix} I_r & 0_{r \times (n-r)} \end{bmatrix}$. Now, define $V_1 \triangleq A^*U_1 B^{-1} \in \mathbb{F}^{m \times r}$, and note that

$$V_1^*V_1 = B^{-1}U_1^*AA^*U_1 B^{-1} = B^{-1}U_1^*U \begin{bmatrix} B^2 & 0 \\ 0 & 0 \end{bmatrix} U^*U_1 B^{-1} = I_r.$$

Next, note that, since $U_2^*U = \begin{bmatrix} 0_{(n-r)\times r} & I_{n-r} \end{bmatrix}$, it follows that

$$U_2^*AA^* = \begin{bmatrix} 0 & I \end{bmatrix} \begin{bmatrix} B^2 & 0 \\ 0 & 0 \end{bmatrix} U^* = 0.$$

However, since $\mathcal{R}(A) = \mathcal{R}(AA^*)$, it follows that $U_2^*A = 0$. Finally, let $V_2 \in \mathbb{F}^{m\times(m-r)}$ be such that $V \triangleq \begin{bmatrix} V_1 & V_2 \end{bmatrix} \in \mathbb{F}^{m\times m}$ is unitary. Hence, we have

$$U \begin{bmatrix} B & 0 \\ 0 & 0 \end{bmatrix} V^* = \begin{bmatrix} U_1 & U_2 \end{bmatrix} \begin{bmatrix} B & 0 \\ 0 & 0 \end{bmatrix} \begin{bmatrix} V_1^* \\ V_2^* \end{bmatrix} = U_1 B V_1^* = U_1 B B^{-1} U_1^* A$$

$$= U_1 U_1^* A = (U_1 U_1^* + U_2 U_2^*)A = UU^*A = A,$$

which yields (5.6.8) with $S_1 = U$ and $S_2 = V^*$. $\qquad\square$

An immediate corollary of the singular value decomposition is the *polar decomposition*.

Corollary 5.6.4. Let $A \in \mathbb{F}^{n\times n}$. Then, there exists a positive-semidefinite matrix $M \in \mathbb{F}^{n\times n}$ and a unitary matrix $S \in \mathbb{F}^{n\times n}$ such that

$$A - MS. \qquad (5.6.9)$$

Proof. It follows from the singular value decomposition that there exist unitary matrices $S_1, S_2 \in \mathbb{F}^{n\times n}$ and a diagonal positive-definite matrix $B \in \mathbb{F}^{r\times r}$, where $r \triangleq \operatorname{rank} A$, such that $A = S_1 \begin{bmatrix} B & 0 \\ 0 & 0 \end{bmatrix} S_2$. Hence,

$$A = S_1 \begin{bmatrix} B & 0 \\ 0 & 0 \end{bmatrix} S_1^* S_1 S_2 = MS,$$

where $M \triangleq S_1 \begin{bmatrix} B & 0 \\ 0 & 0 \end{bmatrix} S_1^*$ is positive semidefinite and $S \triangleq S_1 S_2$ is unitary. $\quad\square$

Proposition 5.6.5. Let $A \in \mathbb{F}^{n\times m}$, let $r \triangleq \operatorname{rank} A$, and define the Hermitian matrix $\mathcal{A} \triangleq \begin{bmatrix} 0 & A \\ A^* & 0 \end{bmatrix} \in \mathbb{F}^{(n+m)\times(n+m)}$. Then, $\operatorname{In} \mathcal{A} = \begin{bmatrix} r & 0 & r \end{bmatrix}^{\mathrm{T}}$, and the $2r$ nonzero eigenvalues of $\mathcal{A}$ are the r positive singular values of A and their negatives.

Proof. Since $\chi_{\mathcal{A}}(s) = \det(s^2 I - A^*A)$, it follows that

$$\operatorname{mspec}(\mathcal{A})\backslash\{0,\ldots,0\}_{\mathrm{m}} = \{\sigma_1(A), -\sigma_1(A), \ldots, \sigma_r(A), -\sigma_r(A)\}_{\mathrm{m}}. \qquad\square$$

5.7 Facts on the Inertia

Fact 5.7.1. Let $A \in \mathbb{F}^{n\times n}$, and assume that A is idempotent. Then,

$$\operatorname{rank} A = \operatorname{sig} A = \operatorname{tr} A$$

and

$$\text{In } A = \left[\begin{array}{c} 0 \\ n - \text{tr } A \\ \text{tr } A \end{array} \right].$$

Fact 5.7.2. Let $A \in \mathbb{F}^{n \times n}$, and assume that A is involutory. Then,

$$\text{rank } A = n,$$
$$\text{sig } A = \text{tr } A,$$

and

$$\text{In } A = \left[\begin{array}{c} \frac{1}{2}(n - \text{tr } A) \\ 0 \\ \frac{1}{2}(n + \text{tr } A) \end{array} \right].$$

Fact 5.7.3. Let $A \in \mathbb{F}^{n \times n}$, and assume that A is tripotent. Then,

$$\text{rank } A = \text{tr } A^2,$$
$$\text{sig } A = \text{tr } A,$$

and

$$\text{In } A = \left[\begin{array}{c} \frac{1}{2}\left(\text{tr } A^2 - \text{tr } A\right) \\ n - \text{tr } A^2 \\ \frac{1}{2}\left(\text{tr } A^2 + \text{tr } A\right) \end{array} \right].$$

Fact 5.7.4. Let $A \in \mathbb{F}^{n \times n}$, and assume that A is either skew Hermitian, skew involutory, or nilpotent. Then,

$$\text{sig } A = \nu_-(A) = \nu_+(A) = 0$$

and

$$\text{In } A = \left[\begin{array}{c} 0 \\ n \\ 0 \end{array} \right].$$

Fact 5.7.5. Let $A \in \mathbb{F}^{n \times n}$, assume that A is group invertible, and assume that $\text{spec}(A) \cap \jmath\mathbb{R} \subseteq \{0\}$. Then,

$$\text{rank } A = \nu_+(A) + \nu_-(A)$$

and

$$\text{def } A = \nu_0(A) = \text{am}_A(0).$$

Fact 5.7.6. Let $A \in \mathbb{F}^{n \times n}$, and assume that A is Hermitian. Then,

$$\text{rank } A = \nu_+(A) + \nu_-(A)$$

and

$$\text{In } A = \begin{bmatrix} \nu_-(A) \\ \nu_0(A) \\ \nu_+(A) \end{bmatrix} = \begin{bmatrix} \frac{1}{2}(\text{rank } A - \text{sig } A) \\ \text{def } A \\ \frac{1}{2}(\text{rank } A + \text{sig } A) \end{bmatrix}.$$

Fact 5.7.7. Let $A \in \mathbb{F}^{n \times n}$, and assume that A is positive semidefinite. Then,

$$\text{rank } A = \text{sig } A = \nu_+(A)$$

and

$$\text{In } A = \begin{bmatrix} 0 \\ \text{def } A \\ \text{rank } A \end{bmatrix}.$$

Fact 5.7.8. Let $A, B \in \mathbb{F}^{n \times n}$, and assume that A and B are Hermitian. Then, In $A = $ In B if and only if rank $A = $ rank B, sig $A = $ sig B, and def $A = $ def B.

Fact 5.7.9. Let $A \in \mathbb{F}^{n \times n}$. Then, the following statements are equivalent:

$i)$ A is an elementary projector.

$ii)$ A is a projector, and tr $A = n - 1$.

$iii)$ A is a projector, and In $A = \begin{bmatrix} 0 \\ 1 \\ n-1 \end{bmatrix}$.

Furthermore, the following statements are equivalent:

$i)$ A is an elementary reflector.

$ii)$ A is a reflector, and tr $A = n - 2$.

$iii)$ A is a reflector, and In $A = \begin{bmatrix} 1 \\ 0 \\ n-1 \end{bmatrix}$.

(Proof: See Proposition 5.5.25.)

Fact 5.7.10. Let $A \in \mathbb{F}^{n \times n}$, and assume that A is positive semidefinite. Then,

$$\text{In } A = \begin{bmatrix} 0 \\ \text{def } A \\ \text{rank } A \end{bmatrix}.$$

If, in addition, A is positive definite, then

$$\text{In } A = \begin{bmatrix} 0 \\ 0 \\ n \end{bmatrix}.$$

Fact 5.7.11. Let $A \in \mathbb{F}^{n \times n}$. Then, there exist a nonsingular matrix $S \in \mathbb{F}^{n \times n}$ and a skew-Hermitian matrix $B \in \mathbb{F}^{n \times n}$ such that

$$A = S \left(\begin{bmatrix} I_{\nu_-(A+A^*)} & 0 & 0 \\ 0 & 0_{\nu_0(A+A^*) \times \nu_0(A+A^*)} & 0 \\ 0 & 0 & -I_{\nu_+(A+A^*)} \end{bmatrix} + B \right) S^*.$$

(Proof: Write $A = \frac{1}{2}(A + A^*) + \frac{1}{2}(A - A^*)$, and apply Proposition 5.4.6 to $\frac{1}{2}(A + A^*)$.)

Fact 5.7.12. Let $A \in \mathbb{F}^{n \times n}$, assume that A is Hermitian, let $S \in \mathbb{F}^{m \times n}$, and assume that $\operatorname{rank} S = n$. Then, $\nu_+(SAS^*) = \nu_+(A)$ and $\nu_-(SAS^*) = \nu_-(A)$. (Proof: See [275, p. 194].)

5.8 Facts on Matrix Transformations for One Matrix

Fact 5.8.1. Let $A \in \mathbb{F}^{n \times n}$, and assume that $\operatorname{spec}(A) = \{1\}$. Then, A^k is similar to A for all $k \in \mathbb{P}$.

Fact 5.8.2. Let $A \in \mathbb{F}^{n \times n}$, and assume there exists a nonsingular matrix $S \in \mathbb{F}^{n \times n}$ such that $S^{-1}AS$ is upper triangular. Then, for all $r = 1, \ldots, n$, $\mathcal{R}\left(S \begin{bmatrix} I_r \\ 0 \end{bmatrix} \right)$ is an invariant subspace of A. (Remark: Analogous results hold for lower triangular matrices and block-triangular matrices.)

Fact 5.8.3. Let $A \in \mathbb{F}^{n \times n}$. Then, there exist unique matrices $B, C \in \mathbb{F}^{n \times n}$ such that the following properties are satisfied:

 i) B is diagonalizable over $\mathbb{F}$.

 ii) C is nilpotent.

 iii) $A = B + C$.

 iv) $BC = CB$.

Furthermore, $\operatorname{mspec}(A) = \operatorname{mspec}(B)$. (Proof: See [357, p. 112] or [381, p. 74]. Existence follows from the real Jordan form. The last statement follows from Fact 5.9.15.) (Remark: This result is the *S-N decomposition* or the *Jordan-Chevalley decomposition.*)

Fact 5.8.4. Let $A \in \mathbb{F}^{n \times n}$, and let $r \triangleq \operatorname{rank} A$. Then, A is group invertible if and only if there exist a nonsingular matrix $B \in \mathbb{F}^{r \times r}$ and a nonsingular matrix $S \in \mathbb{F}^{n \times n}$ such that

$$A = S \begin{bmatrix} B & 0 \\ 0 & 0 \end{bmatrix} S^{-1}.$$

Fact 5.8.5. Let $A \in \mathbb{F}^{n \times n}$, and assume that A is normal. Then, there exists a nonsingular matrix $S \in \mathbb{F}^{n \times n}$ such that

$$A^{\mathrm{T}} = SAS^{-1}$$

and such that $S = S^{\mathrm{T}}$ and $S^{-1} = \overline{S}$. (Remark: If $\mathbb{F} = \mathbb{R}$, then S is a reflector.) (Proof: For $\mathbb{F} = \mathbb{C}$, let $A = UBU^*$, where U is unitary and B is diagonal. Then, $A^{\mathrm{T}} = SA\overline{S}$, where $S \triangleq \overline{U}U^{-1}$. For $\mathbb{F} = \mathbb{R}$, use the real normal form and let $S \triangleq U\tilde{I}U^{\mathrm{T}}$, where U is orthogonal and $\tilde{I} \triangleq \mathrm{diag}(\hat{I}, \dots, \hat{I})$.)

Fact 5.8.6. Let $A \in \mathbb{F}^{n \times n}$. Then, there exists an involutory matrix $S \in \mathbb{F}^{n \times n}$ such that

$$A^{\mathrm{T}} = SAS^{\mathrm{T}}.$$

(Remark: Note A^{T} rather than A^*.) (Proof: See [233] and [303].)

Fact 5.8.7. Let $A \in \mathbb{F}^{n \times n}$. Then, there exists a reverse-symmetric, nonsingular matrix $S \in \mathbb{F}^{n \times n}$ such that $A^{\hat{\mathrm{T}}} = SAS^{-1}$. (Proof: The result follows from Corollary 5.3.8. See [464].)

Fact 5.8.8. Let $A \in \mathbb{F}^{n \times n}$. Then, there exist reverse-symmetric matrices $S_1, S_2 \in \mathbb{F}^{n \times n}$ such that S_2 is nonsingular and $A = S_1 S_2$. (Proof: The result follows from Corollary 5.3.9. See [464].)

Fact 5.8.9. Let $A \in \mathbb{R}^{n \times n}$, and assume that A is not of the form aI, where $a \in \mathbb{R}$. Then, A is similar to a matrix with diagonal entries $0, \dots, 0, \mathrm{tr}\, A$. (Proof: See [592, p. 77].) (Remark: This result is due to Gibson.)

Fact 5.8.10. Let $A \in \mathbb{R}^{n \times n}$, and assume that A is not zero. Then, A is similar to a matrix whose diagonal entries are all nonzero. (Proof: See [592, p. 79].) (Remark: This result is due to Marcus and Purves.)

Fact 5.8.11. Let $A \in \mathbb{R}^{n \times n}$, and assume that A is symmetric. Then, there exists an orthogonal matrix $S \in \mathbb{R}^{n \times n}$ such that $-1 \notin \mathrm{spec}(S)$ and SAS^{T} is diagonal. (Proof: See [592, p. 101].) (Remark: This result is due to Hsu.)

Fact 5.8.12. Let $A \in \mathbb{R}^{n \times n}$, and assume that A is symmetric. Then, there exist a diagonal matrix $B \in \mathbb{R}^{n \times n}$ and a skew-symmetric matrix $C \in \mathbb{R}^{n \times n}$ such that

$$A = [2(I + C)^{-1} - I]B[2(I + C)^{-1} - I]^{\mathrm{T}}.$$

(Proof: Use Fact 5.8.11. See [592, p. 101].)

Fact 5.8.13. Let $A \in \mathbb{F}^{n \times n}$. Then, there exists a unitary matrix $S \in \mathbb{F}^{n \times n}$ such that S^*AS has equal diagonal entries. (Proof: See [265] or [592, p. 78], or use Fact 5.8.14.) (Remark: The diagonal entries are equal to $(1/n)\operatorname{tr} A$.) (Remark: This result is due to Parker. See [284].)

Fact 5.8.14. Let $A \in \mathbb{F}^{n \times n}$. Then, the following statements are equivalent:

$i)$ $\operatorname{tr} A = 0$.

$ii)$ There exist matrices $B, C \in \mathbb{F}^{n \times n}$ such that $A = [B, C]$.

$iii)$ A is unitarily similar to a matrix whose diagonal entries are zero.

(Remark: This result is *Shoda's theorem*. See [7, 284, 415, 427] or [325, p. 146].)

Fact 5.8.15. Let $A \in \mathbb{F}^{n \times n}$, and assume that A is idempotent. Then, A and A^* are unitarily similar. (Proof: The result follows from Fact 5.9.7 and the fact that $\left[\begin{smallmatrix} 1 & a \\ 0 & 0 \end{smallmatrix}\right]$ and $\left[\begin{smallmatrix} 1 & 0 \\ a & 0 \end{smallmatrix}\right]$ are unitarily similar. See [232].)

Fact 5.8.16. Let $A \in \mathbb{F}^{n \times n}$, and assume that A is symmetric. Then, there exists a unitary matrix $S \in \mathbb{F}^{n \times n}$ such that

$$A = SBS^{\mathrm{T}},$$

where

$$B \triangleq \operatorname{diag}[\sigma_1(A), \ldots, \sigma_n(A)].$$

(Proof: See [367, p. 207].) (Remark: A is symmetric, complex, and T-congruent to B.)

Fact 5.8.17. Let $A \in \mathbb{F}^{n \times n}$. Then, $\left[\begin{smallmatrix} A & 0 \\ 0 & -A \end{smallmatrix}\right]$ and $\left[\begin{smallmatrix} 0 & A \\ A & 0 \end{smallmatrix}\right]$ are unitarily similar. (Proof: Use the unitary transformation $\frac{1}{\sqrt{2}}\left[\begin{smallmatrix} I & -I \\ I & I \end{smallmatrix}\right]$.)

Fact 5.8.18. Let $n \in \mathbb{P}$. Then,

$$\hat{I}_n = \begin{cases} S \begin{bmatrix} -I_{n/2} & 0 \\ 0 & -I_{n/2} \end{bmatrix} S^{\mathrm{T}}, & n \text{ even}, \\[2em] S \begin{bmatrix} -I_{n/2} & 0 & 0 \\ 0 & 1 & 0 \\ 0 & 0 & I_{n/2} \end{bmatrix} S^{\mathrm{T}}, & n \text{ odd}, \end{cases}$$

where

$$
S \triangleq \begin{cases} \dfrac{1}{\sqrt{2}} \begin{bmatrix} I_{n/2} & -\hat{I}_{n/2} \\ \hat{I}_{n/2} & I_{n/2} \end{bmatrix}, & n \text{ even}, \\[2em] \dfrac{1}{\sqrt{2}} \begin{bmatrix} I_{n/2} & 0 & -\hat{I}_{n/2} \\ 0 & \sqrt{2} & 0 \\ \hat{I}_{n/2} & 0 & I_{n/2} \end{bmatrix}, & n \text{ odd}. \end{cases}
$$

Therefore,

$$
\mathrm{mspec}\left(\hat{I}_n\right) = \begin{cases} \{-1, 1, \ldots, -1, 1\}_{\mathrm{m}}, & n \text{ even}, \\ \{1, -1, 1, \ldots, -1, 1\}_{\mathrm{m}}, & n \text{ odd}. \end{cases}
$$

(Remark: See [764].)

Fact 5.8.19. Let $A \in \mathbb{F}^{n \times n}$, assume that A is unitary, and let $m \le n/2$. Then, there exist unitary matrices $U, V \in \mathbb{F}^{n \times n}$ such that

$$
A = U \begin{bmatrix} \Gamma & -\Sigma & 0 \\ \Sigma & \Gamma & 0 \\ 0 & 0 & I_{n-2m} \end{bmatrix} V,
$$

where $\Gamma, \Sigma \in \mathbb{R}^{m \times m}$ are diagonal and positive semidefinite and satisfy

$$
\Gamma^2 + \Sigma^2 = I_m.
$$

(Proof: See [667, p. 37].) (Remark: This result is the *CS decomposition*.)

Fact 5.8.20. Let $A \in \mathbb{C}^{n \times n}$. Then, there exists a matrix $B \in \mathbb{R}^{n \times n}$ such that $A\overline{A}$ and B^2 are similar. (Proof: See [228].)

5.9 Facts on Matrix Transformations for Two or More Matrices

Fact 5.9.1. Let $q(s) \triangleq s^2 - \beta_1 s - \beta_0 \in \mathbb{R}[s]$ be irreducible, and let $\lambda = \nu + \jmath\omega$ denote a root of q so that $\beta_1 = 2\nu$ and $\beta_0 = -(\nu^2 + \omega^2)$. Then,

$$
\mathcal{H}_1(q) = \begin{bmatrix} 0 & 1 \\ \beta_0 & \beta_1 \end{bmatrix} = \begin{bmatrix} 1 & 0 \\ \nu & \omega \end{bmatrix} \begin{bmatrix} \nu & \omega \\ -\omega & \nu \end{bmatrix} \begin{bmatrix} 1 & 0 \\ -\nu/\omega & 1/\omega \end{bmatrix} = S\mathcal{J}_1(q)S^{-1}.
$$

The transformation matrix $S = \left[\begin{smallmatrix} 1 & 0 \\ \nu & \omega \end{smallmatrix}\right]$ is not unique; an alternative choice is $S = \left[\begin{smallmatrix} \omega & \nu \\ 0 & \nu^2+\omega^2 \end{smallmatrix}\right]$. Similarly,

$$\mathcal{H}_2(q) = \begin{bmatrix} 0 & 1 & 0 & 0 \\ \beta_0 & \beta_1 & 1 & 0 \\ 0 & 0 & 0 & 1 \\ 0 & 0 & \beta_0 & \beta_1 \end{bmatrix} = S \begin{bmatrix} \nu & \omega & 1 & 0 \\ -\omega & \nu & 0 & 1 \\ 0 & 0 & \nu & \omega \\ 0 & 0 & -\omega & \nu \end{bmatrix} S^{-1} = S\mathcal{J}_2(q)S^{-1},$$

where

$$S \triangleq \begin{bmatrix} \omega & \nu & \omega & \nu \\ 0 & \nu^2 + \omega^2 & \omega & \nu^2 + \omega^2 + \nu \\ 0 & 0 & -2\omega\nu & 2\omega^2 \\ 0 & 0 & -2\omega(\nu^2 + \omega^2) & 0 \end{bmatrix}.$$

Fact 5.9.2. Let $q(s) \triangleq s^2 - 2\nu s + \nu^2 + \omega^2 \in \mathbb{R}[s]$ with roots $\lambda = \nu + \jmath\omega$ and $\overline{\lambda} = \nu - \jmath\omega$. Then,

$$\mathcal{H}_1(q) = \begin{bmatrix} \nu & \omega \\ -\omega & \nu \end{bmatrix} = \frac{1}{\sqrt{2}}\begin{bmatrix} 1 & 1 \\ \jmath & -\jmath \end{bmatrix}\begin{bmatrix} \lambda & 0 \\ 0 & \overline{\lambda} \end{bmatrix}\frac{1}{\sqrt{2}}\begin{bmatrix} 1 & -\jmath \\ 1 & \jmath \end{bmatrix}$$

and

$$\mathcal{H}_2(q) = \begin{bmatrix} \nu & \omega & 1 & 0 \\ -\omega & \nu & 0 & 1 \\ 0 & 0 & \nu & \omega \\ 0 & 0 & -\omega & \nu \end{bmatrix} = S \begin{bmatrix} \lambda & 1 & 0 & 0 \\ 0 & \lambda & 0 & 0 \\ 0 & 0 & \overline{\lambda} & 1 \\ 0 & 0 & 0 & \overline{\lambda} \end{bmatrix} S^{-1},$$

where

$$S \triangleq \frac{1}{\sqrt{2}}\begin{bmatrix} 1 & 0 & 1 & 0 \\ \jmath & 0 & -\jmath & 0 \\ 0 & 1 & 0 & 1 \\ 0 & \jmath & 0 & -\jmath \end{bmatrix}, \qquad S^{-1} = \frac{1}{\sqrt{2}}\begin{bmatrix} 1 & -\jmath & 0 & 0 \\ 0 & 0 & 1 & -\jmath \\ 1 & \jmath & 0 & 0 \\ 0 & 0 & 1 & \jmath \end{bmatrix}.$$

Fact 5.9.3. Left equivalence, right equivalence, biequivalence, unitary left equivalence, unitary right equivalence, and unitary biequivalence are equivalence relations on $\mathbb{F}^{n \times m}$. Similarity, congruence, and unitary similarity are equivalence relations on $\mathbb{F}^{n \times n}$.

Fact 5.9.4. Let $A, B \in \mathbb{F}^{n \times m}$. Then, A and B are in the same equivalence class of $\mathbb{F}^{n \times m}$ induced by biequivalent transformations if and only if A and B are biequivalent to $\begin{bmatrix} I & 0 \\ 0 & 0 \end{bmatrix}$. Now, let $n = m$. Then, A and B are in the same equivalence class of $\mathbb{F}^{n \times n}$ induced by similarity transformations if and only if A and B have the same Jordan form.

Fact 5.9.5. Let $A \in \mathbb{F}^{n \times n}$, and assume that A is normal. Then, A is unitarily similar to its Jordan form.

Fact 5.9.6. Let $A, B \in \mathbb{F}^{n \times n}$, assume that A and B are normal, and assume that A and B are similar. Then, A and B are unitarily similar. (Proof: Since A and B are similar, it follows that $\mathrm{mspec}(A) = \mathrm{mspec}(B)$.

Since A and B are normal, it follows that they are unitarily similar to the same diagonal matrix. See Fact 5.9.5. See [326, p. 104].) (Remark: See [286, p. 8] for related results.)

Fact 5.9.7. Let $A \in \mathbb{F}^{n \times n}$, assume that A is idempotent, and let $r \triangleq \operatorname{rank} A$. Then, there exist a unitary matrix $S \in \mathbb{F}^{n \times n}$ and positive real numbers $a_1, \ldots, a_k$ such that

$$A = S\operatorname{diag}\left(\begin{bmatrix} 1 & a_1 \\ 0 & 0 \end{bmatrix}, \ldots, \begin{bmatrix} 1 & a_k \\ 0 & 0 \end{bmatrix}, I_{r-k}, 0_{(n-r-k) \times (n-r-k)} \right) S^*.$$

(Proof: See [232].) (Remark: This result provides a canonical form for idempotent matrices under unitary similarity.) (Remark: See Fact 5.8.15.)

Fact 5.9.8. Let $A, B \in \mathbb{F}^{n \times n}$, and let $r \triangleq 2n^2$. Then, the following statements are equivalent:

$i)$ A and B are unitarily similar.

$ii)$ For all $k_1, \ldots, k_r, l_1, \ldots, l_r \in \mathbb{N}$ such that $\sum_{i,j=1}^{r}(k_i^2 + l_j^2) \leq r$, $\operatorname{tr} A^{k_1} A^{l_1 *} \cdots A^{k_r} A^{l_r *} = \operatorname{tr} B^{k_1} B^{l_1 *} \cdots B^{k_r} B^{l_r *}$.

(Proof: See [124].) (Remark: This result is due to Pearcy. See [409, pp. 71, 72].)

Fact 5.9.9. Let $A, B \in \mathbb{F}^{n \times n}$, and assume that A and B are idempotent. Then, the following statements are equivalent:

$i)$ A and B are unitarily similar.

$ii)$ For all $i = 1, \ldots, \lfloor n/2 \rfloor$, $\operatorname{tr} A = \operatorname{tr} B$ and $\operatorname{tr} (AA^*)^i = \operatorname{tr} (BB^*)^i$.

$iii)$ $\chi_{AA^*} = \chi_{BB^*}$.

(Proof: The result follows from Fact 5.9.7. See [232].)

Fact 5.9.10. Let $A, B \in \mathbb{F}^{n \times n}$, and assume that either A or B is nonsingular. Then, AB and BA are similar. (Proof: If A is nonsingular, then $AB = A(BA)A^{-1}$.)

Fact 5.9.11. Let $A, B \in \mathbb{F}^{n \times n}$, assume that A and B are Hermitian, and assume that $\operatorname{mspec}(A) = \operatorname{mspec}(B)$. Then, A and B are unitarily similar.

Fact 5.9.12. Let $A, B \in \mathbb{F}^{n \times n}$, assume that A and B are projectors, and assume that $\operatorname{rank} A = \operatorname{rank} B$. Then, A and B are unitarily similar.

Fact 5.9.13. Let $A, B \in \mathbb{F}^{n \times n}$, assume that A and B are reflectors, and assume that $\operatorname{sig} A = \operatorname{sig} B$. Then, A and B are unitarily similar.

Fact 5.9.14. Let $A, B \in \mathbb{F}^{n \times n}$, and assume that A and B are projectors. Then, AB and BA are unitarily similar. (Remark: This result is due to Dixmier. See [602].)

Fact 5.9.15. Let $\mathcal{S} \subset \mathbb{F}^{n \times n}$, and assume that $AB = BA$ for all $A, B \in \mathcal{S}$. Then, there exists a unitary matrix $S \in \mathbb{F}^{n \times n}$ such that, for all $A \in \mathcal{S}$, SAS^* is upper triangular. (Proof: See [367, p. 81] and [601].) (Remark: See Fact 8.13.5.)

Fact 5.9.16. Let $A, B \mathbb{C}^{n \times n}$, and assume that A and B are projectors. Then, there exists a unitary matrix $S \in \mathbb{C}^{n \times n}$ such that SAS^* and SBS^* are upper triangular if and only if $[A, B]$ is nilpotent. (Proof: See [680].) (Remark: See Fact 5.9.15.)

Fact 5.9.17. Let $\mathcal{S} \subset \mathbb{F}^{n \times n}$, and assume that every matrix $A \in \mathcal{S}$ is normal. Then, $AB = BA$ for all $A, B \in \mathcal{S}$ if and only if there exists a unitary matrix $S \in \mathbb{F}^{n \times n}$ such that, for all $A \in \mathcal{S}$, SAS^* is diagonal. (Remark: See Fact 8.13.2 and [367, pp. 103, 172].)

Fact 5.9.18. Let $\mathcal{S} \subset \mathbb{F}^{n \times n}$, and assume that every matrix $A \in \mathcal{S}$ is diagonalizable over $\mathbb{F}$. Then, $AB = BA$ for all $A, B \in \mathcal{S}$ if and only if there exists a nonsingular matrix $S \in \mathbb{F}^{n \times n}$ such that, for all $A \in \mathcal{S}$, SAS^{-1} is diagonal. (Proof: See [367, p. 52].)

Fact 5.9.19. Let $A_1, \ldots, A_r \in \mathbb{F}^{n \times n}$, and assume that $A_i A_j = A_j A_i$ for all $i, j = 1, \ldots, r$. Then,

$$\mathrm{dim\, span} \left\{ \prod_{i=1}^{r} A_i^{n_i}\colon\ 0 \leq n_i \leq n-1, i = 1, \ldots, r \right\} \leq \tfrac{1}{4} n^2 + 1.$$

(Remark: This result gives a bound on the dimension of a commutative subalgebra.) (Remark: This result is due to Schur. See [449].)

Fact 5.9.20. Let $A, B \in \mathbb{F}^{n \times n}$, and assume that $AB = BA$. Then,

$$\mathrm{dim\, span} \{ A^i B^j\colon\ 0 \leq i \leq n-1, 0 \leq j \leq n-1 \} \leq n.$$

(Remark: This result gives a bound on the dimension of a commutative subalgebra generated by two matrices.) (Remark: This result is due to Gerstenhaber. See [81, 449].)

Fact 5.9.21. Let $A, B \in \mathbb{F}^{n \times n}$, and assume that A and B are normal, nonsingular, and congruent. Then, $\mathrm{In}\, A = \mathrm{In}\, B$. (Remark: This result is due to Ando.)

Fact 5.9.22. Let $A, B \in \mathbb{F}^{n \times m}$. Then, the following statements hold:

i) The matrices A and B are unitarily left equivalent if and only if $A^*A = B^*B$.

ii) The matrices A and B are unitarily right equivalent if and only if $AA^* = BB^*$.

iii) The matrices A and B are unitarily biequivalent if and only if A and B have the same singular values with the same multiplicity.

(Proof: See [373] and [613, pp. 372, 373].) (Remark: In [373] A and B need not be the same size.) (Remark: The singular value decomposition provides a canonical form under unitary biequivalence in analogy with the Smith form under biequivalence.) (Remark: Note that $AA^* = BB^*$ implies that $\mathcal{R}(A) = \mathcal{R}(B)$, which implies right equivalence, which is an alternative proof of the immediate fact that unitary right equivalence implies right equivalence.)

Fact 5.9.23. Let $A, B \in \mathbb{F}^{n \times n}$. Then, the following statements hold:

i) $A^*A = B^*B$ if and only if there exists a unitary matrix $S \in \mathbb{F}^{n \times n}$ such that $A = SB$.

ii) $A^*A \leq B^*B$ if and only if there exists a matrix $S \in \mathbb{F}^{n \times n}$ such that $A = SB$ and $S^*S \leq I$.

iii) $A^*B + B^*A = 0$ if and only if there exists a unitary matrix $S \in \mathbb{F}^{n \times n}$ such that $(I - S)A = (I + S)B$.

iv) $A^*B + B^*A \geq 0$ if and only if there exists a matrix $S \in \mathbb{F}^{n \times n}$ such that $(I - S)A = (I + S)B$ and $S^*S \leq I$.

(Proof: See [605].) (Remark: Statements *iii)* and *iv)* follow from *i)* and *ii)* by replacing A and B with $A - B$ and $A + B$, respectively.)

Fact 5.9.24. Let $A \in \mathbb{F}^{n \times n}$, $B \in \mathbb{F}^{m \times m}$, and $C \in \mathbb{F}^{n \times m}$. Then, there exists a matrix $X \in \mathbb{F}^{n \times m}$ satisfying

$$AX + XB + C = 0$$

if and only if the matrices

$$\begin{bmatrix} A & 0 \\ 0 & -B \end{bmatrix}, \qquad \begin{bmatrix} A & C \\ 0 & -B \end{bmatrix}$$

are similar. In this case,

$$\begin{bmatrix} A & C \\ 0 & -B \end{bmatrix} = \begin{bmatrix} I & X \\ 0 & I \end{bmatrix} \begin{bmatrix} A & 0 \\ 0 & -B \end{bmatrix} \begin{bmatrix} I & -X \\ 0 & I \end{bmatrix}.$$

(Proof: For sufficiency, see [455, pp. 422–424] or [592, pp. 194, 195].) (Remark: $AX + XB + C = 0$ is *Sylvester's equation*. See Proposition 7.2.4,

Corollary 7.2.5, and Proposition 11.8.3.) (Remark: This result is due to Roth. See [121].)

Fact 5.9.25. Let $A \in \mathbb{F}^{n \times n}$, $B \in \mathbb{F}^{m \times m}$, and $C \in \mathbb{F}^{n \times m}$. Then, there exist matrices $X, Y \in \mathbb{F}^{n \times m}$ satisfying

$$AX + YB + C = 0$$

if and only if

$$\operatorname{rank} \begin{bmatrix} A & 0 \\ 0 & -B \end{bmatrix} = \operatorname{rank} \begin{bmatrix} A & C \\ 0 & -B \end{bmatrix}.$$

(Proof: See [592, pp. 194, 195].) (Remark: $AX + YB + C = 0$ is a generalization of Sylvester's equation. See Fact 5.9.24.) (Remark: This result is due to Roth.)

Fact 5.9.26. Let $A, B \in \mathbb{F}^{n \times n}$, and assume that A and B are idempotent. Then, the matrices

$$\begin{bmatrix} A + B & A \\ 0 & -A - B \end{bmatrix}, \qquad \begin{bmatrix} A + B & 0 \\ 0 & -A - B \end{bmatrix}$$

are similar. In fact,

$$\begin{bmatrix} A + B & A \\ 0 & -A - B \end{bmatrix} = \begin{bmatrix} I & X \\ 0 & I \end{bmatrix} \begin{bmatrix} A + B & 0 \\ 0 & -A - B \end{bmatrix} \begin{bmatrix} I & -X \\ 0 & I \end{bmatrix},$$

where $X \triangleq \frac{1}{4}(I + A - B)$. (Remark: This result is due to Tian.) (Remark: See Fact 5.9.24.)

5.10 Facts on Eigenvalues and Singular Values for One Matrix

Fact 5.10.1. Let $A \in \mathbb{F}^{n \times n}$, let $\alpha \in \mathbb{F}$, and assume that $A^2 = \alpha A$. Then, $\operatorname{spec}(A) \subseteq \{0, \alpha\}$.

Fact 5.10.2. Let $A \in \mathbb{F}^{n \times n}$, assume that A is Hermitian, and let $\alpha \in \mathbb{R}$. Then, $A^2 = \alpha A$ if and only if $\operatorname{spec}(A) \subseteq \{0, \alpha\}$. (Remark: See Fact 3.5.17.)

Fact 5.10.3. Let $A \in \mathbb{F}^{n \times n}$, and assume that A is Hermitian. Then,

$$\operatorname{spabs}(A) = \lambda_{\max}(A)$$

and

$$\operatorname{sprad}(A) = \sigma_{\max}(A) = \max\{|\lambda_{\min}(A)|, \lambda_{\max}(A)\}.$$

If, in addition, A is positive semidefinite, then

$$\operatorname{sprad}(A) = \sigma_{\max}(A) = \operatorname{spabs}(A) = \lambda_{\max}(A).$$

(Remark: See Fact 5.11.2.)

Fact 5.10.4. Let $A \in \mathbb{F}^{n \times n}$, and assume that A is skew Hermitian. Then, the eigenvalues of A are imaginary. (Proof: Let $\lambda \in \mathrm{spec}(A)$. Since $0 \leq AA^* = -A^2$, it follows that $-\lambda^2 \geq 0$, and thus $\lambda^2 \leq 0$.)

Fact 5.10.5. Let $A \in \mathbb{F}^{n \times n}$, assume that every eigenvalue of A is real, and assume that exactly r eigenvalues of A, including algebraic multiplicity, are nonzero. Then,

$$(\mathrm{tr}\, A)^2 \leq r\, \mathrm{tr}\, A^2.$$

Furthermore, equality holds if and only if the nonzero eigenvalues of A are equal. (Remark: For arbitrary $A \in \mathbb{F}^{n \times n}$ with r nonzero eigenvalues, it is not generally true that $|\mathrm{tr}\, A|^2 \leq r|\mathrm{tr}\, A^2|$. For example, consider $\mathrm{mspec}(A) = \{1, 1, \jmath, -\jmath\}_{\mathrm{m}}$.)

Fact 5.10.6. Let $A \in \mathbb{R}^{n \times n}$, and let $\mathrm{mspec}(A) = \{\lambda_1, \ldots, \lambda_n\}_{\mathrm{m}}$. Then,

$$\sum_{i=1}^{n} (\mathrm{Re}\, \lambda_i)(\mathrm{Im}\, \lambda_i) = 0$$

and

$$\mathrm{tr}\, A^2 = \sum_{i=1}^{n} (\mathrm{Re}\, \lambda_i)^2 - \sum_{i=1}^{n} (\mathrm{Im}\, \lambda_i)^2.$$

Fact 5.10.7. Let $n \geq 2$, let $a_1, \ldots, a_n > 0$, and define the symmetric matrix $A \in \mathbb{R}^{n \times n}$ by $A_{(i,j)} \triangleq a_i + a_j$ for all $i, j = 1, \ldots, n$. Then,

$$\mathrm{rank}\, A \leq 2$$

and

$$\mathrm{mspec}(A) = \{\lambda, \mu, 0, \ldots, 0\}_{\mathrm{m}},$$

where

$$\lambda \triangleq \sum_{i=1}^{n} a_i + \sqrt{n \sum_{i=1}^{n} a_i^2}, \quad \mu \triangleq \sum_{i=1}^{n} a_i - \sqrt{n \sum_{i=1}^{n} a_i^2}.$$

Furthermore, the following statements hold:

  i) $\lambda > 0$.

  ii) $\mu \leq 0$.

Furthermore, the following statements are equivalent:

  iii) $\mu < 0$.

  iv) At least two of the numbers $a_1, \ldots, a_n > 0$ are distinct.

  v) $\mathrm{rank}\, A = 2$.

In this case,

$$\lambda_{\min}(A) = \mu < 0 < \mathrm{tr}\, A = 2\sum_{i=1}^{n} a_i < \lambda_{\max}(A) = \lambda.$$

(Proof: $A = a1_{1 \times n} + 1_{n \times 1}a^{\mathrm{T}}$, where $a \triangleq \begin{bmatrix} a_1 & \cdots & a_n \end{bmatrix}^{\mathrm{T}}$. Then, it follows from Fact 2.10.26 that $\operatorname{rank} A \le \operatorname{rank}(a1_{1 \times n}) + \operatorname{rank}(1_{n \times 1}a^{\mathrm{T}}) = 2$. Furthermore, $\operatorname{mspec}(A)$ follows from Fact 5.10.8, while Fact 1.4.11 implies that $\mu \le 0$.) (Remark: See Fact 8.7.29.)

Fact 5.10.8. Let $x, y \in \mathbb{R}^n$. Then,

$$\operatorname{mspec}(xy^{\mathrm{T}} + yx^{\mathrm{T}}) = \left\{ x^{\mathrm{T}}y + \sqrt{x^{\mathrm{T}}xy^{\mathrm{T}}y}, x^{\mathrm{T}}y - \sqrt{x^{\mathrm{T}}xy^{\mathrm{T}}y}, 0, \ldots, 0 \right\}_{\mathrm{m}},$$

$$\operatorname{sprad}(xy^{\mathrm{T}} + yx^{\mathrm{T}}) = \begin{cases} x^{\mathrm{T}}y + \sqrt{x^{\mathrm{T}}xy^{\mathrm{T}}y}, & x^{\mathrm{T}}y \ge 0, \\ \left| x^{\mathrm{T}}y - \sqrt{x^{\mathrm{T}}xy^{\mathrm{T}}y} \right|, & x^{\mathrm{T}}y \le 0, \end{cases}$$

and

$$\operatorname{spabs}(xy^{\mathrm{T}} + yx^{\mathrm{T}}) = x^{\mathrm{T}}y + \sqrt{x^{\mathrm{T}}xy^{\mathrm{T}}y}.$$

(Problem: Extend this result to $\mathbb{C}$ and $xy^{\mathrm{T}} + zw^{\mathrm{T}}$. See Fact 4.9.13.)

Fact 5.10.9. Let $A \in \mathbb{F}^{n \times n}$, and let $\operatorname{mspec}(A) = \{\lambda_1, \ldots, \lambda_n\}_{\mathrm{m}}$. Then,

$$\operatorname{mspec}\left[(I + A)^2\right] = \left\{(1 + \lambda_1)^2, \ldots, (1 + \lambda_n)^2\right\}_{\mathrm{m}}.$$

If A is nonsingular, then

$$\operatorname{mspec}(A^{-1}) = \left\{\lambda_1^{-1}, \ldots, \lambda_n^{-1}\right\}_{\mathrm{m}}.$$

Finally, if $I + A$ is nonsingular, then

$$\operatorname{mspec}\left[(I + A)^{-1}\right] = \left\{(1 + \lambda_1)^{-1}, \ldots, (1 + \lambda_n)^{-1}\right\}_{\mathrm{m}}$$

and

$$\operatorname{mspec}\left[A(I + A)^{-1}\right] = \left\{\lambda_1(1 + \lambda_1)^{-1}, \ldots, \lambda_n(1 + \lambda_n)^{-1}\right\}_{\mathrm{m}}.$$

(Proof: Use Fact 5.10.10.)

Fact 5.10.10. Let $p, q \in \mathbb{F}[s]$, assume that p and q are coprime, define $g \triangleq p/q \in \mathbb{F}(s)$, let $A \in \mathbb{F}^{n \times n}$, let $\operatorname{mspec}(A) = \{\lambda_1, \ldots, \lambda_n\}_{\mathrm{m}}$, assume that $\operatorname{roots}(q) \cap \operatorname{spec}(A) = \varnothing$, and define $g(A) \triangleq p(A)[q(A)]^{-1}$. Then,

$$\operatorname{mspec}[g(A)] = \{g(\lambda_1), \ldots, g(\lambda_n)\}_{\mathrm{m}}.$$

Fact 5.10.11. Let $x \in \mathbb{F}^n$ and $y \in \mathbb{F}^m$. Then,

$$\sigma_{\max}(xy^*) = \sqrt{x^*xy^*y}.$$

If, in addition, $m = n$, then

$$\operatorname{mspec}(xy^*) = \{x^*y, 0, \ldots, 0\}_{\mathrm{m}},$$

$$\operatorname{mspec}(I + xy^*) = \{1 + x^*y, 1, \ldots, 1\}_{\mathrm{m}},$$

$$\operatorname{sprad}(xy^*) = |x^*y|,$$

$$\text{spabs}(xy^*) = \max\{0, \text{Re}\, x^*y\}.$$

Fact 5.10.12. Let $A \in \mathbb{F}^{n \times n}$, and assume that rank $A = 1$. Then,

$$\sigma_{\max}(A) = (\text{tr}\, AA^*)^{1/2}.$$

Fact 5.10.13. Let $x, y \in \mathbb{F}^n$, and assume that $x^*y \neq 0$. Then,

$$\sigma_{\max}\left[(x^*y)^{-1}xy^*\right] \geq 1.$$

Fact 5.10.14. Let $A \in \mathbb{F}^{n \times n}$, and let $\text{mspec}(A) = \{\lambda_1, \ldots, \lambda_n\}_{\text{m}}$, where $\lambda_1, \ldots, \lambda_n$ are ordered such that $|\lambda_1| \geq \cdots \geq |\lambda_n|$. Then, for all $k = 1, \ldots, n$,

$$\prod_{i=1}^{k} |\lambda_i| \leq \prod_{i=1}^{k} \sigma_i(A)$$

with equality for $k = n$, that is,

$$|\det A| = \prod_{i=1}^{n} |\lambda_i| = \prod_{i=1}^{n} \sigma_i(A).$$

Hence, for all $k = 1, \ldots, n$,

$$\prod_{i=k}^{n} \sigma_i(A) \leq \prod_{i=k}^{n} |\lambda_i|.$$

(Proof: See [111, p. 43], [369, p. 171], or [806, p. 19].) (Remark: This result is due to Weyl.) (Remark: See Fact 8.14.20 and Fact 9.11.19.)

Fact 5.10.15. Let $\beta_0, \ldots, \beta_{n-1} \in \mathbb{F}$, define $A \in \mathbb{F}^{n \times n}$ by

$$A \triangleq \begin{bmatrix} 0 & 1 & 0 & \cdots & 0 & 0 \\ 0 & 0 & 1 & \ddots & 0 & 0 \\ 0 & 0 & 0 & \ddots & 0 & 0 \\ \vdots & \vdots & \vdots & \ddots & \ddots & \vdots \\ 0 & 0 & 0 & \cdots & 0 & 1 \\ -\beta_0 & -\beta_1 & -\beta_2 & \cdots & -\beta_{n-2} & -\beta_{n-1} \end{bmatrix},$$

and define $\alpha \triangleq 1 + \sum_{i=0}^{n-1} |\beta_i|^2$. Then,

$$\sigma_1(A) = \sqrt{\tfrac{1}{2}\left(\alpha + \sqrt{\alpha^2 - 4|\beta_0|^2}\right)},$$

$$\sigma_2(A) = \cdots = \sigma_{n-1}(A) = 1,$$

$$\sigma_n(A) = \sqrt{\tfrac{1}{2}\left(\alpha - \sqrt{\alpha^2 - 4|\beta_0|^2}\right)}.$$

(Proof: See [418, 429] or [354, p. 523].) (Remark: See Fact 11.18.1.)

Fact 5.10.16. Let $\beta \in \mathbb{C}$. Then,

$$\sigma_{\max}\left(\begin{bmatrix} 1 & 2\beta \\ 0 & 1 \end{bmatrix}\right) = |\beta| + \sqrt{1 + |\beta|^2}$$

and

$$\sigma_{\min}\left(\begin{bmatrix} 1 & 2\beta \\ 0 & 1 \end{bmatrix}\right) = \sqrt{1 + |\beta|^2} - |\beta|.$$

(Proof: See [480].) (Remark: Inequalities involving the singular values of block-triangular matrices are given in [480].)

Fact 5.10.17. Let $A \in \mathbb{F}^{n \times m}$. Then,

$$\sigma_{\max}\left(\begin{bmatrix} I & 2A \\ 0 & I \end{bmatrix}\right) = \sigma_{\max}(A) + \sqrt{1 + \sigma_{\max}^2(A)}.$$

(Proof: See [354, p. 116].)

Fact 5.10.18. Let $A \in \mathbb{F}^{n \times m}$, and let $r = \operatorname{rank} A$. Then, for all $i = 1, \ldots, r$,

$$\sigma_i(AA^*) = \sigma_i(A^*A) = \sigma_i^2(A).$$

In particular,

$$\sigma_{\max}(AA^*) = \sigma_{\max}^2(A),$$

and, if $n = m$, then

$$\sigma_{\min}(AA^*) = \sigma_{\min}^2(A).$$

Furthermore, for all $i = 1, \ldots, r$,

$$\sigma_i(AA^*A) = \sigma_i^3(A).$$

Fact 5.10.19. Let $A \in \mathbb{F}^{n \times n}$. Then, $\sigma_{\max}(A) \leq 1$ if and only if $A^*A \leq I$.

Fact 5.10.20. Let $A \in \mathbb{F}^{n \times n}$. Then, for all $i = 1, \ldots, n$,

$$\sigma_i(A^{\mathrm{A}}) = \prod_{\substack{j=1 \\ j \neq n+1-i}}^{n} \sigma_j(A).$$

(Proof: See Fact 4.10.3 and [592, p. 149].)

Fact 5.10.21. Let $A \in \mathbb{F}^{n \times n}$. Then, $\sigma_1(A) = \sigma_n(A)$ if and only if there exist $\lambda \in \mathbb{F}$ and a unitary matrix $B \in \mathbb{F}^{n \times n}$ such that $A = \lambda B$. (Proof: See [592, pp. 149, 165].)

Fact 5.10.22. Let $A \in \mathbb{F}^{n \times n}$, and let $\lambda \in \operatorname{spec}(A)$. Then, the following inequalities hold:

i) $\sigma_{\min}(A) \leq |\lambda| \leq \sigma_{\max}(A)$.

$ii)$ $\lambda_{\min}\left[\frac{1}{2}(A + A^*)\right] \leq \operatorname{Re} \lambda \leq \lambda_{\max}\left[\frac{1}{2}(A + A^*)\right]$.

$iii)$ $\lambda_{\min}\left[\frac{1}{2j}(A - A^*)\right] \leq \operatorname{Im} \lambda \leq \lambda_{\max}\left[\frac{1}{2j}(A - A^*)\right]$.

(Remark: $i)$ is *Browne's theorem*, $ii)$ is *Bendixson's theorem*, and $iii)$ is *Hirsch's theorem*. See [511, pp. 140–144] and [177, p. 17]. See Fact 5.11.3 and Fact 9.10.6.)

Fact 5.10.23. Let $A \in \mathbb{R}^{n \times n}$, where $n \geq 2$, be the tridiagonal matrix

$$A \triangleq \begin{bmatrix} b_1 & c_1 & 0 & \cdots & 0 & 0 \\ a_1 & b_2 & c_2 & \cdots & 0 & 0 \\ 0 & a_2 & b_3 & \ddots & 0 & 0 \\ \vdots & \vdots & \ddots & \ddots & \ddots & \vdots \\ 0 & 0 & 0 & \ddots & b_{n-1} & c_{n-1} \\ 0 & 0 & 0 & \cdots & a_{n-1} & b_n \end{bmatrix},$$

and assume that $a_i c_i > 0$ for all $i = 1, \ldots, n-1$. Then, A is simple, and every eigenvalue of A is real. (Proof: SAS^{-1} is symmetric, where $S \triangleq \operatorname{diag}(d_1, \ldots, d_n)$, $d_1 \triangleq 1$, and $d_{i+1} \triangleq (c_i/a_i)^{1/2}d_i$ for all $i = 1, \ldots, n-1$. For a proof of the fact that A is simple, see [260, p. 198].)

Fact 5.10.24. Let $A \in \mathbb{R}^{n \times n}$ be the tridiagonal matrix

$$A \triangleq \begin{bmatrix} 0 & 1 & 0 & & & & \\ n-1 & 0 & 2 & & & 0 & \\ 0 & n-2 & 0 & \ddots & & & \\ & \ddots & \ddots & \ddots & \ddots & & \\ & & & \ddots & \ddots & 0 & n-2 & 0 \\ & 0 & & & \ddots & 2 & 0 & n-1 \\ & & & & & 0 & 1 & 0 \end{bmatrix}.$$

Then,

$$\chi_A(s) = \prod_{i=1}^{n} [s - (n + 1 - 2i)].$$

Hence,

$$\operatorname{spec}(A) = \begin{cases} \{n-1, -(n-1), \ldots, 1, -1\}, & n \text{ even}, \\ \{n-1, -(n-1), \ldots, 2, -2, 0\}, & n \text{ odd}. \end{cases}$$

(Proof: See [685].)

Fact 5.10.25. Let $A \in \mathbb{R}^{n \times n}$, where $n \geq 1$, be the tridiagonal matrix

$$A \triangleq \begin{bmatrix} b & c & 0 & \cdots & 0 & 0 \\ a & b & c & \cdots & 0 & 0 \\ 0 & a & b & \ddots & 0 & 0 \\ \vdots & \vdots & \ddots & \ddots & \ddots & \vdots \\ 0 & 0 & 0 & \ddots & b & c \\ 0 & 0 & 0 & \cdots & a & b \end{bmatrix},$$

and assume that $ac > 0$. Then,

$$\text{spec}(A) = \{b + \sqrt{ac}\cos[i\pi/(n+1)]: \ i = 1, \ldots, n\}.$$

(Remark: See [354, p. 522].)

Fact 5.10.26. Let $A \in \mathbb{R}^{n \times n}$, and assume that A has real eigenvalues. Then,

$$\lambda_{\min}(A) \leq \tfrac{1}{n}\text{tr}\,A - \sqrt{\tfrac{1}{n^2-n}\big[\text{tr}\,A^2 - \tfrac{1}{n}(\text{tr}\,A)^2\big]}$$

$$\leq \tfrac{1}{n}\text{tr}\,A + \sqrt{\tfrac{1}{n^2-n}\big[\text{tr}\,A^2 - \tfrac{1}{n}(\text{tr}\,A)^2\big]}$$

$$\leq \lambda_{\max}(A)$$

$$\leq \tfrac{1}{n}\text{tr}\,A + \sqrt{\tfrac{n-1}{n}\big[\text{tr}\,A^2 - \tfrac{1}{n}(\text{tr}\,A)^2\big]}.$$

Furthermore, for all $i = 1, \ldots, n$,

$$\big|\lambda_i(A) - \tfrac{1}{n}\text{tr}\,A\big| \leq \sqrt{\tfrac{n-1}{n}\big[\text{tr}\,A^2 - \tfrac{1}{n}(\text{tr}\,A)^2\big]}.$$

(Proof: See [789].)

Fact 5.10.27. Let $A \in \mathbb{R}^{n \times n}$, and assume that $r \triangleq \text{rank}\,A \geq 2$. If $r\,\text{tr}\,A^2 \leq (\text{tr}\,A)^2$, then

$$\text{sprad}(A) \geq \sqrt{\frac{(\text{tr}\,A)^2 - \text{tr}\,A^2}{r(r-1)}}.$$

If $(\text{tr}\,A)^2 \leq r\,\text{tr}\,A^2$, then

$$\text{sprad}(A) \geq \frac{|\text{tr}\,A|}{r} + \sqrt{\frac{r\,\text{tr}\,A^2 - (\text{tr}\,A)^2}{r^2(r-1)}}.$$

If $\text{rank}\,A = 2$, then equality holds in both cases. Finally, if A is skew symmetric, then

$$\text{sprad}(A) \geq \sqrt{\frac{3}{r(r-1)}}\,\|A\|_{\text{F}}.$$

(Proof: See [376].) (Remark: The Frobenius norm $\|\cdot\|_F$ is defined in Section 9.2.)

Fact 5.10.28. Let $A \in \mathbb{F}^{n \times n}$. Then,

$$\text{spabs}(A) \leq \tfrac{1}{2}\lambda_{\max}(A + A^*).$$

Furthermore, equality holds if and only if A is normal. (Proof: See *xii*) and *xiv*) of Fact 9.10.8.)

5.11 Facts on Eigenvalues and Singular Values for Two or More Matrices

Fact 5.11.1. Let $A \in \mathbb{F}^{n \times n}$ and $B \in \mathbb{F}^{n \times m}$, let $r \triangleq \text{rank } B$, and define $\mathcal{A} \triangleq \left[\begin{smallmatrix} A & B \\ B^* & 0 \end{smallmatrix}\right]$. Then, $\nu_-(\mathcal{A}) \geq r$, $\nu_0(\mathcal{A}) \geq 0$, and $\nu_+(\mathcal{A}) \geq r$. If, in addition, $n = m$ and B is nonsingular, then $\text{In } \mathcal{A} = \left[\begin{array}{ccc} n & 0 & n \end{array}\right]^{\mathrm{T}}$. (Proof: See [375].) (Remark: See Proposition 5.6.5.)

Fact 5.11.2. Let $A, B \in \mathbb{F}^{n \times n}$. Then,

$$\text{sprad}(A + B) \leq \sigma_{\max}(A + B) \leq \sigma_{\max}(A) + \sigma_{\max}(B).$$

If, in addition, A and B are Hermitian, then

$$\text{sprad}(A + B) = \sigma_{\max}(A + B) \leq \sigma_{\max}(A) + \sigma_{\max}(B) = \text{sprad}(A) + \text{sprad}(B).$$

(Remark: See Fact 5.10.3.)

Fact 5.11.3. Let $A, B \in \mathbb{F}^{n \times n}$, and let λ be an eigenvalue of $A + B$. Then,

$$\tfrac{1}{2}\lambda_{\min}(A^* + A) + \tfrac{1}{2}\lambda_{\min}(B^* + B) \leq \text{Re } \lambda \leq \tfrac{1}{2}\lambda_{\max}(A^* + A) + \tfrac{1}{2}\lambda_{\max}(B^* + B).$$

(Proof: See [177, p. 18].) (Remark: See Fact 5.10.22.)

Fact 5.11.4. Let $A, B \in \mathbb{F}^{n \times n}$ be normal, and let $\text{mspec}(A) = \{\lambda_1, \ldots, \lambda_n\}$ and $\text{mspec}(B) = \{\mu_1, \ldots, \mu_n\}$. Then,

$$\min \text{Re} \sum_{i=1}^{n} \lambda_i \mu_{\sigma(i)} \leq \text{Re tr } AB \leq \max \text{Re} \sum_{i=1}^{n} \lambda_i \mu_{\sigma(i)},$$

where "max" and "min" are taken over all permutations σ of the eigenvalues of B. If, in addition, A and B are Hermitian, then $\text{tr } AB$ is real, and

$$\sum_{i=1}^{n} \lambda_i(A)\lambda_{n-i}(B) \leq \text{tr } AB \leq \sum_{i=1}^{n} \lambda_i(A)\lambda_i(B).$$

Furthermore, the last inequality is an identity if and only if there exists a unitary matrix $S \in \mathbb{F}^{n \times n}$ such that $A = S\text{diag}[\lambda_1(A), \ldots, \lambda_n(A)]S^*$ and

$B = S\text{diag}[\lambda_1(B), \ldots, \lambda_n(B)]S^*$. (Proof: See [508]. For the last statement, see [473] or [135, p. 10].) (Remark: The upper bound for $\text{tr}\,AB$ is due to Fan.) (Remark: See Proposition 8.4.13 and Fact 8.10.16.)

Fact 5.11.5. Let $A \in \mathbb{F}^{n \times m}$ and $B \in \mathbb{F}^{m \times n}$, and let $r \triangleq \min\{\text{rank}\,A, \text{rank}\,B\}$. Then,

$$|\text{tr}\,AB| \le \sum_{i=1}^{r} \sigma_i(A)\sigma_i(B).$$

(Proof: See [592, p. 148]. Alternatively, applying Fact 5.11.4 to $\left[\begin{smallmatrix} 0 & A \\ A^* & 0 \end{smallmatrix}\right]$ and $\left[\begin{smallmatrix} 0 & B^* \\ B & 0 \end{smallmatrix}\right]$ and using Proposition 5.6.5 yields the weaker result $|\text{Re}\,\text{tr}\,AB| \le \sum_{i=1}^{r} \sigma_i(A)\sigma_i(B)$. See [135, p. 14].) (Remark: This result is due to Mirsky.)

Fact 5.11.6. Let $A, B \in \mathbb{R}^{n \times n}$, assume that B is symmetric, and define $C \triangleq \frac{1}{2}(A + A^{\mathrm{T}})$. Then,

$$\lambda_{\min}(C)\,\text{tr}\,B - \lambda_{\min}(B)[n\lambda_{\min}(C) - \text{tr}\,A]$$

$$\le \text{tr}\,AB \le \lambda_{\max}(C)\,\text{tr}\,B - \lambda_{\max}(B)[n\lambda_{\max}(C) - \text{tr}\,A].$$

(Proof: See [250].) (Remark: See Fact 5.11.4, Proposition 8.4.13, and Fact 8.10.16. Extensions are given in [576].)

Fact 5.11.7. Let $A, B \in \mathbb{R}^{n \times n}$, and assume that $AB = BA$. Then,

$$\text{sprad}(AB) \le \text{sprad}(A)\,\text{sprad}(B),$$

$$\text{sprad}(A + B) \le \text{sprad}(A) + \text{sprad}(B).$$

(Remark: If $AB \ne BA$, then both of these inequalities may be violated. Consider $A = \left[\begin{smallmatrix} 0 & 1 \\ 0 & 0 \end{smallmatrix}\right]$ and $B = \left[\begin{smallmatrix} 0 & 0 \\ 1 & 0 \end{smallmatrix}\right]$.)

Fact 5.11.8. Let $A, B, C \in \mathbb{F}^{n \times n}$, assume that $\text{spec}(A) \cap \text{spec}(B) = \varnothing$, and assume that $[A + B, C] = 0$ and $[AB, C] = 0$. Then, $[A, C] = [B, C] = 0$. (Proof: The result follows from Corollary 7.2.5.) (Remark: This result is due to Embry. See [121].)

Fact 5.11.9. Let $M \in \mathbb{R}^{r \times r}$, assume that M is positive definite, let $C, K \in \mathbb{R}^{r \times r}$, assume that C and K are positive semidefinite, and consider the equation

$$M\ddot{q} + C\dot{q} + Kq = 0.$$

Then, $x(t) \triangleq \begin{bmatrix} q(t) \\ \dot{q}(t) \end{bmatrix}$ satisfies $\dot{x}(t) = Ax(t)$, where A is the $2r \times 2r$ matrix

$$A \triangleq \begin{bmatrix} 0 & I \\ -M^{-1}K & -M^{-1}C \end{bmatrix}.$$

Furthermore, the following statements hold:

i) A, K, and M satisfy

$$\det A = \frac{\det K}{\det M}.$$

ii) A and K satisfy

$$\operatorname{rank} A = r + \operatorname{rank} K.$$

iii) A is nonsingular if and only if K is positive definite. In this case,

$$A^{-1} = \begin{bmatrix} -K^{-1}C & -K^{-1}M \\ I & 0 \end{bmatrix}.$$

iv) Let $\lambda \in \mathbb{C}$. Then, $\lambda \in \operatorname{spec}(A)$ if and only if $\det(\lambda^2 M + \lambda C + K) = 0$.

v) If $\lambda \in \operatorname{spec}(A)$, $\operatorname{Re} \lambda = 0$, and $\operatorname{Im} \lambda \neq 0$, then λ is semisimple.

vi) $\operatorname{mspec}(A) \subset \mathrm{CLHP}$.

vii) If $C = 0$, then $\operatorname{spec}(A) \subset \jmath\mathbb{R}$.

viii) If C and K are positive definite, then $\operatorname{spec}(A) \subset \mathrm{OLHP}$.

ix) If $C = 0$, then $\hat{x}(t) \triangleq \begin{bmatrix} M^{1/2} q(t) \\ M^{1/2} \dot{q}(t) \end{bmatrix}$ satisfies $\dot{x}(t) = \hat{A}x(t)$, where $\hat{A}$ is the Hamiltonian matrix

$$\hat{A} \triangleq \begin{bmatrix} 0 & I \\ -M^{-1/2}KM^{-1/2} & 0 \end{bmatrix}.$$

(Remark: M, C, K are mass, damping, and stiffness matrices. See [103].)
(Remark: See Fact 11.17.36.) (Problem: Prove v).)

Fact 5.11.10. Let $A, B \in \mathbb{R}^{n \times n}$, and assume that A and B are positive semidefinite. Then, every eigenvalue λ of $\begin{bmatrix} 0 & B \\ -A & 0 \end{bmatrix}$ satisfies $\operatorname{Re} \lambda = 0$. (Proof: Square this matrix.) (Problem: What happens if A and B have different dimensions?) In addition, let $C \in \mathbb{R}^{n \times n}$, and assume that C is (positive semidefinite, positive definite). Then, every eigenvalue of $\begin{bmatrix} 0 & A \\ -B & -C \end{bmatrix}$ satisfies ($\operatorname{Re} \lambda \leq 0$, $\operatorname{Re} \lambda < 0$). (Problem: Consider also $\begin{bmatrix} -C & A \\ -B & -C \end{bmatrix}$ and $\begin{bmatrix} -C & A \\ -A & -C \end{bmatrix}$.)

Fact 5.11.11. Let $A, B \in \mathbb{F}^{n \times n}$, and define the matrix polynomial $P_{A,B} \in \mathbb{F}^{n \times n}[s]$, called a *matrix pencil*, by

$$P_{A,B}(s) \triangleq sB - A.$$

Furthermore, define the *characteristic polynomial* $\chi_{A,B} \in \mathbb{F}[s]$ of (A, B) by

$$\chi_{A,B}(s) = \det(sB - A).$$

The pencil $P_{A,B}$ is *regular* if $\chi_{A,B}$ is not the zero polynomial, and *singular* if $\chi_{A,B} = 0$. The roots of $\chi_{A,B}$ are *generalized eigenvalues* of (A, B). Then, the following statements hold:

i) If $\mathcal{N}(A) \cap \mathcal{N}(B) \neq 0$, then $P_{A,B}$ is singular.

ii) $\deg \chi_{A,B} \leq n$.

iii) $\deg \chi_{A,B} = n$ if and only if B is nonsingular.

iv) $m_{\chi_{A,B}}(0) = n - \deg \chi_{B,A}$.

v) If B is nonsingular, then $\chi_{A,B} = \chi_{B^{-1}A}$.

vi) There exist unitary matrices $S_1, S_2 \in \mathbb{C}^{n \times n}$ such that S_1AS_2 and S_1BS_2 are upper triangular.

vii) If $A, B \in \mathbb{R}^{n \times n}$, then there exist orthogonal matrices $S_1, S_2 \in \mathbb{R}^{n \times n}$ such that S_1AS_2 and S_1BS_2 are upper Hessenberg matrices with 1×1 and 2×2 diagonally located blocks.

viii) If $P_{A,B}$ is regular, then there exist nonsingular matrices $S_1, S_2 \in \mathbb{C}^{n \times n}$ such that

$$P_{A,B}(s) = S_1 \left(\begin{bmatrix} A_1 & 0 \\ 0 & I_{n-m} \end{bmatrix} - s \begin{bmatrix} I_m & 0 \\ 0 & B_2 \end{bmatrix} \right) S_2,$$

where $m \triangleq \deg \chi_{A,B}$, $A_1 \in \mathbb{C}^{m \times m}$ is in Jordan form, and $B_2 \in \mathbb{C}^{(n-m) \times (n-m)}$ is nilpotent and in Jordan form. Furthermore,

$$\mathrm{mroots}(\chi_{A,B}) = \mathrm{mspec}(A_1).$$

(Proof: See [456, Chapter 12] and [667, Chapter VI].) (Remark: *vii*) is the *Weierstrass canonical form* for a square, regular matrix pencil. Extensions to singular pencils and nonsquare matrices are given by the *Kronecker canonical form*. For details, see [38, Chapter 2], [286, Chapter XII], [407, pp. 395–398], [456, Chapter 12], and [667, Chapter VI]. Computational algorithms are given in [737]. Applications to linear system theory are discussed in [177, pp, 52–55] and [410].) (Remark: See Fact 2.9.16.)

Fact 5.11.12. Let $A, B \in \mathbb{F}^{n \times n}$, assume that $P_{A,B}$ is a regular pencil, let $\mathcal{S} \subseteq \mathbb{F}^n$, assume that $\mathcal{S}$ is a subspace, let $k \triangleq \dim \mathcal{S}$, let $S \in \mathbb{F}^{n \times k}$, and assume that $\mathcal{R}(S) = \mathcal{S}$. Then, the following statements are equivalent:

i) $\dim(A\mathcal{S} + B\mathcal{S}) = \dim \mathcal{S}$.

ii) There exists a matrix $M \in \mathbb{F}^{k \times k}$ such that $AS = BSM$.

(Proof: See [456, p. 144].) (Remark: $\mathcal{S}$ is a *deflating subspace* of $P_{A,B}$. See Fact 2.9.16.)

5.12 Facts on Matrix Eigenstructure

Fact 5.12.1. Let $A \in \mathbb{F}^{n \times n}$. Then, $\mathcal{R}(A) = \mathcal{R}(A^2)$ if and only if $\mathrm{ind}\, A \leq 1$.

Fact 5.12.2. Let $A \in \mathbb{F}^{n \times n}$, and let $\mathrm{spec}(A) = \{0, \lambda_1, \ldots, \lambda_r\}$. Then, A is group invertible if and only if $\mathrm{rank}\, A = \sum_{i=1}^{r} \mathrm{am}_A(\lambda_i)$.

Fact 5.12.3. Let $A \in \mathbb{F}^{n \times n}$, and assume that A is diagonalizable. Then, A^{A}, A^*, $\overline{A}$, and A^{T} are diagonalizable. If, in addition, A is nonsingular, then A^{-1} is diagonalizable. (Proof: See Fact 2.14.9 and Fact 3.5.6.)

Fact 5.12.4. Let $A \in \mathbb{F}^{n \times n}$, assume that A is diagonalizable over $\mathbb{F}$ with eigenvalues $\lambda_1, \ldots, \lambda_n$, and let $B \triangleq \mathrm{diag}(\lambda_1, \ldots, \lambda_n)$. If, $x_1, \ldots, x_n \in \mathbb{F}^n$ are linearly independent eigenvectors of A associated with $\lambda_1, \ldots, \lambda_n$, respectively, then $A = SBS^{-1}$, where $S \triangleq \begin{bmatrix} x_1 & \cdots & x_n \end{bmatrix}$. Conversely, if $S \in \mathbb{F}^{n \times n}$ is nonsingular and $A = SBS^{-1}$, then, for all $i = 1, \ldots, n$, $\mathrm{col}_i(S)$ is an associated eigenvector.

Fact 5.12.5. Let $A \in \mathbb{F}^{n \times n}$, let $S \in \mathbb{F}^{n \times n}$, assume that S is nonsingular, let $\lambda \in \mathbb{C}$, and assume that $\mathrm{row}_1(S^{-1}AS) = \lambda e_1^{\mathrm{T}}$. Then, $\lambda \in \mathrm{spec}(A)$, and $\mathrm{col}_1(S)$ is an associated eigenvector.

Fact 5.12.6. Let $A \in \mathbb{F}^{n \times n}$. Then, A is cyclic if and only if there exists a vector $x \in \mathbb{F}^n$ such that $\begin{bmatrix} x & Ax & \cdots & A^{n-1}x \end{bmatrix}$ is nonsingular. (Proof: See Fact 12.20.11.)

Fact 5.12.7. Let $A \in \mathbb{R}^{n \times n}$. Then, A is cyclic and semisimple if and only if A is simple.

Fact 5.12.8. Let $A = \mathrm{revdiag}(a_1, \ldots, a_n) \in \mathbb{R}^{n \times n}$. Then, A is semisimple if and only if, for all $i = 1, \ldots, n$, a_i and a_{n+1-i} are either both zero or both nonzero. (Proof: See [325, p. 116], [420], or [592, pp. 68, 86].)

Fact 5.12.9. Let $A \in \mathbb{F}^{n \times n}$. Then, A has at least m real eigenvalues and m associated linearly independent eigenvectors if and only if there exists a positive-semidefinite matrix $S \in \mathbb{F}^{n \times n}$ such that $\mathrm{rank}\, S = m$ and $AS = SA^*$. (Proof: See [592, pp. 68, 86].) (Remark: See Proposition 5.5.18.) (Remark: This result is due to Drazin and Haynsworth.)

Fact 5.12.10. Let $A \in \mathbb{F}^{n \times n}$, assume that A is normal, and let $\mathrm{mspec}(A) = \{\lambda_1, \ldots, \lambda_n\}_{\mathrm{m}}$. Then, there exist vectors $x_1, \ldots, x_n \in \mathbb{C}^n$ such that $x_i^* x_j = \delta_{ij}$ for all $i, j = 1, \ldots, n$ and

$$A = \sum_{i=1}^{n} \lambda_i x_i x_i^*.$$

(Remark: This result is a restatement of Corollary 5.4.4.)

Fact 5.12.11. Let $A \in \mathbb{F}^{n \times n}$, and let $\mathrm{mspec}(A) = \{\lambda_1, \ldots, \lambda_n\}_{\mathrm{m}}$. Then, A is normal if and only if the singular values of A are $|\lambda_1|, \ldots, |\lambda_n|$. (Proof:

See [309].)

Fact 5.12.12. Let $A \in \mathbb{F}^{n \times n}$, and assume that A is either involutory or skew involutory. Then, A is semisimple.

Fact 5.12.13. Let $A \in \mathbb{R}^{n \times n}$, and assume that A is involutory. Then, A is diagonalizable over $\mathbb{R}$.

Fact 5.12.14. Let $A \in \mathbb{F}^{n \times n}$, assume that A is semisimple, and assume that $A^3 = A^2$. Then, A is idempotent.

Fact 5.12.15. Let $A \in \mathbb{F}^{n \times n}$. Then, A is cyclic if and only if every matrix $B \in \mathbb{F}^{n \times n}$ satisfying $AB = BA$ is a polynomial in A. (Proof: See [369, p. 275].)

Fact 5.12.16. Let $A, B \in \mathbb{F}^{n \times n}$. Then, B is a polynomial in A if and only if B commutes with every matrix that commutes with A. (Proof: See [369, p. 276].) (Remark: See Fact 4.8.13.)

Fact 5.12.17. Let $A, B \in \mathbb{C}^{n \times n}$, and assume that $AB = BA$. Then, there exists a nonzero vector $x \in \mathbb{C}^n$ that is an eigenvector of both A and B. (Proof: See [367, p. 51].)

Fact 5.12.18. Let $A, B \in \mathbb{F}^{n \times n}$. Then, the following statements hold:

i) Assume that A and B are Hermitian. Then, AB is Hermitian if and only if $AB = BA$.

ii) Assume that A is normal and $AB = BA$. Then, $A^*B = BA^*$.

iii) Assume that B is Hermitian and $AB = BA$. Then, $A^*B = BA^*$.

iv) Assume that A and B are normal and $AB = BA$. Then, AB is normal.

v) Assume that A, B, and AB are normal. Then, BA is normal.

vi) Assume that A and B are normal and either A or B has the property that distinct eigenvalues have unequal absolute values. Then, AB is normal if and only if $AB = BA$.

(Proof: See [195, 775], [329, p. 157], and [592, p. 102].)

Fact 5.12.19. Let $A, B, C \in \mathbb{F}^{n \times n}$, and assume that A and B are normal and $AC = CB$. Then, $A^*C = CB^*$. (Proof: Consider $\begin{bmatrix} A & 0 \\ 0 & B \end{bmatrix}$ and $\begin{bmatrix} 0 & C \\ 0 & 0 \end{bmatrix}$ in *ii)* of Fact 5.12.18. See [326, p. 104] or [329, p. 321].) (Remark: This result is the *Putnam-Fuglede theorem*.)

Fact 5.12.20. Let $A, B \in \mathbb{R}^{n \times n}$, and assume that A and B are skew symmetric. Then, there exists an orthogonal matrix $S \in \mathbb{R}^{n \times n}$ such that

$$A = S \begin{bmatrix} 0_{(n-l) \times (n-l)} & A_{12} \\ -A_{12}^{\mathrm{T}} & A_{22} \end{bmatrix} S^{\mathrm{T}}$$

and

$$B = S \begin{bmatrix} B_{11} & B_{12} \\ -B_{12}^{\mathrm{T}} & 0_{l \times l} \end{bmatrix} S^{\mathrm{T}},$$

where $l \triangleq \lfloor n/2 \rfloor$. Consequently,

$$\mathrm{mspec}(AB) = \mathrm{mspec}(-A_{12}B_{12}^{\mathrm{T}}) \cup \mathrm{mspec}(-A_{12}^{\mathrm{T}}B_{12}),$$

and thus every nonzero eigenvalue of AB has even algebraic multiplicity. (Proof: See [17].)

Fact 5.12.21. Let $A, B \in \mathbb{R}^{n \times n}$, and assume that A and B are skew symmetric. If n is even, then there exists a monic polynomial p of degree $n/2$ such that $\chi_{AB}(s) = p^2(s)$ and $p(AB) = 0$. If n is odd, then there exists a monic polynomial $p(s)$ of degree $(n-1)/2$ such that $\chi_{AB}(s) = sp^2(s)$ and $ABp(AB) = 0$. Consequently, if n is (even, odd), then χ_{AB} is (even, odd) and (every, every nonzero) eigenvalue of AB has even algebraic multiplicity and geometric multiplicity of at least 2. (Proof: See [231, 304].)

Fact 5.12.22. Let $A, B \in \mathbb{F}^{n \times n}$, and assume that A and B are projectors. Then, $\mathrm{spec}(AB) \subset [0,1]$ and $\mathrm{spec}(A - B) \subset [-1, 1]$. (Proof: See [23] or [592, p. 147].) (Remark: The first result is due to Afriat.)

Fact 5.12.23. Let $q(t)$ denote the displacement of a mass $m > 0$ connected to a spring $k \geq 0$ and dashpot $c \geq 0$ and subject to a force $f(t)$. Then, $q(t)$ satisfies

$$m\ddot{q}(t) + c\dot{q}(t) + kq(t) = f(t)$$

or

$$\ddot{q}(t) + \frac{c}{m}\dot{q}(t) + \frac{k}{m}q(t) = \frac{1}{m}f(t).$$

Now, define the *natural frequency* $\omega_{\mathrm{n}} \triangleq \sqrt{k/m}$ and, if $k > 0$, the *damping ratio* $\zeta \triangleq c/2\sqrt{km}$ to obtain

$$\ddot{q}(t) + 2\zeta\omega_{\mathrm{n}}\dot{q}(t) + \omega_{\mathrm{n}}^2 q(t) = \frac{1}{m}f(t).$$

If $k = 0$, then set $\omega_{\mathrm{n}} = 0$ and $\zeta\omega_{\mathrm{n}} = c/2m$. Next, define $x_1(t) \triangleq q(t)$ and $x_2(t) \triangleq \dot{q}(t)$ so that this equation can be written as

$$\begin{bmatrix} \dot{x}_1(t) \\ \dot{x}_2(t) \end{bmatrix} = \begin{bmatrix} 0 & 1 \\ -\omega_n^2 & -2\zeta\omega_n \end{bmatrix} \begin{bmatrix} x_1(t) \\ x_2(t) \end{bmatrix} + \begin{bmatrix} 0 \\ 1/m \end{bmatrix} f(t).$$

The eigenvalues of the companion matrix $A_c \triangleq \begin{bmatrix} 0 & 1 \\ -\omega_n^2 & -2\zeta\omega_n \end{bmatrix}$ are given by

$$\mathrm{mspec}(A_c) = \begin{cases} \{-\zeta\omega_n - \jmath\omega_d, -\zeta\omega_n + \jmath\omega_d\}_m, & 0 \le \zeta \le 1, \\ \left\{(-\zeta - \sqrt{\zeta^2-1})\omega_n, (-\zeta + \sqrt{\zeta^2-1})\omega_n\right\}, & \zeta > 1, \end{cases}$$

where $\omega_d \triangleq \omega_n\sqrt{1-\zeta^2}$ is the *damped natural frequency*. The matrix A_c has repeated eigenvalues in exactly two cases, namely,

$$\mathrm{mspec}(A_c) = \begin{cases} \{0, 0\}_m, & \omega_n = 0, \\ \{-\omega_n, -\omega_n\}_m, & \zeta = 1. \end{cases}$$

In both of these cases the matrix A_c is defective. In the case $\omega_n = 0$, the matrix A_c is also in Jordan form, while, in the case $\zeta = 1$, it follows that $A_c = SA_J S^{-1}$, where $S \triangleq \begin{bmatrix} -1 & 0 \\ \omega_n & -1 \end{bmatrix}$ and A_J is the Jordan form matrix $A_J \triangleq \begin{bmatrix} -\omega_n & 1 \\ 0 & -\omega_n \end{bmatrix}$. If A_c is not defective, that is, if $\omega_n \ne 0$ and $\zeta \ne 1$, then the Jordan form A_J of A_c is given by

$$A_J \triangleq \begin{cases} \begin{bmatrix} -\zeta\omega_n + \jmath\omega_d & 0 \\ 0 & -\zeta\omega_n - \jmath\omega_d \end{bmatrix}, & 0 \le \zeta < 1, \ \omega_n \ne 0, \\ \begin{bmatrix} \left(-\zeta - \sqrt{\zeta^2-1}\right)\omega_n & 0 \\ 0 & \left(-\zeta + \sqrt{\zeta^2-1}\right)\omega_n \end{bmatrix}, & \zeta > 1, \ \omega_n \ne 0. \end{cases}$$

In the case $0 \le \zeta < 1$ and $\omega_n \ne 0$, define the real normal form

$$A_n \triangleq \begin{bmatrix} -\zeta\omega_n & \omega_d \\ -\omega_d & -\zeta\omega_n \end{bmatrix}.$$

The matrices A_c, A_J, and A_n are related by the similarity transformations

$$A_c = S_1 A_J S_1^{-1} = S_2 A_n S_2^{-1}, \qquad A_J = S_3 A_n S_3^{-1},$$

where

$$S_1 \triangleq \begin{bmatrix} 1 & 1 \\ -\zeta\omega_n + \jmath\omega_d & -\zeta\omega_n - \jmath\omega_d \end{bmatrix}, \quad S_1^{-1} = \frac{\jmath}{2\omega_d}\begin{bmatrix} -\zeta\omega_n - \jmath\omega_d & -1 \\ \zeta\omega_n - \jmath\omega_d & 1 \end{bmatrix},$$

$$S_2 \triangleq \frac{1}{\omega_d}\begin{bmatrix} 1 & 0 \\ -\zeta\omega_n & \omega_d \end{bmatrix}, \qquad S_2^{-1} = \begin{bmatrix} \omega_d & 0 \\ \zeta\omega_n & 1 \end{bmatrix},$$

$$S_3 \triangleq \frac{1}{2\omega_d}\begin{bmatrix} 1 & -\jmath \\ 1 & \jmath \end{bmatrix}, \qquad S_3^{-1} = \omega_d\begin{bmatrix} 1 & 1 \\ \jmath & -\jmath \end{bmatrix}.$$

In the case $\zeta > 1$ and $\omega_n \neq 0$, the matrices A_c and A_J are related by

$$A_c = S_4 A_J S_4^{-1},$$

where

$$S_4 \triangleq \begin{bmatrix} 1 & 1 \\ -\zeta\omega_n + \jmath\omega_d & -\zeta\omega_n - \jmath\omega_d \end{bmatrix}, \quad S_4^{-1} = \frac{\jmath}{2\omega_d} \begin{bmatrix} -\zeta\omega_n - \jmath\omega_d & -1 \\ \zeta\omega_n - \jmath\omega_d & 1 \end{bmatrix}.$$

Finally, define the energy coordinates matrix

$$A_e \triangleq \begin{bmatrix} 0 & \omega_n \\ -\omega_n & -2\zeta\omega_n \end{bmatrix}.$$

Then, $A_e = S_5 A_c S_5^{-1}$, where

$$S_5 \triangleq \sqrt{\frac{m}{2}} \begin{bmatrix} 1 & 0 \\ 0 & 1/\omega_n \end{bmatrix}.$$

5.13 Facts on Companion, Vandermonde, and Circulant Matrices

Fact 5.13.1. Let $p \in \mathbb{F}[s]$, where $p(s) = s^n + \beta_{n-1}s^{n-1} + \cdots + \beta_0$, and define $C_b(p), C_r(p), C_t(p), C_l(p) \in \mathbb{F}^{n \times n}$ by

$$C_b(p) \triangleq \begin{bmatrix} 0 & 1 & 0 & \cdots & 0 & 0 \\ 0 & 0 & 1 & \ddots & 0 & 0 \\ 0 & 0 & 0 & \ddots & 0 & 0 \\ \vdots & \vdots & \vdots & \ddots & \ddots & \vdots \\ 0 & 0 & 0 & \cdots & 0 & 1 \\ -\beta_0 & -\beta_1 & -\beta_2 & \cdots & -\beta_{n-2} & -\beta_{n-1} \end{bmatrix},$$

$$C_r(p) \triangleq \begin{bmatrix} 0 & 0 & 0 & \cdots & 0 & -\beta_0 \\ 1 & 0 & 0 & \cdots & 0 & -\beta_1 \\ 0 & 1 & 0 & \cdots & 0 & -\beta_2 \\ \vdots & \ddots & \ddots & \ddots & \vdots & \vdots \\ 0 & 0 & 0 & \ddots & 0 & -\beta_{n-2} \\ 0 & 0 & 0 & \cdots & 1 & -\beta_{n-1} \end{bmatrix},$$

$$C_{\mathrm{t}}(p) \triangleq \begin{bmatrix} -\beta_{n-1} & -\beta_{n-2} & \cdots & -\beta_2 & -\beta_1 & -\beta_0 \\ 1 & 0 & \cdots & 0 & 0 & 0 \\ \vdots & \ddots & \ddots & \vdots & \vdots & \vdots \\ 0 & 0 & \ddots & 0 & 0 & 0 \\ 0 & 0 & \ddots & 1 & 0 & 0 \\ 0 & 0 & \cdots & 0 & 1 & 0 \end{bmatrix},$$

$$C_{\mathrm{l}}(p) \triangleq \begin{bmatrix} -\beta_{n-1} & 1 & \cdots & 0 & 0 & 0 \\ -\beta_{n-2} & 0 & \ddots & 0 & 0 & 0 \\ \vdots & \vdots & \ddots & \ddots & \ddots & \vdots \\ -\beta_2 & 0 & \cdots & 0 & 1 & 0 \\ -\beta_1 & 0 & \cdots & 0 & 0 & 1 \\ -\beta_0 & 0 & \cdots & 0 & 0 & 0 \end{bmatrix}.$$

Then,

$$C_{\mathrm{r}}(p) = C_{\mathrm{b}}^{\mathrm{T}}(p), \quad C_{\mathrm{l}}(p) = C_{\mathrm{t}}^{\mathrm{T}}(p),$$
$$C_{\mathrm{t}}(p) = \hat{I} C_{\mathrm{b}}(p) \hat{I}, \quad C_{\mathrm{l}}(p) = \hat{I} C_{\mathrm{r}}(p) \hat{I},$$
$$C_{\mathrm{l}}(p) = C_{\mathrm{b}}^{\hat{\mathrm{T}}}(p), \quad C_{\mathrm{t}}(p) = C_{\mathrm{r}}^{\hat{\mathrm{T}}}(p),$$

and

$$\chi_{C_{\mathrm{b}}(p)} = \chi_{C_{\mathrm{r}}(p)} = \chi_{C_{\mathrm{t}}(p)} = \chi_{C_{\mathrm{l}}(p)} = p.$$

Furthermore,

$$C_{\mathrm{r}}(p) = S C_{\mathrm{b}}(p) S^{-1}$$

and

$$C_{\mathrm{l}}(p) = \hat{S} C_{\mathrm{t}}(p) \hat{S}^{-1},$$

where $S, \hat{S} \in \mathbb{F}^{n \times n}$ are the Hankel matrices

$$S \triangleq \begin{bmatrix} \beta_1 & \beta_2 & \cdots & \beta_{n-1} & 1 \\ \beta_2 & \beta_3 & \cdot^{\cdot} & 1 & 0 \\ \vdots & \cdot^{\cdot} & \cdot^{\cdot} & \cdot^{\cdot} & \vdots \\ \beta_{n-1} & 1 & \cdot^{\cdot} & 0 & 0 \\ 1 & 0 & \cdots & 0 & 0 \end{bmatrix}$$

and

$$\hat{S} \triangleq \hat{I}S\hat{I} = \begin{bmatrix} 0 & 0 & \cdots & 0 & 1 \\ 0 & 0 & \cdot{}^{\cdot}{}^{\cdot} & 1 & \beta_{n-1} \\ \vdots & \cdot{}^{\cdot}{}^{\cdot} & \cdot{}^{\cdot}{}^{\cdot} & \cdot{}^{\cdot}{}^{\cdot} & \vdots \\ 0 & 1 & \cdot{}^{\cdot}{}^{\cdot} & \beta_3 & \beta_2 \\ 1 & \beta_{n-1} & \cdots & \beta_2 & \beta_1 \end{bmatrix}.$$

(Remark: $(C_{\mathrm{b}}(p), C_{\mathrm{r}}(p), C_{\mathrm{t}}(p), C_{\mathrm{l}}(p))$ are the (*bottom, right, top, left*) companion matrices. Note that $C_{\mathrm{b}}(p) = C(p)$. See [78, p. 282] and [407, p. 659].) (Remark: $S = B(p,1)$, where $B(p,1)$ is a Bezout matrix. See Fact 4.8.6.)

Fact 5.13.2. Let $p \in \mathbb{F}[s]$, where $p(s) = s^n + \beta_{n-1}s^{n-1} + \cdots + \beta_0$, assume that $\beta_0 \neq 0$, and let

$$C_{\mathrm{b}}(p) \triangleq \begin{bmatrix} 0 & 1 & 0 & \cdots & 0 & 0 \\ 0 & 0 & 1 & \ddots & 0 & 0 \\ 0 & 0 & 0 & \ddots & 0 & 0 \\ \vdots & \vdots & \vdots & \ddots & \ddots & \vdots \\ 0 & 0 & 0 & \cdots & 0 & 1 \\ -\beta_0 & -\beta_1 & -\beta_2 & \cdots & -\beta_{n-2} & -\beta_{n-1} \end{bmatrix}.$$

Then,

$$C_{\mathrm{b}}^{-1}(p) = C_{\mathrm{t}}(\hat{p}) = \begin{bmatrix} -\beta_1/\beta_0 & \cdots & -\beta_{n-2}/\beta_0 & -\beta_{n-1}/\beta_0 & -1/\beta_0 \\ 1 & \cdots & 0 & 0 & 0 \\ \vdots & \ddots & \vdots & \vdots & \vdots \\ 0 & \cdots & 1 & 0 & 0 \\ 0 & \cdots & 0 & 1 & 0 \end{bmatrix},$$

where $\hat{p}(s) \triangleq \beta_0^{-1}s^n p(1/s)$. (Remark: See Fact 4.9.6.)

Fact 5.13.3. Let $\lambda_1, \ldots, \lambda_n \in \mathbb{F}$, and define the *Vandermonde matrix* $V(\lambda_1, \ldots, \lambda_n) \in \mathbb{F}^{n \times n}$ by

$$V(\lambda_1, \ldots, \lambda_n) \triangleq \begin{bmatrix} 1 & 1 & \cdots & 1 \\ \lambda_1 & \lambda_2 & \cdots & \lambda_n \\ \lambda_1^2 & \lambda_2^2 & \cdots & \lambda_n^2 \\ \lambda_1^3 & \lambda_2^3 & \cdots & \lambda_n^3 \\ \vdots & \vdots & \ddots & \vdots \\ \lambda_1^{n-1} & \lambda_2^{n-1} & \cdots & \lambda_n^{n-1} \end{bmatrix}.$$

Then,

$$\det V(\lambda_1, \ldots, \lambda_n) = \prod_{1 \le i < j \le n} (\lambda_i - \lambda_j).$$

Thus, $V(\lambda_1, \ldots, \lambda_n)$ is nonsingular if and only if $\lambda_1, \ldots, \lambda_n$ are distinct. (Remark: This result yields Proposition 4.5.3. Let $x_1, \ldots, x_k$ be eigenvectors of $V(\lambda_1, \ldots, \lambda_n)$ associated with distinct eigenvalues $\lambda_1, \ldots, \lambda_k$ of $V(\lambda_1, \ldots, \lambda_n)$. Assume $\alpha_1 x_1 + \cdots + \alpha_k x_k = 0$ so that $V^i(\lambda_1, \ldots, \lambda_n)(\alpha_1 x_1 + \cdots + \alpha_k x_k) = \alpha_1 \lambda_1^i x_i + \cdots + \alpha_k \lambda_k^i x_k = 0$ for all $i = 0, 1, \ldots, k-1$. Let $X \triangleq \begin{bmatrix} x_1 & \cdots & x_k \end{bmatrix} \in \mathbb{F}^{n \times k}$ and $D \triangleq \operatorname{diag}(\alpha_1, \ldots, \alpha_k)$. Then, $XDV^{\mathrm{T}}(\lambda_1, \ldots, \lambda_k) = 0$, which implies that $XD = 0$. Hence, $\alpha_i x_i = 0$ for all $i = 1, \ldots, k$, and thus $\alpha_1 = \alpha_2 = \cdots = \alpha_k = 0$.) (Proof: Connections between the Vandermonde matrix and the *Pascal matrix, Stirling matrix, Bernoulli matrix, Bernstein matrix*, and companion matrices are discussed in [3]. See also Fact 11.10.4.)

Fact 5.13.4. Let $\lambda_1, \ldots, \lambda_n \in \mathbb{F}$ and, for $i = 1, \ldots, n$, define

$$p_i(s) \triangleq \prod_{\substack{j=1 \\ j \ne i}}^{n} (s - \lambda_j).$$

Furthermore, define $A \in \mathbb{F}^{n \times n}$ by

$$A \triangleq \begin{bmatrix} p_1(0) & \frac{1}{1!} p_1'(0) & \cdots & \frac{1}{(n-1)!} p_1^{(n-1)}(0) \\ \vdots & \ddots & \ddots & \vdots \\ p_n(0) & \frac{1}{1!} p_n'(0) & \cdots & \frac{1}{(n-1)!} p_n^{(n-1)}(0) \end{bmatrix}.$$

Then,

$$\operatorname{diag}[p_1(\lambda_1), \ldots, p_n(\lambda_n)] = AV(\lambda_1, \ldots, \lambda_n).$$

(Proof: See [260, p. 159].)

Fact 5.13.5. Let $p \in \mathbb{F}[s]$, where $p(s) = s^n + \beta_{n-1} s^{n-1} + \cdots + \beta_1 s + \beta_0$, and assume that p has distinct roots $\lambda_1, \ldots, \lambda_n \in \mathbb{C}$. Then,

$$C(p) = V^{-1}(\lambda_1, \ldots, \lambda_n) \operatorname{diag}(\lambda_1, \ldots, \lambda_n) V(\lambda_1, \ldots, \lambda_n).$$

Fact 5.13.6. Let $A \in \mathbb{F}^{n \times n}$. Then, A is cyclic if and only if A is similar to a companion matrix. (Proof: The result follows from Corollary 5.3.4. Alternatively, let $\operatorname{spec}(A) = \{\lambda_1, \ldots, \lambda_r\}$ and $A = SBS^{-1}$, where $S \in \mathbb{C}^{n \times n}$ is nonsingular and $B = \operatorname{diag}(B_1, \ldots, B_r)$ is the Jordan form of A, where, for all $i = 1, \ldots, r$, $B_i \in \mathbb{C}^{n_i \times n_i}$ and $\lambda_i, \ldots, \lambda_i$ are the diagonal entries of B_i. Now, define $R \in \mathbb{C}^{n \times n}$ by $R \triangleq \begin{bmatrix} R_1 & \cdots & R_r \end{bmatrix} \in \mathbb{C}^{n \times n}$, where, for all $i = 1, \ldots, r$, $R_i \in \mathbb{C}^{n \times n_i}$ is the matrix

$$R_i \triangleq \begin{bmatrix} 1 & 0 & \cdots & 0 \\ \lambda_i & 1 & \cdots & 0 \\ \vdots & \vdots & \ddots & \vdots \\ \lambda_i^{n-2} & \binom{n-2}{1}\lambda_i^{n-3} & \cdots & \binom{n-2}{n_i-1}\lambda_i^{n-n_i-1} \\ \lambda_i^{n-1} & \binom{n-1}{1}\lambda_i^{n-2} & \cdots & \binom{n-1}{n_i-1}\lambda_i^{n-n_i} \end{bmatrix}.$$

Then, since $\lambda_1, \ldots, \lambda_r$ are distinct, it follows that R is nonsingular. Furthermore, $C = RBR^{-1}$ is in companion form, and thus $A = SR^{-1}CRS$. If $n_i = 1$ for all $i = 1, \ldots, r$, then R is a Vandermonde matrix. See Fact 5.13.3 and Fact 5.13.5.)

Fact 5.13.7. Let $a_0, \ldots, a_{n-1} \in \mathbb{F}$, and define $\mathrm{circ}(a_0, \ldots, a_{n-1}) \in \mathbb{F}^{n \times n}$ by

$$\mathrm{circ}(a_0, \ldots, a_{n-1}) \triangleq \begin{bmatrix} a_0 & a_1 & a_2 & \cdots & a_{n-2} & a_{n-1} \\ a_{n-1} & a_0 & a_1 & \cdots & a_{n-3} & a_{n-2} \\ a_{n-2} & a_{n-1} & a_0 & \ddots & a_{n-4} & a_{n-3} \\ \vdots & \vdots & \ddots & \ddots & \ddots & \vdots \\ a_2 & a_3 & a_4 & \ddots & a_0 & a_1 \\ a_1 & a_2 & a_3 & \cdots & a_{n-1} & a_0 \end{bmatrix}.$$

A matrix of this form is *circulant*. Furthermore, define the *primary circulant*

$$P \triangleq \mathrm{circ}(0, 1, 0, \ldots, 0) \triangleq \begin{bmatrix} 0 & 1 & 0 & \cdots & 0 & 0 \\ 0 & 0 & 1 & \ddots & 0 & 0 \\ 0 & 0 & 0 & \ddots & 0 & 0 \\ \vdots & \ddots & \ddots & \ddots & \ddots & \vdots \\ 0 & 0 & 0 & \ddots & 0 & 1 \\ 1 & 0 & 0 & \cdots & 0 & 0 \end{bmatrix}.$$

Finally, define $p(s) \triangleq a_{n-1}s^{n-1} + \cdots + a_1 s + a_0$. Then, the following statements hold:

i) $p(P) = \mathrm{circ}(a_0, \ldots, a_{n-1})$.

ii) If $A, B \in \mathbb{F}^{n \times n}$ are circulant, then A and B commute and AB is circulant.

iii) If A is circulant, then A^* is circulant.

iv) If A is circulant and $k \geq 0$, then A^k is circulant.

v) If A is nonsingular and circulant, then A^{-1} is circulant.

vi) $A \in \mathbb{F}^{n \times n}$ is circulant if and only if $A = PAP^{\mathrm{T}}$.

vii) P is an orthogonal matrix, and $P^n = I_n$.

viii) $P = C(p)$, where $p \in \mathbb{F}[s]$ is defined by $p(s) \triangleq s^n - 1$.

ix) If $A \in \mathbb{F}^{n \times n}$ is circulant, then A is reverse symmetric, Toeplitz, and normal.

x) $A \in \mathbb{F}^{n \times n}$ is normal if and only if A is unitarily similar to a normal matrix.

Next, let $\theta \triangleq e^{2\pi j/n}$, and define the *Fourier matrix* $S \in \mathbb{C}^{n \times n}$ by

$$S \triangleq n^{-1/2} V(1, \theta, \dots, \theta^{n-1}) = \frac{1}{\sqrt{n}} \begin{bmatrix} 1 & 1 & 1 & \cdots & 1 \\ 1 & \theta & \theta^2 & \cdots & \theta^{n-1} \\ 1 & \theta^2 & \theta^4 & \cdots & \theta^{n-2} \\ \vdots & \vdots & \vdots & \ddots & \vdots \\ 1 & \theta^{n-1} & \theta^{n-2} & \cdots & \theta \end{bmatrix}.$$

Then, the following statements hold:

i) S is symmetric and unitary.

ii) $S^4 = I_n$.

iii) $\mathrm{spec}(S) \subseteq \{1, -1, j, -j\}$.

iv) $\mathrm{Re}\, S$ and $\mathrm{Im}\, S$ are symmetric, commute, and satisfy

$$(\mathrm{Re}\, S)^2 + (\mathrm{Im}\, S)^2 = I_n.$$

v) $S^{-1}PS = \mathrm{diag}(1, \theta, \dots, \theta^{n-1})$.

vi) $S^{-1}\mathrm{circ}(a_0, \dots, a_{n-1})S = \mathrm{diag}[p(1), p(\theta), \dots, p(\theta^{n-1})]$.

vii) $\mathrm{mspec}[\mathrm{circ}(a_0, \dots, a_{n-1})] = \{p(1), p(\theta), p(\theta^2), \dots, p(\theta^{n-1})\}_{\mathrm{m}}$.

viii) $\mathrm{spec}(P) = \{1, \theta, \theta^2, \dots, \theta^{n-1}\}$.

(Proof: See [10, pp. 81–98], [207, p. 81], and [811, pp. 106–110].) (Remark: Circulant matrices play an important role in digital signal processing, specifically, in the efficient implementation of the *fast Fourier transform*. See [534, pp. 356–380] and [740, pp. 206, 207].) (Remark: If a real Toeplitz matrix is normal, then it must be either symmetric, skew symmetric, circulant, or skew circulant. See [41] and the references therein. A unified treatment of the solutions of quadratic, cubic, and quartic equations using circulant matrices is given in [408].)

5.14 Facts on Matrix Factorizations

Fact 5.14.1. Let $A \in \mathbb{F}^{n \times n}$. Then, A is normal if and only if there exists a unitary matrix $S \in \mathbb{F}^{n \times n}$ such that $A^* = AS$. (Proof: See [592, pp. 102, 113].)

Fact 5.14.2. Let $A \in \mathbb{F}^{n \times n}$, and assume that A is nonsingular. Then, A^{-1} and A^* are similar if and only if there exists a nonsingular matrix $B \in \mathbb{F}^{n \times n}$ such that $A = B^{-1}B^*$. Furthermore, A is unitary if and only if there exists a normal, nonsingular matrix $B \in \mathbb{F}^{n \times n}$ such that $A = B^{-1}B^*$. (Proof: See [217]. Sufficiency in the second statement follows from Fact 3.7.10.)

Fact 5.14.3. Let $A \in \mathbb{F}^{m \times m}$ and $B \in \mathbb{F}^{n \times n}$. Then, there exist matrices $C \in \mathbb{F}^{m \times n}$ and $D \in \mathbb{F}^{n \times m}$ such that $A = CD$ and $B = DC$ if and only if the following statements hold:

i) The Jordan blocks associated with nonzero eigenvalues are identical in A and B.

ii) Let $n_1 \geq n_2 \geq \cdots \geq n_r$ denote the sizes of the Jordan blocks of A associated with $0 \in \mathrm{spec}(A)$, and let $m_1 \geq m_2 \geq \cdots \geq m_r$ denote the sizes of the Jordan blocks of B associated with $0 \in \mathrm{spec}(B)$, where $n_i = 0$ or $m_i = 0$ as needed. Then, $|n_i - m_i| \leq 1$ for all $i = 1, \ldots, r$.

(Proof: See [400].) (Remark: See Fact 5.14.4.)

Fact 5.14.4. Let $A, B \in \mathbb{F}^{n \times n}$, and assume that A and B are nonsingular. Then, A and B are similar if and only if there exist nonsingular matrices $C, D \in \mathbb{F}^{n \times n}$ such that $A = CD$ and $B = DC$. (Proof: Sufficiency follows from Fact 5.9.10. Necessity is a special case of Fact 5.14.3.)

Fact 5.14.5. Let $A, B \in \mathbb{F}^{n \times n}$, and assume that A and B are nonsingular. Then, $\det A = \det B$ if and only if there exist nonsingular matrices $C, D, E \in \mathbb{R}^{n \times n}$ such that $A = CDE$ and $B = EDC$. (Remark: This result is due to Shoda and Taussky-Todd. See [143].)

Fact 5.14.6. Let $A \in \mathbb{F}^{n \times n}$. Then, there exist matrices $B, C \in \mathbb{F}^{n \times n}$ such that B is unitary, C is upper triangular, and $A = BC$. If, in addition, A is nonsingular, then there exist unique matrices $B, C \in \mathbb{F}^{n \times n}$ such that B is unitary, C is upper triangular with positive diagonal entries, and $A = BC$. (Proof: See [367, p. 112] or [613, p. 362].) (Remark: This result is the *QR decomposition*. The orthogonal matrix B is constructed as a product of elementary reflectors.)

Fact 5.14.7. Let $A \in \mathbb{F}^{n \times m}$, and assume that rank $A = m$. Then, there exist a unique matrix $B \in \mathbb{F}^{n \times m}$ and a matrix $C \in \mathbb{F}^{m \times m}$ such that $B^*B = I_m$, C is upper triangular with positive diagonal entries, and $A = BC$. (Proof: See [367, p. 15] or [613, p. 206].) (Remark: $C \in \mathrm{UT}_+(n)$. See Fact 3.13.3.) (Remark: This factorization is a consequence of *Gram-Schmidt orthonormalization.*)

Fact 5.14.8. Let $A \in \mathbb{F}^{n \times n}$, let $r \triangleq$ rank A, and assume that the first r leading principal subdeterminants of A are nonzero. Then, there exist matrices $B, C \in \mathbb{F}^{n \times n}$ such that B is lower triangular, C is upper triangular, and $A = BC$. Either B or C can be chosen to be nonsingular. Furthermore, both B and C are nonsingular if and only if A is nonsingular. (Proof: See [367, p. 160].) (Remark: This result is the *LU decomposition.*)

Fact 5.14.9. Let $A \in \mathbb{F}^{n \times n}$, and let $r \triangleq$ rank A. Then, A is range Hermitian if and only if there exist a nonsingular matrix $S \in \mathbb{F}^{n \times n}$ and a nonsingular matrix $B \in \mathbb{F}^{r \times r}$ such that

$$A = S \begin{bmatrix} B & 0 \\ 0 & 0 \end{bmatrix} S^*.$$

(Remark: S need not be unitary for sufficiency. See Corollary 5.4.4.) (Proof: Use the QR decomposition Fact 5.14.6 to let $S \triangleq \hat{S}R$, where $\hat{S}$ is unitary and R is upper triangular. See [699].)

Fact 5.14.10. Let $A \in \mathbb{F}^{n \times n}$. Then, A is nonsingular if and only if A is the product of elementary matrices. (Problem: How many factors are needed?)

Fact 5.14.11. Let $A \in \mathbb{F}^{n \times n}$, assume that A is a projector, and let $r \triangleq$ rank A. Then, there exist nonzero vectors $x_1, \ldots, x_{n-r} \in \mathbb{F}^n$ such that $x_i^* x_j = 0$ for all $i \neq j$ and such that

$$A = \prod_{i=1}^{n-r} \left[I - (x_i^* x_i)^{-1} x_i x_i^* \right].$$

(Remark: Every projector is the product of mutually orthogonal elementary projectors.) (Proof: A is unitarily similar to $\mathrm{diag}(1, \ldots, 1, 0, \ldots, 0)$, which can be written as the product of elementary projectors.)

Fact 5.14.12. Let $A \in \mathbb{F}^{n \times n}$. Then, A is a reflector if and only if there exist $m \leq n$ nonzero vectors $x_1, \ldots, x_m \in \mathbb{F}^n$ such that $x_i^* x_j = 0$ for all $i \neq j$ and such that

$$A = \prod_{i=1}^{m} \left[I - 2(x_i^* x_i)^{-1} x_i x_i^* \right].$$

In this case, m is the algebraic multiplicity of $-1 \in \mathrm{spec}(A)$. (Remark:

Every reflector is the product of mutually orthogonal elementary reflectors.)
(Proof: A is unitarily similar to $\mathrm{diag}(\pm1,\ldots,\pm1)$, which can be written as
the product of elementary reflectors.)

Fact 5.14.13. Let $A \in \mathbb{R}^{n \times n}$. Then, A is orthogonal if and only if there
exist $m \in \mathbb{P}$ and nonzero vectors $x_1,\ldots,x_m \in \mathbb{F}^n$ such that $\det A = (-1)^m$
and

$$A = \prod_{i=1}^{m} \left[I - 2(x_i^* x_i)^{-1} x_i x_i^* \right].$$

(Remark: Every unitary matrix is the product of elementary reflectors. This
factorization is a result of Cartan and Dieudonné. See [56, p. 24] and
[632, 734]. The minimal number of factors is unsettled; see Fact 3.10.4. See
Fact 3.7.16.)

Fact 5.14.14. Let $A \in \mathbb{R}^{n \times n}$, where $n \geq 2$. Then, A is orthogonal and
$\det A = 1$ if and only if there exist $m \in \mathbb{P}$ such that $1 \leq m \leq n(n-1)/2$,
$\theta_1,\ldots,\theta_m \in \mathbb{R}$, and $j_1,\ldots,j_m, k_1,\ldots,k_m \in \{1,\ldots,n\}$ such that

$$A = \prod_{i=1}^{m} P(\theta_i, j_i, k_i),$$

where

$$P(\theta, j, k) \triangleq I_n + [(\cos\theta) - 1](E_{j,j} + E_{k,k}) + (\sin\theta)(E_{j,k} - E_{k,j}).$$

(Proof: See [253].) (Remark: $P(\theta, j, k)$ is a *plane* or *Givens rotation*. See
Fact 3.7.16.) (Remark: Suppose that $\det A = -1$, and let $B \in \mathbb{R}^{n \times n}$ be an
elementary reflector. Then, $AB \in \mathrm{SO}(n)$. Therefore, the factorization given
above holds with an additional elementary reflector.) (Problem: Generalize
this result to $\mathbb{C}^{n \times n}$.) (Remark: See [469].)

Fact 5.14.15. Let $A \in \mathbb{F}^{n \times n}$. Then, $A^{2*}A = A^*A^2$ if and only if there
exist a projector $B \in \mathbb{F}^{n \times n}$ and a Hermitian matrix $C \in \mathbb{F}^{n \times n}$ such that
$A = BC$. (Proof: See [602].)

Fact 5.14.16. Let $A \in \mathbb{R}^{n \times n}$. Then, $|\det A| = 1$ if and only if A is the
product of $n+2$ or fewer involutory matrices that have exactly one negative
eigenvalue. In addition, the following statements hold:

i) If $n = 2$, then 3 or fewer factors are needed.

ii) If $A \neq \alpha I$ for all $\alpha \in \mathbb{R}$ and $\det A = (-1)^n$, then n or fewer factors
are needed.

iii) If $\det A = (-1)^{n+1}$, then $n+1$ or fewer factors are needed.

(Proof: See [168, 600].) (Remark: The minimal number of factors for a
unitary matrix A is given in [230].)

Fact 5.14.17. Let $A \in \mathbb{F}^{n \times n}$, and define $r_0 \triangleq n$ and $r_k \triangleq \operatorname{rank} A^k$ for all $k = 1, 2, \ldots$. Then, there exists a matrix $B \in \mathbb{C}^{n \times n}$ such that $A = B^2$ if and only if the sequence $\{r_k - r_{k+1}\}_{k=0}^{\infty}$ does not contain two successive elements that are the same odd integer and, if $r_0 - r_1$ is odd, then $r_0 + r_2 \geq 1 + 2r_1$. Now, assume that $A \in \mathbb{R}^{n \times n}$. Then, there exists $B \in \mathbb{R}^{n \times n}$ such that $A = B^2$ if and only if the above condition holds and, for every negative eigenvalue λ of A and for every positive integer k, the Jordan form of A has an even number of $k \times k$ blocks associated with λ. (Proof: See [369, p. 472].) (Remark: See Fact 11.17.34.) (Remark: For all $l \geq 2$, $A \triangleq N_l$ does not have a complex square root.) (Remark: Uniqueness is discussed in [399]. mth roots are considered in [594].)

Fact 5.14.18. Let $A \in \mathbb{C}^{n \times n}$, and assume that A is group invertible. Then, there exists $B \in \mathbb{C}^{n \times n}$ such that $A = B^2$.

Fact 5.14.19. Let $A \in \mathbb{F}^{n \times n}$, assume that A is nonsingular, and define $\{P_k\}_{k=0}^{\infty} \subset \mathbb{F}^{n \times n}$ and $\{Q_k\}_{k=0}^{\infty} \subset \mathbb{F}^{n \times n}$ by

$$P_0 \triangleq A, \qquad Q_0 \triangleq I,$$

and, for all $k \in \mathbb{P}$,

$$P_{k+1} \triangleq \tfrac{1}{2}(P_k + Q_k^{-1}),$$

$$Q_{k+1} \triangleq \tfrac{1}{2}(Q_k + P_k^{-1}).$$

Then,

$$B \triangleq \lim_{k \to \infty} P_k$$

exists and satisfies $B^2 = A$. Furthermore,

$$\lim_{k \to \infty} Q_k = A^{-1}.$$

(Proof: See [216, 351].) (Remark: This sequence is a modified Newton-Raphson algorithm based on the *matrix sign function*. See [419].) (Remark: See Fact 8.7.24.)

Fact 5.14.20. Let $A \in \mathbb{F}^{n \times n}$, assume that A is positive semidefinite, and let $r \triangleq \operatorname{rank} A$. Then, there exists $B \in \mathbb{F}^{n \times r}$ such that $A = BB^*$.

Fact 5.14.21. Let $A \in \mathbb{F}^{n \times n}$, and let $k \in \mathbb{P}$. Then, there exists a unique matrix $B \in \mathbb{F}^{n \times n}$ such that

$$A = B(B^*B)^k.$$

(Proof: See [587].)

Fact 5.14.22. Let $A \in \mathbb{F}^{n \times n}$. Then, there exist symmetric matrices $B, C \in \mathbb{F}^{n \times n}$, one of which is nonsingular, such that $A = BC$. (Proof: See [592, p. 82].) (Remark: Note that

$$\begin{bmatrix} \beta_1 & \beta_2 & 1 \\ \beta_2 & 1 & 0 \\ 1 & 0 & 0 \end{bmatrix} \begin{bmatrix} 0 & 1 & 0 \\ 0 & 0 & 1 \\ -\beta_0 & -\beta_1 & -\beta_2 \end{bmatrix} = \begin{bmatrix} -\beta_0 & 0 & 0 \\ 0 & \beta_2 & 1 \\ 0 & 1 & 0 \end{bmatrix}$$

and use Theorem 5.2.3.) (Remark: This result is due to Frobenius. The identity is a *Bezout matrix factorization*; see Fact 4.8.6. See [136, 137, 327].) (Remark: B and C are symmetric for $\mathbb{F} = \mathbb{C}$.)

Fact 5.14.23. Let $A \in \mathbb{C}^{n \times n}$. Then, det A is real if and only if A is the product of four Hermitian matrices. Furthermore, four is the smallest number of factors in general. (Proof: See [797].)

Fact 5.14.24. Let $A \in \mathbb{R}^{n \times n}$. Then, the following statements hold:

i) A is the product of two positive-semidefinite matrices if and only if A is similar to a positive-semidefinite matrix.

ii) If A is nilpotent, then A is the product of three positive-semidefinite matrices.

iii) If A is singular, then A is the product of four positive-semidefinite matrices.

iv) det $A > 0$ and $A \neq \alpha I$ for all $\alpha \leq 0$ if and only if A is the product of four positive-definite matrices.

v) det $A > 0$ if and only if A is the product of five positive-definite matrices.

(Proof: [61, 327, 796, 797].) (Remark: See [797] for factorizations of complex matrices and operators.) (Example:

$$\begin{bmatrix} -1 & 0 \\ 0 & -1 \end{bmatrix} = \begin{bmatrix} 2 & 0 \\ 0 & 1/2 \end{bmatrix} \begin{bmatrix} 5 & 7 \\ 7 & 10 \end{bmatrix} \begin{bmatrix} 13/2 & -5 \\ -5 & 4 \end{bmatrix} \begin{bmatrix} 8 & 5 \\ 5 & 13/4 \end{bmatrix} \begin{bmatrix} 25/8 & -11/2 \\ -11/2 & 10 \end{bmatrix}.)$$

Fact 5.14.25. Let $A \in \mathbb{R}^{n \times n}$. Then, the following statements hold:

i) $A = BC$, where $B \in \mathbf{S}^n$ and $C \in \mathbf{N}^n$, if and only if A^2 is diagonalizable over $\mathbb{R}$ and spec$(A) \subset [0, \infty)$.

ii) $A = BC$, where $B \in \mathbf{S}^n$ and $C \in \mathbf{P}^n$, if and only if A is diagonalizable over $\mathbb{R}$.

iii) $A = BC$, where $B, C \in \mathbf{N}^n$, if and only if $A = DE$, where $D \in \mathbf{N}^n$ and $E \in \mathbf{P}^n$.

iv) $A = BC$, where $B \in \mathbf{N}^n$ and $C \in \mathbf{P}^n$, if and only if A is diagonalizable over $\mathbb{R}$ and spec$(A) \subset [0, \infty)$.

v) $A = BC$, where $B, C \in \mathbf{P}^n$, if and only if A is diagonalizable over $\mathbb{R}$ and spec$(A) \subset [0, \infty)$.

(Proof: See [366, 793, 796].)

Fact 5.14.26. Let $A \in \mathbb{R}^{n \times n}$, assume that A is singular, and assume that A is not a 2×2 nilpotent matrix. Then, there exist nilpotent matrices $B, C \in \mathbb{R}^{n \times n}$ such that $A = BC$ and $\operatorname{rank} A = \operatorname{rank} B = \operatorname{rank} C$. (Proof: See [795].)

Fact 5.14.27. Let $A \in \mathbb{R}^{n \times n}$, and assume that A is nonsingular. Then, A is similar to A^{-1} if and only if A is the product of two involutory matrices. If, in addition, A is orthogonal, then A is the product of two reflectors. (Proof: See [66, 227, 791, 792] or [592, p. 108].) (Problem: Construct these reflectors for $A = \begin{bmatrix} \cos\theta & \sin\theta \\ -\sin\theta & \cos\theta \end{bmatrix}$.)

Fact 5.14.28. Let $A \in \mathbb{R}^{n \times n}$. Then, $|\det A| = 1$ if and only if A is the product of four or fewer involutory matrices. (Proof: [67, 318, 655].)

Fact 5.14.29. Let $A \in \mathbb{R}^{n \times n}$. Then, A is the identity or singular if and only if A is the product of n or fewer idempotent matrices. Furthermore, $\operatorname{rank}(A - I) \le k\operatorname{def} A$, where $k \in \mathbb{P}$, if and only if A is the product of k idempotent matrices. (Proof: See [68].) (Problem: Explicitly construct the $k = 2$ factors when $\operatorname{rank} A = 1$ and A is not idempotent. Example: $\begin{bmatrix} 2 & 0 \\ 0 & 0 \end{bmatrix} = \begin{bmatrix} 1 & 1 \\ 0 & 0 \end{bmatrix}\begin{bmatrix} 1 & 0 \\ 1 & 0 \end{bmatrix}$.)

Fact 5.14.30. Let $A \in \mathbb{R}^{n \times n}$, where $n \ge 2$. Then, A is the product of two commutators. (Proof: See [797].)

Fact 5.14.31. Let $A \in \mathbb{R}^{n \times n}$, and assume that $\det A = 1$. Then, there exist nonsingular matrices $B, C \in \mathbb{R}^{n \times n}$ such that $A = BCB^{-1}C^{-1}$. (Proof: See [641].) (Remark: The product is a *multiplicative commutator*. This result is due to Shoda.)

Fact 5.14.32. Let $A \in \mathbb{R}^{n \times n}$, assume that A is orthogonal, and assume that $\det A = 1$. Then, there exist reflectors $B, C \in \mathbb{R}^{n \times n}$ such that $A = BCB^{-1}C^{-1}$. (Proof: See [692].)

Fact 5.14.33. Let $A \in \mathbb{F}^{n \times n}$, and assume that A is nonsingular. Then, there exists an involutory matrix $B \in \mathbb{F}^{n \times n}$ and a symmetric matrix $C \in \mathbb{F}^{n \times n}$ such that $A = BC$. (Proof: See [303].)

Fact 5.14.34. Let $A \in \mathbb{F}^{n \times n}$, and assume that n is even. Then, the following statements are equivalent:

  i) A is the product of two skew-symmetric matrices.

  ii) Every elementary divisor of A has even algebraic multiplicity.

  iii) There exists a matrix $B \in \mathbb{F}^{n/2 \times n/2}$ such that A is similar to $\begin{bmatrix} B & 0 \\ 0 & B \end{bmatrix}$.

(Remark: In i) the factors are skew symmetric even when A is complex.)
(Proof: See [304, 797].)

Fact 5.14.35. Let $A \in \mathbb{R}^{n \times n}$, and assume that A is skew symmetric. If $n = 4, 8, 12 \ldots$, then A is the product of five or fewer skew-symmetric matrices. If $n = 6, 10, 14, \ldots$, then A is the product of seven or fewer skew-symmetric matrices. (Proof: See [448].)

Fact 5.14.36. Let $A \in \mathbb{F}^{n \times n}$. Then, there exist a symmetric matrix $B \in \mathbb{F}^{n \times n}$ and a skew-symmetric matrix $C \in \mathbb{F}^{n \times n}$ such that $A = BC$ if and only if A is similar to $-A$. (Proof: See [616].)

Fact 5.14.37. Let $A \in \mathbb{F}^{n \times m}$, and let $r \triangleq \operatorname{rank} A$. Then, there exist matrices $B \in \mathbb{F}^{n \times r}$ and $C \in \mathbb{R}^{r \times m}$ such that $A = BC$. Furthermore, $\operatorname{rank} B = \operatorname{rank} C = r$.

5.15 Facts on the Polar Decomposition

Fact 5.15.1. Let $A \in \mathbb{F}^{n \times m}$. Then,

$$(AA^*)^{1/2}A = A(A^*A)^{1/2}.$$

(Remark: See Fact 5.15.4.) (Remark: The positive-semidefinite square root is defined in (8.5.2).)

Fact 5.15.2. Let $A \in \mathbb{F}^{n \times m}$, where $n \leq m$. Then, there exist $M \in \mathbb{F}^{n \times n}$ and $S \in \mathbb{F}^{n \times m}$ such that M is positive semidefinite, S satisfies $SS^* = I_n$, and $A = MS$. Furthermore, M is given uniquely by $M = (AA^*)^{1/2}$. If, in addition, $\operatorname{rank} A = n$, then S is given uniquely by $S = (AA^*)^{-1/2}A$.

Fact 5.15.3. Let $A \in \mathbb{F}^{n \times m}$, where $m \leq n$. Then, there exist $M \in \mathbb{F}^{m \times m}$ and $S \in \mathbb{F}^{n \times m}$ such that M is positive semidefinite, S satisfies $S^*S = I_m$, and $A = SM$. Furthermore, M is given uniquely by $M = (A^*A)^{1/2}$. If, in addition, $\operatorname{rank} A = m$, then S is given uniquely by $S = A(A^*A)^{-1/2}$.

Fact 5.15.4. Let $A \in \mathbb{F}^{n \times n}$, and assume that A is nonsingular. Then, there exist unique matrices $M, S \in \mathbb{F}^{n \times n}$ such that $A = MS$, M is positive definite, and S is unitary. In particular, $M = (AA^*)^{1/2}$ and $S = (AA^*)^{-1/2}A$. (Remark: See Fact 5.15.1.)

Fact 5.15.5. Let $A \in \mathbb{F}^{n \times n}$, and assume that A is nonsingular. Then, there exist unique matrices $M, S \in \mathbb{F}^{n \times n}$ such that $A = SM$, M is positive definite, and S is unitary. In particular, $M = (A^*A)^{1/2}$ and $S = (AA^*)^{-1/2}A$.

Fact 5.15.6. Let $M_1, M_2 \in \mathbb{F}^{n \times n}$, assume that M_1, M_2 are positive definite, let $S_1, S_2 \in \mathbb{F}^{n \times n}$, assume that S_1, S_2 are unitary, and assume that $M_1 S_1 = S_2 M_2$. Then, $S_1 = S_2$. (Proof: Let $A = M_1 S_1 = S_2 M_2$. Then, $S_1 = \left(S_2 M_2^2 S_2^*\right)^{-1/2} S_2 M_2 = S_2$.)

Fact 5.15.7. Let $A \in \mathbb{F}^{n \times n}$, and assume that A is singular. Then, there exist a matrix $S \in \mathbb{F}^{n \times n}$ and unique matrices $M_1, M_2 \in \mathbb{F}^{n \times n}$ such that $A = M_1 S = S M_2$. In particular, $M_1 = (AA^*)^{1/2}$ and $M_2 = (A^*A)^{1/2}$. (Remark: S is not uniquely determined.)

Fact 5.15.8. Let $A \in \mathbb{F}^{n \times n}$, assume that A is nonsingular, and let $M, S \in \mathbb{F}^{n \times n}$ be such that $A = MS$, M is positive semidefinite, and S is unitary. Then, A is normal if and only if $MS = SM$. (Proof: See [367, p. 414].)

5.16 Notes

It is sometimes useful to define block-companion form matrices in which the scalars are replaced by matrix blocks [294]. The companion form provides only one of many connections between matrices and polynomials. Additional connections are given by the *comrade form*, *Leslie form*, *Schwarz form*, *Routh form*, *confederate form*, and *congenial form*. See [75, 78] and Fact 11.17.25 and Fact 11.17.26 for the Schwarz and Routh forms. The companion matrix is sometimes called a *Frobenius matrix* or the *Frobenius canonical form*, see [3].

The multi-companion form and the elementary multi-companion form are generally known as *rational canonical forms*, while the multi-companion form is traditionally called the *Frobenius canonical form* [80]. The derivation of the Jordan form by means of the elementary multi-companion form and the hypercompanion form follows [581]. Corollary 5.3.8, Corollary 5.3.9, and Proposition 5.5.18 are given in [136, 137, 682, 683, 686]. Corollary 5.3.9 is due to Frobenius. Canonical forms for congruence transformations are given in [466, 697].

Chapter Six

Generalized Inverses

Generalized inverses provide a useful extension of the matrix inverse to singular matrices and to rectangular matrices that are neither left nor right invertible.

6.1 Moore-Penrose Generalized Inverse

Let $A \in \mathbb{F}^{n \times m}$. If A is nonzero, then, by the singular value decomposition Theorem 5.6.3, there exist orthogonal matrices $S_1 \in \mathbb{F}^{n \times n}$ and $S_2 \in \mathbb{F}^{m \times m}$ such that

$$A = S_1 \begin{bmatrix} B & 0 \\ 0 & 0 \end{bmatrix} S_2, \qquad (6.1.1)$$

where $B \triangleq \operatorname{diag}[\sigma_1(A), \ldots, \sigma_r(A)]$, $r \triangleq \operatorname{rank} A$, and $\sigma_1(A) \geq \sigma_2(A) \geq \cdots \geq \sigma_r(A) > 0$ are the positive singular values of A. In (6.1.1), some of the bordering zero matrices may be empty. Then, the (*Moore-Penrose*) *generalized inverse* A^+ of A is the $m \times n$ matrix

$$A^+ \triangleq S_2^* \begin{bmatrix} B^{-1} & 0 \\ 0 & 0 \end{bmatrix} S_1^*. \qquad (6.1.2)$$

If $A = 0_{n \times m}$, then $A^+ \triangleq 0_{m \times n}$, while, if $m = n$ and $\det A \neq 0$, then $A^+ = A^{-1}$. In general, it is helpful to remember that A^+ and A^* are the same size. It is easy to verify that A^+ satisfies

$$AA^+A = A, \qquad (6.1.3)$$

$$A^+AA^+ = A^+, \qquad (6.1.4)$$

$$(AA^+)^* = AA^+, \qquad (6.1.5)$$

$$(A^+A)^* = A^+A. \qquad (6.1.6)$$

Hence, for all $A \in \mathbb{F}^{n \times m}$ there exists a matrix $X \in \mathbb{F}^{m \times n}$ satisfying the four conditions

$$AXA = A, \tag{6.1.7}$$
$$XAX = X, \tag{6.1.8}$$
$$(AX)^* = AX, \tag{6.1.9}$$
$$(XA)^* = XA. \tag{6.1.10}$$

We now show that X is uniquely defined by (6.1.7)–(6.1.10).

Theorem 6.1.1. Let $A \in \mathbb{F}^{n \times m}$. Then, $X = A^+$ is the unique matrix $X \in \mathbb{F}^{m \times n}$ satisfying (6.1.7)–(6.1.10).

Proof. Suppose there exists a matrix $X \in \mathbb{F}^{m \times n}$ satisfying (6.1.7)–(6.1.10). Then,

$$
\begin{aligned}
X &= XAX = X(AX)^* = XX^*A^* = XX^*(AA^+A)^* = XX^*A^*A^{+*}A^* \\
&= X(AX)^*(AA^+)^* = XAXAA^+ = XAA^+ = (XA)^*A^+ = A^*X^*A^+ \\
&= (AA^+A)^*X^*A^+ = A^*A^{+*}A^*X^*A^+ = (A^+A)^*(XA)^*A^+ \\
&= A^+AXAA^+ = A^+AA^+ = A^+. \qquad \square
\end{aligned}
$$

Given $A \in \mathbb{F}^{n \times m}$, $X \in \mathbb{F}^{m \times n}$ is a *(1)-inverse* of A if (6.1.7) holds, a *(1,2)-inverse* of A if (6.1.7) and (6.1.8) hold, and so forth.

Proposition 6.1.2. Let $A \in \mathbb{F}^{n \times m}$, and assume that A is right invertible. Then, $X \in \mathbb{F}^{m \times n}$ is a right inverse of A if and only if X is a (1)-inverse of A. Furthermore, every right inverse (or, equivalently, every (1)-inverse) of A is also a (2,3)-inverse of A.

Proof. Suppose that $AX = I_n$, that is, $X \in \mathbb{F}^{m \times n}$ is a right inverse of A. Then, $AXA = A$, which implies that X is a (1)-inverse of A. Conversely, let X be a (1)-inverse of A, that is, $AXA = A$. Then, letting $\hat{X} \in \mathbb{F}^{m \times n}$ denote a right inverse of A, it follows that $AX = AXA\hat{X} = A\hat{X} = I_n$. Hence, X is a right inverse of A. Finally, if X is a right inverse of A, then it is also a (2,3)-inverse of A. $\qquad \square$

Proposition 6.1.3. Let $A \in \mathbb{F}^{n \times m}$, and assume that A is left invertible. Then, $X \in \mathbb{F}^{m \times n}$ is a left inverse of A if and only if X is a (1)-inverse of A. Furthermore, every left inverse (or, equivalently, every (1)-inverse) of A is also a (2,4)-inverse of A.

It can now be seen that A^+ is a particular (right, left) inverse when A is (right, left) invertible.

Corollary 6.1.4. Let $A \in \mathbb{F}^{n \times m}$. If A is right invertible, then A^+ is a right inverse of A. Furthermore, if A is left invertible, then A^+ is a left inverse of A.

The following result provides an explicit expression for A^+ when A is either right invertible or left invertible. It is helpful to note that A is (right, left) invertible if and only if (AA^*, A^*A) is positive definite.

Proposition 6.1.5. Let $A \in \mathbb{F}^{n \times m}$. If A is right invertible, then

$$A^+ = A^*(AA^*)^{-1}. \qquad (6.1.11)$$

If A is left invertible, then

$$A^+ = (A^*A)^{-1}A^*. \qquad (6.1.12)$$

Proof. It suffices to verify (6.1.7)–(6.1.10) with $X = A^+$. $\qquad \square$

Proposition 6.1.6. Let $A \in \mathbb{F}^{n \times m}$. Then, the following statements hold:

i) $A = 0$ if and only if $A^+ = 0$.

ii) $(A^+)^+ = A$.

iii) $\overline{A}^+ = \overline{A^+}$.

iv) $\left(A^\mathrm{T}\right)^+ = (A^+)^\mathrm{T} = A^{+\mathrm{T}}$.

v) $(A^*)^+ = (A^+)^* \triangleq A^{+*}$.

vi) $\mathcal{R}(A) = \mathcal{R}(AA^*) = \mathcal{R}(AA^+) = \mathcal{R}(A^{+*}) = \mathcal{N}(I - AA^+)$.

vii) $\mathcal{R}(A^*) = \mathcal{R}(A^*A) = \mathcal{R}(A^+) = \mathcal{R}(A^+A)$.

viii) $\mathcal{N}(A) = \mathcal{N}(A^+A) = \mathcal{N}(A^*A) = \mathcal{R}(I - A^+A)$.

ix) $\mathcal{N}(A^*) = \mathcal{N}(A^+) = \mathcal{N}(AA^+) = \mathcal{R}(I - AA^+)$.

x) AA^+ is the projector onto $\mathcal{R}(A)$.

xi) A^+A is the projector onto $\mathcal{R}(A^*)$.

xii) $I - A^+A$ is the projector onto $\mathcal{N}(A)$.

xiii) $I - AA^+$ is the projector onto $\mathcal{N}(A^*)$.

xiv) $x \in \mathcal{R}(A)$ if and only if $x = AA^+x$.

xv) $\operatorname{rank} A = \operatorname{rank} A^+ = \operatorname{rank} AA^+ = \operatorname{rank} A^+A = \operatorname{tr} AA^+ = \operatorname{tr} A^+A$.

xvi) $(A^*A)^+ = A^+A^{+*}$.

xvii) $(AA^*)^+ = A^{+*}A^+$.

xviii) $AA^+ = A(A^*A)^+A^*$.

xix) $A^+A = A^*(AA^*)^+A$.

xx) $A = AA^*A^{*+} = A^{*+}A^*A$.

xxi) $A^* = A^*AA^+ = A^+AA^*$.

xxii) $A^+ = A^*(AA^*)^+ = (A^*A)^+A^* = A^*(A^*AA^*)^+A^*$.

xxiii) $A^{+*} = (AA^*)^+A = A(A^*A)^+$.

xxiv) $A = A(A^*A)^+A^*A = AA^*A(A^*A)^+$.

xxv) $A = AA^*(AA^*)^+A = (AA^*)^+AA^*A$.

xxvi) If $S_1 \in \mathbb{F}^{n \times n}$ and $S_2 \in \mathbb{F}^{m \times m}$ are unitary, then $(S_1AS_2)^+ = S_2^*A^+S_1^*$.

xxvii) A is (range Hermitian, normal, Hermitian, positive semidefinite, positive definite) if and only if A^+ is.

Theorem 2.6.3 showed that the equation $Ax = b$, where $A \in \mathbb{F}^{n \times m}$ and $b \in \mathbb{F}^n$, has a solution $x \in \mathbb{F}^m$ if and only if $\operatorname{rank} A = \operatorname{rank} \begin{bmatrix} A & b \end{bmatrix}$. In particular, $Ax = b$ has a unique solution $x \in \mathbb{F}^m$ if and only if $\operatorname{rank} A = \operatorname{rank} \begin{bmatrix} A & b \end{bmatrix} = m$, while $Ax = b$ has infinitely many solutions if and only if $\operatorname{rank} A = \operatorname{rank} \begin{bmatrix} A & b \end{bmatrix} < m$. We are now prepared to characterize these solutions.

Proposition 6.1.7. Let $A \in \mathbb{F}^{n \times m}$ and $b \in \mathbb{F}^n$. Then, the following statements are equivalent:

i) There exists a vector $x \in \mathbb{F}^m$ satisfying $Ax = b$.

ii) $\operatorname{rank} A = \operatorname{rank} \begin{bmatrix} A & b \end{bmatrix}$.

iii) $b \in \mathcal{R}(A)$.

iv) $AA^+b = b$.

Now, assume that *i*)–*iv*) are satisfied. Then, the following statements hold:

v) If $x \in \mathbb{F}^m$ satisfies $Ax = b$, then

$$x = A^+b + (I - A^+A)x. \tag{6.1.13}$$

vi) For all $y \in \mathbb{F}^m$, $x \in \mathbb{F}^m$ given by

$$x = A^+b + (I - A^+A)y \tag{6.1.14}$$

satisfies $Ax = b$.

vii) Let $x \in \mathbb{F}^m$ be given by (6.1.14), where $y \in \mathbb{F}^m$. Then, $y = 0$ minimizes x^*x.

viii) Assume $\operatorname{rank} A = m$. Then, there exists a unique vector $x \in \mathbb{F}^m$ satisfying $Ax = b$ given by $x = A^+b$. If, in addition, $A^{\mathrm{L}} \in \mathbb{F}^{m \times m}$ is a left inverse of A, then $A^{\mathrm{L}}b = A^+b$.

ix) Assume $\operatorname{rank} A = n$, and let $A^{\mathrm{R}} \in \mathbb{F}^{m \times n}$ be a right inverse of A. Then, $x = A^{\mathrm{R}}b$ satisfies $Ax = b$.

Proof. The equivalence of *i*)–*iii*) is immediate. To prove the equivalence of *iv*), note that, if there exists a vector $x \in \mathbb{F}^n$ satisfying $Ax = b$, then

$b = Ax = AA^{+}Ax = AA^{+}b$. Conversely, if $b = AA^{+}b$, then $x = A^{+}b$ satisfies $Ax = b$.

Now, suppose that $i)$–$iv)$ hold. To prove $v)$, let $x \in \mathbb{F}^m$ satisfy $Ax = b$ so that $A^{+}Ax = A^{+}b$. Hence, $x = x + A^{+}b - A^{+}Ax = A^{+}b + (I - A^{+}A)x$. To prove $vi)$, let $y \in \mathbb{F}^m$, and let $x \in \mathbb{F}^m$ be given by (6.1.14). Then, $Ax = AA^{+}b = b$. To prove $vii)$, let $y \in \mathbb{F}^m$, and let $x \in \mathbb{F}^n$ be given by (6.1.14). Then, $x^{*}x = b^{*}A^{+*}A^{+}b + y^{*}(I - A^{+}A)y$. Therefore, $x^{*}x$ is minimized by $y = 0$. To prove $viii)$, suppose that rank $A = m$. Then, A is left invertible, and it follows from Corollary 6.1.4 that A^{+} is a left inverse of A. Hence, it follows from (6.1.13) that $x = A^{+}b$ is the unique solution of $Ax = b$. In addition, $x = A^{L}b$. To prove $ix)$, let $x = A^{R}b$, and note that $AA^{R}b = b$. $\square$

Definition 6.1.8. Let $A \in \mathbb{F}^{n \times m}$, $B \in \mathbb{F}^{n \times l}$, $C \in \mathbb{F}^{k \times m}$, and $D \in \mathbb{F}^{k \times l}$, and define $\mathcal{A} \triangleq \begin{bmatrix} A & B \\ C & D \end{bmatrix} \in \mathbb{F}^{(n+k) \times (m+l)}$. Then, the *Schur complement* $D|\mathcal{A}$ of D with respect to $\mathcal{A}$ is defined by

$$D|\mathcal{A} \triangleq A - BD^{+}C. \tag{6.1.15}$$

Likewise, the *Schur complement* $A|\mathcal{A}$ of A with respect to $\mathcal{A}$ is defined by

$$A|\mathcal{A} \triangleq D - CA^{+}B. \tag{6.1.16}$$

6.2 Drazin Generalized Inverse

We now introduce a different type of generalized inverse, which applies only to square matrices yet is more useful in certain applications. Let $A \in \mathbb{F}^{n \times n}$. Then, A has a decomposition

$$A = S \begin{bmatrix} J_1 & 0 \\ 0 & J_2 \end{bmatrix} S^{-1}, \tag{6.2.1}$$

where $S \in \mathbb{F}^{n \times n}$ is nonsingular, $J_1 \in \mathbb{F}^{m \times m}$ is nonsingular, and $J_2 \in \mathbb{F}^{(n-m) \times (n-m)}$ is nilpotent. Then, the *Drazin generalized inverse* A^{D} of A is the matrix

$$A^{\mathrm{D}} \triangleq S \begin{bmatrix} J_1^{-1} & 0 \\ 0 & 0 \end{bmatrix} S^{-1}. \tag{6.2.2}$$

Let $A \in \mathbb{F}^{n \times n}$. Then, it follows from Definition 5.5.1 that $\mathrm{ind}\,A = \mathrm{ind}_A(0)$. If A is nonsingular, then $\mathrm{ind}\,A = 0$, whereas $\mathrm{ind}\,A = 1$ if and only if A is singular and the zero eigenvalue of A is semisimple. In particular, $\mathrm{ind}\,0_{n \times n} = 1$. Note that $\mathrm{ind}\,A$ is the size of the largest Jordan block of A associated with the zero eigenvalue of A.

It can be seen that A^D satisfies

$$A^D A A^D = A^D, \tag{6.2.3}$$

$$A A^D = A^D A, \tag{6.2.4}$$

$$A^{k+1} A^D = A^k, \tag{6.2.5}$$

where $k = \operatorname{ind} A$. Hence, for all $A \in \mathbb{F}^{n \times n}$ such that $\operatorname{ind} A = k$ there exists a matrix $X \in \mathbb{F}^{n \times n}$ satisfying the three conditions

$$XAX = X, \tag{6.2.6}$$

$$AX = XA, \tag{6.2.7}$$

$$A^{k+1}X = A^k. \tag{6.2.8}$$

We now show that X is uniquely defined by (6.2.6)–(6.2.8).

Theorem 6.2.1. Let $A \in \mathbb{F}^{n \times n}$, and let $k \triangleq \operatorname{ind} A$. Then, $X = A^D$ is the unique matrix $X \in \mathbb{F}^{n \times n}$ satisfying (6.2.6)–(6.2.8).

Proof. Let $X \in \mathbb{F}^{n \times n}$ satisfy (6.2.6)–(6.2.8). If $k = 0$, then it follows from (6.2.8) that $X = A^{-1}$. Hence, let $A = S \begin{bmatrix} J_1 & 0 \\ 0 & J_2 \end{bmatrix} S^{-1}$, where $k = \operatorname{ind} A \geq 1$, $S \in \mathbb{F}^{n \times n}$ is nonsingular, $J_1 \in \mathbb{F}^{m \times m}$ is nonsingular, and $J_2 \in \mathbb{F}^{(n-m) \times (n-m)}$ is nilpotent. Now, let $\hat{X} \triangleq S^{-1}XS = \begin{bmatrix} \hat{X}_1 & \hat{X}_{12} \\ \hat{X}_{21} & \hat{X}_2 \end{bmatrix}$ be partitioned conformably with $S^{-1}AS = \begin{bmatrix} J_1 & 0 \\ 0 & J_2 \end{bmatrix}$. Since, by (6.2.7), $A\hat{X} = \hat{X}A$, it follows that $J_1 \hat{X}_1 = \hat{X}_1 J_1$, $J_1 \hat{X}_{12} = \hat{X}_{12} J_2$, $J_2 \hat{X}_{21} = \hat{X}_{21} J_1$, and $J_2 \hat{X}_2 = \hat{X}_2 J_2$. Since $J_2^k = 0$, it follows that $J_1 \hat{X}_{12} J_2^{k-1} = 0$, and thus $\hat{X}_{12} J_2^{k-1} = 0$. By repeating this argument, it follows that $J_1 \hat{X}_{12} J_2 = 0$, and thus $\hat{X}_{12} J_2 = 0$, which implies that $J_1 \hat{X}_{12} = 0$, and thus $\hat{X}_{12} = 0$. Similarly, $\hat{X}_{21} = 0$, so that $\hat{X} = \begin{bmatrix} \hat{X}_1 & 0 \\ 0 & \hat{X}_2 \end{bmatrix}$. Now, (6.2.8) implies that $J_1^{k+1} \hat{X}_1 = J_1^k$, and hence $\hat{X}_1 = J_1^{-1}$. Next, (6.2.6) implies that $\hat{X}_2 J_2 \hat{X}_2 = \hat{X}_2$, which, together with $J_2 \hat{X}_2 = \hat{X}_2 J_2$, yields $\hat{X}_2^2 J_2 = \hat{X}_2$. Consequently, $0 = \hat{X}_2^2 J_2^k = \hat{X}_2 J_2^{k-1}$, and thus, by repeating this argument, $\hat{X}_2 = 0$. Therefore, $A^D = S \begin{bmatrix} J_1^{-1} & 0 \\ 0 & 0 \end{bmatrix} S^{-1} = S \begin{bmatrix} \hat{X}_1 & 0 \\ 0 & 0 \end{bmatrix} S^{-1} = S\hat{X}S^{-1} = X$. $\square$

Let $A \in \mathbb{F}^{n \times n}$, and assume that $\operatorname{ind} A \leq 1$ so that, by Corollary 5.5.15, A is group invertible. In this case, the Drazin inverse A^D is denoted by $A^\#$, which is the *group generalized inverse* of A. Therefore, $A^\#$ satisfies

$$A^\# A A^\# = A^\#, \tag{6.2.9}$$

$$A A^\# = A^\# A, \tag{6.2.10}$$

$$A A^\# A = A, \tag{6.2.11}$$

while $A^\#$ is the unique matrix $X \in \mathbb{F}^{n \times n}$ satisfying

$$XAX = X, \qquad\qquad (6.2.12)$$
$$AX = XA, \qquad\qquad (6.2.13)$$
$$AXA = A. \qquad\qquad (6.2.14)$$

Proposition 6.2.2. Let $A \in \mathbb{F}^{n \times n}$, and assume that A is group invertible. Then, the following statements hold:

i) $A = 0$ if and only if $A^{\#} = 0$.

ii) $(A^{\#})^{\#} = A$.

iii) If A is idempotent, then $A^{\#} = A$.

iv) $AA^{\#}$ and $A^{\#}A$ are idempotent.

v) $\left(A^{\mathrm{T}}\right)^{\#} = (A^{\#})^{\mathrm{T}}$.

vi) $\operatorname{rank} A = \operatorname{rank} A^{\#} = \operatorname{rank} AA^{\#} = \operatorname{rank} A^{\#}A$.

vii) $\mathcal{R}(A) = \mathcal{R}(AA^{\#}) = \mathcal{N}(I - AA^{\#}) = \mathcal{R}(AA^{+}) = \mathcal{N}(I - AA^{+})$.

viii) $\mathcal{N}(A) = \mathcal{N}(AA^{\#}) = \mathcal{R}(I - AA^{\#}) = \mathcal{N}(A^{+}A) = \mathcal{R}(I - A^{+}A)$.

ix) $AA^{\#}$ is the idempotent matrix onto $\mathcal{R}(A)$ along $\mathcal{N}(A)$.

An alternative expression for the idempotent matrix onto $\mathcal{R}(A)$ along $\mathcal{N}(A)$ is given by Proposition 5.5.9.

6.3 Facts on the Moore-Penrose Generalized Inverse for One Matrix

Fact 6.3.1. Let $A \in \mathbb{F}^{n \times m}$, and assume that $\operatorname{rank} A = 1$. Then,
$$A^{+} = (\operatorname{tr} AA^{*})^{-1}A^{*}.$$
Consequently, if $x \in \mathbb{F}^{n}$ and $y \in \mathbb{F}^{n}$ are nonzero, then
$$(xy^{*})^{+} = (x^{*}xy^{*}y)^{-1}yx^{*}.$$

Fact 6.3.2. Let $A \in \mathbb{F}^{n \times m}$ and $B \in \mathbb{F}^{m \times n}$. Then, the following statements are equivalent:

i) $B = A^{+}$.

ii) $A^{*}AB = A^{*}$ and $B^{*}BA = B^{*}$.

iii) $BAA^{*} = A^{*}$ and $ABB^{*} = B^{*}$.

(Remark: See [339, pp. 503, 513].)

Fact 6.3.3. Let $A \in \mathbb{F}^{n \times m}$, $B \in \mathbb{F}^{k \times n}$, and $C \in \mathbb{F}^{m \times l}$, and assume that B is left inner and C is right inner. Then,

$$(BAC)^+ = C^*A^+B^*.$$

(Proof: See [339, p. 506].)

Fact 6.3.4. Let $A \in \mathbb{F}^{n \times n}$. Then, the following statements are equivalent:

i) A^+ is idempotent.

ii) $AA^*A = A^2$.

If A is range Hermitian, then the following statements are equivalent:

i) A^+ is idempotent.

ii) $AA^* = A^*A = A$.

The following statements are equivalent:

i) A^+ is a projector.

ii) A is a projector.

(Proof: See [704] and [772].)

Fact 6.3.5. Let $A \in \mathbb{F}^{n \times n}$. Then,

$$\operatorname{rank} [A, A^+] = 2\operatorname{rank} \begin{bmatrix} A & A^* \end{bmatrix} - 2\operatorname{rank} A$$
$$= \operatorname{rank}(A - A^2A^+)$$
$$= \operatorname{rank}(A - A^+A^2).$$

(Proof: See [712].)

Fact 6.3.6. Let $A \in \mathbb{F}^{n \times n}$. Then, the following statements are equivalent:

i) A is range Hermitian.

ii) $A^+A = AA^+$.

iii) $\operatorname{rank} \begin{bmatrix} A & A^* \end{bmatrix} = \operatorname{rank} A$.

iv) $A = A^2A^+$.

v) $A = A^+A^2$.

vi) $(AA^+)^2 = A^2(A^+)^2$.

vii) $(A^+A)^2 = (A^+)^2A^2$.

viii) $\operatorname{ind} A \leq 1$, and $(A^+)^2 = (A^2)^+$.

ix) $\operatorname{ind} A \leq 1$, and $AA^+A^*A = A^*A^2A^+$.

$x)$ $A^2 A^+ + A^* A^{+*} A = 2A.$

$xi)$ $A^2 A^+ + \left(A^2 A^+\right)^* = A + A^*.$

(Proof: See [180, 704].) (Remark: See Fact 6.3.5 and Fact 6.5.9.)

Fact 6.3.7. Let $A \in \mathbb{F}^{n \times n}$. Then, the following statements are equivalent:

$i)$ $A^+ A^* = A^* A^+.$

$ii)$ $A A^+ A^* A = A A^* A^+ A.$

$iii)$ $A A^* A^2 = A^2 A^* A.$

If these conditions hold, then A is *star-dagger*. If A is star-dagger, then $A^2 (A^+)^2$ and $(A^+)^2 A^2$ are positive semidefinite. (Proof: See [338, 704].) (Remark: See Fact 6.3.8.)

Fact 6.3.8. Let $A \in \mathbb{F}^{n \times n}$. Then, the following statements are equivalent:

$i)$ A is normal.

$ii)$ A is range Hermitian, and $A^+ A^* = A^* A^+.$

$iii)$ $A(A A^* A)^+ = (A A^* A)^+ A.$

$iv)$ $A A^+ A^* A^2 A^+ = A A^*.$

$v)$ $A(A^* + A^+) = (A^* + A^+)A.$

$vi)$ $A^* A (A A^*)^+ A^* A = A A^*.$

$vii)$ $2 A A^* (A A^* + A^* A)^+ A A^* = A A^*.$

$viii)$ There exists a matrix $X \in \mathbb{F}^{n \times n}$ such that $A A^* X = A^* A$ and $A^* A X = A A^*.$

$ix)$ There exists a matrix $X \in \mathbb{F}^{n \times n}$ such that $A X = A^*$ and $A^{+*} X = A^+.$

(Proof: See [180].) (Remark: See Fact 3.5.8, Fact 3.7.10, Fact 5.14.2, Fact 6.3.7, and Fact 6.5.10.)

Fact 6.3.9. Let $A \in \mathbb{F}^{n \times m}$, and assume that $\operatorname{rank} A = m$. Then,

$$(A A^*)^+ = A(A^* A)^{-2} A^*.$$

Fact 6.3.10. Let $A \in \mathbb{F}^{n \times m}$. Then,

$$A^+ = \lim_{\alpha \downarrow 0} A^* (A A^* + \alpha I)^{-1} = \lim_{\alpha \downarrow 0} (A^* A + \alpha I)^{-1} A^*.$$

Fact 6.3.11. Let $A \in \mathbb{F}^{n \times m}$, let $\chi_{AA^*}(s) = s^n + \beta_{n-1}s^{n-1} + \cdots + \beta_1 s + \beta_0$, and let $n - k$ denote the smallest integer in $\{0, \ldots, n-1\}$ such that $\beta_k \neq 0$. Then,

$$A^+ = -\beta_{n-k}^{-1} A^* \left[(AA^*)^{k-1} + \beta_{n-1}(AA^*)^{k-2} + \cdots + \beta_{n-k+1} I \right].$$

(Proof: See [214].)

Fact 6.3.12. Let $A \in \mathbb{F}^{n \times n}$, and assume that A is Hermitian. Then, $\text{In } A = \text{In } A^+$.

Fact 6.3.13. Let $A \in \mathbb{F}^{n \times n}$, and assume that A is a projector. Then, $A^+ = A$.

Fact 6.3.14. Let $A \in \mathbb{F}^{n \times n}$. Then, $A^+ = A$ if and only if A is tripotent and A^2 is Hermitian.

Fact 6.3.15. Let $A \in \mathbb{F}^{n \times n}$, and assume that A is idempotent. Then,

$$A^+ A + (I - A)(I - A)^+ = I$$

and

$$AA^+ + (I - A)^+(I - A) = I.$$

(Proof: $\mathcal{N}(A) = \mathcal{R}(I - A^+A) = \mathcal{R}(I - A) = \mathcal{R}[(I - A)(I - A^+)]$.) (Remark: The first identity states that the projector onto the null space of A is the same as the projector onto the range of $I - A$, while the second identity states that the projector onto the range of A is the same as the projector onto the null space of $I - A$.) (Remark: See Fact 3.9.18 and Fact 9.11.14.)

Fact 6.3.16. Let $A \in \mathbb{F}^{n \times n}$, and assume that A is idempotent. Then,

$$A^* A^+ A = A^+ A$$

and

$$AA^+ A^* = AA^+.$$

(Proof: Note that $A^* A^+ A$ is a projector, and $\mathcal{R}(A^* A^+ A) = \mathcal{R}(A^*) = \mathcal{R}(A^+ A)$.)

Fact 6.3.17. Let $A \in \mathbb{F}^{n \times n}$, and assume that A is idempotent. Then, $A + A^* - I$ is nonsingular, and

$$(A + A^* - I)^{-1} = AA^+ + A^+ A - I.$$

(Proof: Use Fact 6.3.16.) (Remark: See [535, p. 457] for a geometric interpretation of this identity. See Fact 9.11.14.)

Fact 6.3.18. Let $A \in \mathbb{F}^{n \times n}$, and assume that A is idempotent. Then, $2A(A + A^*)^+ A^*$ is the projector onto $\mathcal{R}(A) \cap \mathcal{R}(A^*)$. (Proof: See [721].)

Fact 6.3.19. Let $A \in \mathbb{F}^{n \times n}$, and assume that A is idempotent. Then, A is a projector if and only if $[A(I - A^*)]^+$ is idempotent. (Proof: See [720].)

Fact 6.3.20. Let $A \in \mathbb{F}^{n \times n}$, and let $r \triangleq \operatorname{rank} A$. Then, $A^+ = A^*$ if and only if $\sigma_1(A) = \sigma_r(A) = 1$.

Fact 6.3.21. Let $A \in \mathbb{F}^{n \times m}$, assume that A is nonzero, and let $r \triangleq \operatorname{rank} A$. Then, for all $i = 1, \ldots, r$, the singular values of A^+ are given by

$$\sigma_i(A^+) = \sigma_{r+1-i}^{-1}(A).$$

In particular,

$$\sigma_r(A) = 1/\sigma_{\max}(A^+).$$

If, in addition, $A \in \mathbb{F}^{n \times n}$ and A is nonsingular, then

$$\sigma_{\min}(A) = 1/\sigma_{\max}(A^{-1}).$$

Fact 6.3.22. Let $A \in \mathbb{F}^{n \times m}$. Then, $X = A^+$ is the unique matrix satisfying

$$\operatorname{rank} \begin{bmatrix} A & AA^+ \\ A^+A & X \end{bmatrix} = \operatorname{rank} A.$$

(Remark: See Fact 2.15.10 and Fact 6.5.5.) (Proof: See [261].)

Fact 6.3.23. Let $A \in \mathbb{F}^{n \times n}$, and assume that A is centrohermitian. Then, A^+ is centrohermitian. (Proof: See [465].)

Fact 6.3.24. Let $A \in \mathbb{F}^{n \times n}$. Then, the following statements are equivalent:

i) $A^2 = AA^*A$.

ii) A is the product of two projectors.

iii) $A = A(A^+)^2A$.

(Remark: This result is due to Crimmins. See [602].)

Fact 6.3.25. Let $A \in \mathbb{F}^{n \times m}$. Then,

$$A^+ = 4(I + A^+A)^+A^+(I + AA^+)^+.$$

(Proof: Use Fact 6.4.34 with $B = A$.)

Fact 6.3.26. Let $A \in \mathbb{F}^{n \times n}$, and assume that A is unitary. Then,

$$\lim_{k \to \infty} \tfrac{1}{k} \sum_{i=0}^{k-1} A^i = I - (A - I)(A - I)^+.$$

(Remark: $I - (A - I)(A - I)^+$ is the projector onto $\{x: Ax = x\} = \mathcal{N}(A - I)$.)
(Remark: This result is the *ergodic theorem*.) (Proof: Use Fact 11.18.13 and

Fact 11.18.15, and note that $(A - I)^* = (A - I)^+$. See [325, p. 185].)

Fact 6.3.27. Let $A \in \mathbb{F}^{n \times m}$, and define $\{B_i\}_{i=1}^{\infty}$ by

$$B_{i+1} \triangleq 2B_i - B_i A B_i,$$

where $B_0 \triangleq \alpha A^*$ and $\alpha \in (0, 2/\sigma_{\max}^2(A))$. Then,

$$\lim_{i \to \infty} B_i = A^+.$$

(Proof: See [78, p. 259] or [158, p. 250]. This result is due to Ben-Israel.) (Remark: This sequence is a Newton-Raphson algorithm.) (Remark: B_0 satisfies $\mathrm{sprad}(I - B_0 A) < 1$.) (Remark: For the case in which A is square and nonsingular, see Fact 2.14.28.) (Problem: Does convergence hold for all $B_0 \in \mathbb{F}^{n \times n}$ satisfying $\mathrm{sprad}(I - B_0 A) < 1$?)

6.4 Facts on the Moore-Penrose Generalized Inverse for Two or More Matrices

Fact 6.4.1. Let $A \in \mathbb{F}^{n \times m}$ and $B \in \mathbb{F}^{m \times l}$. Then, $AB = 0$ if and only if $B^+ A^+ = 0$.

Fact 6.4.2. Let $A \in \mathbb{F}^{n \times m}$ and $B \in \mathbb{F}^{n \times l}$. Then, $A^+ B = 0$ if and only if $A^* B = 0$.

Fact 6.4.3. Let $A \in \mathbb{F}^{n \times m}$ and $B \in \mathbb{F}^{m \times l}$. Then,

$$(AB)^+ = B_1^+ A_1^+,$$

where $B_1 \triangleq A^+ A B$ and $A_1 \triangleq A B_1 B_1^+$. That is,

$$(AB)^+ = (A^+ A B)^+ \left[A B (A^+ A B)^+ \right]^+.$$

(Proof: See [9, p. 55].) (Remark: This result is due to Cline.)

Fact 6.4.4. Let $A \in \mathbb{F}^{n \times m}$ and $B \in \mathbb{F}^{m \times l}$. Then, the following statements are equivalent:

i) $(AB)^+ = B^+ A^+$.

ii) $\mathcal{R}(A^* A B) \subseteq \mathcal{R}(B)$ and $\mathcal{R}(BB^* A^*) \subseteq \mathcal{R}(A^*)$.

iii) $(AB)(AB)^+ = (AB)B^+ A^+$ and $(AB)^+(AB) = B^+ A^+ AB$.

iv) $A^* A B = BB^+ A^* A B$ and $ABB^* = ABB^* A^+ A$.

v) $AB(AB)^+ A = ABB^+$ and $A^+ A B = B(AB)^+ AB$.

vi) $A^* A B B^+ = BB^+ A^* A$ and $A^+ A B B^* = BB^* A^+ A$.

vii) $(ABB^+)^+ = BB^+ A^+$ and $(A^+ A B)^+ = B^+ A^+ A$.

viii) $B^+(ABB^+)^+ = B^+A^+$ and $(A^+AB)^+A = B^+A^+$.

(Proof: See [9, p. 53] and [710].) (Remark: The equivalence of *i*) and *ii*) is due to Greville.) (Remark: Conditions under which B^+A^+ is a (1)-inverse of AB are given in [710].)

Fact 6.4.5. Let $A \in \mathbb{F}^{n \times m}$ and $B \in \mathbb{F}^{m \times l}$. Then, the following statements are equivalent:

i) $(AB)^+ = B^+A^+ - B^+[(I - BB^+)(I - A^+A)]^+A^+$.

ii) $\mathcal{R}(AA^*AB) = \mathcal{R}(AB)$ and $\mathcal{R}[(ABB^*B)^*] = \mathcal{R}[(AB)^*]$.

(Proof: See [700].)

Fact 6.4.6. Let $A, B \in \mathbb{F}^{n \times n}$, and assume that A and B are projectors. Then,
$$\mathcal{R}[(A - B)^+ - (A - B)] = \mathcal{R}(AB - BA).$$
Consequently, $(A - B)^+ = (A - B)$ if and only if $AB = BA$. (Proof: See [708].)

Fact 6.4.7. Let $A, B \in \mathbb{F}^{n \times n}$, and assume that A and B are projectors. Then,
$$AB = A(AB)^+B,$$
$$(AB)^+ = BA - B(B_\perp A_\perp)^+A,$$
and
$$(AB)^2 = AB + AB(B_\perp A_\perp)^+AB.$$
Furthermore, $(AB)^+$ is idempotent. (Proof: See Fact 6.4.5 and [700]. The last statement follows from Fact 6.3.4. See [772].) (Problem: Use the second identity to prove the last statement.)

Fact 6.4.8. Let $A \in \mathbb{F}^{n \times r}$ and $B \in \mathbb{F}^{r \times m}$, and assume that rank $A =$ rank $B = r$. Then,
$$(AB)^+ = B^+A^+ = B^*(BB^*)^{-1}(A^*A)^{-1}A^*.$$

Fact 6.4.9. Let $A, B \in \mathbb{F}^{n \times n}$, assume that A and B are range Hermitian, and assume that $(AB)^+ = A^+B^+$. Then, AB is range Hermitian. (Proof: See [337].) (Remark: See Fact 8.15.11.)

Fact 6.4.10. Let $A, B \in \mathbb{F}^{n \times n}$, and assume that A is range Hermitian. Then, $AB = BA$ if and only if $A^+B = BA^+$. (Proof: See [703].)

Fact 6.4.11. Let $A, B \in \mathbb{F}^{n \times n}$, and assume that A and B are range Hermitian. Then, the following statements are equivalent:

i) $AB = BA$.

ii) $A^+B = BA^+$.

iii) $AB^+ = B^+A$.

iv) $A^+B^+ = B^+A^+$.

(Proof: See [703].)

Fact 6.4.12. Let $A \in \mathbb{F}^{n \times m}$ and $B \in \mathbb{F}^{m \times l}$, and assume that rank $B = m$. Then,
$$AB(AB)^+ = AA^+.$$

Fact 6.4.13. Let $A \in \mathbb{F}^{n \times n}$ and $B \in \mathbb{F}^{m \times n}$, and assume that A is idempotent. Then,
$$A^*(BA)^+ = (BA)^+.$$

(Proof: See [339, p. 514].)

Fact 6.4.14. Let $A, B \in \mathbb{F}^{n \times n}$, and assume that A and B are projectors. Then, the following statements are equivalent:

i) AB is a projector.

ii) $[(AB)^+]^2 = [(AB)^2]^+$.

(Proof: See [722].) (Remark: See Fact 3.9.20.)

Fact 6.4.15. Let $A \in \mathbb{F}^{n \times m}$, $B \in \mathbb{F}^{m \times n}$, and $C \in \mathbb{F}^{m \times n}$, and assume that $BAA^* = A^*$ and $A^*AC = A^*$. Then,
$$A^+ = BAC.$$

(Proof: See [9, p. 36].) (Remark: This result is due to Decell.)

Fact 6.4.16. Let $A \in \mathbb{F}^{n \times m}$. Then, there exists a matrix $B \in \mathbb{F}^{m \times m}$ satisfying $BAB = B$ if and only if there exist projectors $C \in \mathbb{F}^{n \times n}$ and $D \in \mathbb{F}^{m \times m}$ such that $B = (CAD)^+$. (Proof: See [308].)

Fact 6.4.17. Let $A \in \mathbb{F}^{n \times n}$. Then, A is idempotent if and only if there exist projectors $B, C \in \mathbb{F}^{n \times n}$ such that $A = (BC)^+$. (Proof: Let $A = I$ in Fact 6.4.16.) (Remark: See [310].)

Fact 6.4.18. Let $A, B \in \mathbb{F}^{n \times m}$. Then, the following statements are equivalent:

i) $\mathcal{R}\left(\begin{bmatrix} A \\ A^*A \end{bmatrix}\right) = \mathcal{R}\left(\begin{bmatrix} B \\ B^*B \end{bmatrix}\right)$.

ii) $\mathcal{R}\left(\begin{bmatrix} A \\ A^+A \end{bmatrix}\right) = \mathcal{R}\left(\begin{bmatrix} B \\ B^+B \end{bmatrix}\right)$.

iii) $A = B$.

(Remark: This result is due to Tian.)

Fact 6.4.19. Let $A \in \mathbb{F}^{n \times m}$, $B \in \mathbb{F}^{n \times l}$, $C \in \mathbb{F}^{k \times m}$, and $D \in \mathbb{F}^{k \times l}$. Then,

$$\begin{bmatrix} A & B \\ C & D \end{bmatrix} = \begin{bmatrix} I & 0 \\ CA^+ & I \end{bmatrix} \begin{bmatrix} A & B - AA^+B \\ C - CA^+A & D - CA^+B \end{bmatrix} \begin{bmatrix} I & A^+B \\ 0 & I \end{bmatrix}.$$

(Proof: See [709].)

Fact 6.4.20. Let $A \in \mathbb{F}^{n \times m}$, $B \in \mathbb{F}^{n \times l}$, $C \in \mathbb{F}^{k \times m}$, and $D \in \mathbb{F}^{k \times l}$. Then,

$$\begin{aligned} \operatorname{rank} \begin{bmatrix} A & B \end{bmatrix} &= \operatorname{rank} A + \operatorname{rank}(B - AA^+B) \\ &= \operatorname{rank} B + \operatorname{rank}(A - BB^+A) \\ &= \operatorname{rank} A + \operatorname{rank} B - \dim[\mathcal{R}(A) \cap \mathcal{R}(B)], \end{aligned}$$

$$\begin{aligned} \operatorname{rank} \begin{bmatrix} A \\ C \end{bmatrix} &= \operatorname{rank} A + \operatorname{rank}(C - CA^+A) \\ &= \operatorname{rank} C + \operatorname{rank}(A - AC^+C) \\ &= \operatorname{rank} A + \operatorname{rank} C - \dim[\mathcal{R}(A^*) \cap \mathcal{R}(C^*)], \end{aligned}$$

$$\operatorname{rank} \begin{bmatrix} A & B \\ C & 0 \end{bmatrix} = \operatorname{rank} B + \operatorname{rank} C + \operatorname{rank}\left[(I_n - BB^+)A(I_m - C^+C)\right],$$

and

$$\begin{aligned} \operatorname{rank} \begin{bmatrix} A & B \\ C & D \end{bmatrix} &= \operatorname{rank} A + \operatorname{rank} X + \operatorname{rank} Y \\ &\quad + \operatorname{rank}\left[(I_k - YY^+)(D - CA^+B)(I_l - X^+X)\right], \end{aligned}$$

where $X \triangleq B - AA^+B$ and $Y \triangleq C - CA^+A$. Consequently,

$$\operatorname{rank} A + \operatorname{rank}(D - CA^+B) \leq \operatorname{rank} \begin{bmatrix} A & B \\ C & D \end{bmatrix},$$

and, if $AA^+B = B$ and $CA^+A = C$, then

$$\operatorname{rank} A + \operatorname{rank}(D - CA^+B) = \operatorname{rank} \begin{bmatrix} A & B \\ C & D \end{bmatrix}.$$

Finally, if $n = m$ and A is nonsingular, then

$$n + \operatorname{rank}(D - CA^{-1}B) = \operatorname{rank} \begin{bmatrix} A & B \\ C & D \end{bmatrix}.$$

(Proof: See [163, 514], Fact 2.10.24, and Fact 2.10.25.) (Remark: With certain restrictions the generalized inverses can be replaced by (1)-inverses.) (Remark: See Proposition 2.8.3.)

Fact 6.4.21. Let $A \in \mathbb{F}^{n \times m}$, $B \in \mathbb{F}^{n \times l}$, $C \in \mathbb{F}^{l \times n}$, $D \in \mathbb{F}^{l \times l}$, and assume that D is nonsingular. Then,

$$\operatorname{rank} A = \operatorname{rank}\left(A - BD^{-1}C\right) + \operatorname{rank} BD^{-1}C$$

if and only if there exist matrices $X \in \mathbb{F}^{m \times l}$ and $Y \in \mathbb{F}^{l \times n}$ such that $B = AX$, $C = YA$, and $D = YAX$. (Proof: See [182].)

Fact 6.4.22. Let $A \in \mathbb{F}^{n \times m}$, $B \in \mathbb{F}^{n \times l}$, $C \in \mathbb{F}^{k \times m}$, and $D \in \mathbb{F}^{k \times l}$. Then,

$$\operatorname{rank} A + \operatorname{rank}(D - CA^{+}B) = \operatorname{rank} \begin{bmatrix} A^{*}AA^{*} & A^{*}B \\ CA^{*} & D \end{bmatrix}.$$

(Proof: See [706].)

Fact 6.4.23. Let $A_{11} \in \mathbb{F}^{n \times m}$, $A_{12} \in \mathbb{F}^{n \times l}$, $A_{21} \in \mathbb{F}^{k \times m}$, and $A_{22} \in \mathbb{F}^{k \times l}$, and define $A \triangleq \begin{bmatrix} A_{11} & A_{12} \\ A_{21} & A_{22} \end{bmatrix} \in \mathbb{F}^{(n+k) \times (m+l)}$ and $B \triangleq AA^{+} = \begin{bmatrix} B_{11} & B_{12} \\ B_{12}^{\mathrm{T}} & B_{22} \end{bmatrix}$, where $B_{11} \in \mathbb{F}^{n \times m}$, $B_{12} \in \mathbb{F}^{n \times l}$, $B_{21} \in \mathbb{F}^{k \times m}$, and $B_{22} \in \mathbb{F}^{k \times l}$. Then,

$$\operatorname{rank} B_{12} = \operatorname{rank} \begin{bmatrix} A_{11} & A_{12} \end{bmatrix} + \operatorname{rank} \begin{bmatrix} A_{21} & A_{22} \end{bmatrix} - \operatorname{rank} A.$$

(Proof: See [714].) (Remark: See Fact 3.9.10 and Fact 3.8.17.)

Fact 6.4.24. Let $A, B \in \mathbb{F}^{n \times n}$. Then,

$$\operatorname{rank} \begin{bmatrix} 0 & A \\ B & I \end{bmatrix} = \operatorname{rank} A + \operatorname{rank} \begin{bmatrix} B & I - A^{+}A \end{bmatrix}$$

$$= \operatorname{rank} \begin{bmatrix} A \\ I - BB^{+} \end{bmatrix} + \operatorname{rank} B$$

$$= \operatorname{rank} A + \operatorname{rank} B + \operatorname{rank}\left[(I - BB^{+})(I - A^{+}A)\right]$$

$$= n + \operatorname{rank} AB.$$

Hence, the following statements hold:

i) $\operatorname{rank} AB = \operatorname{rank} A + \operatorname{rank} B - n$ if and only if $(I - BB^{+})(I - A^{+}A) = 0$.

ii) $\operatorname{rank} AB = \operatorname{rank} A$ if and only if $\begin{bmatrix} B & I - A^{+}A \end{bmatrix}$ is right invertible.

iii) $\operatorname{rank} AB = \operatorname{rank} B$ if and only if $\begin{bmatrix} A \\ I - BB^{+} \end{bmatrix}$ is left invertible.

(Proof: See [514].) (Remark: The generalized inverses can be replaced by arbitrary (1)-inverses.)

Fact 6.4.25. Let $A \in \mathbb{F}^{n \times m}$, $B \in \mathbb{F}^{m \times l}$, and $C \in \mathbb{F}^{l \times k}$. Then,

$$\operatorname{rank} \begin{bmatrix} 0 & AB \\ BC & B \end{bmatrix} = \operatorname{rank} B + \operatorname{rank} ABC$$

$$= \operatorname{rank} AB + \operatorname{rank} BC$$

$$+ \operatorname{rank} \left[(I - BC)(BC)^{+}\right]B[(I - (AB)^{+}(AB)].$$

Furthermore, the following statements are equivalent:

i) $\operatorname{rank} \begin{bmatrix} 0 & AB \\ BC & B \end{bmatrix} = \operatorname{rank} AB + \operatorname{rank} BC.$

ii) $\operatorname{rank} ABC = \operatorname{rank} AB + \operatorname{rank} BC - \operatorname{rank} B.$

iii) There exist matrices $X \in \mathbb{F}^{k \times l}$ and $Y \in \mathbb{F}^{m \times n}$ such that

$$BCX + YAB = B.$$

(Proof: See [514, 714] and Fact 5.9.25.) (Remark: See Fact 2.10.28.)

Fact 6.4.26. Let $A \in \mathbb{F}^{n \times m}$ and $b \in \mathbb{F}^n$. Then,

$$\begin{bmatrix} A & b \end{bmatrix}^+ = \begin{bmatrix} A^+(I - bc) \\ c \end{bmatrix},$$

where

$$c \triangleq \begin{cases} (b - AA^+b)^+, & b \neq AA^+b, \\[2mm] \dfrac{b^*(AA^*)^+}{1 + b^*(AA^*)^+b}, & b = AA^+b. \end{cases}$$

(Proof: See [9, p. 44], [260, p. 270], or [639, p. 148].) (Remark: This result is due to Greville.)

Fact 6.4.27. Let $A \in \mathbb{F}^{n \times m}$ and $B \in \mathbb{F}^{n \times l}$. Then,

$$\begin{bmatrix} A & B \end{bmatrix}^+ = \begin{bmatrix} A^+ - A^+B(C^+ + D) \\ C^+ + D \end{bmatrix},$$

where

$$C \triangleq (I - AA^+)B$$

and

$$D \triangleq (I - C^+C)[I + (I - C^+C)B^*(AA^*)^+B(I - C^+C)]^{-1}B^*(AA^*)^+(I - BC^+).$$

Furthermore,

$$\begin{bmatrix} A & B \end{bmatrix}^+ = \begin{cases} \begin{bmatrix} A^*(AA^* + BB^*)^{-1} \\ B^*(AA^* + BB^*)^{-1} \end{bmatrix}, & \operatorname{rank} \begin{bmatrix} A & B \end{bmatrix} = n, \\[4mm] \begin{bmatrix} A^*A & A^*B \\ B^*A & B^*B \end{bmatrix}^{-1} \begin{bmatrix} A^* \\ B^* \end{bmatrix}, & \operatorname{rank} \begin{bmatrix} A & B \end{bmatrix} = m + l, \\[4mm] \begin{bmatrix} A^*(AA^*)^{-1}(I - BE) \\ E \end{bmatrix}, & \operatorname{rank} A = n, \end{cases}$$

where

$$E \triangleq \begin{bmatrix} I + B^*(AA^*)^{-1}B \end{bmatrix}^{-1} B^*(AA^*)^{-1}.$$

(Proof: See [186] or [502, p. 14].) (Remark: If $\begin{bmatrix} A & B \end{bmatrix}$ is square and nonsingular and $A^*B = 0$, then the second expression yields Fact 2.15.7.)

Fact 6.4.28. Let $A \in \mathbb{F}^{n \times m}$ and $B \in \mathbb{F}^{n \times l}$. Then,

$$\mathrm{rank}\left(\begin{bmatrix} A & B \end{bmatrix}^+ - \begin{bmatrix} A^+ \\ B^+ \end{bmatrix}\right) = \mathrm{rank}\begin{bmatrix} AA^*B & BB^*A \end{bmatrix}.$$

Hence, if $A^*B = 0$, then

$$\begin{bmatrix} A & B \end{bmatrix}^+ = \begin{bmatrix} A^+ \\ B^+ \end{bmatrix}.$$

(Proof: See [700].)

Fact 6.4.29. Let $A \in \mathbb{F}^{n \times n}$, assume that A is positive semidefinite, let $B \in \mathbb{F}^{n \times m}$, and define

$$\mathcal{A} \triangleq \begin{bmatrix} A & B \\ B^* & 0 \end{bmatrix}.$$

Then,

$$\mathcal{A}^+ = \begin{bmatrix} C^+ - C^+BD^+B^*C^+ & C^+BD^+ \\ (C^+BD^+)^* & DD^+ - D^+ \end{bmatrix},$$

where

$$C \triangleq A + BB^*, \qquad D \triangleq B^+C^+C.$$

(Proof: See [503, p. 58].) (Remark: Representations for the generalized inverse of a partitioned matrix are given in [59, 70, 94, 155, 158, 335, 383, 533, 534, 536, 537, 607, 618, 701, 769].)

Fact 6.4.30. Let $A \in \mathbb{F}^{n \times n}$, assume that A is Hermitian, let $b \in \mathbb{F}^n$, and define $S \triangleq I - A^+A$. Then,

$(A + bb^*)^+$

$$= \begin{cases} \left[I - (b^*Sb)^{-1}Sbb^*\right]A^+\left[I - (b^*Sb)^{-1}bb^*S\right] + (b^*Sb)^{-2}Sbb^*S, & Sb \neq 0, \\[2mm] A^+ - (1 + b^*A^+b)A^+bb^*A^+, & 1 + b^*A^+b \neq 0, \\[2mm] \left[I - (b^*A^{2+}b)^{-1}A^+bb^*A^+\right]A^+\left[I - (b^*A^{2+}b)^{-1}A^+bb^*A^+\right], & b^*A^+b = 0. \end{cases}$$

(Proof: See [540].) (Remark: Expressions for $(A + BB^*)^+$, where $B \in \mathbb{F}^{n \times l}$, are given in [540].)

Fact 6.4.31. Let $A \in \mathbb{F}^{n \times n}$, assume that A is positive semidefinite, let $C \in \mathbb{F}^{m \times m}$, assume that C is positive definite, and let $B \in \mathbb{F}^{n \times m}$. Then,

$$(A + BCB^*)^+ = A^+ - A^+B(C^{-1} + B^*A^+B)^{-1}B^*A^+$$

if and only if

$$AA^+B = B.$$

(Proof: See [564].) (Remark: $AA^+B = B$ is equivalent to $\mathcal{R}(B) \subseteq \mathcal{R}(A)$.) (Remark: Extensions of the matrix inversion lemma are considered in [264] and [339, pp. 426–428, 447, 448].)

Fact 6.4.32. Let $A, B \in \mathbb{F}^{n \times m}$, and assume that $A^*B = 0$ and $BA^* = 0$. Then,

$$(A + B)^+ = A^+ + B^+.$$

(Proof: Use Fact 2.10.8 and Fact 6.4.33. See [187] and [339, p. 513].) (Remark: This result is due to Penrose.)

Fact 6.4.33. Let $A, B \in \mathbb{F}^{n \times m}$, and assume that $\operatorname{rank}(A + B) = \operatorname{rank} A + \operatorname{rank} B$. Then,

$$(A + B)^+ = (I - C^+B)A^+(I - BC^+) + C^+,$$

where $C \triangleq (I - AA^+)B(I - A^+A)$. (Proof: See [187].)

Fact 6.4.34. Let $A, B \in \mathbb{F}^{n \times m}$. Then,

$$(A + B)^+ = (I + A^+B)^+(A^+ + A^+BA^+)(I + BA^+)^+$$

if and only if $AA^+B = B = BA^+A$. Furthermore, if $n = m$ and A is nonsingular, then

$$(A + B)^+ = (I + A^{-1}B)^+(A^{-1} + A^{-1}BA^{-1})(I + BA^{-1})^+.$$

(Proof: See [187].) (Remark: If A and $A + B$ are nonsingular, then the last statement yields $(A + B)^{-1} = (A + B)^{-1}(A + B)(A + B)^{-1}$ for which the assumption that A is nonsingular is superfluous.)

Fact 6.4.35. Let $A \in \mathbb{F}^{n \times m}$, $B \in \mathbb{F}^{l \times k}$, and $C \in \mathbb{F}^{n \times k}$. Then, there exists a matrix $X \in \mathbb{F}^{m \times l}$ satisfying $AXB = C$ if and only if $AA^+CB^+B = C$. Furthermore, X satisfies $AXB = C$ if and only if there exists a matrix $Y \in \mathbb{F}^{m \times l}$ such that

$$X = A^+CB^+ + Y - A^+AYBB^+.$$

Finally, if $Y = 0$, then $\operatorname{tr} X^*X$ is minimized. (Proof: Use Proposition 6.1.7. See [503, p. 37] and, for Hermitian solutions, see [423].)

Fact 6.4.36. Let $A \in \mathbb{F}^{n \times m}$, and assume that $\operatorname{rank} A = m$. Then, $A^{\mathrm{L}} \in \mathbb{F}^{m \times n}$ is a left inverse of A if and only if there exists a matrix $B \in \mathbb{F}^{m \times n}$ such that

$$A^{\mathrm{L}} = A^+ + B(I - AA^+).$$

(Proof: Use Fact 6.4.30 with $A = C = I_m$.)

Fact 6.4.37. Let $A \in \mathbb{F}^{n \times m}$, and assume that $\operatorname{rank} A = n$. Then, $A^{\mathrm{R}} \in \mathbb{F}^{m \times n}$ is a right inverse of A if and only if there exists a matrix $B \in \mathbb{F}^{m \times n}$ such that

$$A^{\mathrm{R}} = A^+ + (I - A^+ A)B.$$

(Proof: Use Fact 6.4.35 with $B = C = I_n$.)

Fact 6.4.38. Let $A \in \mathbb{F}^{n \times n}$, $x, y \in \mathbb{F}^n$, and $a \in \mathbb{F}$, and assume that $x \in \mathcal{R}(A)$. Then,

$$\begin{bmatrix} A & x \\ y^{\mathrm{T}} & a \end{bmatrix} = \begin{bmatrix} I & 0 \\ y^{\mathrm{T}} & 1 \end{bmatrix} \begin{bmatrix} A & 0 \\ y^{\mathrm{T}} - y^{\mathrm{T}}A & a - y^{\mathrm{T}}A^+ x \end{bmatrix} \begin{bmatrix} I & A^+ x \\ 0 & 1 \end{bmatrix}.$$

(Remark: See Fact 2.13.1 and Fact 2.13.7, and note that $x = AA^+ x$.) (Problem: Obtain a factorization for the case $x \notin \mathcal{R}(A)$.)

Fact 6.4.39. Let $A \in \mathbb{F}^{n \times m}$ and $B \in \mathbb{F}^{n \times l}$. Then,

$$\det \begin{bmatrix} A^*A & B^*A \\ B^*A & B^*B \end{bmatrix} = \det(A^*A)\det[B^*(I - AA^+)B]$$

$$= \det(B^*B)\det[A^*(I - BB^+)A].$$

Fact 6.4.40. Let $A \in \mathbb{F}^{n \times n}$, $B \in \mathbb{F}^{n \times m}$, $C \in \mathbb{F}^{m \times n}$, and $D \in \mathbb{F}^{m \times m}$, assume that either $\operatorname{rank} \begin{bmatrix} A & B \end{bmatrix} = \operatorname{rank} A$ or $\operatorname{rank} \begin{bmatrix} A \\ C \end{bmatrix} = \operatorname{rank} A$, and let $A^- \in \mathbb{F}^{n \times n}$ be a (1)-inverse of A. Then,

$$\det \begin{bmatrix} A & B \\ C & D \end{bmatrix} = (\det A)\det(D - CA^- B).$$

(Proof: See [78, p. 266].)

Fact 6.4.41. Let $A, B \in \mathbb{F}^{n \times n}$, and assume that A and B are projectors. Then,

$$\lim_{k \to \infty} A(BA)^k = 2A(A + B)^+ B.$$

Furthermore, $2A(A + B)^+ B$ is the projector onto $\mathcal{R}(A) \cap \mathcal{R}(B)$. (Proof: See [24].) (Remark: See Fact 6.4.42 and Fact 8.15.10.)

Fact 6.4.42. Let $A \in \mathbb{R}^{n \times m}$ and $B \in \mathbb{R}^{n \times l}$. Then,

$$\mathcal{R}(A) \cap \mathcal{R}(B) = \mathcal{R}[AA^+(AA^+ + BB^+)^+ BB^+].$$

(Remark: See Theorem 2.3.1 and Fact 8.15.10.)

Fact 6.4.43. Let $A \in \mathbb{R}^{n \times m}$ and $B \in \mathbb{R}^{n \times l}$. Then, $\mathcal{R}(A) \subseteq \mathcal{R}(B)$ if and only if $BB^+ A = A$. (Proof: See [9, p. 35].)

Fact 6.4.44. Let $A \in \mathbb{R}^{n \times m}$ and $B \in \mathbb{R}^{n \times l}$. Then,

$$\operatorname{rank} AA^+(AA^+ + BB^+)^+BB^+ = \operatorname{rank} A + \operatorname{rank} B - \operatorname{rank} \begin{bmatrix} A & B \end{bmatrix}.$$

(Proof: Use Fact 6.4.42, Fact 2.10.26, and Fact 2.10.22.)

Fact 6.4.45. Let $A \triangleq \begin{bmatrix} A_{11} & A_{12} \\ A_{21} & A_{22} \end{bmatrix} \in \mathbb{F}^{(n+m) \times (n+m)}$, $B \in \mathbb{F}^{(n+m) \times l}$, $C \in \mathbb{F}^{l \times (n+m)}$, $D \in \mathbb{F}^{l \times l}$, and $\mathcal{A} \triangleq \begin{bmatrix} A & B \\ C & D \end{bmatrix}$, and assume that A and A_{11} are nonsingular. Then,

$$A|\mathcal{A} = (A_{11}|A)|(A_{11}|\mathcal{A}).$$

(Proof: See [592, pp. 18, 19].) (Remark: This result is the *Crabtree-Haynsworth quotient formula*. See [375].) (Remark: Extensions are given in [816].) (Problem: Extend this result to the case in which either A or A_{11} is singular.)

Fact 6.4.46. Let $A \in \mathbb{F}^{n \times m}$ and $b \in \mathbb{F}^n$, and define

$$f(x) \triangleq (Ax - b)^*(Ax - b),$$

where $x \in \mathbb{F}^m$. Then, f has a minimizer. Furthermore, $x \in \mathbb{F}^m$ minimizes f if and only if there exists a vector $y \in \mathbb{F}^m$ such that

$$x = A^+b + (I - A^+A)y.$$

In this case,

$$f(x) = b^*(I - AA^+)b.$$

Finally, f has a unique minimizer if and only if A is left invertible. (Remark: The minimization of f is the *least squares problem*. See [9, 128, 663].) (Remark: This result is a special case of Fact 8.12.23.)

Fact 6.4.47. Let $A \in \mathbb{F}^{n \times m}$, $B \in \mathbb{F}^{n \times l}$, and define

$$f(X) \triangleq \operatorname{tr}[(AX - B)^*(AX - B)],$$

where $X \in \mathbb{F}^{m \times l}$. Then, $X = A^+B$ minimizes f. (Problem: Determine all minimizers.) (Problem: Consider $f(X) = \operatorname{tr}[(AX - B)^*C(AX - B)]$, where $C \in \mathbb{F}^{n \times n}$ is positive definite.)

Fact 6.4.48. Let $A \in \mathbb{F}^{n \times m}$ and $B \in \mathbb{F}^{l \times m}$, and define

$$f(X) \triangleq \operatorname{tr}[(XA - B)^*(XA - B)],$$

where $X \in \mathbb{F}^{l \times n}$. Then, $X = BA^+$ minimizes f.

Fact 6.4.49. Let $A, B \in \mathbb{F}^{n \times m}$, and define

$$f(X) \triangleq \operatorname{tr}[(AX - B)^*(AX - B)],$$

where $X \in \mathbb{F}^{m \times m}$ is unitary. Then, $X = S_1S_2$ minimizes f, where $S_1 \begin{bmatrix} \hat{B} & 0 \\ 0 & 0 \end{bmatrix} S_2$ is the singular value decomposition of A^*B. (Proof: See [78, p. 224].)

6.5 Facts on the Drazin and Group Generalized Inverses

Fact 6.5.1. Let $A \in \mathbb{F}^{n \times n}$. Then, AA^{D} is idempotent.

Fact 6.5.2. Let $A \in \mathbb{F}^{n \times n}$. Then, $A = A^{\mathrm{D}}$ if and only if A is tripotent.

Fact 6.5.3. Let $A \in \mathbb{F}^{n \times n}$. Then,
$$\left(A^*\right)^{\mathrm{D}} = \left(A^{\mathrm{D}}\right)^*.$$

Fact 6.5.4. Let $A \in \mathbb{F}^{n \times n}$, and let $r \in \mathbb{P}$. Then,
$$\left(A^{\mathrm{D}}\right)^r = \left(A^r\right)^{\mathrm{D}}.$$

Fact 6.5.5. Let $A \in \mathbb{F}^{n \times n}$. Then, $X = A^{\mathrm{D}}$ is the unique matrix satisfying
$$\mathrm{rank} \begin{bmatrix} A & AA^{\mathrm{D}} \\ A^{\mathrm{D}}A & X \end{bmatrix} = \mathrm{rank}\, A.$$

(Remark: See Fact 2.15.10 and Fact 6.3.22.) (Proof: See [817].)

Fact 6.5.6. Let $A, B \in \mathbb{F}^{n \times n}$, and assume that $AB = BA$. Then,
$$(AB)^{\mathrm{D}} = B^{\mathrm{D}}A^{\mathrm{D}},$$
$$A^{\mathrm{D}}B = BA^{\mathrm{D}},$$
$$AB^{\mathrm{D}} = B^{\mathrm{D}}A.$$

Fact 6.5.7. Let $A \in \mathbb{F}^{n \times n}$, and assume that $\mathrm{ind}\, A = \mathrm{rank}\, A = 1$. Then,
$$A^{\#} = \left(\mathrm{tr}\, A^2\right)^{-1}A.$$
Consequently, if $x, y \in \mathbb{F}^n$ satisfy $x^*y \neq 0$, then
$$(xy^*)^{\#} = (x^*y)^{-2}xy^*.$$
In particular,
$$1_{n \times n}^{\#} = n^{-2}1_{n \times n}.$$

Fact 6.5.8. Let $A \in \mathbb{F}^{n \times n}$, and let $k \triangleq \mathrm{ind}\, A$. Then,
$$A^{\mathrm{D}} = A^k \left(A^{2k+1}\right)^{+} A^k.$$
If, in particular, $\mathrm{ind}\, A \leq 1$, then
$$A^{\#} = A\left(A^3\right)^{+}A.$$

(Proof: See [96, pp. 165, 174].)

Fact 6.5.9. Let $A \in \mathbb{F}^{n \times n}$. Then, the following statements are equivalent:

i) A is range Hermitian.

ii) $A^+ = A^{\mathrm{D}}$.

iii) $\mathrm{ind}\, A \leq 1$, and $A^+ = A^{\#}$.

iv) $\mathrm{ind}\, A \leq 1$, and $A^*A^{\#}A + AA^{\#}A^* = 2A^*$.

v) $\mathrm{ind}\, A \leq 1$, and $A^+A^{\#}A + AA^{\#}A^+ = 2A^+$.

(Proof: See [180].) (Remark: See Fact 6.3.6.)

Fact 6.5.10. Let $A \in \mathbb{F}^{n \times n}$. Then, the following statements are equivalent:

i) A is normal.

ii) $\mathrm{ind}\, A \leq 1$, and $A^{\#}A^* = A^*A^{\#}$.

(Proof: See [180].) (Remark: See Fact 3.5.8, Fact 3.7.10, Fact 5.14.2, and Fact 6.3.8.)

Fact 6.5.11. Let $A \in \mathbb{F}^{n \times n}$. Then, A is group invertible if and only if $\lim_{\alpha \to 0}(A + \alpha I)^{-1}A$ exists. In this case,

$$\lim_{\alpha \to 0}(A + \alpha I)^{-1}A = AA^{\#}.$$

(Proof: See [158, p. 138].)

Fact 6.5.12. Let $A, B \in \mathbb{F}^{n \times n}$, assume that A and B are group invertible, and consider the following conditions:

i) $ABA = B$.

ii) $BAB = A$.

iii) $A^2 = B^2$.

Then, if two of the above conditions are satisfied, then the third condition is satisfied. Furthermore, if *i)–iii)* are satisfied, then the following statements hold:

i) A and B are group invertible.

ii) $A^{\#} = A^3$ and $B^{\#} = B^3$.

iii) $A^5 = A$ and $B^5 = B$.

iv) $A^4 = B^4 = (AB)^4$.

v) If A and B are nonsingular, then $A^4 = B^4 = (AB)^4 = I$.

(Proof: See [251].)

6.6 Notes

The proof of the uniqueness of A^+ is given in [503]. Most of the results given in this chapter can be found in [158]. The last equality in *xii*) of Proposition 6.1.6 is given in [819]. Reverse-order laws for the generalized inverse of a product are discussed in [768]. Additional books on generalized inverses include [96, 138, 606]. Generalized inverses are widely used in least squares methods; see [134, 158, 459]. Applications to singular differential equations are considered in [157]. Historical remarks are given in [95].

Chapter Seven

Kronecker and Schur Algebra

In this chapter we introduce Kronecker matrix algebra, which is useful for solving linear matrix equations.

7.1 Kronecker Product

For $A \in \mathbb{F}^{n \times m}$ define the *vec* operator as

$$\operatorname{vec} A \triangleq \begin{bmatrix} \operatorname{col}_1(A) \\ \vdots \\ \operatorname{col}_m(A) \end{bmatrix} \in \mathbb{F}^{nm}, \tag{7.1.1}$$

which is the column vector of size $nm \times 1$ obtained by stacking the columns of A. We recover A from $\operatorname{vec} A$ by writing

$$A = \operatorname{vec}^{-1}(\operatorname{vec} A). \tag{7.1.2}$$

Proposition 7.1.1. Let $A \in \mathbb{F}^{n \times m}$ and $B \in \mathbb{F}^{m \times n}$. Then,

$$\operatorname{tr} AB = \left(\operatorname{vec} A^{\mathrm{T}}\right)^{\mathrm{T}} \operatorname{vec} B = \left(\operatorname{vec} B^{\mathrm{T}}\right)^{\mathrm{T}} \operatorname{vec} A. \tag{7.1.3}$$

Proof. Note that

$$\begin{aligned}
\operatorname{tr} AB &= \sum_{i=1}^{n} \operatorname{row}_i(A)\operatorname{col}_i(B) \\
&= \sum_{i=1}^{n} \left[\operatorname{col}_i\left(A^{\mathrm{T}}\right)\right]^{\mathrm{T}}\operatorname{col}_i(B) \\
&= \begin{bmatrix} \operatorname{col}_1^{\mathrm{T}}\left(A^{\mathrm{T}}\right) & \cdots & \operatorname{col}_n^{\mathrm{T}}\left(A^{\mathrm{T}}\right) \end{bmatrix} \begin{bmatrix} \operatorname{col}_1(B) \\ \vdots \\ \operatorname{col}_n(B) \end{bmatrix} \\
&= \left(\operatorname{vec} A^{\mathrm{T}}\right)^{\mathrm{T}}\operatorname{vec} B. \qquad \square
\end{aligned}$$

Next, we introduce the Kronecker product.

Definition 7.1.2. Let $A \in \mathbb{F}^{n \times m}$ and $B \in \mathbb{F}^{l \times k}$. Then, the *Kronecker product* $A \otimes B \in \mathbb{F}^{nl \times mk}$ of A is the partitioned matrix

$$A \otimes B \triangleq \begin{bmatrix} A_{(1,1)}B & A_{(1,2)}B & \cdots & A_{(1,m)}B \\ \vdots & \vdots & \ddots & \vdots \\ A_{(n,1)}B & A_{(n,2)}B & \cdots & A_{(n,m)}B \end{bmatrix}. \tag{7.1.4}$$

Unlike matrix multiplication, the Kronecker product $A \otimes B$ does not entail a restriction on either the size of A or the size of B.

The following results are immediate consequences of the definition of the Kronecker product.

Proposition 7.1.3. Let $\alpha \in \mathbb{F}$, $A \in \mathbb{F}^{n \times m}$, and $B \in \mathbb{F}^{l \times k}$. Then,

$$A \otimes (\alpha B) = (\alpha A) \otimes B = \alpha (A \otimes B), \tag{7.1.5}$$

$$\overline{A \otimes B} = \overline{A} \otimes \overline{B}, \tag{7.1.6}$$

$$(A \otimes B)^{\mathrm{T}} = A^{\mathrm{T}} \otimes B^{\mathrm{T}}, \tag{7.1.7}$$

$$(A \otimes B)^{*} = A^{*} \otimes B^{*}. \tag{7.1.8}$$

Proposition 7.1.4. Let $A, B \in \mathbb{F}^{n \times m}$ and $C \in \mathbb{F}^{l \times k}$. Then,

$$(A + B) \otimes C = A \otimes C + B \otimes C \tag{7.1.9}$$

and

$$C \otimes (A + B) = C \otimes A + C \otimes B. \tag{7.1.10}$$

Proposition 7.1.5. Let $A \in \mathbb{F}^{n \times m}$, $B \in \mathbb{F}^{l \times k}$, and $C \in \mathbb{F}^{p \times q}$. Then,

$$A \otimes (B \otimes C) = (A \otimes B) \otimes C. \tag{7.1.11}$$

Hence, we write $A \otimes B \otimes C$ for $A \otimes (B \otimes C)$ and $(A \otimes B) \otimes C$.

The next result illustrates an important form of compatibility between matrix multiplication and the Kronecker product.

Proposition 7.1.6. Let $A \in \mathbb{F}^{n \times m}$, $B \in \mathbb{F}^{l \times k}$, $C \in \mathbb{F}^{m \times q}$, and $D \in \mathbb{F}^{k \times p}$. Then,

$$(A \otimes B)(C \otimes D) = AC \otimes BD. \tag{7.1.12}$$

Proof. Note that the ij block of $(A \otimes B)(C \otimes D)$ is given by

$$[(A \otimes B)(C \otimes D)]_{ij} = \begin{bmatrix} A_{(i,1)}B & \cdots & A_{(i,m)}B \end{bmatrix} \begin{bmatrix} C_{(1,j)}D \\ \vdots \\ C_{(m,j)}D \end{bmatrix}$$

$$= \sum_{k=1}^{m} A_{(i,k)}C_{(k,j)}BD = (AC)_{(i,j)}BD$$

$$= (AC \otimes BD)_{ij}. \qquad \square$$

Next, we consider the inverse of a Kronecker product.

Proposition 7.1.7. Suppose $A \in \mathbb{F}^{n \times n}$ and $B \in \mathbb{F}^{m \times m}$ are nonsingular. Then,

$$(A \otimes B)^{-1} = A^{-1} \otimes B^{-1}. \qquad (7.1.13)$$

Proof. Note that

$$(A \otimes B)\left(A^{-1} \otimes B^{-1}\right) = AA^{-1} \otimes BB^{-1} = I_n \otimes I_m = I_{nm}. \qquad \square$$

Proposition 7.1.8. Let $x \in \mathbb{F}^n$ and $y \in \mathbb{F}^m$. Then,

$$xy^{\mathrm{T}} = x \otimes y^{\mathrm{T}} = y^{\mathrm{T}} \otimes x \qquad (7.1.14)$$

and

$$\mathrm{vec}\, xy^{\mathrm{T}} = y \otimes x. \qquad (7.1.15)$$

The following result concerns the vec of the product of three matrices.

Proposition 7.1.9. Let $A \in \mathbb{F}^{n \times m}$, $B \in \mathbb{F}^{m \times l}$, and $C \in \mathbb{F}^{l \times k}$. Then,

$$\mathrm{vec}(ABC) = \left(C^{\mathrm{T}} \otimes A\right)\mathrm{vec}\, B. \qquad (7.1.16)$$

Proof. Using (7.1.12) and (7.1.15), it follows that

$$\mathrm{vec}\, ABC = \mathrm{vec} \sum_{i=1}^{l} A\mathrm{col}_i(B)e_i^{\mathrm{T}}C = \sum_{i=1}^{l} \mathrm{vec}\left[A\mathrm{col}_i(B)\left(C^{\mathrm{T}}e_i\right)^{\mathrm{T}}\right]$$

$$= \sum_{i=1}^{l} \left[C^{\mathrm{T}}e_i\right] \otimes \left[A\mathrm{col}_i(B)\right] = \left(C^{\mathrm{T}} \otimes A\right)\sum_{i=1}^{l} e_i \otimes \mathrm{col}_i(B)$$

$$= \left(C^{\mathrm{T}} \otimes A\right)\sum_{i=1}^{l} \mathrm{vec}\left[\mathrm{col}_i(B)e_i^{\mathrm{T}}\right] = \left(C^{\mathrm{T}} \otimes A\right)\mathrm{vec}\, B. \qquad \square$$

The following result concerns the eigenvalues and eigenvectors of the Kronecker product of two matrices.

Proposition 7.1.10. Let $A \in \mathbb{F}^{n \times n}$ and $B \in \mathbb{F}^{m \times m}$. Then,

$$\operatorname{mspec}(A \otimes B) = \{\lambda\mu \colon \ \lambda \in \operatorname{mspec}(A), \mu \in \operatorname{mspec}(B)\}_{\mathrm{m}}. \qquad (7.1.17)$$

If, in addition, $x \in \mathbb{C}^n$ is an eigenvector of A associated with $\lambda \in \operatorname{spec}(A)$ and $y \in \mathbb{C}^n$ is an eigenvector of B associated with $\mu \in \operatorname{spec}(B)$, then $x \otimes y$ is an eigenvector of $A \otimes B$ associated with $\lambda\mu$.

Proof. Using (7.1.12), we have

$$(A \otimes B)(x \otimes y) = (Ax) \otimes (By) \ = (\lambda x) \otimes (\mu y) = \lambda\mu(x \otimes y). \qquad \square$$

Proposition 7.1.10 shows that $\operatorname{mspec}(A \otimes B) = \operatorname{mspec}(B \otimes A)$. Consequently, it follows that $\det(A \otimes B) = \det(B \otimes A)$ and $\operatorname{tr}(A \otimes B) = \operatorname{tr}(B \otimes A)$. The following results are generalizations of these identities.

Proposition 7.1.11. Let $A \in \mathbb{F}^{n \times n}$ and $B \in \mathbb{F}^{m \times m}$. Then,

$$\det(A \otimes B) = \det(B \otimes A) = (\det A)^m (\det B)^n. \qquad (7.1.18)$$

Proof. Let $\operatorname{mspec}(A) = \{\lambda_1, \ldots, \lambda_n\}_{\mathrm{m}}$ and $\operatorname{mspec}(B) = \{\mu_1, \ldots, \mu_m\}_{\mathrm{m}}$. Then, Proposition 7.1.10 implies that

$$\det(A \otimes B) = \prod_{i,j=1}^{n,m} \lambda_i \mu_j = \left(\lambda_1^m \prod_{j=1}^m \mu_j\right) \cdots \left(\lambda_n^m \prod_{j=1}^m \mu_j\right)$$

$$= (\lambda_1 \cdots \lambda_n)^m (\mu_1 \cdots \mu_m)^n = (\det A)^m (\det B)^n. \qquad \square$$

Proposition 7.1.12. Let $A \in \mathbb{F}^{n \times n}$ and $B \in \mathbb{F}^{m \times m}$. Then,

$$\operatorname{tr}(A \otimes B) = \operatorname{tr}(B \otimes A) = (\operatorname{tr} A)(\operatorname{tr} B). \qquad (7.1.19)$$

Proof. Note that

$$\operatorname{tr}(A \otimes B) = \operatorname{tr}(A_{(1,1)} B) + \cdots + \operatorname{tr}(A_{(n,n)} B)$$

$$= [A_{(1,1)} + \cdots + A_{(n,n)}]\operatorname{tr} B = (\operatorname{tr} A)(\operatorname{tr} B). \qquad \square$$

Next, define the *Kronecker permutation matrix* $P_{n,m} \in \mathbb{F}^{nm \times nm}$ by

$$P_{n,m} \triangleq \sum_{i,j=1}^{n,m} E_{i,j,n \times m} \otimes E_{j,i,m \times n}. \qquad (7.1.20)$$

Proposition 7.1.13. Let $A \in \mathbb{F}^{n \times m}$. Then,

$$\operatorname{vec} A^{\mathrm{T}} = P_{n,m} \operatorname{vec} A. \qquad (7.1.21)$$

7.2 Kronecker Sum and Linear Matrix Equations

Next, we define the Kronecker sum of two square matrices.

Definition 7.2.1. Let $A \in \mathbb{F}^{n \times n}$ and $B \in \mathbb{F}^{m \times m}$. Then, the *Kronecker sum* $A \oplus B \in \mathbb{F}^{nm \times nm}$ of A and B is

$$A \oplus B \triangleq A \otimes I_m + I_n \otimes B. \qquad (7.2.1)$$

Proposition 7.2.2. Let $A \in \mathbb{F}^{n \times n}$, $B \in \mathbb{F}^{m \times m}$, and $C \in \mathbb{F}^{l \times l}$. Then,

$$A \oplus (B \oplus C) = (A \oplus B) \oplus C. \qquad (7.2.2)$$

Hence, we write $A \oplus B \oplus C$ for $A \oplus (B \oplus C)$ and $(A \oplus B) \oplus C$.

In Proposition 7.1.10 it was shown that, if $\lambda \in \text{spec}(A)$ and $\mu \in \text{spec}(B)$, then $\lambda\mu \in \text{spec}(A \otimes B)$. Next, we present an analogous result involving Kronecker sums.

Proposition 7.2.3. Let $A \in \mathbb{F}^{n \times n}$ and $B \in \mathbb{F}^{m \times m}$. Then,

$$\text{mspec}(A \oplus B) = \{\lambda + \mu \colon \ \lambda \in \text{mspec}(A), \ \mu \in \text{mspec}(B)\}_{\text{m}}. \qquad (7.2.3)$$

Now, let $x \in \mathbb{C}^n$ be an eigenvector of A associated with $\lambda \in \text{spec}(A)$, and let $y \in \mathbb{C}^m$ be an eigenvector of B associated with $\mu \in \text{spec}(B)$. Then, $x \otimes y$ is an eigenvector of $A \oplus B$ associated with $\lambda + \mu$.

Proof. Note that

$$
\begin{aligned}
(A \oplus B)(x \otimes y) &= (A \otimes I_m)(x \otimes y) + (I_n \otimes B)(x \otimes y) \\
&= (Ax \otimes y) + (x \otimes By) = (\lambda x \otimes y) + (x \otimes \mu y) \\
&= \lambda(x \otimes y) + \mu(x \otimes y) = (\lambda + \mu)(x \otimes y). \qquad \square
\end{aligned}
$$

The next result concerns the existence and uniqueness of solutions to *Sylvester's equation*. See Fact 5.9.24 and Proposition 11.8.3.

Proposition 7.2.4. Let $A \in \mathbb{F}^{n \times n}$, $B \in \mathbb{F}^{m \times m}$, and $C \in \mathbb{F}^{n \times m}$. Then, $X \in \mathbb{F}^{n \times m}$ satisfies

$$AX + XB + C = 0 \qquad (7.2.4)$$

if and only if X satisfies

$$\left(B^{\mathrm{T}} \oplus A\right)\text{vec}\, X + \text{vec}\, C = 0. \qquad (7.2.5)$$

Consequently, $B^{\mathrm{T}} \oplus A$ is nonsingular if and only if there exists a unique matrix $X \in \mathbb{F}^{n \times m}$ satisfying (7.2.4). In this case, X is given by

$$X = -\operatorname{vec}^{-1}\left[\left(B^{\mathrm{T}} \oplus A\right)^{-1} \operatorname{vec} C\right]. \qquad (7.2.6)$$

Furthermore, $B^{\mathrm{T}} \oplus A$ is singular and rank $B^{\mathrm{T}} \oplus A = \operatorname{rank}\left[\ B^{\mathrm{T}} \oplus A \quad \operatorname{vec} C\ \right]$ if and only if there exist infinitely many matrices $X \in \mathbb{F}^{n \times m}$ satisfying (7.5.4). In this case, the set of solutions of (7.2.4) is given by $X + \mathcal{N}(B^{\mathrm{T}} \oplus A)$.

Proof. Note that (7.2.4) is equivalent to

$$0 = \operatorname{vec}(AXI + IXB) + \operatorname{vec} C = (I \otimes A)\operatorname{vec} X + \left(B^{\mathrm{T}} \otimes I\right)\operatorname{vec} X + \operatorname{vec} C$$
$$= \left(B^{\mathrm{T}} \otimes I + I \otimes A\right)\operatorname{vec} X + \operatorname{vec} C = \left(B^{\mathrm{T}} \oplus A\right)\operatorname{vec} X + \operatorname{vec} C,$$

which yields (7.2.5). The remaining results follow from Corollary 2.6.5. □

For the following corollary, note Fact 5.9.24.

Corollary 7.2.5. Let $A \in \mathbb{F}^{n \times n}$, $B \in \mathbb{F}^{m \times m}$, and $C \in \mathbb{F}^{n \times m}$, and assume that $\operatorname{spec}(A)$ and $\operatorname{spec}(-B)$ are disjoint. Then, there exists a unique matrix $X \in \mathbb{F}^{n \times m}$ satisfying (7.2.4). Furthermore, the matrices $\left[\begin{smallmatrix} A & 0 \\ 0 & -B \end{smallmatrix}\right]$ and $\left[\begin{smallmatrix} A & C \\ 0 & -B \end{smallmatrix}\right]$ are similar and satisfy

$$\begin{bmatrix} A & C \\ 0 & -B \end{bmatrix} = \begin{bmatrix} I & X \\ 0 & I \end{bmatrix}\begin{bmatrix} A & 0 \\ 0 & -B \end{bmatrix}\begin{bmatrix} I & -X \\ 0 & I \end{bmatrix}. \qquad (7.2.7)$$

7.3 Schur Product

An alternative form of vector and matrix multiplication is given by the *Schur product*. If $A \in \mathbb{F}^{n \times m}$ and $B \in \mathbb{F}^{n \times m}$, then $A \circ B \in \mathbb{F}^{n \times m}$ is defined by

$$(A \circ B)_{(i,j)} \triangleq A_{(i,j)}B_{(i,j)}, \qquad (7.3.1)$$

that is, $A \circ B$ is formed by means of entry-by-entry multiplication. For matrices $A, B, C \in \mathbb{F}^{n \times m}$, the commutative, associative, and distributive identities

$$A \circ B = B \circ A, \qquad (7.3.2)$$
$$A \circ (B \circ C) = (A \circ B) \circ C, \qquad (7.3.3)$$
$$A \circ (B + C) = A \circ B + A \circ C \qquad (7.3.4)$$

are valid. For a real scalar $\alpha \geq 0$ and $A \in \mathbb{F}^{n \times m}$, the *Schur power* $A^{\{\alpha\}}$ is defined by

$$\left(A^{\{\alpha\}}\right)_{(i,j)} \triangleq \left(A_{(i,j)}\right)^{\alpha}. \qquad (7.3.5)$$

Thus, $A^{\{2\}} = A \circ A$. Note that $A^{\{0\}} = 1_{n \times m}$, while $\alpha < 0$ is allowed if A has no zero entries. Finally, for all $A \in \mathbb{F}^{n \times m}$,

$$A \circ 1_{n \times m} = 1_{n \times m} \circ A = A. \qquad (7.3.6)$$

Proposition 7.3.1. Let $A, B \in \mathbb{F}^{n \times m}$. Then, $A \circ B$ is the submatrix of $A \otimes B$ whose entries lie at the intersections of $\mathrm{row}_1(A \otimes B), \mathrm{row}_{n+2}(A \otimes B), \mathrm{row}_{2n+3}(A \otimes B), \ldots, \mathrm{row}_{n^2}(A \otimes B)$ and columns $\mathrm{col}_1(A \otimes B), \mathrm{col}_{m+2}(A \otimes B), \mathrm{col}_{2m+3}(A \otimes B), \ldots, \mathrm{col}_{m^2}(A \otimes B)$. If, in addition, $n = m$, then $A \circ B$ is a principal submatrix of $A \otimes B$.

Proof. See [510] or [369, p. 304]. $\qquad \square$

7.4 Facts on the Kronecker Product

Fact 7.4.1. Let $x, y \in \mathbb{F}^n$. Then,

$$x \otimes y = (x \otimes I_n)y = (I_n \otimes y)x.$$

Fact 7.4.2. Let $A \in \mathbb{F}^{n \times n}$ and $B \in \mathbb{F}^{m \times m}$, and assume that A and B are (diagonal, upper triangular, lower triangular). Then, so is $A \otimes B$.

Fact 7.4.3. Let $A \in \mathbb{F}^{n \times n}$, $B \in \mathbb{F}^{m \times m}$, and $l \in \mathbb{P}$. Then,

$$(A \otimes B)^l = A^l \otimes B^l.$$

Fact 7.4.4. Let $A \in \mathbb{F}^{n \times m}$. Then,

$$\mathrm{vec}\, A = (I_m \otimes A)\,\mathrm{vec}\, I_m = (A^{\mathrm{T}} \otimes I_n)\,\mathrm{vec}\, I_n.$$

Fact 7.4.5. Let $A \in \mathbb{F}^{n \times m}$ and $B \in \mathbb{F}^{m \times l}$. Then,

$$\mathrm{vec}\, AB = (I_l \otimes A)\,\mathrm{vec}\, B = (B^{\mathrm{T}} \otimes A)\,\mathrm{vec}\, I_m = \sum_{i=1}^{m} \mathrm{col}_i(B^{\mathrm{T}}) \otimes \mathrm{col}_i(A).$$

Fact 7.4.6. Let $A \in \mathbb{F}^{n \times m}$, $B \in \mathbb{F}^{m \times l}$, and $C \in \mathbb{F}^{l \times n}$. Then,

$$\mathrm{tr}\, ABC = (\mathrm{vec}\, A)^{\mathrm{T}}(B \otimes I)\,\mathrm{vec}\, C^{\mathrm{T}}.$$

Fact 7.4.7. Let $A, B, C \in \mathbb{F}^{n \times n}$, and assume that C is symmetric. Then,

$$(\mathrm{vec}\, C)^{\mathrm{T}}(A \otimes B)\,\mathrm{vec}\, C = (\mathrm{vec}\, C)^{\mathrm{T}}(B \otimes A)\,\mathrm{vec}\, C.$$

Fact 7.4.8. Let $A \in \mathbb{F}^{n \times m}$, $B \in \mathbb{F}^{m \times l}$, $C \in \mathbb{F}^{l \times k}$, and $D \in \mathbb{F}^{k \times n}$. Then,

$$\mathrm{tr}\, ABCD = (\mathrm{vec}\, A)^{\mathrm{T}}(B \otimes D^{\mathrm{T}})\,\mathrm{vec}\, C^{\mathrm{T}}.$$

Fact 7.4.9. Let $A \in \mathbb{F}^{n \times m}$, $B \in \mathbb{F}^{m \times l}$, and $k \in \mathbb{P}$. Then,

$$(AB)^{\otimes k} = A^{\otimes k} B^{\otimes k},$$

where $A^{\otimes k} \triangleq A \otimes A \otimes \cdots \otimes A$, with A appearing k times.

Fact 7.4.10. Let $A, C \in \mathbb{F}^{n \times m}$ and $B, D \in \mathbb{F}^{l \times k}$, assume that A is (left equivalent, right equivalent, biequivalent) to C, and assume that B is (left equivalent, right equivalent, biequivalent) to D. Then, $A \otimes B$ is (left equivalent, right equivalent, biequivalent) to $C \otimes D$.

Fact 7.4.11. Let $A, B, C, D \in \mathbb{F}^{n \times n}$, assume that A is (similar, congruent, unitarily similar) to C, and assume that B is (similar, congruent, unitarily similar) to D. Then, $A \otimes B$ is (similar, congruent, unitarily similar) to $C \otimes D$.

Fact 7.4.12. Let $A \in \mathbb{F}^{n \times n}$ and $B \in \mathbb{F}^{m \times m}$, and let $\gamma \in \operatorname{spec}(A \otimes B)$. Then,

$$\sum \operatorname{gm}_A(\lambda) \operatorname{gm}_B(\mu) \leq \operatorname{gm}_{A \otimes B}(\gamma) \leq \operatorname{am}_{A \otimes B}(\gamma) = \sum \operatorname{am}_A(\lambda) \operatorname{am}_B(\mu),$$

where both sums are taken over all $\lambda \in \operatorname{spec}(A)$ and $\mu \in \operatorname{spec}(B)$ such that $\lambda \mu = \gamma$.

Fact 7.4.13. Let $A \in \mathbb{F}^{n \times n}$ and $B \in \mathbb{F}^{m \times m}$, and let $\gamma \in \operatorname{spec}(A \otimes B)$. Then, $\operatorname{ind}_{A \otimes B}(\gamma) = 1$ if and only if $\operatorname{ind}_A(\lambda) = 1$ and $\operatorname{ind}_B(\mu) = 1$ for all $\lambda \in \operatorname{spec}(A)$ and $\mu \in \operatorname{spec}(B)$ such that $\lambda \mu = \gamma$.

Fact 7.4.14. Let $A \in \mathbb{F}^{n \times n}$ and $B \in \mathbb{F}^{m \times m}$. Then,

$$\operatorname{ind} A \otimes B = \max\{\operatorname{ind} A, \operatorname{ind} B\}.$$

Fact 7.4.15. Let $A \in \mathbb{F}^{n \times n}$ and $B \in \mathbb{F}^{n \times n}$, and assume that A and B are (group invertible, range Hermitian, range symmetric, Hermitian, symmetric, normal, positive semidefinite, positive definite, unitary, orthogonal, projectors, reflectors, involutory, idempotent, tripotent, nilpotent, semisimple). Then, so is $A \otimes B$.

Fact 7.4.16. Let $A_1, \ldots, A_l \in \mathbb{F}^{n \times n}$, and assume that $A_1, \ldots, A_l$ are skew Hermitian. If l is (even, odd), then $A_1 \otimes \cdots \otimes A_l$ is (Hermitian, skew Hermitian).

Fact 7.4.17. Let $A_{i,j} \in \mathbb{F}^{n_i \times n_j}$ for all $i = 1, \ldots, k$ and $j = 1, \ldots, l$. Then,

$$
\begin{bmatrix} A_{11} & A_{22} & \cdots \\ A_{21} & A_{22} & \cdots \\ \vdots & \ddots & \ddots \end{bmatrix} \otimes B = \begin{bmatrix} A_{11} \otimes B & A_{22} \otimes B & \cdots \\ A_{21} \otimes B & A_{22} \otimes B & \cdots \\ \vdots & & \ddots & \ddots \end{bmatrix}.
$$

Fact 7.4.18. Let $x \in \mathbb{F}^k$, and let $A_i \in \mathbb{F}^{n \times n_i}$ for all $i = 1, \ldots, l$. Then,

$$x \otimes \begin{bmatrix} A_1 & \cdots & A_l \end{bmatrix} = \begin{bmatrix} x \otimes A_1 & \cdots & x \otimes A_l \end{bmatrix}.$$

Fact 7.4.19. Let $A \in \mathbb{F}^{n \times n}$ and $B \in \mathbb{F}^{m \times m}$. Then, the eigenvalues of $\sum_{i,j=1,1}^{k,l} \gamma_{ij} A^i \otimes B^j$ are of the form $\sum_{i,j=1,1}^{k,l} \gamma_{ij} \lambda^i \mu^j$, where $\lambda \in \operatorname{spec}(A)$ and $\mu \in \operatorname{spec}(B)$ and an associated eigenvector is given by $x \otimes y$, where $x \in \mathbb{F}^n$ is an eigenvector of A associated with $\lambda \in \operatorname{spec}(A)$ and $y \in \mathbb{F}^n$ is an eigenvector of B associated with $\mu \in \operatorname{spec}(B)$. (Remark: This result is due to Stephanos.) (Proof: Let $Ax = \lambda x$ and $By = \mu y$. Then, $\gamma_{ij}(A^i \otimes B^j)(x \otimes y) = \gamma_{ij} \lambda^i \mu^j (x \otimes y)$. See [277], [455, p. 411], or [499, p. 83].)

Fact 7.4.20. Let $A \in \mathbb{F}^{n \times m}$ and $B \in \mathbb{F}^{l \times k}$. Then,

$$\operatorname{rank}(A \otimes B) = (\operatorname{rank} A)(\operatorname{rank} B).$$

(Proof: Use the singular value decomposition of $A \otimes B$.) (Remark: See Fact 8.16.12.)

Fact 7.4.21. Let $A \in \mathbb{F}^{n \times n}$ and $B \in \mathbb{F}^{m \times m}$. Then,

$$\operatorname{rank}(I - A \otimes B) \leq nm - [n - \operatorname{rank}(I - A)][m - \operatorname{rank}(I - B)].$$

(Proof: See [183].)

Fact 7.4.22. Let $A \in \mathbb{F}^{n \times m}$ and $B \in \mathbb{F}^{l \times k}$, and assume that $nl = mk$ and $n \neq m$. Then, $A \otimes B$ is singular. (Proof: See [369, p. 250].)

Fact 7.4.23. Let $A \in \mathbb{F}^{n \times m}$ and $B \in \mathbb{F}^{m \times n}$. Then,

$$|n - m| \min\{n, m\} \leq \operatorname{am}_{A \otimes B}(0).$$

(Proof: See [369, p. 249].)

Fact 7.4.24. The Kronecker permutation matrix has the following properties:

$i)$ $P_{n,m}$ is a permutation matrix.

$ii)$ $P_{n,m}^{\mathrm{T}} = P_{n,m}^{-1} = P_{m,n}$.

$iii)$ $P_{n,m}$ is orthogonal.

$iv)$ $P_{n,m} P_{m,n} = I_{nm}$.

$v)$ $P_{n,n}$ is orthogonal, symmetric, and involutory.

$vi)$ $P_{n,n}$ is a reflector.

$vii)$ $\operatorname{sig} P_{n,n} = \operatorname{tr} P_{n,n} = n$.

viii) The inertia of $P_{n,n}$ is given by

$$\text{In } P_{n,n} = \begin{bmatrix} \frac{1}{2}(n^2 - n) \\ 0 \\ \frac{1}{2}(n^2 + n) \end{bmatrix}.$$

ix) $P_{1,m} = I_m$ and $P_{n,1} = I_n$.

x) If $x \in \mathbb{F}^n$ and $y \in \mathbb{F}^m$, then

$$P_{n,m}(y \otimes x) = x \otimes y.$$

xi) If $A \in \mathbb{F}^{n \times m}$, then

$$P_{n,l}(I_l \otimes A) = (A \otimes I_l)P_{m,l}.$$

xii) If $A \in \mathbb{F}^{n \times m}$ and $B \in \mathbb{F}^{l \times k}$, then

$$P_{n,l}(A \otimes B)P_{m,k} = B \otimes A$$

and

$$\text{vec}(A \otimes B) = (I_m \otimes P_{k,n} \otimes I_l)[(\text{vec } A) \otimes (\text{vec } B)].$$

xiii) If $A \in \mathbb{F}^{n \times m}$ and $B \in \mathbb{F}^{m \times n}$, then

$$\text{tr } AB = \text{tr}[P_{m,n}(A \otimes B)].$$

Fact 7.4.25. Let $A \in \mathbb{F}^{n \times m}$ and $B \in \mathbb{F}^{l \times k}$. Then,

$$(A \otimes B)^+ = A^+ \otimes B^+.$$

7.5 Facts on the Kronecker Sum

Fact 7.5.1. Let $A \in \mathbb{F}^{n \times n}$. Then,

$$(A \oplus A)^2 = A^2 \oplus A^2 + 2A \otimes A.$$

Fact 7.5.2. Let $A \in \mathbb{F}^{n \times n}$ and $B \in \mathbb{F}^{m \times m}$, and let $\gamma \in \text{spec}(A \oplus B)$. Then,

$$\sum \text{gm}_A(\lambda)\text{gm}_B(\mu) \leq \text{gm}_{A \oplus B}(\gamma) \leq \text{am}_{A \oplus B}(\gamma) = \sum \text{am}_A(\lambda)\text{am}_B(\mu),$$

where both sums are taken over all $\lambda \in \text{spec}(A)$ and $\mu \in \text{spec}(B)$ such that $\lambda + \mu = \gamma$.

Fact 7.5.3. Let $A \in \mathbb{F}^{n \times n}$ and $B \in \mathbb{F}^{m \times m}$, and let $\gamma \in \text{spec}(A \oplus B)$. Then, $\text{ind}_{A \oplus B}(\gamma) = 1$ if and only if $\text{ind}_A(\lambda) = 1$ and $\text{ind}_B(\mu) = 1$ for all $\lambda \in \text{spec}(A)$ and $\mu \in \text{spec}(B)$ such that $\lambda + \mu = \gamma$.

Fact 7.5.4. Let $A \in \mathbb{F}^{n \times n}$ and $B \in \mathbb{F}^{m \times m}$, and assume that A and B are (group invertible, range Hermitian, Hermitian, symmetric, skew Hermitian, skew symmetric, normal, positive semidefinite, positive definite, semidissipative, dissipative, nilpotent, semisimple). Then, so is $A \oplus B$.

Fact 7.5.5. Let $A \in \mathbb{F}^{n \times n}$ and $B \in \mathbb{F}^{m \times m}$. Then,

$$n \operatorname{rank} B + m \operatorname{rank} A - 2(\operatorname{rank} A)(\operatorname{rank} B)$$
$$\leq \operatorname{rank}(A \oplus B)$$
$$\leq nm - [n - \operatorname{rank}(I + A)][m - \operatorname{rank}(I - B)].$$

If, in addition, $-A$ and B are idempotent, then

$$\operatorname{rank}(A \oplus B) = n \operatorname{rank} B + m \operatorname{rank} A - 2(\operatorname{rank} A)(\operatorname{rank} B).$$

(Proof: See [183].)

Fact 7.5.6. Let $A \in \mathbb{F}^{n \times n}$, let $B \in \mathbb{F}^{m \times m}$, assume that B is positive semidefinite, and let $\operatorname{mspec}(B) = \{\lambda_1, \ldots, \lambda_m\}_{\mathrm{m}}$. Then,

$$\det(A \oplus B) = \prod_{i=1}^{m} \det(\lambda_i I + A).$$

(Proof: Specialize Fact 7.5.7.)

Fact 7.5.7. Let $A, D \in \mathbb{F}^{n \times n}$, let $C, B \in \mathbb{F}^{m \times m}$, assume that B is positive semidefinite, assume that C is positive definite, define $p(s) \triangleq \det(B - sC)$, and let $\operatorname{mroots}(p) = \{\lambda_1, \ldots, \lambda_m\}_{\mathrm{m}}$. Then,

$$\det(A \otimes C + D \otimes B) = (\det C)^n \prod_{i=1}^{m} \det(\lambda_i D + A).$$

(Proof: See [538, pp. 40, 41].)

Fact 7.5.8. Let $A, D \in \mathbb{F}^{n \times n}$, let $C, B \in \mathbb{F}^{m \times m}$, assume that $\operatorname{rank} B = 1$, and assume that C is nonsingular. Then,

$$\det(A \otimes C + D \otimes B) = (\det C)^n (\det A)^{m-1} \det[A + (\operatorname{tr} BC^{-1})D].$$

(Proof: See [538, p. 41].)

Fact 7.5.9. Let $A \in \mathbb{F}^{n \times n}$ and $B \in \mathbb{F}^{m \times m}$. Then, $\operatorname{spec}(A)$ and $\operatorname{spec}(-B)$ are disjoint if and only if, for all $C \in \mathbb{F}^{n \times m}$, the matrices $\left[\begin{smallmatrix} A & 0 \\ 0 & -B \end{smallmatrix}\right]$ and $\left[\begin{smallmatrix} A & C \\ 0 & -B \end{smallmatrix}\right]$ are similar. (Proof: Sufficiency follows from Fact 5.9.24, while necessity follows from Proposition 7.2.3 and Corollary 2.6.4.)

Fact 7.5.10. Let $A \in \mathbb{F}^{n \times n}$, $B \in \mathbb{F}^{m \times m}$, and $C \in \mathbb{F}^{n \times m}$, and assume that $\det(B^{\mathrm{T}} \oplus A) \neq 0$. Then, $X \in \mathbb{F}^{n \times m}$ satisfies

$$A^2 X + 2AXB + XB^2 + C = 0$$

if and only if

$$X = - \text{vec}^{-1}\Big[(B^{\mathrm{T}} \oplus A)^{-2}\text{vec}\, C\Big].$$

Fact 7.5.11. Let $A \in \mathbb{F}^{n \times m}$, and let $k \in \mathbb{P}$ satisfy $1 \le k \le \min\{n,m\}$. Furthermore, define the kth *compound* $A^{(k)}$ to be the $\binom{n}{k} \times \binom{m}{k}$ matrix whose entries are $k \times k$ subdeterminants of A, ordered lexicographically. (Example: For $n = k = 3$, subsets of the rows and columns of A are chosen in the order $(1,1,1), (1,1,2), (1,1,3), (1,2,1), (1,2,2), \ldots.$) Specifically, $\big(A^{(k)}\big)_{(i,j)}$ is the $k \times k$ subdeterminant of A corresponding to the ith selection of k rows of A and the jth selection of k columns of A. Then, the following statements hold:

i) $\big[A^{(k)}\big]^{\mathrm{T}} = \big[A^{\mathrm{T}}\big]^{(k)}.$

ii) $\det A^{(k)} = (\det A)^{\binom{n-1}{k-1}}.$

iii) If $n = m$ and A is nonsingular, then $\big[A^{(k)}\big]^{-1} = \big[A^{-1}\big]^{(k)}.$

iv) If $B \in \mathbb{F}^{m \times l}$, then $(AB)^{(k)} = A^{(k)}B^{(k)}.$

Now, assume that $n = m$, let $\text{mspec}(A) = \{\lambda_1, \ldots, \lambda_n\}_{\mathrm{m}}$, and, for $i = 0, \ldots, k$, define $A^{(k,i)}$ by

$$(A + sI)^{(k)} = s^k A^{(k,0)} + s^{k-1}A^{(k,1)} + \cdots + s A^{(k,k-1)} + A^{(k,k)}.$$

Then,

$$\text{mspec}\Big(A^{(2,1)}\Big) = \{\lambda_i + \lambda_j:\ i,j = 1, \ldots, n,\ i < j\}_{\mathrm{m}},$$

$$\text{mspec}\Big(A^{(2)}\Big) = \{\lambda_i \lambda_j:\ i,j = 1, \ldots, n,\ i < j\}_{\mathrm{m}},$$

and

$$\text{mspec}\Big(\big[A^{(2,1)}\big]^2 - 4A^{(2)}\Big) = \big\{(\lambda_i - \lambda_j)^2:\ i,j = 1, \ldots, n,\ i < j\big\}_{\mathrm{m}}.$$

(Proof: See [260, pp. 142–155] and [592, p. 124].) (Remark: $\big(A^{(2,1)}\big)^2 - 4A^{(2)}$ is the *discriminant* of A. The discriminant of A is singular if and only if A has a repeated eigenvalue.) (Remark: The compound operation is related to the *bialternate product* since $\text{mspec}(2A \cdot I) = \text{mspec}\big(A^{(2,1)}\big)$ and $\text{mspec}(A \cdot A) = \text{mspec}\big(A^{(2)}\big)$. See [277,302], [405, pp. 313–320], and [499, pp. 84, 85].) (Problem: Express $A \cdot B$ in terms of compounds.)

7.6 Facts on the Schur Product

Fact 7.6.1. Let $x, y, z \in \mathbb{F}^n$. Then,

$$x^{\mathrm{T}}(y \circ z) = z^{\mathrm{T}}(x \circ y) = y^{\mathrm{T}}(x \circ z).$$

Fact 7.6.2. Let $w, y \in \mathbb{F}^n$ and $x, z \in \mathbb{F}^m$. Then,
$$\left(wx^{\mathrm{T}}\right) \circ \left(yz^{\mathrm{T}}\right) = (w \circ y)(x \circ z)^{\mathrm{T}}.$$

Fact 7.6.3. Let $A \in \mathbb{F}^{n \times n}$ and $d \in \mathbb{F}^n$. Then,
$$\mathrm{diag}(d)A = A \circ d1_{1 \times n}.$$

Fact 7.6.4. Let $A, B \in \mathbb{F}^{n \times m}$, $D_1 \in \mathbb{F}^{n \times n}$, and $D_2 \in \mathbb{F}^{m \times m}$, and assume that D_1 and D_2 are diagonal. Then,
$$(D_1 A) \circ (B D_2) = D_1 (A \circ B) D_2.$$

Fact 7.6.5. Let $A, B \in \mathbb{F}^{n \times m}$. Then,
$$\mathrm{rank}(A \circ B) \le \mathrm{rank}(A \otimes B) = (\mathrm{rank}\, A)(\mathrm{rank}\, B).$$
(Proof: Use Proposition 7.3.1.) (Remark: See Fact 8.16.12.)

Fact 7.6.6. Let $A, B \in \mathbb{F}^{n \times m}$. Then,
$$\mathrm{tr}\left[(A \circ B)(A \circ B)^{\mathrm{T}}\right] = \mathrm{tr}\left[(A \circ A)(B \circ B)^{\mathrm{T}}\right].$$

Fact 7.6.7. Let $A \in \mathbb{F}^{n \times m}$, $B \in \mathbb{F}^{m \times n}$, $a \in \mathbb{F}^m$, and $b \in \mathbb{F}^n$. Then,
$$\mathrm{tr}\left[A\left(B \circ ab^{\mathrm{T}}\right)\right] = a^{\mathrm{T}}(A^{\mathrm{T}} \circ B)b.$$

Fact 7.6.8. Let $A, B \in \mathbb{F}^{n \times m}$ and $C \in \mathbb{F}^{m \times n}$. Then,
$$I \circ \left[A\left(B^{\mathrm{T}} \circ C\right)\right] = I \circ \left[(A \circ B)C\right] = I \circ \left[\left(A \circ C^{\mathrm{T}}\right)B^{\mathrm{T}}\right].$$
Hence,
$$\mathrm{tr}\left[A\left(B^{\mathrm{T}} \circ C\right)\right] = \mathrm{tr}[(A \circ B)C] = \mathrm{tr}\left[\left(A \circ C^{\mathrm{T}}\right)B^{\mathrm{T}}\right].$$

Fact 7.6.9. Let $x \in \mathbb{R}^m$ and $A \in \mathbb{R}^{n \times m}$, and define $x^A \in \mathbb{R}^n$ by
$$x^A \triangleq \begin{bmatrix} \prod_{i=1}^m x_{(i)}^{A_{(1,i)}} \\ \vdots \\ \prod_{i=1}^m x_{(i)}^{A_{(n,i)}} \end{bmatrix},$$
where every component of x^A is assumed to exist. Then, the following statements hold:

i) If $a \in \mathbb{R}$, then $a^x = \begin{bmatrix} a^{x_{(1)}} \\ \vdots \\ a^{x_{(m)}} \end{bmatrix}$.

ii) $x^{-A} = \left(x^A\right)^{\{-1\}}$.

iii) If $y \in \mathbb{R}^m$, then $(x \circ y)^A = x^A \circ y^A$.

iv) If $B \in \mathbb{R}^{n \times m}$, then $x^{A+B} = x^A \circ x^B$.

  v) If $B \in \mathbb{R}^{l \times n}$, then $\left(x^A\right)^B = x^{BA}$.

  vi) If $a \in \mathbb{R}$, then $(a^x)^A = a^{Ax}$.

  vii) If $A^{\mathrm{L}} \in \mathbb{R}^{m \times n}$ is a left inverse of A and $y = x^A$, then $x = y^{A^{\mathrm{L}}}$.

  $viii$) If $A \in \mathbb{R}^{n \times n}$ is nonsingular and $y = x^A$, then $x = y^{A^{-1}}$.

  ix) Define $f(x) \triangleq x^A$. Then, $f'(x) = \operatorname{diag}\left(x^A\right) A \operatorname{diag}\left(x^{\{-1\}}\right)$.

(Remark: These operations arise in modeling chemical reaction kinetics. See [474].)

Fact 7.6.10. Let $A \in \mathbb{R}^{n \times n}$, and assume that A is nonsingular. Then,
$$\left(A \circ A^{-\mathrm{T}}\right) 1_{n \times 1} = 1_{n \times 1}$$
and
$$1_{1 \times n}\left(A \circ A^{-\mathrm{T}}\right) = 1_{1 \times n}.$$

(Proof: See [401].)

Fact 7.6.11. Let $A \in \mathbb{R}^{n \times n}$, and assume that $A \geq\geq 0$. Then,
$$\operatorname{sprad}\left[\left(A \circ A^{\mathrm{T}}\right)^{\{1/2\}}\right] \leq \operatorname{sprad}(A) \leq \operatorname{sprad}\left[\tfrac{1}{2}\left(A + A^{\mathrm{T}}\right)\right].$$

(Proof: See [636].)

Fact 7.6.12. Let $A_1, \dots, A_r \in \mathbb{R}^{n \times n}$ and $\alpha_1, \dots, \alpha_r \in \mathbb{R}$, and assume that $A_i \geq\geq 0$ for all $i = 1, \dots, r$, $\alpha_i > 0$ for all $i = 1, \dots, r$, and $\sum_{i=1}^r \alpha_i \geq 1$. Then,
$$\operatorname{sprad}\left(A_1^{\{\alpha_1\}} \circ \cdots \circ A_r^{\{\alpha_r\}}\right) \leq \prod_{i=1}^r [\operatorname{sprad}(A_i)]^{\alpha_i}.$$

In particular, let $A \in \mathbb{R}^{n \times n}$, and assume that $A \geq\geq 0$. Then, for all $\alpha \geq 1$,
$$\operatorname{sprad}\left(A^{\{\alpha\}}\right) \leq [\operatorname{sprad}(A)]^{\alpha}$$
and, for all $\alpha \leq 1$,
$$[\operatorname{sprad}(A)]^{\alpha} \leq \operatorname{sprad}\left(A^{\{\alpha\}}\right).$$

Furthermore,
$$\operatorname{sprad}\left(A^{\{1/2\}} \circ A^{\mathrm{T}\{1/2\}}\right) \leq \operatorname{sprad}(A)$$
and
$$[\operatorname{sprad}(A \circ A)]^{1/2} \leq \operatorname{sprad}(A).$$

If, in addition, $B \in \mathbb{R}^{n \times n}$ is such that $B \geq\geq 0$, then
$$\operatorname{sprad}(A \circ B) \leq [\operatorname{sprad}(A \circ A) \operatorname{sprad}(B \circ B)]^{1/2} \leq \operatorname{sprad}(A) \operatorname{sprad}(B)$$
and
$$\operatorname{sprad}\left(A^{\{1/2\}} \circ B^{\{1/2\}}\right) \leq \sqrt{\operatorname{sprad}(A) \operatorname{sprad}(B)}.$$

If, in addition, $A >> 0$ and $B >> 0$, then

$$\operatorname{sprad}(A \circ B) < \operatorname{sprad}(A) \operatorname{sprad}(B).$$

(Proof: See [242, 411].)

7.7 Notes

A history of the Kronecker product is given in [346]. Kronecker matrix algebra is discussed in [144, 305, 347, 503, 531, 656, 746]. Applications are discussed in [608, 741].

The fact that the Schur product is a principal submatrix of the Kronecker product is noted in [510]. A variation of Kronecker matrix algebra for symmetric matrices can be developed in terms of the half-vectorization operator "vech" and the associated elimination and duplication matrices [347, 502, 727].

Generalizations of the Schur and Kronecker products, known as the block-Kronecker, Khatri-Rao, and Tracy-Singh products, are discussed in [372, 385, 438, 490]. A related operation is the *bialternate product*, which is a variation of the compound operation discussed in Fact 7.5.11. See [277, 302], [405, pp. 313–320], and [499, pp. 84, 85]. The Schur product is also called the Hadamard product.

Chapter Eight

Positive-Semidefinite Matrices

In this chapter we focus on positive-semidefinite and positive-definite matrices. These matrices arise in a variety of applications, such as covariance analysis in signal processing and controllability analysis in linear system theory, and they have many special properties.

8.1 Positive-Semidefinite and Positive-Definite Orderings

Let $A \in \mathbb{F}^{n \times n}$ be a Hermitian matrix. As shown in Corollary 5.4.5, A is unitarily similar to a real diagonal matrix whose diagonal entries are the eigenvalues of A. We denote these eigenvalues by $\lambda_1, \ldots, \lambda_n$ or, for clarity, by $\lambda_1(A), \ldots, \lambda_n(A)$. As in Chapter 4, we employ the convention

$$\lambda_1 \geq \lambda_2 \geq \cdots \geq \lambda_n, \tag{8.1.1}$$

and, for convenience, we define

$$\lambda_{\max}(A) \triangleq \lambda_1, \quad \lambda_{\min}(A) \triangleq \lambda_n. \tag{8.1.2}$$

Then, A is positive semidefinite if and only if $\lambda_{\min}(A) \geq 0$, while A is positive definite if and only if $\lambda_{\min}(A) > 0$.

For convenience, let $\mathbf{H}^n, \mathbf{N}^n$, and $\mathbf{P}^n$ denote, respectively, the Hermitian, positive-semidefinite, and positive-definite matrices in $\mathbb{F}^{n \times n}$. Hence, $\mathbf{P}^n \subset \mathbf{N}^n \subset \mathbf{H}^n$. If $A \in \mathbf{N}^n$, then we write $A \geq 0$, while, if $A \in \mathbf{P}^n$, then we write $A > 0$. If $A, B \in \mathbf{H}^n$, then $A - B \in \mathbf{N}^n$ is possible even if neither A nor B is positive semidefinite. In this case, we write $A \geq B$ or $B \leq A$. Similarly, $A - B \in \mathbf{P}^n$ is denoted by $A > B$ or $B < A$. This notation is consistent with the case $n = 1$, where $\mathbf{H}^1 = \mathbb{R}$, $\mathbf{N}^1 = [0, \infty)$, and $\mathbf{P}^1 = (0, \infty)$.

Since $0 \in \mathbf{N}^n$, it follows that $\mathbf{N}^n$ is a pointed cone. Furthermore, if $A, -A \in \mathbf{N}^n$, then $x^*Ax = 0$ for all $x \in \mathbb{F}^n$, which implies that $A = 0$. Hence, $\mathbf{N}^n$ is a one-sided cone. Finally, $\mathbf{N}^n$ and $\mathbf{P}^n$ are convex cones since, if $A, B \in \mathbf{N}^n$, then $\alpha A + \beta B \in \mathbf{N}^n$ for all $\alpha, \beta > 0$, and likewise for $\mathbf{P}^n$. The following result shows that the relation "$\leq$" is a partial ordering on $\mathbf{H}^n$.

Proposition 8.1.1. The relation "$\leq$" is reflexive, antisymmetric, and transitive on $\mathbf{H}^n$, that is, if $A, B, C \in \mathbf{H}^n$, then the following statements hold:

i) $A \leq A$.

ii) If $A \leq B$ and $B \leq A$, then $A = B$.

iii) If $A \leq B$ and $B \leq C$, then $A \leq C$.

Proof. Since $\mathbf{N}^n$ is a pointed, one-sided, and convex cone, it follows from Proposition 2.3.6 that the relation "$\leq$" is reflexive, antisymmetric, and transitive. $\qquad\square$

Additional properties of "$\leq$" and "$<$" are given by the following result.

Proposition 8.1.2. Let $A, B, C, D \in \mathbf{H}^n$. Then, the following statements hold:

i) If $A \geq 0$, then $\alpha A \geq 0$ for all $\alpha \geq 0$, and $\alpha A \leq 0$ for all $\alpha \leq 0$.

ii) If $A > 0$, then $\alpha A > 0$ for all $\alpha > 0$, and $\alpha A < 0$ for all $\alpha < 0$.

iii) If $A \geq 0$ and $B \geq 0$, then $\alpha A + \beta B \geq 0$ for all $\alpha, \beta \geq 0$.

iv) If $A \geq 0$ and $B > 0$, then $A + B > 0$.

v) $A^2 \geq 0$.

vi) $A^2 > 0$ if and only if $\det A \neq 0$.

vii) If $A \leq B$ and $B < C$, then $A < C$.

$viii$) If $A < B$ and $B \leq C$, then $A < C$.

ix) If $A \leq B$ and $C \leq D$, then $A + C \leq B + D$.

x) If $A \leq B$ and $C < D$, then $A + C < B + D$.

Furthermore, let $S \in \mathbb{F}^{m \times n}$. Then, the following statements hold:

xi) If $A \leq B$, then $SAS^* \leq SBS^*$.

xii) If $A < B$ and $\operatorname{rank} S = m$, then $SAS^* < SBS^*$.

$xiii$) If $SAS^* \leq SBS^*$ and $\operatorname{rank} S = n$, then $A \leq B$.

xiv) If $SAS^* < SBS^*$ and $\operatorname{rank} S = n$, then $m = n$ and $A < B$.

Proof. Results i)–xi) are immediate. To prove xii), note that $A < B$ implies that $(B - A)^{1/2}$ is positive definite. Thus, $\operatorname{rank} S(A - B)^{1/2} = m$, which implies that $S(A - B)S^*$ is positive definite. To prove $xiii$), note that, since $\operatorname{rank} S = n$, it follows that S has a left inverse $S^{\mathrm{L}} \in \mathbb{F}^{n \times m}$. Thus, xi)

implies that $A = S^{\mathrm{L}}SAS^*S^{\mathrm{L}*} \leq S^{\mathrm{L}}SBS^*S^{\mathrm{L}*} = B$. To prove $xiv)$, note that, since $S(B-A)S^*$ is positive definite, it follows that $\operatorname{rank} S = m$. Hence, $m = n$ and S is nonsingular. Thus, $xii)$ implies that $A = S^{-1}SAS^*S^{-*} < S^{-1}SBS^*S^{-*} = B$. $\square$

The following result is an immediate consequence of Corollary 5.4.7.

Corollary 8.1.3. Let $A, B \in \mathbf{H}^n$, and assume that A and B are congruent. Then, A is positive semidefinite if and only if B is positive semidefinite. Furthermore, A is positive definite if and only if B is positive definite.

8.2 Submatrices

We first consider some identities involving a partitioned positive-semidefinite matrix.

Lemma 8.2.1. Let $A = \begin{bmatrix} A_{11} & A_{12} \\ A_{12}^* & A_{22} \end{bmatrix} \in \mathbf{N}^{n+m}$. Then,

$$A_{12} = A_{11}A_{11}^+A_{12}, \tag{8.2.1}$$

$$A_{12} = A_{12}A_{22}A_{22}^+. \tag{8.2.2}$$

Proof. Since $A \geq 0$, it follows from Corollary 5.4.5 that $A = BB^*$, where $B = \begin{bmatrix} B_1 \\ B_2 \end{bmatrix} \in \mathbb{F}^{(n+m) \times r}$ and $r \triangleq \operatorname{rank} A$. Thus, $A_{11} = B_1B_1^*$, $A_{12} = B_1B_2^*$, and $A_{22} = B_2B_2^*$. Since A_{11} is Hermitian, it follows from $xxvii)$ of Proposition 6.1.6 that A_{11}^+ is also Hermitian. Next, defining $S \triangleq B_1 - B_1B_1^*(B_1B_1^*)^+B_1$, it follows that $SS^* = 0$, and thus $\operatorname{tr} SS^* = 0$. Hence, Lemma 2.2.3 implies that $S = 0$, and thus $B_1 = B_1B_1^*(B_1B_1^*)^+B_1$. Consequently, $B_1B_2^* = B_1B_1^*(B_1B_1^*)^+B_1B_2^*$, that is, $A_{12} = A_{11}A_{11}^+A_{12}$. The second result is analogous. $\square$

Corollary 8.2.2. Let $A = \begin{bmatrix} A_{11} & A_{12} \\ A_{12}^* & A_{22} \end{bmatrix} \in \mathbf{N}^{n+m}$. Then, the following statements hold:

$i)$ $\mathcal{R}(A_{12}) \subseteq \mathcal{R}(A_{11})$.

$ii)$ $\mathcal{R}(A_{12}^*) \subseteq \mathcal{R}(A_{22})$.

$iii)$ $\operatorname{rank} \begin{bmatrix} A_{11} & A_{12} \end{bmatrix} = \operatorname{rank} A_{11}$.

$iv)$ $\operatorname{rank} \begin{bmatrix} A_{12}^* & A_{22} \end{bmatrix} = \operatorname{rank} A_{22}$.

Proof. Results $i)$ and $ii)$ follow from (8.2.1) and (8.2.2), while $iii)$ and $iv)$ are consequences of $i)$ and $ii)$. $\square$

Next, if (8.2.1) holds, then the partitioned matrix $A \triangleq \begin{bmatrix} A_{11} & A_{12} \\ A_{12}^* & A_{22} \end{bmatrix}$ can be factored as

$$
\begin{bmatrix} A_{11} & A_{12} \\ A_{12}^* & A_{22} \end{bmatrix} = \begin{bmatrix} I & 0 \\ A_{12}^* A_{11}^+ & I \end{bmatrix} \begin{bmatrix} A_{11} & 0 \\ 0 & A_{11}|A \end{bmatrix} \begin{bmatrix} I & A_{11}^+ A_{12} \\ 0 & I \end{bmatrix}, \quad (8.2.3)
$$

while, if (8.2.2) holds, then

$$
\begin{bmatrix} A_{11} & A_{12} \\ A_{12}^* & A_{22} \end{bmatrix} = \begin{bmatrix} I & A_{12} A_{22}^+ \\ 0 & I \end{bmatrix} \begin{bmatrix} A_{22}|A & 0 \\ 0 & A_{22} \end{bmatrix} \begin{bmatrix} I & 0 \\ A_{22}^+ A_{12}^* & I \end{bmatrix}, \quad (8.2.4)
$$

where

$$
A_{11}|A = A_{22} - A_{12}^* A_{11}^+ A_{12} \tag{8.2.5}
$$

and

$$
A_{22}|A = A_{11} - A_{12} A_{22}^+ A_{12}^*. \tag{8.2.6}
$$

Hence, it follows from Lemma 8.2.1 that, if A is positive semidefinite, then (8.2.3) and (8.2.4) are valid, and, furthermore, the Schur complements (see Definition 6.1.8) $A_{11}|A$ and $A_{22}|A$ are both positive semidefinite. Consequently, we have the following result.

Proposition 8.2.3. Let $A \triangleq \begin{bmatrix} A_{11} & A_{12} \\ A_{12}^* & A_{22} \end{bmatrix} \in \mathbf{H}^{n+m}$. Then, the following statements are equivalent:

   i) $A \geq 0$.

   ii) $A_{11} \geq 0$, $A_{12} = A_{11} A_{11}^+ A_{12}$, and $A_{12}^* A_{11}^+ A_{12} \leq A_{22}$.

   iii) $A_{22} \geq 0$, $A_{12} = A_{12} A_{22} A_{22}^+$, and $A_{12} A_{22}^+ A_{12}^* \leq A_{11}$.

The following statements are also equivalent:

   iv) $A > 0$.

   v) $A_{11} > 0$ and $A_{12}^* A_{11}^{-1} A_{12} < A_{22}$.

   vi) $A_{22} > 0$ and $A_{12} A_{22}^{-1} A_{12}^* < A_{11}$.

The following result follows from (2.8.16) and (2.8.17) or from (8.2.3) and (8.2.4).

Proposition 8.2.4. Let $A \triangleq \begin{bmatrix} A_{11} & A_{12} \\ A_{12}^* & A_{22} \end{bmatrix} \in \mathbf{P}^{n+m}$. Then,

$$
A^{-1} = \begin{bmatrix} A_{11}^{-1} + A_{11}^{-1} A_{12} (A_{11}|A)^{-1} A_{12}^* A_{11}^{-1} & -A_{11}^{-1} A_{12} (A_{11}|A)^{-1} \\ -(A_{11}|A)^{-1} A_{12}^* A_{11}^{-1} & (A_{11}|A)^{-1} \end{bmatrix} \tag{8.2.7}
$$

and

$$
A^{-1} = \begin{bmatrix} (A_{22}|A)^{-1} & -(A_{22}|A)^{-1} A_{12} A_{22}^{-1} \\ -A_{22}^{-1} A_{12}^* (A_{22}|A)^{-1} & A_{22}^{-1} A_{12}^* (A_{22}|A)^{-1} A_{12} A_{22}^{-1} + A_{22}^{-1} \end{bmatrix}, \tag{8.2.8}
$$

where

$$A_{11}|A = A_{22} - A_{12}^* A_{11}^{-1} A_{12} \qquad (8.2.9)$$

and

$$A_{22}|A = A_{11} - A_{12} A_{22}^{-1} A_{12}^*. \qquad (8.2.10)$$

Now, let $A^{-1} = \begin{bmatrix} B_{11} & B_{12} \\ B_{12}^* & B_{22} \end{bmatrix}$. Then,

$$B_{11}|A^{-1} = A_{22}^{-1} \qquad (8.2.11)$$

and

$$B_{22}|A^{-1} = A_{11}^{-1}. \qquad (8.2.12)$$

Lemma 8.2.5. Let $A \in \mathbb{F}^{n \times n}$, $b \in \mathbb{F}^n$, and $a \in \mathbb{R}$. Then, $B \triangleq \begin{bmatrix} A & b \\ b^* & a \end{bmatrix}$ is positive semidefinite if and only if A is positive semidefinite, $b = AA^+b$, and $b^*A^+b \le a$. Furthermore, B is positive definite if and only if A is positive definite and $b^*A^{-1}b < a$. In this case,

$$\det B = (\det A)(a - b^*A^{-1}b). \qquad (8.2.13)$$

For the following result note that a matrix is a principal submatrix of itself, while the determinant of a matrix is also a principal subdeterminant of the matrix.

Proposition 8.2.6. Let $A \in \mathbf{H}^n$. Then, the following statements are equivalent:

i) A is positive semidefinite.

ii) Every principal submatrix of A is positive semidefinite.

iii) Every principal subdeterminant of A is nonnegative.

iv) For all $i = 1, \ldots, n$, the sum of all $i \times i$ principal subdeterminants of A is nonnegative.

Proof. To prove *i)* $\implies$ *ii)*, let $\hat{A} \in \mathbb{F}^{m \times m}$ be the principal submatrix of A obtained from A by retaining rows and columns $i_1, \ldots, i_m$. Then, $\hat{A} = S^{\mathrm{T}}AS$, where $S \triangleq \begin{bmatrix} e_{i_1} & \cdots & e_{i_m} \end{bmatrix} \in \mathbb{R}^{n \times m}$. Now, let $\hat{x} \in \mathbb{F}^m$. Since A is positive semidefinite, it follows that $\hat{x}^*\hat{A}\hat{x} = \hat{x}^*S^{\mathrm{T}}AS\hat{x} \ge 0$, and thus $\hat{A}$ is positive semidefinite.

Next, the implications *ii)* $\implies$ *iii)* $\implies$ *iv)* are immediate. To prove *iv)* $\implies$ *i)*, note that it follows from Proposition 4.4.5 that

$$\chi_A(s) = \sum_{i=0}^{n} \beta_i s^i = \sum_{i=0}^{n} (-1)^{n-i} \gamma_{n-i} s^i = (-1)^n \sum_{i=0}^{n} \gamma_{n-i}(-s)^i, \qquad (8.2.14)$$

where, for all $i = 1, \ldots, n$, γ_i is the sum of all $i \times i$ principal subdeterminants of A, and $\beta_n = \gamma_0 = 1$. By assumption, $\gamma_i \ge 0$ for all $i = 1, \ldots, n$. Now, suppose there exists $\lambda \in \text{spec}(A)$ such that $\lambda < 0$. Then, $0 = (-1)^n \chi_A(\lambda) = \sum_{i=0}^{n} \gamma_{n-i}(-\lambda)^i > 0$, which is a contradiction. $\square$

Proposition 8.2.7. Let $A \in \mathbf{H}^n$. Then, the following statements are equivalent:

i) A is positive definite.

ii) Every principal submatrix of A is positive definite.

iii) Every principal subdeterminant of A is positive.

iv) Every leading principal submatrix of A is positive definite.

v) Every leading principal subdeterminant of A is positive.

Proof. To prove $i) \implies ii)$, let $\hat{A} \in \mathbb{F}^{m \times m}$ and S be as in the proof of Proposition 8.2.6, and let $\hat{x}$ be nonzero so that $S\hat{x}$ is nonzero. Since A is positive definite, it follows that $\hat{x}^* \hat{A} \hat{x} = \hat{x}^* S^{\mathrm{T}} A S \hat{x} > 0$, and hence $\hat{A}$ is positive definite.

Next, the implications $i) \implies ii) \implies iii) \implies v)$ and $ii) \implies iv) \implies v)$ are immediate. To prove $v) \implies i)$, suppose that the leading principal submatrix $A_i \in \mathbb{F}^{i \times i}$ has positive determinant for all $i = 1, \ldots, n$. The result is true for $n = 1$. For $n \geq 2$, we show that, if A_i is positive definite, then so is A_{i+1}. Writing $A_{i+1} = \begin{bmatrix} A_i & b_i \\ b_i^* & a_i \end{bmatrix}$, it follows from Lemma 8.2.5 that $\det A_{i+1} = (\det A_i)(a_i - b_i^* A_i^{-1} b_i) > 0$, and hence $a_i - b_i^* A_i^{-1} b_i = \det A_{i+1}/\det A_i > 0$. Lemma 8.2.5 now implies that A_{i+1} is positive definite. Using this argument for all $i = 2, \ldots, n$ implies that A is positive definite. $\square$

The example $A = \begin{bmatrix} 0 & 0 \\ 0 & -1 \end{bmatrix}$ shows that every principal subdeterminant of A, rather than just the leading principal subdeterminants of A, must be checked to determine whether A is positive semidefinite. A less obvious example is $A = \begin{bmatrix} 1 & 1 & 1 \\ 1 & 1 & 1 \\ 1 & 1 & 0 \end{bmatrix}$, whose eigenvalues are 0, $1 + \sqrt{3}$, and $1 - \sqrt{3}$. In this case, the principal subdeterminant $\det A_{[1,1]} = \det \begin{bmatrix} 1 & 1 \\ 1 & 0 \end{bmatrix} < 0$.

8.3 Simultaneous Diagonalization

This section considers the simultaneous diagonalization of a pair of matrices $A, B \in \mathbf{H}^n$. There are two types of simultaneous diagonalization. *Cogredient diagonalization* involves a nonsingular matrix $S \in \mathbb{F}^{n \times n}$ such that SAS^* and SBS^* are both diagonal, whereas *contragredient diagonalization* involves finding a nonsingular matrix $S \in \mathbb{F}^{n \times n}$ such that SAS^* and $S^{-*}BS^{-1}$ are both diagonal. Both types of simultaneous transformation involve only congruence transformations. We begin by assuming that one of the matrices is positive definite, in which case the results are quite simple to prove. Our first result involves cogredient diagonalization.

Theorem 8.3.1. Let $A, B \in \mathbf{H}^n$, and assume that A is positive definite. Then, there exists a nonsingular matrix $S \in \mathbb{F}^{n \times n}$ such that $SAS^* = I$ and SBS^* is diagonal.

Proof. Setting $S_1 = A^{-1/2}$, it follows that $S_1 A S_1^* = I$. Now, since $S_1 B S_1^*$ is Hermitian, it follows from Corollary 5.4.5 that there exists a unitary matrix $S_2 \in \mathbb{F}^{n \times n}$ such that $SBS^* = S_2 S_1 B S_1^* S_2^*$ is diagonal, where $S = S_2 S_1$. Finally, $SAS^* = S_2 S_1 A S_1^* S_2^* = S_2 I S_2^* = I$. $\square$

An analogous result holds for contragredient diagonalization.

Theorem 8.3.2. Let $A, B \in \mathbf{H}^n$, and assume that A is positive definite. Then, there exists a nonsingular matrix $S \in \mathbb{F}^{n \times n}$ such that $SAS^* = I$ and $S^{-*} B S^{-1}$ is diagonal.

Proof. Setting $S_1 = A^{-1/2}$, it follows that $S_1 A S_1^* = I$. Since $S_1^{-*} B S_1^{-1}$ is Hermitian, it follows that there exists a unitary matrix $S_2 \in \mathbb{F}^{n \times n}$ such that $S^{-*} B S^{-1} = S_2^{-*} S_1^{-*} B S_1^{-1} S_2^{-1} = S_2 \left(S_1^{-*} B S_1^{-1} \right) S_2^*$ is diagonal, where $S = S_2 S_1$. Finally, $SAS^* = S_2 S_1 A S_1^* S_2^* = S_2 I S_2^* = I$. $\square$

Corollary 8.3.3. Let $A, B \in \mathbf{P}^n$. Then, there exists a nonsingular matrix $S \in \mathbb{F}^{n \times n}$ such that SAS^* and $S^{-*} B S^{-1}$ are equal and diagonal.

Proof. By Theorem 8.3.2 there exists a nonsingular matrix $S_1 \in \mathbb{F}^{n \times n}$ such that $S_1 A S_1^* = I$ and $B_1 = S_1^{-*} B S_1^{-1}$ is diagonal. Defining $S \triangleq B_1^{1/4} S_1$ yields $SAS^* = S^{-*} B S^{-1} = B_1^{1/2}$. $\square$

The transformation S of Corollary 8.3.3 is a *balancing transformation*.

Next, we weaken the requirement in Theorem 8.3.1 and Theorem 8.3.2 that A be positive definite by assuming only that A is positive semidefinite. In this case, however, we assume that B is also positive semidefinite.

Theorem 8.3.4. Let $A, B \in \mathbf{N}^n$. Then, there exists a nonsingular matrix $S \in \mathbb{F}^{n \times n}$ such that $SAS^* = \left[\begin{smallmatrix} I & 0 \\ 0 & 0 \end{smallmatrix} \right]$ and SBS^* is diagonal.

Proof. Let the nonsingular matrix $S_1 \in \mathbb{F}^{n \times n}$ be such that $S_1 A S_1^* = \left[\begin{smallmatrix} I & 0 \\ 0 & 0 \end{smallmatrix} \right]$, and similarly partition $S_1 B S_1^* = \left[\begin{smallmatrix} B_{11} & B_{12} \\ B_{12}^* & B_{22} \end{smallmatrix} \right]$, which is positive semidefinite. Letting $S_2 \triangleq \left[\begin{smallmatrix} I & -B_{12} B_{22}^+ \\ 0 & I \end{smallmatrix} \right]$, it follows from Lemma 8.2.1 that

$$S_2 S_1 B S_1^* S_2^* = \begin{bmatrix} B_{11} - B_{12} B_{22}^+ B_{12}^* & 0 \\ 0 & B_{22} \end{bmatrix}.$$

Next, let U_1 and U_2 be unitary matrices such that $U_1(B_{11} - B_{12}B_{22}^+B_{12}^*)U_1^*$ and $U_2B_{22}U_2^*$ are diagonal. Then, defining $S_3 \triangleq \begin{bmatrix} U_1 & 0 \\ 0 & U_2 \end{bmatrix}$ and $S \triangleq S_3S_2S_1$, it follows that $SAS^* = \begin{bmatrix} I & 0 \\ 0 & 0 \end{bmatrix}$ and $SBS^* = S_3S_2S_1BS_1^*S_2^*S_3^*$ is diagonal. $\square$

Theorem 8.3.5. Let $A, B \in \mathbf{N}^n$. Then, there exists a nonsingular matrix $S \in \mathbb{F}^{n \times n}$ such that $SAS^* = \begin{bmatrix} I & 0 \\ 0 & 0 \end{bmatrix}$ and $S^{-*}BS^{-1}$ is diagonal.

Proof. Let $S_1 \in \mathbb{F}^{n \times n}$ be a nonsingular matrix such that $S_1AS_1^* = \begin{bmatrix} I & 0 \\ 0 & 0 \end{bmatrix}$, and similarly partition $S_1^{-*}BS_1^{-1} = \begin{bmatrix} B_{11} & B_{12} \\ B_{12}^* & B_{22} \end{bmatrix}$, which is positive semidefinite. Letting $S_2 \triangleq \begin{bmatrix} I & B_{11}^+B_{12} \\ 0 & I \end{bmatrix}$, it follows that

$$S_2^{-*}S_1^{-*}BS_1^{-1}S_2^{-1} = \begin{bmatrix} B_{11} & 0 \\ 0 & B_{22} - B_{12}^*B_{11}^+B_{12} \end{bmatrix}.$$

Now, let U_1 and U_2 be unitary matrices such that $U_1B_{11}U_1^*$ and $U_2(B_{22} - B_{12}^*B_{11}^+B_{12})U_2^*$ are diagonal. Then, defining $S_3 \triangleq \begin{bmatrix} U_1 & 0 \\ 0 & U_2 \end{bmatrix}$ and $S \triangleq S_3S_2S_1$, it follows that $SAS^* = \begin{bmatrix} I & 0 \\ 0 & 0 \end{bmatrix}$ and $S^{-*}BS^{-1} = S_3^{-*}S_2^{-*}S_1^{-*}BS_1^{-1}S_2^{-1}S_3^{-1}$ is diagonal. $\square$

Corollary 8.3.6. Let $A, B \in \mathbf{N}^n$. Then, AB is semisimple, and every eigenvalue of AB is nonnegative. If, in addition, A and B are positive definite, then every eigenvalue of AB is positive.

Proof. It follows from Theorem 8.3.5 that there exists a nonsingular matrix $S \in \mathbb{R}^{n \times n}$ such that $A_1 = SAS^*$ and $B_1 = S^{-*}BS^{-1}$ are diagonal with nonnegative diagonal entries. Hence, $AB = S^{-1}A_1B_1S$ is semisimple and has nonnegative eigenvalues. $\square$

A more direct approach to showing that AB has nonnegative eigenvalues is to use Corollary 4.4.10 and note that $\lambda_i(AB) = \lambda_i(B^{1/2}AB^{1/2}) \geq 0$.

Corollary 8.3.7. Let $A, B \in \mathbf{N}^n$, and assume that rank A = rank B = rank AB. Then, there exists a nonsingular matrix $S \in \mathbb{F}^{n \times n}$ such that $SAS^* = S^{-*}BS^{-1}$ and such that SAS^* is diagonal.

Proof. By Theorem 8.3.5 there exists a nonsingular matrix $S_1 \in \mathbb{F}^{n \times n}$ such that $S_1AS_1^* = \begin{bmatrix} I_r & 0 \\ 0 & 0 \end{bmatrix}$, where $r \triangleq$ rank A, and such that $B_1 = S_1^{-*}BS_1^{-1}$ is diagonal. Hence, $AB = S_1^{-1}\begin{bmatrix} I_r & 0 \\ 0 & 0 \end{bmatrix}B_1S_1$. Since rank A = rank B = rank AB = r, it follows that $B_1 = \begin{bmatrix} \hat{B}_1 & 0 \\ 0 & 0 \end{bmatrix}$, where $\hat{B}_1 \in \mathbb{F}^{r \times r}$ is positive diagonal. Hence, $S_1^{-*}BS_1^{-1} = \begin{bmatrix} \hat{B}_1 & 0 \\ 0 & 0 \end{bmatrix}$. Now, define $S_2 \triangleq \begin{bmatrix} \hat{B}_1^{1/4} & 0 \\ 0 & I \end{bmatrix}$ and $S \triangleq S_2S_1$. Then, $SAS^* = S_2S_1AS_1^*S_2^* = \begin{bmatrix} \hat{B}_1^{1/2} & 0 \\ 0 & 0 \end{bmatrix} = S_2^{-*}S_1^{-*}BS_1^{-1}S_2^{-1} = S^{-*}BS^{-1}$. $\square$

8.4 Eigenvalue Inequalities

Next, we turn our attention to inequalities involving eigenvalues. We begin with a series of lemmas.

Lemma 8.4.1. Let $A \in \mathbf{H}^n$ and let $\beta \in \mathbb{R}$. Then, the following statements hold:

i) $\beta I \leq A$ if and only if $\beta \leq \lambda_{\min}(A)$.

ii) $\beta I < A$ if and only if $\beta < \lambda_{\min}(A)$.

iii) $A \leq \beta I$ if and only if $\lambda_{\max}(A) \leq \beta$.

iv) $A < \beta I$ if and only if $\lambda_{\max}(A) < \beta$.

Proof. To prove i), assume that $\beta I \leq A$, and let $S \in \mathbb{F}^{n \times n}$ be a unitary matrix such that $B = SAS^*$ is diagonal. Then, $\beta I \leq B$, which yields $\beta \leq \lambda_{\min}(B) = \lambda_{\min}(A)$. Conversely, let $S \in \mathbb{F}^{n \times n}$ be a unitary matrix such that $B = SAS^*$ is diagonal. Since the diagonal entries of B are the eigenvalues of A, it follows that $\lambda_{\min}(A)I \leq B$, which implies that $\beta I \leq \lambda_{\min}(A)I \leq S^*BS = A$. Results ii), iii), and iv) are proved in a similar manner. $\square$

Corollary 8.4.2. Let $A \in \mathbf{H}^n$. Then,

$$\lambda_{\min}(A)I \leq A \leq \lambda_{\max}(A)I. \qquad (8.4.1)$$

Proof. The result follows from i) and iii) of Lemma 8.4.1 with $\beta = \lambda_{\min}(A)$ and $\beta = \lambda_{\max}(A)$, respectively. $\square$

Lemma 8.4.3. Let $A \in \mathbf{H}^n$. Then,

$$\lambda_{\min}(A) = \min_{x \in \mathbb{F}^n \setminus \{0\}} \frac{x^*Ax}{x^*x} \qquad (8.4.2)$$

and

$$\lambda_{\max}(A) = \max_{x \in \mathbb{F}^n \setminus \{0\}} \frac{x^*Ax}{x^*x}. \qquad (8.4.3)$$

Proof. It follows from (8.4.1) that $\lambda_{\min}(A) \leq x^*Ax/x^*x$ for all nonzero $x \in \mathbb{F}^n$. Letting $x \in \mathbb{F}^n$ be an eigenvector of A associated with $\lambda_{\min}(A)$, it follows that this lower bound is attained. This proves (8.4.2). An analogous argument yields (8.4.3). $\square$

The following result is the *Cauchy interlacing theorem*.

Lemma 8.4.4. Let $A \in \mathbf{H}^n$, and let A_0 be an $(n-1) \times (n-1)$ principal submatrix of A. Then, for all $i = 1, \ldots, n-1$,

$$\lambda_{i+1}(A) \leq \lambda_i(A_0) \leq \lambda_i(A). \tag{8.4.4}$$

Proof. Note that (8.4.4) is the chain of inequalities

$$\lambda_n(A) \leq \lambda_{n-1}(A_0) \leq \lambda_{n-1}(A) \leq \cdots \leq \lambda_2(A) \leq \lambda_1(A_0) \leq \lambda_1(A).$$

Suppose that this chain of inequalities does not hold. In particular, first suppose that the right-most inequality that is not true is $\lambda_j(A_0) \leq \lambda_j(A)$, so that $\lambda_j(A) < \lambda_j(A_0)$. Choose δ such that $\lambda_j(A) < \delta < \lambda_j(A_0)$ and such that δ is not an eigenvalue of A_0. If $j = 1$, then $A - \delta I$ is negative definite, while, if $j \geq 2$, then $\lambda_j(A) < \delta < \lambda_j(A_0) \leq \lambda_{j-1}(A_0) \leq \lambda_{j-1}(A)$, so that $A - \delta I$ has $j-1$ positive eigenvalues. Thus, $\nu_+(A - \delta I) = j-1$. Furthermore, since $\delta < \lambda_i(A_0)$, it follows that $\nu_+(A_0 - \delta I) \geq j$.

Now, assume for convenience that the rows and columns of A are ordered so that A_0 is the $(n-1) \times (n-1)$ leading principal submatrix of A. Thus, $A = \begin{bmatrix} A_0 & \beta \\ \beta^* & \gamma \end{bmatrix}$, where $\beta \in \mathbb{F}^{n-1}$ and $\gamma \in \mathbb{F}$. Next, note the identity

$A - \delta I$

$$= \begin{bmatrix} I & 0 \\ \beta^*(A_0 - \delta I)^{-1} & 1 \end{bmatrix} \begin{bmatrix} A_0 - \delta I & 0 \\ 0 & \gamma - \delta - \beta^*(A_0 - \delta I)^{-1}\beta \end{bmatrix} \begin{bmatrix} I & (A_0 - \delta I)^{-1}\beta \\ 0 & 1 \end{bmatrix},$$

where $A_0 - \delta I$ is nonsingular since δ was chosen to not be an eigenvalue of A_0. Since the right-hand side of this identity involves a congruence transformation, and since $\nu_+(A_0 - \delta I) \geq j$, it follows from Corollary 5.4.7 that $\nu_+(A - \delta I) \geq j$. However, this inequality contradicts the fact that $\nu_+(A - \delta I) = j-1$.

Finally, suppose that the right-most inequality in (8.4.4) that is not true is $\lambda_{j+1}(A) \leq \lambda_j(A_0)$, so that $\lambda_j(A_0) < \lambda_{j+1}(A)$. Choose δ such that $\lambda_j(A_0) < \delta < \lambda_{j+1}(A)$ and such that δ is not an eigenvalue of A_0. Then, it follows that $\nu_+(A - \delta I) \geq j+1$ and $\nu_+(A_0 - \delta I) = j-1$. Using the congruence transformation as in the previous case, it follows that $\nu_+(A - \delta I) \leq j$, which contradicts the fact that $\nu_+(A - \delta I) \geq j+1$. $\qquad\square$

The following result is the *inclusion principle*.

Theorem 8.4.5. Let $A \in \mathbf{H}^n$, and let $A_0 \in \mathbf{H}^k$ be a $k \times k$ principal submatrix of A. Then, for all $i = 1, \ldots, k$,

$$\lambda_{i+n-k}(A) \leq \lambda_i(A_0) \leq \lambda_i(A). \tag{8.4.5}$$

Proof. For $k = n-1$, the result is given by Lemma 8.4.4. Hence, let $k = n-2$, and let A_1 denote an $(n-1) \times (n-1)$ principal submatrix of A such that the $(n-2) \times (n-2)$ principal submatrix A_0 of A is also a principal submatrix of A_1. Therefore, Lemma 8.4.4 implies that $\lambda_n(A) \leq \lambda_{n-1}(A_1) \leq \cdots \leq \lambda_2(A_1) \leq \lambda_2(A) \leq \lambda_1(A_1) \leq \lambda_1(A)$ and $\lambda_{n-1}(A_1) \leq \lambda_{n-2}(A_0) \leq \cdots \leq \lambda_2(A_0) \leq \lambda_2(A_1) \leq \lambda_1(A_0) \leq \lambda_1(A_1)$. Combining these inequalities yields $\lambda_{i+2}(A) \leq \lambda_i(A_0) \leq \lambda_i(A)$ for all $i = 1, \ldots, n-2$, while proceeding in a similar manner with $k < n-2$ yields (8.4.5). $\qquad\square$

Corollary 8.4.6. Let $A \in \mathbf{H}^n$, and let $A_0 \in \mathbf{H}^k$ be a $k \times k$ principal submatrix of A. Then,

$$\lambda_{\min}(A) \leq \lambda_{\min}(A_0) \leq \lambda_{\max}(A_0) \leq \lambda_{\max}(A) \tag{8.4.6}$$

and

$$\lambda_{\min}(A_0) \leq \lambda_k(A). \tag{8.4.7}$$

The following result compares the maximum and minimum eigenvalues with the maximum and minimum diagonal entries.

Corollary 8.4.7. Let $A \in \mathbf{H}^n$. Then,

$$\lambda_{\min}(A) \leq \mathrm{d}_{\min}(A) \leq \mathrm{d}_{\max}(A) \leq \lambda_{\max}(A). \tag{8.4.8}$$

Lemma 8.4.8. Let $A, B \in \mathbf{H}^n$, and assume that $A \leq B$ and $\mathrm{mspec}(A) = \mathrm{mspec}(B)$. Then, $A = B$.

Proof. Let $\alpha \geq 0$ be such that $0 < \hat{A} \leq \hat{B}$, where $\hat{A} \triangleq A + \alpha I$ and $\hat{B} \triangleq B + \alpha I$. Note that $\mathrm{mspec}(\hat{A}) = \mathrm{mspec}(\hat{B})$, and thus $\det \hat{A} = \det \hat{B}$. Next, it follows that $I \leq \hat{A}^{-1/2}\hat{B}\hat{A}^{-1/2}$. Hence, it follows from $i)$ of Lemma 8.4.1 that $\lambda_{\min}\left(\hat{A}^{-1/2}\hat{B}\hat{A}^{-1/2}\right) \geq 1$. Furthermore, $\det\left(\hat{A}^{-1/2}\hat{B}\hat{A}^{-1/2}\right) = \det \hat{B}/\det \hat{A} = 1$, which implies that $\lambda_i(\hat{A}^{-1/2}\hat{B}\hat{A}^{-1/2}) = 1$ for all $i = 1, \ldots, n$. Hence, $\hat{A}^{-1/2}\hat{B}\hat{A}^{-1/2} = I$, and thus $\hat{A} = \hat{B}$. Hence, $A = B$. $\qquad\square$

The following result is the *monotonicity theorem* or *Weyl's inequality*.

Theorem 8.4.9. Let $A, B \in \mathbf{H}^n$, and assume that $A \leq B$. Then, for all $i = 1, \ldots, n$,

$$\lambda_i(A) \leq \lambda_i(B). \tag{8.4.9}$$

If $A \neq B$, then there exists $i \in \{1, \ldots, n\}$ such that

$$\lambda_i(A) < \lambda_i(B). \tag{8.4.10}$$

If $A < B$, then (8.4.10) holds for all $i = 1, \ldots, n$.

Proof. Since $A \leq B$, it follows from Corollary 8.4.2 that $\lambda_{\min}(A)I \leq A \leq B \leq \lambda_{\max}(B)I$. Hence, it follows from $iii)$ and $i)$ of Lemma 8.4.1 that $\lambda_{\min}(A) \leq \lambda_{\min}(B)$ and $\lambda_{\max}(A) \leq \lambda_{\max}(B)$. Next, let $S \in \mathbb{F}^{n \times n}$ be a unitary matrix such that $SAS^* = \mathrm{diag}[\lambda_1(A), \ldots, \lambda_n(A)]$. Furthermore, for $2 \leq i \leq n-1$, let $A_0 = \mathrm{diag}[\lambda_1(A), \ldots, \lambda_i(A)]$, and let B_0 denote the $i \times i$

leading principal submatrices of SAS^* and SBS^*, respectively. Since $A \leq B$, it follows that $A_0 \leq B_0$, which implies that $\lambda_{\min}(A_0) \leq \lambda_{\min}(B_0)$. It now follows from (8.4.7) that

$$\lambda_i(A) = \lambda_{\min}(A_0) \leq \lambda_{\min}(B_0) \leq \lambda_i(SBS^*) = \lambda_i(B),$$

which proves (8.4.9). If $A \neq B$, then it follows from Lemma 8.4.8 that $\mathrm{mspec}(A) \neq \mathrm{mspec}(B)$ and thus there exists $i \in \{1, \ldots, n\}$ such that (8.4.10) holds. If $A < B$, then $\lambda_{\min}(A_0) < \lambda_{\min}(B_0)$, which implies that (8.4.10) holds for all $i = 1, \ldots, n$. $\square$

Corollary 8.4.10. Let $A, B \in \mathbf{H}^n$. Then, the following statements hold:

   i) If $A \leq B$, then $\mathrm{tr}\, A \leq \mathrm{tr}\, B$.

  ii) If $A \leq B$ and $\mathrm{tr}\, A = \mathrm{tr}\, B$, then $A = B$.

 iii) If $A < B$, then $\mathrm{tr}\, A < \mathrm{tr}\, B$.

 iv) If $0 \leq A \leq B$, then $0 \leq \det A \leq \det B$.

  v) If $0 \leq A < B$, then $0 \leq \det A < \det B$.

 vi) If $0 < A \leq B$ and $\det A = \det B$, then $A = B$.

Proof. Statements *i*), *iii*), *iv*), and *v*) follow from Theorem 8.4.9. To prove *ii*), note that, since $A \leq B$ and $\mathrm{tr}\, A = \mathrm{tr}\, B$, it follows from Theorem 8.4.9 that $\mathrm{mspec}(A) = \mathrm{mspec}(B)$. Now, Lemma 8.4.8 implies that $A = B$. A similar argument yields *vi*). $\square$

The following result, which is a generalization of Theorem 8.4.9, is due to Weyl.

Theorem 8.4.11. Let $A, B \in \mathbf{H}^n$. If $i + j \geq n + 1$, then

$$\lambda_i(A) + \lambda_j(B) \leq \lambda_{i+j-n}(A + B). \tag{8.4.11}$$

If $i + j \leq n + 1$, then

$$\lambda_{i+j-1}(A + B) \leq \lambda_i(A) + \lambda_j(B). \tag{8.4.12}$$

In particular, for all $i = 1, \ldots, n$,

$$\lambda_i(A) + \lambda_{\min}(B) \leq \lambda_i(A + B) \leq \lambda_i(A) + \lambda_{\max}(B), \tag{8.4.13}$$

$$\lambda_{\min}(A) + \lambda_{\min}(B) \leq \lambda_{\min}(A + B) \leq \lambda_{\min}(A) + \lambda_{\max}(B), \tag{8.4.14}$$

$$\lambda_{\max}(A) + \lambda_{\min}(B) \leq \lambda_{\max}(A + B) \leq \lambda_{\max}(A) + \lambda_{\max}(B). \tag{8.4.15}$$

Proof. See [367, p. 182]. $\square$

Lemma 8.4.12. Let $A, B, C \in \mathbf{H}^n$. If $A \leq B$ and C is positive semidefinite, then

$$\operatorname{tr} AC \leq \operatorname{tr} BC. \tag{8.4.16}$$

If $A < B$ and C is positive definite, then

$$\operatorname{tr} AC < \operatorname{tr} BC. \tag{8.4.17}$$

Proof. Since $C^{1/2}AC^{1/2} \leq C^{1/2}BC^{1/2}$, it follows from $i)$ of Corollary 8.4.10 that
$$\operatorname{tr} AC = \operatorname{tr} C^{1/2}AC^{1/2} \leq \operatorname{tr} C^{1/2}BC^{1/2} = \operatorname{tr} BC.$$

Result (8.4.17) follows from $ii)$ of Corollary 8.4.10 in a similar fashion. $\square$

Proposition 8.4.13. Let $A, B \in \mathbb{F}^{n \times n}$, and assume that B is positive semidefinite. Then,

$$\tfrac{1}{2}\lambda_{\min}(A + A^*)\operatorname{tr} B \leq \operatorname{Re} \operatorname{tr} AB \leq \tfrac{1}{2}\lambda_{\max}(A + A^*)\operatorname{tr} B. \tag{8.4.18}$$

If, in addition, A is Hermitian, then

$$\lambda_{\min}(A)\operatorname{tr} B \leq \operatorname{tr} AB \leq \lambda_{\max}(A)\operatorname{tr} B. \tag{8.4.19}$$

Proof. It follows from Corollary 8.4.2 that $\tfrac{1}{2}\lambda_{\min}(A+A^*)I \leq \tfrac{1}{2}(A+A^*)$, while Lemma 8.4.12 implies that $\tfrac{1}{2}\lambda_{\min}(A + A^*)\operatorname{tr} B = \operatorname{tr} \tfrac{1}{2}\lambda_{\min}(A+A^*)IB \leq \operatorname{tr} \tfrac{1}{2}(A + A^*)B = \operatorname{Re}\operatorname{tr} AB$, which proves the left-hand inequality of (8.4.18). Similarly, the right-hand inequality holds. $\square$

For results relating to Proposition 8.4.13, see Fact 5.11.4 and Fact 5.11.6.

Proposition 8.4.14. Let $A, B \in \mathbf{P}^n$, and assume that $\det B = 1$. Then,

$$(\det A)^{1/n} \leq \tfrac{1}{n}\operatorname{tr} AB. \tag{8.4.20}$$

Furthermore, equality holds if and only if $B = (\det A)^{1/n}A^{-1}$.

Proof. Using the arithmetic-mean-geometric-mean inequality given by Fact 1.4.11, it follows that

$$(\det A)^{1/n} = \left(\det B^{1/2}AB^{1/2}\right)^{1/n} = \left[\prod_{i=1}^{n}\lambda_i\left(B^{1/2}AB^{1/2}\right)\right]^{1/n}$$

$$\leq \tfrac{1}{n}\sum_{i=1}^{n}\lambda_i\left(B^{1/2}AB^{1/2}\right) = \tfrac{1}{n}\operatorname{tr} AB.$$

Equality holds if and only if there exists $\beta > 0$ such that $B^{1/2}AB^{1/2} = \beta I$. In this case, $\beta = (\det A)^{1/n}$ and $B = (\det A)^{1/n}A^{-1}$. $\square$

The following corollary of Proposition 8.4.14 is *Minkowski's determinant theorem*.

Corollary 8.4.15. Let $A, B \in \mathbf{N}^n$, and let $p \in [1, n]$. Then,

$$\det A + \det B \leq \left[(\det A)^{1/p} + (\det B)^{1/p} \right]^p \tag{8.4.21}$$

$$\leq \left[(\det A)^{1/n} + (\det B)^{1/n} \right]^n \tag{8.4.22}$$

$$\leq \det(A + B). \tag{8.4.23}$$

If $A = 0$ or $B = 0$ or $\det(A + B) = 0$, then (8.4.21)–(8.4.23) become identities. If there exists $\alpha \geq 0$ such that $B = \alpha A$, then (8.4.23) becomes an identity. Conversely, if $A + B$ is positive definite and (8.4.23) holds as an identity, then there exists $\alpha \geq 0$ such that either $B = \alpha A$ or $A = \alpha B$. Finally, if $p > 1$ or $p < n$, A is positive definite, and (8.4.21), (8.4.22), and (8.4.23) hold as identities, then $B = 0$.

Proof. Inequalities (8.4.21) and (8.4.22) are consequences of the power-sum inequality Fact 1.4.16. Now, assume that $A + B$ is positive definite, since otherwise (8.4.21)–(8.4.23) are identities. To prove (8.4.23), Proposition 8.4.14 implies that

$$(\det A)^{1/n} + (\det B)^{1/n} \leq \tfrac{1}{n} \mathrm{tr} \left[A[\det(A + B)]^{1/n} (A + B)^{-1} \right]$$

$$+ \tfrac{1}{n} \mathrm{tr} \left[B[\det(A + B)]^{1/n} (A + B)^{-1} \right]$$

$$= [\det(A + B)]^{1/n}.$$

Now, suppose that $A + B$ is positive definite and that (8.4.23) holds as an identity. Then, either A or B is positive definite. Hence, suppose that A is positive definite. Multiplying the identity $(\det A)^{1/n} + (\det B)^{1/n} = [\det(A + B)]^{1/n}$ by $(\det A)^{-1/n}$ yields

$$1 + \left(\det A^{-1/2} B A^{-1/2} \right)^{1/n} = \left[\det \left(I + A^{-1/2} B A^{-1/2} \right) \right]^{1/n}.$$

Letting $\lambda_1, \ldots, \lambda_n$ denote the eigenvalues of $A^{-1/2} B A^{-1/2}$, it follows that $1 + (\lambda_1 \cdots \lambda_n)^{1/n} = [(1 + \lambda_1) \cdots (1 + \lambda_n)]^{1/n}$. It now follows from Fact 1.4.15 that $\lambda_1 = \cdots = \lambda_n$.

To prove the last statement, note that, since $\det A > 0$ and $p > 1$ or $p < n$, it follows from Fact 1.4.16 that $\det B = 0$. Hence, $\det A = \det(A+B)$. Since $A \leq A + B$, it follows from $v)$ of Corollary 8.4.10 that $B = 0$. $\qquad\square$

8.5 Matrix Inequalities

Lemma 8.5.1. Let $A, B \in \mathbf{H}^n$, and assume that $0 \leq A \leq B$. Then, $\mathcal{R}(A) \subseteq \mathcal{R}(B)$.

Proof. Let $x \in \mathcal{N}(B)$. Then, $x^*Bx = 0$, and thus $x^*Ax = 0$, which implies that $Ax = 0$. Hence, $\mathcal{N}(B) \subseteq \mathcal{N}(A)$, and thus $\mathcal{N}(A)^\perp \subseteq \mathcal{N}(B)^\perp$. Since A and B are Hermitian, it follows from Theorem 2.4.3 that $\mathcal{R}(A) = \mathcal{N}(A)^\perp$ and $\mathcal{R}(B) = \mathcal{N}(B)^\perp$. Hence, $\mathcal{R}(A) \subseteq \mathcal{R}(B)$. $\square$

The following result is the *Douglas-Fillmore-Williams lemma.*

Theorem 8.5.2. Let $A \in \mathbb{F}^{n \times m}$ and $B \in \mathbb{F}^{n \times l}$. Then, the following statements are equivalent:

i) There exists a matrix $C \in \mathbb{F}^{l \times m}$ such that $A = BC$.

ii) There exists $\alpha > 0$ such that $AA^* \leq \alpha BB^*$.

iii) $\mathcal{R}(A) \subseteq \mathcal{R}(B)$.

Proof. First we prove that *i)* implies *ii)*. Since $A = BC$, it follows that $AA^* = BCC^*B^*$. Since $CC^* \leq \lambda_{\max}(CC^*)I$, it follows that $AA^* \leq \alpha BB^*$, where $\alpha \triangleq \lambda_{\max}(CC^*)$. To prove that *ii)* implies *iii)*, first note that Lemma 8.5.1 implies that $\mathcal{R}(AA^*) \subseteq \mathcal{R}(\alpha BB^*) = \mathcal{R}(BB^*)$. Since, by Theorem 2.4.3, $\mathcal{R}(AA^*) = \mathcal{R}(A)$ and $\mathcal{R}(BB^*) = \mathcal{R}(B)$, it follows that $\mathcal{R}(A) \subseteq \mathcal{R}(B)$. Finally, to prove that *iii)* implies *i)*, use Theorem 5.6.3 to write $B = S_1 \begin{bmatrix} D & 0 \\ 0 & 0 \end{bmatrix} S_2$, where $S_1 \in \mathbb{F}^{n \times n}$ and $S_2 \in \mathbb{F}^{l \times l}$ are unitary and $D \in \mathbb{R}^{r \times r}$ is diagonal with positive diagonal entries, where $r \triangleq \operatorname{rank} B$. Since $\mathcal{R}(S_1^*A) \subseteq \mathcal{R}(S_1^*B)$ and $S_1^*B = \begin{bmatrix} D & 0 \\ 0 & 0 \end{bmatrix} S_2$, it follows that $S_1^*A = \begin{bmatrix} A_1 \\ 0 \end{bmatrix}$, where $A_1 \in \mathbb{F}^{r \times m}$. Consequently,

$$A = S_1 \begin{bmatrix} A_1 \\ 0 \end{bmatrix} = S_1 \begin{bmatrix} D & 0 \\ 0 & 0 \end{bmatrix} S_2 S_2^* \begin{bmatrix} D^{-1} & 0 \\ 0 & 0 \end{bmatrix} \begin{bmatrix} A_1 \\ 0 \end{bmatrix} = BC,$$

where $C \triangleq S_2^* \begin{bmatrix} D^{-1} & 0 \\ 0 & 0 \end{bmatrix} \begin{bmatrix} A_1 \\ 0 \end{bmatrix} \in \mathbb{F}^{l \times m}$. $\square$

Proposition 8.5.3. Let $\{A_i\}_{i=1}^\infty \subset \mathbf{N}^n$ satisfy $0 \leq A_i \leq A_j$ for all $i \leq j$, and assume there exists $B \in \mathbf{N}^n$ satisfying $A_i \leq B$ for all $i \in \mathbb{P}$. Then, $A \triangleq \lim_{i \to \infty} A_i$ exists and satisfies $0 \leq A \leq B$.

Proof. Let $k \in \{1, \ldots, n\}$. Then, the sequence $\{A_{i(k,k)}\}_{i=1}^\infty$ is nondecreasing and bounded from above. Hence, $A_{(k,k)} \triangleq \lim_{i \to \infty} A_{i(k,k)}$ exists. Now, let $k, l \in \{1, \ldots, n\}$, where $k \neq l$. Since $A_i \leq A_j$ for all $i < j$, it follows that $(e_k + e_l)^*A_i(e_k + e_l) \leq (e_k + e_l)^*A_j(e_k + e_l)$, which implies that $A_{i(k,l)} - A_{j(k,l)} \leq \frac{1}{2}[A_{j(k,k)} - A_{i(k,k)} + A_{j(l,l)} - A_{i(l,l)}]$. Alternatively, replacing $e_k + e_l$ by $e_k - e_l$ yields $A_{j(k,l)} - A_{i(k,l)} \leq \frac{1}{2}[A_{j(k,k)} - A_{i(k,k)} + A_{j(l,l)} - A_{i(l,l)}]$. Thus, $A_{i(k,l)} - A_{j(k,l)} \to 0$ as $i, j \to \infty$, which implies that $A_{(k,l)} \triangleq \lim_{i \to \infty} A_{i(k,l)}$

exists. Hence, $A \triangleq \lim_{i \to \infty} A_i$ exists. Since $A_i \leq B$ for all $i = 1, 2, \ldots$, it follows that $A \leq B$. $\square$

Let $A = SBS^* \in \mathbb{F}^{n \times n}$ be Hermitian, where $S \in \mathbb{F}^{n \times n}$ is unitary, $B \in \mathbb{R}^{n \times n}$ is diagonal, $\mathrm{spec}(A) \subset \mathcal{D}$, and $\mathcal{D} \subseteq \mathbb{R}$. Furthermore, let $f \colon \mathcal{D} \mapsto \mathbb{R}$. Then, we define $f(A) \in \mathbf{H}^n$ by

$$f(A) \triangleq S f(B) S^*, \tag{8.5.1}$$

where $[f(B)]_{(i,i)} \triangleq f[B_{(i,i)}]$. Hence, with an obvious extension of notation, $f \colon \{X \in \mathbf{H}^n \colon \mathrm{spec}(X) \subset \mathcal{D}\} \mapsto \mathbf{H}^n$. If f is one-to-one, then its inverse $f^{-1} \colon \{X \in \mathbf{H}^n \colon \mathrm{spec}(X) \subset f(\mathcal{D})\} \mapsto \mathbf{H}^n$ exists.

Suppose that $A \in \mathbb{F}^{n \times n}$ is positive semidefinite. Then, for all $r \geq 0$ (not necessarily an integer), $A^r = SB^rS^*$ is positive semidefinite, where, for all $i = 1, \ldots, n$, $(B^r)_{(i,i)} = [B_{(i,i)}]^r$. Note that $A^0 \triangleq I$. In particular,

$$A^{1/2} = SB^{1/2}S^* \tag{8.5.2}$$

is the unique positive-semidefinite *square root* of A since

$$A^{1/2}A^{1/2} = SB^{1/2}S^*SB^{1/2}S^* = SBS^* = A. \tag{8.5.3}$$

The uniqueness of $A^{1/2}$ follows from the fact that the square root function is one-to-one and onto the range $(0, \infty)$; see [460] and [367, p. 405]. Hence, if $C \in \mathbb{F}^{n \times m}$, then C^*C is positive semidefinite, and we define

$$\langle C \rangle \triangleq (C^*C)^{1/2}. \tag{8.5.4}$$

Now, suppose that A is positive definite. Then, A^r is positive definite for all $r \in \mathbb{R}$, and, if $r \neq 0$, then $(A^r)^{1/r} = A$. Furthermore,

$$\log A = S(\log B)S^* \in \mathbf{H}^n, \tag{8.5.5}$$

where, for all $i = 1, \ldots, n$, $(\log B)_{(i,i)} = \log B_{(i,i)}$.

Lemma 8.5.4. Let $A \in \mathbf{P}^n$. If $A \leq I$, then $I \leq A^{-1}$. Furthermore, if $A < I$, then $I < A^{-1}$.

Proof. Since $A \leq I$, it follows from $xi)$ of Proposition 8.1.2 that $I = A^{-1/2}AA^{-1/2} \leq A^{-1/2}IA^{-1/2} = A^{-1}$. Similarly, $A < I$ implies that $I = A^{-1/2}AA^{-1/2} < A^{-1/2}IA^{-1/2} = A^{-1}$. $\square$

Proposition 8.5.5. Let $A, B \in \mathbf{H}^n$, and assume that either A and B are positive definite or A and B are negative definite. If $A \leq B$, then $B^{-1} \leq A^{-1}$. If, in addition, $A < B$, then $B^{-1} < A^{-1}$.

Proof. Assume that A and B are positive definite. Since $A \leq B$, it follows that $B^{-1/2}AB^{-1/2} \leq I$. Now, Lemma 8.5.4 implies that $I \leq B^{1/2}A^{-1}B^{1/2}$, which implies that $B^{-1} \leq A^{-1}$. If A and B are negative definite, then $A \leq B$

is equivalent to $-B \leq -A$. The case $A < B$ is proved in a similar manner. □

The following result is *Furuta's inequality.*

Proposition 8.5.6. Let $A, B \in \mathbf{N}^n$, and assume that $0 \leq A \leq B$. Furthermore, let $p, q, r \in \mathbb{R}$ satisfy $p \geq 0$, $q \geq 1$, $r \geq 0$, and $p+2r \leq (1+2r)q$. Then,

$$A^{(p+2r)/q} \leq (A^r B^p A^r)^{1/q} \tag{8.5.6}$$

and

$$(B^r A^p B^r)^{1/q} \leq B^{(p+2r)/q}. \tag{8.5.7}$$

Proof. See [279] or [282, pp. 129, 130]. □

Corollary 8.5.7. Let $A, B \in \mathbf{N}^n$, and assume that $0 \leq A \leq B$. Then,

$$A^2 \leq \left(AB^2A\right)^{1/2} \tag{8.5.8}$$

and

$$\left(BA^2B\right)^{1/2} \leq B^2. \tag{8.5.9}$$

Proof. In Proposition 8.5.6 set $r = 1$, $p = 2$, and $q = 2$. □

Corollary 8.5.8. Let $A, B, C \in \mathbf{N}^n$, and assume that $0 \leq A \leq C \leq B$. Then,

$$\left(CA^2C\right)^{1/2} \leq C^2 \leq \left(CB^2C\right)^{1/2}. \tag{8.5.10}$$

Proof. The result follows from Corollary 8.5.7. See also [756]. □

The following result provides representations for A^r, where $r \in [0, 1)$.

Proposition 8.5.9. Let $A \in \mathbf{P}^n$ and $r \in (0, 1)$. Then,

$$A^r = \left(\cos \frac{r\pi}{2}\right)I + \frac{\sin r\pi}{\pi} \int_0^\infty \left[\frac{x^{r+1}}{1+x^2}I - (A + xI)^{-1}x^r\right] dx \tag{8.5.11}$$

and

$$A^r = \frac{\sin r\pi}{\pi} \int_0^\infty (A + xI)^{-1}Ax^{r-1} \, dx. \tag{8.5.12}$$

Proof. Let $t \geq 0$. As shown in [108], [111, p. 143],

$$\int_0^\infty \left[\frac{x^{r+1}}{1+x^2} - \frac{x^r}{t+x}\right] dx = \frac{\pi}{\sin r\pi}\left(t^r - \cos \frac{r\pi}{2}\right).$$

Solving for t^r and replacing t by A yields (8.5.11). Likewise, it follows from [820, p. 448, formula 589] that

$$\int\limits_0^\infty \frac{tx^{r-1}}{t+x}\,\mathrm{d}x = \frac{t^r\pi}{\sin r\pi}.$$

Replacing t by A yields (8.5.12). □

The following result is the *Löwner-Heinz inequality*.

Corollary 8.5.10. Let $A, B \in \mathbf{N}^n$, assume that $0 \le A \le B$, and let $r \in [0,1]$. Then, $A^r \le B^r$. If, in addition, $A < B$ and $r \in (0,1]$, then $A^r < B^r$.

Proof. Let $0 < A \le B$, and let $r \in (0,1)$. In Proposition 8.5.6, replace p, q, r with $r, 1, 0$. The first result now follows from (8.5.6). Alternatively, it follows from (8.5.11) of Proposition 8.5.9 that

$$B^r - A^r = \frac{\sin r\pi}{\pi}\int\limits_0^\infty [(A+xI)^{-1} - (B+xI)^{-1}]x^r\,\mathrm{d}x.$$

Since $A \le B$, it follows from Proposition 8.5.5 that, for all $x \ge 0$, $(B + xI)^{-1} \le (A + xI)^{-1}$. Hence, $A^r \le B^r$. By continuity, the result holds for $A, B \in \mathbf{N}^n$ and $r \in [0,1]$. In the case $A < B$, it follows from Proposition 8.5.5 that, for all $x \ge 0$, $(B + xI)^{-1} < (A + xI)^{-1}$, so that $A^r < B^r$.

Alternatively, it follows from (8.5.12) of Proposition 8.5.9 that

$$B^r - A^r = \frac{\sin r\pi}{\pi}\int\limits_0^\infty [(A+xI)^{-1}A - (B+xI)^{-1}B]x^{r-1}\,\mathrm{d}x.$$

Since $A \le B$, it follows that, for all $x \ge 0$, $(B+xI)^{-1}B \le (A+xI)^{-1}A$. Hence, $A^r \le B^r$. Alternative proofs are given in [806, p. 2] and [282, p. 127].

For the case $r = 1/2$, let $\lambda \in \mathbb{R}$ be an eigenvalue of $B^{1/2} - A^{1/2}$, and let $x \in \mathbb{F}^n$ be an associated eigenvector. Then,

$$\lambda x^*\left(B^{1/2} + A^{1/2}\right)x = x^*\left(B^{1/2} + A^{1/2}\right)\left(B^{1/2} - A^{1/2}\right)x$$
$$= x^*\left(B - B^{1/2}A^{1/2} + A^{1/2}B^{1/2} - A\right)$$
$$= x^*(B-A)x \ge 0.$$

Since $B^{1/2} + A^{1/2}$ is positive semidefinite, it follows that either $\lambda \ge 0$ or $x^*\left(B^{1/2} + A^{1/2}\right)x = 0$. In the latter case, $B^{1/2}x = A^{1/2}x = 0$, which implies that $\lambda = 0$. □

The Löwner-Heinz inequality does not extend to $r > 1$. In fact, $A \triangleq \begin{bmatrix} 2 & 1 \\ 1 & 1 \end{bmatrix}$ and $B \triangleq \begin{bmatrix} 1 & 0 \\ 0 & 0 \end{bmatrix}$ satisfy $A \geq B \geq 0$, whereas, for all $r > 1$, $A^r \not\geq B^r$. For details, see [282, pp. 127, 128].

Many of the results already given involve functions that are nondecreasing or increasing on suitable sets of matrices.

Definition 8.5.11. Let $\mathcal{D} \subseteq \mathbf{H}^n$, and let $\phi \colon \mathcal{D} \mapsto \mathbf{H}^m$. The function ϕ is *nondecreasing* if $\phi(A) \leq \phi(B)$ for all $A, B \in \mathcal{D}$ such that $A \leq B$, it is *increasing* if it is nondecreasing and $\phi(A) < \phi(B)$ for all $A, B \in \mathcal{D}$ such that $A < B$, and it is *strongly increasing* if it is nondecreasing and $\phi(A) < \phi(B)$ for all $A, B \in \mathcal{D}$ such that $A \leq B$ and $A \neq B$. The function ϕ is (*nonincreasing, decreasing, strongly decreasing*) if $-\phi$ is (nondecreasing, increasing, strongly increasing).

Proposition 8.5.12. The following functions are nondecreasing:

  i) $\phi \colon \mathbf{H}^n \mapsto \mathbf{H}^m$ defined by $\phi(A) \triangleq BAB^*$, where $B \in \mathbb{F}^{m \times n}$.

  ii) $\phi \colon \mathbf{H}^n \mapsto \mathbb{R}$ defined by $\phi(A) \triangleq \operatorname{tr} AB$, where $B \in \mathbf{N}^n$.

  iii) $\phi \colon \mathbf{N}^{n+m} \mapsto \mathbf{N}^n$ defined by $\phi(A) \triangleq A_{22}|A$, where $A \triangleq \begin{bmatrix} A_{11} & A_{12} \\ A_{12}^* & A_{22} \end{bmatrix}$.

The following functions are increasing:

  iv) $\phi \colon \mathbf{H}^n \mapsto \mathbb{R}$ defined by $\phi(A) \triangleq \lambda_i(A)$, where $i \in \{1, \ldots, n\}$.

  v) $\phi \colon \mathbf{N}^n \mapsto \mathbf{N}^n$ defined by $\phi(A) \triangleq A^r$, where $r \in [0, 1]$.

  vi) $\phi \colon \mathbf{N}^n \mapsto \mathbf{N}^n$ defined by $\phi(A) \triangleq A^{1/2}$.

  vii) $\phi \colon \mathbf{P}^n \mapsto -\mathbf{P}^n$ defined by $\phi(A) \triangleq -A^{-r}$, where $r \in [0, 1]$.

  viii) $\phi \colon \mathbf{P}^n \mapsto -\mathbf{P}^n$ defined by $\phi(A) \triangleq -A^{-1}$.

  ix) $\phi \colon \mathbf{P}^n \mapsto -\mathbf{P}^n$ defined by $\phi(A) \triangleq -A^{-1/2}$.

  x) $\phi \colon -\mathbf{P}^n \mapsto \mathbf{P}^n$ defined by $\phi(A) \triangleq (-A)^{-r}$, where $r \in [0, 1]$.

  xi) $\phi \colon -\mathbf{P}^n \mapsto \mathbf{P}^n$ defined by $\phi(A) \triangleq -A^{-1}$.

  xii) $\phi \colon -\mathbf{P}^n \mapsto \mathbf{P}^n$ defined by $\phi(A) \triangleq -A^{-1/2}$.

  xiii) $\phi \colon \mathbf{H}^n \mapsto \mathbf{H}^m$ defined by $\phi(A) \triangleq BAB^*$, where $B \in \mathbb{F}^{m \times n}$ and $\operatorname{rank} B = m$.

  xiv) $\phi \colon \mathbf{P}^{n+m} \mapsto \mathbf{P}^n$ defined by $\phi(A) \triangleq A_{22}|A$, where $A \triangleq \begin{bmatrix} A_{11} & A_{12} \\ A_{12}^* & A_{22} \end{bmatrix}$.

  xv) $\phi \colon \mathbf{P}^{n+m} \mapsto \mathbf{P}^n$ defined by $\phi(A) \triangleq -(A_{22}|A)^{-1}$, where $A \triangleq \begin{bmatrix} A_{11} & A_{12} \\ A_{12}^* & A_{22} \end{bmatrix}$.

  xvi) $\phi \colon \mathbf{P}^n \mapsto \mathbf{H}^n$ defined by $\phi(A) \triangleq \log A$.

The following functions are strongly increasing:

xvii) ϕ: $\mathbf{H}^n \mapsto [0, \infty)$ defined by $\phi(A) \triangleq \operatorname{tr} BAB^*$, where $B \in \mathbb{F}^{m \times n}$ and rank $B = m$.

xviii) ϕ: $\mathbf{H}^n \mapsto \mathbb{R}$ defined by $\phi(A) \triangleq \operatorname{tr} AB$, where $B \in \mathbf{P}^n$.

xix) ϕ: $\mathbf{N}^n \mapsto [0, \infty)$ defined by $\phi(A) \triangleq \det A$.

Proof. For the proof of *iii)*, see [479]. To prove *xvi)*, let $A, B \in \mathbf{P}^n$, and assume that $A \leq B$. Then, for all $r \in [0, 1]$, it follows from *v)* that $r^{-1}(A^r - I) \leq r^{-1}(B^r - I)$. Letting $r \to 0^+$ yields $\log A \leq \log B$, which proves that log is nondecreasing. See [282, p. 139]. To prove that log is increasing, assume that $A < B$, and let $\varepsilon > 0$ be such that $A + \varepsilon I < B$. Then, it follows that $\log A < \log(A + \varepsilon I) \leq \log B$. $\square$

Finally, we consider convex functions defined with respect to matrix inequalities.

Definition 8.5.13. Let $\mathcal{D} \subseteq \mathbb{F}^{n \times m}$ be a convex set, and let ϕ: $\mathcal{D} \mapsto \mathbf{H}^p$. The function ϕ is *convex* if

$$\phi[\alpha A_1 + (1 - \alpha)A_2] \leq \alpha\phi(A_1) + (1 - \alpha)\phi(A_2) \tag{8.5.13}$$

for all $\alpha \in [0, 1]$ and $A_1, A_2 \in \mathcal{D}$. The function ϕ is *concave* if $-\phi$ is convex.

Lemma 8.5.14. Let $\mathcal{D} \subseteq \mathbb{F}^{n \times m}$ and $\mathcal{S} \subseteq \mathbf{H}^p$ be convex sets, and let ϕ_1: $\mathcal{D} \mapsto \mathcal{S}$ and ϕ_2: $\mathcal{S} \mapsto \mathbf{H}^q$. Then, the following statements hold:

i) If ϕ_1 is convex and ϕ_2 is nondecreasing and convex, then $\phi_2 \bullet \phi_1$: $\mathcal{D} \mapsto \mathbf{H}^q$ is convex.

ii) If ϕ_1 is concave and ϕ_2 is nonincreasing and convex, then $\phi_2 \bullet \phi_1$: $\mathcal{D} \mapsto \mathbf{H}^q$ is convex.

iii) If $\mathcal{S}$ is symmetric, $\phi_2(-A) = -\phi_2(A)$ for all $A \in \mathcal{S}$, ϕ_1 is concave, and ϕ_2 is nonincreasing and concave, then $\phi_2 \bullet \phi_1$: $\mathcal{D} \mapsto \mathbf{H}^q$ is convex.

iv) If $\mathcal{S}$ is symmetric, $\phi_2(-A) = -\phi_2(A)$ for all $A \in \mathcal{S}$, ϕ_1 is convex, and ϕ_2 is nondecreasing and concave, then $\phi_2 \bullet \phi_1$: $\mathcal{D} \mapsto \mathbf{H}^q$ is convex.

Proof. To prove *i)* and *ii)*, let $\alpha \in [0, 1]$ and $A_1, A_2 \in \mathcal{D}$. In both cases it follows that

$$\phi_2(\phi_1[\alpha A_1 + (1 - \alpha)A_2]) \leq \phi_2[\alpha\phi_1(A_1) + (1 - \alpha)\phi_1(A_2)]$$
$$\leq \alpha\phi_2[\phi_1(A_1)] + (1 - \alpha)\phi_2[\phi_1(A_2)].$$

Statements *iii)* and *iv)* follow from *i)* and *ii)*, respectively. $\square$

Proposition 8.5.15. The following functions are convex:

$i)$ ϕ: $\mathbf{N}^n \mapsto \mathbf{N}^n$ defined by $\phi(A) \triangleq A^r$, where $r \in [1, 2]$.

$ii)$ ϕ: $\mathbf{N}^n \mapsto \mathbf{N}^n$ defined by $\phi(A) \triangleq A^2$.

$iii)$ ϕ: $\mathbf{P}^n \mapsto \mathbf{P}^n$ defined by $\phi(A) \triangleq A^{-r}$, where $r \in [0, 1]$.

$iv)$ ϕ: $\mathbf{P}^n \mapsto \mathbf{P}^n$ defined by $\phi(A) \triangleq A^{-1}$.

$v)$ ϕ: $\mathbf{P}^n \mapsto \mathbf{P}^n$ defined by $\phi(A) \triangleq A^{-1/2}$.

$vi)$ ϕ: $\mathbf{N}^n \mapsto -\mathbf{N}^n$ defined by $\phi(A) \triangleq -A^r$, where $r \in [0, 1]$.

$vii)$ ϕ: $\mathbf{N}^n \mapsto -\mathbf{N}^n$ defined by $\phi(A) \triangleq -A^{1/2}$.

$viii)$ ϕ: $\mathbf{N}^n \mapsto \mathbf{H}^m$ defined by $\phi(A) \triangleq \gamma BAB^*$, where $\gamma \in \mathbb{R}$ and $B \in \mathbb{F}^{m \times n}$.

$ix)$ ϕ: $\mathbf{N}^n \mapsto \mathbf{N}^m$ defined by $\phi(A) \triangleq BA^r B^*$, where $B \in \mathbb{F}^{m \times n}$ and $r \in [1, 2]$.

$x)$ ϕ: $\mathbf{P}^n \mapsto \mathbf{N}^m$ defined by $\phi(A) \triangleq BA^{-r}B^*$, where $B \in \mathbb{F}^{m \times n}$ and $r \in [0, 1]$.

$xi)$ ϕ: $\mathbf{N}^n \mapsto -\mathbf{N}^m$ defined by $\phi(A) \triangleq -BA^r B^*$, where $B \in \mathbb{F}^{m \times n}$ and $r \in [0, 1]$.

$xii)$ ϕ: $\mathbf{P}^n \mapsto -\mathbf{P}^m$ defined by $\phi(A) \triangleq -(BA^{-r}B^*)^{-p}$, where $B \in \mathbb{F}^{m \times n}$ has rank m and $r, p \in [0, 1]$.

$xiii)$ ϕ: $\mathbb{F}^{n \times m} \mapsto \mathbf{N}^n$ defined by $\phi(A) \triangleq ABA^*$, where $B \in \mathbf{N}^m$.

$xiv)$ ϕ: $\mathbf{P}^n \times \mathbb{F}^{m \times n} \mapsto \mathbf{N}^m$ defined by $\phi(A, B) \triangleq BA^{-1}B^*$.

$xv)$ ϕ: $\mathbf{N}^{n+m} \mapsto \mathbf{N}^n$ defined by $\phi(A) \triangleq -A_{22}|A$, where $A \triangleq \begin{bmatrix} A_{11} & A_{12} \\ A_{12}^* & A_{22} \end{bmatrix}$.

$xvi)$ ϕ: $\mathbf{P}^{n+m} \mapsto \mathbf{P}^n$ defined by $\phi(A) \triangleq (A_{22}|A)^{-1}$, where $A \triangleq \begin{bmatrix} A_{11} & A_{12} \\ A_{12}^* & A_{22} \end{bmatrix}$.

$xvii)$ ϕ: $\mathbf{H}^n \mapsto [0, \infty)$ defined by $\phi(A) \triangleq \operatorname{tr} A^k$, where k is a nonnegative even integer.

$xviii)$ ϕ: $\mathbf{P}^n \mapsto (0, \infty)$ defined by $\phi(A) \triangleq \operatorname{tr} A^{-r}$, where $r > 0$.

$xix)$ ϕ: $\mathbf{P}^n \mapsto (-\infty, 0)$ defined by $\phi(A) \triangleq -(\operatorname{tr} A^{-r})^{-p}$, where $r, p \in [0, 1]$.

$xx)$ ϕ: $\mathbf{N}^n \times \mathbf{N}^n \mapsto (-\infty, 0]$ defined by $\phi(A, B) \triangleq -\operatorname{tr}(A^r + B^r)^{1/r}$, where $r \in [0, 1]$.

$xxi)$ ϕ: $\mathbf{N}^n \times \mathbf{N}^n \mapsto [0, \infty)$ defined by $\phi(A, B) \triangleq \operatorname{tr}(A^2 + B^2)^{1/2}$.

$xxii)$ ϕ: $\mathbf{N}^n \times \mathbf{N}^m \mapsto \mathbb{R}$ defined by $\phi(A, B) \triangleq -\operatorname{tr} A^r X B^p X^*$, where $X \in \mathbb{F}^{n \times m}$, $r, p \geq 0$, and $r + p \leq 1$.

xxiii) ϕ: $\mathbf{N}^n \mapsto (-\infty, 0)$ defined by $\phi(A) \triangleq -\operatorname{tr} A^r X A^p X^*$, where $X \in \mathbb{F}^{n \times n}$, $r, p \geq 0$, and $r + p \leq 1$.

xxiv) ϕ: $\mathbf{P}^n \times \mathbf{P}^m \times \mathbb{F}^{m \times n} \mapsto \mathbb{R}$ defined by $\phi(A, B, X) \triangleq (\operatorname{tr} A^{-p} X B^{-r} X^*)^q$, where $r, p \geq 0$, $r + p \leq 1$, and $q \geq (2 - r - p)^{-1}$.

xxv) ϕ: $\mathbf{P}^n \times \mathbb{F}^{n \times n} \mapsto [0, \infty)$ defined by $\phi(A, X) \triangleq \operatorname{tr} A^{-p} X A^{-r} X^*$, where $r, p \geq 0$ and $r + p \leq 1$.

xxvi) ϕ: $\mathbf{P}^n \times \mathbb{F}^{n \times n} \mapsto [0, \infty)$ defined by $\phi(A) \triangleq \operatorname{tr} A^{-p} X A^{-r} X^*$, where $r, p \in [0, 1]$ and $X \in \mathbb{F}^{n \times n}$.

xxvii) ϕ: $\mathbf{P}^n \mapsto \mathbb{R}$ defined by $\phi(A) \triangleq \operatorname{tr}([A^r, X][A^{1-r}, X])$, where $X \in \mathbf{H}^n$.

xxviii) ϕ: $\mathbf{P}^n \mapsto \mathbf{H}^m$ defined by $\phi(A) \triangleq A \log A$.

xxix) ϕ: $\mathbf{N}^n \backslash \{0\} \mapsto \mathbb{R}$ defined by $\phi(A) \triangleq -\log \operatorname{tr} A^r$, where $r \in [0, 1]$.

xxx) ϕ: $\mathbf{P}^n \times \mathbf{P}^n \mapsto (0, \infty)$ defined by $\phi(A, B) \triangleq \operatorname{tr}[A(\log A - \log B)]$.

xxxi) ϕ: $\mathbf{N}^n \mapsto (-\infty, 0]$ defined by $\phi(A) \triangleq -(\det A)^{1/n}$.

xxxii) ϕ: $\mathbf{P}^n \mapsto \mathbb{R}$ defined by $\phi(A) \triangleq -\log \det A$.

xxxiii) ϕ: $\mathbf{P}^n \mapsto (0, \infty)$ defined by $\phi(A) \triangleq \det A^{-1}$.

xxxiv) ϕ: $\mathbf{N}^n \times \mathbf{N}^m \mapsto -\mathbf{N}^{nm}$ defined by $\phi(A, B) \triangleq -A^r \otimes B^{1-r}$, where $r \in [0, 1]$.

xxxv) ϕ: $\mathbf{N}^n \times \mathbf{N}^n \mapsto -\mathbf{N}^n$ defined by $\phi(A, B) \triangleq -A^r \circ B^{1-r}$, where $r \in [0, 1]$.

xxxvi) ϕ: $\mathbf{H}^n \mapsto \mathbb{R}$ defined by $\phi(A) \triangleq \sum_{i=1}^k \lambda_i(A)$, where $k \in \{1, \ldots, n\}$.

xxxvii) ϕ: $\mathbf{H}^n \mapsto \mathbb{R}$ defined by $\phi(A) \triangleq -\sum_{i=k}^n \lambda_i(A)$, where $k \in \{1, \ldots, n\}$.

Proof. Statements *i)* and *iii)* are proved in [27] and [111, p. 123].

Let $\alpha \in [0, 1]$ for the remainder of the proof.

To prove *ii)* directly, let $A_1, A_2 \in \mathbf{H}^n$. Since

$$\alpha(1 - \alpha) = (\alpha - \alpha^2)^{1/2}[(1 - \alpha) - (1 - \alpha)^2]^{1/2},$$

it follows that

$$0 \leq \left[(\alpha - \alpha^2)^{1/2} A_1 - [(1 - \alpha) - (1 - \alpha)^2]^{1/2} A_2\right]^2$$

$$= (\alpha - \alpha^2) A_1^2 + [(1 - \alpha) - (1 - \alpha)^2] A_2^2 - \alpha(1 - \alpha)(A_1 A_2 + A_2 A_1).$$

Hence,

$$[\alpha A_1 + (1 - \alpha) A_2]^2 \leq \alpha A_1^2 + (1 - \alpha) A_2^2,$$

which shows that $\phi(A) = A^2$ is convex.

To prove iv) directly, let $A_1, A_2 \in \mathbf{P}^n$. Then, $\begin{bmatrix} A_1^{-1} & I \\ I & A_1 \end{bmatrix}$ and $\begin{bmatrix} A_2^{-1} & I \\ I & A_2 \end{bmatrix}$ are positive semidefinite, and thus

$$\alpha \begin{bmatrix} A_1^{-1} & I \\ I & A_1 \end{bmatrix} + (1 - \alpha) \begin{bmatrix} A_2^{-1} & I \\ I & A_2 \end{bmatrix}$$

$$= \begin{bmatrix} \alpha A_1^{-1} + (1 - \alpha) A_2^{-1} & I \\ I & \alpha A_1 + (1 - \alpha) A_2 \end{bmatrix}$$

is positive semidefinite. It now follows from Proposition 8.2.3 that $[\alpha A_1 + (1 - \alpha) A_2]^{-1} \leq \alpha A_1^{-1} + (1 - \alpha) A_2^{-1}$, which shows that $\phi(A) = A^{-1}$ is convex.

To prove v) directly, note that $\phi(A) = A^{-1/2} = \phi_2[\phi_1(A)]$, where $\phi_1(A) \triangleq A^{1/2}$ and $\phi_2(B) \triangleq B^{-1}$. It follows from vii) that ϕ_1 is concave, while it follows from iv) that ϕ_2 is convex. Furthermore, $viii$) of Proposition 8.5.12 implies that ϕ_2 is nonincreasing. It thus follows from ii) of Lemma 8.5.14 that $\phi(A) = A^{-1/2}$ is convex.

To prove vi), let $A \in \mathbf{P}^n$ and note that $\phi(A) = -A^r = \phi_2[\phi_1(A)]$, where $\phi_1(A) \triangleq A^{-r}$ and $\phi_2(B) \triangleq -B^{-1}$. It follows from iii) that ϕ_1 is convex, while it follows from iv) that ϕ_2 is concave. Furthermore, $viii$) of Proposition 8.5.12 implies that ϕ_2 is nondecreasing. It thus follows from iv) of Lemma 8.5.14 that $\phi(A) = A^r$ is convex on $\mathbf{P}^n$. Continuity implies that $\phi(A) = A^r$ is convex on $\mathbf{N}^n$.

To prove vii) directly, let $A_1, A_2 \in \mathbf{N}^n$. Then,

$$0 \leq \alpha(1 - \alpha)\left(A_1^{1/2} - A_2^{1/2}\right)^2,$$

which is equivalent to

$$\left[\alpha A_1^{1/2} + (1 - \alpha) A_2^{1/2}\right]^2 \leq \alpha A_1 + (1 - \alpha) A_2.$$

Using vi) of Proposition 8.5.12 yields

$$\alpha A_1^{1/2} + (1 - \alpha) A_2^{1/2} \leq [\alpha A_1 + (1 - \alpha) A_2]^{1/2}.$$

Finally, multiplying by -1 shows that $\phi(A) = -A^{1/2}$ is convex.

The proof of $viii$) is immediate. Statements ix), x), and xi) follow from i), iii), and vi), respectively.

To prove xii), note that $\phi(A) = -(BA^{-r}B^*)^{-p} = \phi_2[\phi_1(A)]$, where $\phi_1(A) = -BA^{-r}B^*$ and $\phi_2(C) = C^{-p}$. Statement x) implies that ϕ_1 is concave, while iii) implies that ϕ_2 is convex. Furthermore, vii) of Proposition 8.5.12 implies that ϕ_2 is nonincreasing. It thus follows from ii) of Lemma 8.5.14 that $\phi(A) = -(BA^{-r}B^*)^{-p}$ is convex.

To prove $xiii)$, let $A_1, A_2 \in \mathbb{F}^{n \times m}$, and let $B \in \mathbf{N}^m$. Then,

$$0 \leq \alpha(1-\alpha)(A_1 - A_2)B(A_1 - A_2)^*$$
$$= \alpha A_1 B A_1^* + (1-\alpha)A_2 B A_2^* - [\alpha A_1 + (1-\alpha)A_2]B[\alpha A_1 + (1-\alpha)A_2]^*.$$

Thus,

$$[\alpha A_1 + (1-\alpha)A_2]B[\alpha A_1 + (1-\alpha)A_2]^* \leq \alpha A_1 B A_1^* + (1-\alpha)A_2 B A_2^*,$$

which shows that $\phi(A) = ABA^*$ is convex.

To prove $xiv)$, let $A_1, A_2 \in \mathbf{P}^n$ and $B_1, B_2 \in \mathbb{F}^{m \times n}$. Then, it follows from Proposition 8.2.3 that $\begin{bmatrix} B_1 A_1^{-1} B_1^* & B_1 \\ B_1^* & A_1 \end{bmatrix}$ and $\begin{bmatrix} B_2 A_2^{-1} B_2^* & B_2 \\ B_2^* & A_2 \end{bmatrix}$ are positive semidefinite, and thus

$$\alpha \begin{bmatrix} B_1 A_1^{-1} B_1^* & B_1 \\ B_1^* & A_1 \end{bmatrix} + (1-\alpha) \begin{bmatrix} B_2 A_2^{-1} B_2^* & B_2 \\ B_2^* & A_2 \end{bmatrix}$$
$$= \begin{bmatrix} \alpha B_1 A_1^{-1} B_1^* + (1-\alpha)B_2 A_2^{-1} B_2^* & \alpha B_1 + (1-\alpha)B_2 \\ \alpha B_1^* + (1-\alpha)B_2^* & \alpha A_1 + (1-\alpha)A_2 \end{bmatrix}$$

is positive semidefinite. It thus follows from Proposition 8.2.3 that

$$[\alpha B_1 + (1-\alpha)B_2][\alpha A_1 + (1-\alpha)A_2]^{-1}[\alpha B_1 + (1-\alpha)B_2]^*$$
$$\leq \alpha B_1 A_1^{-1} B_1^* + (1-\alpha)B_2 A_2^{-1} B_2^*,$$

which shows that $\phi(A, B) = BA^{-1}B^*$ is convex.

To prove $xv)$, let $A \triangleq \begin{bmatrix} A_{11} & A_{12} \\ A_{12}^* & A_{22} \end{bmatrix} \in \mathbf{P}^{n+m}$ and $B \triangleq \begin{bmatrix} B_{11} & B_{12} \\ B_{12}^* & B_{22} \end{bmatrix} \in \mathbf{P}^{n+m}$. Then, it follows from $xiv)$ with A_1, B_1, A_2, B_2 replaced by $A_{22}, A_{12}, B_{22}, B_{12}$, respectively, that

$$[\alpha A_{12} + (1-\alpha)B_{12}][\alpha A_{22} + (1-\alpha)B_{22}]^{-1}[\alpha A_{12} + (1-\alpha)B_{12}]^*$$
$$\leq \alpha A_{12}A_{22}^{-1}A_{12}^* + (1-\alpha)B_{12}B_{22}^{-1}B_{12}^*.$$

Hence,

$$-[\alpha A_{22} + (1-\alpha)B_{22}]|[\alpha A + (1-\alpha)B]$$
$$= [\alpha A_{12} + (1-\alpha)B_{12}][\alpha A_{22} + (1-\alpha)B_{22}]^{-1}[\alpha A_{12} + (1-\alpha)B_{12}]^*$$
$$\quad - [\alpha A_{11} + (1-\alpha)B_{11}]$$
$$\leq \alpha(A_{12}A_{22}^{-1}A_{12}^* - A_{11}) + (1-\alpha)(B_{12}B_{22}^{-1}B_{12}^* - B_{11})$$
$$= \alpha(-A_{22}|A) + (1-\alpha)(-B_{22}|B),$$

which shows that $\phi(A) \triangleq -A_{22}|A$ is convex. By continuity, the result holds for $A \in \mathbf{N}^{n+m}$.

To prove xvi), note that $\phi(A) = (A_{22}|A)^{-1} = \phi_2[\phi_1(A)]$, where $\phi_1(A) = A_{22}|A$ and $\phi_2(B) = B^{-1}$. It follows from xv) that ϕ_1 is concave, while it follows from iv) that ϕ_2 is convex. Furthermore, $viii$) of Proposition 8.5.12 implies that ϕ_2 is nonincreasing. It thus follows from Lemma 8.5.14 that $\phi(A) \triangleq (A_{22}|A)^{-1}$ is convex.

Result $xvii$) is given in [135, p. 106].

Result $xviii$) is given in by Theorem 9 of [483].

To prove xix), note that $\phi(A) = -(\operatorname{tr} A^{-r})^{-p} = \phi_2[\phi_1(A)]$, where $\phi_1(A) = \operatorname{tr} A^{-r}$ and $\phi_2(B) = -B^{-p}$. Statement iii) implies that ϕ_1 is convex and that ϕ_2 is concave. Furthermore, vii) of Proposition 8.5.12 implies that ϕ_2 is nondecreasing. It thus follows from iv) of Lemma 8.5.14 that $\phi(A) = -(\operatorname{tr} A^{-r})^{-p}$ is convex.

Results xx) and xxi) are proved in [160].

Results $xxii$)–$xxvi$) are given by Corollary 1.1, Theorem 1, Corollary 2.1, Theorem 2, and Theorem 8, respectively, of [160]. A proof of $xxii$) in the case $p = 1 - r$ is given in [111, p. 273].

Result $xxvii$) is proved in [111, p. 274] and [160].

Result $xxviii$) is given in [111, p. 123] and [281].

To prove $xxix$), note that $\phi(A) = -\log \operatorname{tr} A^r = \phi_2[\phi_1(A)]$, where $\phi_1(A) = \operatorname{tr} A^r$ and $\phi_2(x) = -\log x$. Statement vi) implies that ϕ_1 is concave. Furthermore, ϕ_2 is convex and nonincreasing. It thus follows from ii) of Lemma 8.5.14 that $\phi(A) = -\log \operatorname{tr} A^r$ is convex.

Result xxx) is given in [111, p. 275].

To prove $xxxi$), let $A_1, A_2 \in \mathbf{N}^n$. From Corollary 8.4.15 it follows that $(\det A_1)^{1/n} + (\det A_2)^{1/n} \leq [\det(A_1 + A_2)]^{1/n}$. Replacing A_1 and A_2 by αA_1 and $(1 - \alpha)A_2$, respectively, and multiplying by -1 shows that $\phi(A) = -(\det A)^{1/n}$ is convex.

To prove $xxxii$), note that $\phi(A) = -n\log[(\det A)^{1/n}] = \phi_2[\phi_1(A)]$, where $\phi_1(A) = (\det A)^{1/n}$ and $\phi_2(x) = -n\log x$. It follows from xix) that ϕ_1 is concave. Since ϕ_2 is nonincreasing and convex, it follows from ii) of Lemma 8.5.14 that $\phi(A) = -\log \det A$ is convex.

To prove $xxxiii$), note that $\phi(A) = \det A^{-1} = \phi_2[\phi_1(A)]$, where $\phi_1(A) = \log \det A^{-1}$ and $\phi_2(x) = e^x$. It follows from xx) that ϕ_1 is convex. Since ϕ_2 is nondecreasing and convex, it follows from i) of Lemma 8.5.14 that

$\phi(A) = \det A^{-1}$ is convex.

Next, $xxxiv$) is given in [111, p. 273] and [806, p. 9]. Statement $xxxv$) is given in [806, p. 9].

Finally, $xxxvi$) is given in [516, p. 478]. Statement $xxxvii$) follows immediately from $xxxvi$).   □

The following result is a corollary of xv) of Proposition 8.5.15 for the case $\alpha = 1/2$. Versions of this result appear in [163,342,479,488] and [592, p. 152].

Corollary 8.5.16. Let $A \triangleq \begin{bmatrix} A_{11} & A_{12} \\ A_{12}^* & A_{22} \end{bmatrix} \in \mathbb{F}^{n+m}$ and $B \triangleq \begin{bmatrix} B_{11} & B_{12} \\ B_{12}^* & B_{22} \end{bmatrix} \in \mathbb{F}^{n+m}$, and assume that A and B are positive semidefinite. Then,

$$A_{11}|A + B_{11}|B \le (A_{11} + B_{11})|(A + B).$$

The following corollary of $xxxvi$) and $xxxvii$) gives a strong majorization condition for the eigenvalues of a pair of Hermitian matrices.

Corollary 8.5.17. Let $A, B \in \mathbf{H}^n$. Then, for all $k = 1, \ldots, n$,

$$\sum_{i=1}^{k} \lambda_i(A + B) \le \sum_{i=1}^{k} [\lambda_i(A) + \lambda_i(B)] \qquad (8.5.14)$$

with equality for $k = n$. Furthermore, for all $k = 1, \ldots, n$,

$$\sum_{i=k}^{n} [\lambda_i(A) + \lambda_i(B)] \le \sum_{i=k}^{n} \lambda_i(A + B) \qquad (8.5.15)$$

with equality for $k = 1$.

Proof. See [111, p. 69], [369, p. 201], or [516, p. 478].   □

8.6 Facts on Range and Rank

Fact 8.6.1. Let $A, B \in \mathbb{F}^{n \times n}$, and assume that A and B are positive semidefinite. Then, there exists $\alpha > 0$ such that $A \le \alpha B$ if and only if $\mathcal{R}(A) \subseteq \mathcal{R}(B)$. In this case, $\operatorname{rank} A \le \operatorname{rank} B$. (Proof: Use Theorem 8.5.2 and Corollary 8.5.10.)

Fact 8.6.2. Let $A, B \in \mathbb{F}^{n \times n}$, and assume that A is positive semidefinite and B is either positive semidefinite or skew Hermitian. Then, the following identities hold:

   $i)$ $\mathcal{N}(A + B) = \mathcal{N}(A) \cap \mathcal{N}(B)$.

ii) $\mathcal{R}(A + B) = \mathcal{R}(A) + \mathcal{R}(B)$.

(Proof: Use $[(\mathcal{N}(A) \cap \mathcal{N}(B)]^{\perp} = \mathcal{R}(A) + \mathcal{R}(B)$.)

Fact 8.6.3. Let $A \in \mathbb{F}^{n \times n}$, and assume that $A + A^* \geq 0$. Then, the following identities hold:

i) $\mathcal{N}(A) = \mathcal{N}(A + A^*) \cap \mathcal{N}(A - A^*)$.

ii) $\mathcal{R}(A) = \mathcal{R}(A + A^*) + \mathcal{R}(A - A^*)$.

iii) $\operatorname{rank} A = \operatorname{rank} \begin{bmatrix} A + A^* & A - A^* \end{bmatrix}$.

Fact 8.6.4. Let $A, B \in \mathbb{F}^{n \times n}$, and assume that A and B are positive semidefinite. Then,

$$\operatorname{rank}(A + B) = \operatorname{rank} \begin{bmatrix} A & B \end{bmatrix} = \operatorname{rank} \begin{bmatrix} A \\ B \end{bmatrix}$$

and

$$\operatorname{rank} \begin{bmatrix} A & B \\ 0 & A \end{bmatrix} = \operatorname{rank} A + \operatorname{rank}(A + B).$$

(Proof: Using Fact 8.6.2,

$$\mathcal{R}\left(\begin{bmatrix} A & B \end{bmatrix} \right) = \mathcal{R}\left(\begin{bmatrix} A & B \end{bmatrix} \begin{bmatrix} A \\ B \end{bmatrix} \right) = \mathcal{R}(A^2 + B^2) = \mathcal{R}(A^2) + \mathcal{R}(B^2)$$
$$= \mathcal{R}(A) + \mathcal{R}(B) = \mathcal{R}(A + B).$$

Alternatively, it follows from Fact 6.4.20 that

$$\operatorname{rank} \begin{bmatrix} A & B \end{bmatrix} = \operatorname{rank} \begin{bmatrix} A + B & B \end{bmatrix}$$
$$= \operatorname{rank}(A + B) + \operatorname{rank}[B - (A + B)(A + B)^+ B].$$

Next, note that

$$\operatorname{rank}[B - (A + B)(A + B)^+ B] = \operatorname{rank}\left(B^{1/2}[I - (A + B)(A + B)^+] B^{1/2} \right)$$
$$\leq \operatorname{rank}\left(B^{1/2}[I - BB^+] B^{1/2} \right) = 0.$$

For the second result use Theorem 8.3.4 to simultaneously diagonalize A and B.)

8.7 Facts on Identities and Inequalities for One Matrix

Fact 8.7.1. Let $A \in \mathbb{F}^{n \times n}$, and assume that A is Hermitian. Then,

$$\operatorname{rank} A = \nu_+(A) + \nu_-(A)$$

and

$$\operatorname{def} A = \nu_0(A).$$

Fact 8.7.2. Let $A \in \mathbb{F}^{n \times n}$, assume that A is positive semidefinite, and assume there exists $i \in \{1, \ldots, n\}$ such that $A_{(i,i)} = 0$. Then, $\mathrm{row}_i(A) = 0$ and $\mathrm{col}_i(A) = 0$.

Fact 8.7.3. Let $A \in \mathbb{F}^{n \times n}$, and assume that A is positive semidefinite. Then, $A_{(i,i)} \geq 0$ for all $i = 1, \ldots, n$, and $|A_{(i,j)}|^2 \leq A_{(i,i)} A_{(j,j)}$ for all $i, j = 1, \ldots, n$.

Fact 8.7.4. Let $A \in \mathbb{F}^{n \times n}$. Then, $A \geq 0$ if and only if $A \geq -A$.

Fact 8.7.5. Let $A \in \mathbb{F}^{n \times n}$, and assume that A is Hermitian. Then, $A^2 \geq 0$.

Fact 8.7.6. Let $A \in \mathbb{F}^{n \times n}$, and assume that A is skew Hermitian. Then, $A^2 \leq 0$.

Fact 8.7.7. Let $A \in \mathbb{F}^{n \times n}$. Then,

$$(A + A^*)^2 \geq 0$$

and

$$(A - A^*)^2 \leq 0.$$

Fact 8.7.8. Let $A \in \mathbb{F}^{n \times n}$. Then,

$$A^2 + A^{2*} \leq AA^* + A^*A.$$

Equality holds if and only if $A = A^*$.

Fact 8.7.9. Let $A \in \mathbb{F}^{n \times n}$, and let $\alpha > 0$. Then,

$$A + A^* \leq \alpha I + \alpha^{-1} AA^*.$$

Equality holds if and only if $A = \alpha I$.

Fact 8.7.10. Let $A \in \mathbb{F}^{n \times n}$, and assume that A is positive definite. Then,

$$2I \leq A + A^{-1}.$$

Equality holds if and only if $A = I$. Furthermore,

$$2n \leq \mathrm{tr}\, A + \mathrm{tr}\, A^{-1}.$$

Fact 8.7.11. Let $A \in \mathbb{F}^{n \times n}$, and assume that A is positive definite. Then,

$$\left(1_{1 \times n} A^{-1} 1_{n \times 1}\right)^{-1} 1_{n \times n} \leq A.$$

(Proof: Set $B = 1_{n \times n}$ in Fact 8.16.10. See [813].)

Fact 8.7.12. Let $A \in \mathbb{F}^{n \times n}$, and assume that A is positive definite. Then, $\begin{bmatrix} A & I \\ I & A^{-1} \end{bmatrix}$ is positive semidefinite.

Fact 8.7.13. Let $A \in \mathbb{F}^{n \times n}$, and assume that A is Hermitian. Then, $A^2 \leq A$ if and only if $0 \leq A \leq I$.

Fact 8.7.14. Let $A \in \mathbb{F}^{n \times n}$, and assume that A is Hermitian. Then, $\alpha I + A \geq 0$ if and only if $\alpha \geq -\lambda_{\min}(A)$. Furthermore,

$$A^2 + A + \tfrac{1}{4}I \geq 0.$$

Fact 8.7.15. Let $A \in \mathbb{F}^{n \times m}$. Then, $AA^* \leq I_n$ if and only if $A^*A \leq I_m$.

Fact 8.7.16. Let $A \in \mathbb{F}^{n \times n}$, and assume that either $AA^* \leq A^*A$ or $A^*A \leq AA^*$. Then, A is normal. (Proof: Use ii) of Corollary 8.4.10.)

Fact 8.7.17. Let $A \in \mathbb{F}^{n \times n}$, and assume that A is a projector. Then,

$$0 \leq A \leq I.$$

Fact 8.7.18. Let $A \in \mathbb{F}^{n \times n}$, and assume that A is nonsingular. Then,

$$\langle A^{-1} \rangle = \langle A^* \rangle^{-1}.$$

Fact 8.7.19. Let $A \in \mathbb{F}^{n \times m}$, and assume that A^*A is nonsingular. Then,

$$\langle A^* \rangle = A \langle A \rangle^{-1/2} A^*.$$

Fact 8.7.20. Let $A \in \mathbb{F}^{n \times n}$, and assume that A is nonsingular. Then, $\langle A^* \rangle^{-1/2} A$ is unitary.

Fact 8.7.21. Let $A \in \mathbb{F}^{n \times n}$. Then, A is normal if and only if $\langle A \rangle = \langle A^* \rangle$. (Remark: See Fact 3.5.8.)

Fact 8.7.22. Let $A \in \mathbb{F}^{n \times n}$, assume that A is normal, and let $\alpha, \beta \in (0, \infty)$. Then,

$$\langle \alpha A + \beta A^* \rangle \leq \alpha \langle A \rangle + \beta \langle A^* \rangle.$$

In particular,

$$\langle A + A^* \rangle \leq \langle A \rangle + \langle A^* \rangle.$$

(Proof: See [468, 815].) (Remark: See Fact 8.9.5.)

Fact 8.7.23. Let $A \in \mathbb{F}^{n \times n}$. Then, A is positive definite if and only if $I + A$ is nonsingular and the matrices $I - B$ and $I + B$ are positive definite, where $B \triangleq (I + A)^{-1}(I - A)$. (Proof: See [246].) (Remark: For additional results on the Cayley transform, see Fact 3.7.22, Fact 3.7.23, Fact 3.7.24, Fact 3.12.9, and Fact 11.18.10.)

Fact 8.7.24. Let $A \in \mathbb{R}^{n \times n}$, assume that A is positive definite, assume that $A \leq I$, and define $\{B_k\}_{k=0}^{\infty}$ by $B_0 \triangleq 0$ and

$$B_{k+1} \triangleq B_k + \tfrac{1}{2}(A - B_k^2).$$

Then,
$$\lim_{k\to\infty} B_k = A^{1/2}.$$

(Proof: See [92, p. 181].) (Remark: See Fact 5.14.19.)

Fact 8.7.25. Let $A \in \mathbb{R}^{n\times n}$, assume that A is nonsingular, and define $\{B_k\}_{k=0}^{\infty}$ by $B_0 \triangleq A$ and
$$B_{k+1} \triangleq \tfrac{1}{2}\left(B_k + B_k^{-\mathrm{T}}\right).$$

Then,
$$\lim_{k\to\infty} B_k = \left(AA^{\mathrm{T}}\right)^{-1/2}A.$$

(Remark: The limit is unitary. See Fact 8.7.20. See [78, p. 224].)

Fact 8.7.26. Let $0 \le \alpha_1 \le \cdots \le \alpha_n$, and define $A \in \mathbb{R}^{n\times n}$ by $A_{(i,j)} \triangleq \min\{\alpha_i, \alpha_j\}$ for all $i, j = 1, \ldots, n$. Then, A is positive semidefinite. (Problem: Determine rank A. When is A positive definite?) (Remark: When $\alpha_i = i$ for all $i = 1, \ldots, n$, the matrix A is a covariance matrix arising in the theory of Brownian motion.)

Fact 8.7.27. Let $\lambda_1, \ldots, \lambda_n \in \mathbb{C}$, assume that $\operatorname{Re}\lambda_i < 0$ for all $i = 1, \ldots, n$, and, for all $i, j = 1, \ldots, n$, define $A \in \mathbb{C}^{n\times n}$ by
$$A_{(i,j)} \triangleq \frac{-1}{\lambda_i + \overline{\lambda}_j}.$$

Then, A is positive semidefinite. (Proof: Note that $A = 2B \circ (1_{n\times n} - C)^{\{-1\}}$, where $B_{(i,j)} = \frac{1}{(\lambda_i-1)(\overline{\lambda}_j-1)}$ and $C_{(i,j)} = \frac{(\lambda_i+1)(\overline{\lambda}_j+1)}{(\lambda_i-1)(\overline{\lambda}_j-1)}$. Then, note that B is positive semidefinite and that $(1_{n\times n} - C)^{\{-1\}} = 1_{n\times n} + C + C^{\{2\}} + C^{\{3\}} + \cdots$. Alternatively, A satisfies a Lyapunov equation with coefficient $\operatorname{diag}(\lambda_1, \ldots, \lambda_n)$. See [369, p. 348].) (Remark: A is a Cauchy matrix. See Fact 3.15.13.)

Fact 8.7.28. Let $a_1, \ldots, a_n \ge 0$ and $p \in \mathbb{R}$, assume that either $a_1, \ldots, a_n$ are positive or p is positive, and, for all $i, j = 1, \ldots, n$, define $A \in \mathbb{R}^{n\times n}$ by
$$A_{(i,j)} \triangleq (a_i a_j)^p.$$

Then, A is positive semidefinite. (Proof: Let $a \triangleq \begin{bmatrix} a_1 & \cdots & a_n \end{bmatrix}^{\mathrm{T}}$ and $A \triangleq a^{\{p\}}a^{\{p\}\mathrm{T}}$.)

Fact 8.7.29. Let $a_1, \ldots, a_n > 0$, let $\alpha > 0$, and, for all $i, j = 1, \ldots, n$, define $A \in \mathbb{R}^{n\times n}$ by
$$A_{(i,j)} \triangleq \frac{1}{(a_i + a_j)^\alpha}.$$

Then, A is positive semidefinite. (Proof: See [588].) (Remark: See Fact 5.10.7.)

Fact 8.7.30. Let $a_1, \ldots, a_n > 0$, let $r \in [-1, 1]$, and, for all $i, j = 1, \ldots, n$, define $A \in \mathbb{R}^{n \times n}$ by

$$A_{(i,j)} \triangleq \frac{a_i^r + a_j^r}{a_i + a_j}.$$

Then, A is positive semidefinite. (Proof: See [806, p. 74].)

Fact 8.7.31. Let $a_1, \ldots, a_n > 0$, let $q > 0$, let $p \in [-q, q]$, and, for all $i, j = 1, \ldots, n$, define $A \in \mathbb{R}^{n \times n}$ by

$$A_{(i,j)} \triangleq \frac{a_i^p + a_j^p}{a_i^q + a_j^q}.$$

Then, A is positive semidefinite. (Proof: Let $r = p/q$ and $b_i = a_i^q$. Then, $A_{(i,j)} = (b_i^r + b_j^r)/(b_i + b_j)$. Now, use Fact 8.7.30. See [521] for the case $q \geq p \geq 0$.) (Remark: The case $q = 1$ and $p = 0$ yields a Cauchy matrix. In the case $n = 2$, $A \geq 0$ yields Fact 1.4.8.) (Problem: When is A positive definite?)

Fact 8.7.32. Let $a_1, \ldots, a_n > 0$, let $p \in [-1, 1]$ and $q \in (-2, 2)$, and, for all $i, j = 1, \ldots, n$, define $A \in \mathbb{R}^{n \times n}$ by

$$A_{(i,j)} \triangleq \frac{a_i^p + a_j^p}{a_i^2 + qa_ia_j + a_j^2}.$$

Then, A is positive semidefinite. (Proof: See [805] or [806, p. 76].)

Fact 8.7.33. Let $\lambda_1, \ldots, \lambda_n \in \text{OUD}$, and let $w_1, \ldots, w_n \in \mathbb{C}$. Then, there exists a holomorphic function $\phi : \text{OUD} \mapsto \text{OUD}$ such that $\phi(\lambda_i) = w_i$ for all $i = 1, \ldots, n$ if and only if $A \in \mathbb{C}^{n \times n}$ is positive semidefinite, where, for all $i, j = 1, \ldots, n$,

$$A_{(i,j)} \triangleq \frac{1 - \overline{w_i}w_j}{1 - \overline{\lambda_i}\lambda_j}.$$

(Proof: See [523].) (Remark: A is a *Pick matrix*.)

Fact 8.7.34. Let $A \in \mathbb{F}^{n \times n}$, and assume that A is Hermitian, $A_{(i,i)} > 0$ for all $i = 1, \ldots, n$, and, for all $i, j = 1, \ldots, n$,

$$|A_{(i,j)}| < \tfrac{1}{n-1}\sqrt{A_{(i,i)}A_{(j,j)}}.$$

Then, A is positive definite. (Proof: Note that

$$x^*Ax = \sum_{i=1}^{n-1} \sum_{j=i+1}^{n} \begin{bmatrix} x_{(i)} \\ x_{(j)} \end{bmatrix}^* \begin{bmatrix} \frac{1}{n-1}A_{(i,i)} & A_{(i,j)} \\ \overline{A_{(i,j)}} & \frac{1}{n-1}A_{(j,j)} \end{bmatrix} \begin{bmatrix} x_{(i)} \\ x_{(j)} \end{bmatrix}.)$$

(Remark: This result is due to Roup.)

Fact 8.7.35. Let $\alpha_0, \ldots, \alpha_n > 0$, and define the tridiagonal matrix $A \in \mathbb{R}^{n \times n}$ by

$$
A \triangleq
\begin{bmatrix}
\alpha_0 + \alpha_1 & -\alpha_1 & 0 & 0 & \cdots & 0 \\
-\alpha_1 & \alpha_1 + \alpha_2 & -\alpha_2 & 0 & \cdots & 0 \\
0 & -\alpha_2 & \alpha_2 + \alpha_3 & -\alpha_3 & \cdots & 0 \\
\vdots & \vdots & \vdots & \vdots & & \vdots \\
0 & 0 & 0 & 0 & \cdots & \alpha_{n-1} + \alpha_n
\end{bmatrix}.
$$

Then, A is positive definite. (Proof: For $k = 2, \ldots, n$, the $k \times k$ leading principal subdeterminant of A is given by $\left[\sum_{i=0}^{k} \alpha_i^{-1} \right] \alpha_0 \alpha_1 \cdots \alpha_k$. See [80, p. 115].) (Remark: A is a stiffness matrix arising in structural analysis.)

Fact 8.7.36. Let $x_1, \ldots, x_n \in \mathbb{F}^n$, and define $A \in \mathbb{F}^{n \times n}$ by $A_{(i,j)} \triangleq x_i^* x_j$ for all $i, j = 1, \ldots, n$, and $B \triangleq \begin{bmatrix} x_1 & \cdots & x_n \end{bmatrix}$. Then, $A = B^* B$. Consequently, A is positive semidefinite and $\operatorname{rank} A = \operatorname{rank} B$. Conversely, let $A \in \mathbb{F}^{n \times n}$, and assume that A is positive semidefinite. Then, there exist $x_1, \ldots, x_n \in \mathbb{F}^n$ such that $A = B^* B$, where $B = \begin{bmatrix} x_1 & \cdots & x_n \end{bmatrix}$. (Proof: The converse is an immediate consequence of Corollary 5.4.5.) (Remark: A is the *Gram matrix* of $x_1, \ldots, x_n$.)

Fact 8.7.37. Let $A \in \mathbb{F}^{n \times n}$, and assume that A is positive semidefinite. Then, there exists a matrix $B \in \mathbb{F}^{n \times n}$ such that B is upper triangular, B has nonnegative diagonal entries, and $A = BB^*$. If, in addition, A is positive definite, then B is unique and has positive diagonal entries. (Remark: This result is the *Cholesky decomposition*.)

Fact 8.7.38. Let $x \in \mathbb{F}^n$. Then,

$$
xx^* \leq x^* x I.
$$

Fact 8.7.39. Let $A \in \mathbb{F}^{n \times m}$, and assume that $\operatorname{rank} A = m$. Then,

$$
0 \leq A(A^* A)^{-1} A^* \leq I.
$$

Fact 8.7.40. Let $A \in \mathbb{F}^{n \times n}$, and assume that A is positive definite. Then,

$$
A^{-1} \leq \frac{\alpha + \beta}{\alpha \beta} I - \frac{1}{\alpha \beta} A \leq \frac{(\alpha + \beta)^2}{4 \alpha \beta} A^{-1},
$$

where $\alpha \triangleq \lambda_{\max}(A)$ and $\beta \triangleq \lambda_{\min}(A)$. (Proof: See [517].)

Fact 8.7.41. Let $A \in \mathbb{F}^{n \times n}$, and assume that A is positive semidefinite. If $\alpha \in [0, 1]$, then

$$
A^\alpha \leq \alpha A + (1 - \alpha) I.
$$

Furthermore, if $\alpha > 1$ or if $\alpha < 0$ and A is positive definite, then

$$\alpha A + (1 - \alpha)I \leq A^\alpha.$$

(Proof: See [282, pp. 122, 123].) (Remark: This result is a special case of the Young inequality. See Fact 8.8.23 and Fact 1.4.1.)

Fact 8.7.42. Let $A \in \mathbb{F}^{n \times n}$, and assume that A is positive definite. Then,

$$I - A^{-1} \leq \log A \leq A - I.$$

Furthermore, if $A \geq I$, then $\log A$ is positive semidefinite, and, if $A > I$, then $\log A$ is positive definite. (Proof: See Fact 1.4.2.)

8.8 Facts on Identities and Inequalities for Two or More Matrices

Fact 8.8.1. Let $A, B \in \mathbb{F}^{n \times n}$, and assume that A and B are positive semidefinite. Then, $0 < A \leq B$ if and only if $\operatorname{sprad}(AB^{-1}) < 1$.

Fact 8.8.2. Let $A, B \in \mathbb{F}^{n \times n}$, and assume that A and B are positive definite. Then,

$$(A^{-1} + B^{-1})^{-1} = A(A + B)^{-1}B.$$

Fact 8.8.3. Let $A, B \in \mathbb{F}^{n \times n}$, and assume that A is positive definite, B is Hermitian, and $A + B$ is nonsingular. Then,

$$(A + B)^{-1} + (A + B)^{-1}B(A + B)^{-1} \leq A^{-1}.$$

If, in addition, B is nonsingular, the inequality is strict. (Proof: The inequality is equivalent to $BA^{-1}B \geq 0$. See [565].)

Fact 8.8.4. Let $A \in \mathbb{F}^{n \times m}$ and $B \in \mathbb{F}^{m \times m}$, and assume that B is positive semidefinite. Then, $ABA^* = 0$ if and only if $AB = 0$.

Fact 8.8.5. Let $A, B \in \mathbb{F}^{n \times n}$, and assume that A and B are positive semidefinite. Then, AB is positive semidefinite if and only if AB is normal.

Fact 8.8.6. Let $A, B \in \mathbb{F}^{n \times n}$, assume that A and B are Hermitian, and assume that either i) A and B are positive semidefinite or ii) either A or B is positive definite. Then, AB is group invertible. (Proof: Use Theorem 8.3.2 and Theorem 8.3.5.)

Fact 8.8.7. Let $A, B \in \mathbb{F}^{n \times n}$, assume that A and B are Hermitian, and assume that A and $AB + BA$ are positive definite. Then, B is positive definite. (Proof: See [461, p. 120] or [777]. Alternatively, the result follows from Corollary 11.8.4.)

Fact 8.8.8. Let $A, B \in \mathbb{F}^{n \times n}$, and assume that A and B are Hermitian and nonsingular. Then, the following statements hold:

   i) If every eigenvalue of AB is positive, then $\operatorname{In} A = \operatorname{In} B$.

   ii) If $\operatorname{In} A = \operatorname{In} B$ and $A \leq B$, then $B^{-1} \leq A^{-1}$.

(Proof: See [30].)

Fact 8.8.9. Let $A, B \in \mathbb{F}^{n \times n}$, assume that A and B are Hermitian, and assume that $A \leq B$. Then, $A_{(i,i)} \leq B_{(i,i)}$ for all $i = 1, \ldots, n$.

Fact 8.8.10. Let $A, B \in \mathbb{F}^{n \times n}$, assume that A and B are Hermitian, and assume that $\langle A \rangle \leq B$. Then, either $A \leq B$ or $-A \leq B$. (Proof: See [814].)

Fact 8.8.11. Let $A, B \in \mathbb{F}^{n \times n}$, and assume that A is positive semidefinite and B is positive definite. Then, $A \leq B$ if and only if $AB^{-1}A \leq A$.

Fact 8.8.12. Let $A, B, C, D \in \mathbb{F}^{n \times n}$, assume that A, B, C, D are positive semidefinite, and assume that $0 < D \leq C$ and $BCB \leq ADA$. Then, $B \leq A$. (Proof: See [49, 169].)

Fact 8.8.13. Let $A, B \in \mathbb{F}^{n \times n}$, and assume that A and B are projectors. Then,

$$(A + B)^{1/2} \leq A^{1/2} + B^{1/2}$$

if and only if $AB = BA$. (Proof: See [718, p. 30].)

Fact 8.8.14. Let $A, B \in \mathbb{F}^{n \times n}$, assume that A and B are positive semidefinite, and assume that $0 \leq A \leq B$. Then,

$$\left(A + \tfrac{1}{4}A^2\right)^{1/2} \leq \left(B + \tfrac{1}{4}B^2\right)^{1/2}.$$

(Proof: See [544].)

Fact 8.8.15. Let $A \in \mathbb{F}^{n \times n}$, assume that A is positive semidefinite, and let $B \in \mathbb{F}^{l \times n}$. Then, BAB^* is positive definite if and only if $B(A + A^2)B^*$ is positive definite. (Proof: Diagonalize A using a unitary transformation and note that $BA^{1/2}$ and $B(A + A^2)^{1/2}$ have the same rank.)

Fact 8.8.16. Let $A_1, \ldots, A_k \in \mathbb{F}^{n \times n}$, assume that $A_1, \ldots, A_k$ are positive semidefinite, and let $p, q \in \mathbb{R}$ satisfy $1 \leq p \leq q$. Then,

$$\left(\frac{1}{k}\sum_{i=1}^{k} A_i^p\right)^{1/p} \leq \left(\frac{1}{k}\sum_{i=1}^{k} A_i^q\right)^{1/q}.$$

(Proof: See [108].)

Fact 8.8.17. Let $A, B \in \mathbb{F}^{n \times n}$, and assume that A and B are Hermitian. Then, there exists a Hermitian matrix $C \in \mathbb{F}^{n \times n}$ that is a least upper bound for A and B in the sense that $A \leq C$, $B \leq C$, and, if $D \in \mathbb{F}^{n \times n}$ is a Hermitian matrix satisfying $A \leq D$ and $B \leq D$, then $C \leq D$. (Proof: First consider the case in which A and B are both positive semidefinite.) (Problem: Generalize to three or more matrices.)

Fact 8.8.18. Let $A, B \in \mathbb{F}^{n \times n}$, assume that A and B are positive semidefinite, and let $p, q \in \mathbb{R}$ satisfy $p \geq q \geq 0$. Then,

$$\left[\tfrac{1}{2}(A^q + B^q) \right]^{1/q} \leq \left[\tfrac{1}{2}(A^p + B^p) \right]^{1/p}.$$

Furthermore,

$$\mu(A, B) \triangleq \lim_{p \to \infty} \left[\tfrac{1}{2}(A^p + B^p) \right]^{1/p}$$

exists and satisfies

$$A \leq \mu(A, B), \quad B \leq \mu(A, B).$$

(Proof: See [93].) (Problem: If $A \leq C$ and $B \leq C$, then does it follow that $\mu(A, B) \leq C$? See [32, 413].)

Fact 8.8.19. Let $A, B \in \mathbb{F}^{n \times n}$, assume that A and B are positive semidefinite, let $p \in (1, \infty)$, and let $\alpha \in [0, 1]$. Then,

$$\alpha^{1-1/p}A + (1 - \alpha)^{1-1/p}B \leq (A^p + B^p)^{1/p}.$$

(Proof: See [32].)

Fact 8.8.20. Let $A, B, C \in \mathbb{F}^{n \times n}$. Then,

$$A^*A + B^*B = (B + CA)^*(I + CC^*)^{-1}(B + CA) + (A - C^*B)(I + C^*C)^{-1}(A - C^*B).$$

(Proof: See [375].) (Remark: See Fact 8.11.22.)

Fact 8.8.21. Let $A, B \in \mathbb{F}^{n \times n}$, assume that A and B are positive definite, let $C \in \mathbb{F}^{n \times n}$ satisfy $B = C^*C$, and let $\alpha \in [0, 1]$. Then,

$$C^*\left(C^{-*}AC^{-1}\right)^\alpha C \leq \alpha A + (1 - \alpha)B.$$

If, in addition, $\alpha \in (0, 1)$, then equality holds if and only if $A = B$. (Proof: See [532].)

Fact 8.8.22. Let $A, B \in \mathbb{F}^{n \times n}$, assume that A is positive semidefinite, and let $p \in \mathbb{R}$. Furthermore, assume that either A and B are nonsingular or $p > 1$. Then,

$$(BAB^*)^p = BA^{1/2}(A^{1/2}B^*BA^{1/2})^{p-1}A^{1/2}B^*.$$

(Proof: See [278].)

Fact 8.8.23. Let $A, B \in \mathbb{F}^{n \times n}$, and assume that A and B are positive definite. Then,

$$A^{1/2}\left(A^{-1/2}BA^{-1/2}\right)^{1/2}A^{1/2} = A\left(A^{-1}B\right)^{1/2}$$

$$= (A + B)\left[(A + B)^{-1}A(A + B)^{-1}B\right]^{1/2},$$

where $\left(A^{-1}B\right)^{1/2}$ has positive eigenvalues and satisfies $\left[(A^{-1}B)^{1/2}\right]^2 = A^{-1}B$. Denote the above quantity by $A\#B$. Then, $A\#B$ is the unique positive-definite solution of $XA^{-1}X = B$, and satisfies

$$A\#A = A,$$

$$A\#B = B\#A,$$

$$\det A\#B = \sqrt{(\det A)\det B},$$

$$(A\#B)^{-1} = A^{-1}\#B^{-1},$$

$$S^*AS\#S^*BS = S^*(B\#A)S,$$

$$2\left(A^{-1} + B^{-1}\right)^{-1} \leq A\#B \leq \tfrac{1}{2}(A + B),$$

$$(A\#B)B^{-1}(A\#B) = A,$$

$$\begin{bmatrix} A & A\#B \\ A\#B & B \end{bmatrix} \geq 0,$$

$$\operatorname{rank}\begin{bmatrix} A & A\#B \\ A\#B & B \end{bmatrix} = n,$$

where $S \in \mathbb{F}^{n \times n}$ is nonsingular. If $C, D \in \mathbf{P}^n$, $C \leq A$, and $D \leq B$, then

$$C\#D \leq A\#B.$$

Furthermore, if $X \in \mathbf{H}^n$, then $\begin{bmatrix} A & X \\ X & B \end{bmatrix}$ is positive semidefinite if and only if $XA^{-1}X \leq B$. In this case, $-A\#B \leq X \leq A\#B$. Let $A_0 \triangleq A$ and $B_0 \triangleq B$, and, for all $k \in \mathbb{N}$, define $A_{k+1} \triangleq 2(A_k^{-1} + B_k^{-1})^{-1}$ and $B_{k+1} \triangleq \tfrac{1}{2}(A_k + B_k)$. Then, for all $k \in \mathbb{N}$,

$$A_k \leq A_{k+1} \leq A\#B \leq B_{k+1} \leq B_k$$

and

$$\lim_{k \to \infty} A_k = \lim_{k \to \infty} B_k = A\#B.$$

Finally, if $\alpha \in [0, 1]$, then

$$\left[\alpha A^{-1} + (1 - \alpha)B^{-1}\right]^{-1} \leq A^{1/2}\left(A^{-1/2}BA^{-1/2}\right)^{1-\alpha}A^{1/2} \leq \alpha A + (1 - \alpha)B,$$

or, equivalently,

$$\left[\alpha A + (1 - \alpha)B\right]^{-1} \leq A^{-1/2}\left(A^{-1/2}BA^{-1/2}\right)^{\alpha-1}A^{-1/2} \leq \alpha A^{-1} + (1 - \alpha)B^{-1}.$$

Hence,

$$\text{tr}\left[\alpha A + (1-\alpha)B\right]^{-1} \leq \text{tr}\left[A^{-1}\left(A^{-1/2}BA^{-1/2}\right)^{\alpha-1}\right] \leq \text{tr}\left[\alpha A^{-1} + (1-\alpha)B^{-1}\right].$$

(Proof: See [263, 460, 716].) (Remark: The inequalities involving α improve iv) of Proposition 8.5.15. Alternative means and their differences are considered in [11]. $A\#B$ is the *geometric mean* of A and B. A related mean is defined in [263].) (Remark: A geometric mean for an arbitrary number of positive-definite matrices is given in [33].) (Remark: The penultimate string of inequalities is the *Young inequality*. See [282, p. 122] and Fact 8.7.41.)

Fact 8.8.24. Let $\{x_i\}_{i=1}^{\infty} \subset \mathbb{R}^n$, assume that $\sum_{i=1}^{\infty} x_i$ exists, and let $\{A_i\}_{i=1}^{\infty} \subset \mathbf{N}^n$ be such that $A_i \leq A_{i+1}$ for all $i \in \mathbb{P}$ and $\lim_{i\to\infty} \text{tr}\, A_i = \infty$. Then,

$$\lim_{k\to\infty} (\text{tr}\, A_k)^{-1}\sum_{i=1}^{k} A_i x_i = 0.$$

If, in addition A_i is positive definite for all $i \in \mathbb{P}$ and $\{\lambda_{\max}(A_i)/\lambda_{\min}(A_i)\}_{i=1}^{\infty}$ is bounded, then

$$\lim_{k\to\infty} A_k^{-1}\sum_{i=1}^{k} A_i x_i = 0.$$

(Proof: See [20].) (Remark: These identities are matrix versions of the *Kronecker lemma*.)

Fact 8.8.25. Let $A, B \in \mathbb{F}^{n\times n}$, and assume that A and B are positive definite. Then, $B \leq A$ if and only if, for all $p, q, r, t \in \mathbb{R}$ such that $p \geq 1$, $q \geq 1$, $t \in [0, 1]$, and $r \geq t$,

$$\left[A^{r/2}\left(A^{-t/2}B^pA^{-t/2}\right)^q A^{r/2}\right]^{\frac{r-t}{(p-t)q+r}} \leq A^{r-t}.$$

(Proof: See [283].) (Remark: This result is due to Fujii and Nakamoto.)

Fact 8.8.26. Let $A, B \in \mathbb{F}^{n\times n}$, assume that A and B are positive definite, assume that $A \leq B$, and let $r \geq 1$. Then,

$$\left[\frac{\lambda_{\min}(A)}{\lambda_{\max}(A)}\right]^r A^r \leq B^r.$$

(Proof: See [282, p. 196].) (Remark: A more general version of this result is given in [282, pp. 193, 194].)

Fact 8.8.27. Let $A, B \in \mathbb{F}^{n\times n}$, and assume that A is positive definite and B is positive semidefinite. Then, the following statements are equivalent:

i) For all $k \in \mathbb{N}$, $B^k \leq A^k$.

ii) For all $p \in (0, \infty)$, $B^p \leq A^p$.

iii) For all $p, r \in \mathbb{R}$ such that $p > r \geq 0$,

$$\left(A^{-r/2}B^pA^{-r/2}\right)^{(2p-r)/(p-r)} \leq A^{2p-r}.$$

(Proof: See [283].) (Remark: A and B are related by the *spectral order*.)

Fact 8.8.28. Let $A, B \in \mathbb{F}^{n \times n}$, and assume that A and B are positive definite. Then, the following statements are equivalent:

i) $\log B \leq \log A$.

ii) For all $p, r \in [0, \infty)$, $B^r \leq \left(B^{r/2}A^pB^{r/2}\right)^{1/2}$.

iii) For all $p, r \in [0, \infty)$, $\left(A^{r/2}B^pA^{r/2}\right)^{r/(p+r)} \leq A^r$.

(Proof: See [800] and [282, pp. 139, 200].)

Fact 8.8.29. Let $A, B \in \mathbb{F}^{n \times n}$, assume that A and B are positive definite, let $p, q \in (0, \infty)$, and consider the following conditions:

i) $B \leq A$.

ii) $\log B \leq \log A$.

iii) $\left(A^{q/2}B^pA^{q/2}\right)^{q/(p+q)} \leq A^q$.

iv) $\log\left(A^{q/2}B^pA^{q/2}\right) \leq \log A^{p+q}$.

Then, *i*) $\implies$ *ii*) $\implies$ *iii*) $\implies$ *iv*). (Remark: This result is due to Yanagida, Ito, and Yamazaki.)

Fact 8.8.30. Let $A, B \in \mathbb{F}^{n \times n}$, assume that A and B are positive definite, and let $\alpha, \beta \in \mathbb{R}$ satisfy either $\alpha I \leq A \leq \beta I$ or $\alpha I \leq B \leq \beta I$. If $\log B \leq \log A$, then, for all $p > 0$,

$$B^p \leq S\left(p, e^{\beta - \alpha}\right)A^p,$$

where, for $p > 0$ and $h > 0$, $S(p, h)$ is defined by

$$S(p, h) \triangleq \begin{cases} \dfrac{(h^p - 1)h^{p/(h^p - 1)}}{ep \log h}, & h \neq 1, \\ 1, & h = 1. \end{cases}$$

(Proof: See [404].) (Remark: $S(p, h)$ is *Specht's ratio*.) (Remark: See Fact 11.14.9.)

8.9 Facts on Identities and Inequalities for Partitioned Matrices

Fact 8.9.1. Let $A \in \mathbb{F}^{n \times n}$, and assume that A is positive semidefinite. Then, $\begin{bmatrix} A & A \\ A & A \end{bmatrix}$ and $\begin{bmatrix} A & -A \\ -A & A \end{bmatrix}$ are positive semidefinite. Furthermore, if $\begin{bmatrix} \alpha & \beta \\ \beta & \gamma \end{bmatrix} \in \mathbb{F}^{2 \times 2}$ is positive semidefinite, then so is $\begin{bmatrix} \alpha A & \beta A \\ \beta A & \gamma A \end{bmatrix}$. Finally, if A and $\begin{bmatrix} \alpha & \beta \\ \beta & \gamma \end{bmatrix}$ are positive definite, then $\begin{bmatrix} \alpha A & \beta A \\ \beta A & \gamma A \end{bmatrix}$ is positive definite. (Proof: Use Fact 7.4.15.)

Fact 8.9.2. Let $A, B, C \in \mathbb{F}^{n \times n}$, assume that $\begin{bmatrix} A & B \\ B^* & C \end{bmatrix} \in \mathbb{F}^{2n \times 2n}$ is positive semidefinite, and assume that $\begin{bmatrix} \alpha & \beta \\ \beta & \gamma \end{bmatrix} \in \mathbb{F}^{2 \times 2}$ is positive semidefinite. Then, $\begin{bmatrix} \alpha A & \beta B \\ \beta B^* & \gamma C \end{bmatrix}$ is positive semidefinite. If, in addition, $\begin{bmatrix} A & B \\ B^* & C \end{bmatrix}$ is positive definite and α and γ are positive, then $\begin{bmatrix} \alpha A & \beta B \\ \beta B^* & \gamma C \end{bmatrix}$ is positive definite. (Proof: Note that $\begin{bmatrix} \alpha A & \beta B \\ \beta B^* & \gamma C \end{bmatrix} = \left(\begin{bmatrix} \alpha & \beta \\ \beta & \gamma \end{bmatrix} \otimes 1_{n \times n} \right) \circ \begin{bmatrix} A & B \\ B^* & C \end{bmatrix}$, and use Fact 7.4.15 and Fact 8.16.8.) (Problem: Extend this result to nonsquare B.)

Fact 8.9.3. Let $A, B, C \in \mathbb{F}^{n \times m}$, and define
$$\mathcal{A} \triangleq \begin{bmatrix} A & B \\ B^* & C \end{bmatrix}.$$
If $\mathcal{A}$ is positive semidefinite, then
$$0 \leq BC^+ B^* \leq A.$$
If $\mathcal{A}$ is positive definite, then
$$0 \leq BC^{-1} B^* < A.$$
Now, assume that $n = m$. If $\mathcal{A}$ is positive semidefinite, then
$$-A - C \leq B + B^* \leq A + C.$$
If $\mathcal{A}$ is positive definite, then
$$-A - C < B + B^* < A + C.$$
(Proof: Consider $S \mathcal{A} S^{\mathrm{T}}$, where $S \triangleq \begin{bmatrix} I & I \end{bmatrix}$ and $S \triangleq \begin{bmatrix} I & -I \end{bmatrix}$.) (Remark: See Fact 8.16.27.)

Fact 8.9.4. Let $A, B, C \in \mathbb{F}^{n \times m}$, define
$$\mathcal{A} \triangleq \begin{bmatrix} A & B \\ B^* & C \end{bmatrix},$$
and assume that $B = AA^+ B$. Then,

$$\operatorname{In}\mathcal{A} = \operatorname{In} A + \operatorname{In}(A|\mathcal{A}).$$

(Remark: This result is the *Haynsworth inertia additivity formula.* See [596].) (Remark: If $\mathcal{A}$ is positive semidefinite, then $B = AA^+B$.)

Fact 8.9.5. Let $A \in \mathbb{F}^{n \times m}$, and define

$$\mathcal{A} \triangleq \begin{bmatrix} \langle A^* \rangle & A \\ A^* & \langle A \rangle \end{bmatrix}.$$

Then, $\mathcal{A}$ is positive semidefinite. If, in addition, $n = m$, then

$$-\langle A^* \rangle - \langle A \rangle \le A + A^* \le \langle A^* \rangle + \langle A \rangle.$$

(Proof: Use Fact 8.9.3.) (Remark: See Fact 8.7.22 and Fact 8.15.2.)

Fact 8.9.6. Let $A \in \mathbb{F}^{n \times m}$ and $B \in \mathbb{F}^{n \times l}$, and define

$$\mathcal{A} \triangleq \begin{bmatrix} A^*A & A^*B \\ B^*A & B^*B \end{bmatrix}.$$

Then, $\mathcal{A}$ is positive semidefinite, and

$$0 \le A^*B(B^*B)^+B^*A \le A^*A.$$

If $m = l$, then

$$-A^*A - B^*B \le A^*B + B^*A \le A^*A + B^*B.$$

If, in addition, $m = l = 1$ and $B^*B \ne 0$, then

$$|A^*B|^2 \le A^*AB^*B.$$

(Remark: This result is the Cauchy-Schwarz inequality. See Fact 8.11.15.) (Remark: See Fact 8.16.28.)

Fact 8.9.7. Let $A, B \in \mathbb{F}^{n \times m}$, and define

$$\mathcal{A} \triangleq \begin{bmatrix} I + A^*A & I + A^*B \\ I + B^*A & I + B^*B \end{bmatrix}.$$

Then, $\mathcal{A}$ is positive semidefinite, and

$$0 \le (I + A^*B)(I + B^*B)^{-1}(I + B^*A) \le I + A^*A.$$

(Remark: See Fact 8.11.17.)

Fact 8.9.8. Let $A \in \mathbb{F}^{n \times n}$ and $B \in \mathbb{F}^{n \times m}$, assume that A is positive semidefinite, and define

$$\mathcal{A} \triangleq \begin{bmatrix} A & AB \\ B^*A & B^*AB \end{bmatrix}.$$

Then,

$$\mathcal{A} = \left[\begin{array}{c} A^{1/2} \\ B^*A^{1/2} \end{array} \right] \left[\begin{array}{cc} A^{1/2} & A^{1/2}B \end{array} \right],$$

and thus $\mathcal{A}$ is positive semidefinite. Furthermore,

$$0 \le AB(B^*AB)^+B^*A \le A.$$

Now, assume that $n = m$. Then,

$$-A - B^*AB \le AB + B^*A \le A + B^*AB.$$

Fact 8.9.9. Let $A \in \mathbb{F}^{n \times n}$ and $B \in \mathbb{F}^{n \times m}$, assume that A is positive definite, and define

$$\mathcal{A} \triangleq \left[\begin{array}{cc} A & B \\ B^* & B^*A^{-1}B \end{array} \right].$$

Then,

$$\mathcal{A} = \left[\begin{array}{c} A^{1/2} \\ B^*A^{-1/2} \end{array} \right] \left[\begin{array}{cc} A^{1/2} & A^{-1/2}B \end{array} \right],$$

and thus $\mathcal{A}$ is positive semidefinite. Furthermore,

$$0 \le B(B^*A^{-1}B)^+B^* \le A.$$

Furthermore, if $\operatorname{rank} B = m$, then

$$\operatorname{rank}\left[A - B^*(BA^{-1}B^*)^{-1}B\right] = n - m.$$

Now, assume that $n = m$. Then,

$$-A - B^*A^{-1}B \le B + B^* \le A + B^*A^{-1}B.$$

(Proof: Use Fact 8.9.3.) (Remark: See Fact 8.16.29.) (Remark: The matrix $I - A^{-1/2}B^*(BA^{-1}B^*)^+BA^{-1/2}$ is a projector.)

Fact 8.9.10. Let $A \in \mathbb{F}^{n \times n}$ and $B \in \mathbb{F}^{m \times n}$, assume that A is positive definite, and define

$$\mathcal{A} \triangleq \left[\begin{array}{cc} BAB^* & BB^* \\ BB^* & BA^{-1}B^* \end{array} \right].$$

Then,

$$\mathcal{A} = \left[\begin{array}{c} BA^{1/2} \\ BA^{-1/2} \end{array} \right] \left[\begin{array}{cc} A^{1/2}B^* & A^{-1/2}B^* \end{array} \right],$$

and thus $\mathcal{A}$ is positive semidefinite. Furthermore,

$$0 \le BB^*(BA^{-1}B^*)^+BB^* \le BAB^*.$$

Now, assume that $n = m$. Then,

$$-BAB^* - BA^{-1}B^* \le 2BB^* \le BAB^* + BA^{-1}B^*.$$

(Proof: Use Fact 8.9.3.) (Remark: See Fact 8.11.16 and Fact 8.16.29.)

Fact 8.9.11. Let $A, B \in \mathbb{F}^{n \times m}$, let $\alpha, \beta \in (0, \infty)$, and define

$$\mathcal{A} \triangleq \begin{bmatrix} \beta^{-1}I + \alpha A^*A & (A+B)^* \\ A+B & \alpha^{-1}I + \beta BB^* \end{bmatrix}.$$

Then,

$$\mathcal{A} = \begin{bmatrix} \beta^{-1/2}I & \alpha^{1/2}A^* \\ \beta^{1/2}B & \alpha^{-1/2}I \end{bmatrix} \begin{bmatrix} \beta^{-1/2}I & \beta^{1/2}B^* \\ \alpha^{1/2}A & \alpha^{-1/2}I \end{bmatrix}$$

$$= \begin{bmatrix} \alpha A^*A & A^* \\ A & \alpha^{-1}I \end{bmatrix} + \begin{bmatrix} \beta^{-1}I & B^* \\ B & \beta BB^* \end{bmatrix},$$

and thus $\mathcal{A}$ is positive semidefinite. Furthermore,

$$(A+B)^*(\alpha^{-1}I + \beta BB^*)^{-1}(A+B) \leq \beta^{-1}I + \alpha A^*A.$$

Now, assume that $n = m$. Then,

$$-\left(\beta^{-1/2} + \alpha^{-1/2}\right)I - \alpha A^*A - \beta BB^* \leq A + B + (A+B)^*$$

$$\leq \left(\beta^{-1/2} + \alpha^{-1/2}\right)I + \alpha A^*A + \beta BB^*.$$

(Remark: See Fact 8.11.18 and Fact 8.16.30.)

Fact 8.9.12. Let $A, B \in \mathbb{F}^{n \times m}$, assume that $I - A^*A$ and $I - B^*B$ are positive definite, and define

$$\mathcal{A} \triangleq \begin{bmatrix} (I - A^*A)^{-1} & (I - B^*A)^{-1} \\ (I - A^*B)^{-1} & (I - B^*B)^{-1} \end{bmatrix}.$$

Then, $\mathcal{A}$ is positive semidefinite, and

$$I - B^*B \leq (I - B^*A)(I - A^*A)^{-1}(I - A^*B).$$

Furthermore,

$$-(I - A^*A)^{-1} - (I - B^*B)^{-1} \leq (I - B^*A)^{-1} + (I - A^*B)^{-1}$$

$$\leq (I - A^*A)^{-1} + (I - B^*B)^{-1}.$$

(Remark: These results and Fact 8.11.19 are *Hua's inequalities*. See [28].)

Fact 8.9.13. Let $A, B \in \mathbb{F}^{n \times m}$ and $C, D \in \mathbb{F}^{m \times m}$, assume that C and D are positive definite, and define

$$\mathcal{A} \triangleq \begin{bmatrix} AC^{-1}A^* + BD^{-1}B^* & A+B \\ (A+B)^* & C+D \end{bmatrix}.$$

Then, $\mathcal{A}$ is positive semidefinite, and

$$(A+B)(C+D)^{-1}(A+B)^* \leq AC^{-1}A^* + BD^{-1}B^*.$$

Now, assume that $n = m$. Then,

$$-AC^{-1}A^* - BD^{-1}B^* - C - D \leq A + B + (A + B)^*$$

$$\leq AC^{-1}A^* + BD^{-1}B^* + C + D.$$

(Proof: See [342, 484] or [592, p. 151].) (Remark: Replacing A, B, C, D by $\alpha B_1, (1 - \alpha)B_2, \alpha A_1, (1 - \alpha)A_2$ yields xiv) of Proposition 8.5.15.)

Fact 8.9.14. Let $A, B, C \in \mathbb{F}^{n \times n}$, assume that $\left[\begin{smallmatrix} A & B \\ B^* & C \end{smallmatrix}\right] \in \mathbb{F}^{2n \times 2n}$ is positive semidefinite, and assume that $AB = BA$. Then,

$$B^*B \leq A^{1/2}CA^{1/2}.$$

(Proof: See [813].)

Fact 8.9.15. Let $A, B \in \mathbb{F}^{n \times n}$, and assume that A and B are Hermitian. Then, $-A \leq B \leq A$ if and only if $\left[\begin{smallmatrix} A & B \\ B & A \end{smallmatrix}\right]$ is positive semidefinite. Furthermore, $-A < B < A$ if and only if $\left[\begin{smallmatrix} A & B \\ B & A \end{smallmatrix}\right]$ is positive definite. (Proof: Note that

$$\frac{1}{\sqrt{2}}\begin{bmatrix} I & -I \\ I & I \end{bmatrix}\begin{bmatrix} A & B \\ B & A \end{bmatrix}\frac{1}{\sqrt{2}}\begin{bmatrix} I & I \\ -I & I \end{bmatrix} = \begin{bmatrix} A - B & 0 \\ 0 & A + B \end{bmatrix}.)$$

Fact 8.9.16. Let $A \in \mathbb{F}^{n \times n}$, $B \in \mathbb{F}^{n \times m}$, and $C \in \mathbb{F}^{m \times m}$, assume that $\left[\begin{smallmatrix} A & B \\ B^* & C \end{smallmatrix}\right]$ is positive semidefinite, and let $r \triangleq \operatorname{rank} B$. Then, for all $k = 1, \ldots, r$,

$$\prod_{i=1}^{k} \sigma_i(B) \leq \prod_{i=1}^{k} \max\{\lambda_i(A), \lambda_i(C)\}.$$

(Proof: See [813].)

Fact 8.9.17. Let $A \in \mathbb{R}^{n \times n}$, assume that A is positive definite, let $S \subseteq \{1, \ldots, n\}$, and let $A_{[S]}$ denote the principal submatrix of A obtained by deleting $\operatorname{row}_i(A)$ and $\operatorname{col}_i(A)$ for all $i \in S$. Then,

$$\left(A_{[S]}\right)^{-1} \leq \left(A^{-1}\right)_{[S]}.$$

(Proof: See [367, p. 474].) (Remark: Generalizations of this result are given in [181].)

Fact 8.9.18. Let $A_{ij} \in \mathbb{F}^{n_i \times n_j}$ for all $i, j = 1, \ldots, k$, define

$$A \triangleq \begin{bmatrix} A_{11} & \cdots & A_{1k} \\ \vdots & \ddots & \vdots \\ A_{1k} & \cdots & A_{kk} \end{bmatrix},$$

and assume that A is square and positive definite. Furthermore, define

$$\hat{A} \triangleq \begin{bmatrix} \hat{A}_{11} & \cdots & \hat{A}_{1k} \\ \vdots & \ddots & \vdots \\ \hat{A}_{1k} & \cdots & \hat{A}_{kk} \end{bmatrix},$$

where $\hat{A}_{ij} = 1_{1\times n_i} A_{ij} 1_{n_j \times 1}$ is the sum of the entries of A_{ij} for all $i, j = 1, \ldots, k$. Then, $\hat{A}$ is positive definite. (Proof: $\hat{A} = BAB^{\mathrm{T}}$, where the entries of $B \in \mathbb{R}^{k \times \sum_{i=1}^{k} n_i}$ are zeros and ones. See [26].)

Fact 8.9.19. Let $A, D \in \mathbb{F}^{n \times n}$, $B \in \mathbb{F}^{n \times m}$, and $C \in \mathbb{F}^{m \times m}$, and assume that $\begin{bmatrix} A & B \\ B^* & C \end{bmatrix} \in \mathbb{F}^{n \times n}$ is positive semidefinite, C is positive definite, and D is positive definite. Then, $\begin{bmatrix} A+D & B \\ B^* & C \end{bmatrix}$ is positive definite.

8.10 Facts on the Trace

Fact 8.10.1. Let $A, B \in \mathbb{F}^{n \times n}$, and assume that either A and B are Hermitian or A and B are skew Hermitian. Then, $\mathrm{tr}\, AB$ is real.

Fact 8.10.2. Let $A, B \in \mathbb{F}^{n \times n}$, assume that A and B are Hermitian, and assume that $-A \le B \le A$. Then,

$$\mathrm{tr}\, B^2 \le \mathrm{tr}\, A^2.$$

(Proof: $0 \le \mathrm{tr}[(A-B)(A+B)] = \mathrm{tr}\, A^2 - \mathrm{tr}\, B^2$. See [719].)

Fact 8.10.3. Let $A, B \in \mathbb{F}^{n \times n}$, and assume that A and B are positive semidefinite. Then, $AB = 0$ if and only if $\mathrm{tr}\, AB = 0$.

Fact 8.10.4. Let $A, B \in \mathbb{F}^{n \times n}$, assume that A and B are positive semidefinite, and let $p, q \ge 1$ satisfy $1/p + 1/q = 1$. Then,

$$\mathrm{tr}\, AB \le (\mathrm{tr}\, A^p)^{1/p} (\mathrm{tr}\, B^q)^{1/q}.$$

Furthermore, equality holds if and only if A^{p-1} and B are linearly dependent. (Remark: This result is a matrix version of Hölder's inequality.)

Fact 8.10.5. Let $A, B \in \mathbb{F}^{n \times n}$, and let $k \in \mathbb{N}$. Then,

$$|\mathrm{tr}\, (AB)^{2k}| \le \mathrm{tr}\, (A^*ABB^*)^k \le \mathrm{tr}\, (A^*A)^k (BB^*)^k.$$

(Proof: See [803].)

Fact 8.10.6. Let $A, B \in \mathbb{F}^{n \times n}$, assume that A and B are Hermitian, and let $k \in \mathbb{P}$. Then,

$$|\mathrm{tr}\, (AB)^{2k}| \le \mathrm{tr}\, (A^2 B^2)^k \le \begin{Bmatrix} \mathrm{tr}\, A^{2k} B^{2k} \\ (\mathrm{tr}\, A^2 B^2)^k \end{Bmatrix}.$$

(Proof: See [803].)

Fact 8.10.7. Let $A, B \in \mathbb{F}^{n \times n}$, and assume that A and B are positive semidefinite. Then,

$$\operatorname{tr} AB \leq \left[\operatorname{tr}\left(A^{1/2}BA^{1/2}\right)^{1/2}\right]^2 \leq (\operatorname{tr} A)(\operatorname{tr} B) \leq \tfrac{1}{4}(\operatorname{tr} A + \operatorname{tr} B)^2,$$

(Remark: Note that

$$\operatorname{tr}\left(A^{1/2}BA^{1/2}\right)^{1/2} = \sum_{i=1}^{n} \lambda_i^{1/2}(AB)$$

and

$$\operatorname{tr} AB = \operatorname{tr} A^{1/2}BA^{1/2} = \operatorname{tr}\left[\left(A^{1/2}BA^{1/2}\right)^{1/2}\left(A^{1/2}BA^{1/2}\right)^{1/2}\right].$$

The second inequality follows from Proposition 9.3.6 with $p = q = 2$, $r = 1$, and A and B replaced by $A^{1/2}$ and $B^{1/2}$.)

Fact 8.10.8. Let $A, B \in \mathbb{F}^{n \times n}$, assume that A and B are positive semidefinite, and let $p \geq 0$ and $r \geq 1$. Then,

$$\operatorname{tr}\left(A^{1/2}BA^{1/2}\right)^{pr} \leq \operatorname{tr}\left(A^{r/2}B^rA^{r/2}\right)^p.$$

In particular,

$$\operatorname{tr}\left(A^{1/2}BA^{1/2}\right)^{2p} \leq \operatorname{tr}\left(AB^2A\right)^p$$

and

$$\operatorname{tr} AB \leq \operatorname{tr}\left(AB^2A\right)^{1/2}.$$

(Proof: Use Fact 8.14.10 and Fact 8.14.11.) (Remark: This inequality is due to Araki. See [40] and [111, p. 258].) (Problem: Compare the upper bounds

$$\operatorname{tr} AB \leq \left[\operatorname{tr}\left(A^{1/2}BA^{1/2}\right)^{1/2}\right]^2$$

and

$$\operatorname{tr} AB \leq \operatorname{tr}\left(AB^2A\right)^{1/2}.)$$

Fact 8.10.9. Let $A, B \in \mathbb{F}^{n \times n}$, assume that A and B are positive semidefinite, and let $k, m \in \mathbb{P}$, where $m \geq k$. Then,

$$\operatorname{tr}\left(A^kB^k\right)^m \leq \operatorname{tr}\left(A^mB^m\right)^k.$$

In particular,

$$\operatorname{tr}(AB)^m \leq \operatorname{tr} A^mB^m.$$

If, in addition, m is even, then

$$\operatorname{tr}(AB)^m \leq \operatorname{tr}\left(A^2B^2\right)^{m/2} \leq \operatorname{tr} A^mB^m.$$

(Proof: Use Fact 8.14.10 and Fact 8.14.11.) (Remark: The result $\operatorname{tr}(AB)^m$ $\leq \operatorname{tr} A^m B^m$ is the *Lieb-Thirring inequality*. See [111, p. 279]. The inequality $\operatorname{tr}(AB)^m \leq \operatorname{tr}(A^2 B^2)^{m/2}$ follows from Fact 8.10.8. See [803].) (Problem: Compare the upper bounds

$$\operatorname{tr} AB \leq \left[\operatorname{tr} \left(A^{1/2} B A^{1/2} \right)^{1/2} \right]^2$$

and

$$\operatorname{tr} AB \leq \operatorname{tr} \left(AB^2 A \right)^{1/2}.)$$

Fact 8.10.10. Let $A, B \in \mathbb{F}^{n \times n}$, assume that A and B are positive semidefinite, and let $p \geq r \geq 0$. Then,

$$\left[\operatorname{tr} \left(A^{1/2} B A^{1/2} \right)^p \right]^{1/p} \leq \left[\operatorname{tr} \left(A^{1/2} B A^{1/2} \right)^r \right]^{1/r}.$$

In particular,

$$\left[\operatorname{tr} \left(A^{1/2} B A^{1/2} \right)^2 \right]^{1/2} \leq \operatorname{tr} AB \leq \left\{ \begin{array}{c} \operatorname{tr} \left(AB^2 A \right)^{1/2} \\[2mm] \left[\operatorname{tr} \left(A^{1/2} B A^{1/2} \right)^{1/2} \right]^2 \end{array} \right\}.$$

(Proof: The result follows from the power-sum inequality Fact 1.4.16. See [202].)

Fact 8.10.11. Let $A \in \mathbb{F}^{n \times n}$, and let $p, q \in [0, \infty)$. Then,

$$\operatorname{tr}(A^{*p} A^p)^q \leq \operatorname{tr}(A^* A)^{pq}.$$

Furthermore, equality holds if and only if $\operatorname{tr} A^{*p} A^p = \operatorname{tr}(A^* A)^p$. (Proof: See [650].)

Fact 8.10.12. Let $A \in \mathbb{F}^{n \times n}$, $p \in [2, \infty)$, and $q \in [1, \infty)$. Then, A is normal if and only if
$$\operatorname{tr}(A^{*p} A^p)^q = \operatorname{tr}(A^* A)^{pq}.$$

(Proof: See [650].)

Fact 8.10.13. Let $A, B \in \mathbb{F}^{n \times n}$, assume that A and B are positive semidefinite, assume that $A \leq B$, and let $p, q \geq 0$. Then,

$$\operatorname{tr} A^p B^q \leq \operatorname{tr} B^{p+q}.$$

If, in addition, A and B are positive definite, then this inequality holds for all $p, q \in \mathbb{R}$ satisfying $q \geq -1$ and $p + q \geq 0$. (Proof: See [139].)

Fact 8.10.14. Let $A, B \in \mathbb{F}^{n \times n}$, assume that A and B are positive semidefinite, and let $\alpha \in [0, 1]$. Then,

$$\operatorname{tr} A^\alpha B^{1-\alpha} \leq (\operatorname{tr} A)^\alpha (\operatorname{tr} B)^{1-\alpha} \leq \operatorname{tr}[\alpha A + (1 - \alpha) B].$$

Furthermore, the first inequality is an equality if and only if A and B are linearly dependent, while the second inequality is an equality if and only if $A = B$. (Remark: See Fact 1.4.5 and Fact 8.10.15.)

Fact 8.10.15. Let $A, B \in \mathbb{F}^{n \times n}$, assume that A and B are positive definite, and let $\alpha \in [0, 1]$. Then,

$$\operatorname{tr} A^{-\alpha} B^{\alpha-1} \le \left(\operatorname{tr} A^{-1}\right)^{\alpha}\left(\operatorname{tr} B^{-1}\right)^{1-\alpha} \le \operatorname{tr}\left[\alpha A^{-1} + (1-\alpha)B^{-1}\right]$$

and

$$\operatorname{tr}\left[\alpha A + (1-\alpha)B\right]^{-1} \le \left(\operatorname{tr} A^{-1}\right)^{\alpha}\left(\operatorname{tr} B^{-1}\right)^{1-\alpha} \le \operatorname{tr}\left[\alpha A^{-1} + (1-\alpha)B^{-1}\right].$$

(Remark: The lower inequalities refine the convexity of $\phi(A) = \operatorname{tr} A^{-1}$. See Fact 1.4.5 and Fact 8.10.14.) (Problem: Compare this result to Fact 8.8.23.)

Fact 8.10.16. Let $A, B \in \mathbb{F}^{n \times n}$, and assume that B is positive semidefinite. Then,

$$|\operatorname{tr} AB| \le \sigma_{\max}(A)\operatorname{tr} B.$$

(Proof: Use Proposition 8.4.13 and $\sigma_{\max}(A + A^*) \le 2\sigma_{\max}(A)$.) (Remark: See Fact 5.11.4.)

Fact 8.10.17. Let $A, B \in \mathbb{F}^{n \times n}$, assume that A and B are positive semidefinite, and let $p \ge 1$. Then,

$$[\operatorname{tr}(A^p + B^p)]^{1/p} \le [\operatorname{tr}(A + B)^p]^{1/p} \le (\operatorname{tr} A^p)^{1/p} + (\operatorname{tr} B^p)^{1/p}.$$

(Proof: See [139].) (Remark: The first inequality is the *McCarthy inequality*. The second inequality is a special case of the triangle inequality for the norm $\|\cdot\|_{\sigma p}$ and a matrix version of Minkowski's inequality.)

Fact 8.10.18. Let $A, B \in \mathbb{F}^{n \times n}$, assume that B is positive semidefinite, and assume that $A^*A \le B$. Then,

$$|\operatorname{tr} A| \le \operatorname{tr} B^{1/2}.$$

(Proof: Corollary 8.5.10 with $r = 2$ implies that $(A^*A)^{1/2} \le \operatorname{tr} B^{1/2}$. Letting $\operatorname{mspec}(A) = \{\lambda_1, \ldots, \lambda_n\}_{\mathrm{m}}$, it follows from Fact 9.10.1 that $|\operatorname{tr} A| \le \sum_{i=1}^{n}|\lambda_i| \le \sum_{i=1}^{n}\sigma_i(A) = \operatorname{tr}(A^*A)^{1/2} \le \operatorname{tr} B^{1/2}$. See [89].)

Fact 8.10.19. Let $A, P \in \mathbb{F}^{n \times n}$, $B, Q \in \mathbb{F}^{n \times m}$, and $C, R \in \mathbb{F}^{m \times m}$, and assume that $\begin{bmatrix} A & B \\ B^* & C \end{bmatrix}, \begin{bmatrix} P & Q \\ Q^* & R \end{bmatrix} \in \mathbb{F}^{(n+m) \times (n+m)}$ are positive semidefinite. Then,

$$|\operatorname{tr} BQ^*|^2 \le (\operatorname{tr} AP)(\operatorname{tr} CR).$$

(Proof: See [468, 815].)

Fact 8.10.20. Let $A, B \in \mathbb{F}^{n \times m}$, let $X \in \mathbb{F}^{n \times n}$, and assume that X is positive definite. Then,

$$|\operatorname{tr} A^*B|^2 \le (\operatorname{tr} A^*XA)(\operatorname{tr} B^*X^{-1}A).$$

(Proof: Use Fact 8.10.19 with $\left[\begin{smallmatrix} X & I \\ I & X^{-1} \end{smallmatrix}\right]$ and $\left[\begin{smallmatrix} AA^* & AB^* \\ BA^* & BB^* \end{smallmatrix}\right]$. See [468, 815].)

Fact 8.10.21. Let $A \in \mathbb{F}^{n \times n}$, $B \in \mathbb{F}^{n \times m}$, and $C \in \mathbb{F}^{m \times m}$, and assume that $\left[\begin{smallmatrix} A & B \\ B^* & C \end{smallmatrix}\right] \in \mathbb{F}^{(n+m) \times (n+m)}$ is positive semidefinite. Then,

$$\operatorname{tr} B^* B \le \sqrt{(\operatorname{tr} A^2)(\operatorname{tr} C^2)} \le (\operatorname{tr} A)(\operatorname{tr} C).$$

(Proof: Use Fact 8.10.19 with $P = A$, $Q = B$, and $R = C$.) (Remark: The inequality $\operatorname{tr} B^* B \le (\operatorname{tr} A)(\operatorname{tr} C)$ is given in [579].)

Fact 8.10.22. Let $A, B, Q, S_1, S_2 \in \mathbb{R}^{n \times n}$, assume that A and B are symmetric, assume that Q, S_1, and S_2 are orthogonal, assume that $S_1^{\mathrm{T}} A S_1$ and $S_2^{\mathrm{T}} B S_2$ are diagonal with the diagonal entries ordered from largest to smallest, and define the orthogonal matrices $Q_1, Q_2 \in \mathbb{R}^{n \times n}$ by $Q_1 \triangleq S_1 \operatorname{revdiag}(\pm 1, \dots, \pm 1) S_1^{\mathrm{T}}$ and $Q_2 \triangleq S_2 \operatorname{diag}(\pm 1, \dots, \pm 1) S_2^{\mathrm{T}}$. Then,

$$\operatorname{tr} A Q_1 B Q_1^{\mathrm{T}} \le \operatorname{tr} A Q B Q^{\mathrm{T}} \le \operatorname{tr} A Q_2 B Q_2^{\mathrm{T}}.$$

(Proof: See [83, 473].)

Fact 8.10.23. Let $A_{11} \in \mathbb{R}^{n \times n}$, $A_{12} \in \mathbb{R}^{n \times m}$, and $A_{22} \in \mathbb{R}^{m \times m}$, define $A \triangleq \left[\begin{smallmatrix} A_{11} & A_{12} \\ A_{12}^{\mathrm{T}} & A_{22} \end{smallmatrix}\right] \in \mathbb{R}^{(n+m) \times (n+m)}$, and assume that A is symmetric. Then, A is positive semidefinite if and only if, for all $B \in \mathbb{R}^{n \times m}$,

$$\operatorname{tr} B A_{12}^{\mathrm{T}} \le \operatorname{tr} \left(A_{11}^{1/2} B A_{22} B^{\mathrm{T}} A_{11}^{1/2} \right)^{1/2}.$$

(Proof: See [89].)

Fact 8.10.24. Let $A, B \in \mathbb{F}^{n \times n}$, and assume that A and B are positive definite. Then,

$$\operatorname{tr}(A - B) \le \operatorname{tr}[A(\log A - \log B)]$$

and

$$(\log \operatorname{tr} A - \log \operatorname{tr} B) \operatorname{tr} A \le \operatorname{tr}[A(\log A - \log B)].$$

(Proof: See [111, p. 281] and [85].) (Remark: The second inequality is equivalent to the thermodynamic inequality. See Fact 11.12.22.) (Remark: $\operatorname{tr}[A(\log A - \log B)]$ is the *relative entropy of Umegaki*.)

Fact 8.10.25. Let $A, B \in \mathbb{F}^{n \times n}$, assume that A and B are positive definite, and let $p > 0$. Then,

$$\operatorname{tr}[A(\log A + \log B)] \le \tfrac{1}{p} \operatorname{tr}[A^{p/2} B^p A^{p/2}].$$

Furthermore, the right-hand side is a nonincreasing function of p, and

$$\operatorname{tr}[A(\log A + \log B)] = \lim_{p \downarrow 0} \tfrac{1}{p} \operatorname{tr}[A^{p/2} B^p A^{p/2}].$$

(Proof: See [31, 86, 278, 349].) (Remark: This inequality has applications to quantum information theory.)

8.11 Facts on the Determinant

Fact 8.11.1. Let $A \in \mathbb{F}^{n \times n}$, and assume that $A + A^*$ is positive semidefinite. Then,

$$\det \tfrac{1}{2}(A + A^*) \leq |\det A|.$$

Furthermore, if $A + A^*$ is positive definite, then equality holds if and only if A is Hermitian. (Remark: This result is the *Ostrowski-Taussky inequality*.) (Remark: This result is equivalent to Fact 8.11.3.)

Fact 8.11.2. Let $A \in \mathbb{F}^{n \times n}$, and assume that A is positive definite. Then,

$$n + \log \det A \leq n(\det A)^{1/n} \leq \operatorname{tr} A \leq \left(n \operatorname{tr} A^2\right)^{1/2},$$

with equality if and only if $A = I$.

Fact 8.11.3. Let $A, B \in \mathbb{F}^{n \times n}$, and assume that A is positive semidefinite and B is skew Hermitian. Then,

$$\det A \leq |\det(A + B)|.$$

Furthermore, if A is positive definite, then equality holds if and only if $B = 0$. (Proof: See [592, pp. 146, 163] and [339, p. 447].) (Remark: This result is equivalent to Fact 8.11.1.) (Remark: Suppose that A and B are real. If A is positive definite, then $\det(A + B) = (\det A)\det\left(I + A^{-1/2}BA^{-1/2}\right)$ is real. If A is positive semidefinite, then a continuity argument implies that $\det(A + B)$ is real. Thus, $\det A \leq \det(A + B)$. (Remark: Extensions of this result are given in [123].)

Fact 8.11.4. Let $A, B \in \mathbb{F}^{n \times n}$, assume that A and B are positive semidefinite, and assume that $B \leq A$. Then,

$$\det A + n \det B \leq \det(A + B).$$

(Proof: See [592, pp. 154, 166].) (Remark: Under weaker conditions, Corollary 8.4.15 implies that $\det A + \det B \leq \det(A + B)$.)

Fact 8.11.5. Let $A \in \mathbb{F}^{n \times n}$, and assume that $\frac{1}{2\jmath}(A - A^*)$ is positive definite. Then,

$$B \triangleq \left[\tfrac{1}{2}(A + A^*)\right]^{1/2} A^{-1} A^* \left[\tfrac{1}{2}(A + A^*)\right]^{-1/2}$$

is unitary. (Proof: See [249].) (Remark: A is *strictly dissipative* if $\frac{1}{2\jmath}(A - A^*)$ is negative definite. A is strictly dissipative if and only if $-\jmath A$ is dissipative. See [247, 248].) (Remark: $A^{-1}A^*$ is similar to a unitary matrix. See Fact 3.7.10.)

Fact 8.11.6. Let $A \in \mathbb{R}^{n \times n}$, and assume that $A + A^{\mathrm{T}}$ is positive semidefinite. Then,

$$\left[\tfrac{1}{2}(A + A^{\mathrm{T}})\right]^{\mathrm{A}} \leq \tfrac{1}{2}(A^{\mathrm{A}} + A^{\mathrm{AT}}).$$

Now, assume that $A + A^{\mathrm{T}}$ is positive definite. Then,

$$\left[\det \tfrac{1}{2}(A + A^{\mathrm{T}})\right]\left[\tfrac{1}{2}(A + A^{\mathrm{T}})\right]^{-1} \leq (\det A)\left[\tfrac{1}{2}(A^{-1} + A^{-\mathrm{T}})\right].$$

Furthermore,

$$\left[\det \tfrac{1}{2}(A + A^{\mathrm{T}})\right]\left[\tfrac{1}{2}(A + A^{\mathrm{T}})\right]^{-1} < (\det A)\left[\tfrac{1}{2}(A^{-1} + A^{-\mathrm{T}})\right]$$

if and only if $\mathrm{rank}(A - A^{\mathrm{T}}) \geq 4$. Finally, if $n \geq 4$ and $A - A^{\mathrm{T}}$ is nonsingular, then

$$(\det A)\left[\tfrac{1}{2}(A^{-1} + A^{-\mathrm{T}})\right] < \left[\det A - \det \tfrac{1}{2}(A - A^{\mathrm{T}})\right]\left[\tfrac{1}{2}(A + A^{\mathrm{T}})\right]^{-1}.$$

(Proof: See [248, 394].) (Remark: This result does not hold for complex matrices.)

Fact 8.11.7. Let $A, B \in \mathbb{F}^{n \times n}$, assume that B is Hermitian, and assume that $A^*BA < A + A^*$. Then, $\det A \neq 0$.

Fact 8.11.8. Let $A, B \in \mathbb{F}^{n \times n}$, assume that A and B are positive definite, and let $\alpha \in [0, 1]$. Then,

$$(\det A)^{\alpha}(\det B)^{1-\alpha} \leq \det[\alpha A + (1 - \alpha)B].$$

Furthermore, equality holds if and only if $A = B$. (Remark: This result is due to Bergstrom.)

Fact 8.11.9. Let $A, B \in \mathbb{F}^{n \times n}$, assume that A and B are positive semidefinite, assume that $0 \leq A \leq B$, and let $\alpha \in [0, 1]$. Then,

$$\det[\alpha A + (1 - \alpha)B] \leq \alpha \det A + (1 - \alpha)\det B.$$

(Proof: See [762].)

Fact 8.11.10. Let $A, B \in \mathbb{F}^{n \times n}$, and assume that A and B are positive definite. Then,

$$\frac{\det A}{\det A_{[1,1]}} + \frac{\det B}{\det B_{[1,1]}} \leq \frac{\det(A + B)}{\det(A_{[1,1]} + B_{[1,1]})}.$$

(Proof: See [592, p. 145].)

Fact 8.11.11. Let $A_1, \ldots, A_k \in \mathbb{F}^{n \times n}$, assume that $A_1, \ldots, A_k$ are positive semidefinite, and let $\lambda_1, \ldots, \lambda_k \in \mathbb{C}$. Then,

$$\det\left(\sum_{i=1}^{k} \lambda_i A_i\right) \leq \det\left(\sum_{i=1}^{k} |\lambda_i| A_i\right).$$

(Proof: See [592, p. 144].)

Fact 8.11.12. Let $A, B, C \in \mathbb{R}^{n \times n}$, let $D \triangleq A + \jmath B$, and assume that $CB + B^{\mathrm{T}}C^{\mathrm{T}} < D + D^*$. Then, $\det A \neq 0$.

Fact 8.11.13. Let $A, B \in \mathbb{F}^{n \times n}$, assume that A and B are positive semidefinite, and let $m \in \mathbb{P}$. Then,

$$n^{1/m}(\det AB)^{1/n} \leq (\operatorname{tr} A^m B^m)^{1/m}.$$

(Proof: See [202].) (Remark: Assuming $\det B = 1$ and setting $m = 1$ yields Proposition 8.4.14.)

Fact 8.11.14. Let $A, B, C \subset \mathbb{F}^{n \times n}$, define

$$\mathcal{A} \triangleq \left[\begin{array}{cc} A & B \\ B^* & C \end{array} \right],$$

and assume that $\mathcal{A}$ is positive semidefinite. Then,

$$|\det(B + B^*)| \leq \det(A + C).$$

If, in addition, $\mathcal{A}$ is positive definite, then

$$|\det(B + B^*)| < \det(A + C).$$

(Remark: Use Fact 8.9.3.)

Fact 8.11.15. Let $A, B \in \mathbb{F}^{n \times m}$. Then,

$$|\det A^* B|^2 \leq (\det A^* A)(\det B^* B).$$

(Proof: Use Fact 8.9.6 or apply Fact 8.11.31 to $\left[\begin{smallmatrix} A^*A & B^*A \\ A^*B & B^*B \end{smallmatrix} \right]$.)

Fact 8.11.16. Let $A \in \mathbb{F}^{n \times n}$, assume that A is positive definite, and let $B \in \mathbb{F}^{m \times n}$, where $\operatorname{rank} B = m$. Then,

$$(\det BB^*)^2 \leq (\det BAB^*) \det BA^{-1}B^*.$$

(Proof: Use Fact 8.9.10.)

Fact 8.11.17. Let $A, B \in \mathbb{F}^{n \times m}$. Then,

$$|\det(I + AB^*)|^2 \leq \det(I + AA^*) \det(I + BB^*).$$

(Proof: Use Fact 8.9.7.)

Fact 8.11.18. Let $A, B \in \mathbb{F}^{n \times n}$, and let $\alpha, \beta \in (0, \infty)$. Then,

$$|\det(A + B)|^2 \leq \det(\beta^{-1}I + \alpha A^* A) \det(\alpha^{-1}I + \beta BB^*).$$

(Proof: Use Fact 8.9.11. See [812].)

Fact 8.11.19. Let $A, B \in \mathbb{F}^{n \times m}$, and assume that $I - A^* A$ and $I - B^* B$ are positive definite. Then,

$$0 < \det(I - A^* A) \det(I - B^* B) \leq [\det(I - A^* B)]^2.$$

(Remark: These results and Fact 8.9.12 are *Hua's inequalities*. See [28].)

Fact 8.11.20. Let $A \in \mathbb{F}^{n \times m}$, $B \in \mathbb{F}^{n \times l}$, $C \in \mathbb{F}^{n \times m}$, and $D \in \mathbb{F}^{n \times l}$. Then,
$$|\det(AC^* + BD^*)|^2 \leq \det(AA^* + BB^*)\det(CC^* + DD^*).$$

(Proof: Use Fact 8.11.27 and $\mathcal{A}\mathcal{A}^* \geq 0$, where $\mathcal{A} \triangleq \left[\begin{smallmatrix} A & B \\ C & D \end{smallmatrix}\right]$.) (Remark: See Fact 2.13.19.)

Fact 8.11.21. Let $A \in \mathbb{F}^{n \times m}$, $B \in \mathbb{F}^{n \times m}$, $C \in \mathbb{F}^{k \times m}$, and $D \in \mathbb{F}^{k \times m}$. Then,
$$|\det(A^*B + C^*D)|^2 \leq \det(A^*A + C^*C)\det(B^*B + D^*D).$$

(Proof: Use Fact 8.11.27 and $\mathcal{A}^*\mathcal{A} \geq 0$, where $\mathcal{A} \triangleq \left[\begin{smallmatrix} A & B \\ C & D \end{smallmatrix}\right]$.) (Remark: See Fact 2.13.15.)

Fact 8.11.22. Let $A, B, C \in \mathbb{F}^{n \times n}$. Then,
$$|\det(B + CA)|^2 \leq \det(A^*A + B^*B)\det(I + CC^*).$$

(Proof: See [375].) (Remark: See Fact 8.8.20.)

Fact 8.11.23. Let $A, B \in \mathbb{F}^{n \times m}$. Then, there exist unitary matrices $S_1, S_2 \in \mathbb{F}^{n \times n}$ such that
$$I + \langle A + B \rangle \leq S_1(I + \langle A \rangle)^{1/2} S_2(I + \langle B \rangle) S_2^*(I + \langle A \rangle)^{1/2} S_1^*.$$

Therefore,
$$\det(I + \langle A + B \rangle) \leq \det(I + \langle A \rangle)\det(I + \langle B \rangle).$$

(Proof: See [28, 693].) (Remark: This result is due to Seiler and Simon.)

Fact 8.11.24. Let $A, B \in \mathbb{F}^{n \times n}$, assume that $A + A^* > 0$ and $B + B^* \geq 0$, and let $\alpha > 0$. Then, $\alpha I + AB$ is nonsingular and has no negative eigenvalues. Hence,
$$\det(\alpha I + AB) > 0.$$

(Proof: See [319].) (Remark: Equivalently, $-A$ is dissipative and $-B$ is semidissipative.) (Problem: Find a positive lower bound for $\det(\alpha I + AB)$ in terms of α, A, and B.)

Fact 8.11.25. Let $A \in \mathbb{F}^{n \times n}$. Then,
$$|\det A| \leq \prod_{i=1}^{n} \left(\sum_{j=1}^{n} |A_{(i,j)}|^2 \right)^{1/2}.$$

Furthermore, equality holds if and only if AA^* is diagonal. Now, let $\alpha > 0$ be such that, for all $i, j = 1, \ldots, n$, $|A_{(i,j)}| \leq \alpha$. Then,
$$|\det A| \leq \alpha^n n^{n/2}.$$

If, in addition, at least one entry of A has absolute value less than α, then
$$|\det A| < \alpha^n n^{n/2}.$$

(Remark: Replace A with AA^* in Fact 8.14.8.)

Fact 8.11.26. Let $A \in \mathbb{F}^{n \times n}$, $B \in \mathbb{F}^{n \times m}$, and $C \in \mathbb{F}^{m \times m}$, define $\mathcal{A} \triangleq \left[\begin{smallmatrix} A & B \\ B^* & C \end{smallmatrix}\right] \in \mathbb{F}^{(n+m) \times (n+m)}$, and assume that $\mathcal{A}$ is positive definite. Then,

$$\det \mathcal{A} = (\det A)\det\left(C - B^* C^+ B\right) \leq (\det A)\det C \leq \prod_{i=1}^{n+m} \mathcal{A}_{(i,i)}.$$

(Proof: The second inequality is obtained by successive application of the first inequality.) (Remark: $\det \mathcal{A} \leq (\det A)\det C$ is *Fischer's inequality*.)

Fact 8.11.27. Let $A, B, C \in \mathbb{F}^{n \times n}$, define $\mathcal{A} \triangleq \left[\begin{smallmatrix} A & B \\ B^* & C \end{smallmatrix}\right] \in \mathbb{F}^{2n \times 2n}$, and assume that $\mathcal{A}$ is positive semidefinite. Then,

$$0 \leq (\det A)\det C - |\det B|^2 \leq \det \mathcal{A} \leq (\det A)\det C.$$

Hence,

$$|\det B|^2 \leq (\det A)\det C.$$

Furthermore, $\mathcal{A}$ is positive definite if and only if

$$|\det B|^2 < (\det A)\det C.$$

(Proof: Assuming that A is positive definite, it follows that $0 \leq B^* A^{-1} B \leq C$, which implies that $|\det B|^2/\det A \leq \det C$. Then, use continuity for the case in which A is singular. For an alternative proof, see [592, p. 142]. For the case in which $\mathcal{A}$ is positive definite, note that $0 \leq B^* A^{-1} B < C$, and thus $|\det B|^2/\det A < \det C$.) (Remark: This result is due to Everitt.) (Remark: See Fact 8.11.31.) (Remark: It is not generally true that $|\det(B^* B)|^2 < (\det A)\det C$ for nonsquare B. See [813].)

Fact 8.11.28. Let $A \in \mathbb{F}^{n \times n}$, $B \in \mathbb{F}^{n \times m}$, and $C \in \mathbb{F}^{m \times m}$, define $\mathcal{A} \triangleq \left[\begin{smallmatrix} A & B \\ B^* & C \end{smallmatrix}\right] \in \mathbb{F}^{(n+m) \times (n+m)}$, and assume that $\mathcal{A}$ is positive semidefinite and A is positive definite. Then,

$$B^* A^{-1} B \leq \left[\frac{\lambda_{\max}(\mathcal{A}) - \lambda_{\min}(\mathcal{A})}{\lambda_{\max}(\mathcal{A}) + \lambda_{\min}(\mathcal{A})}\right]^2 C.$$

(Proof: See [468, 815].)

Fact 8.11.29. Let $A, B, C \in \mathbb{F}^{n \times n}$, define $\mathcal{A} \triangleq \left[\begin{smallmatrix} A & B \\ B^* & C \end{smallmatrix}\right] \in \mathbb{F}^{2n \times 2n}$, and assume that $\mathcal{A}$ is positive semidefinite. Then,

$$|\det B|^2 \leq \left[\frac{\lambda_{\max}(\mathcal{A}) - \lambda_{\min}(\mathcal{A})}{\lambda_{\max}(\mathcal{A}) + \lambda_{\min}(\mathcal{A})}\right]^{2n} (\det A)\det C.$$

Hence,

$$|\det B|^2 \leq \left[\frac{\lambda_{\max}(\mathcal{A}) - \lambda_{\min}(\mathcal{A})}{\lambda_{\max}(\mathcal{A}) + \lambda_{\min}(\mathcal{A})}\right]^2 (\det A)\det C.$$

Now, define $\hat{\mathcal{A}} \triangleq \left[\begin{smallmatrix} \det A & \det B \\ \det B^* & \det C \end{smallmatrix}\right] \in \mathbb{F}^{2 \times 2}$. Then,

$$|\det B|^2 \le \left[\frac{\lambda_{\max}(\hat{A}) - \lambda_{\min}(\hat{A})}{\lambda_{\max}(\hat{A}) + \lambda_{\min}(\hat{A})}\right]^2 (\det A)\det C.$$

(Proof: See [468, 815].) (Remark: The second and third bounds are not comparable. See [468, 815].)

Fact 8.11.30. Let $A \in \mathbb{F}^{n\times n}$, $B \in \mathbb{F}^{n\times m}$, and $C \in \mathbb{F}^{m\times m}$, define $\mathcal{A} \triangleq \left[\begin{smallmatrix} A & B \\ B^* & C \end{smallmatrix}\right] \in \mathbb{F}^{(n+m)\times(n+m)}$, assume that $\mathcal{A}$ is positive semidefinite, and assume that A and C are positive definite. Then,

$$\det(A|\mathcal{A})\det(C|\mathcal{A}) \le \det \mathcal{A}.$$

(Proof: See [375].) (Remark: This result is the *reversed Fischer inequality*.)

Fact 8.11.31. Let $A_{ij} \in \mathbb{F}^{n\times n}$ for all $i, j = 1, \ldots, k$, define $A \triangleq \begin{bmatrix} A_{11} & \cdots & A_{1k} \\ \vdots & \ddots & \vdots \\ A_{1k}^* & \cdots & A_{kk} \end{bmatrix}$, and assume that A is positive semidefinite. Then,

$$\det \begin{bmatrix} \det A_{11} & \cdots & \det A_{1k} \\ \vdots & \ddots & \vdots \\ \det A_{1k}^* & \cdots & \det A_{kk} \end{bmatrix} \le \det A$$

and

$$\begin{bmatrix} \operatorname{tr} A_{11} & \cdots & \operatorname{tr} A_{1k} \\ \vdots & \ddots & \vdots \\ \operatorname{tr} A_{1k}^* & \cdots & \operatorname{tr} A_{kk} \end{bmatrix} \ge 0.$$

(Remark: The matrix whose (i, j) entry is $\det A_{ij}$ is a *determinantal compression* of A. See [211, 212, 579, 691].)

8.12 Facts on Quadratic Forms

Fact 8.12.1. Let $x, y \in \mathbb{F}^n$. Then, $xx^* \le yy^*$ if and only if there exists $\alpha \in \mathbb{F}$ such that $|\alpha| \in [0, 1]$ and $x = \alpha y$.

Fact 8.12.2. Let $x, y \in \mathbb{F}^n$. Then, $xy^* + yx^* \ge 0$ if and only if x and y are linearly dependent. (Proof: Evaluate the product of the nonzero eigenvalues of $xy^* + yx^*$, and use the Cauchy-Schwarz inequality $|x^*y|^2 \le x^*x y^*y$.)

Fact 8.12.3. Let $A \in \mathbb{F}^{n\times n}$, assume that A is positive definite, and let $x, y \in \mathbb{F}^n$. Then,

$$2\operatorname{Re} x^*y \le x^*Ax + y^*A^{-1}y.$$

Furthermore, if $y = Ax$, then equality holds. Therefore,

$$x^*Ax = \max_{z \in \mathbb{F}^n} [2\mathrm{Re}\, x^*z - z^*Az].$$

(Proof: $\left(A^{1/2}x - A^{-1/2}y\right)^* \left(A^{1/2}x - A^{-1/2}y\right) \geq 0$.) (Remark: This result is due to Bellman. See [468, 815].)

Fact 8.12.4. Let $A \in \mathbb{F}^{n \times n}$, assume that A is positive definite, and let $x, y \in \mathbb{F}^n$. Then,

$$|x^*y|^2 \leq (x^*Ax)(y^*A^{-1}y).$$

(Proof: Use Fact 8.9.6 with A replaced by $A^{1/2}x$ and B replaced by $A^{-1/2}y$.)

Fact 8.12.5. Let $A \in \mathbb{F}^{n \times n}$, assume that A is positive definite, and let $x \in \mathbb{F}^n$. Then,

$$(x^*x)^2 \leq (x^*Ax)(x^*A^{-1}x) \leq \frac{(\alpha + \beta)^2}{4\alpha\beta}(x^*x)^2,$$

where $\alpha \triangleq \lambda_{\min}(A)$ and $\beta \triangleq \lambda_{\max}(A)$. (Remark: The second inequality is the *Kantorovich inequality*. See Fact 1.4.18 and [13]. See also [491].)

Fact 8.12.6. Let $A \in \mathbb{F}^{n \times n}$, assume that A is positive definite, let $y \in \mathbb{F}^n$, let $\alpha > 0$, and define $f: \mathbb{F}^n \mapsto \mathbb{R}$ by $f(x) \triangleq |x^*y|^2$. Then,

$$x_0 = \sqrt{\frac{\alpha}{y^*A^{-1}y}} A^{-1}y$$

minimizes $f(x)$ subject to $x^*Ax \leq \alpha$. Furthermore, $f(x_0) = \alpha y^*A^{-1}y$. (Proof: See [18].)

Fact 8.12.7. Let $A \in \mathbb{F}^{n \times n}$, assume that A is positive semidefinite, and let $x \in \mathbb{F}^n$. Then,

$$\left(x^*A^2x\right)^2 \leq (x^*Ax)(x^*A^3x)$$

and

$$(x^*Ax)^2 \leq (x^*x)(x^*A^2x).$$

(Proof: Apply the Cauchy-Schwarz inequality Corollary 9.1.7.)

Fact 8.12.8. Let $A \in \mathbb{F}^{n \times n}$, assume that A is positive semidefinite, and let $x \in \mathbb{F}^n$. If $\alpha \in [0, 1]$, then

$$x^*A^\alpha x \leq (x^*x)^{1-\alpha}(x^*Ax)^\alpha.$$

Furthermore, if $\alpha > 1$, then

$$(x^*Ax)^\alpha \leq (x^*x)^{\alpha-1}x^*A^\alpha x.$$

(Remark: The first inequality is the *Hölder-McCarthy inequality*, which is equivalent to the Young inequality. See Fact 8.8.23 and [282, p. 125]. Matrix versions of the second inequality are given in [362].)

Fact 8.12.9. Let $A, B \in \mathbb{R}^n$, and assume that A is Hermitian and B is positive definite. Then,

$$\lambda_{\max}(AB^{-1}) = \max\{\lambda \in \mathbb{R} \colon \ \det(A - \lambda B) = 0\} = \min_{x \in \mathbb{F}^n \setminus \{0\}} \frac{x^* A x}{x^* B x}.$$

(Proof: Use Lemma 8.4.3.)

Fact 8.12.10. Let $A, B \in \mathbb{F}^{n \times n}$, and assume that A is positive definite and B is positive semidefinite. Then,

$$4(x^* x)(x^* B x) < (x^* A x)^2$$

for all nonzero $x \in \mathbb{F}^n$ if and only if there exists $\alpha > 0$ such that

$$\alpha I + \alpha^{-1} B < A.$$

In this case, $4B < A^2$, and hence $2B^{1/2} < A$. (Proof: Sufficiency follows from $\alpha x^* x + \alpha^{-1} x^* B x < x^* A x$. Necessity follows from Fact 8.12.11. The last result follows from $(A - 2\alpha I)^2 \geq 0$ or $2B^{1/2} \leq \alpha I + \alpha^{-1} B$.)

Fact 8.12.11. Let $A, B, C \in \mathbb{F}^{n \times n}$, assume that A, B, C are positive semidefinite, and assume that

$$4(x^* C x)(x^* B x) < (x^* A x)^2$$

for all nonzero $x \in \mathbb{F}^n$. Then, there exists $\alpha > 0$ such that

$$\alpha C + \alpha^{-1} B < A.$$

(Proof: See [582].)

Fact 8.12.12. Let $A, B \in \mathbb{F}^{n \times n}$, and assume that A is Hermitian and B is positive semidefinite. Then, $x^* A x < 0$ for all $x \in \mathbb{F}^n$ such that $Bx = 0$ and $x \neq 0$ if and only if there exists $\alpha > 0$ such that $A < \alpha B$. (Proof: Suppose that, for every $\alpha > 0$, there exists a nonzero vector x such that $x^* A x \geq \alpha x^* B x$. Now, $Bx = 0$ implies that $x^* A x \geq 0$.)

Fact 8.12.13. Let $A, B \in \mathbb{C}^{n \times n}$, and assume that A and B are Hermitian. Then, the following statements are equivalent:

i) There exist $\alpha, \beta \in \mathbb{R}$ such that $\alpha A + \beta B$ is positive definite.

ii) $\{x \in \mathbb{C}^n \colon \ x^* A x = x^* B x = 0\} = \{0\}$.

(Remark: This result is *Finsler's lemma*. See [48, 733].) (Remark: See Fact 8.13.6.)

Fact 8.12.14. Let $A, B \in \mathbb{R}^{n \times n}$, and assume that A and B are symmetric. Then, the following statements are equivalent:

i) There exist $\alpha, \beta \in \mathbb{R}$ such that $\alpha A + \beta B$ is positive definite.

ii) Either $x^{\mathrm{T}}Ax \geq 0$ for all $x \in \{y \in \mathbb{F}^n \colon y^{\mathrm{T}}By = 0\}$ or $x^{\mathrm{T}}Ax \leq 0$ for all $x \in \{y \in \mathbb{F}^n \colon y^{\mathrm{T}}By = 0\}$.

Now, assume that $n \geq 3$. Then, the following statement is equivalent to *i*) and *ii*):

iii) $\{x \in \mathbb{R}^n \colon x^{\mathrm{T}}Ax = x^{\mathrm{T}}Bx = 0\} = \{0\}$.

(Remark: This result is *Finsler's lemma.* See [48, 733].) (Remark: See Fact 8.13.6.)

Fact 8.12.15. Let $A, B \in \mathbb{C}^{n \times n}$, assume that A and B are Hermitian, and assume that $x^*(A + \jmath B)x$ is nonzero for all nonzero $x \in \mathbb{C}^n$. Then, there exists $t \in \mathbb{R}$ such that $(\sin t)A + (\cos t)B$ is positive definite. (Proof: See [194].) (Remark: This result is due to Stewart.)

Fact 8.12.16. Let $A \in \mathbb{R}^{n \times n}$, assume that A is symmetric, let $B \in \mathbb{R}^{m \times n}$, and assume that B has full row rank. Then, $x^{\mathrm{T}}Ax$ is positive for all $x \in \mathcal{N}(B)$ if and only if the last $n - m$ leading principal subdeterminants of the matrix

$$\begin{bmatrix} 0_{m \times m} & B \\ B^{\mathrm{T}} & A \end{bmatrix}$$

have sign $(-1)^m$. (Proof: See [53, p. 20] or [496, p. 312].) (Remark: This result is due to Mann.)

Fact 8.12.17. Let $A \in \mathbb{F}^{n \times n}$, assume that A is positive semidefinite and nonzero, let $x, y \in \mathbb{F}^n$, and assume that $x^*y = 0$. Then,

$$|x^*Ay|^2 \leq \left[\frac{\lambda_{\max}(A) - \lambda_{\min}(A)}{\lambda_{\max}(A) + \lambda_{\min}(A)} \right]^2 (x^*Ax)(y^*Ay).$$

Furthermore, there exist vectors $x, y \in \mathbb{F}^n$ satisfying $x^*y = 0$ for which equality holds. (Proof: See [369, p. 443] or [468, 815].) (Remark: This result is the *Wielandt inequality.*)

Fact 8.12.18. Let $A \in \mathbb{F}^{n \times n}$, $B \in \mathbb{F}^{n \times m}$, and $C \in \mathbb{F}^{m \times m}$, define $\mathcal{A} \triangleq \begin{bmatrix} A & B \\ B^* & C \end{bmatrix}$, and assume that A and C are positive semidefinite. Then, the following statements are equivalent:

i) $\mathcal{A}$ is positive semidefinite.

ii) $|x^*By|^2 \leq (x^*Ax)(y^*Cy)$ for all $x \in \mathbb{F}^n$ and $y \in \mathbb{F}^m$.

iii) $2|x^*By| \leq x^*Ax + y^*Cy$ for all $x \in \mathbb{F}^n$ and $y \in \mathbb{F}^m$.

If, in addition, A and C are positive definite, then the following statement is equivalent to *i*)–*iii*):

iv) $\mathrm{sprad}(B^*A^{-1}BC^{-1}) \leq 1$.

Finally, if A is positive semidefinite and nonzero, then, for all $x \in \mathbb{F}^n$ and $y \in \mathbb{F}^m$,

$$|x^*By|^2 \leq \left[\frac{\lambda_{\max}(A) - \lambda_{\min}(A)}{\lambda_{\max}(A) + \lambda_{\min}(A)}\right]^2 (x^*Ax)(y^*Cy).$$

(Proof: See [367, p. 473] and [468, 815].)

Fact 8.12.19. Let $A \in \mathbb{F}^{n \times n}$, assume that A is Hermitian, let $x, y \in \mathbb{F}^n$, and assume that $x^*x = y^*y = 1$ and $x^*y = 0$. Then,

$$2|x^*Ay| \leq \lambda_{\max}(A) - \lambda_{\min}(A).$$

Furthermore, there exist vectors $x, y \in \mathbb{F}^n$ satisfying $x^*x = y^*y = 1$ and $x^*y = 0$ for which equality holds. (Proof: See [468, 815].) (Remark: $\lambda_{\max}(A) - \lambda_{\min}(A)$ is the *spread* of A.)

Fact 8.12.20. Let $A \in \mathbb{R}^{n \times n}$, and assume that A is positive definite. Then,

$$\int_{\mathbb{R}^n} e^{-x^{\mathrm{T}}Ax} \, \mathrm{d}x = \frac{\pi^{n/2}}{\sqrt{\det A}}.$$

Fact 8.12.21. Let $A, B \in \mathbb{R}^{n \times n}$, assume that A and B are positive definite, and, for $k = 0, 1, 2, 3$, define

$$\mathfrak{I}_k \triangleq \frac{1}{(2\pi)^{n/2}\sqrt{\det A}} \int_{\mathbb{R}^n} \left(x^{\mathrm{T}}Bx\right)^k e^{-\frac{1}{2}x^{\mathrm{T}}A^{-1}x} \, \mathrm{d}x.$$

Then,

$$\mathfrak{I}_0 = 1,$$
$$\mathfrak{I}_1 = \operatorname{tr} AB,$$
$$\mathfrak{I}_2 = (\operatorname{tr} AB)^2 + 2\operatorname{tr}(AB)^2,$$
$$\mathfrak{I}_3 = (\operatorname{tr} AB)^3 + 6(\operatorname{tr} AB)\left[\operatorname{tr}(AB)^2\right] + 8\operatorname{tr}(AB)^3.$$

(Proof: See [538, p. 80].) (Remark: These identities are *Lancaster's formulas.*)

Fact 8.12.22. Let $A \in \mathbb{R}^{n \times n}$, assume that A is positive definite, let $B \in \mathbb{R}^{n \times n}$, let $a, b \in \mathbb{R}^n$, and let $\alpha, \beta \in \mathbb{R}$. Then,

$$\int_{\mathbb{R}^n} \left(x^{\mathrm{T}}Bx + b^{\mathrm{T}}x + \beta\right)e^{-(x^{\mathrm{T}}Ax + a^{\mathrm{T}}x + \alpha)} \, \mathrm{d}x$$

$$= \frac{\pi^{n/2}}{2\sqrt{\det A}}\left[2\beta + \operatorname{tr}\left(A^{-1}B\right) - b^{\mathrm{T}}A^{-1}a + \tfrac{1}{2}a^{\mathrm{T}}A^{-1}BA^{-1}a\right]e^{\frac{1}{4}a^{\mathrm{T}}A^{-1}a - \alpha}.$$

(Proof: See [339, p. 322].)

Fact 8.12.23. Let $A \in \mathbb{F}^{n \times n}$, assume that A is Hermitian, let $b \in \mathbb{F}^n$ and $c \in \mathbb{R}$, and define $f\colon \mathbb{F}^n \mapsto \mathbb{R}$ by $f(x) \triangleq x^*Ax + \operatorname{Re}(b^*x) + c$. Then, f

is convex if and only if A is positive semidefinite, while f is strictly convex if and only if A is positive definite. (Remark: *Strictly convex* means that $f(\alpha x_1 + (1 - \alpha)x_2) < \alpha f(x_1) + (1 - \alpha)f(x_2)$ for all $\alpha \in (0,1)$ and for all $x_1, x_2 \in \mathbb{F}^n$ such that $x_1 \neq x_2$.) Now, assume that A is positive semidefinite. Then, f has a minimizer if and only if $b \in \mathcal{R}(A)$. In this case, the following statements hold. The vector $x_0 \in \mathbb{F}^n$ is a minimizer of f if and only if x_0 satisfies $Ax_0 = -\frac{1}{2}b$. Hence, $x_0 \in \mathbb{F}^m$ minimizes f if and only if there exists a vector $y \in \mathbb{F}^m$ such that

$$x_0 = -\tfrac{1}{2}A^+b + (I - A^+A)y.$$

The minimum of f is given by

$$f(x_0) = c - x_0^*Ax_0 = c - \tfrac{1}{4}b^*A^+b.$$

Furthermore, if A is positive definite, then $x_0 = -\frac{1}{2}A^{-1}b$ is the unique minimizer of f, and the minimum of f is given by

$$f(x_0) = c - x_0^*Ax_0 = c - \tfrac{1}{4}b^*A^{-1}b.$$

(Proof: Use Proposition 6.1.7 and note that, for every x_0 satisfying $Ax_0 = -\frac{1}{2}b$, it follows that

$$f(x_0) = (x - x_0)^*A(x - x_0) + c - x_0^*Ax_0$$
$$= (x - x_0)^*A(x - x_0) + c - \tfrac{1}{4}b^*A^+b.)$$

(Remark: This result is the *quadratic minimization lemma*.) (Remark: See Fact 6.4.46.)

Fact 8.12.24. Let $p, q \in \mathbb{R}$, and define $\phi \colon \mathbf{P}^n \times \mathbf{P}^n \to (0, \infty)$ by

$$\phi(A, B) \triangleq \operatorname{tr} A^pB^q.$$

Then, the following statements hold:

i) If $p, q \in (0, 1)$ and $p + q \leq 1$, then ϕ is concave.

ii) If either $p, q \in [-1, 0)$ or $p \in [-1, 0)$, $q \in [1, 2]$, and $p + q \geq 1$, or $p \in [1, 2]$, $q \in [-1, 0]$, and $p + q \geq 1$, then ϕ is convex.

iii) If p, q do not satisfy the hypotheses of either *i)* or *ii)*, then ϕ is neither convex nor concave.

(Proof: See [88].)

8.13 Facts on Matrix Transformations

Fact 8.13.1. Let $A \in \mathbb{F}^{n \times n}$. Then, AA^* and A^*A are unitarily similar.

Fact 8.13.2. Let $A, B \in \mathbb{F}^{n \times n}$, assume that A and B are Hermitian, and assume that A is nonsingular. Then, the following statements are equivalent:

  i) There exists a unitary matrix $S \in \mathbb{F}^{n \times n}$ such that SAS^* and SBS^* are diagonal.

  ii) $AB = BA$.

  iii) $A^{-1}B$ is Hermitian.

(Proof: See [367, p. 229].) (Remark: The equivalence of *i)* and *ii)* is given by Fact 5.9.17.)

Fact 8.13.3. Let $A, B \in \mathbb{F}^{n \times n}$, assume that A and B are Hermitian, and assume that A is nonsingular. Then, there exists a nonsingular matrix $S \in \mathbb{F}^{n \times n}$ such that SAS^* and SBS^* are diagonal if and only if $A^{-1}B$ is diagonalizable over $\mathbb{R}$. (Proof: See [367, p. 229] or [592, p. 95].)

Fact 8.13.4. Let $A, B \in \mathbb{F}^{n \times n}$, assume that A and B are symmetric, and assume that A is nonsingular. Then, there exists a nonsingular matrix $S \in \mathbb{F}^{n \times n}$ such that SAS^{T} and SBS^{T} are diagonal if and only if $A^{-1}B$ is diagonalizable. (Proof: See [367, p. 229] and [733].) (Remark: A and B are complex symmetric.)

Fact 8.13.5. Let $A, B \in \mathbb{F}^{n \times n}$, and assume that $\{x \in \mathbb{F}^n: \ x^*Ax = x^*Bx = 0\} = \{0\}$. Then, there exists a nonsingular matrix $S \in \mathbb{F}^{n \times n}$ such that SAS^* and SBS^* are upper triangular. (Proof: See [592, p. 96].) (Remark: A and B need not be Hermitian.) (Remark: See Fact 8.13.6 and Fact 5.9.15.)

Fact 8.13.6. Let $A, B \in \mathbb{F}^{n \times n}$, assume that A and B are Hermitian, assume that $\{x \in \mathbb{F}^n: \ x^*Ax = x^*Bx = 0\} = \{0\}$, and, if $\mathbb{F} = \mathbb{R}$, assume that $n \geq 3$. Then, there exists a nonsingular matrix $S \in \mathbb{F}^{n \times n}$ such that SAS^* and SBS^* are diagonal. (Proof: The result follows from Fact 8.13.5. See [505] or [592, p. 96].) (Remark: For $\mathbb{F} = \mathbb{R}$, this result is due to Pesonen and Milnor. See [733].) (Remark: See Fact 8.12.13 and Fact 8.12.14.)

Fact 8.13.7. Let $A, B \in \mathbb{F}^{n \times n}$, and assume that A and B are Hermitian. Then, there exists a nonsingular matrix $S \in \mathbb{F}^{n \times n}$ such that SAS^* and SBS^* are diagonal if and only if there exists a positive-definite matrix $M \in \mathbb{F}^{n \times n}$ such that $AMB = BMA$. (Proof: See [48].)

Fact 8.13.8. Let $A, B \in \mathbb{R}^{n \times n}$, assume that A and B are symmetric and nonsingular, and assume there exist $\alpha, \beta \in \mathbb{R}$ such that $\alpha A + \beta B$ is positive definite. Then, there exists a nonsingular matrix $S \in \mathbb{R}^{n \times n}$ such that SAS^{T} and SBS^{T} are diagonal. (Remark: This result is due to Weierstrass. See [733].) (Remark: Suppose that B is positive definite. Then, by necessity of Fact 8.13.3, it follows that $A^{-1}B$ is diagonalizable over $\mathbb{R}$, which proves *iii)* $\implies$ *i)* of Proposition 5.5.18.)

Fact 8.13.9. Let $A \in \mathbb{F}^{n \times n}$. Then, A is diagonalizable over $\mathbb{F}$ with (nonnegative, positive) eigenvalues if and only if there exist (positive-semidefinite, positive-definite) matrices $B, C \in \mathbb{F}^{n \times n}$ such that $A = BC$. (Proof: To prove sufficiency, use Theorem 8.3.5 and note that

$$A = S^{-1}(SBS^*)(S^{-*}CS^{-1})S.)$$

8.14 Facts on Eigenvalues and Singular Values

Fact 8.14.1. Let $A = \begin{bmatrix} a & b \\ \bar{b} & c \end{bmatrix} \in \mathbb{F}^{2 \times 2}$, and assume that A is Hermitian. Then,

$$2|b| \leq \lambda_1 - \lambda_2.$$

Now, assume that A is positive semidefinite. Then,

$$\sqrt{2}|b| \leq \left(\sqrt{\lambda_1} - \sqrt{\lambda_2} \right) \sqrt{\lambda_1 + \lambda_2}.$$

If $c > 0$, then

$$\frac{|b|}{\sqrt{c}} \leq \sqrt{\lambda_1} - \sqrt{\lambda_2}.$$

If $a > 0$ and $c > 0$, then

$$\frac{|b|}{\sqrt{ac}} \leq \frac{\lambda_1 - \lambda_2}{\lambda_1 + \lambda_2}.$$

Finally, if A is positive definite, then

$$\frac{|b|}{a} \leq \frac{\lambda_1 - \lambda_2}{2\sqrt{\lambda_1 \lambda_2}}$$

and

$$4|b| \leq \frac{\lambda_1^2 - \lambda_2^2}{\sqrt{\lambda_1 \lambda_2}}.$$

(Proof: See [468, 815].) (Remark: These inequalities are useful for deriving inequalities involving quadratic forms. See Fact 8.12.18 and Fact 8.12.17.)

Fact 8.14.2. Let $A \in \mathbb{F}^{n \times m}$. Then,

$$\operatorname{tr} \langle A \rangle = \sum_{i=1}^{\min\{n,m\}} \sigma_i(A).$$

Fact 8.14.3. Let $A \in \mathbb{F}^{n \times n}$, and define

$$\mathcal{A} \triangleq \begin{bmatrix} \sigma_{\max}(A)I & A^* \\ A & \sigma_{\max}(A)I \end{bmatrix}.$$

Then, $\mathcal{A}$ is positive semidefinite. Furthermore,

$$\langle A + A^* \rangle \leq \left\{ \begin{array}{c} \langle A \rangle + \langle A^* \rangle \leq 2\sigma_{\max}(A)I \\ A^*A + I \end{array} \right\} \leq \left[\sigma_{\max}^2(A) + 1 \right]I.$$

(Proof: See [813].)

Fact 8.14.4. Let $A \in \mathbb{F}^{n \times n}$. Then, for all $i = 1, \ldots, n$,

$$\lambda_i\left[\tfrac{1}{2}(A + A^*) \right] \leq \sigma_i(A).$$

Hence,

$$|\operatorname{tr} A| \leq \operatorname{tr} \langle A \rangle.$$

(Proof: See [369, p. 151] or [653].)

Fact 8.14.5. Let $A \in \mathbb{F}^{n \times n}$, and let $\operatorname{mspec}(A) = \{\lambda_1, \ldots, \lambda_n\}_{\mathrm{m}}$, where $\lambda_1, \ldots, \lambda_n$ are ordered such that $|\lambda_1| \geq \cdots \geq |\lambda_n|$. If $r > 0$, then, for all $k = 1, \ldots, n$,

$$\sum_{i=1}^{k} |\lambda_i|^r \leq \sum_{i=1}^{k} \sigma_i^r(A).$$

In particular, for all $k = 1, \ldots, n$,

$$\sum_{i=1}^{k} |\lambda_i| \leq \sum_{i=1}^{k} \sigma_i(A).$$

Hence,

$$|\operatorname{tr} A| \leq \sum_{i=1}^{n} |\lambda_i| \leq \sum_{i=1}^{n} \sigma_i(A) = \operatorname{tr} \langle A \rangle.$$

Furthermore, for all $k = 1, \ldots, n$,

$$\sum_{i=1}^{k} |\lambda_i|^2 \leq \sum_{i=1}^{k} \sigma_i^2(A).$$

Hence,

$$|\operatorname{tr} A^2| \leq \sum_{i=1}^{n} |\lambda_i|^2 \leq \sum_{i=1}^{n} \sigma_i(A^2) = \operatorname{tr} \langle A^2 \rangle \leq \sum_{i=1}^{n} \sigma_i^2(A) = \operatorname{tr} A^*A.$$

(Proof: The result follows from Fact 8.17.6 and Fact 5.10.14. See [111, p. 42], [369, p. 176], or [806, p. 19]. See Fact 9.11.18 for the inequality $\operatorname{tr} \langle A^2 \rangle = \operatorname{tr} \left(A^{2*}A^2 \right)^{1/2} \leq \operatorname{tr} A^*A$.) Finally,

$$\sum_{i=1}^{n} |\lambda_i|^2 = \operatorname{tr} A^*A$$

if and only if A is normal. (Proof: See [592, p. 146].) (Remark: $\sum_{i=1}^{n} |\lambda_i|^2 \leq \operatorname{tr} A^*A$ is *Schur's inequality*. See Fact 9.10.2.) (Problem: Determine when equality holds for the remaining inequalities.)

Fact 8.14.6. Let $A \in \mathbb{F}^{n \times n}$. Then,

$$|\operatorname{tr} A^2| \leq \operatorname{tr}(\langle A \rangle \langle A^* \rangle).$$

(Proof: See [468, 815].) (Problem: Compare the bounds

$$|\operatorname{tr} A^2| \leq \operatorname{tr}(\langle A \rangle \langle A^* \rangle)$$

and

$$|\operatorname{tr} A^2| \leq \operatorname{tr} A^* A.)$$

Fact 8.14.7. Let $A \in \mathbb{F}^{n \times n}$, and assume that A is Hermitian. Then, for all $k = 1, \ldots, n$,

$$\sum_{i=1}^{k} \mathrm{d}_i(A) \leq \sum_{i=1}^{k} \lambda_i(A)$$

with equality for $k = n$, that is,

$$\operatorname{tr} A = \sum_{i=1}^{n} \mathrm{d}_i(A) = \sum_{i=1}^{n} \lambda_i(A).$$

Hence, for all $k = 1, \ldots, n$,

$$\sum_{i=k}^{n} \lambda_i(A) \leq \sum_{i=k}^{n} \mathrm{d}_i(A).$$

Furthermore, the vector $\begin{bmatrix} \mathrm{d}_1(A) \cdots \mathrm{d}_n(A) \end{bmatrix}^{\mathrm{T}}$ is an element of the convex hull of the $n!$ vectors obtaining by permuting the components of $\begin{bmatrix} \lambda_1(A) \cdots \lambda_n(A) \end{bmatrix}^{\mathrm{T}}$. (Proof: See [111, p. 35], [367, p. 193], or [806, p. 18]. The last statement follows from Fact 8.17.1.) (Remark: This result is *Schur's theorem*.)

Fact 8.14.8. Let $A \in \mathbb{F}^{n \times n}$, and assume that A is positive semidefinite. Then, for all $k = 1, \ldots, n$,

$$\prod_{i=k}^{n} \lambda_i(A) \leq \prod_{i=k}^{n} d_i(A).$$

In particular,

$$\det A \leq \prod_{i=1}^{n} A_{(i,i)}.$$

Now, assume that A is positive definite. Then, equality holds if and only if A is diagonal. (Proof: See [367, p. 200], [806, p. 18], [367, p. 477], or [282, pp. 21–24].) (Remark: The case $k = 1$ is *Hadamard's inequality*.) (Remark: See Fact 9.11.8.)

Fact 8.14.9. Let $A, B \in \mathbf{H}^n$, and define $\gamma \triangleq \left[\begin{array}{ccc} \gamma_1 & \cdots & \gamma_n \end{array}\right]$, where the components of γ are the components of $\left[\begin{array}{ccc} \lambda_1(A) & \cdots & \lambda_n(A) \end{array}\right] + \left[\begin{array}{ccc} \lambda_n(B) & \cdots & \lambda_1(B) \end{array}\right]$ arranged in decreasing order. Then, for all $k = 1, \ldots, n$,

$$\sum_{i=1}^{k} \gamma_i \leq \sum_{i=1}^{k} \lambda_i(A + B).$$

(Proof: The result follows from the Lidskii-Wielandt inequalities. See [111, p. 71] or [112].) (Remark: This result provides a lower bound to complement (8.5.14).)

Fact 8.14.10. Let $A, B \in \mathbb{F}^{n \times n}$, and assume that A and B are positive semidefinite. If $p \geq 1$, then

$$\sum_{i=1}^{n} \lambda_i^p(A)\lambda_{n-i+1}^p(B) \leq \operatorname{tr}\left(B^{1/2}AB^{1/2}\right)^p \leq \operatorname{tr} A^p B^p \leq \sum_{i=1}^{n} \lambda_i^p(A)\lambda_i^p(B).$$

If $0 \leq p \leq 1$, then

$$\sum_{i=1}^{n} \lambda_i^p(A)\lambda_{n-i+1}^p(B) \leq \operatorname{tr} A^p B^p \leq \operatorname{tr}\left(B^{1/2}AB^{1/2}\right)^p \leq \sum_{i=1}^{n} \lambda_i^p(A)\lambda_i^p(B).$$

Now, suppose that A and B are positive definite. If $p \leq -1$, then

$$\sum_{i=1}^{n} \lambda_i^p(A)\lambda_{n-i+1}^p(B) \leq \operatorname{tr}\left(B^{1/2}AB^{1/2}\right)^p \leq \operatorname{tr} A^p B^p \leq \sum_{i=1}^{n} \lambda_i^p(A)\lambda_i^p(B).$$

If $-1 \leq p \leq 0$, then

$$\sum_{i=1}^{n} \lambda_i^p(A)\lambda_{n-i+1}^p(B) \leq \operatorname{tr} A^p B^p \leq \operatorname{tr}\left(B^{1/2}AB^{1/2}\right)^p \leq \sum_{i=1}^{n} \lambda_i^p(A)\lambda_i^p(B).$$

(Proof: See [751]. See also [156, 463, 485, 754].) (Remark: See Fact 8.10.8. See Fact 8.10.5 for the indefinite case.)

Fact 8.14.11. Let $A, B \in \mathbb{F}^{n \times n}$, assume that A and B are positive semidefinite, and let $p \geq r \geq 0$. Then,

$$\left[\begin{array}{ccc} \lambda_1^{1/p}(A^p B^p) & \cdots & \lambda_n^{1/p}(A^p B^p) \end{array}\right]$$

weakly log majorizes and thus weakly majorizes

$$\left[\begin{array}{ccc} \lambda_1^{1/r}(A^r B^r) & \cdots & \lambda_n^{1/r}(A^r B^r) \end{array}\right].$$

(Proof: See [111, p. 257] or [806, p. 20] and Fact 8.17.6.)

Fact 8.14.12. Let $A, B \in \mathbb{F}^{n \times n}$, and assume that A and B are positive semidefinite. Then,

$$\lambda_{\max}(A + B) \leq \max\{\lambda_{\max}(A), \lambda_{\max}(B)\} + \sigma_{\max}^{1/2}\left(A^{1/2}BA^{1/2}\right).$$

(Proof: See [430].) (Remark: See Fact 9.12.11 and Fact 9.12.13.)

Fact 8.14.13. Let $A, B \in \mathbb{F}^{n \times n}$, and assume that A and B are positive semidefinite. Then,

$$\lambda_{\max}(A + B)$$
$$\leq \tfrac{1}{2}\left[\lambda_{\max}(A) + \lambda_{\max}(B) + \sqrt{[\lambda_{\max}(A) - \lambda_{\max}(B)]^2 + 4\sigma_{\max}^2(A^{1/2}B^{1/2})}\right].$$

(Proof: See [433].)

Fact 8.14.14. Let $f\colon \mathbb{R} \mapsto \mathbb{R}$ be convex, define $f\colon \mathbf{H}^n \mapsto \mathbb{R}$ by (8.5.1), let $A, B \in \mathbb{F}^{n \times n}$, and assume that A and B are Hermitian. Then, for all $\alpha \in [0, 1]$,

$$\left[\ \alpha\lambda_1[f(A)] + (1 - \alpha)\lambda_1[f(B)] \quad \cdots \quad \alpha\lambda_n[f(A)] + (1 - \alpha)\lambda_n[f(B)]\ \right]$$

weakly majorizes

$$\left[\ \lambda_1[f(\alpha A + (1 - \alpha)B)] \quad \cdots \quad \lambda_n[f(\alpha A + (1 - \alpha)B)]\ \right].$$

If, in addition, f is either nonincreasing or nondecreasing, then, for all $i = 1, \ldots, n$,

$$\lambda_i[f(\alpha A + (1 - \alpha)B)] \leq \alpha\lambda_i[f(A)] + (1 - \alpha)\lambda_i[f(B)].$$

(Proof: See [51].) (Remark: Convexity of $f\colon \mathbb{R} \mapsto \mathbb{R}$ does not imply convexity of $f\colon \mathbf{H}^n \mapsto \mathbb{R}$.)

Fact 8.14.15. Let $A, B \in \mathbb{F}^{n \times n}$, and assume that A and B are positive semidefinite. If $r \in [0, 1]$, then

$$\left[\ \lambda_1(A^r + B^r) \quad \cdots \quad \lambda_n(A^r + B^r)\ \right]$$

weakly majorizes

$$\left[\ \lambda_1[(A + B)^r] \quad \cdots \quad \lambda_n[(A + B)^r]\ \right],$$

and, for all $i = 1, \ldots, n$,

$$2^{1-r}\lambda_i[(A + B)^r] \leq \lambda_i(A^r + B^r).$$

If $r \geq 1$, then

$$\left[\ \lambda_1[(A + B)^r] \quad \cdots \quad \lambda_n[(A + B)^r]\ \right]$$

weakly majorizes

$$\left[\ \lambda_1(A^r + B^r) \quad \cdots \quad \lambda_n(A^r + B^r)\ \right],$$

and, for all $i = 1, \ldots, n$,

$$\lambda_i(A^r + B^r) \leq 2^{r-1}\lambda_i[(A + B)^r].$$

(Proof: The result follows from Fact 8.14.14. See [34, 50, 51].)

Fact 8.14.16. Let $A \in \mathbb{F}^{n \times n}$, assume that A is Hermitian, and let $S \in \mathbb{R}^{k \times n}$ satisfy $SS^* = I_k$. Then, for all $i = 1, \ldots, k$,

$$\lambda_{i+n-k}(A) \leq \lambda_i(SAS^*) \leq \lambda_i(A).$$

Consequently,

$$\sum_{i=1}^{k} \lambda_{i+n-k}(A) \leq \operatorname{tr} SAS^* \leq \sum_{i=1}^{k} \lambda_i(A)$$

and

$$\prod_{i=1}^{k} \lambda_{i+n-k}(A) \leq \det SAS^* \leq \prod_{i=1}^{k} \lambda_i(A).$$

(Proof: See [367, p. 190].) (Remark: This result is the *Poincaré separation theorem*.)

Fact 8.14.17. Let $A \in \mathbb{F}^{n \times n}$, and assume that A is Hermitian. Then, for all $k = 1, \ldots, n$,

$$\sum_{i=1}^{k} \lambda_i = \max\{\operatorname{tr} S^*AS \colon S \in \mathbb{F}^{n \times k} \text{ and } S^*S = I_k\}$$

and

$$\sum_{i=n+1-k}^{n} \lambda_i = \min\{\operatorname{tr} S^*AS \colon S \in \mathbb{F}^{n \times k} \text{ and } S^*S = I_k\}.$$

(Proof: See [367, p. 191].) (Remark: This result is the *minimum principle*.)

Fact 8.14.18. Let $A \in \mathbb{F}^{n \times n}$. Then, $\left[\begin{smallmatrix} I & A \\ A^* & I \end{smallmatrix}\right]$ is positive semidefinite if and only if $\sigma_{\max}(A) \leq 1$. Furthermore, $\left[\begin{smallmatrix} I & A \\ A^* & I \end{smallmatrix}\right]$ is positive definite if and only if $\sigma_{\max}(A) < 1$. (Proof: Note that

$$\begin{bmatrix} I & A \\ A^* & I \end{bmatrix} = \begin{bmatrix} I & 0 \\ A^* & I \end{bmatrix} \begin{bmatrix} I & 0 \\ 0 & I - A^*A \end{bmatrix} \begin{bmatrix} I & A \\ 0 & I \end{bmatrix}.)$$

Fact 8.14.19. Let $A_{11} \in \mathbb{F}^{n \times n}$, $A_{12} \in \mathbb{F}^{n \times m}$, and $A_{22} \in \mathbb{F}^{m \times m}$, and assume that $\left[\begin{smallmatrix} A_{11} & A_{12} \\ A_{12}^* & A_{22} \end{smallmatrix}\right] \in \mathbb{F}^{(n+m) \times (n+m)}$ is positive semidefinite. Then,

$$\sigma_{\max}^2(A_{12}) \leq \sigma_{\max}(A_{11})\sigma_{\max}(A_{22}).$$

(Proof: Use $A_{22} \geq A_{12}^* A_{11}^+ A_{12} \geq 0$, factor $A_{11}^+ = MM^*$, where M has full column rank, and recall that $\sigma_{\max}(SS^*) = \sigma_{\max}^2(S)$.) (Problem: Consider alternative norms.)

Fact 8.14.20. Let $A, B \in \mathbb{F}^{n \times n}$, and assume that A and B are positive semidefinite. Then, for all $k = 1, \ldots, n$,

$$\prod_{i=1}^{k} \lambda_i(AB) \leq \prod_{i=1}^{k} \sigma_i(AB) \leq \prod_{i=1}^{k} \lambda_i(A)\lambda_i(B)$$

with equality for $k = n$. Furthermore, for all $k = 1, \ldots, n$,

$$\prod_{i=k}^{n} \lambda_i(A)\lambda_i(B) \leq \prod_{i=k}^{n} \sigma_i(AB) \leq \prod_{i=k}^{n} \lambda_i(AB).$$

(Proof: Use Fact 5.10.14 and Fact 9.11.19.)

Fact 8.14.21. Let $A, B \in \mathbb{F}^{n \times n}$, assume that A and B are positive semidefinite, and let $p \in [0, 1]$. Then,

$$\sigma_{\max}(A^p - B^p) \leq \sigma_{\max}^p(A - B).$$

(Proof: See [428].)

Fact 8.14.22. Let $A, B \in \mathbb{F}^{n \times n}$, and assume that A and B are positive semidefinite. Then, the following statements hold:

i) If $q \in [0, 1]$, then

$$\sigma_{\max}(A^q B^q) \leq \sigma_{\max}^q(AB)$$

and

$$\sigma_{\max}(B^q A^q B^q) \leq \sigma_{\max}^q(BAB).$$

ii) If $q \in [0, 1]$, then

$$\lambda_{\max}(A^q B^q) \leq \lambda_{\max}^q(AB).$$

iii) If $q \geq 1$, then

$$\sigma_{\max}^q(AB) \leq \sigma_{\max}(A^q B^q).$$

iv) If $q \geq 1$, then

$$\lambda_{\max}^q(AB) \leq \lambda_{\max}(A^q B^q).$$

v) If $p \geq q > 0$, then

$$\sigma_{\max}^{1/q}(A^q B^q) \leq \sigma_{\max}^{1/p}(A^p B^p).$$

(Proof: See [111, pp. 255–258] and [280].)

Fact 8.14.23. Let $A, B \in \mathbb{F}^{n \times n}$, and assume that A and B are positive semidefinite. Then, the following statements hold:

i) $\sigma_{\max}[\log(I + A)\log(I + B)] \leq \left(\log\left[1 + \sigma_{\max}^{1/2}(AB)\right]\right)^2.$

ii) $\sigma_{\max}[\log(I + B)\log(I + A)\log(I + B)] \leq \left(\log\left[1 + \sigma_{\max}^{1/3}(BAB)\right]\right)^3.$

iii) $\det[\log(I + A)\log(I + B)] \leq \det\left[\log\left(I + \langle AB \rangle^{1/2}\right)\right]^2.$

iv) $\det[\log(I + B)\log(I + A)\log(I + B)] \leq \det\left(\log\left[I + (BAB)^{1/3}\right]\right)^3.$

(Proof: See [730].) (Remark: See Fact 11.14.4.)

Fact 8.14.24. Let $A, B \in \mathbb{F}^{n \times n}$, and assume that A and B are positive semidefinite. Then,

$$\sigma_{\max}\left[(I + A)^{-1}AB(I + B)^{-1}\right] \leq \frac{\sigma_{\max}(AB)}{\left[1 + \sigma_{\max}^{1/2}(AB)\right]^2}.$$

(Proof: See [730].)

8.15 Facts on Generalized Inverses

Fact 8.15.1. Let $A \in \mathbb{F}^{n \times n}$, and assume that A is positive semidefinite. Then, the following statements hold:

i) $A^+ = A^{\mathrm{D}} = A^{\#} \geq 0.$

ii) $\operatorname{rank} A = \operatorname{rank} A^+.$

iii) $A^{+1/2} \triangleq \left(A^{1/2}\right)^+ = (A^+)^{1/2}.$

iv) $A^{1/2} = A(A^+)^{1/2} = (A^+)^{1/2}A.$

v) $AA^+ = A^{1/2}\left(A^{1/2}\right)^+.$

vi) $\left[\begin{smallmatrix} A & AA^+ \\ A^+A & A^+ \end{smallmatrix}\right]$ is positive semidefinite.

vii) $A^+A + AA^+ \leq A + A^+.$

viii) $A^+A \circ AA^+ \leq A \circ A^+.$

(Proof: See [813] or Fact 8.9.3 and Fact 8.16.27 for *vi*)–*viii*).)

Fact 8.15.2. Let $A \in \mathbb{F}^{n \times m}$. Then,

$$\langle A^* \rangle = A\langle A \rangle^{+1/2}A^*.$$

(Remark: See Fact 8.9.5.)

Fact 8.15.3. Let $A, B \in \mathbb{F}^{n \times n}$, and assume that A and B are positive semidefinite. Then,

$$A = (A + B)(A + B)^+A.$$

Fact 8.15.4. Let $A, B \in \mathbb{F}^{n \times n}$, and assume that A and B are positive semidefinite. Then, $A \leq B$ if and only if $\mathcal{R}(A) \subseteq \mathcal{R}(B)$ and $\operatorname{sprad}(B^+A) \leq 1$. (Proof: See [660].)

Fact 8.15.5. Let $A, B \in \mathbb{R}^{n \times n}$, assume that A and B are positive semidefinite, and assume that $A \leq B$. Then, the following statements are equivalent:

  i) $B^+ \leq A^+$.

  ii) $\mathcal{R}(A) = \mathcal{R}(B)$.

  iii) $\operatorname{rank} A = \operatorname{rank} B$.

Furthermore, the following statements are equivalent:

  iv) $A^+ \leq B^+$.

  v) $A^2 = AB$.

(Proof: See [336, 539].)

Fact 8.15.6. Let $A, B \in \mathbb{F}^{n \times n}$, assume that A and B are positive semidefinite, and assume that $A \leq B$. Then,

$$0 \leq AA^+ \leq BB^+.$$

If, in addition, $\operatorname{rank} A = \operatorname{rank} B$, then

$$AA^+ = BB^+.$$

Fact 8.15.7. Let $A, B \in \mathbb{F}^{n \times n}$, assume that A and B are positive semidefinite, and assume that $A \leq B$. Then,

$$0 \leq AB^+A \leq A \leq A + B\left[(I - AA^+)B(I - AA^+)\right]^+B \leq B.$$

(Proof: See [336].)

Fact 8.15.8. Let $A, B \in \mathbb{F}^{n \times n}$, and assume that A and B are positive semidefinite. Then,

$$\operatorname{spec}\left[(A + B)^+A\right] \subset [0, 1].$$

(Proof: Let C be positive definite and satisfy $B \leq C$. Then,

$$(A + C)^{-1/2}C(A + C)^{-1/2} \leq I.$$

The result now follows from Fact 8.15.9.)

Fact 8.15.9. Let $A, B, C \in \mathbb{F}^{n \times n}$, assume that A, B, C are positive semidefinite, and assume that $B \leq C$. Then, for all $i = 1, \ldots, n$,

$$\lambda_i\left[(A + B)^+B\right] \leq \lambda_i\left[(A + C)^+C\right].$$

Consequently,

$$\operatorname{tr}\left[(A + B)^+B\right] \leq \operatorname{tr}\left[(A + C)^+C\right].$$

(Proof: See [752].) (Remark: See Fact 8.15.8.)

Fact 8.15.10. Let $A, B \in \mathbb{F}^{n \times n}$, assume that A and B are positive semidefinite, and define
$$A : B \triangleq A(A + B)^+ B.$$
Then, the following statements hold:

i) $A(A + B)^+ B = ACB$ for every (1)-inverse C of $A + B$.

ii) $A : B = B : A = B - B(A + B)^+ B = A - A(A + B)^+ A$.

iii) $\mathcal{R}(A : B) = \mathcal{R}(A) \cap \mathcal{R}(B)$.

iv) $\mathcal{N}(A : B) = \mathcal{N}(A) + \mathcal{N}(B)$.

v) For all $\alpha, \beta > 0$,
$$\left(\alpha^{-1} A \right) : \left(\beta^{-1} B \right) \leq \alpha A + \beta B.$$

vi) $A : B \geq X$ for all positive-semidefinite matrices $X \in \mathbb{F}^{n \times n}$ such that
$$\begin{bmatrix} A + B & A \\ A & A - X \end{bmatrix} \geq 0.$$

vii) $\operatorname{rank}(A : B) = \operatorname{rank} A + \operatorname{rank} B - \operatorname{rank}(A + B)$.

viii) $\phi \colon \mathbf{N}^n \times \mathbf{N}^n \mapsto -\mathbf{N}^n$ defined by $\phi(A, B) \triangleq -A : B$ is convex.

ix) If A and B are projectors, then
$$A : B = (A^+ + B^+)^+,$$
and $2(A : B)$ is the projector onto $\mathcal{R}(A) \cap \mathcal{R}(B)$.

x) If $A + B$ is positive definite, then
$$A : B = A(A + B)^{-1} B.$$

xi) If A and B are positive definite, then
$$A : B = \left(A^{-1} + B^{-1} \right)^{-1}.$$

xii) If $C, D \in \mathbb{F}^{n \times n}$ are positive semidefinite, then
$$(A : B) : C = A : (B : C)$$
and
$$A : C + B : D \leq (A + B) : (C + D).$$

(Proof: See [21, 22, 25, 440], [606, p. 189], and [806, p. 9].) (Remark: $A : B$ is the *parallel sum* of A and B.) (Remark: See Fact 6.4.41 and Fact 6.4.42.) (Remark: A symmetric expression for the parallel sum of three or more positive-semidefinite matrices is given in [705].)

Fact 8.15.11. Let $A, B \in \mathbb{F}^{n \times n}$, and assume that A and B are positive semidefinite. If $(AB)^+ = B^+ A^+$, then AB is range Hermitian. Furthermore, the following statements are equivalent:

   i) AB is range Hermitian.

   ii) $(AB)^{\#} = B^{+}A^{+}$.

   iii) $(AB)^{+} = B^{+}A^{+}$.

(Proof: See [526].) (Remark: See Fact 6.4.9.)

Fact 8.15.12. Let $A \in \mathbb{F}^{n \times n}$ and $C \in \mathbb{F}^{m \times m}$, assume that A and C are positive semidefinite, let $B \in \mathbb{F}^{n \times m}$, and define $X \triangleq A^{+1/2}BC^{+1/2}$. Then, the following statements are equivalent:

   i) $\begin{bmatrix} A & B \\ B^{*} & C \end{bmatrix}$ is positive semidefinite.

   ii) $AA^{+}B = B$ and $X^{*}X \leq I_{m}$.

   iii) $BC^{+}C = B$ and $X^{*}X \leq I_{m}$.

   iv) $B = A^{1/2}XC^{1/2}$ and $X^{*}X \leq I_{m}$.

   v) There exists a matrix $Y \in \mathbb{F}^{n \times m}$ such that $B = A^{1/2}YC^{1/2}$ and $Y^{*}Y \leq I_{m}$.

(Proof: See [806, p. 15].)

8.16 Facts on the Kronecker and Schur Products

Fact 8.16.1. Let $A \in \mathbb{F}^{n \times n}$, assume that A is positive semidefinite, and assume that every entry of A is nonzero. Then, $A^{\{-1\}}$ is positive semidefinite if and only if rank $A = 1$. (Proof: See [471].)

Fact 8.16.2. Let $A \in \mathbb{F}^{n \times n}$, assume that A is positive semidefinite, and let $k \in \mathbb{P}$. If $r \in [0, 1]$, then

$$(A^{r})^{\{k\}} \leq \left(A^{\{k\}}\right)^{r}.$$

If $r \in [1, 2]$, then

$$\left(A^{\{k\}}\right)^{r} \leq (A^{r})^{\{k\}}.$$

If A is positive definite and $r \in [0, 1]$, then

$$\left(A^{\{k\}}\right)^{-r} \leq (A^{-r})^{\{k\}}.$$

(Proof: See [806, p. 8].)

Fact 8.16.3. Let $A \in \mathbb{F}^{n \times n}$, and assume that A is positive semidefinite. Then,

$$(I \circ A)^{2} \leq \tfrac{1}{2}(I \circ A^{2} + A \circ A) \leq I \circ A^{2}$$

and

$$A \circ A \leq I \circ A^{2}.$$

Now, assume that A is positive definite. Then,

$$(A \circ A)^{-1} \leq A^{-1} \circ A^{-1}$$

and

$$1 \in \mathrm{spec}(A \circ A^{-1}).$$

Next, let $\alpha \triangleq \lambda_{\min}(A)$ and $\beta \triangleq \lambda_{\max}(A)$. Then,

$$\left(A \circ A^{-1}\right)^{-1} \leq I \leq \left(A^{1/2} \circ A^{-1/2}\right)^2 \leq \tfrac{1}{2}\left(I + A \circ A^{-1}\right) \leq A \circ A^{-1}$$

and

$$\frac{2\alpha\beta}{\alpha^2 + \beta^2} I \leq \frac{2\alpha\beta}{\alpha^2 + \beta^2}\left(A^2 \circ A^{-2}\right)^{1/2} \leq \frac{\alpha\beta}{\alpha^2 + \beta^2}\left(I + A^2 \circ A^{-2}\right) \leq A \circ A^{-1}.$$

Define $\Phi(A) \triangleq A \circ A^{-1}$, and, for all $k \in \mathbb{P}$, define

$$\Phi^{(k+1)}(A) \triangleq \Phi\left[\Phi^{(k)}(A)\right],$$

where $\Phi^{(1)}(A) \triangleq \Phi(A)$. Then, for all $k \in \mathbb{P}$,

$$\Phi^{(k)}(A) \geq I$$

and

$$\lim_{k \to \infty} \Phi^{(k)}(A) = I.$$

(Proof: See [259, 401, 749, 750], [367, p. 475], and set $B = A^{-1}$ in Fact 8.16.23.) (Remark: The convergence result also holds if A is an *H-matrix* [401]. $A \circ A^{-1}$ is the *relative gain array*.) (Remark: See Fact 8.16.25.)

Fact 8.16.4. Let $A \in \mathbb{F}^{n \times n}$, assume that A is positive semidefinite, and assume that $I_n \circ A = I_n$. Then,

$$\det A \leq \lambda_{\min}(A \circ A).$$

(Proof: See [763].)

Fact 8.16.5. Let $A \in \mathbb{F}^{n \times n}$. Then,

$$-A^*A \circ I \leq A^* \circ A \leq A^*A \circ I.$$

(Proof: Use Fact 8.16.28 with $B = I$.)

Fact 8.16.6. Let $A \in \mathbb{F}^{n \times n}$. Then,

$$\langle A \circ A^* \rangle \leq \left\{ \begin{array}{c} A^*A \circ I \\ \langle A \rangle \circ \langle A^* \rangle \end{array} \right\} \leq \sigma_{\max}^2(A)I.$$

(Proof: See [813] and Fact 8.16.16.)

Fact 8.16.7. Let $A \triangleq \left[\begin{smallmatrix} A_{11} & A_{12} \\ A_{12}^* & A_{22} \end{smallmatrix} \right] \in \mathbb{F}^{(n+m) \times (n+m)}$ and $B \triangleq \left[\begin{smallmatrix} B_{11} & B_{12} \\ B_{12}^* & B_{22} \end{smallmatrix} \right] \in \mathbb{F}^{(n+m) \times (n+m)}$, and assume that A and B are positive semidefinite. Then,

$$(A_{11}|A) \circ (B_{11}|B) \leq (A_{11}|A) \circ B_{22} \leq (A_{11} \circ B_{11})|(A \circ B).$$

(Proof: See [479].)

Fact 8.16.8. Let $A, B \in \mathbb{F}^{n \times n}$, and assume that A and B are positive semidefinite. Then, $A \circ B$ is positive semidefinite. If, in addition, B is positive definite and every diagonal entry of A is positive, then $A \circ B$ is positive definite. (Proof: By Fact 7.4.15, $A \otimes B$ is positive semidefinite, and the Schur product $A \circ B$ is a principal submatrix of the Kronecker product. If A is positive definite, use Fact 8.16.15 to obtain $\det(A \circ B) > 0$.) (Remark: The first result is *Schur's theorem*.)

Fact 8.16.9. Let $A \in \mathbb{F}^{n \times n}$, and assume that A is positive definite. Then, there exist positive-definite matrices $B, C \in \mathbb{F}^{n \times n}$ such that $A = B \circ C$. (Remark: See [592, pp. 154, 166].) (Remark: This result is due to Djokovic.)

Fact 8.16.10. Let $A, B \in \mathbb{F}^{n \times n}$, and assume that A is positive definite and B is positive semidefinite. Then,

$$\left(1_{1 \times n} A^{-1} 1_{n \times 1}\right)^{-1} B \le A \circ B.$$

(Proof: See [262].) (Remark: Setting $B = 1_{n \times n}$ yields Fact 8.7.11.)

Fact 8.16.11. Let $A, B \in \mathbb{F}^{n \times n}$, and assume that A and B are positive definite. Then,

$$\left(1_{1 \times n} A^{-1} 1_{n \times 1} 1_{1 \times n} B^{-1} 1_{n \times 1}\right)^{-1} 1_{n \times n} \le A \circ B.$$

(Proof: See [813].)

Fact 8.16.12. Let $A \in \mathbb{F}^{n \times n}$, assume that A is positive definite, let $B \in \mathbb{F}^{n \times n}$, and assume that B is positive semidefinite. Then,

$$\operatorname{rank} B \le \operatorname{rank}(A \circ B) \le \operatorname{rank}(A \otimes B) = (\operatorname{rank} A)(\operatorname{rank} B).$$

(Remark: See Fact 7.4.20, Fact 7.6.5, and Fact 8.16.10.) (Remark: The first inequality is due to Djokovic. See [592, pp. 154, 166].)

Fact 8.16.13. Let $A, B \in \mathbb{F}^{n \times n}$, and assume that A and B are positive semidefinite. If $p \ge 1$, then

$$\operatorname{tr} (A \circ B)^p \le \operatorname{tr} A^p \circ B^p.$$

If $0 \le p \le 1$, then

$$\operatorname{tr} A^p \circ B^p \le \operatorname{tr} (A \circ B)^p.$$

Now, assume that A and B are positive definite. If $p \le 0$, then

$$\operatorname{tr} (A \circ B)^p \le \operatorname{tr} A^p \circ B^p.$$

(Proof: See [754].)

Fact 8.16.14. Let $A, B \in \mathbb{F}^{n \times n}$, and assume that A and B are positive semidefinite. Then, for all $k = 1, \ldots, n$,

$$\prod_{i=k}^{n} \lambda_i(A)\lambda_i(B) \leq \prod_{i=k}^{n} \sigma_i(AB) \leq \prod_{i=k}^{n} \lambda_i(AB) \leq \prod_{i=k}^{n} \lambda_i^2(A\#B) \leq \prod_{i=k}^{n} \lambda_i(A \circ B).$$

Consequently,

$$\lambda_{\min}(AB)I \leq A \circ B$$

and

$$\det AB = [\det(A\#B)]^2 \leq \det(A \circ B).$$

(Proof: See [29, 259], [806, p. 21], and Fact 8.14.20.)

Fact 8.16.15. Let $A, B \in \mathbb{F}^{n \times n}$, and assume that A and B are positive semidefinite. Then,

$$\det AB \leq \left(\prod_{i=1}^{n} A_{(i,i)} \right) \det B \leq \det(A \circ B).$$

If, in addition, A and B are positive definite, then the right-hand inequality is an equality if and only if B is diagonal. (Proof: See [513].) (Remark: The left-hand inequality follows from Hadamard's inequality Fact 8.14.8. The right-hand inequality is *Oppenheim's inequality*.) (Problem: Compare $\left(\prod_{i=1}^{n} A_{(i,i)} \right) \det B$ and $[\det(A\#B)]^2$.)

Fact 8.16.16. Let $A_1, A_2, B_1, B_2 \in \mathbb{F}^{n \times n}$, assume that A_1, A_2, B_1, B_2 are positive semidefinite, and assume that $0 \leq A_1 \leq B_1$ and $0 \leq A_2 \leq B_2$. Then,

$$0 \leq A_1 \otimes A_2 \leq B_1 \otimes B_2$$

and

$$0 \leq A_1 \circ A_2 \leq B_1 \circ B_2.$$

(Proof: See [27].) (Problem: Under which conditions are these inequalities strict?)

Fact 8.16.17. Let $A, B \in \mathbb{F}^{n \times n}$, assume that A and B are positive semidefinite, assume that $0 \leq A \leq B$, and let $k \in \mathbb{P}$. Then,

$$A^{\{k\}} \leq B^{\{k\}}.$$

(Proof: $0 \leq (B - A) \circ (B + A)$ implies that $A \circ A \leq B \circ B$.)

Fact 8.16.18. Let $A_1, \ldots, A_k, B_1, \ldots, B_k \in \mathbb{F}^{n \times n}$, and assume that $A_1, \ldots, A_k, B_1, \ldots, B_k$ are positive semidefinite. Then,

$$(A_1 + B_1) \otimes \cdots \otimes (A_k + B_k) \leq A_1 \otimes \cdots \otimes A_k + B_1 \otimes \cdots \otimes B_k.$$

(Proof: See [531, p. 143].)

Fact 8.16.19. Let $A_1, A_2, B_1, B_2 \in \mathbb{F}^{n \times n}$, assume that A_1, A_2, B_1, B_2 are positive semidefinite, assume that $0 \leq A_1 \leq B_1$ and $0 \leq A_2 \leq B_2$, and let $\alpha \in [0, 1]$. Then,

$$[\alpha A_1 + (1 - \alpha)B_1] \otimes [\alpha A_2 + (1 - \alpha)B_2] \leq \alpha(A_1 \otimes A_2) + (1 - \alpha)(B_1 \otimes B_2).$$

(Proof: See [762].)

Fact 8.16.20. Let $A, B \in \mathbb{F}^{n \times n}$, and assume that A and B are Hermitian. Then, for all $i = 1, \ldots, n$,

$$\lambda_n(A)\lambda_n(B) \leq \lambda_{i+n^2-n}(A \otimes B) \leq \lambda_i(A \circ B) \leq \lambda_i(A \otimes B) \leq \lambda_1(A)\lambda_1(B).$$

(Proof: The result follows from Proposition 7.3.1 and Theorem 8.4.5. For A, B positive semidefinite, the result is given in [510].)

Fact 8.16.21. Let $A \in \mathbb{F}^{n \times n}$ and $B \in \mathbb{F}^{m \times m}$, assume that A and B are positive semidefinite, and let $r \in (0, \infty)$. Then,

$$(A \otimes B)^r = A^r \otimes B^r.$$

Fact 8.16.22. Let $A \in \mathbb{F}^{n \times m}$ and $B \in \mathbb{F}^{k \times l}$. Then,

$$\langle A \otimes B \rangle = \langle A \rangle \otimes \langle B \rangle.$$

Fact 8.16.23. Let $A, B \in \mathbb{F}^{n \times n}$, and assume that A and B are positive semidefinite. If $r \in [0, 1]$, then

$$A^r \circ B^r \leq (A \circ B)^r.$$

If $r \in [1, 2]$, then

$$(A \circ B)^r \leq A^r \circ B^r.$$

If A and B are positive definite and $r \in [0, 1]$, then

$$(A \circ B)^{-r} \leq A^{-r} \circ B^{-r}.$$

Therefore,

$$(A \circ B)^2 \leq A^2 \circ B^2,$$

$$A \circ B \leq (A^2 \circ B^2)^{1/2},$$

$$A^{1/2} \circ B^{1/2} \leq (A \circ B)^{1/2}.$$

Furthermore,

$$A^2 \circ B^2 - \tfrac{1}{4}(\beta - \alpha)^2 I \leq (A \circ B)^2 \leq \tfrac{1}{2}\left[A^2 \circ B^2 + (AB)^{\{2\}}\right] \leq A^2 \circ B^2$$

and

$$A \circ B \leq (A^2 \circ B^2)^{1/2} \leq \frac{\alpha + \beta}{2\sqrt{\alpha\beta}} A \circ B,$$

where $\alpha \triangleq \lambda_{\min}(A \otimes B)$ and $\beta \triangleq \lambda_{\max}(A \otimes B)$. Hence,

$$A \circ B - \tfrac{1}{4}\left(\sqrt{\beta} - \sqrt{\alpha}\right)^2 I \leq \left(A^{1/2} \circ B^{1/2}\right)^2$$

$$\leq \tfrac{1}{2}\left[A \circ B + \left(A^{1/2}B^{1/2}\right)^{\{2\}}\right]$$

$$\leq A \circ B$$

$$\leq \left(A^2 \circ B^2\right)^{1/2}$$

$$\leq \frac{\alpha + \beta}{2\sqrt{\alpha\beta}} A \circ B.$$

(Proof: See [27, 546, 749], [367, p. 475], and [806, p. 8].)

Fact 8.16.24. Let $A, B \in \mathbb{F}^{n \times n}$, assume that A and B are positive definite, and let k, l be nonzero integers such that $k \leq l$. Then,

$$\left(A^k \circ B^k\right)^{1/k} \leq \left(A^l \circ B^l\right)^{1/l}.$$

In particular,

$$\left(A^{-1} \circ B^{-1}\right)^{-1} \leq A \circ B,$$

$$(A \circ B)^{-1} \leq A^{-1} \circ B^{-1},$$

and, for all $k > 0$,

$$A \circ B \leq (A^k \circ B^k)^{1/k},$$

$$A^{1/k} \circ B^{1/k} \leq (A \circ B)^{1/k}.$$

Furthermore,

$$(A \circ B)^{-1} \leq A^{-1} \circ B^{-1} \leq \frac{(\alpha + \beta)^2}{4\alpha\beta}(A \circ B)^{-1},$$

where $\alpha \triangleq \lambda_{\min}(A \otimes B)$ and $\beta \triangleq \lambda_{\max}(A \otimes B)$. (Proof: See [546].) (Problem: Consider real numbers $p \leq q \leq -1$ to unify this result with Fact 8.16.33.)

Fact 8.16.25. Let $A, B \in \mathbb{F}^{n \times n}$, and assume that A and B are positive definite. Then,

$$2I \leq A \circ B^{-1} + B \circ A^{-1}.$$

(Proof: See [749, 813].) (Remark: Setting $B = A$ yields an inequality of Fact 8.16.3.)

Fact 8.16.26. Let $A, B \in \mathbb{F}^{n \times m}$, and define

$$\mathcal{A} \triangleq \begin{bmatrix} A^*A \circ B^*B & (A \circ B)^* \\ A \circ B & I \end{bmatrix}.$$

Then, $\mathcal{A}$ is positive semidefinite. Furthermore,

$$(A \circ B)^*(A \circ B) \leq \tfrac{1}{2}(A^*A \circ B^*B + A^*B \circ B^*A) \leq A^*A \circ B^*B.$$

(Proof: See [371, 749, 813].)

Fact 8.16.27. Let $A, B, C \in \mathbb{F}^{n \times n}$, define

$$\mathcal{A} \triangleq \left[\begin{array}{cc} A & B \\ B^* & C \end{array} \right],$$

and assume that $\mathcal{A}$ is positive semidefinite. Then,

$$-A \circ C \leq B \circ B^* \leq A \circ C$$

and

$$|\det(B \circ B^*)| \leq \det(A \circ C).$$

If, in addition, $\mathcal{A}$ is positive definite, then

$$-A \circ C < B \circ B^* < A \circ C$$

and

$$|\det(B \circ B^*)| < \det(A \circ C).$$

(Proof: See [813].) (Remark: See Fact 8.9.3.)

Fact 8.16.28. Let $A, B \in \mathbb{F}^{n \times m}$. Then,

$$-A^*A \circ B^*B \leq A^*B \circ B^*A \leq A^*A \circ B^*B$$

and

$$|\det(A^*B \circ B^*A)| \leq \det(A^*A \circ B^*B).$$

(Proof: Apply Fact 8.16.27 to $\left[\begin{smallmatrix} A^*A & A^*B \\ B^*A & B^*B \end{smallmatrix} \right]$.) (Remark: See Fact 8.16.5 and Fact 8.9.6.)

Fact 8.16.29. Let $A, B \in \mathbb{F}^{n \times n}$, and assume that A is positive definite. Then,

$$-A \circ B^*A^{-1}B \leq B \circ B^* \leq A \circ B^*A^{-1}B$$

and

$$|\det(B \circ B^*)| \leq \det(A \circ B^*A^{-1}B).$$

(Proof: Use Fact 8.16.27 and Fact 8.9.10.)

Fact 8.16.30. Let $A, B \in \mathbb{F}^{n \times n}$, and let $\alpha, \beta \in (0, \infty)$.

$$-\left(\beta^{-1/2}I + \alpha A^*A \right) \circ \left(\alpha^{-1/2}I + \beta BB^* \right) \leq (A + B) \circ (A + B)^*$$

$$\leq \left(\beta^{-1/2}I + \alpha A^*A \right) \circ \left(\alpha^{-1/2}I + \beta BB^* \right).$$

(Remark: See Fact 8.9.11.)

Fact 8.16.31. Let $A, B \in \mathbb{F}^{n \times m}$, and define

$$\mathcal{A} \triangleq \left[\begin{array}{cc} A^*A \circ I & (A \circ B)^* \\ A \circ B & BB^* \circ I \end{array} \right].$$

Then, $\mathcal{A}$ is positive semidefinite. Now, assume that $n = m$. Then,

$$-A^*A \circ I - BB^* \circ I \leq A \circ B + (A \circ B)^* \leq A^*A \circ I + BB^* \circ I$$

and

$$-A^*A \circ BB^* \circ I \leq A \circ A^* \circ B \circ B^* \leq A^*A \circ BB^* \circ I.$$

(Remark: See Fact 8.16.27.)

Fact 8.16.32. Let $A, B \in \mathbb{F}^{n \times n}$, and assume that A and B are positive semidefinite. Then,
$$A \circ B \leq \tfrac{1}{2}\left(A^2 + B^2\right) \circ I.$$

(Proof: Use Fact 8.16.31.)

Fact 8.16.33. Let $A, B \in \mathbb{F}^{n \times n}$, assume that A and B are positive semidefinite, and let $p, q \in [1, \infty)$ satisfy $p \leq q$. Then,

$$(A^p \circ B^p)^{1/p} \leq (A^q \circ B^q)^{1/q}.$$

(Proof: Since $p/q \leq 1$, it follows from Fact 8.16.23 that $A^p \circ B^p = (A^q)^{p/q} \circ (A^q)^{p/q} \leq (A^q \circ B^q)^{p/q}$. Then, use Corollary 8.5.10 with p replaced by $1/p$. See [806, p. 8].)

Fact 8.16.34. Let $A, B \in \mathbb{F}^{n \times n}$, assume that A and B are positive definite, and let $p, q \in (0, \infty)$ satisfy $p \leq q$. Then,

$$I \circ (\log A + \log B) \leq \log \left(A^p \circ B^p\right)^{1/p} \leq \log \left(A^q \circ B^q\right)^{1/q}$$

and

$$I \circ (\log A + \log B) = \lim_{p \downarrow 0} \log \left(A^p \circ B^p\right)^{1/p}.$$

(Proof: See [748].)

Fact 8.16.35. Let $A, B \in \mathbb{F}^{n \times n}$, and assume that A and B are positive definite. Then,
$$I \circ (\log A + \log B) \leq \log(A \circ B).$$

(Proof: Set $p = 1$ in Fact 8.16.34. See [27] and [806, p. 8].) (Remark: See Fact 11.12.20.)

Fact 8.16.36. Let $A, B \in \mathbb{F}^{n \times n}$, assume that A and B are positive definite, and let $C, D \in \mathbb{F}^{m \times n}$. Then,

$$(C \circ D)(A \circ B)^{-1}(C \circ D)^* \leq \left(CA^{-1}C^*\right) \circ \left(DB^{-1}D^*\right).$$

In particular,

$$(A \circ B)^{-1} \leq A^{-1} \circ B^{-1}$$

and

$$(C \circ D)(C \circ D)^* \leq (CC^*) \circ (DD^*).$$

(Proof: Form the Schur complement of the lower right block of the Schur product of the positive-semidefinite matrices $\begin{bmatrix} A & C^* \\ C & CA^{-1}C^* \end{bmatrix}$ and $\begin{bmatrix} B & D^* \\ D & DB^{-1}D^* \end{bmatrix}$. See [512, 755] or [806, p. 13].)

Fact 8.16.37. Let $A, B \in \mathbb{F}^{n \times n}$, assume that A and B are positive semidefinite, and let $p, q \in (1, \infty)$ satisfy $1/p + 1/q = 1$. Then,

$$(A \circ B) + (C \circ D) \le (A^p + C^p)^{1/p} \circ (B^q + D^q)^{1/q}.$$

(Proof: Use *xxiv*) of Proposition 8.5.15 with $r = 1/p$. See [806, p. 10].) (Remark: Note the relationship between the *conjugate parameters* p, q and the *barycentric coordinates* $\alpha, 1 - \alpha$. See Fact 1.4.20.)

8.17 Facts on Majorization

Fact 8.17.1. Let $x, y \in \mathbb{R}^n$, where $x_{(1)} \ge \cdots \ge x_{(n)}$ and $y_{(1)} \ge \cdots \ge y_{(n)}$. Then, y strongly majorizes x if and only if x is an element of the convex hull of the vectors $y_1, \ldots, y_{n!} \in \mathbb{R}^n$, where each vector is formed by permuting the components of y. (Proof: See [516, p. 113].) (Remark: This result is due to Rado.) (Remark: See Fact 8.14.7.)

Fact 8.17.2. Let $x, y \in \mathbb{R}^n$, where $x_{(1)} \ge \cdots \ge x_{(n)}$ and $y_{(1)} \ge \cdots \ge y_{(n)}$, assume that y strongly majorizes x, let $f \colon [\min\{x_{(n)}, y_{(n)}\}, y_{(1)}] \mapsto \mathbb{R}$, and assume that f is convex. Then, $\begin{bmatrix} f(y_{(1)}) & \cdots & f(y_{(n)}) \end{bmatrix}^{\mathrm{T}}$ weakly majorizes $\begin{bmatrix} f(x_{(1)}) & \cdots & f(x_{(n)}) \end{bmatrix}^{\mathrm{T}}$. (Proof: See [111, p. 42], [369, p. 173], or [516, p. 116].)

Fact 8.17.3. Let $x, y \in \mathbb{R}^n$, where $x_{(1)} \ge \cdots \ge x_{(n)} \ge 0$ and $y_{(1)} \ge \cdots \ge y_{(n)} \ge 0$, assume that y strongly log majorizes x, let $f \colon [0, \infty) \mapsto \mathbb{R}$, and assume that $g(z) \triangleq f(e^z)$ is convex. Then, $\begin{bmatrix} f(y_{(1)}) & \cdots & f(y_{(n)}) \end{bmatrix}^{\mathrm{T}}$ weakly majorizes $\begin{bmatrix} f(x_{(1)}) & \cdots & f(x_{(n)}) \end{bmatrix}^{\mathrm{T}}$. (Proof: Apply Fact 8.17.2.)

Fact 8.17.4. Let $x, y \in \mathbb{R}^n$, where $x_{(1)} \ge \cdots \ge x_{(n)}$ and $y_{(1)} \ge \cdots \ge y_{(n)}$, assume that y weakly majorizes x, let $f \colon [\min\{x_{(n)}, y_{(n)}\}, y_{(1)}] \mapsto \mathbb{R}$, and assume that f is convex and increasing. Then, $\begin{bmatrix} f(y_{(1)}) & \cdots & f(y_{(n)}) \end{bmatrix}^{\mathrm{T}}$ weakly majorizes $\begin{bmatrix} f(x_{(1)}) & \cdots & f(x_{(n)}) \end{bmatrix}^{\mathrm{T}}$. (Proof: See [111, p. 42], [369, p. 173], or [516, p. 116].)

Fact 8.17.5. Let $x, y \in \mathbb{R}^n$, where $x_{(1)} \ge \cdots \ge x_{(n)} \ge 0$ and $y_{(1)} \ge \cdots \ge y_{(n)} \ge 0$, assume that y log majorizes x, let $f \colon [0, \infty) \mapsto \mathbb{R}$, and assume that $g(z) \triangleq f(e^z)$ is convex and increasing. Then, $\begin{bmatrix} f(y_{(1)}) & \cdots & f(y_{(n)}) \end{bmatrix}^{\mathrm{T}}$ weakly majorizes $\begin{bmatrix} f(x_{(1)}) & \cdots & f(x_{(n)}) \end{bmatrix}^{\mathrm{T}}$. (Proof: Use Fact 8.17.4.)

Fact 8.17.6. Let $x, y \in \mathbb{R}^n$, where $x_{(1)} \geq \cdots \geq x_{(n)} \geq 0$ and $y_{(1)} \geq \cdots \geq y_{(n)} \geq 0$, and assume that y weakly log majorizes x. Then, y weakly majorizes x. (Proof: Use Fact 8.17.4 with $f(t) = e^t$. See [806, p. 19].)

Fact 8.17.7. Let $x, y \in \mathbb{R}^n$, where $x_{(1)} \geq \cdots \geq x_{(n)} \geq 0$ and $y_{(1)} \geq \cdots \geq y_{(n)} \geq 0$, assume that y weakly majorizes x, and let $p \in [1, \infty)$. Then, for all $k = 1, \ldots, n$,

$$\left(\sum_{i=1}^{k} x_{(i)}^p \right)^{1/p} \leq \left(\sum_{i=1}^{k} y_{(i)}^p \right)^{1/p}.$$

(Proof: Use Fact 8.17.4. See [516, p. 96].) (Remark: $\phi(x) \triangleq \left(\sum_{i=1}^{k} x_{(i)}^p \right)^{1/p}$ is a *symmetric gauge function*. See Fact 9.8.33.)

8.18 Notes

The ordering $A \leq B$ is traditionally called the *Löwner ordering*. Proposition 8.2.3 is given in [8] and [442] with extensions in [89]. The proof of Proposition 8.2.6 is based on [146, p. 120], as suggested in [679]. The proof given in [285, p. 307] is incomplete.

Theorem 8.3.4 is due to Newcomb [559]. Proposition 8.4.13 is given in [363,548]. Special cases such as Fact 8.10.16 appear in numerous papers. The proofs of Lemma 8.4.4 and Theorem 8.4.5 are based on [667]. Theorem 8.4.9 can also be obtained as a corollary of the *Fischer minimax theorem* given in [367,516], which provides a geometric characterization of the eigenvalues of a symmetric matrix. Theorem 8.3.5 appears in [606, p. 121]. Theorem 8.5.2 is given in [25]. Additional inequalities appear in [541].

Functions that are nondecreasing on $\mathbf{P}^n$ are characterized by the theory of *monotone matrix functions* [111,234]. See [544] for a summary of the principal results.

The literature on convex maps is extensive. Result *xiv*) of Proposition 8.5.15 is due to Lieb and Ruskai [484]. Result *xxii*) is the *Lieb concavity theorem* [483]. Result *xxxiv*) is due to Ando. Results *xxxvi*) and *xxxvii*) are due to Fan. Some extensions to strict convexity are considered in [516]. See also [27,530,550].

Products of positive-definite matrices are studied in [61–64,796]. Alternative orderings for positive-semidefinite matrices are considered in [58,336].

Essays on the legacy of Issai Schur appear in [403]. Schur complements are discussed in [162, 163, 342, 479, 488]. Majorization and eigenvalue inequalities for sums and products of matrices are discussed in [112].

Chapter Nine

Norms

Norms are used to quantify vectors and matrices, and they play a basic role in convergence analysis. This chapter introduces vector and matrix norms and their numerous properties.

9.1 Vector Norms

For $\alpha \in \mathbb{F}$, let $|\alpha|$ denote the absolute value of α. For $x \in \mathbb{F}^n$ and $A \in \mathbb{F}^{n \times m}$, every component of x and every entry of A can be replaced by its absolute value to obtain $|x| \in \mathbb{R}^n$ and $|A| \in \mathbb{R}^{n \times m}$ defined by

$$|x|_{(i)} \triangleq |x_{(i)}| \tag{9.1.1}$$

for all $i = 1, \ldots, n$ and

$$|A|_{(i,j)} \triangleq |A_{(i,j)}| \tag{9.1.2}$$

for all $i = 1, \ldots, n$ and $j = 1, \ldots, m$. For many applications it is useful to have a scalar measure of the magnitude of x or A. *Norms* provide such measures.

Definition 9.1.1. A *norm* $\| \cdot \|$ on $\mathbb{F}^n$ is a function $\| \cdot \| \colon \mathbb{F}^n \mapsto [0, \infty)$ that satisfies the following conditions:

i) $\|x\| \geq 0$ for all $x \in \mathbb{F}^n$.

ii) $\|x\| = 0$ if and only if $x = 0$.

iii) $\|\alpha x\| = |\alpha| \|x\|$ for all $\alpha \in \mathbb{F}$ and $x \in \mathbb{F}^n$.

iv) $\|x + y\| \leq \|x\| + \|y\|$ for all $x, y \in \mathbb{F}^n$.

Condition *iv)* is the *triangle inequality*.

A norm $\| \cdot \|$ on $\mathbb{F}^n$ is *monotone* if $|x| \leq\leq |y|$ implies that $\|x\| \leq \|y\|$ for all $x, y \in \mathbb{F}^n$, while $\| \cdot \|$ is *absolute* if $\| |x| \| = \|x\|$ for all $x \in \mathbb{F}^n$.

Proposition 9.1.2. Let $\| \cdot \|$ be a norm on $\mathbb{F}^n$. Then, $\| \cdot \|$ is monotone if and only if $\| \cdot \|$ is absolute.

Proof. First, suppose that $\| \cdot \|$ is monotone. Let $x \in \mathbb{F}^n$, and define $y \triangleq |x|$. Then, $|y| = |x|$, and thus $|y| \leq\leq |x|$ and $|y| \leq\leq |x|$. Hence, $\|x\| \leq \|y\|$ and $\|y\| \leq \|x\|$, which implies that $\|x\| = \|y\|$. Thus, $\||x|\| = \|y\| = \|x\|$, which proves that $\| \cdot \|$ is absolute.

Conversely, suppose that $\| \cdot \|$ is absolute and, for convenience, let $n = 2$. Now, let $x, y \in \mathbb{F}^2$ be such that $|x| \leq\leq |y|$. Then, there exist $\alpha_1, \alpha_2 \in [0, 1]$ and $\theta_1, \theta_2 \in \mathbb{R}$ such that $x_{(i)} = \alpha_i e^{\jmath \theta_i} y_{(i)}$ for $i = 1, 2$. Since $\| \cdot \|$ is absolute, it follows that

$$
\begin{aligned}
\|x\| &= \left\| \begin{bmatrix} \alpha_1 e^{\jmath \theta_1} y_{(1)} \\ \alpha_2 e^{\jmath \theta_2} y_{(2)} \end{bmatrix} \right\| \\
&= \left\| \begin{bmatrix} \alpha_1 \ |y_{(1)}| \\ \alpha_2 \ |y_{(2)}| \end{bmatrix} \right\| \\
&= \left\| \tfrac{1}{2}(1 - \alpha_1) \begin{bmatrix} -|y_{(1)}| \\ |\alpha_2||y_{(2)}| \end{bmatrix} + \tfrac{1}{2}(1 - \alpha_1) \begin{bmatrix} |y_{(1)}| \\ \alpha_2|y_{(2)}| \end{bmatrix} + \alpha_1 \begin{bmatrix} |y_{(1)}| \\ \alpha_2|y_{(2)}| \end{bmatrix} \right\| \\
&\leq \left[\tfrac{1}{2}(1 - \alpha_1) + \tfrac{1}{2}(1 - \alpha_1) + \alpha_1 \right] \left\| \begin{bmatrix} |y_{(1)}| \\ \alpha_2|y_{(2)}| \end{bmatrix} \right\| \\
&= \left\| \begin{bmatrix} |y_{(1)}| \\ \alpha_2|y_{(2)}| \end{bmatrix} \right\| \\
&= \left\| \tfrac{1}{2}(1 - \alpha_2) \begin{bmatrix} |y_{(1)}| \\ -|y_{(2)}| \end{bmatrix} + \tfrac{1}{2}(1 - \alpha_2) \begin{bmatrix} |y_{(1)}| \\ |y_{(2)}| \end{bmatrix} + \alpha_2 \begin{bmatrix} |y_{(1)}| \\ |y_{(2)}| \end{bmatrix} \right\| \\
&\leq \left\| \begin{bmatrix} |y_{(1)}| \\ |y_{(2)}| \end{bmatrix} \right\| \\
&= \|y\|.
\end{aligned}
$$

Thus, $\| \cdot \|$ is monotone. $\square$

As we shall see, there are many different norms. For $x \in \mathbb{F}^n$, a useful class of norms consists of the *Hölder norms* defined by

$$
\|x\|_p \triangleq \begin{cases} \left(\displaystyle\sum_{i=1}^{n} |x_{(i)}|^p \right)^{1/p}, & 1 \leq p < \infty, \\[2mm] \displaystyle\max_{i \in \{1, \ldots, n\}} |x_{(i)}|, & p = \infty. \end{cases} \tag{9.1.3}
$$

Note that, for all $x \in \mathbb{C}^n$ and $p \in [1, \infty]$, $\|\overline{x}\|_p = \|x\|_p$. These norms depend on *Minkowski's inequality* given by the following result.

Lemma 9.1.3. Let $p \in [1, \infty]$, and let $x, y \in \mathbb{F}^n$. Then,

$$\|x + y\|_p \leq \|x\|_p + \|y\|_p. \tag{9.1.4}$$

If $p = 1$, then equality holds if and only if, for all $i = 1, \ldots, n$, there exists $\alpha_i \geq 0$ such that either $x_{(i)} = \alpha_i y_{(i)}$ or $y_{(i)} = \alpha_i x_{(i)}$. If $p \in (1, \infty)$, then equality holds if and only if there exists $\alpha \geq 0$ such that either $x = \alpha y$ or $y = \alpha x$.

Proof. See [87, 511] and Fact 1.4.21. $\qquad \square$

Proposition 9.1.4. Let $p \in [1, \infty]$. Then, $\| \cdot \|_p$ is a norm on $\mathbb{F}^n$.

For $p = 1$,

$$\|x\|_1 = \sum_{i=1}^n |x_{(i)}| \tag{9.1.5}$$

is the *absolute sum norm*; for $p = 2$,

$$\|x\|_2 = \left(\sum_{i=1}^n |x_{(i)}|^2 \right)^{1/2} = \sqrt{x^* x} \tag{9.1.6}$$

is the *Euclidean norm*; and, for $p = \infty$,

$$\|x\|_\infty = \max_{i \in \{1, \ldots, n\}} |x_{(i)}| \tag{9.1.7}$$

is the *infinity norm*.

The Hölder norms satisfy the following monotonicity property, which is related to the power-sum inequality given by Fact 1.4.16.

Proposition 9.1.5. Let $1 \leq p \leq q \leq \infty$, and let $x \in \mathbb{F}^n$. Then,

$$\|x\|_\infty \leq \|x\|_q \leq \|x\|_p \leq \|x\|_1. \tag{9.1.8}$$

Assume, in addition, that $1 \leq p < q \leq \infty$. Then, x has at least two nonzero components if and only if

$$\|x\|_\infty < \|x\|_q < \|x\|_p < \|x\|_1. \tag{9.1.9}$$

Proof. If either $p = q$ or $x = 0$ or x has exactly one nonzero component, then $\|x\|_q = \|x\|_p$. Hence, to prove both (9.1.8) and (9.1.9), it suffices to prove (9.1.9) in the case that $1 < p < q < \infty$ and x has at least two nonzero components. Thus, let $n \geq 2$, let $x \in \mathbb{F}^n$ have at least two nonzero components, and define $f \colon [1, \infty) \to [0, \infty)$ by $f(\beta) \triangleq \|x\|_\beta$. Hence,

$$f'(\beta) = \tfrac{1}{\beta} \|x\|_\beta^{1-\beta} \sum_{i=1}^n \gamma_i,$$

where, for all $i = 1, \ldots, n$,

$$\gamma_i \triangleq \begin{cases} |x_i|^\beta (\log |x_{(i)}| - \log \|x\|_\beta), & x_{(i)} \neq 0, \\ 0, & x_{(i)} = 0. \end{cases}$$

If $x_{(i)} \neq 0$, then $\log |x_{(i)}| < \log \|x\|_\beta$. It thus follows that $f'(\beta) < 0$, which implies that f is decreasing on $[1, \infty)$. Hence, (9.1.9) holds. $\square$

The following result is *Hölder's inequality*. For this result we interpret $1/\infty = 0$. Note that, for all $x, y \in \mathbb{F}^n$, $|x^\mathrm{T} y| \leq |x|^\mathrm{T} |y| = \|x \circ y\|_1$. See Fact 1.4.23.

Proposition 9.1.6. Let $p, q \in [1, \infty]$ satisfy $1/p + 1/q = 1$, and let $x, y \in \mathbb{F}^n$. Then,

$$|x^\mathrm{T} y| \leq \|x\|_p \|y\|_q. \tag{9.1.10}$$

Furthermore, equality holds if and only if $|x^\mathrm{T} y| = |x|^\mathrm{T} |y|$ and

$$\begin{cases} |x| \circ |y| = \|y\|_\infty |x|, & p = 1, \\ |x|^{\{p\}} \text{ and } |y|^{\{q\}} \text{ are linearly dependent}, & 1 < p < \infty, \\ |x| \circ |y| = \|x\|_\infty |y|, & p = \infty. \end{cases} \tag{9.1.11}$$

Proof. See [151, p. 127], [367, p. 536], [416, p. 71], and Fact 1.4.20. $\square$

The case $p = q = 2$ is the *Cauchy-Schwarz inequality*.

Corollary 9.1.7. Let $x, y \in \mathbb{F}^n$. Then,

$$|x^\mathrm{T} y| \leq \|x\|_2 \|y\|_2. \tag{9.1.12}$$

Furthermore, equality holds if and only if x and y are linearly dependent.

Proof. Assume $y \neq 0$, and define $M \triangleq \begin{bmatrix} \sqrt{y^* y} I & (y^* y)^{-1/2} y \end{bmatrix}$. Since $M^* M = \begin{bmatrix} y^* y I & y \\ y^* & 1 \end{bmatrix}$ is positive semidefinite, it follows from *iii*) of Proposition 8.2.3 that $yy^* \leq y^* y I$. Therefore, $x^* yy^* x \leq x^* x y^* y$, which is equivalent to (9.1.12) with x replaced by $\overline{x}$.

Now, suppose that x and y are linearly dependent. Then, there exists $\beta \in \mathbb{F}$ such that either $x = \beta y$ or $y = \beta x$. In both cases it follows that $|x^* y| = \|x\|_2 \|y\|_2$. Conversely, define $f \colon \mathbb{F}^n \times \mathbb{F}^n \to [0, \infty)$ by $f(\mu, \nu) \triangleq \mu^* \mu \nu^* \nu - |\mu^* \nu|^2$. Now, suppose that $f(x, y) = 0$ so that (x, y) minimizes f. Then, it follows that $f_\mu(x, y) = 0$, which implies that $y^* y x = y^* x y$. Hence, x and y are linearly dependent. $\square$

The norms $\| \cdot \|$ and $\| \cdot \|'$ on $\mathbb{F}^n$ are *equivalent* if there exist $\alpha, \beta > 0$ such that

$$\alpha \|x\| \le \|x\|' \le \beta \|x\| \tag{9.1.13}$$

for all $x \in \mathbb{F}^n$. Note that these inequalities can be written as

$$\tfrac{1}{\beta} \|x\|' \le \|x\| \le \tfrac{1}{\alpha} \|x\|'. \tag{9.1.14}$$

Hence, the word "equivalent" is justified.

Theorem 9.1.8. Let $\| \cdot \|$ and $\| \cdot \|'$ be norms on $\mathbb{F}^n$. Then, $\| \cdot \|$ and $\| \cdot \|'$ are equivalent.

Proof. See [367, p. 272].   □

9.2 Matrix Norms

One way to define norms for matrices is by viewing a matrix $A \in \mathbb{F}^{n \times m}$ as a vector in $\mathbb{F}^{nm}$, for example, as vec A.

Definition 9.2.1. A *norm* $\| \cdot \|$ on $\mathbb{F}^{n \times m}$ is a function $\| \cdot \|$: $\mathbb{F}^{n \times m} \mapsto [0, \infty)$ that satisfies the following conditions:

i) $\|A\| \ge 0$ for all $A \in \mathbb{F}^{n \times m}$.

ii) $\|A\| = 0$ if and only if $A = 0$.

iii) $\|\alpha A\| = |\alpha| \|A\|$ for all $\alpha \in \mathbb{F}$.

iv) $\|A + B\| \le \|A\| + \|B\|$ for all $A, B \in \mathbb{F}^{n \times m}$.

If $\| \cdot \|$ is a norm on $\mathbb{F}^{nm}$, then $\| \cdot \|'$ defined by $\|A\|' \triangleq \| \text{vec } A \|$ is a norm on $\mathbb{F}^{n \times m}$. For example, Hölder norms can be defined for matrices by choosing $\| \cdot \| = \| \cdot \|_p$. Hence, for all $A \in \mathbb{F}^{n \times m}$ define

$$\|A\|_p \triangleq \begin{cases} \left(\displaystyle\sum_{i=1}^{n} \sum_{j=1}^{m} |A_{(i,j)}|^p \right)^{1/p}, & 1 \le p < \infty, \\[2mm] \displaystyle\max_{\substack{i \in \{1,\dots,n\} \\ j \in \{1,\dots,m\}}} |A_{(i,j)}|, & p = \infty. \end{cases} \tag{9.2.1}$$

Note that the same symbol $\| \cdot \|_p$ is used to denote the Hölder norm for both vectors and matrices. This notation is consistent since, if $A \in \mathbb{F}^{n \times 1}$, then $\|A\|_p$ coincides with the vector Hölder norm. Furthermore, if $A \in \mathbb{F}^{n \times m}$ and $1 \le p \le \infty$, then

$$\|A\|_p = \| \text{vec } A \|_p. \tag{9.2.2}$$

It follows from (9.1.8) that, if $A \in \mathbb{F}^{n \times m}$ and $1 \leq p \leq q$, then

$$\|A\|_\infty \leq \|A\|_q \leq \|A\|_p \leq \|A\|_1. \tag{9.2.3}$$

If, in addition, $1 < p < q < \infty$ and A has at least two nonzero entries, then

$$\|A\|_\infty < \|A\|_q < \|A\|_p < \|A\|_1. \tag{9.2.4}$$

The Hölder norms in the cases $p = 1, 2, \infty$ are the most commonly used. Let $A \in \mathbb{F}^{n \times m}$. For $p = 2$ we define the *Frobenius norm* $\| \cdot \|_\mathrm{F}$ by

$$\|A\|_\mathrm{F} \triangleq \|A\|_2. \tag{9.2.5}$$

Since $\|A\|_2 = \|\operatorname{vec} A\|_2$, it follows that

$$\|A\|_\mathrm{F} = \|A\|_2 = \|\operatorname{vec} A\|_2 = \|\operatorname{vec} A\|_\mathrm{F}. \tag{9.2.6}$$

It is easy to see that

$$\|A\|_\mathrm{F} = \sqrt{\operatorname{tr} A^*A}. \tag{9.2.7}$$

Let $\| \cdot \|$ be a norm on $\mathbb{F}^{n \times m}$. If $\|S_1 A S_2\| = \|A\|$ for all $A \in \mathbb{F}^{n \times m}$ and for all unitary matrices $S_1 \in \mathbb{F}^{n \times n}$ and $S_2 \in \mathbb{F}^{m \times m}$, then $\| \cdot \|$ is *unitarily invariant*. Now, let $m = n$. If $\|A\| = \|A^*\|$ for all $A \in \mathbb{F}^{n \times n}$, then $\| \cdot \|$ is *self-adjoint*. If $\|I_n\| = 1$, then $\| \cdot \|$ is *normalized*. Note that the Frobenius norm is not normalized since $\|I_n\|_\mathrm{F} = \sqrt{n}$. If $\|SAS^*\| = \|A\|$ for all $A \in \mathbb{F}^{n \times n}$ and for all unitary $S \in \mathbb{F}^{n \times n}$, then $\| \cdot \|$ is *weakly unitarily invariant*.

An important class of norms can be defined in terms of singular values. Let $\sigma_1(A) \geq \sigma_2(A) \geq \cdots$ denote the singular values of $A \in \mathbb{F}^{n \times m}$. The following result gives a weak majorization condition for singular values.

Proposition 9.2.2. Let $A, B \in \mathbb{F}^{n \times m}$. Then, for all $k = 1, \ldots, \min\{n, m\}$,

$$\sum_{i=1}^{k} [\sigma_i(A) - \sigma_i(B)] \leq \sum_{i=1}^{k} \sigma_i(A + B) \leq \sum_{i=1}^{k} [\sigma_i(A) + \sigma_i(B)]. \tag{9.2.8}$$

In particular,

$$\sigma_{\max}(A + B) \leq \sigma_{\max}(A) + \sigma_{\max}(B) \tag{9.2.9}$$

and

$$\operatorname{tr} \langle A + B \rangle \leq \operatorname{tr} \langle A \rangle + \operatorname{tr} \langle B \rangle. \tag{9.2.10}$$

Proof. Define $\mathcal{A}, \mathcal{B} \in \mathbf{H}^{n+m}$ by $\mathcal{A} \triangleq \left[\begin{smallmatrix} 0 & A \\ A^* & 0 \end{smallmatrix} \right]$ and $\mathcal{B} \triangleq \left[\begin{smallmatrix} 0 & B \\ B^* & 0 \end{smallmatrix} \right]$. Then, Corollary 8.5.17 implies that, for all $k = 1, \ldots, n + m$,

$$\sum_{i=1}^{k} \lambda_i(\mathcal{A} + \mathcal{B}) \leq \sum_{i=1}^{k} [\lambda_i(\mathcal{A}) + \lambda_i(\mathcal{B})].$$

Now, consider $k \leq \min\{n, m\}$. Then, it follows from Proposition 5.6.5 that, for all $i = 1, \ldots, k$, $\lambda_i(\mathcal{A}) = \sigma_i(A)$. Setting $k = 1$ yields (9.2.9), while setting $k = \min\{n, m\}$ and using Fact 8.14.2 yields (9.2.10). $\qquad\square$

Proposition 9.2.3. Let $p \in [1, \infty]$, and let $A \in \mathbb{F}^{n \times m}$. Then, $\|\cdot\|_{\sigma p}$ defined by

$$\|A\|_{\sigma p} \triangleq \begin{cases} \left(\displaystyle\sum_{i=1}^{\min\{n,m\}} \sigma_i^p(A) \right)^{1/p}, & 1 \leq p < \infty, \\ \sigma_{\max}(A), & p = \infty, \end{cases} \tag{9.2.11}$$

is a norm on $\mathbb{F}^{n \times m}$.

Proof. Let $p \in [1, \infty]$. Then, it follows from Proposition 9.2.2 and Minkowski's inequality Fact 1.4.21 that

$$\|A + B\|_{\sigma p} = \left(\sum_{i=1}^{\min\{n,m\}} \sigma_i^p(A + B) \right)^{1/p}$$

$$\leq \left(\sum_{i=1}^{\min\{n,m\}} [\sigma_i(A) + \sigma_i(B)]^p \right)^{1/p}$$

$$\leq \left(\sum_{i=1}^{\min\{n,m\}} \sigma_i^p(A) \right)^{1/p} + \left(\sum_{i=1}^{\min\{n,m\}} \sigma_i^p(B) \right)^{1/p}$$

$$= \|A\|_{\sigma p} + \|B\|_{\sigma p}. \qquad\qquad\square$$

The norm $\|\cdot\|_{\sigma p}$ is a *Schatten norm*. Let $A \in \mathbb{F}^{n \times m}$. Then, for all $p \in [1, \infty)$,

$$\|A\|_{\sigma p} = (\operatorname{tr} \langle A \rangle^p)^{1/p}. \tag{9.2.12}$$

Important special cases are

$$\|A\|_{\sigma 1} = \sigma_1(A) + \cdots + \sigma_{\min\{n,m\}}(A) = \operatorname{tr} \langle A \rangle, \tag{9.2.13}$$

$$\|A\|_{\sigma 2} = \left[\sigma_1^2(A) + \cdots + \sigma_{\min\{n,m\}}^2(A) \right]^{1/2} = (\operatorname{tr} A^*A)^{1/2} = \|A\|_{\mathrm{F}}, \tag{9.2.14}$$

and

$$\|A\|_{\sigma\infty} = \sigma_1(A) = \sigma_{\max}(A), \tag{9.2.15}$$

which are the *trace norm*, Frobenius norm, and *spectral norm*, respectively.

By applying Proposition 9.1.5 to the vector $\begin{bmatrix} \sigma_1(A) & \cdots & \sigma_{\min\{n,m\}}(A) \end{bmatrix}^{\mathrm{T}}$, we obtain the following result.

Proposition 9.2.4. Let $p, q \in [1, \infty)$, where $p \leq q$, and let $A \in \mathbb{F}^{n \times m}$. Then,

$$\|A\|_{\sigma\infty} \leq \|A\|_{\sigma q} \leq \|A\|_{\sigma p} \leq \|A\|_{\sigma 1}. \tag{9.2.16}$$

Assume, in addition, that $1 < p < q < \infty$ and $\operatorname{rank} A \geq 2$. Then,

$$\|A\|_{\infty} < \|A\|_q < \|A\|_p < \|A\|_1. \tag{9.2.17}$$

The norms $\|\cdot\|_{\sigma p}$ are not very interesting when applied to vectors. Let $x \in \mathbb{F}^n = \mathbb{F}^{n \times 1}$. Then, $\sigma_{\max}(x) = (x^* x)^{1/2} = \|x\|_2$, and, since $\operatorname{rank} x \leq 1$, it follows that, for all $p \in [1, \infty]$,

$$\|x\|_{\sigma p} = \|x\|_2. \tag{9.2.18}$$

Proposition 9.2.5. Let $A \in \mathbb{F}^{n \times m}$. If $p \in (0, 2]$, then

$$\|A\|_{\sigma p} \leq \|A\|_p. \tag{9.2.19}$$

If $p \geq 2$, then

$$\|A\|_p \leq \|A\|_{\sigma p}. \tag{9.2.20}$$

Proof. See [806, p. 50]. $\qquad\square$

Proposition 9.2.6. Let $\|\cdot\|$ be a norm on $\mathbb{F}^{n \times n}$, and let $A \in \mathbb{F}^{n \times n}$. Then,

$$\operatorname{sprad}(A) = \lim_{k \to \infty} \|A^k\|^{1/k}. \tag{9.2.21}$$

Proof. See [367, p. 322]. $\qquad\square$

9.3 Compatible Norms

The norms $\|\cdot\|$, $\|\cdot\|'$, and $\|\cdot\|''$ on $\mathbb{F}^{n \times l}$, $\mathbb{F}^{n \times m}$, and $\mathbb{F}^{m \times l}$, respectively, are *compatible* if, for all $A \in \mathbb{F}^{n \times m}$ and $B \in \mathbb{F}^{m \times l}$,

$$\|AB\| \leq \|A\|'\|B\|''. \tag{9.3.1}$$

For $l = 1$, the norms $\|\cdot\|$, $\|\cdot\|'$, and $\|\cdot\|''$ on $\mathbb{F}^n$, $\mathbb{F}^{n \times m}$, and $\mathbb{F}^m$, respectively, are compatible if, for all $A \in \mathbb{F}^{n \times m}$ and $x \in \mathbb{F}^m$,

$$\|Ax\| \leq \|A\|'\|x\|''. \tag{9.3.2}$$

Furthermore, the norm $\|\cdot\|$ on $\mathbb{F}^n$ is *compatible* with the norm $\|\cdot\|'$ on $\mathbb{F}^{n \times n}$ if, for all $A \in \mathbb{F}^{n \times n}$ and $x \in \mathbb{F}^n$,

$$\|Ax\| \leq \|A\|'\|x\|. \tag{9.3.3}$$

Note that $\|I_n\|' \geq 1$. The norm $\|\cdot\|$ on $\mathbb{F}^{n \times n}$ is *submultiplicative* if, for all $A, B \in \mathbb{F}^{n \times n}$,

$$\|AB\| \leq \|A\|\|B\|. \tag{9.3.4}$$

Hence, the norm $\| \cdot \|$ on $\mathbb{F}^{n \times n}$ is submultiplicative if and only if $\| \cdot \|$, $\| \cdot \|$, and $\| \cdot \|$ are compatible. In this case, $\|I_n\| \geq 1$.

Proposition 9.3.1. Let $\| \cdot \|$ be a submultiplicative norm on $\mathbb{F}^{n \times n}$, and let $y \in \mathbb{F}^n$ be nonzero. Then, $\|x\|' \triangleq \|xy^*\|$ is a norm on $\mathbb{F}^n$, and $\| \cdot \|'$ is compatible with $\| \cdot \|$.

Proposition 9.3.2. Let $\| \cdot \|$ be a submultiplicative norm on $\mathbb{F}^{n \times n}$, and let $A \in \mathbb{F}^{n \times n}$. Then,

$$\mathrm{sprad}(A) \leq \|A\|. \tag{9.3.5}$$

Proof. Use Proposition 9.3.1 to construct a norm $\| \cdot \|'$ on $\mathbb{F}^n$ that is compatible with $\| \cdot \|$. Furthermore, let $A \in \mathbb{F}^{n \times n}$, let $\lambda \in \mathrm{spec}(A)$, and let $x \in \mathbb{C}^n$ be an eigenvector of A associated with λ. Then, $Ax = \lambda x$ implies that $|\lambda| \|x\|' = \|Ax\|' \leq \|A\| \|x\|'$, and thus $|\lambda| \leq \|A\|$, which implies (9.3.5). Alternatively, it follows from Proposition 9.2.6 that

$$\mathrm{sprad}(A) = \lim_{k \to \infty} \|A^k\|^{1/k} \leq \lim_{k \to \infty} \|A\|^{k/k} = \|A\|. \qquad \square$$

Proposition 9.3.3. Let $A \in \mathbb{F}^{n \times n}$, and let $\varepsilon > 0$. Then, there exists a submultiplicative norm $\| \cdot \|$ on $\mathbb{F}^{n \times n}$ such that

$$\mathrm{sprad}(A) \leq \|A\| \leq \mathrm{sprad}(A) + \varepsilon. \tag{9.3.6}$$

Proof. See [367, p. 297]. $\square$

Corollary 9.3.4. Let $A \in \mathbb{F}^{n \times n}$, and assume that $\mathrm{sprad}(A) < 1$. Then, there exists a submultiplicative norm $\| \cdot \|$ on $\mathbb{F}^{n \times n}$ such that $\|A\| < 1$.

We now identify some compatible norms. We begin with the Hölder norms.

Proposition 9.3.5. Let $A \in \mathbb{F}^{n \times m}$ and $B \in \mathbb{F}^{m \times l}$. If $p \in [1, 2]$, then

$$\|AB\|_p \leq \|A\|_p \|B\|_p. \tag{9.3.7}$$

If $p \in [2, \infty]$ and q satisfies $1/p + 1/q = 1$, then

$$\|AB\|_p \leq \|A\|_p \|B\|_q \tag{9.3.8}$$

and

$$\|AB\|_p \leq \|A\|_q \|B\|_p. \tag{9.3.9}$$

Proof. First let $1 \leq p \leq 2$ so that $q \triangleq p/(p-1) \geq 2$. Using Hölder's inequality (9.1.10) and (9.1.8) with $p \leq q$ yields

$$\|AB\|_p = \left(\sum_{i,j=1}^{n,l} |\mathrm{row}_i(A)\mathrm{col}_j(B)|^p \right)^{1/p}$$

$$\leq \left(\sum_{i,j=1}^{n,l} \|\text{row}_i(A)\|_p^p \|\text{col}_j(B)\|_q^p \right)^{1/p}$$

$$= \left(\sum_{i=1}^{n} \|\text{row}_i(A)\|_p^p \right)^{1/p} \left(\sum_{j=1}^{l} \|\text{col}_j(B)\|_q^p \right)^{1/p}$$

$$\leq \left(\sum_{i=1}^{n} \|\text{row}_i(A)\|_p^p \right)^{1/p} \left(\sum_{j=1}^{l} \|\text{col}_j(B)\|_p^p \right)^{1/p}$$

$$= \|A\|_p \|B\|_p.$$

Next, let $2 \leq p \leq \infty$ so that $q \triangleq p/(p-1) \leq 2$. Using Hölder's inequality (9.1.10) and (9.1.8) with $q \leq p$ yields

$$\|AB\|_p \leq \left(\sum_{i=1}^{n} \|\text{row}_i(A)\|_p^p \right)^{1/p} \left(\sum_{j=1}^{l} \|\text{col}_j(B)\|_q^p \right)^{1/p}$$

$$\leq \left(\sum_{i=1}^{n} \|\text{row}_i(A)\|_p^p \right)^{1/p} \left(\sum_{j=1}^{l} \|\text{col}_j(B)\|_q^q \right)^{1/q}$$

$$= \|A\|_p \|B\|_q.$$

Similarly, it can be shown that (9.3.9) holds. $\square$

Proposition 9.3.6. Let $A \in \mathbb{F}^{n \times m}$, $B \in \mathbb{F}^{m \times l}$, and $p, q \in [1, \infty]$, and define $r \triangleq 1/(1/p + 1/q) \geq 1$. Then,

$$\|AB\|_{\sigma r} \leq \|A\|_{\sigma p} \|B\|_{\sigma q}. \tag{9.3.10}$$

Proof. Using Proposition 9.6.3 and Hölder's inequality with $1/(p/r) + 1/(q/r) = 1$, it follows that

$$\|AB\|_{\sigma r} = \left(\sum_{i=1}^{\min\{n,m,l\}} \sigma_i^r(AB) \right)^{1/r}$$

$$\leq \left(\sum_{i=1}^{\min\{n,m,l\}} \sigma_i^r(A) \sigma_i^r(B) \right)^{1/r}$$

$$\leq \left[\left(\sum_{i=1}^{\min\{n,m,l\}} \sigma_i^p(A) \right)^{r/p} \left(\sum_{i=1}^{\min\{n,m,l\}} \sigma_i^q(B) \right)^{r/q} \right]^{1/r}$$

$$= \|A\|_{\sigma p}\|B\|_{\sigma q}. \qquad\qquad \square$$

Let $A, B \in \mathbb{F}^{n \times m}$. Using (9.2.16) and (9.3.10), it follows that

$$\|AB\|_{\sigma\infty} \leq \|AB\|_{\sigma 2} \leq \left\{ \begin{array}{c} \|A\|_{\sigma\infty}\|B\|_{\sigma 2} \\ \|A\|_{\sigma 2}\|B\|_{\sigma\infty} \\ \|AB\|_{\sigma 1} \end{array} \right\} \leq \|A\|_{\sigma 2}\|B\|_{\sigma 2} \qquad (9.3.11)$$

or, equivalently,

$$\sigma_{\max}(AB) \leq \|AB\|_{\mathrm{F}} \leq \left\{ \begin{array}{c} \sigma_{\max}(A)\|B\|_{\mathrm{F}} \\ \|A\|_{\mathrm{F}}\sigma_{\max}(B) \\ \mathrm{tr}\,\langle AB \rangle \end{array} \right\} \leq \|A\|_{\mathrm{F}}\|B\|_{\mathrm{F}}. \qquad (9.3.12)$$

Also, for all $r \in [1, \infty]$,

$$\|AB\|_{\sigma r} \leq \left\{ \begin{array}{l} \|A\|_{\sigma r}\sigma_{\max}(B) \\ \sigma_{\max}(A)\|B\|_{\sigma r} \end{array} \right. . \qquad (9.3.13)$$

In particular, setting $r = \infty$ yields

$$\sigma_{\max}(AB) \leq \sigma_{\max}(A)\sigma_{\max}(B). \qquad (9.3.14)$$

Note that the inequality $\|AB\|_{\mathrm{F}} \leq \|A\|_{\mathrm{F}}\|B\|_{\mathrm{F}}$ in (9.3.12) is equivalent to (9.3.7) with $p = 2$ as well as (9.3.8) and (9.3.9) with $p = q = 2$. Finally, it follows from the Cauchy-Schwarz inequality Corollary 9.1.7 that

$$|\mathrm{tr}\, A^*B| \leq \|A\|_{\mathrm{F}}\|B\|_{\mathrm{F}}. \qquad (9.3.15)$$

9.4 Induced Norms

In this section we consider the case in which there exists a nonzero vector $x \in \mathbb{F}^m$ such that (9.3.3) holds as an equality. This condition characterizes a special class of norms on $\mathbb{F}^{n \times n}$, namely, the *induced norms*.

Definition 9.4.1. Let $\|\cdot\|''$ and $\|\cdot\|$ be norms on $\mathbb{F}^m$ and $\mathbb{F}^n$, respectively. Then, $\|\cdot\|'\colon \mathbb{F}^{n \times m} \mapsto \mathbb{F}$ defined by

$$\|A\|' = \max_{x \in \mathbb{F}^m \backslash \{0\}} \frac{\|Ax\|}{\|x\|''} \qquad (9.4.1)$$

is an *induced norm* on $\mathbb{F}^{n \times m}$. In this case, $\|\cdot\|'$ is *induced by* $\|\cdot\|''$ and $\|\cdot\|$. If $m = n$ and $\|\cdot\|'' = \|\cdot\|$, then $\|\cdot\|'$ is *induced by* $\|\cdot\|$, and $\|\cdot\|'$ is an *equi-induced norm*.

The next result confirms that $\|\cdot\|'$ defined by (9.4.1) is indeed a norm.

Theorem 9.4.2. Every induced norm is a norm. Furthermore, every equi-induced norm is normalized.

Proof. See [367, p. 293]. $\qquad\square$

Let $A \in \mathbb{F}^{n \times m}$. It can be seen that (9.4.1) is equivalent to

$$\|A\|' = \max_{x \in \{y \in \mathbb{F}^m \colon \|y\|''=1\}} \|Ax\|. \tag{9.4.2}$$

Theorem 10.3.7 implies that the maximum in (9.4.2) exists. Since, for all $x \neq 0$,

$$\|A\|' = \max_{x \in \mathbb{F}^m \setminus \{0\}} \frac{\|Ax\|}{\|x\|''} \geq \frac{\|Ax\|}{\|x\|''}, \tag{9.4.3}$$

it follows that, for all $x \in \mathbb{F}^m$,

$$\|Ax\| \leq \|A\|'\|x\|'' \tag{9.4.4}$$

so that $\|\cdot\|$, $\|\cdot\|'$, and $\|\cdot\|''$ are compatible. If $m = n$ and $\|\cdot\|'' = \|\cdot\|$, then the norm $\|\cdot\|$ is compatible with the induced norm $\|\cdot\|'$. The next result shows that compatible norms can be obtained from induced norms.

Proposition 9.4.3. Let $\|\cdot\|$, $\|\cdot\|'$, and $\|\cdot\|''$ be norms on $\mathbb{F}^l$, $\mathbb{F}^m$, and $\mathbb{F}^n$, respectively. Furthermore, let $\|\cdot\|'''$ be the norm on $\mathbb{F}^{m \times l}$ induced by $\|\cdot\|$ and $\|\cdot\|'$, let $\|\cdot\|''''$ be the norm on $\mathbb{F}^{n \times m}$ induced by $\|\cdot\|'$ and $\|\cdot\|''$, and let $\|\cdot\|'''''$ be the norm on $\mathbb{F}^{n \times l}$ induced by $\|\cdot\|$ and $\|\cdot\|''$. If $A \in \mathbb{F}^{n \times m}$ and $B \in \mathbb{F}^{m \times l}$, then

$$\|AB\|''''' \leq \|A\|''''\|B\|'''. \tag{9.4.5}$$

Proof. Note that, for all $x \in \mathbb{F}^l$, $\|Bx\|' \leq \|B\|'''\|x\|$, and, for all $y \in \mathbb{F}^m$, $\|Ay\|'' \leq \|A\|''''\|y\|'$. Hence, for all $x \in \mathbb{F}^l$, $\|ABx\|'' \leq \|A\|''''\|Bx\|' \leq \|A\|''''\|B\|'''\|x\|$, which implies that

$$\|AB\|''''' = \max_{x \in \mathbb{F}^l \setminus \{0\}} \frac{\|ABx\|''}{\|x\|} \leq \|A\|''''\|B\|'''. \qquad\square$$

Corollary 9.4.4. Every equi-induced norm is submultiplicative.

The following result is a consequence of Corollary 9.4.4 and Proposition 9.3.2.

Corollary 9.4.5. Let $\|\cdot\|$ be an equi-induced norm on $\mathbb{F}^{n \times n}$, and let $A \in \mathbb{F}^{n \times n}$. Then,

$$\mathrm{sprad}(A) \leq \|A\|. \tag{9.4.6}$$

By assigning $\|\cdot\|_p$ to $\mathbb{F}^m$ and $\|\cdot\|_q$ to $\mathbb{F}^n$, the *Hölder-induced norm* on $\mathbb{F}^{n \times m}$ is defined by

$$\|A\|_{q,p} \triangleq \max_{x \in \mathbb{F}^m \setminus \{0\}} \frac{\|Ax\|_q}{\|x\|_p}. \tag{9.4.7}$$

Proposition 9.4.6. Let $p, q, p', q' \in [1, \infty]$, where $p \leq p'$ and $q \leq q'$, and let $A \in \mathbb{F}^{n \times m}$. Then,

$$\|A\|_{q',p} \leq \|A\|_{q,p} \leq \|A\|_{q,p'}. \tag{9.4.8}$$

Proof. The result follows from Proposition 9.1.5. $\qquad\square$

The following result gives explicit expressions for several Hölder-induced norms.

Proposition 9.4.7. Let $A \in \mathbb{F}^{n \times m}$. Then,

$$\|A\|_{2,2} = \sigma_{\max}(A). \tag{9.4.9}$$

Now, let $p \in [1, \infty]$. Then,

$$\|A\|_{p,1} = \max_{i \in \{1,\ldots,m\}} \|\mathrm{col}_i(A)\|_p. \tag{9.4.10}$$

Finally, let $q \in [1, \infty]$ satisfy $1/p + 1/q = 1$. Then,

$$\|A\|_{\infty,p} = \max_{i \in \{1,\ldots,n\}} \|\mathrm{row}_i(A)\|_q. \tag{9.4.11}$$

Proof. Since A^*A is Hermitian, it follows from Corollary 8.4.2 that, for all $x \in \mathbb{F}^m$,

$$x^*A^*Ax \leq \lambda_{\max}(A^*A)x^*x,$$

which implies that, for all $x \in \mathbb{F}^m$, $\|Ax\|_2 \leq \sigma_{\max}(A)\|x\|_2$, and thus $\|A\|_{2,2} \leq \sigma_{\max}(A)$. Now, let $x \in \mathbb{F}^{n \times n}$ be an eigenvector associated with $\lambda_{\max}(A^*A)$ so that $\|Ax\|_2 = \sigma_{\max}(A)\|x\|_2$, which implies that $\sigma_{\max}(A) \leq \|A\|_{2,2}$. Hence, (9.4.9) holds.

Next, note that, for all $x \in \mathbb{F}^m$,

$$\|Ax\|_p = \left\| \sum_{i=1}^m x_{(i)} \mathrm{col}_i(A) \right\|_p \leq \sum_{i=1}^m |x_{(i)}| \|\mathrm{col}_i(A)\|_p \leq \max_{i \in \{1,\ldots,m\}} \|\mathrm{col}_i(A)\|_p \|x\|_1,$$

and hence $\|A\|_{p,1} \leq \max_{i \in \{1,\ldots,m\}} \|\mathrm{col}_i(A)\|_p$. Next, let $j \in \{1,\ldots,m\}$ be such that $\|\mathrm{col}_j(A)\|_p = \max_{i \in \{1,\ldots,m\}} \|\mathrm{col}_i(A)\|_p$. Now, since $\|e_j\|_1 = 1$, it follows that $\|Ae_j\|_p = \|\mathrm{col}_j(A)\|_p \|e_j\|_1$, which implies that

$$\max_{i \in \{1,\ldots,n\}} \|\mathrm{col}_i(A)\|_p = \|\mathrm{col}_j(A)\|_p \leq \|A\|_{p,1},$$

and hence (9.4.10) holds.

Next, for all $x \in \mathbb{F}^m$, it follows from Hölder's inequality (9.1.10) that

$$\|Ax\|_\infty = \max_{i \in \{1,\ldots,n\}} |\mathrm{row}_i(A)x| \leq \max_{i \in \{1,\ldots,n\}} \|\mathrm{row}_i(A)\|_q \|x\|_p,$$

which implies that $\|A\|_{\infty,p} \leq \max_{i \in \{1,\ldots,n\}} \|\mathrm{row}_i(A)\|_q$. Next, let $j \in \{1,\ldots,n\}$ be such that $\|\mathrm{row}_j(A)\|_q = \max_{i \in \{1,\ldots,n\}} \|\mathrm{row}_i(A)\|_q$, and let nonzero $x \in \mathbb{F}^m$

be such that $|\mathrm{row}_j(A)x| = \|\mathrm{row}_j(A)\|_q\|x\|_p$. Hence,

$$\|Ax\|_\infty = \max_{i\in\{1,...,n\}}|\mathrm{row}_i(A)x| \geq |\mathrm{row}_j(A)x| = \|\mathrm{row}_j(A)\|_q\|x\|_p,$$

which implies that

$$\max_{i\in\{1,...,n\}}\|\mathrm{row}_i(A)\|_q = \|\mathrm{row}_j(A)\|_q \leq \|A\|_{\infty,p},$$

and thus (9.4.11) holds.   □

Note that

$$\max_{i\in\{1,...,m\}}\|\mathrm{col}_i(A)\|_2 = \mathrm{d}_{\max}^{1/2}(A^*A) \tag{9.4.12}$$

and

$$\max_{i\in\{1,...,n\}}\|\mathrm{row}_i(A)\|_2 = \mathrm{d}_{\max}^{1/2}(AA^*). \tag{9.4.13}$$

Therefore, it follows from Proposition 9.4.7 that

$$\|A\|_{1,1} = \max_{i\in\{1,...,m\}}\|\mathrm{col}_i(A)\|_1, \tag{9.4.14}$$

$$\|A\|_{2,1} = \mathrm{d}_{\max}^{1/2}(A^*A), \tag{9.4.15}$$

$$\|A\|_{\infty,1} = \|A\|_\infty = \max_{\substack{i\in\{1,...,n\}\\ j\in\{1,...,m\}}}|A_{(i,j)}|, \tag{9.4.16}$$

$$\|A\|_{\infty,2} = \mathrm{d}_{\max}^{1/2}(AA^*), \tag{9.4.17}$$

$$\|A\|_{\infty,\infty} = \max_{i\in\{1,...,n\}}\|\mathrm{row}_i(A)\|_1. \tag{9.4.18}$$

For convenience, we define the *column norm*

$$\|A\|_{\mathrm{col}} \triangleq \|A\|_{1,1} \tag{9.4.19}$$

and the *row norm*

$$\|A\|_{\mathrm{row}} \triangleq \|A\|_{\infty,\infty}. \tag{9.4.20}$$

Proposition 9.4.8. Let $p, q \in [1,\infty]$ be such that $1/p + 1/q = 1$, and let $A \in \mathbb{F}^{n\times m}$. Then,

$$\|A\|_{q,p} \leq \|A\|_q. \tag{9.4.21}$$

Proof. For $p = 1$ and $q = \infty$, (9.4.21) follows from (9.4.16). For $q < \infty$ and $x \in \mathbb{F}^n$, it follows from Hölder's inequality (9.1.10) that

$$\|Ax\|_q = \left(\sum_{i=1}^n |\mathrm{row}_i(A)x|^q\right)^{1/q} \leq \left(\sum_{i=1}^n \|\mathrm{row}_i(A)\|_q^q\|x\|_p^q\right)^{1/q}$$

$$= \left(\sum_{i=1}^n \sum_{j=1}^m |A_{(i,j)}|^q\right)^{1/q}\|x\|_p = \|A\|_q\|x\|_p,$$

which implies (9.4.21). □

Next, we specialize Proposition 9.4.3 to the Hölder-induced norms.

Corollary 9.4.9. Let $1 \le p, q, r \le \infty$, and let $A \in \mathbb{F}^{n \times m}$ and $A \in \mathbb{F}^{m \times l}$. Then,

$$\|AB\|_{r,p} \le \|A\|_{r,q}\|B\|_{q,p}. \tag{9.4.22}$$

In particular,

$$\|AB\|_{\mathrm{col}} \le \|A\|_{\mathrm{col}}\|B\|_{\mathrm{col}}, \tag{9.4.23}$$

$$\sigma_{\max}(AB) \le \sigma_{\max}(A)\sigma_{\max}(B), \tag{9.4.24}$$

$$\|AB\|_{\mathrm{row}} \le \|A\|_{\mathrm{row}}\|B\|_{\mathrm{row}}, \tag{9.4.25}$$

$$\|AB\|_{\infty} \le \|A\|_{\infty}\|B\|_{\mathrm{col}}, \tag{9.4.26}$$

$$\|AB\|_{\infty} \le \|A\|_{\mathrm{row}}\|B\|_{\infty}, \tag{9.4.27}$$

$$\mathrm{d}_{\max}^{1/2}(B^*A^*AB) \le \mathrm{d}_{\max}^{1/2}(A^*A)\|B\|_{\mathrm{col}}, \tag{9.4.28}$$

$$\mathrm{d}_{\max}^{1/2}(B^*A^*AB) \le \sigma_{\max}(A)\mathrm{d}_{\max}^{1/2}(B^*B), \tag{9.4.29}$$

$$\mathrm{d}_{\max}^{1/2}(ABB^*A^*) \le \mathrm{d}_{\max}^{1/2}(AA^*)\sigma_{\max}(B), \tag{9.4.30}$$

$$\mathrm{d}_{\max}^{1/2}(ABB^*A^*) \le \|B\|_{\mathrm{row}}\mathrm{d}_{\max}^{1/2}(BB^*). \tag{9.4.31}$$

The following result is often useful.

Proposition 9.4.10. Let $A \in \mathbb{F}^{n \times n}$, and assume that $\mathrm{sprad}(A) < 1$. Then, there exists a submultiplicative norm $\|\cdot\|$ on $\mathbb{F}^{n \times n}$ such that $\|A\| < 1$. Furthermore, the series $\sum_{k=0}^{\infty} A^k$ converges absolutely, and

$$(I - A)^{-1} = \sum_{k=0}^{\infty} A^k. \tag{9.4.32}$$

Finally,

$$\frac{1}{1 + \|A\|} \le \left\|(I - A)^{-1}\right\| \le \frac{1}{1 - \|A\|} + \|I\| - 1. \tag{9.4.33}$$

If, in addition, $\|\cdot\|$ is normalized, then

$$\frac{1}{1 + \|A\|} \le \left\|(I - A)^{-1}\right\| \le \frac{1}{1 - \|A\|}. \tag{9.4.34}$$

Proof. Corollary 9.3.4 implies that there exists a submultiplicative norm $\|\cdot\|$ on $\mathbb{F}^{n \times n}$ such that $\|A\| < 1$. It thus follows that

$$\left\|\sum_{k=0}^{\infty} A^k\right\| \le \sum_{k=0}^{\infty} \|A^k\| \le \|I\| - 1 + \sum_{k=0}^{\infty} \|A\|^k = \frac{1}{1 - \|A\|} + \|I\| - 1,$$

which proves that the series $\sum_{k=0}^{\infty} A^k$ converges absolutely.

Next, we show that $I - A$ is nonsingular. If $I - A$ is singular, then there exists a nonzero vector $x \in \mathbb{C}^n$ such that $Ax = x$. Hence, $1 \in \mathrm{spec}(A)$, which contradicts $\mathrm{sprad}(A) < 1$. Next, to verify (9.4.32), note that

$$(I - A)\sum_{k=0}^{\infty} A^k = \sum_{k=0}^{\infty} A^k - \sum_{k=1}^{\infty} A^k = I + \sum_{k=1}^{\infty} A^k - \sum_{k=1}^{\infty} A^k = I,$$

which implies (9.4.32) and thus the right-hand inequality in (9.4.33). Furthermore,

$$\begin{aligned}
1 &\leq \|I\| \\
&= \left\|(I - A)(I - A)^{-1}\right\| \\
&\leq \|I - A\| \left\|(I - A)^{-1}\right\| \\
&\leq (1 + \|A\|) \left\|(I - A)^{-1}\right\|,
\end{aligned}$$

which yields the left-hand inequality in (9.4.33). $\square$

9.5 Induced Lower Bound

We now consider a variation of the induced norm.

Definition 9.5.1. Let $\| \cdot \|$ and $\| \cdot \|'$ denote norms on $\mathbb{F}^m$ and $\mathbb{F}^n$, respectively. Then, for $A \in \mathbb{F}^{n \times m}$, $\ell \colon \mathbb{F}^{n \times m} \mapsto \mathbb{R}$ defined by

$$\ell(A) \triangleq \begin{cases} \min\limits_{y \in \mathcal{R}(A) \setminus \{0\}} \max\limits_{x \in \{z \in \mathbb{F}^m \colon Az = y\}} \dfrac{\|y\|'}{\|x\|}, & A \neq 0, \\ 0, & A = 0, \end{cases} \tag{9.5.1}$$

is the *lower bound induced by* $\| \cdot \|$ and $\| \cdot \|'$. Equivalently,

$$\ell(A) \triangleq \begin{cases} \min\limits_{x \in \mathbb{F}^m \setminus \mathcal{N}(A)} \max\limits_{z \in \mathcal{N}(A)} \dfrac{\|Ax\|'}{\|x + z\|}, & A \neq 0, \\ 0, & A = 0. \end{cases} \tag{9.5.2}$$

Proposition 9.5.2. Let $\| \cdot \|$ and $\| \cdot \|'$ be norms on $\mathbb{F}^m$ and $\mathbb{F}^n$, respectively, let $\| \cdot \|''$ be the norm induced by $\| \cdot \|$ and $\| \cdot \|'$, let $\| \cdot \|'''$ be the norm induced by $\| \cdot \|'$ and $\| \cdot \|$, and let ℓ be the lower bound induced by $\| \cdot \|$ and $\| \cdot \|'$. Then, the following statements hold:

  i) $\ell(A)$ exists for all $A \in \mathbb{F}^{n \times m}$, that is, the minimum in (9.5.1) is attained.

 ii) If $A \in \mathbb{F}^{n \times m}$, then $\ell(A) = 0$ if and only if $A = 0$.

iii) For all $A \in \mathbb{F}^{n \times m}$ there exists a vector $x \in \mathbb{F}^m$ such that

$$\ell(A)\|x\| = \|Ax\|'. \tag{9.5.3}$$

iv) For all $A \in \mathbb{F}^{n \times m}$,

$$\ell(A) \leq \|A\|''. \tag{9.5.4}$$

v) If $A \neq 0$ and B is a (1)-inverse of A, then

$$1/\|B\|''' \leq \ell(A) \leq \|B\|'''. \tag{9.5.5}$$

vi) If $A, B \in \mathbb{F}^{n \times m}$ and either $\mathcal{R}(A) \subseteq \mathcal{R}(A+B)$ or $\mathcal{N}(A) \subseteq \mathcal{N}(A+B)$, then

$$\ell(A) - \|B\|''' \leq \ell(A+B). \tag{9.5.6}$$

vii) If $A, B \in \mathbb{F}^{n \times m}$ and either $\mathcal{R}(A+B) \subseteq \mathcal{R}(A)$ or $\mathcal{N}(A+B) \subseteq \mathcal{N}(A)$, then

$$\ell(A+B) \leq \ell(A) + \|B\|'''. \tag{9.5.7}$$

viii) If $n = m$ and $A \in \mathbb{F}^{n \times n}$ is nonsingular, then

$$\ell(A) = 1/\|A^{-1}\|'''. \tag{9.5.8}$$

Proof. See [306]. $\qquad\qquad\qquad\qquad\qquad\qquad\qquad\qquad\qquad\qquad\square$

Proposition 9.5.3. Let $\|\cdot\|$, $\|\cdot\|'$, and $\|\cdot\|''$ be norms on $\mathbb{F}^l$, $\mathbb{F}^m$, and $\mathbb{F}^n$, respectively, let $\|\cdot\|'''$ denote the norm on $\mathbb{F}^{m \times l}$ induced by $\|\cdot\|$ and $\|\cdot\|'$, let $\|\cdot\|''''$ denote the norm on $\mathbb{F}^{n \times m}$ induced by $\|\cdot\|'$ and $\|\cdot\|''$, and let $\|\cdot\|'''''$ denote the norm on $\mathbb{F}^{n \times l}$ induced by $\|\cdot\|$ and $\|\cdot\|''$. If $A \in \mathbb{F}^{n \times m}$ and $B \in \mathbb{F}^{m \times l}$, then

$$\ell(A)\ell'(B) \leq \ell''(AB). \tag{9.5.9}$$

In addition, the following statements hold:

i) If either $\operatorname{rank} B = \operatorname{rank} AB$ or $\operatorname{def} B = \operatorname{def} AB$, then

$$\ell''(AB) \leq \|A\|''\ell(B). \tag{9.5.10}$$

ii) If $\operatorname{rank} A = \operatorname{rank} AB$, then

$$\ell''(AB) \leq \ell(A)\|B\|''''. \tag{9.5.11}$$

iii) If $\operatorname{rank} B = m$, then

$$\|A\|''\ell(B) \leq \|AB\|'''''. \tag{9.5.12}$$

iv) If $\operatorname{rank} A = m$, then

$$\ell(A)\|B\|'''' \leq \|AB\|'''''. \tag{9.5.13}$$

Proof. See [306]. $\qquad\qquad\qquad\qquad\qquad\qquad\qquad\qquad\qquad\qquad\square$

By assigning $\|\cdot\|_p$ to $\mathbb{F}^m$ and $\|\cdot\|_q$ to $\mathbb{F}^n$, the *Hölder-induced lower bound* on $\mathbb{F}^{n \times m}$ is defined by

$$\ell_{q,p}(A) \triangleq \begin{cases} \min_{y \in \mathcal{R}(A) \backslash \{0\}} \max_{x \in \{z \in \mathbb{F}^m : Az=y\}} \frac{\|y\|_q}{\|x\|_p}, & A \neq 0, \\ 0, & A = 0. \end{cases} \tag{9.5.14}$$

The following result shows that $\ell_{2,2}(A)$ is the smallest positive singular value of A.

Proposition 9.5.4. Let $A \in \mathbb{F}^{n \times m}$, assume that A is nonzero, and let $r \triangleq \operatorname{rank} A$. Then,

$$\ell_{2,2}(A) = \sigma_r(A). \tag{9.5.15}$$

Proof. The result follows from the singular value decomposition. $\quad\square$

Corollary 9.5.5. Let $A \in \mathbb{F}^{n \times m}$. If A is right invertible, then

$$\ell_{2,2}(A) = \sigma_n(A). \tag{9.5.16}$$

If A is left invertible, then

$$\ell_{2,2}(A) = \sigma_m(A). \tag{9.5.17}$$

Finally, if $n = m$ and A is nonsingular, then

$$\ell_{2,2}(A^{-1}) = \sigma_{\min}(A^{-1}) = \frac{1}{\sigma_{\max}(A)}. \tag{9.5.18}$$

Proof. Use Proposition 5.6.2 and Fact 6.3.21. $\quad\square$

In contrast to the submultiplicativity condition (9.4.4) satisfied by the induced norm, the induced lower bound satisfies a supermultiplicativity condition. The following result is analogous to Proposition 9.4.3.

Proposition 9.5.6. Let $\|\cdot\|$, $\|\cdot\|'$, and $\|\cdot\|''$ be norms on $\mathbb{F}^l$, $\mathbb{F}^m$, and $\mathbb{F}^n$, respectively. Let $\ell(\cdot)$ be the lower bound induced by $\|\cdot\|$ and $\|\cdot\|'$, let $\ell'(\cdot)$ be the lower bound induced by $\|\cdot\|'$ and $\|\cdot\|''$, let $\ell''(\cdot)$ be the lower bound induced by $\|\cdot\|$ and $\|\cdot\|''$, let $A \in \mathbb{F}^{n \times m}$ and $B \in \mathbb{F}^{m \times l}$, and assume that either A or B is right invertible. Then,

$$\ell'(A)\ell(B) \leq \ell''(AB). \tag{9.5.19}$$

Furthermore, if $1 \leq p, q, r \leq \infty$, then

$$\ell_{r,q}(A)\ell_{q,p}(B) \leq \ell_{r,p}(AB). \tag{9.5.20}$$

In particular,

$$\sigma_m(A)\sigma_l(B) \leq \sigma_l(AB). \tag{9.5.21}$$

Proof. See [306] and [455, pp. 369, 370]. $\quad\square$

9.6 Singular Value Inequalities

Proposition 9.6.1. Let $A \in \mathbb{F}^{n \times m}$ and $B \in \mathbb{F}^{m \times l}$. Then, for all $i \in \{1, \ldots, \min\{n, m\}\}$ and $j \in \{1, \ldots, \min\{m, l\}\}$ such that $i + j \leq \min\{n, l\} + 1$,

$$\sigma_{i+j-1}(AB) \leq \sigma_i(A)\sigma_j(B). \tag{9.6.1}$$

In particular, for all $j = 1, \ldots, \min\{n, m, l\}$,

$$\sigma_j(AB) \leq \sigma_{\max}(A)\sigma_j(B), \tag{9.6.2}$$

and, for all $i = 1, \ldots, \min\{n, m, l\}$,

$$\sigma_i(AB) \leq \sigma_i(A)\sigma_{\max}(B). \tag{9.6.3}$$

Proof. See [369, p. 178]. □

Proposition 9.6.2. Let $A \in \mathbb{F}^{n \times m}$ and $B \in \mathbb{F}^{m \times l}$. Then, for all $k = 1, \ldots, \min\{n, m, l\}$,

$$\prod_{i=1}^{k} \sigma_i(AB) \leq \prod_{i=1}^{k} \sigma_i(A)\sigma_i(B).$$

If, in addition, $n = m = l$, then

$$\prod_{i=1}^{n} \sigma_i(AB) = \prod_{i=1}^{n} \sigma_i(A)\sigma_i(B).$$

Proof. See [369, p. 172]. □

Proposition 9.6.3. Let $A \in \mathbb{F}^{n \times m}$ and $B \in \mathbb{F}^{m \times l}$. If $r \geq 0$, then, for all $k = 1, \ldots, \min\{n, m, l\}$,

$$\sum_{i=1}^{k} \sigma_i^r(AB) \leq \sum_{i=1}^{k} \sigma_i^r(A)\sigma_i^r(B). \tag{9.6.4}$$

In particular, for all $k = 1, \ldots, \min\{n, m, l\}$,

$$\sum_{i=1}^{k} \sigma_i(AB) \leq \sum_{i=1}^{k} \sigma_i(A)\sigma_i(B). \tag{9.6.5}$$

If $r < 0$, $n = m = l$, and A and B are nonsingular, then

$$\sum_{i=1}^{n} \sigma_i^r(AB) \leq \sum_{i=1}^{n} \sigma_i^r(A)\sigma_i^r(B). \tag{9.6.6}$$

Proof. The first statement follows from Proposition 9.6.2 and Fact 8.17.3. For the case $r < 0$, use Fact 8.17.5. See [369, p. 177] or [111, p. 94]. □

Proposition 9.6.4. Let $A \in \mathbb{F}^{n \times m}$ and $B \in \mathbb{F}^{m \times l}$. If $m \leq n$, then, for all $i = 1, \ldots, \min\{n, m, l\}$,

$$\sigma_m(A)\sigma_i(B) \leq \sigma_i(AB). \tag{9.6.7}$$

If $m \leq l$, then, for all $i = 1, \ldots, \min\{n, m, l\}$,

$$\sigma_i(A)\sigma_m(B) \leq \sigma_i(AB). \tag{9.6.8}$$

Proof. Corollary 8.4.2 implies that $\sigma_m^2(A)I_m = \lambda_{\min}(A^*A)I_m \leq A^*A$, which implies that $\sigma_m^2(A)B^*B \leq B^*A^*AB$. Hence, it follows from the monotonicity theorem Theorem 8.4.9 that, for all $i = 1, \ldots, \min\{n, m, l\}$,

$$\sigma_m(A)\sigma_i(B) = \lambda_i[\sigma_m^2(A)B^*B]^{1/2} \leq \lambda_i^{1/2}(B^*A^*AB) = \sigma_i(AB),$$

which proves the left-hand inequality in (9.6.7). Similarly, for all $i = 1, \ldots, \min\{n, m, l\}$,

$$\sigma_i(A)\sigma_m(B) = \lambda_i[\sigma_m^2(B)AA^*]^{1/2} \leq \lambda_i^{1/2}(ABB^*A^*) = \sigma_i(AB). \qquad \square$$

Corollary 9.6.5. Let $A \in \mathbb{F}^{n \times m}$ and $B \in \mathbb{F}^{m \times l}$. Then,

$$\sigma_m(A)\sigma_{\min\{n,m,l\}}(B) \leq \sigma_{\min\{n,m,l\}}(AB) \leq \sigma_{\max}(A)\sigma_{\min\{n,m,l\}}(B), \quad (9.6.9)$$

$$\sigma_m(A)\sigma_{\max}(B) \leq \sigma_{\max}(AB) \leq \sigma_{\max}(A)\sigma_{\max}(B), \tag{9.6.10}$$

$$\sigma_{\min\{n,m,l\}}(A)\sigma_m(B) \leq \sigma_{\min\{n,m,l\}}(AB) \leq \sigma_{\min\{n,m,l\}}(A)\sigma_{\max}(B), \quad (9.6.11)$$

$$\sigma_{\max}(A)\sigma_m(B) \leq \sigma_{\max}(AB) \leq \sigma_{\max}(A)\sigma_{\max}(B). \tag{9.6.12}$$

Specializing Corollary 9.6.5 to the case in which A or B is square yields the following result.

Corollary 9.6.6. Let $A \in \mathbb{F}^{n \times n}$ and $B \in \mathbb{F}^{n \times l}$. Then, for all $i = 1, \ldots, \min\{n, l\}\}$,

$$\sigma_{\min}(A)\sigma_i(B) \leq \sigma_i(AB) \leq \sigma_{\max}(A)\sigma_i(B). \tag{9.6.13}$$

In particular,

$$\sigma_{\min}(A)\sigma_{\max}(B) \leq \sigma_{\max}(AB) \leq \sigma_{\max}(A)\sigma_{\max}(B). \tag{9.6.14}$$

If $A \in \mathbb{F}^{n \times m}$ and $B \in \mathbb{F}^{m \times m}$, then, for all $i = 1, \ldots, \min\{n, m\}\}$,

$$\sigma_i(A)\sigma_{\min}(B) \leq \sigma_i(AB) \leq \sigma_i(A)\sigma_{\max}(B). \tag{9.6.15}$$

In particular,

$$\sigma_{\max}(A)\sigma_{\min}(B) \leq \sigma_{\max}(AB) \leq \sigma_{\max}(A)\sigma_{\max}(B). \tag{9.6.16}$$

Corollary 9.6.7. Let $A \in \mathbb{F}^{n \times m}$ and $B \in \mathbb{F}^{m \times l}$. If $m \leq n$, then

$$\sigma_m(A)\|B\|_{\mathrm{F}} \leq \|AB\|_{\mathrm{F}}. \tag{9.6.17}$$

If $m \le l$, then

$$\|A\|_{\mathrm{F}} \sigma_m(B) \le \|AB\|_{\mathrm{F}}. \tag{9.6.18}$$

Proposition 9.6.8. Let $A, B \in \mathbb{F}^{n \times m}$. Then, for all $i, j \in \{1, \ldots, \min\{n, m\}\}$ such that $i + j \le \min\{n, m\} + 1$,

$$\sigma_{i+j-1}(A + B) \le \sigma_i(A) + \sigma_j(B) \tag{9.6.19}$$

and

$$\sigma_{i+j-1}(A) - \sigma_j(B) \le \sigma_i(A + B). \tag{9.6.20}$$

Proof. See [369, p. 178]. $\square$

Corollary 9.6.9. Let $A, B \in \mathbb{F}^{n \times m}$. Then,

$$\sigma_n(A) - \sigma_{\max}(B) \le \sigma_n(A + B) \le \sigma_n(A) + \sigma_{\max}(B). \tag{9.6.21}$$

Proof. The result follows from Proposition 9.6.8. Alternatively, it follows from Lemma 8.4.3 and the Cauchy-Schwarz inequality Corollary 9.1.7 that, for all nonzero $x \in \mathbb{F}^n$,

$$\lambda_{\min}[(A + B)(A + B)^*] \le \frac{x^*(AA^* + BB^* + AB^* + BA^*)x}{x^*x}$$

$$= \frac{x^*AA^*x}{\|x\|_2^2} + \frac{x^*BB^*x}{\|x\|_2^2} + \mathrm{Re} \frac{2x^*AB^*x}{\|x\|_2^2}$$

$$\le \frac{x^*AA^*x}{\|x\|_2^2} + \sigma_{\max}^2(B) + 2\frac{(x^*AA^*x)^{1/2}}{\|x\|_2}\sigma_{\max}(B).$$

Minimizing with respect to x and using Lemma 8.4.3 yields

$$\sigma_n^2(A + B) = \lambda_{\min}[(A + B)(A + B)^*]$$

$$\le \lambda_{\min}(AA^*) + \sigma_{\max}^2(B) + 2\lambda_{\min}^{1/2}(AA^*)\sigma_{\max}(B)$$

$$= [\sigma_n(A) + \sigma_{\max}(B)]^2,$$

which proves the right-hand inequality of (9.6.21). Finally, the left-hand inequality follows from the right-hand inequality with A and B replaced by $A + B$ and $-B$, respectively. $\square$

9.7 Facts on Vector Norms

Fact 9.7.1. Let $x, y \in \mathbb{F}^n$. Then, x and y are linearly dependent if and only if $|x|^{\{2\}}$ and $|y|^{\{2\}}$ are linearly dependent and $|x^*y| = |x|^{\mathrm{T}}|y|$. (Remark: This equivalence clarifies the relationship between (9.1.11) with $p = 2$ and Corollary 9.1.7.)

Fact 9.7.2. Let $x, y \in \mathbb{F}^n$, and let $\| \cdot \|$ be a norm on $\mathbb{F}^n$. Then,

$$| \|x\| - \|y\| | \leq \|x + y\|$$

and

$$| \|x\| - \|y\| | \leq \|x - y\|.$$

Fact 9.7.3. Let $x, y \in \mathbb{F}^n$, and let $\| \cdot \|$ be a norm on $\mathbb{F}^n$. Then, the following statements hold:

i) If there exists $\beta \geq 0$ such that either $x = \beta y$ or $y = \beta x$, then $\|x + y\| = \|x\| + \|y\|$.

ii) If $\|x + y\| = \|x\| + \|y\|$ and x and y are linearly dependent, then there exists $\beta \geq 0$ such that either $x = \beta y$ or $y = \beta x$.

iii) If $\|x + y\|_2 = \|x\|_2 + \|y\|_2$, then there exists $\beta \geq 0$ such that either $x = \beta y$ or $y = \beta x$.

(Proof: For iii), use v) of Fact 9.7.5.) (Problem: Consider iii) with alternative norms.) (Problem: If x and y are linearly independent, then does it follow that $\|x + y\| < \|x\| + \|y\|$?)

Fact 9.7.4. Let $x, y \in \mathbb{F}^n$, and let $\| \cdot \|$ be a norm on $\mathbb{F}^n$. Then,

$$\|x\|^2 + \|y\|^2 \leq \|x + y\|^2 + \|x - y\|^2 \leq 4(\|x\|^2 + \|y\|^2).$$

(Proof: See [282, pp. 9–10].)

Fact 9.7.5. Let $x, y \in \mathbb{F}^n$. Then, the following statements hold:

i) $\frac{1}{2}(\|x + y\|_2^2 + \|x - y\|_2^2) = \|x\|_2^2 + \|y\|_2^2$.

ii) $\operatorname{Re} x^* y = \frac{1}{4}(\|x + y\|_2^2 - \|x - y\|_2^2) = \frac{1}{2}(\|x + y\|_2^2 - \|x\|_2^2 - \|y\|_2^2)$.

iii) $\|x - y\|_2 = \sqrt{\|x\|_2^2 + \|y\|_2^2 - 2\operatorname{Re} x^* y}$.

iv) $\|x + y\|_2 \|x - y\|_2 \leq \|x\|_2^2 + \|y\|_2^2$.

v) If $\|x + y\|_2 = \|x\|_2 + \|y\|_2$, then $\operatorname{Im} x^* y = 0$ and $\operatorname{Re} x^* y \geq 0$.

Furthermore, the following statements are equivalent:

vi) $\|x + y\|_2^2 = \|x\|_2^2 + \|y\|_2^2$.

vii) $\|x - y\|_2 = \|x + y\|_2$.

viii) $\operatorname{Re} x^* y = 0$.

(Remark: i) is the *parallelogram law*, which relates the diagonals and the sides of a parallelogram, ii) is the *polarization identity*, iii) is the *cosine law*, and the equivalence of vi) and viii) is the *Pythagorean theorem*.)

Fact 9.7.6. Let $x, y \in \mathbb{F}^n$, and assume that x and y are nonzero. Then,

$$\|x\|_2 + \|y\|_2 \le \frac{2\|x - y\|}{\left\|\dfrac{x}{\|x\|} - \dfrac{y}{\|y\|}\right\|}.$$

(Proof: See [811, p. 28].) (Problem: Interpret this inequality geometrically.)

Fact 9.7.7. Let $x \in \mathbb{F}^n$, and let $p, q \in [1, \infty]$ satisfy $1/p + 1/q = 1$. Then,

$$\|x\|_2 \le \sqrt{\|x\|_p \|x\|_q}.$$

Fact 9.7.8. Let $x, y \in \mathbb{F}^n$, let $p \in (0, 1]$, and define $\|\cdot\|_p$ as in (9.1.3). Then,

$$\|x\|_p + \|y\|_p \le \|x + y\|_p.$$

(Remark: This result is a *reverse triangle inequality*.)

Fact 9.7.9. Let $y \in \mathbb{F}^n$, let $\|\cdot\|$ be a norm on $\mathbb{F}^n$, let $\|\cdot\|'$ be the norm on $\mathbb{F}^{n \times n}$ induced by $\|\cdot\|$, and define

$$\|y\|_{\mathrm{D}} \triangleq \max_{x \in \{z \in \mathbb{F}^n: \|z\|=1\}} |y^* x|.$$

Then, $\|\cdot\|_{\mathrm{D}}$ is a norm on $\mathbb{F}^n$. Furthermore,

$$\|y\| = \max_{x \in \{z \in \mathbb{F}^n: \|z\|_{\mathrm{D}}=1\}} |y^* x|.$$

Hence, for all $x \in \mathbb{F}^n$,

$$|x^* y| \le \|x\| \|y\|_{\mathrm{D}}.$$

In addition,

$$\|xy^*\|' = \|x\| \|y\|_{\mathrm{D}}.$$

Finally, let $p \in [1, \infty]$, and let $1/p + 1/q = 1$. Then,

$$\|\cdot\|_{p\mathrm{D}} = \|\cdot\|_q.$$

Hence, for all $x \in \mathbb{F}^n$,

$$|x^* y| \le \|x\|_p \|y\|_q$$

and

$$\|xy^*\|_{p,p} = \|x\|_p \|y\|_q.$$

(Proof: See [667, p. 57].) (Remark: $\|\cdot\|_{\mathrm{D}}$ is the *dual norm* of $\|\cdot\|$.)

Fact 9.7.10. Let $\|\cdot\|$ be a norm on $\mathbb{F}^n$, and let $\alpha > 0$. Then, $\{x \in \mathbb{F}^n: \|x\| \le \alpha\}$ is convex.

Fact 9.7.11. Let $x \in \mathbb{R}^n$, and let $\|\cdot\|$ be a norm on $\mathbb{R}^n$. Then, $x^{\mathrm{T}} y > 0$ for all $y \in \mathbb{B}_{\|x\|}(x) = \{z \in \mathbb{R}^n: \|z - x\| < \|x\|\}$.

Fact 9.7.12. Let $x, y \in \mathbb{R}^n$, assume that x and y are nonzero, assume that $x^\mathrm{T}y = 0$, and let $\|\cdot\|$ be a norm on $\mathbb{R}^n$. Then, $\|x\| \leq \|x+y\|$. (Proof: If $\|x+y\| < \|x\|$, then $x+y \in \mathbb{B}_{\|x\|}(0)$, and thus $y \in \mathbb{B}_{\|x\|}(-x)$. By Fact 9.7.11, $x^\mathrm{T}y < 0$.) (Remark: See [122, 482] for related results concerning matrices.)

Fact 9.7.13. Let $x \in \mathbb{F}^n$ and $y \in \mathbb{F}^m$. Then,

$$\sigma_{\max}(xy^*) = \|xy^*\|_\mathrm{F} = \|x\|_2\|y\|_2$$

and

$$\sigma_{\max}(xx^*) = \|xx^*\|_\mathrm{F} = \|x\|_2^2.$$

Fact 9.7.14. Let $x \in \mathbb{F}^n$ and $y \in \mathbb{F}^m$. Then,

$$\|x \otimes y\|_2 = \left\|\mathrm{vec}\left(x \otimes y^\mathrm{T}\right)\right\|_2 = \left\|\mathrm{vec}\left(yx^\mathrm{T}\right)\right\|_2 = \left\|yx^\mathrm{T}\right\|_2 = \|x\|_2\|y\|_2.$$

Fact 9.7.15. Let $x \in \mathbb{F}^n$, and let $1 \leq p, q \leq \infty$. Then,

$$\|x\|_p = \|x\|_{p,q}.$$

Fact 9.7.16. Let $x \in \mathbb{F}^n$, and let $p, q \in [1, \infty)$, where $p \leq q$. Then,

$$\|x\|_q \leq \|x\|_p \leq n^{1/p-1/q}\|x\|_q.$$

(Proof: See [353], [354, p. 107].) (Remark: See Fact 9.8.13.)

Fact 9.7.17. Let $A \in \mathbb{F}^{n \times n}$, and assume that A is positive definite. Then,
$$\|x\|_A \triangleq (x^*Ax)^{1/2}$$

is a norm on $\mathbb{F}^n$.

Fact 9.7.18. Let $\|\cdot\|$ and $\|\cdot\|'$ be norms on $\mathbb{F}^n$, and let $\alpha, \beta > 0$. Then, $\alpha\|\cdot\| + \beta\|\cdot\|'$ is also a norm on $\mathbb{F}^n$. Furthermore, $\max\{\|\cdot\|, \|\cdot\|'\}$ is a norm on $\mathbb{F}^n$. (Remark: $\min\{\|\cdot\|, \|\cdot\|'\}$ is not generally a norm.)

Fact 9.7.19. Let $A \in \mathbb{F}^{n \times n}$, assume that A is nonsingular, and let $\|\cdot\|$ be a norm on $\mathbb{F}^n$. Then, $\|x\|' \triangleq \|Ax\|$ is a norm on $\mathbb{F}^n$.

Fact 9.7.20. Let $x \in \mathbb{F}^n$, and let $p \in [1, \infty]$. Then,

$$\|\overline{x}\|_p = \|x\|_p.$$

9.8 Facts on Matrix Norms for One Matrix

Fact 9.8.1. Let $A \in \mathbb{F}^{n \times n}$, and assume that $\mathrm{sprad}(A) < 1$. Then, there exists a submultiplicative matrix norm $\|\cdot\|$ on $\mathbb{F}^{n \times n}$ such that $\|A\| < 1$. Furthermore,

$$\lim_{k \to \infty} A^k = 0.$$

Fact 9.8.2. Let $A \in \mathbb{F}^{n \times n}$, assume that A is nonsingular, and let $\| \cdot \|$ be a submultiplicative norm on $\mathbb{F}^{n \times n}$. Then,

$$\|A^{-1}\| \geq \|I_n\| / \|A\|.$$

Fact 9.8.3. Let $A \in \mathbb{F}^{n \times n}$, assume that A is nonzero and idempotent, and let $\| \cdot \|$ be a submultiplicative norm on $\mathbb{F}^{n \times n}$. Then,

$$\|A\| \geq 1.$$

Fact 9.8.4. Let $\| \cdot \|$ be a unitarily invariant norm on $\mathbb{F}^{n \times n}$. Then, $\| \cdot \|$ is self-adjoint.

Fact 9.8.5. Let $A \in \mathbb{F}^{n \times m}$, let $\| \cdot \|$ be a norm on $\mathbb{F}^{n \times m}$, and define $\|A\|' \triangleq \|A^*\|$. Then, $\| \cdot \|'$ is a norm on $\mathbb{F}^{m \times n}$. If, in addition, $n = m$ and $\| \cdot \|$ is induced by $\| \cdot \|''$, then $\| \cdot \|'$ is induced by $\| \cdot \|''_{\mathrm{D}}$. (Proof: See [367, p. 309] and Fact 9.8.8.) (Remark: See Fact 9.7.9 for the definition of the dual norm. $\| \cdot \|'$ is the *adjoint norm* of $\| \cdot \|$.) (Problem: Generalize this result to matrices that are not square and norms that are not equi-induced.)

Fact 9.8.6. Let $1 \leq p \leq \infty$. Then, $\| \cdot \|_{\sigma p}$ is unitarily invariant.

Fact 9.8.7. Let $A \in \mathbb{F}^{n \times n}$, and assume that A is positive semidefinite. Then,

$$\|A\|_{1,\infty} = \max_{x \in \{z \in \mathbb{F}^n : \|z\|_\infty = 1\}} x^* A x.$$

(Remark: This result is due to Tao. See [619] and [354, p. 116].)

Fact 9.8.8. Let $A \in \mathbb{F}^{n \times m}$, and let $p, q \in [1, \infty]$ satisfy $1/p + 1/q = 1$. Then,

$$\|A^*\|_{p,p} = \|A\|_{q,q}.$$

In particular,

$$\|A^*\|_{\mathrm{col}} = \|A\|_{\mathrm{row}}.$$

(Proof: See Fact 9.8.5.)

Fact 9.8.9. Let $A \in \mathbb{F}^{n \times m}$, and let $p, q \in [1, \infty]$ satisfy $1/p + 1/q = 1$. Then,

$$\left\| \begin{bmatrix} 0 & A \\ A^* & 0 \end{bmatrix} \right\|_{p,p} = \max\{\|A\|_{p,p}, \|A\|_{q,q}\}.$$

In particular,

$$\left\| \begin{bmatrix} 0 & A \\ A^* & 0 \end{bmatrix} \right\|_{\mathrm{col}} = \left\| \begin{bmatrix} 0 & A \\ A^* & 0 \end{bmatrix} \right\|_{\mathrm{row}} = \max\{\|A\|_{\mathrm{col}}, \|A\|_{\mathrm{row}}\}.$$

Fact 9.8.10. Let $A \in \mathbb{F}^{n \times m}$. Then, the following inequalities hold:

i) $\|A\|_F \leq \|A\|_1 \leq \sqrt{mn}\|A\|_F$.

ii) $\|A\|_\infty \leq \|A\|_1 \leq mn\|A\|_\infty$.

iii) $\|A\|_{\mathrm{col}} \leq \|A\|_1 \leq m\|A\|_{\mathrm{col}}$.

iv) $\|A\|_{\mathrm{row}} \leq \|A\|_1 \leq n\|A\|_{\mathrm{row}}$.

v) $\sigma_{\max}(A) \leq \|A\|_1 \leq \sqrt{mn\,\mathrm{rank}\,A}\,\sigma_{\max}(A)$.

vi) $\|A\|_\infty \leq \|A\|_F \leq \sqrt{mn}\|A\|_\infty$.

vii) $\frac{1}{\sqrt{n}}\|A\|_{\mathrm{col}} \leq \|A\|_F \leq \sqrt{m}\|A\|_{\mathrm{col}}$.

viii) $\frac{1}{\sqrt{m}}\|A\|_{\mathrm{row}} \leq \|A\|_F \leq \sqrt{n}\|A\|_{\mathrm{row}}$.

ix) $\sigma_{\max}(A) \leq \|A\|_F \leq \sqrt{\mathrm{rank}\,A}\,\sigma_{\max}(A)$.

x) $\frac{1}{n}\|A\|_{\mathrm{col}} \leq \|A\|_\infty \leq \|A\|_{\mathrm{col}}$.

xi) $\frac{1}{m}\|A\|_{\mathrm{row}} \leq \|A\|_\infty \leq \|A\|_{\mathrm{row}}$.

xii) $\frac{1}{\sqrt{mn}}\sigma_{\max}(A) \leq \|A\|_\infty \leq \sigma_{\max}(A)$.

xiii) $\frac{1}{m}\|A\|_{\mathrm{row}} \leq \|A\|_{\mathrm{col}} \leq n\|A\|_{\mathrm{row}}$.

xiv) $\frac{1}{\sqrt{m}}\sigma_{\max}(A) \leq \|A\|_{\mathrm{col}} \leq \sqrt{n}\sigma_{\max}(A)$.

xv) $\frac{1}{\sqrt{n}}\sigma_{\max}(A) \leq \|A\|_{\mathrm{row}} \leq \sqrt{m}\sigma_{\max}(A)$.

(Remark: See [354, p. 115] for matrices that attain these bounds.)

Fact 9.8.11. Let $A \in \mathbb{F}^{n \times n}$. Then,

$$\|A^{\mathrm{A}}\|_F \leq n^{(2-n)/2}\|A\|_F^{n-1}.$$

(Proof: See [592, pp. 151, 165].)

Fact 9.8.12. Let $A \in \mathbb{F}^{n \times n}$, let $\|\cdot\|$ and $\|\cdot\|'$ be norms on $\mathbb{F}^n$, and define the induced norms

$$\|A\|'' \triangleq \max_{x \in \{y \in \mathbb{F}^m: \|y\|=1\}} \|Ax\|$$

and

$$\|A\|''' \triangleq \max_{x \in \{y \in \mathbb{F}^m: \|y\|'=1\}} \|Ax\|'.$$

Then,

$$\max_{A \in \{X \in \mathbb{F}^{n \times n}: \; X \neq 0\}} \frac{\|A\|''}{\|A\|'''} = \max_{A \in \{X \in \mathbb{F}^{n \times n}: \; X \neq 0\}} \frac{\|A\|'''}{\|A\|''}$$

$$= \max_{x \in \{y \in \mathbb{F}^{n}: \; y \neq 0\}} \frac{\|x\|}{\|x\|'} \max_{x \in \{y \in \mathbb{F}^{n}: \; y \neq 0\}} \frac{\|x\|'}{\|x\|}.$$

(Proof: See [367, p. 303].) (Remark: This symmetry property is evident in Fact 9.8.10.)

Fact 9.8.13. Let $A \in \mathbb{F}^{n \times n}$, and let $p, q \in [1, \infty]$. Then,

$$\|A\|_{p,p} \leq \begin{cases} n^{1/p - 1/q} \|A\|_{q,q}, & p \leq q, \\ n^{1/q - 1/p} \|A\|_{q,q}, & q \leq p. \end{cases}$$

Consequently,

$$n^{1/p - 1} \|A\|_{\mathrm{col}} \leq \|A\|_{p,p} \leq n^{1 - 1/p} \|A\|_{\mathrm{col}},$$

$$n^{-|1/p - 1/2|} \sigma_{\max}(A) \leq \|A\|_{p,p} \leq n^{|1/p - 1/2|} \sigma_{\max}(A),$$

$$n^{-1/p} \|A\|_{\mathrm{col}} \leq \|A\|_{p,p} \leq n^{1/p} \|A\|_{\mathrm{row}}.$$

(Proof: See [353] and [354, p. 112].) (Remark: See Fact 9.7.16.) (Problem: Extend these inequalities to matrices that are not square.)

Fact 9.8.14. Let $A \in \mathbb{F}^{n \times m}$, $p, q \in [1, \infty]$, and $\alpha \in [0, 1]$, and let $r \triangleq pq/[(1 - \alpha)p + \alpha q]$. Then,

$$\|A\|_{r,r} \leq \|A\|_{p,p}^{\alpha} \|A\|_{q,q}^{1 - \alpha}.$$

(Proof: See [353] or [354, p. 113].)

Fact 9.8.15. Let $A \in \mathbb{F}^{n \times m}$, and let $p \in [1, \infty]$. Then,

$$\|A\|_{p,p} \leq \|A\|_{\mathrm{col}}^{1/p} \|A\|_{\mathrm{row}}^{1 - 1/p}.$$

In particular,

$$\sigma_{\max}(A) \leq \sqrt{\|A\|_{\mathrm{col}} \|A\|_{\mathrm{row}}}.$$

(Proof: Set $\alpha = 1/p$, $p = 1$, and $q = \infty$ in Fact 9.8.14. See [354, p. 113]. To prove the special case $p = 2$ directly, note that $\lambda_{\max}(A^*A) \leq \|A^*A\|_{\mathrm{col}} \leq \|A^*\|_{\mathrm{col}} \|A\|_{\mathrm{col}} = \|A\|_{\mathrm{row}} \|A\|_{\mathrm{col}}$.)

Fact 9.8.16. Let $A \in \mathbb{F}^{n \times m}$, and let $p \in [1, 2]$. Then,

$$\|A\|_{p,p} \leq \|A\|_{\mathrm{col}}^{2/p - 1} \sigma_{\max}^{2 - 2/p}(A).$$

(Proof: Let $\alpha = 2/p - 1$, $p = 1$, and $q = 2$ in Fact 9.8.14. See [354, p. 113].)

Fact 9.8.17. Let $A \in \mathbb{F}^{n \times n}$, and let $p \in [1, \infty]$. Then,

$$\|A\|_{p,p} \le \|\,|A|\,\|_{p,p} \le n^{\min\{1/p, 1-1/p\}} \|A\|_{p,p} \le \sqrt{n}\|A\|_{p,p}.$$

(Remark: See [354, p. 117].)

Fact 9.8.18. Let $A \in \mathbb{F}^{n \times m}$, and let $p, q \in [1, \infty]$. Then,

$$\|\overline{A}\|_{q,p} = \|A\|_{q,p}.$$

Fact 9.8.19. Let $A \in \mathbb{F}^{n \times m}$, and let $p, q \in [1, \infty]$. Then,

$$\|A^*\|_{q,p} = \|A\|_{p/(p-1), q/(q-1)}.$$

Fact 9.8.20. Let $A \in \mathbb{F}^{n \times m}$, and let $p, q \in [1, \infty]$. Then,

$$\|A\|_{q,p} \le \begin{cases} \|A\|_{p/(p-1)}, & 1/p + 1/q \le 1, \\ \|A\|_q, & 1/p + 1/q \ge 1. \end{cases}$$

Fact 9.8.21. Let $A \in \mathbb{F}^{n \times n}$, and let $\|\cdot\|$ be a unitarily invariant norm on $\mathbb{F}^{n \times n}$. Then,

$$\|\langle A \rangle\| = \|A\|.$$

Fact 9.8.22. Let $A, S \in \mathbb{F}^{n \times n}$, assume that S is nonsingular, and let $\|\cdot\|$ be a unitarily invariant norm on $\mathbb{F}^{n \times n}$. Then,

$$\|A\| \le \tfrac{1}{2}\|SAS^{-1} + S^{-*}AS^*\|.$$

(Proof: See [35, 139].)

Fact 9.8.23. Let $A \in \mathbb{F}^{n \times n}$, assume that A is positive semidefinite, and let $\|\cdot\|$ be a submultiplicative norm on $\mathbb{F}^{n \times n}$. Then,

$$\|A\|^{1/2} \le \left\|A^{1/2}\right\|.$$

In particular,

$$\sigma_{\max}^{1/2}(A) = \sigma_{\max}\left(A^{1/2}\right).$$

Fact 9.8.24. Let $A_{11} \in \mathbb{F}^{n \times n}$, $A_{12} \in \mathbb{F}^{n \times m}$, and $A_{22} \in \mathbb{F}^{m \times m}$, assume that $\begin{bmatrix} A_{11} & A_{12} \\ A_{12}^* & A_{22} \end{bmatrix} \in \mathbb{F}^{(n+m) \times (n+m)}$ is positive semidefinite, let $\|\cdot\|$ and $\|\cdot\|'$ be unitarily invariant norms on $\mathbb{F}^{n \times n}$ and $\mathbb{F}^{m \times m}$, respectively, and let $p > 0$. Then,

$$\|\langle A_{12}\rangle^p\|'^2 \le \|A_{11}^p\| \|A_{22}^p\|'.$$

(Proof: See [371].)

Fact 9.8.25. Let $A \in \mathbb{F}^{n \times n}$, let $\|\cdot\|$ be a norm on $\mathbb{F}^n$, let $\|\cdot\|_{\mathrm{D}}$ denote the dual norm on $\mathbb{F}^n$, and let $\|\cdot\|'$ denote the norm induced by $\|\cdot\|$ on $\mathbb{F}^{n \times n}$.

Then,

$$\|A\|' = \max_{\substack{x,y\in\mathbb{F}^n \\ x,y\neq 0}} \frac{\mathrm{Re}\, y^*Ax}{\|y\|_{\mathrm{D}}\|x\|}.$$

(Proof: See [354, p. 115].) (Remark: See Fact 9.7.9 for the definition of the dual norm.) (Problem: Generalize this result to obtain Fact 9.8.26 as a special case.)

Fact 9.8.26. Let $A \in \mathbb{F}^{n\times m}$, and let $p, q \in [1,\infty]$. Then,

$$\|A\|_{q,p} = \max_{\substack{x\in\mathbb{F}^m, y\in\mathbb{F}^n \\ x,y\neq 0}} \frac{|y^*Ax|}{\|y\|_{q/(q-1)}\|x\|_p}.$$

Fact 9.8.27. Let $A \in \mathbb{F}^{n\times m}$, and let $p, q \in [1,\infty]$ satisfy $1/p + 1/q = 1$. Then,

$$\|A\|_{p,p} = \max_{\substack{x\in\mathbb{F}^m, y\in\mathbb{F}^n \\ x,y\neq 0}} \frac{|y^*Ax|}{\|y\|_q\|x\|_p} = \max_{\substack{x\in\mathbb{F}^m, y\in\mathbb{F}^n \\ x,y\neq 0}} \frac{|y^*Ax|}{\|y\|_{p/(p-1)}\|x\|_p}.$$

(Remark: See Fact 9.11.2 for the case $p = 2$.)

Fact 9.8.28. Let $A \in \mathbb{F}^{n\times n}$, and assume that A is positive definite. Then,

$$\min_{x\in\mathbb{F}^n\backslash\{0\}} \frac{x^*Ax}{\|Ax\|_2\|x\|_2} = \frac{2\sqrt{\alpha\beta}}{\alpha + \beta}$$

and

$$\min_{\alpha\geq 0} \sigma_{\max}(\alpha A - I) = \frac{\alpha - \beta}{\alpha + \beta},$$

where $\alpha \triangleq \lambda_{\max}(A)$ and $\beta \triangleq \lambda_{\min}(A)$. (Proof: See [316].) (Remark: These quantities are *antieigenvalues*.)

Fact 9.8.29. Let $A \in \mathbb{F}^{n\times n}$, and define

$$\mathrm{nrad}(A) \triangleq \max\{|x^*Ax|:\ x \in \mathbb{C}^n \text{ and } x^*x \leq 1\}.$$

Then, the following statements hold:

  i) $\mathrm{nrad}(A) = \max\{|z|:\ z \in \Theta(A)\}$.

  ii) $\mathrm{sprad}(A) \leq \mathrm{nrad}(A) \leq \mathrm{nrad}(|A|) = \frac{1}{2}\mathrm{sprad}(|A| + |A|^\mathrm{T})$.

  iii) $\frac{1}{2}\sigma_{\max}(A) \leq \mathrm{nrad}(A) \leq \frac{1}{2}\left[\sigma_{\max}(A) + \sigma_{\max}^{1/2}(A^2)\right] \leq \sigma_{\max}(A)$.

  iv) If $A^2 = 0$, then $\mathrm{nrad}(A) = \sigma_{\max}(A)$.

  v) If $\mathrm{nrad}(A) = \sigma_{\max}(A)$, then $\sigma_{\max}(A^2) = \sigma_{\max}^2(A)$.

  vi) If A is normal, then $\mathrm{nrad}(A) = \mathrm{sprad}(A)$.

  vii) $\mathrm{nrad}(A^k) \leq [\mathrm{nrad}(A)]^k$ for all $k \in \mathbb{N}$.

viii) nrad($\cdot$) is a weakly unitarily invariant norm on $\mathbb{F}^{n \times n}$.

ix) nrad($\cdot$) is not a submultiplicative norm on $\mathbb{F}^{n \times n}$.

x) $\| \cdot \| \triangleq \alpha \text{nrad}(\cdot)$ is a submultiplicative norm on $\mathbb{F}^{n \times n}$ if and only if $\alpha \geq 4$.

xi) $\text{nrad}(AB) \leq \text{nrad}(A)\text{nrad}(B)$ for all $A, B \in \mathbb{F}^{n \times n}$ such that A and B are normal.

xii) $\text{nrad}(A \circ B) \leq \alpha \text{nrad}(A)\text{nrad}(B)$ for all $A, B \in \mathbb{F}^{n \times n}$ if and only if $\alpha \geq 2$.

xiii) $\text{nrad}(A \oplus B) = \max\{\text{nrad}(A), \text{nrad}(B)\}$ for all $A \in \mathbb{F}^{n \times n}$ and $B \in \mathbb{F}^{m \times m}$.

(Proof: See [367, p. 331] and [369, pp. 43, 44]. For *iii*), see [434].) (Remark: nrad(A) is the *numerical radius* of A. $\Theta(A)$ is the numerical range. See Fact 4.10.18.) (Remark: nrad($\cdot$) is not submultiplicative. The example $A = \begin{bmatrix} 0 & 1 \\ 0 & 0 \end{bmatrix}$, $B = \begin{bmatrix} 0 & 2 \\ 2 & 0 \end{bmatrix}$, where B is normal, $\text{nrad}(A) = 1/2$, $\text{nrad}(B) = 2$, and $\text{nrad}(AB) = 2$, shows that *xi*) is not valid if only one of the matrices A and B is normal, which corrects [369, pp. 43, 73].) (Remark: *vii*) is the *power inequality*.)

Fact 9.8.30. Let $A \in \mathbb{F}^{n \times m}$, let $\gamma > \sigma_{\max}(A)$, and define $\beta \triangleq \sigma_{\max}(A)/\gamma$. Then,

$$\|A\|_{\mathrm{F}} \leq \sqrt{-[\gamma^2/(2\pi)]\log \det(I - \gamma^{-2}A^*A)} \leq \beta^{-1}\sqrt{-\log(1 - \beta^2)}\|A\|_{\mathrm{F}}.$$

(Proof: See [140].)

Fact 9.8.31. Let $\| \cdot \|$ be a unitarily invariant norm on $\mathbb{F}^{n \times n}$. Then, $\|A\| = 1$ for all $A \in \mathbb{F}^{n \times n}$ such that $\text{rank}\, A = 1$ if and only if $\|E_{1,1}\| = 1$. (Proof: $\|A\| = \|E_{1,1}\|\sigma_{\max}(A)$.) (Remark: These equivalent normalizations are used in [667, p. 74] and [111], respectively.)

Fact 9.8.32. Let $\| \cdot \|$ be a unitarily invariant norm on $\mathbb{F}^{n \times n}$. Then, the following statements are equivalent:

i) $\sigma_{\max}(A) \leq \|A\|$ for all $A \in \mathbb{F}^{n \times n}$.

ii) $\| \cdot \|$ is submultiplicative.

iii) $\|A^2\| \leq \|A\|^2$ for all $A \in \mathbb{F}^{n \times n}$.

iv) $\|A^k\| \leq \|A\|^k$ for all $k \in \mathbb{P}$ and $A \in \mathbb{F}^{n \times n}$.

v) $\|A \circ B\| \leq \|A\|\|B\|$ for all $A, B \in \mathbb{F}^{n \times n}$.

vi) $\text{sprad}(A) \leq \|A\|$ for all $A \in \mathbb{F}^{n \times n}$.

vii) $\|Ax\|_2 \leq \|A\|\|x\|_2$ for all $A \in \mathbb{F}^{n \times n}$ and $x \in \mathbb{F}^n$.

viii) $\|A\|_\infty \le \|A\|$ for all $A \in \mathbb{F}^{n \times n}$.

ix) $\|E_{1,1}\| \ge 1$.

x) $\sigma_{\max}(A) \le \|A\|$ for all $A \in \mathbb{F}^{n \times n}$ such that rank $A = 1$.

(Proof: The equivalence of *i)*–*vii)* is given in [368] and [369, p. 211]. Since $\|A\| = \|E_{1,1}\|\sigma_{\max}(A)$ for all $A \in \mathbb{F}^{n \times n}$ such that rank $A = 1$, it follows that *vii)* and *viii)* are equivalent. To prove *ix)* $\implies$ *x)*, let $A \in \mathbb{F}^{n \times n}$ satisfy rank $A = 1$. Then, $\|A\| = \sigma_{\max}(A)\|E_{1,1}\| \ge \sigma_{\max}(A)$. To show *x)* $\implies$ *ii)*, define $\|\cdot\|' \triangleq \|E_{1,1}\|^{-1}\|\cdot\|$. Since $\|E_{1,1}\|' = 1$, it follows from [111, p. 94] that $\|\cdot\|'$ is submultiplicative. Since $\|E_{1,1}\|^{-1} \le 1$, it follows that $\|\cdot\|$ is also submultiplicative. Alternatively, $\|A\|' = \sigma_{\max}(A)$ for all $A \in \mathbb{F}^{n \times n}$ having rank 1. Then, Corollary 3.10 of [667, p. 80] implies that $\|\cdot\|'$, and thus $\|\cdot\|$, is submultiplicative.)

Fact 9.8.33. Let $\Phi\colon \mathbb{F}^n \mapsto [0, \infty)$ satisfy the following conditions:

i) If $x \ne 0$, then $\Phi(x) > 0$.

ii) $\Phi(\alpha x) = |\alpha|\Phi(x)$ for all $\alpha \in \mathbb{R}$.

iii) $\Phi(x + y) \le \Phi(x) + \Phi(y)$ for all $x, y \in \mathbb{F}^n$.

iv) If $A \in \mathbb{F}^{n \times n}$ is a permutation matrix, then $\Phi(Ax) = \Phi(x)$ for all $x \in \mathbb{F}^n$.

v) $\Phi(|x|) = \Phi(x)$ for all $x \in \mathbb{F}^n$.

Furthermore, for $A \in \mathbb{F}^{n \times m}$, define

$$\|A\| \triangleq \Phi[\sigma_1(A), \ldots, \sigma_n(A)].$$

Then, $\|\cdot\|$ is a unitarily invariant norm. Conversely, if $\|\cdot\|$ is a unitarily invariant norm on $\mathbb{F}^{n \times m}$, where $n \le m$, then $\Phi\colon \mathbb{F}^n \mapsto [0, \infty)$ defined by

$$\Phi(x) \triangleq \left\| \begin{bmatrix} x_{(1)} & & & & 0 \\ & \ddots & & & \\ & & x_{(n)} & & \\ 0 & & & 0_{n \times (m-n)} \end{bmatrix} \right\|$$

satisfies *i)*–*v)*. (Proof: See [667, pp. 75–76].) (Remark: Φ is a *symmetric gauge function*. This result is due to von Neumann. See Fact 8.17.7.)

Fact 9.8.34. Let $\|\cdot\|$ and $\|\cdot\|'$ denote norms on $\mathbb{F}^m$ and $\mathbb{F}^n$, respectively, and define $\hat{\ell}\colon \mathbb{F}^{n \times m} \mapsto \mathbb{R}$ by

$$\hat{\ell}(A) \triangleq \min_{x \in \mathbb{F}^m \setminus \{0\}} \frac{\|Ax\|'}{\|x\|},$$

or, equivalently,

$$\hat{\ell}(A) \triangleq \min_{x \in \{y \in \mathbb{F}^m\,:\, \|y\|=1\}} \|Ax\|'.$$

Then, for $A \in \mathbb{F}^{n \times m}$, the following statements hold:

i) $\hat{\ell}(A) \geq 0$.

ii) $\hat{\ell}(A) > 0$ if and only if $\operatorname{rank} A = m$.

iii) $\hat{\ell}(A) = \ell(A)$ if and only if either $A = 0$ or $\operatorname{rank} A = m$.

(Proof: See [455, pp. 369, 370].) (Remark: $\hat{\ell}$ is a weaker version of ℓ.)

Fact 9.8.35. Let $A \in \mathbb{F}^{n \times n}$, let $\|\cdot\|$ be a normalized, submultiplicative norm on $\mathbb{F}^{n \times n}$, and assume that $\|I - A\| < 1$. Then, A is nonsingular. (Remark: See Fact 9.9.41.)

Fact 9.8.36. Let $\|\cdot\|$ be a normalized, submultiplicative norm on $\mathbb{F}^{n \times n}$. Then, $\|\cdot\|$ is equi-induced if and only if $\|A\| \leq \|A\|'$ for all $A \in \mathbb{F}^{n \times n}$ and for all normalized submultiplicative norms $\|\cdot\|'$ on $\mathbb{F}^{n \times n}$. (Proof: See [671].) (Remark: As shown in [174, 210], not every normalized submultiplicative norm on $\mathbb{F}^{n \times n}$ is equi-induced or induced.)

9.9 Facts on Matrix Norms for Two or More Matrices

Fact 9.9.1. $\|\cdot\|'_\infty \triangleq n\|\cdot\|_\infty$ is submultiplicative on $\mathbb{F}^{n \times n}$. (Remark: It is not generally true that $\|AB\|_\infty \leq \|A\|_\infty \|B\|_\infty$. For example, let $A = B = \left[\begin{smallmatrix} 1 & 1 \\ 1 & 1 \end{smallmatrix}\right]$.)

Fact 9.9.2. Let $A \in \mathbb{F}^{n \times m}$ and $B \in \mathbb{F}^{m \times l}$. Then,

$$\|AB\|_\infty \leq m\|A\|_\infty \|B\|_\infty.$$

Furthermore, if $A = 1_{n \times m}$ and $B = 1_{m \times l}$, then $\|AB\|_\infty = m\|A\|_\infty \|B\|_\infty$.

Fact 9.9.3. Let $A, B \in \mathbb{F}^{n \times n}$, and let $\|\cdot\|$ be a submultiplicative norm on $\mathbb{F}^{n \times n}$. Then, $\|AB\| \leq \|A\|\|B\|$. Hence, if $\|A\| \leq 1$ and $\|B\| \leq 1$, then $\|AB\| \leq 1$, and if either $\|A\| < 1$ or $\|B\| < 1$, then $\|AB\| < 1$. (Remark: $\operatorname{sprad}(A) < 1$ and $\operatorname{sprad}(B) < 1$ do not imply that $\operatorname{sprad}(AB) < 1$. Let $A = B^{\mathrm{T}} = \left[\begin{smallmatrix} 0 & 2 \\ 0 & 0 \end{smallmatrix}\right]$.)

Fact 9.9.4. Let $\|\cdot\|$ be a norm on $\mathbb{F}^{m \times m}$, and let

$$\delta > \sup\left\{\frac{\|AB\|}{\|A\|\|B\|}: \ A, B \in \mathbb{F}^{m \times m}, A, B \neq 0\right\}.$$

Then, $\|\cdot\|' \triangleq \delta\|\cdot\|$ is a submultiplicative norm on $\mathbb{F}^{m \times m}$. (Proof: See [367, p. 323].)

Fact 9.9.5. Let $A, B \in \mathbb{F}^{n \times n}$, assume that A and B are positive semidefinite and nonzero, and let $\|\cdot\|$ be a submultiplicative unitarily invariant norm on $\mathbb{F}^{n \times n}$. Then,

$$\frac{\|AB\|}{\|A\|\|B\|} \leq \frac{\|A+B\|}{\|A\|+\|B\|}$$

and

$$\frac{\|A \circ B\|}{\|A\|\|B\|} \leq \frac{\|A+B\|}{\|A\|+\|B\|}.$$

(Proof: See [350].) (Remark: See Fact 9.8.32.)

Fact 9.9.6. Let $A, B \in \mathbb{F}^{n \times n}$, and let $\|\cdot\|$ be a submultiplicative norm on $\mathbb{F}^{n \times n}$. Then, $\|\cdot\|' \triangleq 2\|\cdot\|$ is submultiplicative and satisfies

$$\|[A,B]\|' \leq \|A\|'\|B\|'.$$

Fact 9.9.7. Let $A, B \in \mathbb{F}^{n \times n}$, and let $\|\cdot\|$ be a unitarily invariant norm on $\mathbb{F}^{n \times n}$. Then,

$$\|AB\| \leq \sigma_{\max}(A)\|B\|$$

and

$$\|AB\| \leq \|A\|\sigma_{\max}(B).$$

(Proof: See [432].)

Fact 9.9.8. Let $A, B \in \mathbb{F}^{n \times m}$, and let $\|\cdot\|$ be a unitarily invariant norm on $\mathbb{F}^{n \times n}$. If $p > 0$, then

$$\|\langle B^*A \rangle^p\|^2 \leq \|(A^*A)^p\|\|(B^*B)^p\|.$$

In particular,

$$\|(A^*BB^*A)^{1/4}\|^2 \leq \|A\|\|B\|$$

and

$$\|A^*B\|^2 \leq \|A^*A\|\|B^*B\|.$$

Furthermore,

$$\operatorname{tr}\langle B^*A \rangle \leq \|A\|_{\mathrm{F}}\|B\|_{\mathrm{F}}$$

and

$$\left[\operatorname{tr}(A^*BB^*A)^{1/4}\right]^2 \leq (\operatorname{tr}\langle A \rangle)(\operatorname{tr}\langle B \rangle).$$

(Proof: See [371].) (Problem: Noting Fact 9.10.5, compare the lower bounds for $\|A\|_{\mathrm{F}}^2\|B\|_{\mathrm{F}}^2$ given by

$$|\operatorname{tr}(A^*B)^2| \leq \operatorname{tr} AA^*BB^* \leq \|A\|_{\mathrm{F}}^2\|B\|_{\mathrm{F}}^2$$

and

$$\left[\operatorname{tr}(A^*BB^*A)^{1/2}\right]^2 \leq \|A\|_{\mathrm{F}}^2\|B\|_{\mathrm{F}}^2.)$$

Fact 9.9.9. Let $A, B \in \mathbb{F}^{n \times n}$, and assume that A and B are positive semidefinite. Then,

$$(2\|A\|_{\mathrm{F}}\|B\|_{\mathrm{F}})^{1/2} \le (\|A\|_{\mathrm{F}}^2 + \|B\|_{\mathrm{F}}^2)^{1/2}$$
$$= \|(A^2 + B^2)^{1/2}\|_{\mathrm{F}}$$
$$\le \|A + B\|_{\mathrm{F}}$$
$$\le \sqrt{2}(\|A\|_{\mathrm{F}}^2 + \|B\|_{\mathrm{F}}^2)^{1/2}.$$

Fact 9.9.10. Let $A, B \in \mathbb{F}^{n \times n}$, assume that A and B are positive semidefinite, and let $\|\cdot\|$ be a unitarily invariant norm on $\mathbb{F}^{n \times n}$. Then,

$$\|AB\| \le \tfrac{1}{4}\|(A + B)^2\|.$$

In particular,

$$\mathrm{tr}\, AB \le \mathrm{tr}\, (AB^2A)^{1/2} \le \tfrac{1}{4}\mathrm{tr}\, (A + B)^2,$$
$$\mathrm{tr}\, (AB)^2 \le \mathrm{tr}\, A^2B^2 \le \tfrac{1}{16}\mathrm{tr}\, (A + B)^4,$$
$$\sigma_{\max}(AB) \le \tfrac{1}{4}\sigma_{\max}[(A + B)^2].$$

(Proof: See [806, p. 77] or [118]. The inequalities $\mathrm{tr}\, AB \le \mathrm{tr}\, (AB^2A)^{1/2}$ and $\mathrm{tr}\, (AB)^2 \le \mathrm{tr}\, A^2B^2$ follow from Fact 8.10.8.) (Problem: Noting Fact 9.9.9, compare the lower bounds for $\|A + B\|_{\mathrm{F}}$ given by

$$(2\|A\|_{\mathrm{F}}\|B\|_{\mathrm{F}})^{1/2} \le \|(A^2 + B^2)^{1/2}\|_{\mathrm{F}} \le \|A + B\|_{\mathrm{F}}$$

and

$$2\|AB\|_{\mathrm{F}}^{1/2} \le \|(A + B)^2\|_{\mathrm{F}}^{1/2} \le \|A + B\|_{\mathrm{F}}.)$$

Fact 9.9.11. Let $A, B \in \mathbb{F}^{n \times n}$, assume that A and B are positive semidefinite, and let $\|\cdot\|$ be a unitarily invariant norm on $\mathbb{F}^{n \times n}$. If $p \in [0, 1]$, then

$$\|B^pA^pB^p\| \le \|(BAB)^p\|.$$

Furthermore, if $p \ge 1$, then

$$\|(BAB)^p\| \le \|B^pA^pB^p\|.$$

(Proof: See [111, p. 258].) (Remark: See Fact 8.14.22.)

Fact 9.9.12. Let $A \in \mathbb{F}^{n \times m}$, $B \in \mathbb{F}^{m \times l}$, and $p, q, q', r \in [1, \infty]$, and assume that $1/q + 1/q' = 1$. Then,

$$\|AB\|_p \le \varepsilon_{pq}(n)\varepsilon_{pr}(l)\varepsilon_{q'r}(m)\|A\|_q\|B\|_r,$$

where

$$\varepsilon_{pq}(n) \triangleq \begin{cases} 1, & p \ge q, \\ n^{1/p - 1/q}, & q \ge p. \end{cases}$$

Furthermore, there exist matrices $A \in \mathbb{F}^{n \times m}$ and $B \in \mathbb{F}^{m \times l}$ such that equality holds. (Proof: See [296].) (Remark: Related results are given in [256, 296–298, 475, 715].)

Fact 9.9.13. Let $A, B \in \mathbb{C}^{n \times m}$. Then, there exist unitary matrices $S_1, S_2 \in \mathbb{C}^{m \times m}$ such that

$$\langle A + B \rangle \leq S_1 \langle A \rangle S_1^* + S_2 \langle B \rangle S_2^*.$$

(Remark: This result is a matrix version of the triangle inequality. See [28, 694].)

Fact 9.9.14. Let $A, B \in \mathbb{F}^{n \times n}$, and assume that A and B are positive semidefinite. Then,

$$\|AB - BA\|_{\mathrm{F}}^2 + \|(A - B)^2\|_{\mathrm{F}}^2 \leq \|A^2 - B^2\|_{\mathrm{F}}^2.$$

(Proof: See [432].)

Fact 9.9.15. Let $A, B \in \mathbb{F}^{n \times n}$, assume that A and B are positive semidefinite, and let $p \in [1, \infty]$. Then,

$$\|A - B\|_{\sigma 2p}^2 \leq \|A^2 - B^2\|_{\sigma p}.$$

(Proof: See [426].) (Remark: The case $p = 1$ is due to Powers and Stormer.)

Fact 9.9.16. Let $A, B \in \mathbb{F}^{n \times n}$, and let $p \in [1, \infty]$. Then,

$$\|\langle A \rangle - \langle B \rangle\|_{\sigma p}^2 \leq \|A + B\|_{\sigma 2p} \|A - B\|_{\sigma 2p}.$$

(Proof: See [436].)

Fact 9.9.17. Let $A, B \in \mathbb{F}^{n \times n}$. Then,

$$\|\langle A \rangle - \langle B \rangle\|_{\sigma 1}^2 \leq 2 \|A + B\|_{\sigma 1} \|A - B\|_{\sigma 1}.$$

(Proof: See [436].) (Remark: This result is due to Borchers and Kosaki. See [436].)

Fact 9.9.18. Let $A, B \in \mathbb{F}^{n \times n}$. Then,

$$\|\langle A \rangle - \langle B \rangle\|_{\mathrm{F}} \leq \sqrt{2} \|A - B\|_{\mathrm{F}}$$

and

$$\|\langle A \rangle - \langle B \rangle\|_{\mathrm{F}}^2 + \|\langle A^* \rangle - \langle B^* \rangle\|_{\mathrm{F}}^2 \leq 2 \|A - B\|_{\mathrm{F}}^2.$$

If, in addition, A and B are normal, then

$$\|\langle A \rangle - \langle B \rangle\|_{\mathrm{F}} \leq \|A - B\|_{\mathrm{F}}.$$

(Proof: See [28, 425, 436].)

Fact 9.9.19. Let $A, B \in \mathbb{F}^{n \times n}$. Then,

$$\sigma_{\max}(\langle A \rangle - \langle B \rangle) \le \frac{2}{\pi}\left[2 + \log \frac{\sigma_{\max}(A) + \sigma_{\max}(B)}{\sigma_{\max}(A - B)}\right]\sigma_{\max}(A - B).$$

(Remark: This result is due to Kato. See [436].)

Fact 9.9.20. Let $A, B \in \mathbb{F}^{n \times n}$. If $p \in (0, 2]$, then

$$2^{p-1}(\|A\|_{\sigma p}^p + \|B\|_{\sigma p}^p) \le \|A + B\|_{\sigma p}^p + \|A - B\|_{\sigma p}^p \le 2(\|A\|_{\sigma p}^p + \|B\|_{\sigma p}^p).$$

If $p \in [2, \infty)$, then

$$2(\|A\|_{\sigma p}^p + \|B\|_{\sigma p}^p) \le \|A + B\|_{\sigma p}^p + \|A - B\|_{\sigma p}^p \le 2^{p-1}(\|A\|_{\sigma p}^p + \|B\|_{\sigma p}^p).$$

If $p \in (1, 2]$ and $1/p + 1/q = 1$, then

$$\|A + B\|_{\sigma p}^q + \|A - B\|_{\sigma p}^q \le 2(\|A\|_{\sigma p}^p + \|B\|_{\sigma p}^p)^{q/p}.$$

If $p \in [2, \infty)$ and $1/p + 1/q = 1$, then

$$2(\|A\|_{\sigma p}^p + \|B\|_{\sigma p}^p)^{q/p} \le \|A + B\|_{\sigma p}^q + \|A - B\|_{\sigma p}^q.$$

(Proof: See [361].) (Remark: These inequalities are versions of the *Clarkson inequalities*.) (Remark: See [361] for extensions to unitarily invariant norms. See [119] for additional extensions.)

Fact 9.9.21. Let $A, B \in \mathbb{F}^{n \times n}$. If $p \in [1, 4/3]$ or if $p \in (4/3, 2]$ and $A + B$ and $A - B$ are positive semidefinite, then

$$(\|A\|_{\sigma p} + \|B\|_{\sigma p})^p + |\|A\|_{\sigma p} - \|B\|_{\sigma p}|^p \le \|A + B\|_{\sigma p}^p + \|A - B\|_{\sigma p}^p.$$

Furthermore, if $p \in [4, \infty)$ or if $p \in [2, 4)$ and A and B are positive semidefinite, then

$$\|A + B\|_{\sigma p}^p + \|A - B\|_{\sigma p}^p \le (\|A\|_{\sigma p} + \|B\|_{\sigma p})^p + |\|A\|_{\sigma p} - \|B\|_{\sigma p}|^p.$$

(Proof: See [60, 424].) (Remark: These inequalities are versions of *Hanner's inequality*.)

Fact 9.9.22. Let $A, B \in \mathbb{F}^{n \times n}$, and assume that A and B are Hermitian. If $p \in [1, 2]$, then

$$2^{1/2-1/p}\|(A^2 + B^2)^{1/2}\|_p \le \|A + \jmath B\|_{\sigma p} \le \|(A^2 + B^2)^{1/2}\|_p$$

and

$$2^{1-2/p}(\|A\|_{\sigma p}^2 + \|B\|_{\sigma p}^2) \le \|A + \jmath B\|_{\sigma p}^2 \le 2^{2/p-1}(\|A\|_{\sigma p}^2 + \|B\|_{\sigma p}^2).$$

Furthermore, if $p \in [2, \infty)$, then

$$\|(A^2 + B^2)^{1/2}\|_p \le \|A + \jmath B\|_{\sigma p} \le 2^{1/2-1/p}\|(A^2 + B^2)^{1/2}\|_p$$

and

$$2^{2/p-1}(\|A\|_{\sigma p}^2 + \|B\|_{\sigma p}^2) \le \|A + \jmath B\|_{\sigma p}^2 \le 2^{1-2/p}(\|A\|_{\sigma p}^2 + \|B\|_{\sigma p}^2).$$

(Proof: See [117].)

Fact 9.9.23. Let $A, B \in \mathbb{F}^{n \times n}$, and assume that A and B are Hermitian. If $p \in [1, 2]$, then

$$2^{1-2/p}(\|A\|_{\sigma p}^p + \|B\|_{\sigma p}^p) \le \|A + \jmath B\|_{\sigma p}^p.$$

If $p \in [2, \infty]$, then

$$\|A + \jmath B\|_{\sigma p}^p \le 2^{1-2/p}(\|A\|_{\sigma p}^p + \|B\|_{\sigma p}^p).$$

In particular,

$$\|A + \jmath B\|_{\mathrm{F}}^2 = \|A\|_{\mathrm{F}}^2 + \|B\|_{\mathrm{F}}^2 = \|(A^2 + B^2)^{1/2}\|_{\mathrm{F}}^2.$$

(Proof: See [117, 123].)

Fact 9.9.24. Let $A, B \in \mathbb{F}^{n \times n}$, and assume that A is positive semidefinite and B is Hermitian. If $p \in [1, 2]$, then

$$\|A\|_{\sigma p}^2 + 2^{1-2/p}\|B\|_{\sigma p}^2 \le \|A + \jmath B\|_{\sigma p}^2.$$

If $p \in [2, \infty]$, then

$$\|A + \jmath B\|_{\sigma p}^2 \le \|A\|_{\sigma p}^2 + 2^{1-2/p}\|B\|_{\sigma p}^2.$$

In particular,

$$\|A\|_{\sigma 1}^2 + \tfrac{1}{2}\|B\|_{\sigma 1}^2 \le \|A + \jmath B\|_{\sigma 1}^2,$$

$$\|A + \jmath B\|_{\mathrm{F}}^2 = \|A\|_{\mathrm{F}}^2 + \|B\|_{\mathrm{F}}^2,$$

and

$$\sigma_{\max}^2(A + \jmath B) \le \sigma_{\max}^2(A) + 2\sigma_{\max}^2(B).$$

In fact,

$$\|A\|_{\sigma 1}^2 + \|B\|_{\sigma 1}^2 \le \|A + \jmath B\|_{\sigma 1}^2.$$

(Proof: See [123].)

Fact 9.9.25. Let $A, B \in \mathbb{F}^{n \times n}$, and assume that A and B are positive semidefinite. If $p \in [1, 2]$, then

$$\|A\|_{\sigma p}^2 + \|B\|_{\sigma p}^2 \le \|A + \jmath B\|_{\sigma p}^2.$$

If $p \in [2, \infty]$, then

$$\|A + \jmath B\|_{\sigma p}^2 \le \|A\|_{\sigma p}^2 + \|B\|_{\sigma p}^2.$$

(Proof: See [123].)

Fact 9.9.26. Let $A \in \mathbb{F}^{n \times n}$, let $B \in \mathbb{F}^{n \times n}$, assume that B is Hermitian, and let $\|\cdot\|$ be a unitarily invariant norm on $\mathbb{F}^{n \times n}$. Then,

$$\|A - \tfrac{1}{2}(A + A^*)\| \le \|A - B\|.$$

In particular,

$$\|A - \tfrac{1}{2}(A + A^*)\|_{\mathrm{F}} \le \|A - B\|_{\mathrm{F}}$$

and
$$\sigma_{\max}\left[A - \tfrac{1}{2}(A + A^*)\right] \le \sigma_{\max}(A - B).$$
(Proof: See [111, p. 275] and [592, p. 150].)

Fact 9.9.27. Let $A, M, S, B \in \mathbb{F}^{n \times n}$, and assume that $A = MS$, M is positive semidefinite, and S and B are unitary, and let $\|\cdot\|$ be a unitarily invariant norm on $\mathbb{F}^{n \times n}$. Then,
$$\|A - S\| \le \|A - B\|.$$
In particular,
$$\|A - S\|_{\mathrm{F}} \le \|A - B\|_{\mathrm{F}}.$$
(Proof: See [111, p. 276] and [592, p. 150].) (Remark: $A = MS$ is the polar decomposition of A. See Corollary 5.6.4.)

Fact 9.9.28. Let $A, B \in \mathbb{F}^{n \times n}$, assume that A and B are Hermitian, let $\|\cdot\|$ be a unitarily invariant norm on $\mathbb{F}^{n \times n}$, and let $k \in \mathbb{N}$. Then,
$$\|(A - B)^{2k+1}\| \le 2^{2k}\|A^{2k+1} - B^{2k+1}\|.$$
(Proof: See [111, p. 294].)

Fact 9.9.29. Let $A, B \in \mathbb{F}^{n \times n}$, and let $\|\cdot\|$ be a unitarily invariant norm on $\mathbb{F}^{n \times n}$. Then,
$$\|\langle A \rangle - \langle B \rangle\| \le \sqrt{2\|A + B\|\|A - B\|}.$$
(Proof: See [28].) (Remark: This result is due to Kosaki and Bhatia.)

Fact 9.9.30. Let $A, B \in \mathbb{F}^{n \times n}$, and let $p \ge 1$. Then,
$$\|\langle A \rangle - \langle B \rangle\|_{\sigma p} \le \max\left\{2^{1/p - 1/2}, 1\right\}\sqrt{\|A + B\|_{\sigma p}\|A - B\|_{\sigma p}}.$$
(Proof: See [28].) (Remark: This result is due to Kittaneh, Kosaki, and Bhatia.)

Fact 9.9.31. Let $A, B \in \mathbb{F}^{n \times n}$, assume that A and B are positive semidefinite, and let $p \in [1, \infty)$. Then,
$$\|A + B\|_{\sigma p} \le \left(\|A\|_{\sigma p}^p + \|B\|_{\sigma p}^p\right)^{1/p} + 2^{1/p}\|A^{1/2}B^{1/2}\|_{\sigma p}.$$
(Proof: See [430].)

Fact 9.9.32. Let $A, X, B \in \mathbb{F}^{n \times n}$, and let $\|\cdot\|$ be a unitarily invariant norm on $\mathbb{F}^{n \times n}$. Then,
$$\|A^*XB\| \le \tfrac{1}{2}\|AA^*X + XBB^*\|.$$
In particular,
$$\|A^*B\| \le \tfrac{1}{2}\|AA^* + BB^*\|.$$

(Proof: See [113, 115].) (Remark: See Fact 9.12.24.)

Fact 9.9.33. Let $A, X, B \in \mathbb{F}^{n \times n}$, assume that X is positive semidefinite, and let $\| \cdot \|$ be a unitarily invariant norm on $\mathbb{F}^{n \times n}$. Then,

$$\|A^*XB + B^*XA\| \leq \|A^*XA + B^*XB\|.$$

In particular,

$$\|A^*B + B^*A\| \leq \|A^*A + B^*B\|.$$

(Proof: See [431].) (Remark: See [431] for extensions to the case in which X is not necessarily positive semidefinite.)

Fact 9.9.34. Let $A, X, B \in \mathbb{F}^{n \times n}$, assume that A and B are positive semidefinite, let $p \in [0, 1]$, and let $\| \cdot \|$ be a unitarily invariant norm on $\mathbb{F}^{n \times n}$. Then,

$$\|A^pXB^{1-p} + A^{1-p}XB^p\| \leq \|AX + XB\|$$

and

$$\|A^pXB^{1-p} - A^{1-p}XB^p\| \leq |2p - 1| \|AX - XB\|.$$

(Proof: See [35, 120, 276].) (Remark: These results are the *Heinz inequalities*.)

Fact 9.9.35. Let $A, B \in \mathbb{F}^{n \times n}$, assume that A and B are positive semidefinite, and let $\| \cdot \|$ be a unitarily invariant norm on $\mathbb{F}^{n \times n}$. Then,

$$\|A^{1/2}B^{1/2}\| \leq \tfrac{1}{2}\|A + B\|.$$

(Proof: Let $p = 1/2$ and $X = I$ in Fact 9.9.34.) (Remark: See Fact 9.12.12.)

Fact 9.9.36. Let $A, B \in \mathbb{F}^{n \times n}$, assume that A and B are positive semidefinite, and let $\| \cdot \|$ be a unitarily invariant norm on $\mathbb{F}^{n \times n}$. If $r \in [0, 1]$, then

$$\|A^r - B^r\| \leq \|\langle A - B \rangle^r\|.$$

Furthermore, if $r \in [1, \infty)$, then

$$\|\langle A - B \rangle^r\| \leq \|A^r - B^r\|.$$

In particular,

$$\|(A - B)^2\| \leq \|A^2 - B^2\|.$$

(Proof: See [111, pp. 293, 294] and [432].)

Fact 9.9.37. Let $A, B \in \mathbb{F}^{n \times n}$, assume that A and B are positive semidefinite, let $\| \cdot \|$ be a unitarily invariant norm on $\mathbb{F}^{n \times n}$, and let $z \in \mathbb{C}$. Then,

$$\|A - |z|B\| \leq \|A + zB\| \leq \|A + |z|B\|.$$

(Proof: See [116].)

Fact 9.9.38. Let $A, B \in \mathbb{F}^{n \times n}$, assume that A and B are positive semidefinite, and let $\| \cdot \|$ be a unitarily invariant norm on $\mathbb{F}^{n \times n}$. If $r \in [0, 1]$, then

$$\|(A + B)^r\| \le \|A^r + B^r\|.$$

Furthermore, if $r \in [1, \infty)$, then

$$\|A^r + B^r\| \le \|(A + B)^r\|.$$

In particular, if $k \in \mathbb{P}$, then

$$\|A^k + B^k\| \le \|(A + B)^k\|.$$

(Proof: See [34].)

Fact 9.9.39. Let $A, B \in \mathbb{F}^{n \times n}$, assume that A and B are positive semidefinite, and let $\| \cdot \|$ be a unitarily invariant norm on $\mathbb{F}^{n \times n}$. Then,

$$\|\log(I + A) - \log(I + B)\| \le \|\log(I + \langle A - B \rangle)\|$$

and

$$\|\log(I + A + B)\| \le \|\log(I + A) + \log(I + B)\|.$$

(Proof: See [34] and [111, p. 293].) (Remark: See Fact 11.14.12.)

Fact 9.9.40. Let $A, X, B \in \mathbb{F}^{n \times n}$, assume that A and B are positive definite, and let $\| \cdot \|$ be a unitarily invariant norm on $\mathbb{F}^{n \times n}$. Then,

$$\|(\log A)X - X(\log B)\| \le \|A^{1/2}XB^{-1/2} - A^{-1/2}XB^{1/2}\|.$$

(Proof: See [120].) (Remark: See Fact 11.14.13.)

Fact 9.9.41. Let $A, B \in \mathbb{F}^{n \times n}$, assume that A is nonsingular, let $\| \cdot \|$ be a normalized submultiplicative norm on $\mathbb{F}^{n \times n}$, and assume that $\|A - B\| < 1/\|A^{-1}\|$. Then, B is nonsingular. (Remark: See Fact 9.8.35.)

Fact 9.9.42. Let $A, B \in \mathbb{F}^{n \times n}$, assume that A and $A + B$ are nonsingular, and let $\| \cdot \|$ be a normalized submultiplicative norm on $\mathbb{F}^{n \times n}$. Then,

$$\|A^{-1} - (A + B)^{-1}\| \le \|A^{-1}\| \, \|(A + B)^{-1}\| \, \|B\|.$$

If, in addition, $\|A^{-1}B\| < 1$, then

$$\|A^{-1} + (A + B)^{-1}\| \le \frac{\|A^{-1}\| \, \|A^{-1}B\|}{1 - \|A^{-1}B\|}.$$

Furthermore, if $\|A^{-1}B\| < 1$ and $\|B\| < 1/\|A^{-1}\|$, then

$$\|A^{-1} - (A + B)^{-1}\| \le \frac{\|A^{-1}\|^2 \, \|B\|}{1 - \|A^{-1}\| \, \|B\|}.$$

Fact 9.9.43. Let $A \in \mathbb{F}^{n \times n}$, assume that A is nonsingular, let $E \in \mathbb{F}^{n \times n}$, and let $\| \cdot \|$ be a normalized norm on $\mathbb{F}^{n \times n}$. Then,

$$(A + E)^{-1} = A^{-1}(I + EA^{-1})^{-1}$$
$$= A^{-1} - A^{-1}EA^{-1} + O(\|E\|^2).$$

Fact 9.9.44. Let $A \in \mathbb{F}^{n \times m}$ and $B \in \mathbb{F}^{l \times k}$. Then,

$$\|A \otimes B\|_{\mathrm{col}} = \|A\|_{\mathrm{col}}\|B\|_{\mathrm{col}},$$

$$\|A \otimes B\|_{\infty} = \|A\|_{\infty}\|B\|_{\infty},$$

$$\|A \otimes B\|_{\mathrm{row}} = \|A\|_{\mathrm{row}}\|B\|_{\mathrm{row}}.$$

Furthermore, if $p \in [1, \infty]$, then

$$\|A \otimes B\|_p = \|A\|_p\|B\|_p.$$

Fact 9.9.45. Let $A, B \in \mathbb{F}^{n \times n}$, and let $\| \cdot \|$ be a unitarily invariant norm on $\mathbb{F}^{n \times m}$. Then,

$$\|A \circ B\|^2 \leq \|A^*A\|\|B^*B\|.$$

(Proof: See [370].)

Fact 9.9.46. Let $A \in \mathbb{R}^{n \times n}$, assume that A is nonsingular, let $b \in \mathbb{R}^n$, and let $\hat{x} \in \mathbb{R}^n$. Then,

$$\frac{1}{\kappa(A)} \frac{\|A\hat{x} - b\|}{\|b\|} \leq \frac{\|\hat{x} - A^{-1}b\|}{\|A^{-1}b\|} \leq \kappa(A) \frac{\|A\hat{x} - b\|}{\|b\|},$$

where $\kappa(A) \triangleq \|A\|\|A^{-1}\|$ and the vector and matrix norms are compatible. Equivalently, letting $\hat{b} \triangleq A\hat{x} - b$ and $x \triangleq A^{-1}b$, it follows that

$$\frac{1}{\kappa(A)} \frac{\|\hat{b}\|}{\|b\|} \leq \frac{\|\hat{x} - x\|}{\|x\|} \leq \kappa(A) \frac{\|\hat{b}\|}{\|b\|}.$$

(Remark: This result estimates the accuracy of an approximate solution $\hat{x}$ to $Ax = b$. $\kappa(A)$ is the *condition number* of A.)

Fact 9.9.47. Let $A \in \mathbb{R}^{n \times n}$, assume that A is nonsingular, let $\hat{A} \in \mathbb{R}^{n \times n}$, assume that $\|A^{-1}\hat{A}\| < 1$, and let $b, \hat{b} \in \mathbb{R}^n$. Furthermore, let $x \in \mathbb{R}^n$ satisfy $Ax = b$, and let $\hat{x} \in \mathbb{R}^n$ satisfy $(A + \hat{A})\hat{x} = b + \hat{b}$. Then,

$$\frac{\|\hat{x} - x\|}{\|x\|} \leq \frac{\kappa(A)}{1 - \|A^{-1}\hat{A}\|}\left(\frac{\|\hat{b}\|}{\|b\|} + \frac{\|\hat{A}\|}{\|A\|}\right),$$

where $\kappa(A) \triangleq \|A\|\|A^{-1}\|$ and the vector and matrix norms are compatible.

If, in addition, $\|A^{-1}\|\|\hat{A}\| < 1$, then

$$\frac{1}{\kappa(A)+1}\frac{\|\hat{b}-\hat{A}x\|}{\|b\|} \leq \frac{\|\hat{x}-x\|}{\|x\|} \leq \frac{\kappa(A)}{1-\|A^{-1}\hat{A}\|}\frac{\|\hat{b}-\hat{A}x\|}{\|b\|}.$$

(Proof: See [221, 222].)

Fact 9.9.48. Let $A, \hat{A} \in \mathbb{R}^{n \times n}$ satisfy $\|A^+\hat{A}\| < 1$, let $b \in \mathcal{R}(A)$, let $\hat{b} \in \mathbb{R}^n$, and assume that $b + \hat{b} \in \mathcal{R}(A + \hat{A})$. Furthermore, let $\hat{x} \in \mathbb{R}^n$ satisfy $(A + \hat{A})\hat{x} = b + \hat{b}$. Then, $x \triangleq A^+ b + (I - A^+ A)\hat{x}$ satisfies $Ax = b$ and

$$\frac{\|\hat{x}-x\|}{\|x\|} \leq \frac{\kappa(A)}{1-\|A^+\hat{A}\|}\left(\frac{\|\hat{b}\|}{\|b\|} + \frac{\|\hat{A}\|}{\|A\|}\right),$$

where $\kappa(A) \triangleq \|A\|\|A^{-1}\|$ and the vector and matrix norms are compatible. (Proof: See [221].) (Remark: See [222] for a lower bound.)

Fact 9.9.49. Let $A \in \mathbb{F}^{n \times m}$ be the partitioned matrix

$$A = \begin{bmatrix} A_{11} & A_{12} & \cdots & A_{1k} \\ A_{21} & A_{22} & \cdots & A_{2k} \\ \vdots & \vdots & \ddots & \vdots \\ A_{k1} & A_{k2} & \cdots & A_{kk} \end{bmatrix},$$

where $A_{ij} \in \mathbb{F}^{n_i \times n_j}$ for all $i, j = 1, \ldots, k$. Then, the following statements hold:

i) If $p \in [1, 2]$, then

$$\sum_{i,j=1}^{k} \|A_{ij}\|_{\sigma p}^2 \leq \|A\|_{\sigma p}^2 \leq k^{4/p-2}\sum_{i,j=1}^{k} \|A_{ij}\|_{\sigma p}^2.$$

ii) If $p \in [2, \infty]$, then

$$k^{4/p-2}\sum_{i,j=1}^{k} \|A_{ij}\|_{\sigma p}^2 \leq \|A\|_{\sigma p}^2 \leq \sum_{i,j=1}^{k} \|A_{ij}\|_{\sigma p}^2.$$

iii) If $p \in [1, 2]$, then

$$\|A\|_{\sigma p}^p \leq \sum_{i,j=1}^{k} \|A_{ij}\|_{\sigma p}^p \leq k^{2-p}\|A\|_{\sigma p}^p.$$

iv) If $p \in [2, \infty)$, then

$$k^{2-p}\|A\|_{\sigma p}^p \leq \sum_{i,j=1}^{k} \|A_{ij}\|_{\sigma p}^p \leq \|A\|_{\sigma p}^p.$$

(Proof: See [114].) (Remark: Equality holds for $p = 1$.)

9.10 Facts on Matrix Norms and Eigenvalues

Fact 9.10.1. Let $A \in \mathbb{F}^{n \times n}$, and let $\mathrm{mspec}(A) = \{\lambda_1, \ldots, \lambda_n\}_{\mathrm{m}}$. Then,

$$|\mathrm{tr}\, A| \leq \sum_{i=1}^{n} |\lambda_i| \leq \|A\|_{\sigma 1} = \mathrm{tr}\, \langle A \rangle = \sum_{i=1}^{n} \sigma_i(A).$$

If, in addition, A is positive semidefinite, then

$$\|A\|_{\sigma 1} = \mathrm{tr}\, A.$$

Fact 9.10.2. Let $A \in \mathbb{F}^{n \times n}$, and let $\mathrm{mspec}(A) = \{\lambda_1, \ldots, \lambda_n\}_{\mathrm{m}}$. Then,

$$|\mathrm{tr}\, A^2| \leq \sum_{i=1}^{n} |\lambda_i|^2 \leq \|A\|_{\sigma 2}^2 = \|A\|_{\mathrm{F}}^2 = \mathrm{tr}\, A^*A = \sum_{i=1}^{n} \sigma_i^2(A).$$

If, in addition, A is Hermitian, then

$$\|A\|_{\sigma 2} = \sqrt{\mathrm{tr}\, A^2}.$$

(Proof: $\mathrm{tr}\,(A + A^*)^2 \geq 0$ and $\mathrm{tr}\,(A - A^*)^2 \leq 0$ yield $|\mathrm{tr}\, A^2| \leq \|A\|_{\mathrm{F}}^2$. Use Fact 8.14.5.)

Fact 9.10.3. Let $A \in \mathbb{F}^{n \times n}$, let $\mathrm{mspec}(A) = \{\lambda_1, \ldots, \lambda_n\}_{\mathrm{m}}$, and let $p \in (0, 2]$. Then,

$$\sum_{i=1}^{n} |\lambda_i|^p \leq \|A\|_{\sigma p}^p \leq \|A\|_p^p.$$

In particular,

$$|\mathrm{tr}\, A| \leq \sum_{i=1}^{n} |\lambda_i| \leq \|A\|_{\sigma 1} = \mathrm{tr}\, \langle A \rangle = \sum_{i=1}^{n} \sigma_i(A) \leq \|A\|_1.$$

If, in addition, A is positive semidefinite, then

$$|\mathrm{tr}\, A^p| \leq \sum_{i=1}^{n} |\lambda_i|^p.$$

(Proof: See Fact 8.14.5 and Proposition 9.2.5.)

Fact 9.10.4. Let $A, B \in \mathbb{F}^{n \times m}$, let $\mathrm{mspec}(A^*B) = \{\lambda_1, \ldots, \lambda_m\}_{\mathrm{m}}$, and let $p, q \in [1, \infty]$ satisfy $1/p + 1/q = 1$. Then,

$$|\mathrm{tr}\, A^*B| \leq \sum_{i=1}^{m} |\lambda_i| \leq \sum_{i=1}^{m} \sigma_i(A^*B) = \|AB\|_{\sigma 1} \leq \|A\|_{\sigma p} \|B\|_{\sigma q}.$$

In particular,

$$|\mathrm{tr}\, A^*B| \leq \|A\|_{\mathrm{F}} \|B\|_{\mathrm{F}}.$$

(Proof: Use Proposition 9.3.6.)

Fact 9.10.5. Let $A, B \in \mathbb{F}^{n \times m}$, and let $\mathrm{mspec}(A^*B) = \{\lambda_1, \ldots, \lambda_m\}_{\mathrm{m}}$. Then,

$$|\mathrm{tr}\,(A^*B)^2| \leq \sum_{i=1}^{m} |\lambda_i|^2 \leq \sum_{i=1}^{m} \sigma_i^2(A^*B) = \mathrm{tr}\,AA^*BB^* = \|A^*B\|_{\mathrm{F}}^2 \leq \|A\|_{\mathrm{F}}^2 \|B\|_{\mathrm{F}}^2.$$

(Proof: Use Fact 8.14.5.)

Fact 9.10.6. Let $A \in \mathbb{R}^{n \times n}$, and let $\lambda \in \mathrm{spec}(A)$. Then, the following inequalities hold:

i) $|\lambda| \leq n\|A\|_\infty$.

ii) $|\mathrm{Re}\,\lambda| \leq \frac{n}{2}\|A + A^{\mathrm{T}}\|_\infty$.

iii) $|\mathrm{Im}\,\lambda| \leq \frac{\sqrt{n^2-n}}{2\sqrt{2}}\|A - A^{\mathrm{T}}\|_\infty$.

(Proof: See [511, p. 140].) (Remark: i) and ii) are *Hirsch's theorems*, while iii) is *Bendixson's theorem*. See Fact 5.10.22.)

Fact 9.10.7. Let $A, B \in \mathbb{F}^{n \times n}$, assume that A and B are Hermitian, and let $\mathrm{mspec}(A + \jmath B) = \{\lambda_1, \ldots, \lambda_n\}_{\mathrm{m}}$. Then,

$$\sum_{i=1}^{n} |\mathrm{Re}\,\lambda_i|^2 \leq \|A\|_{\mathrm{F}}^2$$

and

$$\sum_{i=1}^{n} |\mathrm{Im}\,\lambda_i|^2 \leq \|B\|_{\mathrm{F}}^2.$$

(Proof: See [592, p. 146].)

Fact 9.10.8. Let $A \in \mathbb{F}^{n \times n}$, let $\|\cdot\|$ be the norm on $\mathbb{F}^{n \times n}$ induced by the norm $\|\cdot\|'$ on $\mathbb{F}^n$, and define

$$\mu(A) \triangleq \lim_{\varepsilon \to 0^+} \frac{\|I + \varepsilon A\| - 1}{\varepsilon},$$

and let $A, B \in \mathbb{F}^{n \times n}$. Then, the following statements hold:

i) $\mu(A) = \mathrm{D}_+ f(A; I)$, where $f \colon \mathbb{F}^{n \times n} \mapsto \mathbb{R}$ is defined by $f(A) \triangleq \|A\|$.

ii) $\mu(A) = \lim_{\varepsilon \to 0^+} \varepsilon^{-1} \log \|e^{\varepsilon A}\|$.

iii) $\mu(I) = 1$, $\mu(-I) = -1$, and $\mu(0) = 0$.

iv) $-\|A\| \leq -\mu(-A) \leq \mathrm{Re}\,\lambda_i(A) \leq \mu(A) \leq \|A\|$ for all $i = 1, \ldots, n$.

v) $\mu(\alpha A) = |\alpha|\mu[(\mathrm{sign}\,\alpha)A]$ for all $\alpha \in \mathbb{R}$.

vi) $\mu(A + \alpha I) = \mu(A) + \mathrm{Re}\,\alpha$ for all $\alpha \in \mathbb{F}$.

vii) $\max\{\mu(A) - \mu(-B), -\mu(-A) + \mu(B)\} \leq \mu(A + B) \leq \mu(A) + \mu(B)$.

viii) $\mu(\alpha A + (1 - \alpha)B) \leq \alpha\mu(A) + (1 - \alpha)\mu(B)$ for all $\alpha \in [0, 1]$.

ix) $|\mu(A) - \mu(B)| \le \max\{|\mu(A-B)|, |\mu(B-A)|\} \le \|A-B\|$.

x) $\max\{-\mu(-A), -\mu(A)\}\|x\|' \le \|Ax\|'$ for all $x \in \mathbb{F}^n$.

xi) If A is nonsingular, then $\max\{-\mu(-A), -\mu(A)\} \le 1/\|A^{-1}\|$.

xii) $\mathrm{spabs}(A) \le \mu(A)$.

xiii) $\|e^A\| \le e^{\mu(A)}$.

xiv) If $\|\cdot\| = \sigma_{\max}(\cdot)$, then
$$\mu(A) = \tfrac{1}{2}\lambda_{\max}(A + A^*).$$

xv) If $\|\cdot\|' = \|\cdot\|_1$, and thus $\|\cdot\| = \|\cdot\|_{\mathrm{col}}$, then
$$\mu(A) = \max_{j \in \{1,\ldots,n\}} \left(\mathrm{Re}\, a_{jj} + \sum_{\substack{i=1 \\ i \ne j}}^{n} |a_{ij}| \right).$$

xvi) If $\|\cdot\|' = \|\cdot\|_\infty$, and thus $\|\cdot\| = \|\cdot\|_{\mathrm{row}}$, then
$$\mu(A) = \max_{i \in \{1,\ldots,n\}} \left(\mathrm{Re}\, a_{ii} + \sum_{\substack{j=1 \\ j \ne i}}^{n} |a_{ij}| \right).$$

(Proof: See [218, 219, 573, 677].) (Remark: $\mu(\cdot)$ is the *matrix measure* or *logarithmic derivative*. For applications, see [747]. See Fact 9.10.8 for the logarithmic derivative of an asymptotically stable matrix.) (Remark: The directional derivative $D_+ f(A; I)$ is defined in (10.4.2).)

Fact 9.10.9. Let $A, B \in \mathbb{F}^{n \times n}$, assume that A and B are Hermitian, and let $\|\cdot\|$ be a weakly unitarily invariant norm on $\mathbb{F}^{n \times n}$. Then,

$$\left\| \begin{bmatrix} \lambda_1(A) & & 0 \\ & \ddots & \\ 0 & & \lambda_n(A) \end{bmatrix} - \begin{bmatrix} \lambda_1(B) & & 0 \\ & \ddots & \\ 0 & & \lambda_n(B) \end{bmatrix} \right\| \le \|A - B\|$$

$$\le \left\| \begin{bmatrix} \lambda_1(A) & & 0 \\ & \ddots & \\ 0 & & \lambda_n(A) \end{bmatrix} - \begin{bmatrix} \lambda_n(B) & & 0 \\ & \ddots & \\ 0 & & \lambda_1(B) \end{bmatrix} \right\|.$$

In particular,

$$\max_{i \in \{1,\ldots,n\}} |\lambda_i(A) - \lambda_i(B)| \le \sigma_{\max}(A - B) \le \max_{i \in \{1,\ldots,n\}} |\lambda_i(A) - \lambda_{n-i+1}(B)|$$

and

$$\sum_{i=1}^{n} [\lambda_i(A) - \lambda_i(B)]^2 \le \|A - B\|_{\mathrm{F}}^2 \le \sum_{i=1}^{n} [\lambda_i(A) - \lambda_{n-i+1}(B)]^2.$$

(Proof: See [28], [110, p. 38], [111, pp. 63, 69], [414, p. 126], [461, p. 134], [478], or [667, p. 202].) (Remark: The first inequality is the *Lidskii-Mirsky-Wielandt theorem*. The result can be stated without norms using Fact 9.8.33. See [478].) (Remark: See Fact 9.12.19.)

Fact 9.10.10. Let $A, B \in \mathbb{F}^{n \times n}$, and assume that A and B are normal. Then, there exists a permutation σ of $1, \ldots, n$ such that

$$\sum_{i=1}^{n} |\lambda_{\sigma(i)}(A) - \lambda_i(B)|^2 \leq \|A - B\|_{\mathrm{F}}^2.$$

(Proof: See [367, p. 368] or [592, pp. 160–161].) (Remark: This inequality is the *Hoffman-Wielandt theorem*.)

Fact 9.10.11. Let $A, B \in \mathbb{F}^{n \times n}$, and assume that A is Hermitian and B is normal. Furthermore, let $\mathrm{mspec}(B) = \{\lambda_1(B), \ldots, \lambda_n(B)\}_{\mathrm{m}}$, where $\mathrm{Re}\, \lambda_1(B) \geq \cdots \geq \mathrm{Re}\, \lambda_n(B)$. Then,

$$\sum_{i=1}^{n} |\lambda_i(A) - \lambda_i(B)|^2 \leq \|A - B\|_{\mathrm{F}}^2.$$

(Proof: See [367, p. 370].) (Remark: This result is a special case of Fact 9.10.10.)

9.11 Facts on Singular Values for One Matrix

Fact 9.11.1. Let $A \in \mathbb{F}^{n \times m}$. Then,

$$\sigma_{\max}(A) = \max_{x \in \mathbb{F}^n \setminus \{0\}} \left(\frac{x^* A^* A x}{x^* x} \right)^{1/2},$$

and thus

$$\|Ax\|_2 \leq \sigma_{\max}(A) \|x\|_2.$$

Furthermore,

$$\lambda_{\min}^{1/2}(A^* A) = \min_{x \in \mathbb{F}^n \setminus \{0\}} \left(\frac{x^* A^* A x}{x^* x} \right)^{1/2},$$

and thus

$$\lambda_{\min}^{1/2}(A^* A) \|x\|_2 \leq \|Ax\|_2.$$

If, in addition, $m \leq n$, then

$$\sigma_m(A) = \min_{x \in \mathbb{F}^n \setminus \{0\}} \left(\frac{x^* A^* A x}{x^* x} \right)^{1/2},$$

and thus

$$\sigma_m(A) \|x\|_2 \leq \|Ax\|_2.$$

Finally, if $m = n$, then

$$\sigma_{\min}(A) = \min_{x \in \mathbb{F}^n \setminus \{0\}} \left(\frac{x^* A^* A x}{x^* x} \right)^{1/2},$$

and thus

$$\sigma_{\min}(A) \|x\|_2 \leq \|Ax\|_2.$$

(Proof: See Lemma 8.4.3.)

Fact 9.11.2. Let $A \in \mathbb{F}^{n \times m}$. Then,

$$\sigma_{\max}(A) = \max\{|y^* A x|: \ x \in \mathbb{F}^m, \ y \in \mathbb{F}^n, \ \|x\|_2 = \|y\|_2 = 1\}$$

$$= \max\{|y^* A x|: \ x \in \mathbb{F}^m, \ y \in \mathbb{F}^n, \ \|x\|_2 \leq 1, \ \|y\|_2 \leq 1\}.$$

(Remark: See Fact 9.8.27.)

Fact 9.11.3. Let $x \in \mathbb{F}^n$ and $y \in \mathbb{F}^m$, and define $\mathcal{S} \triangleq \{A \in \mathbb{F}^{n \times m}: \sigma_{\max}(A) \leq 1\}$. Then,

$$\max_{A \in \mathcal{S}} x^* A y = \sqrt{x^* x y^* y}.$$

Fact 9.11.4. Let $\|\cdot\|$ be an equi-induced unitarily invariant norm on $\mathbb{F}^{n \times n}$. Then, $\|\cdot\| = \sigma_{\max}(\cdot)$.

Fact 9.11.5. Let $\|\cdot\|$ be an equi-induced self-adjoint norm on $\mathbb{F}^{n \times n}$. Then, $\|\cdot\| = \sigma_{\max}(\cdot)$.

Fact 9.11.6. Let $A \in \mathbb{F}^{n \times n}$, and let $\lambda \in \text{spec}(A)$. Then,

$$\sigma_{\min}(A) \leq |\lambda| \leq \sigma_{\max}(A).$$

Hence,

$$[\sigma_{\min}(A)]^n \leq |\det A| \leq [\sigma_{\max}(A)]^n.$$

(Proof: The second inequality follows from $|\lambda| \|x\|_2 \leq \sigma_{\max}(A) \|x\|_2$ or Proposition 9.3.2.) (Remark: See Fact 11.18.1.)

Fact 9.11.7. Let $A \in \mathbb{F}^{n \times n}$. Then,

$$|\det A| \leq \sigma_{\min}(A) \sigma_{\max}^{n-1}(A).$$

(Proof: Use $|\det A| = \prod_{i=1}^n \sigma_i(A)$ given by Fact 5.10.14.)

Fact 9.11.8. Let $A \in \mathbb{F}^{n \times n}$. Then,

$$|\det A| \leq \prod_{i=1}^n \left[(A^* A)_{(i,i)} \right]^{1/2}.$$

(Proof: The result follows from Hadamard's inequality. See Fact 8.14.8.)

Fact 9.11.9. Let $A \in \mathbb{R}^{n \times n}$, and, for $i = 1, \ldots, n$, let α_i denote the sum of the positive entries of A in $\text{row}_i(A)$ and let β_i denote the sum of the positive entries of A in $\text{row}_i(-A)$. Then,

$$|\det A| \leq \prod_{i=1}^{n} \max\{\alpha_i, \beta_i\} - \prod_{i=1}^{n} \min\{\alpha_i, \beta_i\}.$$

(Proof: See [397].) (Remark: This result is an extension of a result due to Schinzel.)

Fact 9.11.10. Let $A \in \mathbb{F}^{n \times n}$. Then,

$$\sigma_{\min}(A) - 1 \leq \sigma_{\min}(A + I) \leq \sigma_{\min}(A) + 1.$$

(Proof: Use Proposition 9.6.8.)

Fact 9.11.11. Let $A \in \mathbb{F}^{n \times n}$, assume that A is normal, and let $r \in \mathbb{N}$. Then,

$$\sigma_{\max}(A^r) = \sigma_{\max}^r(A).$$

(Remark: Matrices that are not normal might also satisfy these conditions. Consider $\begin{bmatrix} 1 & 0 & 0 \\ 0 & 0 & 0 \\ 0 & 1 & 0 \end{bmatrix}$.)

Fact 9.11.12. Let $A \in \mathbb{F}^{n \times n}$. Then,

$$\sigma_{\max}^2(A) - \sigma_{\max}(A^2) \leq \sigma_{\max}(A^*A - AA^*) \leq \sigma_{\max}^2(A) - \sigma_{\min}^2(A)$$

and

$$\sigma_{\max}^2(A) + \sigma_{\min}^2(A) \leq \sigma_{\max}(A^*A + AA^*) \leq \sigma_{\max}^2(A) + \sigma_{\max}(A^2).$$

If $A^2 = 0$, then

$$\sigma_{\max}(A^*A - AA^*) = \sigma_{\max}^2(A).$$

(Proof: See [432, 435].) (Remark: See Fact 9.12.11.) (Remark: If A is normal, then it follows that $\sigma_{\max}^2(A) \leq \sigma_{\max}(A^2)$, although Fact 9.11.11 implies that equality holds.)

Fact 9.11.13. Let $A \in \mathbb{F}^{n \times n}$. Then, the following statements are equivalent:

i) $\text{sprad}(A) = \sigma_{\max}(A)$.

ii) $\sigma_{\max}(A^i) = \sigma_{\max}^i(A)$ for all $i \in \mathbb{P}$.

iii) $\sigma_{\max}(A^n) = \sigma_{\max}^n(A)$.

(Proof: See [267] and [369, p. 44].) (Remark: The result *iii)* $\implies$ *i)* is due to Ptak.)

Fact 9.11.14. Let $A \in \mathbb{F}^{n \times n}$, assume that A is idempotent, and define the projectors $P, Q \in \mathbb{F}^{n \times n}$ by $P \triangleq AA^+$ and $Q \triangleq I - A^+A$. Then, $P - Q$ is nonsingular,

$$(P - Q)^{-1} = A + A^* - I,$$

$$\sigma_{\max}(PQ) < 1,$$

$$\sigma_{\max}(A) = \frac{1}{\sqrt{1 - \sigma_{\max}^2(PQ)}},$$

and

$$\sigma_{\max}(PQ) = \sigma_{\max}(QP) = \sigma_{\max}(P + Q - I).$$

Furthermore,

$$\sigma_{\max}(A) = \sigma_{\max}(I - A).$$

(Proof: See [603]. The first identity follows from Fact 6.3.17. The last identity follows from the second identity using Fact 6.3.15.) (Remark: See Fact 3.9.18.)

Fact 9.11.15. Let $A \in \mathbb{F}^{n \times n}$. Then,

$$\sigma_{\max}(A) \le \sigma_{\max}(|A|) \le \sqrt{\operatorname{rank} A}\, \sigma_{\max}(A).$$

(Proof: See [354, p. 111].)

Fact 9.11.16. Let $A \in \mathbb{R}^{n \times n}$. Then,

$$\sqrt{\tfrac{1}{2(n^2 - n)} \left(\|A\|_{\mathrm{F}}^2 + \operatorname{tr} A^2 \right)} \le \sigma_{\max}(A).$$

Furthermore, if $\|A\|_{\mathrm{F}} \le \operatorname{tr} A$, then

$$\sigma_{\max}(A) \le \tfrac{1}{n} \operatorname{tr} A + \sqrt{\tfrac{n-1}{n} \left[\|A\|_{\mathrm{F}}^2 - \tfrac{1}{n}(\operatorname{tr} A)^2 \right]}.$$

(Proof: See [529], which considers the complex case.)

Fact 9.11.17. Let $A \in \mathbb{F}^{n \times n}$. Then, the polynomial $p \in \mathbb{R}[s]$ defined by

$$p(s) \triangleq s^n - \|A\|_{\mathrm{F}}^2 s + (n-1)|\det A|^{2/(n-1)}$$

has either exactly one or exactly two positive roots $0 < \alpha \le \beta$. Furthermore, α and β satisfy

$$\alpha^{(n-1)/2} \le \sigma_{\min}(A) \le \sigma_{\max}(A) \le \beta^{(n-1)/2}.$$

(Proof: See [620].)

Fact 9.11.18. Let $A \in \mathbb{F}^{n \times n}$. Then, for all $k = 1, \ldots, n$,

$$\sum_{i=1}^{k} \sigma_i(A^2) \le \sum_{i=1}^{k} \sigma_i^2(A).$$

Hence,

$$\operatorname{tr} \left(A^{2*} A^2 \right)^{1/2} \le \operatorname{tr} A^* A,$$

that is,

$$\text{tr} \langle A^2 \rangle \le \text{tr} \langle A \rangle^2.$$

(Proof: Let $B = A$ and $r = 1$ in Proposition 9.6.3.)

Fact 9.11.19. Let $A \in \mathbb{F}^{n \times n}$, and let $\text{mspec}(A) = \{\lambda_1, \ldots, \lambda_n\}_{\text{m}}$, where $\lambda_1, \ldots, \lambda_n$ are ordered such that $|\lambda_1| \ge \cdots \ge |\lambda_n|$. Then, for all $k = 1, \ldots, n$,

$$\prod_{i=1}^{k} |\lambda_i(A)|^2 \le \prod_{i=1}^{k} \sigma_i(A^2) \le \prod_{i=1}^{k} \sigma_i^2(A)$$

and

$$\prod_{i=1}^{n} |\lambda_i(A)|^2 = \prod_{i=1}^{n} \sigma_i(A^2) = \prod_{i=1}^{n} \sigma_i^2(A) = |\det A|^2.$$

(Proof: See [369, p. 172], and use Fact 5.10.14.) (Remark: See Fact 5.10.14 and Fact 8.14.20.)

Fact 9.11.20. Let $A \in \mathbb{F}^{n \times n}$, and let $\text{mspec}(A) = \{\lambda_1, \cdots, \lambda_n\}_{\text{m}}$, where $\lambda_1, \ldots, \lambda_n$ are ordered such that $|\lambda_1| \ge \cdots \ge |\lambda_n|$. Then, for all $i = 1, \ldots, n$,

$$\lim_{k \to \infty} \sigma_i^{1/k}\left(A^k\right) = |\lambda_i(A)|.$$

In particular,

$$\lim_{k \to \infty} \sigma_{\max}^{1/k}\left(A^k\right) = \text{sprad}(A).$$

(Proof: See [369, p. 180].) (Remark: This identity is due to Yamamoto.) (Remark: The expression for $\text{sprad}(A)$ is a special case of Proposition 9.2.6.)

9.12 Facts on Singular Values for Two or More Matrices

Fact 9.12.1. Let $a_1, \ldots, a_n \in \mathbb{F}^n$ be linearly independent, and, for all $i = 1, \ldots, n$, define

$$A_i \triangleq I - (a_i^* a_i)^{-1} a_i a_i^*.$$

Then,

$$\sigma_{\max}(A_n A_{n-1} \cdots A_1) < 1.$$

(Proof: Define $A \triangleq A_n A_{n-1} \cdots A_1$. Since $\sigma_{\max}(A_i) \le 1$ for all $i = 1, \ldots, n$, it follows that $\sigma_{\max}(A) \le 1$. Suppose that $\sigma_{\max}(A) = 1$, and let $x \in \mathbb{F}^n$ satisfy $x^* x = 1$ and $\|Ax\|_2 = 1$. Then, for all $i = 1, \ldots, n$, $\|A_i A_{i-1} \cdots A_1 x\|_2 = 1$. Consequently, $\|A_1 x\|_2 = 1$, which implies that $a_1^* x = 0$, and thus $A_1 x = x$. Hence, $\|A_i A_{i-1} \cdots A_2 x\|_2 = 1$. Repeating this argument implies that, for all $i = 1, \ldots, n$, $a_i^* x = 0$. Since $a_1, \ldots, a_n$ are linearly independent, it follows that $x = 0$, which is a contradiction.) (Remark: This result is due to Akers and Djokovic.)

Fact 9.12.2. Let $A \in \mathbb{F}^{n \times m}$, $B \in \mathbb{F}^{m \times n}$, and $p \in [1, \infty)$, and assume that AB is normal. Then,

$$\|AB\|_{\sigma p} \leq \|BA\|_{\sigma p}.$$

In particular,

$$\mathrm{tr}\,\langle AB \rangle \leq \mathrm{tr}\,\langle BA \rangle,$$

$$\|AB\|_{\mathrm{F}} \leq \|BA\|_{\mathrm{F}},$$

$$\sigma_{\max}(AB) \leq \sigma_{\max}(BA).$$

(Proof: This result is due to Simon. See [139].)

Fact 9.12.3. Let $A, B \in \mathbb{R}^{n \times n}$, assume that A is nonsingular, and assume that B is singular. Then,

$$\sigma_{\min}(A) \leq \sigma_{\max}(A - B).$$

Furthermore, if $\sigma_{\max}(A^{-1}) = \mathrm{sprad}(A^{-1})$, then there exists a singular matrix $C \in \mathbb{R}^{n \times n}$ such that $\sigma_{\max}(A - C) = \sigma_{\min}(A)$. (Proof: See [592, p. 151].) (Remark: This result is due to Franck.)

Fact 9.12.4. Let $A \in \mathbb{C}^{n \times n}$, assume that A is nonsingular, let $\|\cdot\|$ and $\|\cdot\|'$ be norms on $\mathbb{C}^n$, let $\|\cdot\|''$ be the norm on $\mathbb{C}^{n \times n}$ induced by $\|\cdot\|$ and $\|\cdot\|'$, and let $\|\cdot\|'''$ be the norm on $\mathbb{C}^{n \times n}$ induced by $\|\cdot\|'$ and $\|\cdot\|$. Then,

$$\min\{\|B\|'' \colon\ B \in \mathbb{C}^{n \times n} \text{ and } A + B \text{ is nonsingular}\} = 1/\|A^{-1}\|'''.$$

In particular,

$$\min\{\|B\|_{\mathrm{col}}^* \colon\ B \in \mathbb{C}^{n \times n} \text{ and } A + B \text{ is singular}\} = 1/\|A^{-1}\|_{\mathrm{col}},$$

$$\min\{\sigma_{\max}(B) \colon\ B \in \mathbb{C}^{n \times n} \text{ and } A + B \text{ is singular}\} = \sigma_{\min}(A),$$

$$\min\{\|B\|_{\mathrm{row}} \colon\ B \in \mathbb{C}^{n \times n} \text{ and } A + B \text{ is singular}\} = 1/\|A^{-1}\|_{\mathrm{row}}.$$

(Proof: See [354, p. 111] and [352].) (Remark: This result is due to Gastinel. See [352].) (Remark: The result involving $\sigma_{\max}(B)$ is equivalent to the inequality in Fact 9.12.3.)

Fact 9.12.5. Let $A, B \in \mathbb{F}^{n \times m}$, and assume that $\mathrm{rank}\, A = \mathrm{rank}\, B$ and $\alpha \triangleq \sigma_{\max}(A^+)\sigma_{\max}(A - B) < 1$. Then,

$$\sigma_{\max}(B^+) < \frac{1}{1 - \alpha}\sigma_{\max}(A^+).$$

If, in addition, $n = m$, A and B are nonsingular, and $\sigma_{\max}(A - B) < \sigma_{\min}(A)$, then

$$\sigma_{\max}(B^{-1}) < \frac{\sigma_{\min}(A)}{\sigma_{\min}(A) - \sigma_{\max}(A - B)}\sigma_{\max}(A^{-1}).$$

(Proof: See [354, p. 400].)

Fact 9.12.6. Let $A, B \in \mathbb{F}^{n \times n}$. Then,

$$\sigma_{\max}(I - [A, B]) \geq 1.$$

(Proof: Since $\operatorname{tr}[A, B] = 0$, it follows that there exists $\lambda \in \operatorname{spec}(I - [A, B])$ such that $\operatorname{Re} \lambda \geq 1$, and thus $|\lambda| \geq 1$. Hence, Corollary 9.4.5 implies that $\sigma_{\max}(I - [A, B]) \geq \operatorname{sprad}(I - [A, B]) \geq |\lambda| \geq 1$.)

Fact 9.12.7. Let $A \in \mathbb{F}^{n \times m}$, $B \in \mathbb{F}^{n \times l}$, $C \in \mathbb{F}^{k \times m}$, and $D \in \mathbb{F}^{k \times l}$. Then,

$$\sigma_{\max}\left(\begin{bmatrix} A & B \\ C & D \end{bmatrix}\right) \leq \sigma_{\max}\left(\begin{bmatrix} \sigma_{\max}(A) & \sigma_{\max}(B) \\ \sigma_{\max}(C) & \sigma_{\max}(D) \end{bmatrix}\right).$$

(Proof: See [433] and references given therein.) (Remark: This result is due to Tomiyama.)

Fact 9.12.8. Let $A \in \mathbb{F}^{n \times m}$, $B \in \mathbb{F}^{n \times l}$, and $C \in \mathbb{F}^{k \times m}$. Then, for all $X \in \mathbb{F}^{k \times l}$,

$$\max\left\{\sigma_{\max}([\begin{matrix} A & B \end{matrix}]), \sigma_{\max}\left(\begin{bmatrix} A \\ C \end{bmatrix}\right)\right\} \leq \sigma_{\max}\left(\begin{bmatrix} A & B \\ C & X \end{bmatrix}\right).$$

Furthermore, there exists a matrix $X \in \mathbb{F}^{k \times l}$ such that equality holds. (Remark: This result is *Parrott's theorem*. See [200] and [818, pp. 40–42].)

Fact 9.12.9. Let $A \in \mathbb{F}^{n \times m}$ and $B \in \mathbb{F}^{n \times l}$. Then,

$$\max\{\sigma_{\max}(A), \sigma_{\max}(B)\} \leq \sigma_{\max}([\begin{matrix} A & B \end{matrix}])$$

$$\leq \left[\sigma_{\max}^2(A) + \sigma_{\max}^2(B)\right]^{1/2}$$

$$\leq \sqrt{2}\max\{\sigma_{\max}(A), \sigma_{\max}(B)\}$$

and

$$\left[\sigma_n^2(A) + \sigma_n^2(B)\right]^{1/2} \leq \sigma_n([\begin{matrix} A & B \end{matrix}]) \leq \begin{cases} \left[\sigma_n^2(A) + \sigma_{\max}^2(B)\right]^{1/2} \\ \left[\sigma_{\max}^2(A) + \sigma_n^2(B)\right]^{1/2} \end{cases}.$$

Fact 9.12.10. Let $A, B \in \mathbb{F}^{n \times n}$, and let $\alpha > 0$. Then,

$$\sigma_{\max}(A + B) \leq \left[(1 + \alpha^2)\sigma_{\max}^2(A) + (1 + \alpha^{-2})\sigma_{\max}^2(B)\right]^{1/2}$$

and

$$\sigma_{\min}(A + B) \leq \left[(1 + \alpha^2)\sigma_{\min}^2(A) + (1 + \alpha^{-2})\sigma_{\max}^2(B)\right]^{1/2}.$$

Fact 9.12.11. Let $A, B \in \mathbb{F}^{n \times n}$. Then,

$$\max\{\sigma_{\max}^2(A), \sigma_{\max}^2(B)\} - \sigma_{\max}(AB) \le \sigma_{\max}(A^*A - BB^*)$$

and

$$\sigma_{\max}(A^*A - BB^*) \le \max\{\sigma_{\max}^2(A), \sigma_{\max}^2(B)\} - \min\{\sigma_{\min}^2(A), \sigma_{\min}^2(B)\}.$$

Furthermore,

$$\max\{\sigma_{\max}^2(A), \sigma_{\max}^2(B)\} + \min\{\sigma_{\min}^2(A), \sigma_{\min}^2(B)\} \le \sigma_{\max}(A^*A + BB^*)$$

and

$$\sigma_{\max}(A^*A + BB^*) \le \max\{\sigma_{\max}^2(A), \sigma_{\max}^2(B)\} + \sigma_{\max}(AB).$$

Now, assume that A and B are positive semidefinite. Then,

$$\max\{\lambda_{\max}(A), \lambda_{\max}(B)\} - \sigma_{\max}\left(A^{1/2}B^{1/2}\right) \le \sigma_{\max}(A - B)$$

and

$$\sigma_{\max}(A - B) \le \max\{\lambda_{\max}(A), \lambda_{\max}(B)\} - \min\{\lambda_{\min}(A), \lambda_{\min}(B)\}.$$

Furthermore,

$$\max\{\lambda_{\max}(A), \lambda_{\max}(B)\} + \min\{\lambda_{\min}(A), \lambda_{\min}(B)\} \le \lambda_{\max}(A + B)$$

and

$$\lambda_{\max}(A + B) \le \max\{\lambda_{\max}(A), \lambda_{\max}(B)\} + \sigma_{\max}\left(A^{1/2}B^{1/2}\right).$$

(Proof: See [435, 807].) (Remark: See Fact 8.14.12 and Fact 9.11.12.)

Fact 9.12.12. Let $A, B \in \mathbb{F}^{n \times n}$, and assume that A and B are positive semidefinite. Then,

$$2\sigma_{\max}\left(A^{1/2}B^{1/2}\right) \le \lambda_{\max}(A + B)$$

and

$$\sigma_{\max}\left(A^{1/2}B^{1/2}\right) \le \max\{\lambda_{\max}(A), \lambda_{\max}(B)\}.$$

(Proof: See Fact 9.9.35 and Fact 9.12.11.)

Fact 9.12.13. Let $A, B \in \mathbb{F}^{n \times n}$, and assume that A and B are positive semidefinite. Then,

$$\begin{aligned}
\max\{\sigma_{\max}(A), \sigma_{\max}(B)\} &- \sigma_{\max}\left(A^{1/2}B^{1/2}\right) \\
&\le \sigma_{\max}(A - B) \\
&\le \max\{\sigma_{\max}(A), \sigma_{\max}(B)\} \\
&\le \sigma_{\max}(A + B) \\
&\le \max\{\sigma_{\max}(A), \sigma_{\max}(B)\} + \sigma_{\max}\left(A^{1/2}B^{1/2}\right).
\end{aligned}$$

(Proof: See [430, 435].) (Remark: See Fact 8.14.12.)

Fact 9.12.14. Let $A, B \in \mathbb{F}^{n \times n}$, and assume that A and B are positive semidefinite. Then,

$$\sigma_{\max}\left(A^{1/2}B^{1/2}\right) \leq \sigma_{\max}^{1/2}(AB).$$

Equivalently,

$$\lambda_{\max}\left(A^{1/2}BA^{1/2}\right) \leq \lambda_{\max}^{1/2}(AB^2A).$$

Furthermore, $AB = 0$ if and only if $A^{1/2}B^{1/2} = 0$. (Proof: See [430] and [435].)

Fact 9.12.15. Let $A, B \in \mathbb{F}^{n \times n}$. Then,

$$\sigma_{\min}(A) - \sigma_{\max}(B) \leq |\det(A + B)|^{1/n}$$

$$\leq \prod_{i=1}^{n} |\sigma_i(A) + \sigma_{n-i+1}(B)|^{1/n}$$

$$\leq \sigma_{\max}(A) + \sigma_{\max}(B).$$

(Proof: See [378, p. 63] and [477].)

Fact 9.12.16. Let $A, B \in \mathbb{F}^{n \times n}$, and assume that $\sigma_{\max}(B) \leq \sigma_{\min}(A)$. Then,

$$0 \leq [\sigma_{\min}(A) - \sigma_{\max}(B)]^n$$

$$\leq \prod_{i=1}^{n} |\sigma_i(A) - \sigma_{n-i+1}(B)|$$

$$\leq |\det(A + B)|$$

$$\leq \prod_{i=1}^{n} |\sigma_i(A) + \sigma_{n-i+1}(B)|$$

$$\leq [\sigma_{\max}(A) + \sigma_{\max}(B)]^n.$$

Hence, if $\sigma_{\max}(B) < \sigma_{\min}(A)$, then A is nonsingular and $A + \alpha B$ is nonsingular for all $-1 \leq \alpha \leq 1$. (Proof: See [477].) (Remark: See Fact 11.17.17.)

Fact 9.12.17. Let $A, B \in \mathbb{F}^{n \times m}$. Then,

$$\left[\sigma_1(A + B) \quad \cdots \quad \sigma_{\min\{n,m\}}(A + B)\right]$$

weakly majorizes

$$\left[\sigma_1(A) + \sigma_{\min\{n,m\}}(B) \quad \cdots \quad \sigma_{\min\{n,m\}}(A) + \sigma_1(B)\right].$$

Furthermore, if either $\sigma_{\max}(A) < \sigma_{\min}(B)$ or $\sigma_{\max}(B) < \sigma_{\min}(A)$, then

$$\left[|\sigma_1(A) - \sigma_{\min\{n,m\}}(B)| \quad \cdots \quad |\sigma_{\min\{n,m\}}(A) - \sigma_1(B)|\right]$$

weakly majorizes

$$\left[\sigma_1(A + B) \quad \cdots \quad \sigma_{\min\{n,m\}}(A + B)\right].$$

(Proof: See [477].)

Fact 9.12.18. Let $A \in \mathbb{F}^{n \times n}$, let $k \in \mathbb{P}$ satisfy $k < \operatorname{rank} A$, and let $\|\cdot\|$ be a unitarily invariant norm on $\mathbb{F}^{n \times n}$. Then,

$$\min_{B \in \{X \in \mathbb{F}^{n \times n}: \ \operatorname{rank} X = k\}} \|A - B\| = \|A - B_0\|,$$

where B_0 is formed by replacing the $(\operatorname{rank} A) - k$ smallest singular values in the singular value decomposition of A by zeros. Furthermore,

$$\sigma_{\max}(A - B_0) = \sigma_{k+1}(A)$$

and

$$\|A - B_0\|_{\mathrm{F}} = \sqrt{\sum_{i=k+1}^{r} \sigma_i^2(A)}.$$

(Proof: The result follows from Fact 9.12.19 with $B_\sigma \triangleq \operatorname{diag}[\sigma_1(A), \ldots, \sigma_k(A), 0_{(n-k) \times (m-k)}]$, $S_1 = I_n$, and $S_2 = I_m$. See [299] and [667, p. 208].) (Remark: This result is due to Schmidt and Mirsky.)

Fact 9.12.19. Let $A, B \in \mathbb{F}^{n \times m}$, define $A_\sigma, B_\sigma \in \mathbb{F}^{n \times m}$ by

$$A_\sigma \triangleq \begin{bmatrix} \sigma_1(A) & & & \\ & \ddots & & \\ & & \sigma_r(A) & \\ & & & 0_{(n-r) \times (m-r)} \end{bmatrix},$$

where $r \triangleq \operatorname{rank} A$, and

$$B_\sigma \triangleq \begin{bmatrix} \sigma_1(B) & & & \\ & \ddots & & \\ & & \sigma_l(B) & \\ & & & 0_{(n-l) \times (m-l)} \end{bmatrix},$$

where $l \triangleq \operatorname{rank} B$, let $S_1 \in \mathbb{F}^{n \times n}$ and $S_2 \in \mathbb{F}^{m \times m}$ be unitary matrices, and let $\|\cdot\|$ be a unitarily invariant norm on $\mathbb{F}^{n \times m}$. Then,

$$\|A_\sigma - B_\sigma\| \le \|A - S_1 B S_2\| \le \|A_\sigma + B_\sigma\|.$$

In particular,

$$\max_{i \in \{1, \ldots, \max\{r, l\}\}} |\sigma_i(A) - \sigma_i(B)| \le \sigma_{\max}(A - B) \le \sigma_{\max}(A) + \sigma_{\max}(B).$$

(Proof: See [752].) (Remark: In the case $S_1 = I_n$ and $S_2 = I_m$, the left-hand inequality is *Mirsky's theorem*. See [667, p. 204].) (Remark: See Fact 9.10.9.)

Fact 9.12.20. Let $A, B \in \mathbb{F}^{n \times m}$, and assume that $\operatorname{rank} A = \operatorname{rank} B$. Then,

$$\sigma_{\max}[AA^+(I - BB^+)] = \sigma_{\max}[BB^+(I - AA^+)]$$
$$\leq \min\{\sigma_{\max}(A^+), \sigma_{\max}(B^+)\}\sigma_{\max}(A - B).$$

(Proof: See [354, p. 400] and [667, p. 141].)

Fact 9.12.21. Let $A, B \in \mathbb{F}^{n \times m}$. Then, for all $k = 1, \ldots, \min\{n, m\}$,

$$\sum_{i=1}^{k} \sigma_i(A \circ B) \leq \sum_{i=1}^{k} \sigma_i(A)\sigma_i(B).$$

In particular,

$$\sigma_{\max}(A \circ B) \leq \sigma_{\max}(A)\sigma_{\max}(B).$$

(Proof: See [369, p. 334].)

Fact 9.12.22. Let $A \in \mathbb{F}^{n \times m}$, $B \in \mathbb{F}^{l \times k}$, and $p \in [1, \infty]$. Then,

$$\|A \otimes B\|_{\sigma p} = \|A\|_{\sigma p}\|B\|_{\sigma p}.$$

In particular,

$$\sigma_{\max}(A \otimes B) = \sigma_{\max}(A)\sigma_{\max}(B)$$

and

$$\|A \otimes B\|_{\mathrm{F}} = \|A\|_{\mathrm{F}}\|B\|_{\mathrm{F}}.$$

Fact 9.12.23. Let $A \in \mathbb{F}^{n \times m}$ and $B \in \mathbb{F}^{l \times m}$, and let $p, q > 1$ satisfy $1/p + 1/q = 1$. Then, for all $i = 1, \ldots, \min\{n, m, l\}$,

$$\sigma_i(AB^*) \leq \sigma_i\left(\tfrac{1}{p}\langle A\rangle^p + \tfrac{1}{q}\langle B\rangle^q\right).$$

Equivalently, there exists a unitary matrix $S \in \mathbb{F}^{m \times m}$ such that

$$\langle AB^*\rangle^{1/2} \leq S^*\left(\tfrac{1}{p}\langle A\rangle^p + \tfrac{1}{q}\langle B\rangle^q\right)S.$$

(Proof: See [28] or [806, p. 28].) (Remark: This result is a matrix version of Young's inequality. See Fact 1.4.7 and [359].)

Fact 9.12.24. Let $A \in \mathbb{F}^{n \times m}$ and $B \in \mathbb{F}^{l \times m}$. Then, for all $i = 1, \ldots, \min\{n, m, l\}$,

$$\sigma_i(AB^*) \leq \tfrac{1}{2}\sigma_i(A^*A + B^*B).$$

(Proof: Set $p = q = 2$ in Fact 9.12.23. See [115].) (Remark: See Fact 9.9.32.)

9.13 Notes

The equivalence of absolute and monotone norms given by Proposition 9.1.2 is due to [82]. More general monotonicity conditions are considered in [398]. Induced lower bounds are treated in [455, pp. 369, 370]; see also [667, pp. 33, 80]. The induced norms (9.4.10) and (9.4.11) are given in [354, p. 116] and [176]. Alternative norms for the convolution operator are given in [176, 781]. Proposition 9.3.6 is given in [611, p. 97]. Norm-related topics are discussed in [91]. Spectral perturbation theory in finite and infinite dimensions is treated in [414], where the emphasis is on the regularity of the spectrum as a function of the perturbation rather than on bounds for finite perturbations.

Chapter Ten

Functions of Matrices and Their Derivatives

The norms discussed in Chapter 9 provide the foundation for the development in this chapter of some basic results in topology and analysis.

10.1 Open Sets and Closed Sets

Let $\| \cdot \|$ be a norm on $\mathbb{F}^n$, let $x \in \mathbb{F}^n$, and let $\varepsilon > 0$. Then, define the *open ball of radius ε centered at x* by

$$\mathbb{B}_\varepsilon(x) \triangleq \{y \in \mathbb{F}^n \colon \|x - y\| < \varepsilon\} \tag{10.1.1}$$

and the *sphere of radius ε centered at x* by

$$\mathbb{S}_\varepsilon(x) \triangleq \{y \in \mathbb{F}^n \colon \|x - y\| = \varepsilon\}. \tag{10.1.2}$$

Definition 10.1.1. Let $\mathcal{S} \subseteq \mathbb{F}^n$. The vector $x \in \mathcal{S}$ is an *interior point* of $\mathcal{S}$ if there exists $\varepsilon > 0$ such that $\mathbb{B}_\varepsilon(x) \subseteq \mathcal{S}$. The *interior* of $\mathcal{S}$ is the set

$$\operatorname{int} \mathcal{S} \triangleq \{x \in \mathcal{S} \colon \ x \text{ is an interior point of } \mathcal{S}\}. \tag{10.1.3}$$

Finally, $\mathcal{S}$ is *open* if every element of $\mathcal{S}$ is an interior point, that is, if $\mathcal{S} = \operatorname{int} \mathcal{S}$.

Definition 10.1.2. Let $\mathcal{S} \subseteq \mathcal{S}' \subseteq \mathbb{F}^n$. The vector $x \in \mathcal{S}$ is an *interior point* of $\mathcal{S}$ *relative* to $\mathcal{S}'$ if there exists $\varepsilon > 0$ such that $\mathbb{B}_\varepsilon(x) \cap \mathcal{S}' \subseteq \mathcal{S}$ or, equivalently, $\mathbb{B}_\varepsilon(x) \cap \mathcal{S} = \mathbb{B}_\varepsilon(x) \cap \mathcal{S}'$. The *interior* of $\mathcal{S}$ *relative* to $\mathcal{S}'$ is the set

$$\operatorname{int}_{\mathcal{S}'} \mathcal{S} \triangleq \{x \in \mathcal{S} \colon \ x \text{ is an interior point of } \mathcal{S} \text{ relative to } \mathcal{S}'\}. \tag{10.1.4}$$

Finally, $\mathcal{S}$ is *open relative* to $\mathcal{S}'$ if $\mathcal{S} = \operatorname{int}_{\mathcal{S}'} \mathcal{S}$.

Definition 10.1.3. Let $\mathcal{S} \subseteq \mathbb{F}^n$. The vector $x \in \mathbb{F}^n$ is a *closure point* of $\mathcal{S}$ if, for all $\varepsilon > 0$, the set $\mathcal{S} \cap \mathbb{B}_\varepsilon(x)$ is not empty. The *closure* of $\mathcal{S}$ is the set

$$\operatorname{cl} \mathcal{S} \triangleq \{x \in \mathbb{F}^n \colon \ x \text{ is a closure point of } \mathcal{S}\}. \tag{10.1.5}$$

Finally, the set $\mathcal{S}$ is *closed* if every closure point of $\mathcal{S}$ is an element of $\mathcal{S}$, that is, if $\mathcal{S} = \mathrm{cl}\,\mathcal{S}$.

Definition 10.1.4. Let $\mathcal{S} \subseteq \mathcal{S}' \subseteq \mathbb{F}^n$. The vector $x \in \mathcal{S}'$ is a *closure point* of $\mathcal{S}$ relative to $\mathcal{S}'$ if, for all $\varepsilon > 0$, the set $\mathcal{S} \cap \mathbb{B}_\varepsilon(x)$ is not empty. The *closure* of $\mathcal{S}$ *relative* to $\mathcal{S}'$ is the set

$$\mathrm{cl}_{\mathcal{S}'}\,\mathcal{S} \triangleq \{x \in \mathbb{F}^n \colon x \text{ is a closure point of } \mathcal{S} \text{ relative to } \mathcal{S}'\}. \qquad (10.1.6)$$

Finally, $\mathcal{S}$ is *closed relative* to $\mathcal{S}'$ if $\mathcal{S} = \mathrm{cl}_{\mathcal{S}'}\,\mathcal{S}$.

It follows from Theorem 9.1.8 on the equivalence of norms on $\mathbb{F}^n$ that these definitions are independent of the norm assigned to $\mathbb{F}^n$.

Let $\mathcal{S} \subseteq \mathcal{S}' \subseteq \mathbb{F}^n$. Then,

$$\mathrm{cl}_{\mathcal{S}'}\,\mathcal{S} = (\mathrm{cl}\,\mathcal{S}) \cap \mathcal{S}', \qquad (10.1.7)$$

$$\mathrm{int}_{\mathcal{S}'}\,\mathcal{S} = \mathcal{S}'\backslash\mathrm{cl}(\mathcal{S}'\backslash\mathcal{S}), \qquad (10.1.8)$$

and

$$\mathrm{int}\,\mathcal{S} \subseteq \mathrm{int}_{\mathcal{S}'}\,\mathcal{S} \subseteq \mathcal{S} \subseteq \mathrm{cl}_{\mathcal{S}'}\,\mathcal{S} \subseteq \mathrm{cl}\,\mathcal{S}. \qquad (10.1.9)$$

The set $\mathcal{S}$ is *solid* if $\mathrm{int}\,\mathcal{S}$ is not empty, while $\mathcal{S}$ is *completely solid* if $\mathrm{cl\,int}\,\mathcal{S} = \mathrm{cl}\,\mathcal{S}$. If $\mathcal{S}$ is completely solid, then $\mathcal{S}$ is solid. The *boundary* of $\mathcal{S}$ is the set

$$\mathrm{bd}\,\mathcal{S} \triangleq \mathrm{cl}\,\mathcal{S}\backslash\mathrm{int}\,\mathcal{S}, \qquad (10.1.10)$$

while the *boundary* of $\mathcal{S}$ *relative to* $\mathcal{S}'$ is the set

$$\mathrm{bd}_{\mathcal{S}'}\,\mathcal{S} \triangleq \mathrm{cl}_{\mathcal{S}'}\,\mathcal{S}\backslash\mathrm{int}_{\mathcal{S}'}\,\mathcal{S}. \qquad (10.1.11)$$

Note that the empty set is both open and closed, although it is not solid.

The set $\mathcal{S} \subset \mathbb{F}^n$ is *bounded* if there exists $\delta > 0$ such that, for all $x, y \in \mathcal{S}$,

$$\|x - y\| < \delta. \qquad (10.1.12)$$

The set $\mathcal{S} \subset \mathbb{F}^n$ is *compact* if it is both closed and bounded.

10.2 Limits

Definition 10.2.1. A *sequence* $\{x_1, x_2, \ldots\}_{\mathrm{m}}$ is an ordered multiset with a countably infinite number of elements. We write $\{x_i\}_{i=1}^\infty$ for $\{x_1, x_2, \ldots\}_{\mathrm{m}}$.

Definition 10.2.2. The sequence $\{\alpha_i\}_{i=1}^\infty \subset \mathbb{F}$ *converges* to $\alpha \in \mathbb{F}$ if, for all $\varepsilon > 0$, there exists a positive integer $p \in \mathbb{P}$ such that $|\alpha_i - \alpha| < \varepsilon$ for all $i > p$. In this case, we write $\alpha = \lim_{i\to\infty} \alpha_i$ or $\alpha_i \to \alpha$ as $i \to \infty$, where $i \in \mathbb{P}$.

Definition 10.2.3. The sequence $\{x_i\}_{i=1}^{\infty} \subset \mathbb{F}^n$ *converges* to $x \in \mathbb{F}^n$ if $\lim_{i\to\infty} \|x - x_i\| = 0$, where $\|\cdot\|$ is a norm on $\mathbb{F}^n$. In this case, we write $x = \lim_{i\to\infty} x_i$ or $x_i \to x$ as $i \to \infty$, where $i \in \mathbb{P}$. Similarly, $\{A_i\}_{i=1}^{\infty} \subset \mathbb{F}^{n\times m}$ *converges* to $A \in \mathbb{F}^{n\times m}$ if $\lim_{i\to\infty} \|A - A_i\| = 0$, where $\|\cdot\|$ is a norm on $\mathbb{F}^{n\times m}$. In this case, we write $A = \lim_{i\to\infty} A_i$ or $A_i \to A$ as $i \to \infty$, where $i \in \mathbb{P}$.

It follows from Theorem 9.1.8 that convergence of a sequence is independent of the choice of norm.

Proposition 10.2.4. Let $\mathcal{S} \subseteq \mathbb{F}^n$. The vector $x \in \mathbb{F}^n$ is a closure point of $\mathcal{S}$ if and only if there exists a sequence $\{x_i\}_{i=1}^{\infty} \subseteq \mathcal{S}$ such that $x = \lim_{i\to\infty} x_i$.

Proof. Suppose that $x \in \mathbb{F}^n$ is a closure point of $\mathcal{S}$. Then, for all $i \in \mathbb{P}$, there exists a vector $x_i \in \mathcal{S}$ such that $\|x - x_i\| < 1/i$. Hence, $x - x_i \to 0$ as $i \to \infty$. Conversely, suppose that $\{x_i\}_{i=1}^{\infty} \subseteq \mathcal{S}$ is such that $x_i \to x$ as $i \to \infty$, and let $\varepsilon > 0$. Then, there exists a positive integer $p \in \mathbb{P}$ such that $\|x - x_i\| < \varepsilon$ for all $i > p$. Therefore, $x_{p+1} \in \mathcal{S} \cap \mathbb{B}_\varepsilon(x)$, and thus $\mathcal{S} \cap \mathbb{B}_\varepsilon(x)$ is not empty. Hence, x is a closure point of $\mathcal{S}$. $\square$

Theorem 10.2.5. Let $\mathcal{S} \subset \mathbb{F}^n$ be compact, and let $\{x_i\}_{i=1}^{\infty} \subseteq \mathcal{S}$. Then, there exists a convergent subsequence $\{x_{i_j}\}_{j=1}^{\infty} \subseteq \{x_i\}_{i=1}^{\infty}$ such that $\lim_{j\to\infty} x_{i_j}$ exists and $\lim_{j\to\infty} x_{i_j} \in \mathcal{S}$.

Proof. See [555, p. 145]. $\square$

Next, we define convergence for the *series* $\sum_{i=1}^{\infty} x_i$ in terms of the *partial sums* $\sum_{i=1}^{k} x_i$.

Definition 10.2.6. The series $\sum_{i=1}^{\infty} x_i$, where $\{x_i\}_{i=1}^{\infty} \subset \mathbb{F}^n$, *converges* to $x \in \mathbb{F}^n$ if

$$x = \lim_{k\to\infty} \sum_{i=1}^{k} x_i. \tag{10.2.1}$$

Furthermore, $\sum_{i=1}^{\infty} x_i$ *converges absolutely* if $\sum_{i=1}^{\infty} \|x_i\|$ converges, where $\|\cdot\|$ is a norm on $\mathbb{F}^n$. Similarly, the series $\sum_{i=1}^{\infty} A_i$, where $\{A_i\}_{i=1}^{\infty} \subset \mathbb{F}^{n\times m}$, *converges* to $A \in \mathbb{F}^{n\times m}$ if

$$A = \lim_{k\to\infty} \sum_{i=1}^{k} A_i. \tag{10.2.2}$$

Finally, $\sum_{i=1}^{\infty} A_i$ *converges absolutely* if $\sum_{i=1}^{\infty} \|A_i\|$ converges, where $\|\cdot\|$ is a norm on $\mathbb{F}^{n\times m}$.

10.3 Continuity

Definition 10.3.1. Let $\mathcal{D} \subseteq \mathbb{F}^m$, $f\colon \mathcal{D} \mapsto \mathbb{F}^n$, and $x \in \mathcal{D}$. Then, f is *continuous* at x if, for every convergent sequence $\{x_i\}_{i=1}^{\infty} \subseteq \mathcal{D}$ such that $\lim_{i \to \infty} x_i = x$, it follows that $\lim_{i \to \infty} f(x_i) = f(x)$. Furthermore, let $\mathcal{D}_0 \subseteq \mathcal{D}$. Then, f is *continuous* on $\mathcal{D}_0$ if f is continuous at x for all $x \in \mathcal{D}_0$. Finally, f is *continuous* if it is continuous on $\mathcal{D}$.

Theorem 10.3.2. Let $\mathcal{D} \subseteq \mathbb{F}^n$ be convex, and let $f\colon \mathcal{D} \to \mathbb{F}$ be convex. Then, f is continuous on $\text{int}_{\text{aff } \mathcal{D}}\, \mathcal{D}$.

Proof. See [84, p. 81] and [614, p. 82]. □

Corollary 10.3.3. Let $A \in \mathbb{F}^{n \times m}$, and define $f\colon \mathbb{F}^m \to \mathbb{F}^n$ by $f(x) \triangleq Ax$. Then, f is continuous.

Proof. The result is a consequence of Theorem 10.3.2. Alternatively, let $x \in \mathbb{F}^m$, and let $\{x_i\}_{i=1}^{\infty} \subset \mathbb{F}^m$ be such that $x_i \to x$ as $i \to \infty$. Furthermore, let $\|\cdot\|$ and $\|\cdot\|'$ be compatible norms on $\mathbb{F}^m$ and $\mathbb{F}^{m \times n}$, respectively. Since $\|Ax - Ax_i\| \leq \|A\|'\|x - x_i\|$, it follows that $Ax_i \to Ax$ as $i \to \infty$. □

Theorem 10.3.4. Let $\mathcal{D} \subseteq \mathbb{F}^m$, and let $f\colon \mathcal{D} \mapsto \mathbb{F}^n$. Then, the following statements are equivalent:

i) f is continuous.

ii) For all open $\mathcal{S} \subseteq \mathbb{F}^n$, the set $f^{-1}(\mathcal{S})$ is open relative to $\mathcal{D}$.

iii) For all closed $\mathcal{S} \subseteq \mathbb{F}^n$, the set $f^{-1}(\mathcal{S})$ is closed relative to $\mathcal{D}$.

Proof. See [555, pp. 87, 110]. □

Corollary 10.3.5. Let $A \in \mathbb{F}^{n \times m}$ and $\mathcal{S} \subseteq \mathbb{F}^n$, and define $\mathcal{S}' \triangleq \{x \in \mathbb{F}^m\colon Ax \in \mathcal{S}\}$. If $\mathcal{S}$ is open, then $\mathcal{S}'$ is open. If $\mathcal{S}$ is closed, then $\mathcal{S}'$ is closed.

The following result is the *open mapping theorem*.

Theorem 10.3.6. Let $A \in \mathbb{F}^{n \times m}$ be right invertible, and let $\mathcal{D} \subseteq \mathbb{F}^m$ be open. Then, $A\mathcal{D}$ is open.

Theorem 10.3.7. Let $\mathcal{D} \subset \mathbb{F}^m$ be compact, and let $f\colon \mathcal{D} \mapsto \mathbb{F}^n$ be continuous. Then, $f(\mathcal{D})$ is compact.

Proof. See [555, p. 146]. □

Corollary 10.3.8. Let $\mathcal{D} \subset \mathbb{F}^m$ be compact, and let $f\colon \mathcal{D} \mapsto \mathbb{R}$ be continuous. Then, there exists a vector $x_0 \in \mathcal{D}$ such that $f(x_0) \leq f(x)$ for all $x \in \mathcal{D}$.

The following result provides a statement of the *Brouwer fixed-point theorem*.

Theorem 10.3.9. Let $\mathcal{D} \subseteq \mathbb{F}^m$, assume that $\mathcal{D}$ is convex, let $f \colon \mathcal{D} \to \mathcal{D}$, and assume that f is continuous. Furthermore, assume that at least one of the following statements holds:

  i) $\mathcal{D}$ is bounded.

  ii) $\mathcal{D}$ is closed, and $f(\mathcal{D})$ is bounded.

Then, there exists a vector $x \in \mathcal{D}$ such that $f(x) = x$.

Proof. See [760, pp. 163, 167]. $\square$

10.4 Derivatives

Let $\mathcal{D} \subseteq \mathbb{F}^m$, and let $x_0 \in \mathcal{D}$. Then, the *variational cone of $\mathcal{D}$ with respect to x_0* is the set

$$\text{vcone}(\mathcal{D}, x_0) \triangleq \{\xi \in \mathbb{F}^m \colon \quad \text{there exists } \alpha_0 > 0 \text{ such that}$$
$$x_0 + \alpha\xi \in \mathcal{D}, \alpha \in [0, \alpha_0)\}. \qquad (10.4.1)$$

Note that $\text{vcone}(\mathcal{D}, x_0)$ is a pointed cone, although it may consist of only the origin as can be seen from the example $x_0 = 0$ and

$$\mathcal{D} = \left\{ x \in \mathbb{R}^2 \colon 0 \leq x_{(1)} \leq 1, x_{(1)}^3 \leq x_{(2)} \leq x_{(1)}^2 \right\}.$$

Now, let $\mathcal{D} \subseteq \mathbb{F}^m$ and $f \colon \mathcal{D} \to \mathbb{F}^n$. If $\xi \in \text{vcone}(\mathcal{D}, x_0)$, then the *one-sided directional differential of f at x_0 in the direction ξ* is defined by

$$\mathrm{D}_+ f(x_0; \xi) \triangleq \lim_{\alpha \downarrow 0} \tfrac{1}{\alpha} [f(x_0 + \alpha\xi) - f(x_0)] \qquad (10.4.2)$$

if the limit exists. Similarly, if $\xi \in \text{vcone}(\mathcal{D}, x_0)$ and $-\xi \in \text{vcone}(\mathcal{D}, x_0)$, then the *two-sided directional differential* $\mathrm{D}f(x_0; \xi)$ of f at x_0 in the direction ξ is defined by replacing "$\alpha\downarrow 0$" in (10.4.2) by "$\alpha \to 0$." If $\xi = e_i$ so that the direction ξ is one of the coordinate axes, then the *partial derivative of f with respect to $x_{(i)}$ at x_0*, denoted by $\frac{\partial f(x_0)}{\partial x_{(i)}}$, is given by

$$\frac{\partial f(x_0)}{\partial x_{(i)}} \triangleq \lim_{\alpha \to 0} \tfrac{1}{\alpha} [f(x_0 + \alpha e_i) - f(x_0)], \qquad (10.4.3)$$

that is,

$$\frac{\partial f(x_0)}{\partial x_{(i)}} = \mathrm{D}f(x_0; e_i), \qquad (10.4.4)$$

when the two-sided directional differential $\mathrm{D}f(x_0; e_i)$ exists.

Proposition 10.4.1. Let $\mathcal{D} \subseteq \mathbb{F}^m$ be a convex set, let $f\colon \mathcal{D} \to \mathbb{F}^n$ be convex, and let $x_0 \in \mathrm{int}\,\mathcal{D}$. Then, $\mathrm{D}_+ f(x_0; \xi)$ exists for all $\xi \in \mathbb{F}^m$.

Proof. See [84, p. 83]. $\square$

Note that $\mathrm{D}_+ f(x_0; \xi) = \pm\infty$ is possible if x_0 is an element of the boundary of $\mathcal{D}$ even if f is continuous at x_0. For example, consider $f\colon [0, \infty) \mapsto \mathbb{R}$ given by $f(x) = 1 - \sqrt{x}$.

Next, we consider a stronger form of differentiation.

Proposition 10.4.2. Let $\mathcal{D} \subseteq \mathbb{F}^m$ be solid and convex, let $f\colon \mathcal{D} \to \mathbb{F}^n$, and let $x_0 \in \mathcal{D}$. Then, there exists at most one matrix $F \in \mathbb{F}^{n \times m}$ satisfying

$$\lim_{\substack{x \to x_0 \\ x \in \mathcal{D} \setminus \{x_0\}}} \|x - x_0\|^{-1}[f(x) - f(x_0) - F(x - x_0)] = 0. \tag{10.4.5}$$

Proof. See [760, p. 170]. $\square$

In (10.4.5) the limit is taken over all sequences that are contained in $\mathcal{D}$, do not include x_0, and converge to x_0.

Definition 10.4.3. Let $\mathcal{D} \subseteq \mathbb{F}^m$ be solid and convex, let $f\colon \mathcal{D} \to \mathbb{F}^n$, let $x_0 \in \mathcal{D}$, and assume there exists a matrix $F \in \mathbb{F}^{n \times m}$ satisfying (10.4.5). Then, f is *differentiable at x_0*, and the matrix F is the *(Fréchet) derivative of f at x_0*. In this case, we write $f'(x_0) = F$ and

$$\lim_{\substack{x \to x_0 \\ x \in \mathcal{D} \setminus \{x_0\}}} \|x - x_0\|^{-1}\big[f(x) - f(x_0) - f'(x_0)(x - x_0)\big] = 0. \tag{10.4.6}$$

Note that Proposition 10.4.2 and Definition 10.4.3 do not require that x_0 lie in the interior of $\mathcal{D}$. Sometimes we write $\frac{\mathrm{d}f(x_0)}{\mathrm{d}x}$ for $f'(x_0)$.

Proposition 10.4.4. Let $\mathcal{D} \subseteq \mathbb{F}^m$ be solid and convex, let $f\colon \mathcal{D} \to \mathbb{F}^n$, let $x \in \mathcal{D}$, and assume that f is differentiable at x_0. Then, f is continuous at x_0.

Let $\mathcal{D} \subseteq \mathbb{F}^m$ be solid and convex, and let $f\colon \mathcal{D} \mapsto \mathbb{F}^n$. In terms of its scalar components, f can be written as $f = \begin{bmatrix} f_1 & \cdots & f_n \end{bmatrix}^{\mathrm{T}}$, where $f_i\colon \mathcal{D} \mapsto \mathbb{F}$ for all $i = 1, \ldots, n$ and $f(x) = \begin{bmatrix} f_1(x) & \cdots & f_n(x) \end{bmatrix}^{\mathrm{T}}$ for all $x \in \mathcal{D}$. With this notation, $f'(x_0)$ can be written as

$$f'(x_0) = \begin{bmatrix} f'_1(x_0) \\ \vdots \\ f'_n(x_0) \end{bmatrix}, \qquad (10.4.7)$$

where $f'_i(x_0) \in \mathbb{F}^{1 \times m}$ is the *gradient of f_i at x_0* and $f'(x_0)$ is the *Jacobian of f at x_0*. Furthermore, if $x \in \text{int } \mathcal{D}$, then $f'(x_0)$ is related to the partial derivatives of f by

$$f'(x_0) = \begin{bmatrix} \dfrac{\partial f(x_0)}{\partial x_{(1)}} & \cdots & \dfrac{\partial f(x_0)}{\partial x_{(m)}} \end{bmatrix}, \qquad (10.4.8)$$

where $\frac{\partial f(x_0)}{\partial x_{(i)}} \in \mathbb{F}^{n \times 1}$ for all $i = 1, \ldots, m$. Note that the existence of the partial derivatives of f at x_0 does not imply that f is differentiable at x_0, that is, $f'(x_0)$ given by (10.4.8) may not satisfy (10.4.6). Finally, note that the (i, j) entry of the $n \times m$ matrix $f'(x_0)$ is $\frac{\partial f_i(x_0)}{\partial x_{(j)}}$. For example, if $x \in \mathbb{F}^n$ and $A \in \mathbb{F}^{n \times n}$, then

$$\frac{\mathrm{d}}{\mathrm{d}x} Ax = A. \qquad (10.4.9)$$

Let $\mathcal{D} \subseteq \mathbb{F}^m$ and $f \colon \mathcal{D} \mapsto \mathbb{F}^n$. If $f'(x)$ exists for all $x \in \mathcal{D}$ and $f' \colon \mathcal{D} \mapsto \mathbb{F}^{n \times n}$ is continuous, then f is *continuously differentiable*, or C^1. The *second derivative* of f at $x_0 \in \mathcal{D}$, denoted by $f''(x_0)$, is the derivative of $f' \colon \mathcal{D} \mapsto \mathbb{F}^{n \times n}$ at $x_0 \in \mathcal{D}$. For $x_0 \in \mathcal{D}$ it can be seen that $f''(x_0) \colon \mathbb{F}^m \times \mathbb{F}^m \mapsto \mathbb{F}^n$ is *bilinear*, that is, for all $\hat{\eta} \in \mathbb{F}^m$, the mapping $\eta \mapsto f''(x_0)(\eta, \hat{\eta})$ is linear and, for all $\eta \in \mathbb{F}^m$, the mapping $\hat{\eta} \mapsto f''(x_0)(\eta, \hat{\eta})$ is linear. Letting $f = \begin{bmatrix} f_1 & \cdots & f_n \end{bmatrix}^{\mathrm{T}}$, it follows that

$$f''(x_0)(\eta, \hat{\eta}) = \begin{bmatrix} \eta^{\mathrm{T}} f''_1(x_0) \hat{\eta} \\ \vdots \\ \eta^{\mathrm{T}} f''_n(x_0) \hat{\eta} \end{bmatrix}, \qquad (10.4.10)$$

where, for all $i = 1, \ldots, n$, the matrix $f''_i(x_0)$ is the $m \times m$ *Hessian of f_i at x_0*. We write $f^{(2)}(x_0)$ for $f''(x_0)$ and $f^{(k)}(x_0)$ for the kth derivative of f at x_0. f is C^k if $f^{(k)}(x)$ exists and is continuous on $\mathcal{D}$.

The following result is the *inverse function theorem*.

Theorem 10.4.5. Let $\mathcal{D} \subseteq \mathbb{F}^n$ be open, let $f \colon \mathcal{D} \mapsto \mathbb{F}^n$, and assume that f is C^k. Furthermore, let $x_0 \in \mathcal{D}$ be such that $\det f'(x_0) \neq 0$. Then, there exists an open set $\mathcal{N} \subset \mathbb{F}^n$ containing $f(x_0)$ and a C^k function $g \colon \mathcal{N} \mapsto \mathcal{D}$ such that $f[g(y)] = y$ for all $y \in \mathcal{N}$.

Let $S \colon [t_0, t_1] \mapsto \mathbb{F}^{n \times m}$, and assume that every entry of $S(t)$ is differentiable. Then, define $\dot{S}(t) \triangleq \frac{\mathrm{d}S(t)}{\mathrm{d}t} \in \mathbb{F}^{n \times m}$ for all $t \in [t_0, t_1]$ entrywise, that is, for all $i = 1, \ldots, n$ and $j = 1, \ldots, m$,

$$[\dot{S}(t)]_{(i,j)} \triangleq \frac{\mathrm{d}}{\mathrm{d}t}S_{(i,j)}(t). \tag{10.4.11}$$

If $t = t_0$ or $t = t_1$, then "d/dt" denotes a one-sided derivative. Similarly, define $\int_{t_0}^{t_1} S(t)\,\mathrm{d}t$ entrywise, that is, for all $i = 1, \ldots, n$ and $j = 1, \ldots, m$,

$$\left[\int_{t_0}^{t_1} S(t)\,\mathrm{d}t\right]_{(i,j)} \triangleq \int_{t_0}^{t_1}[S(t)]_{(i,j)}\,\mathrm{d}t. \tag{10.4.12}$$

10.5 Functions of a Matrix

Consider the function $f\colon \mathcal{D} \subseteq \mathbb{C} \mapsto \mathbb{C}$ defined by the power series

$$f(s) = \sum_{i=0}^{\infty} \beta_i s^i, \tag{10.5.1}$$

where $\beta_i \in \mathbb{C}$ for all $i \in \mathbb{N}$, and assume that this series converges for all $|s| < \gamma$. Then, for $A \in \mathbb{C}^{n \times n}$, we define

$$f(A) \triangleq \sum_{i=1}^{\infty} \beta_i A^i, \tag{10.5.2}$$

which converges for all $A \in \mathbb{C}^{n \times n}$ such that $\mathrm{sprad}(A) < \gamma$. Now, assume that $A = SBS^{-1}$, where $S \in \mathbb{C}^{n \times n}$ is nonsingular, $B \in \mathbb{C}^{n \times n}$, and $\mathrm{sprad}(B) < \gamma$. Then,
$$f(A) = Sf(B)S^{-1}. \tag{10.5.3}$$

If, in addition, $B = \mathrm{diag}(J_1, \ldots, J_r)$ is the Jordan form of A, then

$$f(A) = S\mathrm{diag}[f(J_1), \ldots, f(J_r)]S^{-1}. \tag{10.5.4}$$

Letting $J = \lambda I_k + N_k$ denote a Jordan block, $f(J)$ is the upper triangular Toeplitz matrix

$$f(J) = f(\lambda)N_k + f'(\lambda)N_k + \tfrac{1}{2}f''(\lambda)N_k^2 + \cdots + \frac{1}{(k-1)!}f^{(k-1)}(\lambda)N_k^{k-1}$$

$$= \begin{bmatrix} f(\lambda) & f'(\lambda) & \tfrac{1}{2}f''(\lambda) & \cdots & \frac{1}{(k-1)!}f^{(k-1)}(\lambda) \\ 0 & f(\lambda) & f'(\lambda) & \cdots & \frac{1}{(k-2)!}f^{(k-2)}(\lambda) \\ 0 & 0 & f(\lambda) & \cdots & \frac{1}{(k-3)!}f^{(k-3)}(\lambda) \\ \vdots & \vdots & \ddots & \ddots & \vdots \\ 0 & 0 & 0 & \cdots & f(\lambda) \end{bmatrix}. \tag{10.5.5}$$

Next, we extend the definition $f(A)$ to functions $f: \mathcal{D} \subseteq \mathbb{C} \mapsto \mathbb{C}$ that are not necessarily of the form (10.5.1). To do this, let $A \in \mathbb{C}^{n \times n}$, where $\text{spec}(A) \subset \mathcal{D}$, and assume that, for all $\lambda_i \in \text{spec}(A)$, f is $k_i - 1$ times differentiable at λ_i, where $k_i \triangleq \text{ind}_A(\lambda_i)$ is the order of the largest Jordan block associated with λ_i as given by Theorem 5.3.3. In this case, f is *defined* at A, and $f(A)$ is given by (10.5.3) and (10.5.4) with $f(J_i)$ defined as in (10.5.5).

Theorem 10.5.1. Let $A \in \mathbb{F}^{n \times n}$, let $\text{spec}(A) = \{\lambda_1, \ldots, \lambda_r\}$, and, for $i = 1, \ldots, r$, let $k_i \triangleq \text{ind}_A(\lambda_i)$. Furthermore, suppose that $f: \mathcal{D} \subseteq \mathbb{C} \mapsto \mathbb{C}$ is defined at A. Then, there exists a polynomial $p \in \mathbb{F}[s]$ such that $f(A) = p(A)$. Furthermore, there exists a unique polynomial p of minimal degree $\sum_{i=1}^{r} k_i$ satisfying $f(A) = p(A)$ and such that, for all $i = 1, \ldots, r$ and $j = 0, 1, \ldots, k_i - 1$,

$$f^{(j)}(\lambda_i) = p^{(j)}(\lambda_i). \tag{10.5.6}$$

This polynomial is given by

$$p(s) = \sum_{i=1}^{r} \left(\left[\prod_{\substack{j=1 \\ j \neq i}}^{r} (s - \lambda_j)^{n_j} \right] \sum_{k=0}^{k_i - 1} \frac{1}{k!} \frac{\mathrm{d}^k}{\mathrm{d}s^k} \left. \frac{f(s)}{\prod_{\substack{l=1 \\ l \neq i}}^{r} (s - \lambda_l)^{k_l}} \right|_{s=\lambda_i} (s - \lambda_i)^k \right). \tag{10.5.7}$$

If, in addition, A is diagonalizable, then p is given by

$$p(s) = \sum_{i=1}^{r} f(\lambda_i) \prod_{\substack{j=1 \\ j \neq i}}^{r} \frac{s - \lambda_j}{\lambda_i - \lambda_j}. \tag{10.5.8}$$

Proof. See [196, pp. 263, 264]. $\square$

The polynomial (10.5.7) is the *Lagrange-Hermite interpolation polynomial* for f.

The following result, which is known as the *identity theorem*, is a special case of Theorem 10.5.1.

Theorem 10.5.2. Let $A \in \mathbb{F}^{n \times n}$, let $\text{spec}(A) = \{\lambda_1, \ldots, \lambda_r\}$, and, for $i = 1, \ldots, r$, let $k_i \triangleq \text{ind}_A(\lambda_i)$. Furthermore, let $f: \mathcal{D} \subseteq \mathbb{C} \mapsto \mathbb{C}$ and $g: \mathcal{D} \subseteq \mathbb{C} \mapsto \mathbb{C}$ be analytic on a neighborhood of $\text{spec}(A)$. Then, $f(A) = g(A)$ if and only if, for all $i = 1, \ldots, r$ and $j = 0, 1, \ldots, k_i - 1$,

$$f^{(j)}(\lambda_i) = g^{(j)}(\lambda_i). \tag{10.5.9}$$

Corollary 10.5.3. Let $A \in \mathbb{F}^{n \times n}$, and let $f \colon \mathcal{D} \subset \mathbb{C} \to \mathbb{C}$ be analytic on a neighborhood of $\mathrm{mspec}(A)$. Then,

$$\mathrm{mspec}[f(A)] = f[\mathrm{mspec}(A)]. \tag{10.5.10}$$

10.6 Matrix Derivatives

In this section we consider derivatives of differentiable scalar-valued functions with matrix arguments. Consider the linear function $f \colon \mathbb{F}^{m \times n} \mapsto \mathbb{F}$ given by $f(X) = \mathrm{tr}\, AX$, where $A \in \mathbb{F}^{n \times m}$ and $X \in \mathbb{F}^{m \times n}$. In terms of vectors $x \in \mathbb{F}^{mn}$, we can define the linear function $\hat{f}(x) \triangleq (\mathrm{vec}\, A)^{\mathrm{T}}x$ so that $\hat{f}(\mathrm{vec}\, X) = f(X) = (\mathrm{vec}\, A)^{\mathrm{T}}\mathrm{vec}\, X$. Consequently, for all $Y \in \mathbb{F}^{m \times n}$, $f'(X_0)$ can be represented by $f'(X_0)Y = \mathrm{tr}\, AY$.

These observations suggest that a convenient representation of the derivative $\frac{\mathrm{d}}{\mathrm{d}X}f(X)$ of a differentiable scalar-valued differentiable function $f(X)$ of a matrix argument $X \in \mathbb{F}^{m \times n}$ is the $n \times m$ matrix whose (i,j) entry is $\frac{\partial f(X)}{\partial X_{(j,i)}}$. Note the order of indices.

Proposition 10.6.1. Let $x \in \mathbb{F}^n$. Then, the following statements hold:

i) If $A \in \mathbb{F}^{n \times n}$, then

$$\frac{\mathrm{d}}{\mathrm{d}x}x^{\mathrm{T}}Ax = x^{\mathrm{T}}(A + A^{\mathrm{T}}). \tag{10.6.1}$$

ii) If $A \in \mathbb{F}^{n \times n}$ is symmetric, then

$$\frac{\mathrm{d}}{\mathrm{d}x}x^{\mathrm{T}}Ax = 2x^{\mathrm{T}}A. \tag{10.6.2}$$

iii) If $A \in \mathbb{F}^{n \times n}$ is Hermitian, then

$$\frac{\mathrm{d}}{\mathrm{d}x}x^{*}Ax = 2x^{*}A. \tag{10.6.3}$$

Proposition 10.6.2. Let $A \in \mathbb{F}^{n \times m}$ and $B \in \mathbb{F}^{l \times n}$. Then, the following statements hold:

i) For all $X \in \mathbb{F}^{m \times n}$,

$$\frac{\mathrm{d}}{\mathrm{d}X}\,\mathrm{tr}\, AX = A. \tag{10.6.4}$$

ii) For all $X \in \mathbb{F}^{m \times l}$,

$$\frac{\mathrm{d}}{\mathrm{d}X}\,\mathrm{tr}\, AXB = BA. \tag{10.6.5}$$

iii) For all $X \in \mathbb{F}^{l \times m}$,

$$\frac{\mathrm{d}}{\mathrm{d}X}\,\mathrm{tr}\, AX^{\mathrm{T}}B = A^{\mathrm{T}}B^{\mathrm{T}}. \tag{10.6.6}$$

iv) For all $X \in \mathbb{F}^{m \times l}$ and $k \in \mathbb{P}$,

$$\frac{\mathrm{d}}{\mathrm{d}X} \operatorname{tr} (AXB)^k = kB(AXB)^{k-1}A. \qquad (10.6.7)$$

v) For all $X \in \mathbb{F}^{m \times l}$,

$$\frac{\mathrm{d}}{\mathrm{d}X} \det AXB = B(AXB)^{\mathrm{A}}A. \qquad (10.6.8)$$

vi) For all $X \in \mathbb{F}^{m \times l}$ such that AXB is nonsingular,

$$\frac{\mathrm{d}}{\mathrm{d}X} \log \det AXB = B(AXB)^{-1}A. \qquad (10.6.9)$$

Proposition 10.6.3. Let $A \in \mathbb{F}^{n \times m}$ and $B \in \mathbb{F}^{m \times n}$. Then, the following statements hold:

i) For all $X \in \mathbb{F}^{m \times m}$ and $k \in \mathbb{P}$,

$$\frac{\mathrm{d}}{\mathrm{d}X} \operatorname{tr} AX^kB = \sum_{i=0}^{k-1} X^{k-1-i}BAX^i. \qquad (10.6.10)$$

ii) For all nonsingular $X \in \mathbb{F}^{m \times m}$,

$$\frac{\mathrm{d}}{\mathrm{d}X} \operatorname{tr} AX^{-1}B = -X^{-1}BAX^{-1}. \qquad (10.6.11)$$

iii) For all nonsingular $X \in \mathbb{F}^{m \times m}$,

$$\frac{\mathrm{d}}{\mathrm{d}X} \det AX^{-1}B = -X^{-1}B(AX^{-1}B)^{\mathrm{A}}AX^{-1}. \qquad (10.6.12)$$

iv) For all nonsingular $X \in \mathbb{F}^{m \times m}$,

$$\frac{\mathrm{d}}{\mathrm{d}X} \log \det AX^{-1}B = -X^{-1}B(AX^{-1}B)^{-1}AX^{-1}. \qquad (10.6.13)$$

Proposition 10.6.4. The following statements hold:

i) Let $A, B \in \mathbb{F}^{n \times m}$. Then, for all $X \in \mathbb{F}^{m \times n}$,

$$\frac{\mathrm{d}}{\mathrm{d}X} \operatorname{tr} AXBX = AXB + BXA. \qquad (10.6.14)$$

ii) Let $A \in \mathbb{F}^{n \times n}$ and $B \in \mathbb{F}^{m \times m}$. Then, for all $X \in \mathbb{F}^{n \times m}$,

$$\frac{\mathrm{d}}{\mathrm{d}X} \operatorname{tr} AXBX^{\mathrm{T}} = BX^{\mathrm{T}}A + B^{\mathrm{T}}X^{\mathrm{T}}A^{\mathrm{T}}. \qquad (10.6.15)$$

iii) Let $A \in \mathbb{F}^{k \times l}$, $B \in \mathbb{F}^{l \times m}$, $C \in \mathbb{F}^{n \times l}$, $D \in \mathbb{F}^{l \times l}$, and $E \in \mathbb{F}^{l \times k}$. Then, for all $X \in \mathbb{F}^{m \times n}$,

$$\frac{\mathrm{d}}{\mathrm{d}X} \operatorname{tr} A(D + BXC)^{-1}E = -C(D + BXC)^{-1}EA(D + BXC)^{-1}B. \qquad (10.6.16)$$

iv) Let $A \in \mathbb{F}^{k \times l}$, $B \in \mathbb{F}^{l \times m}$, $C \in \mathbb{F}^{n \times l}$, $D \in \mathbb{F}^{l \times l}$, and $E \in \mathbb{F}^{l \times k}$. Then, for all $X \in \mathbb{F}^{n \times m}$,

$$\frac{\mathrm{d}}{\mathrm{d}X} \operatorname{tr} A(D + BX^{\mathrm{T}}C)^{-1}E$$
$$= -B^{\mathrm{T}}(D + BX^{\mathrm{T}}C)^{-\mathrm{T}}A^{\mathrm{T}}E^{\mathrm{T}}(D + BX^{\mathrm{T}}C)^{-\mathrm{T}}C^{\mathrm{T}}. \quad (10.6.17)$$

10.7 Facts on Open, Closed, and Convex Sets

Fact 10.7.1. Let $x \in \mathbb{F}^n$, and let $\varepsilon > 0$. Then, $\mathbb{B}_{\varepsilon}(x)$ is completely solid and convex.

Fact 10.7.2. Let $\mathcal{S} \subset \mathbb{F}^n$, assume that $\mathcal{S}$ is bounded, let $\delta > 0$ satisfy $\|x - y\| < \delta$ for all $x, y \in \mathcal{S}$, and let $x_0 \in \mathcal{S}$. Then, $\mathcal{S} \subseteq \mathbb{B}_{\delta}(x_0)$.

Fact 10.7.3. Let $\mathcal{S}_1 \subseteq \mathcal{S}_2 \subseteq \mathbb{F}^n$. Then,

$$\operatorname{cl} \mathcal{S}_1 \subseteq \operatorname{cl} \mathcal{S}_2$$

and

$$\operatorname{int} \mathcal{S}_1 \subseteq \operatorname{int} \mathcal{S}_2.$$

Fact 10.7.4. Let $\mathcal{S} \subseteq \mathbb{F}^n$. Then, $\operatorname{cl} \mathcal{S}$ is the smallest closed set containing $\mathcal{S}$, and $\operatorname{int} \mathcal{S}$ is the largest open set contained in $\mathcal{S}$.

Fact 10.7.5. Let $\mathcal{S} \subseteq \mathbb{F}^n$. Then,

$$(\operatorname{int} \mathcal{S})^{\sim} = \operatorname{cl}(\mathcal{S}^{\sim})$$

and

$$\operatorname{bd} \mathcal{S} = (\operatorname{cl} \mathcal{S}) \cap (\operatorname{cl} \mathcal{S}^{\sim}) = [(\operatorname{int} \mathcal{S}) \cup \operatorname{int}(\mathcal{S}^{\sim})]^{\sim}.$$

Fact 10.7.6. Let $\mathcal{S} \subseteq \mathbb{F}^n$, and assume that $\mathcal{S}$ is convex. Then, $\operatorname{cl} \mathcal{S}$, $\operatorname{int} \mathcal{S}$, and $\operatorname{int}_{\operatorname{aff} \mathcal{S}} \mathcal{S}$ are also convex. (Proof: See [614, p. 45] and [615, p. 64].)

Fact 10.7.7. Let $\mathcal{S} \subseteq \mathbb{F}^n$, and assume that $\mathcal{S}$ is convex. Then, $\mathcal{S}$ is solid if and only if $\mathcal{S}$ is completely solid.

Fact 10.7.8. Let $\mathcal{S} \subseteq \mathbb{F}^n$, and assume that $\mathcal{S}$ is solid. Then, $\operatorname{co} \mathcal{S}$ is solid and completely solid.

Fact 10.7.9. Let $\mathcal{S} \subseteq \mathbb{F}^n$. Then, $\operatorname{co} \operatorname{cl} \mathcal{S} \subseteq \operatorname{cl} \operatorname{co} \mathcal{S}$. (Remark: Equality does not generally hold. Consider

$$\mathcal{S} = \left\{ x \in \mathbb{R}^2 : \ x_{(1)}^2 x_{(2)}^2 = 1 \text{ for all } x_{(1)} > 0 \right\}.$$

Hence, if $\mathcal{S}$ is closed, then it does not necessarily follow that $\operatorname{co} \mathcal{S}$ is closed.)

Fact 10.7.10. Let $\mathcal{S} \subseteq \mathbb{F}^n$, and assume that $\mathcal{S}$ is either bounded or convex. Then,

$$\mathrm{co}\,\mathrm{cl}\,\mathcal{S} = \mathrm{cl}\,\mathrm{co}\,\mathcal{S}.$$

(Proof: Use Fact 10.7.6 and Fact 10.7.9.)

Fact 10.7.11. Let $\mathcal{S} \subseteq \mathbb{F}^n$, and assume that $\mathcal{S}$ is open. Then, $\mathrm{co}\,\mathcal{S}$ is also open.

Fact 10.7.12. Let $\mathcal{S} \subseteq \mathbb{F}^n$, and assume that $\mathcal{S}$ is compact. Then, $\mathrm{co}\,\mathcal{S}$ is also compact.

Fact 10.7.13. Let $\mathcal{S} \subset \mathbb{F}^n$, assume that $\mathcal{S}$ is symmetric, solid, convex, closed, and bounded, and, for all $x \in \mathbb{F}^n$, define

$$\|x\| \triangleq \min\{\alpha \geq 0 \colon x \in \alpha\mathcal{S}\} = \max\{\alpha \geq 0 \colon \alpha x \in \mathcal{S}\}.$$

Then, $\|\cdot\|$ is a norm on $\mathbb{F}^n$, and $\mathbb{B}_1(0) = \mathrm{int}\,\mathcal{S}$. Conversely, let $\|\cdot\|$ be a norm on $\mathbb{F}^n$. Then, $\mathbb{B}_1(0)$ is convex, bounded, symmetric, and solid. (Proof: See [378, pp. 38, 39].) (Remark: In all cases, $\mathbb{B}_1(0)$ is defined with respect to $\|\cdot\|$. This result is due to Minkowski.)

Fact 10.7.14. Let $\mathcal{S} \subseteq \mathbb{F}^n$, and assume that $\mathcal{S}$ is solid. Then, $\dim \mathcal{S} = n$.

Fact 10.7.15. Let $\mathcal{S} \subseteq \mathbb{F}^m$, assume that $\mathcal{S}$ is solid, let $A \in \mathbb{F}^{n \times m}$, and assume that A is right invertible. Then, $A\mathcal{S}$ is solid. (Proof: Use Theorem 10.3.6.) (Remark: See Fact 2.10.15.)

Fact 10.7.16. Let $\mathcal{S} \subseteq \mathbb{F}^n$, and assume that $\mathcal{S}$ is a subspace. Then, $\mathcal{S}$ is closed.

Fact 10.7.17. $\mathbf{N}^n$ is a closed and completely solid subset of $\mathbb{F}^{n(n+1)/2}$. Furthermore,

$$\mathrm{int}\,\mathbf{N}^n = \mathbf{P}^n.$$

Fact 10.7.18. Let $\mathcal{S} \subseteq \mathbb{F}^n$, and assume that $\mathcal{S}$ is convex. Then,

$$\mathrm{int}\,\mathrm{cl}\,\mathcal{S} = \mathrm{int}\,\mathcal{S}.$$

Fact 10.7.19. Let $\mathcal{D} \subseteq \mathbb{F}^n$, and let x_0 belong to a solid, convex subset of $\mathcal{D}$. Then,

$$\dim \mathrm{vcone}(\mathcal{D}, x_0) = n.$$

Fact 10.7.20. Let $\|\cdot\|$ be a norm on $\mathbb{F}^n$, let $\mathcal{S} \subset \mathbb{F}^n$, assume that $\mathcal{S}$ is a subspace, let $y \in \mathbb{F}^n$, and define

$$\mu \triangleq \max_{x \in \{z \in \mathcal{S} \colon \|z\|=1\}} |y^*x|.$$

Then, there exists a vector $z \in \mathcal{S}^\perp$ such that

$$\max_{x \in \{z \in \mathbb{F}^n \colon \|z\|=1\}} |(y+z)^*x| = \mu.$$

(Proof: See [667, p. 57].) (Remark: This result is a version of the *Hahn-Banach theorem*.) (Problem: Find a simple interpretation in $\mathbb{R}^2$.)

Fact 10.7.21. Let $\mathcal{S} \subset \mathbb{R}^n$, assume that $\mathcal{S}$ is a convex cone, let $x \in \mathbb{R}^n$, and assume that $x \notin \operatorname{int} \mathcal{S}$. Then, there exists a nonzero vector $\lambda \in \mathbb{R}^n$ such that $\lambda^\mathrm{T} x \le 0$ and $\lambda^\mathrm{T} z \ge 0$ for all $z \in \mathcal{S}$. (Remark: This result is a *separation theorem*. See [462, p. 37], [591, p. 443], [614, pp. 95–101], and [672, pp. 96–100].)

Fact 10.7.22. Let $\mathcal{S}_1, \mathcal{S}_2 \subset \mathbb{R}^n$, and assume that $\mathcal{S}_1$ and $\mathcal{S}_2$ are convex. Then, the following statements are equivalent:

i) There exist a nonzero vector $\lambda \in \mathbb{R}^n$ and $\alpha \in \mathbb{R}$ such that $\lambda^\mathrm{T} x \le \alpha$ for all $x \in \mathcal{S}_1$, $\lambda^\mathrm{T} x \ge \alpha$ for all $x \in \mathcal{S}_2$, and either $\mathcal{S}_1$ or $\mathcal{S}_2$ is not contained in the affine hyperplane $\{x \in \mathbb{R}^n \colon \lambda^\mathrm{T} x = \alpha\}$.

ii) $\operatorname{int}_{\operatorname{aff} \mathcal{S}_1} \mathcal{S}_1$ and $\operatorname{int}_{\operatorname{aff} \mathcal{S}_2} \mathcal{S}_2$ are disjoint.

(Proof: See [98, p. 82].) (Remark: This result is a *proper separation theorem*.)

Fact 10.7.23. Let $\mathcal{S}_1, \mathcal{S}_2, \mathcal{S}_3 \subseteq \mathbb{F}^n$, assume that $\mathcal{S}_1$ and $\mathcal{S}_2$ are convex, $\mathcal{S}_2$ is closed, and $\mathcal{S}_3$ is bounded, and assume that $\mathcal{S}_1 + \mathcal{S}_3 \subseteq \mathcal{S}_2 + \mathcal{S}_3$. Then, $\mathcal{S}_1 \subseteq \mathcal{S}_2$. (Proof: See [135, p. 5].) (Remark: This result is due to Radstrom.)

10.8 Facts on Functions and Derivatives

Fact 10.8.1. Let $\{x_i\}_{i=1}^\infty \subset \mathbb{F}^n$. Then, $\lim_{i \to \infty} x_i = x$ if and only if $\lim_{i \to \infty} x_{i(j)} = x_{(j)}$ for all $j = 1, \ldots, n$.

Fact 10.8.2. Let $\mathcal{S}_1 \subseteq \mathbb{F}^n$, assume that $\mathcal{S}_1$ is compact, let $\mathcal{S}_2 \subset \mathbb{F}^m$, let $f \colon \mathcal{S}_1 \times \mathcal{S}_2 \to \mathbb{R}$, and assume that f is continuous. Then, $g \colon \mathcal{S}_2 \to \mathbb{R}$ defined by $g(y) \triangleq \max_{x \in \mathcal{S}_1} f(x, y)$ is continuous.

Fact 10.8.3. Let $f \colon [0, \infty) \to \mathbb{R}$, and assume that $\lim_{t \to \infty} f(t)$ exists. Then,

$$\lim_{t \to \infty} \tfrac{1}{t} \int_0^t f(\tau) \, \mathrm{d}\tau = \lim_{t \to \infty} f(t).$$

Fact 10.8.4. Let $f\colon \mathbb{R}^2 \to \mathbb{R}$, $g\colon \mathbb{R} \to \mathbb{R}$, and $h\colon \mathbb{R} \to \mathbb{R}$. Then, assuming each of the following integrals exists,

$$\frac{\mathrm{d}}{\mathrm{d}\alpha} \int_{g(\alpha)}^{h(\alpha)} f(t, \alpha)\, \mathrm{d}t = f(h(\alpha), \alpha) h'(\alpha) - f(g(\alpha), \alpha) g'(\alpha) + \int_{g(\alpha)}^{h(\alpha)} \frac{\partial}{\partial \alpha} f(t, \alpha)\, \mathrm{d}t.$$

(Remark: This identity is *Leibniz' rule*.)

Fact 10.8.5. Let $\mathcal{D} \subseteq \mathbb{R}^m$, assume that $\mathcal{D}$ is a convex set, and let $f\colon \mathcal{D} \to \mathbb{R}$. Then, f is convex if and only if the set $\{(x, y) \in \mathbb{R}^n \times \mathbb{R}\colon y \geq f(x)\}$ is convex.

Fact 10.8.6. Let $\mathcal{D} \subseteq \mathbb{R}^m$, assume that $\mathcal{D}$ is a convex set, let $f\colon \mathcal{D} \to \mathbb{R}$, and assume that f is convex. Then, $f^{-1}((-\infty, \alpha]) = \{x \in \mathcal{D}\colon f(x) \leq \alpha\}$ is convex.

Fact 10.8.7. Let $f\colon \mathcal{D} \subseteq \mathbb{F}^m \mapsto \mathbb{F}^n$, and assume that $\mathrm{D}_+ f(0; \xi)$ exists. Then, for all $\beta > 0$,

$$\mathrm{D}_+ f(0; \beta\xi) = \beta \mathrm{D}_+ f(0; \xi).$$

Fact 10.8.8. Define $f\colon \mathbb{R} \to \mathbb{R}$ by $f(x) \triangleq |x|$. Then, for all $\xi \in \mathbb{R}$,

$$\mathrm{D}_+ f(0; \xi) = |\xi|.$$

Now, define $f\colon \mathbb{R}^n \to \mathbb{R}^n$ by $f(x) \triangleq \sqrt{x^{\mathrm{T}}x}$. Then, for all $\xi \in \mathbb{R}^n$,

$$\mathrm{D}_+ f(0; \xi) = \sqrt{\xi^{\mathrm{T}}\xi}.$$

Fact 10.8.9. Let $A, B \in \mathbb{F}^{n \times n}$. Then, for all $s \in \mathbb{F}$,

$$\frac{\mathrm{d}}{\mathrm{d}s}(A + sB)^2 = AB + BA + 2sB.$$

Hence,

$$\left.\frac{\mathrm{d}}{\mathrm{d}s}(A + sB)^2\right|_{s=0} = AB + BA.$$

Fact 10.8.10. Let $A, B \in \mathbb{F}^{n \times n}$, and let $\mathcal{D} \triangleq \{s \in \mathbb{F}\colon \det(A + sB) \neq 0\}$. Then, for all $s \in \mathcal{D}$,

$$\frac{\mathrm{d}}{\mathrm{d}s}(A + sB)^{-1} = -(A + sB)^{-1}B(A + sB)^{-1}.$$

Hence, if A is nonsingular, then

$$\left.\frac{\mathrm{d}}{\mathrm{d}s}(A + sB)^{-1}\right|_{s=0} = -A^{-1}BA^{-1}.$$

Fact 10.8.11. Let $\mathcal{D} \subseteq \mathbb{F}$, let $A \colon \mathcal{D} \longrightarrow \mathbb{F}^{n \times n}$, and assume that A is differentiable. Then,

$$\frac{\mathrm{d}}{\mathrm{d}s} \det A(s) = \mathrm{tr}\left[A^{\mathrm{A}}(s) \frac{\mathrm{d}}{\mathrm{d}s} A(s) \right] = \frac{1}{n-1} \mathrm{tr}\left[A(s) \frac{\mathrm{d}}{\mathrm{d}s} A^{\mathrm{A}}(s) \right] = \sum_{i=1}^{n} \det A_i(s),$$

where $A_i(s)$ is obtained by differentiating the entries of the ith row of $A(s)$. (Proof: See [196, p. 267], [592, pp. 199, 212], and [613, p. 430].)

Fact 10.8.12. Let $\mathcal{D} \subseteq \mathbb{F}$, let $A \colon \mathcal{D} \longrightarrow \mathbb{F}^{n \times n}$, assume that A is differentiable, and assume that $A(s)$ is nonsingular for all $x \in \mathcal{D}$. Then,

$$\frac{\mathrm{d}}{\mathrm{d}s} A^{-1}(s) = -A^{-1}(s)\left[\frac{\mathrm{d}}{\mathrm{d}s} A(s) \right] A^{-1}(s)$$

and

$$\mathrm{tr}\left[A^{-1}(s) \frac{\mathrm{d}}{\mathrm{d}s} A(s) \right] = -\mathrm{tr}\left[A(s) \frac{\mathrm{d}}{\mathrm{d}s} A^{-1}(s) \right].$$

(Proof: See [592, pp. 198, 212] and [369, p. 491].)

Fact 10.8.13. Let $A, B \in \mathbb{F}^{n \times n}$. Then, for all $s \in \mathbb{F}$,

$$\frac{\mathrm{d}}{\mathrm{d}s} \det(A + sB) = \mathrm{tr}\left[B(A + sB)^{\mathrm{A}} \right].$$

Hence,

$$\frac{\mathrm{d}}{\mathrm{d}s} \det(A + sB)\bigg|_{s=0} = \mathrm{tr}\, BA^{\mathrm{A}} = \sum_{i=1}^{n} \det\left[A \overset{i}{\leftarrow} \mathrm{col}_i(B) \right].$$

(Proof: Use Fact 10.8.11 and Fact 2.14.8.) (Remark: This result generalizes Lemma 4.4.7.)

Fact 10.8.14. Let $A \in \mathbb{F}^{n \times n}$, $r \in \mathbb{R}$, and $k \in \mathbb{P}$. Then, for all $s \in \mathbb{C}$,

$$\frac{\mathrm{d}^k}{\mathrm{d}s^k} [\det(I + sA)]^r = (r\,\mathrm{tr}\, A)^k [\det(I + sA)]^r.$$

Hence,

$$\frac{\mathrm{d}^k}{\mathrm{d}s^k} [\det(I + sA)]^r \bigg|_{s=0} = (r\,\mathrm{tr}\, A)^k.$$

Fact 10.8.15. Let $A \in \mathbb{R}^{n \times n}$, assume that A is symmetric, let $X \in \mathbb{R}^{m \times n}$, and assume that XAX^{T} is nonsingular. Then,

$$\left(\frac{\mathrm{d}}{\mathrm{d}X} \det XAX^{\mathrm{T}} \right) = 2(\det XAX^{\mathrm{T}})A^{\mathrm{T}}X^{\mathrm{T}}(XAX^{\mathrm{T}})^{-1}.$$

(Proof: See [193].)

10.9 Notes

An introductory treatment of limits and continuity is given in [555]. Fréchet and directional derivatives are discussed in [268], while differentiation of matrix functions is considered in [339, 503, 519, 586, 617, 638]. In [614, 615] the set $\text{int}_{\text{aff}\, \mathcal{S}}\, \mathcal{S}$ is called the relative interior of $\mathcal{S}$. An extensive treatment of matrix functions is given in Chapter 6 of [369]; see also [374]. The identity theorem is discussed in [387]. The chain rule for matrix functions is considered in [503, 522]. Differentiation with respect to complex matrices is discussed in [402].

Chapter Eleven

The Matrix Exponential and Stability Theory

The matrix exponential function is fundamental to the study of linear ordinary differential equations. This chapter focuses on the properties of the matrix exponential as well as on stability theory.

11.1 Definition of the Matrix Exponential

The scalar initial value problem

$$\dot{x}(t) = ax(t), \tag{11.1.1}$$

$$x(0) = x_0, \tag{11.1.2}$$

where $t \in [0, \infty)$ and $a, x(t) \in \mathbb{R}$, has the solution

$$x(t) = e^{at}x_0, \tag{11.1.3}$$

where $t \in [0, \infty)$. We are interested in systems of linear differential equations of the form

$$\dot{x}(t) = Ax(t), \tag{11.1.4}$$

$$x(0) = x_0, \tag{11.1.5}$$

where $t \in [0, \infty)$, $x(t) \in \mathbb{R}^n$, and $A \in \mathbb{R}^{n \times n}$. Here $\dot{x}(t)$ denotes $\frac{\mathrm{d}x(t)}{\mathrm{d}t}$, where the derivative is one sided for $t = 0$ and two sided for $t > 0$. The solution of (11.1.4), (11.1.5) is given by

$$x(t) = e^{tA}x_0, \tag{11.1.6}$$

where $t \in [0, \infty)$ and e^{tA} is the *matrix exponential*. The following definition is based on (10.5.2).

Definition 11.1.1. Let $A \in \mathbb{F}^{n \times n}$. Then, the *matrix exponential* $e^A \in \mathbb{F}^{n \times n}$ or $\exp(A) \in \mathbb{F}^{n \times n}$ is the matrix

$$e^A \triangleq \sum_{k=0}^{\infty} \tfrac{1}{k!}A^k. \tag{11.1.7}$$

Note that $0! \triangleq 1$ and $e^{0_{n \times n}} = I_n$.

Proposition 11.1.2. The series (11.1.7) converges absolutely for all $A \in \mathbb{F}^{n \times n}$. Furthermore, let $\| \cdot \|$ be a normalized submultiplicative norm on $\mathbb{F}^{n \times n}$. Then,

$$\|e^A\| \leq e^{\|A\|}. \tag{11.1.8}$$

Proof. Defining the partial sum $S_r \triangleq \sum_{k=0}^{r} \frac{1}{k!} A^k$, we need to show that $\lim_{r \to \infty} S_r = e^A$. We thus have, as $r \to \infty$,

$$\|e^A - S_r\| = \left\| \sum_{k=r+1}^{\infty} \frac{1}{k!} A^k \right\| \leq \sum_{k=r+1}^{\infty} \frac{1}{k!} \|A\|^k$$

$$= e^{\|A\|} - \sum_{k=0}^{r} \frac{1}{k!} \|A\|^k \to 0.$$

Furthermore, note that

$$\|e^A\| = \left\| \sum_{k=0}^{\infty} \frac{1}{k!} A^k \right\| \leq \sum_{k=0}^{\infty} \frac{1}{k!} \|A^k\| \leq \sum_{k=0}^{\infty} \frac{1}{k!} \|A\|^k = e^{\|A\|},$$

which verifies (11.1.8). $\qquad\qquad\square$

The following result generalizes the well-known scalar result.

Proposition 11.1.3. Let $A \in \mathbb{F}^{n \times n}$. Then,

$$e^A = \lim_{k \to \infty} \left(I + \tfrac{1}{k}A\right)^k. \tag{11.1.9}$$

Proof. It follows from the binomial theorem that

$$\left(I + \tfrac{1}{k}A\right)^k = \sum_{i=0}^{k} \alpha_i(k) A^i,$$

where

$$\alpha_i(k) \triangleq \frac{1}{k^i} \binom{k}{i} = \frac{1}{k^i} \frac{k!}{i!(k-i)!}.$$

For all $i \in \mathbb{P}$, it follows that $\alpha_i(k) \to 1/i!$ as $k \to \infty$. Hence,

$$\lim_{k \to \infty} \left(I + \tfrac{1}{k}A\right)^k = \lim_{k \to \infty} \sum_{i=0}^{k} \alpha_i(k) A^i = \sum_{i=0}^{\infty} \frac{1}{i!} A^i = e^A. \qquad\square$$

The following results are immediate consequences of Definition 11.1.1.

Proposition 11.1.4. Let $A \in \mathbb{F}^{n \times n}$. Then, the following statements hold:

i) $\left(e^A\right)^{\mathrm{T}} = e^{A^{\mathrm{T}}}$.

ii) e^A is nonsingular, and $\left(e^A\right)^{-1} = e^{-A}$.

iii) If $A = \mathrm{diag}(A_1, \ldots, A_k)$, where $A_i \in \mathbb{F}^{n_i \times n_i}$ for all $i = 1, \ldots, k$, then $e^A = \mathrm{diag}\left(e^{A_1}, \ldots, e^{A_k}\right)$.

iv) If $S \in \mathbb{F}^{n \times n}$ is nonsingular, then $e^{SAS^{-1}} = Se^A S^{-1}$.

v) If A and $B \in \mathbb{F}^{n \times n}$ are similar, then e^A and e^B are similar.

vi) If A and $B \in \mathbb{F}^{n \times n}$ are unitarily similar, then e^A and e^B are unitarily similar.

vii) If A is Hermitian, then e^A is positive definite.

viii) If A is skew Hermitian, then e^A is unitary.

ix) If A is normal, then e^A is normal.

The converse of v) is false. For example, $A \triangleq \left[\begin{smallmatrix} 0 & 0 \\ 0 & 0 \end{smallmatrix}\right]$ and $B \triangleq \left[\begin{smallmatrix} 0 & 2\pi \\ -2\pi & 0 \end{smallmatrix}\right]$ satisfy $e^A = e^B = I$, although A and B are not similar. The converses of vii) and viii) are given by v) and vi) of Proposition 11.4.7. The converse of ix) is false. For example, the matrix $A \triangleq \left[\begin{smallmatrix} -2\pi & 4\pi \\ -2\pi & 2\pi \end{smallmatrix}\right]$ is not normal but satisfies $e^A = I$.

Proposition 11.1.5. Let $A \in \mathbb{F}^{n \times n}$. Then, for all $t \in \mathbb{R}$,

$$e^{tA} - I = \int_0^t Ae^{\tau A} \, \mathrm{d}\tau \tag{11.1.10}$$

and

$$\frac{\mathrm{d}}{\mathrm{d}t}e^{tA} = Ae^{tA}. \tag{11.1.11}$$

Proof. Note that

$$\int_0^t Ae^{\tau A} \, \mathrm{d}\tau = \int_0^t \sum_{k=0}^{\infty} \frac{1}{k!}\tau^k A^{k+1} \, \mathrm{d}\tau = \sum_{k=0}^{\infty} \frac{1}{k!}\frac{t^{k+1}}{k+1}A^{k+1} = e^{tA} - I,$$

which yields (11.1.10), while differentiating (11.1.10) with respect to t yields (11.1.11). $\qquad\square$

Proposition 11.1.6. Let $A, B \in \mathbb{F}^{n \times n}$. Then, $AB = BA$ if and only if, for all $t \in [0, \infty)$,

$$e^{tA}e^{tB} = e^{t(A+B)}. \tag{11.1.12}$$

Proof. Suppose $AB = BA$. By expanding e^{tA}, e^{tB}, and $e^{t(A+B)}$, it can be seen that the expansions of $e^{tA}e^{tB}$ and $e^{t(A+B)}$ are identical. Conversely, differentiating (11.1.12) twice with respect to t and setting $t = 0$ yields $AB = BA$. $\qquad\square$

Corollary 11.1.7. Let $A, B \in \mathbb{F}^{n \times n}$, and assume that $AB = BA$. Then,

$$e^A e^B = e^B e^A = e^{A+B}. \tag{11.1.13}$$

The converse of Corollary 11.1.7 is not true. For example, if $A \triangleq \begin{bmatrix} 0 & \pi \\ -\pi & 0 \end{bmatrix}$ and $B \triangleq \begin{bmatrix} 0 & (7+4\sqrt{3})\pi \\ (-7+4\sqrt{3})\pi & 0 \end{bmatrix}$, then $e^A = e^B = -I$ and $e^{A+B} = I$, although $AB \neq BA$. A partial converse is given by Fact 11.12.2.

Proposition 11.1.8. Let $A \in \mathbb{F}^{n \times n}$ and $B \in \mathbb{F}^{m \times m}$. Then,

$$e^{A \otimes I_m} = e^A \otimes I_m, \tag{11.1.14}$$

$$e^{I_n \otimes B} = I_n \otimes e^B, \tag{11.1.15}$$

$$e^{A \oplus B} = e^A \otimes e^B. \tag{11.1.16}$$

Proof. Note that

$$
\begin{aligned}
e^{A \otimes I_m} &= I_{nm} + A \otimes I_m + \tfrac{1}{2!}(A \otimes I_m)^2 + \cdots \\
&= I_n \otimes I_m + A \otimes I_m + \tfrac{1}{2!}(A^2 \otimes I_m) + \cdots \\
&= (I_n + A + \tfrac{1}{2!}A^2 + \cdots) \otimes I_m \\
&= e^A \otimes I_m
\end{aligned}
$$

and similarly for (11.1.15). To prove (11.1.16), note that $(A \otimes I_m)(I_n \otimes B) = A \otimes B$ and $(I_n \otimes B)(A \otimes I_m) = A \otimes B$, which shows that $A \otimes I_m$ and $I_n \otimes B$ commute. Thus, by Corollary 11.1.7,

$$e^{A \oplus B} = e^{A \otimes I_m + I_n \otimes B} = e^{A \otimes I_m} e^{I_n \otimes B} = (e^A \otimes I_m)(I_n \otimes e^B) = e^A \otimes e^B. \quad\square$$

11.2 Structure of the Matrix Exponential

To elucidate the structure of the matrix exponential, recall that, by Theorem 4.6.1, every term A^k in (11.1.7) for $k > r \triangleq \deg \mu_A$ can be expressed as a linear combination of $I, A, \ldots, A^{r-1}$. The following result provides an expression for e^{tA} in terms of $I, A, \ldots, A^{r-1}$.

Proposition 11.2.1. Let $A \in \mathbb{F}^{n \times n}$. Then, for all $t \in \mathbb{R}$,

$$e^{tA} = \frac{1}{2\pi\jmath} \oint_{\mathbb{C}} (zI - A)^{-1} e^{tz} \, \mathrm{d}z = \sum_{i=0}^{n-1} \psi_i(t) A^i, \tag{11.2.1}$$

where, for all $i = 0, \ldots, n-1$, $\psi_i(t)$ is given by

$$\psi_i(t) \triangleq \frac{1}{2\pi j} \oint_{\mathcal{C}} \frac{\chi_A^{[i+1]}(z)}{\chi_A(z)} e^{tz} \, dz, \tag{11.2.2}$$

where $\mathcal{C}$ is a simple, closed contour in the complex plane enclosing $\operatorname{spec}(A)$,

$$\chi_A(s) = s^n + \beta_{n-1} s^{n-1} + \cdots + \beta_1 s + \beta_0, \tag{11.2.3}$$

and the polynomials $\chi_A^{[1]}, \ldots, \chi_A^{[n]}$ are defined by the recursion

$$s\chi_A^{[i+1]}(s) = \chi_A^{[i]}(s) - \beta_i, \quad i = 0, \ldots, n-1,$$

where $\chi_A^{[0]} \triangleq \chi_A$ and $\chi_A^{[n]}(s) = 1$. Furthermore, for all $i = 0, \ldots, n-1$ and $t \geq 0$, $\psi_i(t)$ satisfies

$$\psi_i^{(n)}(t) + \beta_{n-1} \psi_i^{(n-1)}(t) + \cdots + \beta_1 \psi_i'(t) + \beta_0 \psi_i(t) = 0, \tag{11.2.4}$$

where, for all $i, j = 0, \ldots, n-1$,

$$\psi_i^{(j)}(0) = \delta_{ij}. \tag{11.2.5}$$

Proof. See [794, p. 31], [299, p. 381], [470, 492], and Fact 4.9.8. □

The coefficient $\psi_i(t)$ of A^i in (11.2.1) can be further characterized in terms of the Laplace transform. Define

$$\hat{x}(s) \triangleq \mathcal{L}\{x(t)\} \triangleq \int_0^\infty e^{-st} x(t) \, dt. \tag{11.2.6}$$

Note that

$$\mathcal{L}\{\dot{x}(t)\} = s\hat{x}(s) - x(0) \tag{11.2.7}$$

and

$$\mathcal{L}\{\ddot{x}(t)\} = s^2 \hat{x}(s) - sx(0) - \dot{x}(0). \tag{11.2.8}$$

The following result shows that the resolvent of A is the Laplace transform of the exponential of A. See (4.4.20).

Proposition 11.2.2. Let $A \in \mathbb{F}^{n \times n}$, and define $\psi_0, \ldots, \psi_{n-1}$ as in Proposition 11.2.1. Then, for all $s \in \mathbb{C} \backslash \operatorname{spec}(A)$,

$$\mathcal{L}\{e^{tA}\} = \int_0^\infty e^{-st} e^{tA} \, dt = (sI - A)^{-1}. \tag{11.2.9}$$

Furthermore, for all $i = 0, \ldots, n-1$, the Laplace transform $\hat{\psi}_i(s)$ of $\psi_i(t)$ is given by

$$\hat{\psi}_i(s) = \frac{\chi_A^{[i+1]}(s)}{\chi_A(s)} \tag{11.2.10}$$

and

$$(sI - A)^{-1} = \sum_{i=0}^{n-1} \hat{\psi}_i(s) A^i. \tag{11.2.11}$$

Proof. Let $s \in \mathbb{C}$ satisfy $\text{Re}\, s > \text{spabs}(A)$ so that $A - sI$ is asymptotically stable. Thus, it follows from Lemma 11.8.2 that

$$\mathcal{L}\{e^{tA}\} = \int_0^\infty e^{-st} e^{tA}\, dt = \int_0^\infty e^{t(A-sI)}\, dt = (sI - A)^{-1}.$$

By analytic continuation, the expression $\mathcal{L}\{e^{tA}\}$ is given by (11.2.9) for all $s \in \mathbb{C}\backslash\text{spec}(A)$. $\qquad\square$

Comparing (11.2.11) with (4.4.20) implies that

$$\sum_{i=0}^{n-1} \hat{\psi}_i(s) A^i = \frac{s^{n-1}}{\chi_A(s)} I + \frac{s^{n-2}}{\chi_A(s)} B_{n-2} + \cdots + \frac{s}{\chi_A(s)} B_1 + B_0. \tag{11.2.12}$$

To further illustrate the structure of e^{tA}, where $A \in \mathbb{F}^{n \times n}$, let $A = SBS^{-1}$, where $B = \text{diag}(B_1, \ldots, B_k)$ is the Jordan form of A. Hence, by Proposition 11.1.4,

$$e^{tA} = Se^{tB}S^{-1}, \tag{11.2.13}$$

where

$$e^{tB} = \text{diag}(e^{tB_1}, \ldots, e^{tB_k}). \tag{11.2.14}$$

The structure of e^{tB} can thus be determined by considering the block $B_i \in \mathbb{F}^{\alpha_i \times \alpha_i}$, which, for all $i = 1, \ldots, k$, has the form

$$B_i = \lambda_i I_{\alpha_i} + N_{\alpha_i}. \tag{11.2.15}$$

Since $\lambda_i I_{\alpha_i}$ and N_{α_i} commute, it follows from Proposition 11.1.6 that

$$e^{tB_i} = e^{t(\lambda_i I_{\alpha_i} + N_{\alpha_i})} = e^{\lambda_i t I_{\alpha_i}} e^{tN_{\alpha_i}} = e^{\lambda_i t} e^{tN_{\alpha_i}}. \tag{11.2.16}$$

Since $N_{\alpha_i}^{\alpha_i} = 0$, it follows that $e^{tN_{\alpha_i}}$ is a finite sum of powers of tN_{α_i}. Specifically,

$$e^{tN_{\alpha_i}} = I_{\alpha_i} + tN_{\alpha_i} + \tfrac{1}{2}t^2 N_{\alpha_i}^2 + \cdots + \frac{1}{(\alpha_i - 1)!} t^{\alpha_i - 1} N_{\alpha_i}^{\alpha_i - 1}, \tag{11.2.17}$$

and thus

$$e^{tN_{\alpha_i}} = \begin{bmatrix} 1 & t & \frac{t^2}{2} & \cdots & \frac{t^{\alpha_i-1}}{(\alpha_i-1)!} \\ 0 & 1 & t & \ddots & \frac{t^{\alpha_i-2}}{(\alpha_i-2)!} \\ 0 & 0 & 1 & \ddots & \frac{t^{\alpha_i-3}}{(\alpha_i-3)!} \\ \vdots & \vdots & \ddots & \ddots & \vdots \\ 0 & 0 & 0 & \cdots & 1 \end{bmatrix}, \tag{11.2.18}$$

which is a Toeplitz matrix (see Fact 11.11.1).

Note that (11.2.16) follows from (10.5.5) with $f(\lambda) = e^{\lambda t}$. Furthermore, every entry of e^{tB_i} is of the form $\frac{1}{r!}t^r e^{\lambda_i t}$, where $r \in \{0, \alpha_i -1\}$ and λ_i is an eigenvalue of A. Reconstructing A by means of $A = SBS^{-1}$ shows that every entry of A is a linear combination of the entries of the blocks e^{tB_i}. If A is real, then e^{tA} is also real. Thus, the term $e^{\lambda_i t}$ for complex $\lambda_i = \nu_i + \jmath\omega_i \in \text{spec}(A)$, where ν_i and ω_i are real, yields terms of the form $e^{\nu_i t}\cos \omega_i t$ and $e^{\nu_i t}\sin \omega_i t$.

The following result follows from (11.2.18) or Corollary 10.5.3.

Proposition 11.2.3. Let $A \in \mathbb{F}^{n \times n}$. Then,

$$\text{mspec}(e^A) = \left\{ e^\lambda : \ \lambda \in \text{mspec}(A) \right\}_{\text{m}}. \tag{11.2.19}$$

Proof. It can be seen that every diagonal entry of the Jordan form of e^A is of the form e^λ, where $\lambda \in \text{spec}(A)$. $\qquad\square$

Corollary 11.2.4. Let $A \in \mathbb{F}^{n \times n}$. Then,

$$\det e^A = e^{\text{tr}\, A}. \tag{11.2.20}$$

Corollary 11.2.5. Let $A \in \mathbb{F}^{n \times n}$, and assume that $\text{tr}\, A = 0$. Then, $\det e^A = 1$.

11.3 Explicit Expressions

In this section we present explicit expressions for the exponential of a general 2×2 real matrix A. Expressions are given in terms of both the entries of A and the eigenvalues of A.

Lemma 11.3.1. Let $A \triangleq \begin{bmatrix} a & b \\ 0 & d \end{bmatrix} \in \mathbb{C}^{2 \times 2}$. Then,

$$
e^A = \begin{cases} e^a \begin{bmatrix} 1 & b \\ 0 & 1 \end{bmatrix}, & a = d, \\[2em] \begin{bmatrix} e^a & b\frac{e^a - e^d}{a-d} \\ 0 & e^d \end{bmatrix}, & a \neq d. \end{cases} \tag{11.3.1}
$$

The following result gives an expression for e^A in terms of the eigenvalues of A.

Proposition 11.3.2. Let $A \in \mathbb{C}^{2 \times 2}$, and let $\mathrm{mspec}(A) = \{\lambda, \mu\}_{\mathrm{m}}$. Then,

$$
e^A = \begin{cases} e^\lambda[(1-\lambda)I + A], & \lambda = \mu, \\[1em] \frac{\mu e^\lambda - \lambda e^\mu}{\mu - \lambda}I + \frac{e^\mu - e^\lambda}{\mu - \lambda}A, & \lambda \neq \mu. \end{cases} \tag{11.3.2}
$$

Proof. The result follows from Theorem 10.5.1. Alternatively, suppose that $\lambda = \mu$. Then, there exists a nonsingular matrix $S \in \mathbb{C}^{2 \times 2}$ such that $A = S\begin{bmatrix} \lambda & \alpha \\ 0 & \lambda \end{bmatrix}S^{-1}$, where $\alpha \in \mathbb{C}$. Hence, $e^A = e^\lambda S\begin{bmatrix} 1 & \alpha \\ 0 & 1 \end{bmatrix}S^{-1} = e^\lambda[(1-\lambda)I + A]$. Now, suppose that $\lambda \neq \mu$. Then, there exists a nonsingular matrix $S \in \mathbb{C}^{2 \times 2}$ such that $A = S\begin{bmatrix} \lambda & 0 \\ 0 & \mu \end{bmatrix}S^{-1}$. Hence, $e^A = S\begin{bmatrix} e^\lambda & 0 \\ 0 & e^\mu \end{bmatrix}S^{-1}$. Then, the identity $\begin{bmatrix} e^\lambda & 0 \\ 0 & e^\mu \end{bmatrix} = \frac{\mu e^\lambda - \lambda e^\mu}{\mu - \lambda}I + \frac{e^\mu - e^\lambda}{\mu - \lambda}\begin{bmatrix} \lambda & 0 \\ 0 & \mu \end{bmatrix}$ yields the desired result. $\square$

Next, we give an expression for e^A in terms of the entries of $A \in \mathbb{R}^{2 \times 2}$.

Corollary 11.3.3. Let $A \triangleq \begin{bmatrix} a & b \\ c & d \end{bmatrix} \in \mathbb{R}^{2 \times 2}$, and define $\gamma \triangleq (a-d)^2 + 4bc$ and $\delta \triangleq \frac{1}{2}|\gamma|^{1/2}$. Then,

$$
e^A = \begin{cases} e^{\frac{a+d}{2}} \begin{bmatrix} \cos\delta + \frac{a-d}{2\delta}\sin\delta & \frac{b}{\delta}\sin\delta \\[0.8em] \frac{c}{\delta}\sin\delta & \cos\delta - \frac{a-d}{2\delta}\sin\delta \end{bmatrix}, & \gamma < 0, \\[2em] e^{\frac{a+d}{2}} \begin{bmatrix} 1 + \frac{a-d}{2} & b \\[0.8em] c & 1 - \frac{a-d}{2} \end{bmatrix}, & \gamma = 0, \\[2em] e^{\frac{a+d}{2}} \begin{bmatrix} \cosh\delta + \frac{a-d}{2\delta}\sinh\delta & \frac{b}{\delta}\sinh\delta \\[0.8em] \frac{c}{\delta}\sinh\delta & \cosh\delta - \frac{a-d}{2\delta}\sinh\delta \end{bmatrix}, & \gamma > 0. \end{cases} \tag{11.3.3}
$$

Proof. The eigenvalues of A are $\lambda \triangleq \frac{1}{2}(a + d - \sqrt{\gamma})$ and $\mu \triangleq \frac{1}{2}(a + d + \sqrt{\gamma})$. Hence, $\lambda = \mu$ if and only if $\gamma = 0$. The result now follows from Proposition 11.3.2. $\square$

Example 11.3.4. Let $A \triangleq \begin{bmatrix} \nu & \omega \\ -\omega & \nu \end{bmatrix} \in \mathbb{R}^{2 \times 2}$. Then,

$$e^{tA} = e^{\nu t} \begin{bmatrix} \cos \omega t & \sin \omega t \\ -\sin \omega t & \cos \omega t \end{bmatrix}. \tag{11.3.4}$$

On the other hand, if $A \triangleq \begin{bmatrix} \nu & \omega \\ \omega & -\nu \end{bmatrix}$, then

$$e^{tA} = \begin{bmatrix} \cosh \delta t + \frac{\nu}{\delta} \sinh \delta t & \frac{\omega}{\delta} \sinh \delta t \\ \frac{\omega}{\delta} \sinh \delta t & \cosh \delta t - \frac{\nu}{\delta} \sinh \delta t \end{bmatrix}, \tag{11.3.5}$$

where $\delta \triangleq \sqrt{\omega^2 + \nu^2}$.

Example 11.3.5. Let $\alpha \in \mathbb{F}$, and define $A \triangleq \begin{bmatrix} 0 & 1 \\ 0 & \alpha \end{bmatrix}$. Then,

$$e^{tA} = \begin{cases} \begin{bmatrix} 1 & \alpha^{-1}(e^{\alpha t} - 1) \\ 0 & e^{\alpha t} \end{bmatrix}, & \alpha \neq 0, \\ \\ \begin{bmatrix} 1 & t \\ 0 & 1 \end{bmatrix}, & \alpha = 0. \end{cases}$$

Example 11.3.6. Let $\theta \in \mathbb{R}$, and define $A \triangleq \begin{bmatrix} 0 & \theta \\ -\theta & 0 \end{bmatrix}$. Then,

$$e^A = \begin{bmatrix} \cos \theta & \sin \theta \\ -\sin \theta & \cos \theta \end{bmatrix}.$$

Furthermore, define $B \triangleq \begin{bmatrix} 0 & \frac{\pi}{2} - \theta \\ \frac{-\pi}{2} + \theta & 0 \end{bmatrix}$. Then,

$$e^B = \begin{bmatrix} \sin \theta & \cos \theta \\ -\cos \theta & \sin \theta \end{bmatrix}.$$

Example 11.3.7. Consider the second-order mechanical vibration equation

$$m\ddot{q} + c\dot{q} + kq = 0, \tag{11.3.6}$$

where m is positive and c and k are nonnegative. Here m, c, and k denote mass, damping, and stiffness parameters, respectively. Equation (11.3.6) can be written in companion form as the system

$$\dot{x} = Ax, \tag{11.3.7}$$

where

$$x \triangleq \begin{bmatrix} q \\ \dot{q} \end{bmatrix}, \qquad A \triangleq \begin{bmatrix} 0 & 1 \\ -k/m & -c/m \end{bmatrix}. \tag{11.3.8}$$

The inelastic case $k = 0$ is the simplest one since A is upper triangular. In this case,

$$e^{tA} = \begin{cases} \begin{bmatrix} 1 & t \\ 0 & 1 \end{bmatrix}, & k = c = 0, \\[2em] \begin{bmatrix} 1 & \frac{m}{c}(1 - e^{-ct/m}) \\ 0 & e^{-ct/m} \end{bmatrix}, & k = 0, \ c > 0, \end{cases} \tag{11.3.9}$$

where $c = 0$ and $c > 0$ correspond to a rigid body and a damped rigid body, respectively.

Next, we consider the elastic case $c \geq 0$ and $k > 0$. In this case, we define

$$\omega_n \triangleq \sqrt{\frac{k}{m}}, \qquad \zeta \triangleq \frac{c}{2\sqrt{mk}}, \tag{11.3.10}$$

where $\omega_n > 0$ denotes the (undamped) *natural frequency* of vibration and $\zeta \geq 0$ denotes the *damping ratio*. Now, A can be written as

$$A = \begin{bmatrix} 0 & 1 \\ -\omega_n^2 & -2\zeta\omega_n \end{bmatrix}, \tag{11.3.11}$$

and Corollary 11.3.3 yields

$$e^{tA} \tag{11.3.12}$$

$$= \begin{cases} \begin{bmatrix} \cos\omega_n t & \frac{1}{\omega_n}\sin\omega_n t \\[0.8em] -\omega_n\sin\omega_n t & \cos\omega_n t \end{bmatrix}, & \zeta = 0, \\[2.5em] e^{-\zeta\omega_n t}\begin{bmatrix} \cos\omega_d t + \frac{\zeta}{\sqrt{1-\zeta^2}}\sin\omega_d t & \frac{1}{\omega_d}\sin\omega_d t \\[1em] \frac{-\omega_d}{1-\zeta^2}\sin\omega_d t & \cos\omega_d t - \frac{\zeta}{\sqrt{1-\zeta^2}}\sin\omega_d t \end{bmatrix}, & 0 < \zeta < 1, \\[2.5em] e^{-\omega_n t}\begin{bmatrix} 1 + \omega_n t & t \\[0.8em] -\omega_n^2 t & 1 - \omega_n t \end{bmatrix}, & \zeta = 1, \\[2.5em] e^{-\zeta\omega_n t}\begin{bmatrix} \cosh\omega_d t + \frac{\zeta}{\sqrt{\zeta^2-1}}\sinh\omega_d t & \frac{1}{\omega_d}\sinh\omega_d t \\[1em] \frac{-\omega_d}{\zeta^2-1}\sinh\omega_d t & \cosh\omega_d t - \frac{\zeta}{\sqrt{\zeta^2-1}}\sinh\omega_d t \end{bmatrix}, & \zeta > 1, \end{cases}$$

where $\zeta = 0$, $0 < \zeta < 1$, $\zeta = 1$, and $\zeta > 1$ correspond to *undamped*, *underdamped*, *critically damped*, and *overdamped oscillators*, respectively, and where the *damped natural frequency* ω_d is the positive number

$$\omega_d \triangleq \begin{cases} \omega_n\sqrt{1 - \zeta^2}, & 0 < \zeta < 1, \\[1em] \omega_n\sqrt{\zeta^2 - 1}, & \zeta > 1. \end{cases} \tag{11.3.13}$$

11.4 Logarithms

Proposition 11.4.1. Define $\mathcal{D} \triangleq \{z: |z - 1| < 1\}$. Then, the series

$$\log z \triangleq \sum_{i=1}^{\infty} \frac{(-1)^{i+1}}{i} (z - 1)^i \qquad (11.4.1)$$

is analytic on $\mathcal{D}$, and, for all $z \in \mathcal{D}$,

$$e^{\log z} = z. \qquad (11.4.2)$$

Furthermore, if $s \in \mathbb{C}$ and $|s| < \log 2$, then

$$|e^s - 1| < 1 \qquad (11.4.3)$$

and

$$\log e^s = s. \qquad (11.4.4)$$

Proof. For $x \in \mathbb{R}$ such that $|x| < 1$, it follows that

$$\frac{\mathrm{d}}{\mathrm{d}x} \log(1 - x) = \frac{-1}{1 - x} = -\left(1 + x + x^2 + \cdots\right).$$

Integrating implies that

$$\log(1 - x) = -\left(x + \frac{x^2}{2} + \frac{x^3}{3} + \cdots\right).$$

Now, letting $z = 1 - x$ yields

$$\log z = -\left[(1 - z) + \frac{(1 - z)^2}{2} + \frac{(1 - z)^3}{3} + \cdots\right],$$

that is,

$$\log z = \sum_{i=1}^{\infty} \frac{(-1)^{i+1}}{i} (z - 1)^i.$$

This series is analytic and can be extended to an analytic function on $\mathcal{D}$. Finally, for $s \in \mathbb{C}$ such that $|s| < \log 2$, it follows that

$$|e^s - 1| = \left| s + \frac{s^2}{2!} + \frac{s^3}{3!} + \cdots \right| = e^{|s|} - 1 < 1. \qquad \square$$

Let $A \in \mathbb{F}^{n \times n}$ be positive definite so that $A = SBS^* \in \mathbb{F}^{n \times n}$, where $S \in \mathbb{F}^{n \times n}$ is unitary and $B \in \mathbb{R}^{n \times n}$ is diagonal with positive diagonal entries. In Section 8.5, $\log A$ is defined as $\log A = S(\log B)S^* \in \mathbf{H}^n$, where $(\log B)_{(i,i)} \triangleq \log B_{(i,i)}$. It can be seen that $\log A$ satisfies $A = e^{\log A}$. The following definition is not restricted to positive-definite matrices A.

Definition 11.4.2. Let $A \in \mathbb{F}^{n \times n}$. Then, $B \in \mathbb{F}^{n \times n}$ is a *logarithm* of A if $e^B = A$.

The following result, based on Proposition 11.4.1, gives an explicit expression for a logarithm of $A \in \mathbb{F}^{n \times n}$ as well as properties of $\log A$.

Proposition 11.4.3. Let $\| \cdot \|$ be a normalized submultiplicative norm on $\mathbb{F}^{n \times n}$, and, for $A \in \mathbb{F}^{n \times n}$ such that the series below converges, define

$$\log A \triangleq \sum_{i=1}^{\infty} \frac{(-1)^{i+1}}{i} (A - I)^i. \tag{11.4.5}$$

Then, the following statements hold:

i) The series (11.4.5) converges absolutely for all $A \in \mathbb{F}^{n \times n}$ such that $\|A - I\| < 1$.

ii) If $\|A - I\| < 1$, then $\|\log A\| \leq \log(1 + \|A - I\|)$.

iii) The function log: $\{A \in \mathbb{F}^{n \times n}: \|A - I\| < 1\} \mapsto \mathbb{F}^{n \times n}$ defined by (11.4.5) is continuous.

iv) If $\|A - I\| < 1$, then $\log A$ is a logarithm of A, that is, $e^{\log A} = A$.

v) If $B \in \mathbb{F}^{n \times n}$ and $\|B\| < \log 2$, then $\|e^B - I\| < 1$ and $\log e^B = B$.

vi) exp: $\mathbb{B}_{\log 2}(0) \mapsto \mathbb{F}^{n \times n}$ is one-to-one.

Proof. For $\alpha \triangleq \|A - I\| < 1$ it follows from (11.4.5) that $\|\log A\| \leq \sum_{i=1}^{\infty} (-1)^{i+1} \alpha^i / i = \log(1 + \alpha)$, which proves i) and ii). Continuity in iii) follows from uniform convergence. Statements iv) and v) are proved in [324, pp. 34–35]. To prove vi), let $B \in \mathbb{B}_{\log 2}(0)$, so that $e^{\|B\|} < 2$, and thus

$$\|e^B - I\| \leq \sum_{i=1}^{\infty} (i!)^{-1} \|B\|^i = e^{\|B\|} - 1 < 1.$$

Now, let $B_1, B_2 \in \mathbb{B}_{\log 2}(0)$, and assume that $e^{B_1} = e^{B_2}$. Then, it follows from v) that $B_1 = \log e^{B_1} = \log e^{B_2} = B_2$. $\square$

The following result shows that every complex, nonsingular matrix has a complex logarithm.

Proposition 11.4.4. Let $A \in \mathbb{C}^{n \times n}$. Then, there exists a matrix $B \in \mathbb{C}^{n \times n}$ such that $A = e^B$ if and only if A is nonsingular.

Proof. See [324, pp. 35, 60] or [369, p. 474]. $\square$

However, only certain real matrices have a real logarithm.

Proposition 11.4.5. Let $A \in \mathbb{R}^{n \times n}$. Then, there exists a matrix $B \in \mathbb{R}^{n \times n}$ such that $A = e^B$ if and only if A is nonsingular and, for every negative eigenvalue λ of A and for every positive integer k, the Jordan form of A has an even number of $k \times k$ blocks associated with λ.

Proof. See [369, p. 475]. □

Replacing A and B in Proposition 11.4.5 by e^A and A, respectively, yields the following result.

Corollary 11.4.6. Let $A \in \mathbb{R}^{n \times n}$. Then, for every negative eigenvalue λ of e^A and for every positive integer k, the Jordan form of e^A has an even number of $k \times k$ blocks associated with λ.

Since the matrix $A \triangleq \begin{bmatrix} -2\pi & 4\pi \\ -2\pi & 2\pi \end{bmatrix}$ satisfies $e^A = I$, it follows that a positive-definite matrix can have a logarithm that is not normal. However, the following result shows that every positive-definite matrix has at least one Hermitian logarithm.

Proposition 11.4.7. The function exp: $\mathbf{H}^n \mapsto \mathbf{P}^n$ is onto.

Let $A \in \mathbb{R}^{n \times n}$. If there exists a matrix $B \in \mathbb{R}^{n \times n}$ such that $A = e^B$, then Corollary 11.2.4 implies that $\det A = \det e^B = e^{\operatorname{tr} B} > 0$. However, the converse is not true. Consider, for example, $A \triangleq \begin{bmatrix} -1 & 0 \\ 0 & -2 \end{bmatrix}$, which satisfies $\det A > 0$. However, Proposition 11.4.5 implies that there does not exist a matrix $B \in \mathbb{R}^{2 \times 2}$ such that $A = e^B$. On the other hand, note that $A = e^B e^C$, where $B \triangleq \begin{bmatrix} 0 & \pi \\ -\pi & 0 \end{bmatrix}$ and $C \triangleq \begin{bmatrix} 0 & 0 \\ 0 & \log 2 \end{bmatrix}$. While the product of two exponentials of real matrices has positive determinant, the following result shows that the converse is also true.

Proposition 11.4.8. Let $A \in \mathbb{R}^{n \times n}$. Then, there exist matrices $B, C \in \mathbb{R}^{n \times n}$ such that $A = e^B e^C$ if and only if $\det A > 0$.

Proof. Suppose that there exist $B, C \in \mathbb{R}^{n \times n}$ such that $A = e^B e^C$. Then, $\det A = (\det e^B)(\det e^C) > 0$. Conversely, suppose that $\det A > 0$. If A has no negative eigenvalues, then it follows from Proposition 11.4.5 that there exists $B \in \mathbb{R}^{n \times n}$ such that $A = e^B$. Hence, $A = e^B e^{0_{n \times n}}$. Now, suppose that A has at least one negative eigenvalue. Then, Theorem 5.3.5 on the real Jordan form implies that there exist a nonsingular matrix $S \in \mathbb{R}^{n \times n}$ and matrices $A_1 \in \mathbb{R}^{n_1 \times n_1}$ and $A_2 \in \mathbb{R}^{n_2 \times n_2}$ such that $A = S \begin{bmatrix} A_1 & 0 \\ 0 & A_2 \end{bmatrix} S^{-1}$, where every eigenvalue of A_1 is negative and where none of the eigenvalues of A_2 are negative. Since $\det A$ and $\det A_2$ are positive, it follows that n_1 is even. Now, write $A = S \begin{bmatrix} -I_{n_1} & 0 \\ 0 & I_{n_2} \end{bmatrix} \begin{bmatrix} -A_1 & 0 \\ 0 & A_2 \end{bmatrix} S^{-1}$. Since the eigenvalue -1 of $\begin{bmatrix} -I_{n_1} & 0 \\ 0 & I_{n_2} \end{bmatrix}$ appears in an even number of 1×1 Jordan blocks, it follows from Proposition 11.4.5 that there exists a matrix $\hat{B} \in \mathbb{R}^{n \times n}$ such that $\begin{bmatrix} -I_{n_1} & 0 \\ 0 & I_{n_2} \end{bmatrix} = e^{\hat{B}}$. Furthermore, since $\begin{bmatrix} -A_1 & 0 \\ 0 & A_2 \end{bmatrix}$ has no negative eigenvalues, it follows that there exists a matrix $\hat{C} \in \mathbb{R}^{n \times n}$ such that $\begin{bmatrix} -A_1 & 0 \\ 0 & A_2 \end{bmatrix} = e^{\hat{C}}$. Hence, $e^A = S e^{\hat{B}} e^{\hat{C}} S^{-1} = e^{S \hat{B} S^{-1}} e^{S \hat{C} S^{-1}}$. □

Although $e^A e^B$ is generally different from e^{A+B}, the following result, known as the *Baker-Campbell-Hausdorff series*, provides an expansion for a matrix function $C(t)$ that satisfies $e^{C(t)} = e^{tA} e^{tB}$.

Proposition 11.4.9. Let $A_1, \ldots, A_l \in \mathbb{F}^{n \times n}$. Then, there exists $\varepsilon > 0$ such that, for all $t \in (-\varepsilon, \varepsilon)$,

$$e^{tA_1} \cdots e^{tA_l} = e^{C(t)}, \tag{11.4.6}$$

where

$$C(t) \triangleq \sum_{i=1}^{l} t A_i + \sum_{1 \leq i < j \leq l} \tfrac{1}{2} t^2 [A_i, A_j] + O(t^3). \tag{11.4.7}$$

Proof. See [324, chapter 3], [628, p. 35], or [742, p. 97]. $\qquad\square$

To illustrate (11.4.6), let $l = 2$, $A = A_1$, and $B = A_2$. Then, the first few terms of the series are given by

$$e^{tA} e^{tB} = e^{tA + tB + (t^2/2)[A,B] + (t^3/12)[[B,A],A+B] + \cdots}. \tag{11.4.8}$$

The radius of convergence of this series is discussed in [208, 560].

Corollary 11.4.10. Let $A, B \in \mathbb{F}^{n \times n}$. Then,

$$e^{A+B} = \lim_{k \to \infty} \left[e^{(1/k)A} e^{(1/k)B} \right]^k. \tag{11.4.9}$$

Proof. Setting $l = 2$ and $t = 1/k$ in (11.4.6) yields, as $k \to \infty$,

$$\left[e^{(1/k)A} e^{(1/k)B} \right]^k = \left[e^{(1/k)(A+B) + O(1/k^2)} \right]^k = e^{A+B+O(1/k)} \to e^{A+B}. \qquad\square$$

11.5 Lie Groups

Definition 11.5.1. Let $\mathcal{S} \subset \mathbb{F}^{n \times n}$, and assume that $\mathcal{S}$ is a group. Then, $\mathcal{S}$ is a *Lie group* if $\mathcal{S}$ is closed relative to $\mathrm{GL}_{\mathbb{F}}(n)$.

Proposition 11.5.2. Let $\mathcal{S} \subset \mathbb{F}^{n \times n}$, and assume that $\mathcal{S}$ is a group. Then, $\mathcal{S}$ is a Lie group if and only if the limit of every convergent sequence in $\mathcal{S}$ is either an element of $\mathcal{S}$ or is singular.

The groups $\mathrm{SL}_{\mathbb{F}}(n)$, $\mathrm{U}(n)$, $\mathrm{O}(n)$, $\mathrm{SU}(n)$, $\mathrm{SO}(n)$, $\mathrm{U}(n,m)$, $\mathrm{O}(n,m)$, $\mathrm{SU}(n,m)$, $\mathrm{SO}(n,m)$, $\mathrm{Sp}_{\mathbb{F}}(n)$, $\mathrm{Aff}_{\mathbb{F}}(n)$, $\mathrm{SE}_{\mathbb{F}}(n)$, and $\mathrm{Trans}_{\mathbb{F}}(n)$ defined in Proposition 3.3.4 are closed sets, and thus are Lie groups. Although the groups $\mathrm{GL}_{\mathbb{F}}(n)$ and $\mathrm{PL}_{\mathbb{F}}(n)$ are not closed sets, they are closed relative to $\mathrm{GL}_{\mathbb{F}}(n)$, and thus they are Lie groups. The group $\mathcal{S} \subset \mathbb{C}^{2 \times 2}$ defined by

$$S \triangleq \left\{ \begin{bmatrix} e^{\jmath t} & 0 \\ 0 & e^{\pi \jmath t} \end{bmatrix} : t \in \mathbb{R} \right\} \tag{11.5.1}$$

is not closed relative to $\mathrm{GL}_{\mathbb{C}}(2)$, and thus is not a Lie group. For details, see [324, p. 4].

Proposition 11.5.3. Let $S \subset \mathbb{F}^{n \times n}$, and assume that S is a Lie group. Furthermore, define

$$S_0 \triangleq \left\{ A \in \mathbb{F}^{n \times n} : e^{tA} \in S \text{ for all } t \in \mathbb{R} \right\}. \tag{11.5.2}$$

Then, S_0 is a Lie algebra.

Proof. See [324, pp. 43, 44]. $\square$

The Lie algebra S_0 defined by (11.5.2) is the *Lie algebra of* S.

Proposition 11.5.4. Let $S \subset \mathbb{F}^{n \times n}$, assume that S is a Lie group, and let $S_0 \subseteq \mathbb{F}^{n \times n}$ be the Lie algebra of S. Furthermore, let $S \in S$ and $A \in S_0$. Then, $SAS^{-1} \in S$.

Proof. For all $t \in \mathbb{R}$, $e^{tA} \in S$, and thus $e^{tSAS^{-1}} = Se^{tA}S^{-1} \in S$. Hence, $SAS^{-1} \in S$. $\square$

Proposition 11.5.5. The following statements hold:

i) $\mathrm{gl}_{\mathbb{F}}(n)$ is the Lie algebra of $\mathrm{GL}_{\mathbb{F}}(n)$.

ii) $\mathrm{gl}_{\mathbb{R}}(n) = \mathrm{pl}_{\mathbb{R}}(n)$ is the Lie algebra of $\mathrm{PL}_{\mathbb{R}}(n)$.

iii) $\mathrm{pl}_{\mathbb{C}}(n)$ is the Lie algebra of $\mathrm{PL}_{\mathbb{C}}(n)$.

iv) $\mathrm{sl}_{\mathbb{F}}(n)$ is the Lie algebra of $\mathrm{SL}_{\mathbb{F}}(n)$.

v) $\mathrm{u}(n)$ is the Lie algebra of $\mathrm{U}(n)$.

vi) $\mathrm{so}(n)$ is the Lie algebra of $\mathrm{O}(n)$.

vii) $\mathrm{su}(n)$ is the Lie algebra of $\mathrm{SU}(n)$.

viii) $\mathrm{so}(n)$ is the Lie algebra of $\mathrm{SO}(n)$.

ix) $\mathrm{su}(n, m)$ is the Lie algebra of $\mathrm{U}(n, m)$.

x) $\mathrm{so}(n, m)$ is the Lie algebra of $\mathrm{O}(n, m)$.

xi) $\mathrm{su}(n, m)$ is the Lie algebra of $\mathrm{SU}(n, m)$.

xii) $\mathrm{so}(n, m)$ is the Lie algebra of $\mathrm{SO}(n, m)$.

xiii) $\mathrm{sp}_{\mathbb{F}}(n)$ is the Lie algebra of $\mathrm{Sp}_{\mathbb{F}}(n)$.

xiv) $\mathrm{aff}_{\mathbb{F}}(n)$ is the Lie algebra of $\mathrm{Aff}_{\mathbb{F}}(n)$.

xv) $\mathrm{se}_{\mathbb{C}}(n)$ is the Lie algebra of $\mathrm{SE}_{\mathbb{C}}(n)$.

xvi) $\mathrm{se}_{\mathbb{R}}(n)$ is the Lie algebra of $\mathrm{SE}_{\mathbb{R}}(n)$.

xvii) $\mathrm{trans}_{\mathbb{F}}(n)$ is the Lie algebra of $\mathrm{Trans}_{\mathbb{F}}(n)$.

Proof. See [324, pp. 38–41]. ☐

Definition 11.5.6. Let $\mathcal{S} \subseteq \mathbb{F}^{n \times n}$. Then, $\mathcal{S}$ is *pathwise connected* if, for all $B_1, B_2 \in \mathcal{S}$, there exists a continuous function $f \colon [0, 1] \mapsto \mathcal{S}$ such that $f(0) = B_1$ and $f(1) = B_2$.

Proposition 11.5.7. Let $\mathcal{S} \subset \mathbb{F}^{n \times n}$, assume that $\mathcal{S}$ is a Lie group, and let $\mathcal{S}_0 \subseteq \mathbb{F}^{n \times n}$ be the Lie algebra of $\mathcal{S}$. Then, exp: $\mathcal{S}_0 \mapsto \mathcal{S}$. Furthermore, if exp is onto, then $\mathcal{S}$ is pathwise connected.

Proof. Let $A \in \mathcal{S}_0$ so that $e^{tA} \in \mathcal{S}$ for all $t \in \mathbb{R}$. Hence, setting $t = 1$ implies that exp: $\mathcal{S}_0 \mapsto \mathcal{S}$. Now, suppose that exp is onto, let $B \in \mathcal{S}$, and let $A \in \mathcal{S}_0$ be such that $e^A = B$. Then, $f(t) \triangleq e^{tA}$ satisfies $f(0) = I$ and $f(1) = B$, which implies that $\mathcal{S}$ is pathwise connected. ☐

A Lie group can consist of multiple pathwise-connected components.

Proposition 11.5.8. Let $n \geq 1$. Then, the following functions are onto:

i) exp: $\mathrm{gl}_{\mathbb{C}}(n) \mapsto \mathrm{GL}_{\mathbb{C}}(n)$.

ii) exp: $\mathrm{gl}_{\mathbb{R}}(1) \mapsto \mathrm{PL}_{\mathbb{R}}(1)$.

iii) exp: $\mathrm{pl}_{\mathbb{C}}(n) \mapsto \mathrm{PL}_{\mathbb{C}}(n)$.

iv) exp: $\mathrm{sl}_{\mathbb{C}}(n) \mapsto \mathrm{SL}_{\mathbb{C}}(n)$.

v) exp: $\mathrm{u}(n) \mapsto \mathrm{U}(n)$.

vi) exp: $\mathrm{su}(n) \mapsto \mathrm{SU}(n)$.

vii) exp: $\mathrm{so}(n) \mapsto \mathrm{SO}(n)$.

Furthermore, the following functions are not onto:

viii) exp: $\mathrm{gl}_{\mathbb{R}}(n) \mapsto \mathrm{PL}_{\mathbb{R}}(n)$, where $n \geq 2$.

ix) exp: $\mathrm{sl}_{\mathbb{R}}(n) \mapsto \mathrm{SL}_{\mathbb{R}}(n)$.

x) exp: $\mathrm{so}(n) \mapsto \mathrm{O}(n)$.

xi) exp: $\mathrm{sp}(n) \mapsto \mathrm{Sp}(n)$.

Proof. Statement i) follows from Proposition 11.4.4, while ii) is immediate. Statements iii)–vii) can be verified by construction; see [592, pp. 199, 212] for the proof of v) and vii). The example $A \triangleq \begin{bmatrix} -1 & 0 \\ 0 & -2 \end{bmatrix}$ and Proposition 11.4.5 show that $viii$) is not onto. For $\lambda < 0$, $\lambda \neq -1$, Proposition 11.4.5 and the example $\begin{bmatrix} \lambda & 0 \\ 0 & 1/\lambda \end{bmatrix}$ given in [628, p. 39] show that ix) is not onto. See also [56, pp. 84, 85]. Statement $viii$) shows that x) is not onto. For xi), see [220]. $\qquad\square$

Proposition 11.5.9. The Lie groups $\mathrm{GL}_{\mathbb{C}}(n), \mathrm{SL}_{\mathbb{F}}(n), \mathrm{U}(n), \mathrm{SU}(n)$, and $\mathrm{SO}(n)$ are pathwise connected. The Lie groups $\mathrm{GL}_{\mathbb{R}}(n), \mathrm{O}(n), \mathrm{O}(n,1)$, and $\mathrm{SO}(n,1)$ are not pathwise connected.

Proof. See [324, p. 15]. $\qquad\square$

Proposition 11.5.9 and ix) of Proposition 11.5.8 show that the converse of the Proposition 11.5.7 does not hold, that is, pathwise connectedness does not imply that exp is onto. See [628, p. 39].

11.6 Lyapunov Stability Theory

Consider the dynamical system

$$\dot{x}(t) = f[x(t)], \tag{11.6.1}$$

where $t \geq 0$, $x(t) \in \mathcal{D} \subseteq \mathbb{R}^n$, and $f \colon \mathcal{D} \to \mathbb{R}^n$ is continuous. We assume that, for all $x_0 \in \mathcal{D}$ and for all $T > 0$, there exists a unique C^1 solution $x \colon [0,T] \mapsto \mathcal{D}$ satisfying (11.6.1). If $x_e \in \mathcal{D}$ satisfies $f(x_e) = 0$, then $x(t) \equiv x_e$ is an *equilibrium* of (11.6.1). The following definition concerns the stability of an equilibrium of (11.6.1). Throughout this section, let $\| \cdot \|$ denote a norm on $\mathbb{R}^n$.

Definition 11.6.1. Let $x_e \in \mathcal{D}$ be an equilibrium of (11.6.1). Then, x_e is *Lyapunov stable* if, for all $\varepsilon > 0$, there exists $\delta > 0$ such that, if $\|x(0) - x_e\| < \delta$, then $\|x(t) - x_e\| < \varepsilon$ for all $t \geq 0$. Furthermore, x_e is *asymptotically stable* if it is Lyapunov stable and there exists $\varepsilon > 0$ such that, if $\|x(0) - x_e\| < \varepsilon$, then $\lim_{t\to\infty} x(t) = x_e$. In addition, x_e is *globally asymptotically stable* if it is Lyapunov stable, $\mathcal{D} = \mathbb{R}^n$, and, for all $x(0) \in \mathbb{R}^n$, $\lim_{t\to\infty} x(t) = x_e$. Finally, x_e is *unstable* if it is not Lyapunov stable.

Note that, if $x_e \in \mathbb{R}^n$ is a globally asymptotically stable equilibrium, then x_e is the only equilibrium of (11.6.1).

The following result, known as *Lyapunov's direct method*, gives sufficient conditions for Lyapunov stability and asymptotic stability of an equilibrium of (11.6.1).

Theorem 11.6.2. Let $x_e \in \mathcal{D}$ be an equilibrium of the dynamical system (11.6.1), and assume there exists a C^1 function $V \colon \mathcal{D} \mapsto \mathbb{R}$ such that

$$V(x_e) = 0, \tag{11.6.2}$$

such that, for all $x \in \mathcal{D}\backslash\{x_e\}$,

$$V(x) > 0, \tag{11.6.3}$$

and such that, for all $x \in \mathcal{D}$,

$$V'(x)f(x) \le 0. \tag{11.6.4}$$

Then, x_e is Lyapunov stable. If, in addition, for all $x \in \mathcal{D}\backslash\{x_e\}$,

$$V'(x)f(x) < 0, \tag{11.6.5}$$

then x_e is asymptotically stable. Finally, if $\mathcal{D} = \mathbb{R}^n$ and

$$\lim_{\|x\|\to\infty} V(x) = \infty, \tag{11.6.6}$$

then x_e is globally asymptotically stable.

Proof. For convenience, let $x_e = 0$. To prove Lyapunov stability, let $\varepsilon > 0$ be such that $\mathbb{B}_\varepsilon(0) \subseteq \mathcal{D}$. Since $\mathbb{S}_\varepsilon(0)$ is compact and $V(x)$ is continuous, it follows from Theorem 10.3.7 that $V[\mathbb{S}_\varepsilon(0)]$ is compact. Since $0 \notin \mathbb{S}_\varepsilon(0)$, $V(x) > 0$ for all $x \in \mathcal{D}\backslash\{0\}$, and $V[\mathbb{S}_\varepsilon(0)]$ is compact, it follows that $\alpha \triangleq \min V[\mathbb{S}_\varepsilon(0)]$ is positive. Next, since V is continuous, it follows that there exists $\delta \in (0, \varepsilon]$ such that $V(x) < \alpha$ for all $x \in \mathbb{B}_\delta(0)$. Now, let $x(t)$ for all $t \ge 0$ satisfy (11.6.1), where $\|x(0)\| < \delta$. Hence, $V[x(0)] < \alpha$. It thus follows from (11.6.4) that, for all $t \ge 0$,

$$V[x(t)] - V[x(0)] = \int_0^t V'[x(s)]f[x(s)]\,\mathrm{d}s \le 0,$$

and hence, for all $t \ge 0$,

$$V[x(t)] \le V[x(0)] < \alpha.$$

Now, since $V(x) \ge \alpha$ for all $x \in \mathbb{S}_\varepsilon(0)$, it follows that $x(t) \notin \mathbb{S}_\varepsilon(0)$ for all $t \ge 0$. Hence, $\|x(t)\| < \varepsilon$ for all $t \ge 0$, which proves that $x_e = 0$ is Lyapunov stable.

To prove that $x_e = 0$ is asymptotically stable, let $\varepsilon > 0$ be such that $\mathbb{B}_\varepsilon(0) \subseteq \mathcal{D}$. Since (11.6.5) implies (11.6.4), it follows that there exists $\delta > 0$ such that, if $\|x(0)\| < \delta$, then $\|x(t)\| < \varepsilon$ for all $t \ge 0$. Furthermore, $\frac{\mathrm{d}}{\mathrm{d}t}V[x(t)] = V'[x(t)]f[x(t)] < 0$ for all $t \ge 0$, and thus $V[x(t)]$ is decreasing and bounded from below by zero. Now, suppose that $V[x(t)]$ does not converge to zero. Therefore, there exists $L > 0$ such that $V[x(t)] \ge L$ for all $t \ge 0$. Now, let $\delta' > 0$ be such that $V(x) < L$ for all $x \in \mathbb{B}_{\delta'}(0)$. Therefore, $\|x(t)\| \ge \delta'$ for all $t \ge 0$. Next, define $\gamma < 0$ by $\gamma \triangleq \max_{\delta' \le \|x\| \le \varepsilon} V'(x)f(x)$.

Therefore, since $\|x(t)\| < \varepsilon$ for all $t \geq 0$, it follows that

$$V[x(t)] - V[x(0)] = \int_0^t V'[x(\tau)]f[x(\tau)]\,\mathrm{d}\tau \leq \gamma t,$$

and hence

$$V(x(t)) \leq V[x(0)] + \gamma t.$$

However, $t > -V[x(0)]/\gamma$ implies that $V[x(t)] < 0$, which is a contradiction.

To prove that $x_e = 0$ is globally asymptotically stable, let $x(0) \in \mathbb{R}^n$, and let $\beta \triangleq V[x(0)]$. It follows from (11.6.6) that there exists $\varepsilon > 0$ such that $V(x) > \beta$ for all $x \in \mathbb{R}^n$ such that $\|x\| > \varepsilon$. Therefore, $\|x(0)\| \leq \varepsilon$, and, since $V[x(t)]$ is decreasing, it follows that $\|x(t)\| < \varepsilon$ for all $t > 0$. The remainder of the proof is identical to the proof of asymptotic stability. $\square$

11.7 Linear Stability Theory

We now specialize Definition 11.6.1 to the linear system

$$\dot{x}(t) = Ax(t), \qquad (11.7.1)$$

where $t \geq 0$, $x(t) \in \mathbb{R}^n$, and $A \in \mathbb{R}^{n \times n}$. Note that $x_e = 0$ is an equilibrium of (11.7.1), and that $x_e \in \mathbb{R}^n$ is an equilibrium of (11.7.1) if and only if $x_e \in \mathcal{N}(A)$. Hence, if x_e is the globally asymptotically stable equilibrium of (11.7.1), then A is nonsingular and $x_e = 0$.

We consider three types of stability for the linear system (11.7.1). Unlike Definition 11.6.1, these definitions are stated in terms of the dynamics rather than the equilibrium.

Definition 11.7.1. For $A \in \mathbb{F}^{n \times n}$, define the following classes of matrices:

i) A is *Lyapunov stable* if $\mathrm{spec}(A) \subset \mathrm{CLHP}$ and, if $\lambda \in \mathrm{spec}(A)$ and $\mathrm{Re}\,\lambda = 0$, then λ is semisimple.

ii) A is *semistable* if $\mathrm{spec}(A) \subset \mathrm{OLHP} \cup \{0\}$ and, if $0 \in \mathrm{spec}(A)$, then 0 is semisimple.

iii) A is *asymptotically stable* if $\mathrm{spec}(A) \subset \mathrm{OLHP}$.

The following result concerns Lyapunov stability, semistability, and asymptotic stability for (11.7.1).

Proposition 11.7.2. Let $A \in \mathbb{R}^{n \times n}$. Then, the following statements are equivalent:

i) $x_{\mathrm{e}} = 0$ is a Lyapunov-stable equilibrium of (11.7.1).

ii) At least one equilibrium of (11.7.1) is Lyapunov stable.

iii) Every equilibrium of (11.7.1) is Lyapunov stable.

iv) A is Lyapunov stable.

v) For every initial condition $x(0) \in \mathbb{R}^n$, $x(t)$ is bounded for all $t \geq 0$.

vi) $\|e^{tA}\|$ is bounded for all $t \geq 0$, where $\| \cdot \|$ is a norm on $\mathbb{R}^{n \times n}$.

vii) For every initial condition $x(0) \in \mathbb{R}^n$, $e^{tA}x(0)$ is bounded for all $t \geq 0$.

The following statements are equivalent:

viii) A is semistable.

ix) $\lim_{t \to \infty} e^{tA}$ exists. In fact, $\lim_{t \to \infty} e^{tA} = I - AA^{\#}$.

x) For every initial condition $x(0)$, $\lim_{t \to \infty} x(t)$ exists.

The following statements are equivalent:

xi) $x_{\mathrm{e}} = 0$ is an asymptotically stable equilibrium of (11.7.1).

xii) A is asymptotically stable.

xiii) $\mathrm{spabs}(A) < 0$.

xiv) For every initial condition $x(0) \in \mathbb{R}^n$, $\lim_{t \to \infty} x(t) = 0$.

xv) For every initial condition $x(0) \in \mathbb{R}^n$, $e^{tA}x(0) \to 0$ as $t \to \infty$.

xvi) $e^{tA} \to 0$ as $t \to \infty$.

The following definition concerns the stability of a polynomial.

Definition 11.7.3. Let $p \in \mathbb{R}[s]$. Then, define the following terminology:

i) p is *Lyapunov stable* if $\mathrm{roots}(p) \subset \mathrm{CLHP}$ and, if λ is an imaginary root of p, then $\mathrm{m}_p(\lambda) = 1$.

ii) p is *semistable* if $\mathrm{roots}(p) \subset \mathrm{OLHP} \cup \{0\}$ and, if $0 \in \mathrm{roots}(p)$, then $\mathrm{m}_p(0) = 1$.

iii) p is *asymptotically stable* if $\mathrm{roots}(p) \subset \mathrm{OLHP}$.

For the following result, recall Definition 11.7.1.

Proposition 11.7.4. Let $A \in \mathbb{R}^{n \times n}$. Then, the following statements hold:

i) A is Lyapunov stable if and only if μ_A is Lyapunov stable.

ii) A is semistable if and only if μ_A is semistable.

Furthermore, the following statements are equivalent:

iii) A is asymptotically stable

iv) μ_A is asymptotically stable.

v) χ_A is asymptotically stable.

Next, consider the factorization of the minimal polynomial μ_A of A given by

$$\mu_A = \mu_A^{\mathrm{s}} \mu_A^{\mathrm{u}}, \tag{11.7.2}$$

where μ_A^{s} and μ_A^{u} are monic polynomials such that

$$\mathrm{roots}(\mu_A^{\mathrm{s}}) \subset \mathrm{OLHP} \tag{11.7.3}$$

and

$$\mathrm{roots}(\mu_A^{\mathrm{u}}) \subset \mathrm{CRHP}. \tag{11.7.4}$$

Proposition 11.7.5. Let $A \in \mathbb{R}^{n \times n}$, and let $S \in \mathbb{R}^{n \times n}$ be a nonsingular matrix such that

$$A = S \begin{bmatrix} A_1 & A_{12} \\ 0 & A_2 \end{bmatrix} S^{-1}, \tag{11.7.5}$$

where $A_1 \in \mathbb{R}^{r \times r}$ is asymptotically stable, $A_{12} \in \mathbb{R}^{r \times (n-r)}$, and $A_2 \in \mathbb{R}^{(n-r) \times (n-r)}$ satisfies $\mathrm{spec}(A_2) \subset \mathrm{CRHP}$. Then,

$$\mu_A^{\mathrm{s}}(A) = S \begin{bmatrix} 0 & C_{12\mathrm{s}} \\ 0 & \mu_A^{\mathrm{s}}(A_2) \end{bmatrix} S^{-1}, \tag{11.7.6}$$

where $C_{12\mathrm{s}} \in \mathbb{R}^{r \times (n-r)}$ and $\mu_A^{\mathrm{s}}(A_2)$ is nonsingular, and

$$\mu_A^{\mathrm{u}}(A) = S \begin{bmatrix} \mu_A^{\mathrm{u}}(A_1) & C_{12\mathrm{u}} \\ 0 & 0 \end{bmatrix} S^{-1}, \tag{11.7.7}$$

where $C_{12\mathrm{u}} \in \mathbb{R}^{r \times (n-r)}$ and $\mu_A^{\mathrm{u}}(A_1)$ is nonsingular. Consequently,

$$\mathcal{N}[\mu_A^{\mathrm{s}}(A)] = \mathcal{R}[\mu_A^{\mathrm{u}}(A)] = \mathcal{R}\left(S \begin{bmatrix} I_r \\ 0 \end{bmatrix}\right). \tag{11.7.8}$$

If, in addition, $A_{12} = 0$, then

$$\mu_A^{\mathrm{s}}(A) = S \begin{bmatrix} 0 & 0 \\ 0 & \mu_A^{\mathrm{s}}(A_2) \end{bmatrix} S^{-1} \tag{11.7.9}$$

and

$$\mu_A^{\mathrm{u}}(A) = S \begin{bmatrix} \mu_A^{\mathrm{u}}(A_1) & 0 \\ 0 & 0 \end{bmatrix} S^{-1}. \tag{11.7.10}$$

Consequently,

$$\mathcal{R}[\mu_A^{\rm s}(A)] = \mathcal{N}[\mu_A^{\rm u}(A)] = \mathcal{R}\left(S\begin{bmatrix} 0 \\ I_{n-r} \end{bmatrix}\right). \tag{11.7.11}$$

Corollary 11.7.6. Let $A \in \mathbb{R}^{n \times n}$. Then,

$$\mathcal{N}[\mu_A^{\rm s}(A)] = \mathcal{R}[\mu_A^{\rm u}(A)] \tag{11.7.12}$$

and

$$\mathcal{N}[\mu_A^{\rm u}(A)] = \mathcal{R}[\mu_A^{\rm s}(A)]. \tag{11.7.13}$$

Proof. It follows from Theorem 5.3.5 that there exists a nonsingular matrix $S \in \mathbb{R}^{n \times n}$ such that (11.7.5) is satisfied, where $A_1 \in \mathbb{R}^{r \times r}$ is asymptotically stable, $A_{12} = 0$, and $A_2 \in \mathbb{R}^{(n-r) \times (n-r)}$ satisfies $\operatorname{spec}(A_2) \subset \mathrm{CRHP}$. The result now follows from Proposition 11.7.5. $\square$

In view of Corollary 11.7.6, we define the *asymptotically stable subspace* $\mathcal{S}_{\rm s}(A)$ of A by

$$\mathcal{S}_{\rm s}(A) \triangleq \mathcal{N}[\mu_A^{\rm s}(A)] = \mathcal{R}[\mu_A^{\rm u}(A)] \tag{11.7.14}$$

and the *unstable subspace* $\mathcal{S}_{\rm u}(A)$ of A by

$$\mathcal{S}_{\rm u}(A) \triangleq \mathcal{N}[\mu_A^{\rm u}(A)] = \mathcal{R}[\mu_A^{\rm s}(A)]. \tag{11.7.15}$$

Note that

$$\dim \mathcal{S}_{\rm s}(A) = \operatorname{def} \mu_A^{\rm s}(A) = \operatorname{rank} \mu_A^{\rm u}(A) = \sum_{\substack{\lambda \in \operatorname{spec}(A) \\ \operatorname{Re}\lambda < 0}} \operatorname{am}_A(\lambda) \tag{11.7.16}$$

and

$$\dim \mathcal{S}_{\rm u}(A) = \operatorname{def} \mu_A^{\rm u}(A) = \operatorname{rank} \mu_A^{\rm s}(A) = \sum_{\substack{\lambda \in \operatorname{spec}(A) \\ \operatorname{Re}\lambda \geq 0}} \operatorname{am}_A(\lambda). \tag{11.7.17}$$

Lemma 11.7.7. Let $A \in \mathbb{R}^{n \times n}$, assume that $\operatorname{spec}(A) \subset \mathrm{CRHP}$, let $x \in \mathbb{R}^n$, and assume that $\lim_{t \to \infty} e^{tA}x = 0$. Then, $x = 0$.

For the following result, note Proposition 11.7.2, Proposition 5.5.8, Fact 3.8.2, Fact 11.17.3, and Proposition 6.1.7.

Proposition 11.7.8. Let $A \in \mathbb{R}^{n \times n}$. Then, the following statements hold:

i) $\mathcal{S}_{\rm s}(A) = \{x \in \mathbb{R}^n : \lim_{t \to \infty} e^{tA}x = 0\}$.

ii) $\mu_A^{\rm s}(A)$ and $\mu_A^{\rm u}(A)$ are group invertible.

iii) $P_{\rm s} \triangleq I - \mu_A^{\rm s}(A)[\mu_A^{\rm s}(A)]^{\#}$ and $P_{\rm u} \triangleq I - \mu_A^{\rm u}(A)[\mu_A^{\rm u}(A)]^{\#}$ are idempotent.

iv) $P_{\rm s} + P_{\rm u} = I$.

v) $P_{\rm s\perp} = P_{\rm u}$ and $P_{\rm u\perp} = P_{\rm s}$.

vi) $\mathcal{S}_\mathrm{s}(A) = \mathcal{R}(P_\mathrm{s}) = \mathcal{N}(P_\mathrm{u})$.

vii) $\mathcal{S}_\mathrm{u}(A) = \mathcal{R}(P_\mathrm{u}) = \mathcal{N}(P_\mathrm{s})$.

viii) $\mathcal{S}_\mathrm{s}(A)$ and $\mathcal{S}_\mathrm{u}(A)$ are invariant subspaces of A.

ix) $\mathcal{S}_\mathrm{s}(A)$ and $\mathcal{S}_\mathrm{u}(A)$ are complementary subspaces.

x) P_s is the idempotent matrix onto $\mathcal{S}_\mathrm{s}(A)$ along $\mathcal{S}_\mathrm{u}(A)$.

xi) P_u is the idempotent matrix onto $\mathcal{S}_\mathrm{u}(A)$ along $\mathcal{S}_\mathrm{s}(A)$.

Proof. To prove *i)*, let $S \in \mathbb{R}^{n \times n}$ be a nonsingular matrix such that

$$A = S \begin{bmatrix} A_1 & 0 \\ 0 & A_2 \end{bmatrix} S^{-1},$$

where $A_1 \in \mathbb{R}^{r \times r}$ is asymptotically stable and $\mathrm{spec}(A_2) \subset \mathrm{CRHP}$. It then follows from Proposition 11.7.5 that

$$\mathcal{S}_\mathrm{s}(A) = \mathcal{N}[\mu_A^\mathrm{s}(A)] = \mathcal{R}\left(S \begin{bmatrix} I_r \\ 0 \end{bmatrix} \right).$$

Furthermore,

$$e^{tA} = S \begin{bmatrix} e^{tA_1} & 0 \\ 0 & e^{tA_2} \end{bmatrix} S^{-1}.$$

To prove $\mathcal{S}_\mathrm{s}(A) \subseteq \{z \in \mathbb{R}^n\colon\ \lim_{t\to\infty} e^{tA}z = 0\}$, let $x \triangleq S\begin{bmatrix} x_1 \\ 0 \end{bmatrix} \in \mathcal{S}_\mathrm{s}(A)$, where $x_1 \in \mathbb{R}^r$. Then, $e^{tA}x = S\begin{bmatrix} e^{tA_1}x_1 \\ 0 \end{bmatrix} \to 0$ as $t \to \infty$. Hence, $x \in \{z \in \mathbb{R}^n\colon\ \lim_{t\to\infty} e^{tA}z = 0\}$. Conversely, to prove $\{z \in \mathbb{R}^n\colon\ \lim_{t\to\infty} e^{tA}z = 0\} \subseteq \mathcal{S}_\mathrm{s}(A)$, let $x \triangleq S\begin{bmatrix} x_1 \\ x_2 \end{bmatrix} \in \mathbb{R}^n$ satisfy $\lim_{t\to\infty} e^{tA}x = 0$. Hence, $e^{tA_2}x_2 \to 0$ as $t \to \infty$. Since $\mathrm{spec}(A_2) \subset \mathrm{CRHP}$, it follows from Lemma 11.7.7 that $x_2 = 0$. Hence, $x \in \mathcal{R}\left(S\begin{bmatrix} I_r \\ 0 \end{bmatrix}\right) = \mathcal{S}_\mathrm{s}(A)$.

The remaining statements follow directly from Proposition 11.7.5. $\square$

11.8 The Lyapunov Equation

In this section we specialize Theorem 11.6.2 to the linear system (11.7.1).

Corollary 11.8.1. Let $A \in \mathbb{R}^{n \times n}$, and assume there exist a positive-semidefinite matrix $R \in \mathbb{R}^{n \times n}$ and a positive-definite matrix $P \in \mathbb{R}^{n \times n}$ satisfying

$$A^\mathrm{T}P + PA + R = 0. \tag{11.8.1}$$

Then, A is Lyapunov stable. If, in addition, for all nonzero $\omega \in \mathbb{R}$,

$$\mathrm{rank} \begin{bmatrix} \jmath\omega I - A \\ R \end{bmatrix} = n, \tag{11.8.2}$$

then A is semistable. Finally, if R is positive definite, then A is asymptotically stable.

Proof. Define $V(x) \triangleq x^\mathrm{T}Px$, which satisfies (11.6.2) with $x_\mathrm{e} = 0$ and satisfies (11.6.3) for all nonzero $x \in \mathcal{D} = \mathbb{R}^n$. Furthermore, Theorem 11.6.2 implies that $V'(x)f(x) = 2x^\mathrm{T}PAx = x^\mathrm{T}(A^\mathrm{T}P + PA)x = -x^\mathrm{T}Rx$, which satisfies (11.6.4) for all $x \in \mathbb{R}^n$. Thus, Theorem 11.6.2 implies that A is Lyapunov stable. If, in addition, R is positive definite, then (11.6.5) is satisfied for all $x \neq 0$, and thus A is asymptotically stable.

Alternatively, we shall prove the first and third statements without using Theorem 11.6.2. Letting $\lambda \in \mathrm{spec}(A)$, and letting $x \in \mathbb{C}^n$ be an associated eigenvector, it follows that $0 \geq -x^*Rx = x^*(A^\mathrm{T}P + PA)x = (\overline{\lambda} + \lambda)x^*Px$. Therefore, $\mathrm{spec}(A) \subset \mathrm{CLHP}$. Now, suppose that $\jmath\omega \in \mathrm{spec}(A)$, where $\omega \in \mathbb{R}$, and let $x \in \mathcal{N}[(\jmath\omega I - A)^2]$. Defining $y \triangleq (\jmath\omega I - A)x$, it follows that $(\jmath\omega I - A)y = 0$, and thus $Ay = \jmath\omega y$. Therefore, $-y^*Ry = y^*(A^\mathrm{T}P + PA)y = -\jmath\omega y^*Py + \jmath\omega y^*Py = 0$, and thus $Ry = 0$. Hence, $0 = x^*Ry = -x^*(A^\mathrm{T}P + PA)y = -x^*(A^\mathrm{T} + \jmath\omega I)Py = y^*Py$. Since P is positive definite, it follows that $y = 0$, that is, $(\jmath\omega I - A)x = 0$. Therefore, $x \in \mathcal{N}(\jmath\omega I - A)$. Now, Proposition 5.5.14 implies that $\jmath\omega$ is semisimple. Therefore, A is Lyapunov stable.

Next, to prove that A is asymptotically stable, let $\lambda \in \mathrm{spec}(A)$, and let $x \in \mathbb{C}^n$ be an associated eigenvector. Thus, $0 > -x^*Rx = (\overline{\lambda} + \lambda)x^*Px$, which implies that A is asymptotically stable.

Finally, to prove that A is semistable, let $\jmath\omega \in \mathrm{spec}(A)$, where $\omega \in \mathbb{R}$ is nonzero, and let $x \in \mathbb{C}^n$ be an associated eigenvector. Then,

$$-x^*Rx = x^*(A^\mathrm{T}P + PA)x = x^*[(\jmath\omega I - A)^*P + P(\jmath\omega I - A]x = 0.$$

Therefore, $Rx = 0$, and thus

$$\begin{bmatrix} \jmath\omega I - A \\ R \end{bmatrix} x = 0,$$

which implies that $x = 0$, which contradicts $x \neq 0$. Consequently, $\jmath\omega \notin \mathrm{spec}(A)$ for all nonzero $\omega \in \mathbb{R}$, and thus A is semistable. $\quad\square$

Equation (11.8.1) is a *Lyapunov equation*. Converse results for Corollary 11.8.1 are given by Corollary 11.8.4, Proposition 11.8.5, Proposition 11.8.6, Proposition 11.8.7, and Proposition 12.8.3. The following lemma will be useful for analyzing (11.8.1).

Lemma 11.8.2. Assume that $A \in \mathbb{F}^{n \times n}$ is asymptotically stable. Then,

$$\int_0^\infty e^{tA} \, \mathrm{d}t = -A^{-1}. \tag{11.8.3}$$

Proof. Proposition 11.1.5 implies that $\int_0^t e^{\tau A} \, \mathrm{d}\tau = A^{-1}(e^{tA} - I)$. Letting $t \to \infty$ yields (11.8.3). $\qquad \square$

The following result concerns Sylvester's equation. See Fact 5.9.24 and Proposition 7.2.4.

Proposition 11.8.3. Let $A, B, C \in \mathbb{R}^{n \times n}$. Then, there exists a unique matrix $X \in \mathbb{R}^{n \times n}$ satisfying

$$AX + XB + C = 0 \tag{11.8.4}$$

if and only if $B^{\mathrm{T}} \oplus A$ is nonsingular. In this case, X is given by

$$X = -\operatorname{vec}^{-1}\left[\left(B^{\mathrm{T}} \oplus A\right)^{-1} \operatorname{vec} C\right]. \tag{11.8.5}$$

If, in addition, $B^{\mathrm{T}} \oplus A$ is asymptotically stable, then X is given by

$$X = \int_0^\infty e^{tA} C e^{tB} \, \mathrm{d}t. \tag{11.8.6}$$

Proof. The first two statements follow from Proposition 7.2.4. If $B^{\mathrm{T}} \oplus A$ is asymptotically stable, then it follows from (11.8.5) using Lemma 11.8.2 and Proposition 11.1.8 that

$$X = \int_0^\infty \operatorname{vec}^{-1}\left(e^{t(B^{\mathrm{T}} \oplus A)} \operatorname{vec} C\right) \mathrm{d}t = \int_0^\infty \operatorname{vec}^{-1}\left(e^{tB^{\mathrm{T}}} \otimes e^{tA}\right) \operatorname{vec} C \, \mathrm{d}t$$

$$= \int_0^\infty \operatorname{vec}^{-1} \operatorname{vec}\left(e^{tA} C e^{tB}\right) \mathrm{d}t = \int_0^\infty e^{tA} C e^{tB} \, \mathrm{d}t. \qquad \square$$

The following result provides a converse to Corollary 11.8.1 for the case of asymptotic stability.

Corollary 11.8.4. Let $A \in \mathbb{R}^{n \times n}$, and let $R \in \mathbb{R}^{n \times n}$. Then, there exists a unique matrix $P \in \mathbb{R}^{n \times n}$ satisfying (11.8.1) if and only if $A \oplus A$ is nonsingular. In this case, if R is symmetric, then P is symmetric. Now, assume that A is asymptotically stable. Then, $P \in \mathbf{S}^n$ is given by

$$P = \int_0^\infty e^{tA^{\mathrm{T}}} R e^{tA}\, \mathrm{d}t. \tag{11.8.7}$$

Finally, if R is (positive semidefinite, positive definite), then P is (positive semidefinite, positive definite).

Proof. First note that $A \oplus A$ is nonsingular if and only if $(A \oplus A)^{\mathrm{T}} = A^{\mathrm{T}} \oplus A^{\mathrm{T}}$ is nonsingular. Now, the first statement follows from Proposition 11.8.3. To prove the second statement, note that $A^{\mathrm{T}}(P - P^{\mathrm{T}}) + (P - P^{\mathrm{T}})A = 0$, which implies that P is symmetric. Now, suppose that A is asymptotically stable. Then, Fact 11.17.31 implies that $A \oplus A$ is asymptotically stable. Consequently, (11.8.7) follows from (11.8.6). $\qquad\square$

The following result provides a converse to Corollary 11.8.1 for the case of Lyapunov stability.

Proposition 11.8.5. Let $A \in \mathbb{R}^{n \times n}$, and assume that A is Lyapunov stable. Then, there exist a positive-definite matrix P and a positive-semidefinite matrix R satisfying (11.8.1).

Proof. Let $S \in \mathbb{R}^{n \times n}$ be a nonsingular matrix such that $SAS^{-1} = \begin{bmatrix} A_1 & 0 \\ 0 & A_2 \end{bmatrix}$ is in real Jordan form, where $A_1 \in \mathbb{R}^{n_1 \times n_1}$, $\mathrm{spec}(A_1) \subset \mathrm{OLHP}$, $\mathrm{spec}(A_2) \subset \jmath\mathbb{R}$, and A_2 is skew symmetric. Let $R_1 \in \mathbb{R}^{n_1 \times n_1}$ be positive definite, and let $P_1 \in \mathbb{R}^{n_1 \times n_1}$ be the positive-definite solution of $A_1^{\mathrm{T}}P_1 + P_1 A_1 + R_1 = 0$. Since $A_2 + A_2^{\mathrm{T}} = 0$, it follows that $(SAS^{-1})^{\mathrm{T}}\hat{P} + \hat{P}SAS^{-1} + \hat{R} = 0$, where $\hat{P} \triangleq \begin{bmatrix} P_1 & 0 \\ 0 & 0 \end{bmatrix}$ and $\hat{R} \triangleq \begin{bmatrix} R_1 & 0 \\ 0 & 0 \end{bmatrix}$. Therefore, (11.8.1) is satisfied with $P \triangleq S^{\mathrm{T}}\hat{P}S$ and $R \triangleq S^{\mathrm{T}}\hat{R}S$. $\qquad\square$

The following results also include converse statements. We first consider asymptotic stability.

Consider the Lyapunov equation

$$A^{\mathrm{T}}P + PA + R = 0. \tag{11.8.8}$$

Proposition 11.8.6. Let $A \in \mathbb{R}^{n \times n}$. The following statements are equivalent:

i) A is asymptotically stable.

ii) For every positive-definite matrix $R \in \mathbb{R}^{n \times n}$ there exists a positive-definite matrix $P \in \mathbb{R}^{n \times n}$ such that (11.8.8) is satisfied.

iii) There exist a positive-definite matrix $R \in \mathbb{R}^{n \times n}$ and a positive-definite matrix $P \in \mathbb{R}^{n \times n}$ such that (11.8.8) is satisfied.

Proof. The result $i)$ $\implies$ $ii)$ follows from Corollary 11.8.1. The implications $ii)$ $\implies$ $iii)$ and $iii)$ $\implies$ $iv)$ are immediate. To prove $iv)$ $\implies$ $i)$, note that, since there exists a positive-semidefinite matrix P satisfying (11.8.8), it follows from Proposition 12.4.3 that (A, C) is observably asymptotically stable. Thus, there exists a nonsingular matrix $S \in \mathbb{R}^{n \times n}$ such that $A = S\begin{bmatrix} A_1 & 0 \\ A_{21} & A_2 \end{bmatrix}S^{-1}$ and $C = \begin{bmatrix} C_1 & 0 \end{bmatrix}S^{-1}$, where (C_1, A_1) is observable and A_1 is asymptotically stable. Furthermore, since $(S^{-1}AS, CS)$ is detectable, it follows that A_2 is also asymptotically stable. Consequently, A is asymptotically stable. $\qquad\square$

Next, we consider the case of Lyapunov stability.

Proposition 11.8.7. Let $A \in \mathbb{R}^{n \times n}$. Then, A is Lyapunov stable if and only if there exist a positive-semidefinite matrix $R \in \mathbb{R}^{n \times n}$ and a positive-definite matrix $P \in \mathbb{R}^{n \times n}$ such that (11.8.8) is satisfied.

Proof. Suppose that A is Lyapunov stable. Then, there exists a nonsingular matrix $S \in \mathbb{R}^{n \times n}$ such that $A = S\begin{bmatrix} A_1 & 0 \\ 0 & A_2 \end{bmatrix}S^{-1}$, $A_1 \in \mathbb{R}^{n_1 \times n_1}$, $A_2 \in \mathbb{R}^{n_2 \times n_2}$, $\mathrm{spec}(A_1) \subset \jmath\mathbb{R}$, and $\mathrm{spec}(A_2) \subset \mathrm{OLHP}$. Let $S_1 \in \mathbb{R}^{n_1 \times n_1}$ be such that $A_1 = S_1 J_1 S_1^{-1}$, where $J_1 \in \mathbb{R}^{n_1 \times n_1}$ is skew symmetric. Then, it follows that $A_1^{\mathrm{T}}P_1 + P_1 A_1 = 0$, where $P_1 = S_1^{-\mathrm{T}}S_1^{-1}$ is positive definite. Next, let $R_2 \in \mathbb{R}^{n_2 \times n_2}$ be positive definite, and let $P_2 \in \mathbb{R}^{n_2 \times n_2}$ be the positive-definite solution of $A_2^{\mathrm{T}}P_2 + P_2 A_2 + R_2 = 0$. Hence, (11.8.8) is satisfied with $P \triangleq S^{-\mathrm{T}}\begin{bmatrix} P_1 & 0 \\ 0 & P_2 \end{bmatrix}S^{-1}$ and $R \triangleq S^{-\mathrm{T}}\begin{bmatrix} 0 & 0 \\ 0 & R_2 \end{bmatrix}S^{-1}$.

Conversely, suppose there exist a positive-semidefinite matrix $R \in \mathbb{R}^{n \times n}$ and a positive-definite matrix $P \in \mathbb{R}^{n \times n}$ such that (11.8.8) is satisfied. Let $\lambda \in \mathrm{spec}(A)$, and let $x \in \mathbb{R}^n$ be an eigenvector of A associated with λ. It thus follows from (11.8.8) that $0 = x^*A^{\mathrm{T}}Px + x^*PAx + x^*Rx = (\lambda + \bar{\lambda})x^*Px + x^*Rx$. Therefore, $\mathrm{Re}\,\lambda = -x^*Rx/(2x^*Px)$, which shows that $\mathrm{spec}(A) \subset \mathrm{CLHP}$. Now, let $\jmath\omega \in \mathrm{spec}(A)$, and suppose that $x \in \mathbb{R}^n$ satisfies $(\jmath\omega I - A)^2 x = 0$. Then, $(\jmath\omega I - A)y = 0$, where $y = (\jmath\omega I - A)x$. Computing $0 = y^*(A^{\mathrm{T}}P + PA)y + y^*Ry$ yields $y^*Ry = 0$ and thus $Ry = 0$. Therefore, $(A^{\mathrm{T}}P + PA)y = 0$ and thus $y^*Py = (A^{\mathrm{T}} + \jmath\omega I)Py = 0$. Since P is positive definite, it follows that $y = (\jmath\omega I - A)x = 0$. Therefore, $\mathcal{N}(\jmath\omega I - A) = \mathcal{N}[(\jmath\omega I - A)^2]$. Hence, it follows from Proposition 5.5.14 that $\jmath\omega$ is semisimple. $\qquad\square$

Corollary 11.8.8. Let $A \in \mathbb{R}^{n \times n}$. Then, the following statements hold:

$i)$ A is Lyapunov stable if and only if there exists a positive-definite matrix $P \in \mathbb{R}^{n \times n}$ such that $A^{\mathrm{T}}P + PA$ is negative semidefinite.

$ii)$ A is asymptotically stable if and only if there exists a positive-definite matrix $P \in \mathbb{R}^{n \times n}$ such that $A^{\mathrm{T}}P + PA$ is negative definite.

11.9 Discrete-Time Stability Theory

The theory of difference equations is concerned with the behavior of discrete-time dynamical systems of the form

$$x_{k+1} = f(x_k), \tag{11.9.1}$$

where $f\colon \mathbb{R}^n \to \mathbb{R}^n$, $k \in \mathbb{N}$, $x_k \in \mathbb{R}^n$, and x_0 is the initial condition. The solution $x_k \equiv x_e$ is an equilibrium of (11.9.1) if $x_e = f(x_e)$.

A linear discrete-time system has the form

$$x_{k+1} = Ax_k, \tag{11.9.2}$$

where $A \in \mathbb{R}^{n \times n}$. For $k \in \mathbb{N}$, x_k is given by

$$x_k = A^k x_0. \tag{11.9.3}$$

The behavior of $\{x_k\}_{k=0}^\infty$ is determined by the stability of A. To study the stability of discrete-time systems it is helpful to define the *open unit disk* (OUD) and the *closed unit disk* (CUD) by

$$\text{OUD} \triangleq \{x \in \mathbb{C}\colon |x| < 1\} \tag{11.9.4}$$

and

$$\text{CUD} \triangleq \{x \in \mathbb{C}\colon |x| \le 1\}. \tag{11.9.5}$$

Definition 11.9.1. For $A \in \mathbb{F}^{n \times n}$, define the following classes of matrices:

 i) A is *discrete-time Lyapunov stable* if $\text{spec}(A) \subset \text{CUD}$ and, if $\lambda \in \text{spec}(A)$ and $|\lambda| = 1$, then λ is semisimple.

 ii) A is *discrete-time semistable* if $\text{spec}(A) \subset \text{OUD} \cup \{1\}$ and, if $1 \in \text{spec}(A)$, then 1 is semisimple.

 iii) A is *discrete-time asymptotically stable* if $\text{spec}(A) \subset \text{OUD}$.

Proposition 11.9.2. Let $A \in \mathbb{R}^{n \times n}$ and consider the linear discrete-time system (11.9.2). Then, the following statements are equivalent:

 i) A is discrete-time Lyapunov stable.

 ii) For every initial condition $x_0 \in \mathbb{R}^n$, the sequence $\{\|x_k\|\}_{k=1}^\infty$ is bounded, where $\|\cdot\|$ is a norm on $\mathbb{R}^n$.

 iii) For every initial condition $x_0 \in \mathbb{R}^n$, the sequence $\{\|A^k x_0\|\}_{k=1}^\infty$ is bounded, where $\|\cdot\|$ is a norm on $\mathbb{R}^n$.

 iv) The sequence $\{\|A^k\|\}_{k=1}^\infty$ is bounded, where $\|\cdot\|$ is a norm on $\mathbb{R}^{n \times n}$.

The following statements are equivalent:

 v) A is discrete-time semistable.

vi) $\lim_{k\to\infty} A^k$ exists. In fact, $\lim_{k\to\infty} A^k = I - (I - A)^\#(I - A)$.

vii) For every initial condition $x_0 \in \mathbb{R}^n$, $\lim_{k\to\infty} x_k$ exists.

The following statements are equivalent:

viii) A is discrete-time asymptotically stable.

ix) $\mathrm{sprad}(A) < 1$.

x) For every initial condition $x_0 \in \mathbb{R}^n$, $\lim_{k\to\infty} x_k = 0$.

xi) For every initial condition $x_0 \in \mathbb{R}^n$, $A^k x_0 \to 0$ as $k \to \infty$.

xii) $A^k \to 0$ as $k \to \infty$.

The following definition concerns the discrete-time stability of a polynomial.

Definition 11.9.3. Let $p \in \mathbb{R}[s]$. Then, define the following terminology:

i) p is *discrete-time Lyapunov stable* if $\mathrm{roots}(p) \subset \mathrm{CUD}$ and, if λ is an imaginary root of p, then $\mathrm{m}_p(\lambda) = 1$.

ii) p is *discrete-time semistable* if $\mathrm{roots}(p) \subset \mathrm{OUD} \cup \{1\}$ and, if $1 \in \mathrm{roots}(p)$, then $\mathrm{m}_p(1) = 1$.

iii) p is *discrete-time asymptotically stable* if $\mathrm{roots}(p) \subset \mathrm{OUD}$.

Proposition 11.9.4. Let $A \in \mathbb{R}^{n\times n}$. Then, the following statements hold:

i) A is discrete-time Lyapunov stable if and only if μ_A is discrete-time Lyapunov stable.

ii) A is discrete-time semistable if and only if μ_A is discrete-time semistable.

Furthermore, the following statements are equivalent:

i) A is discrete-time asymptotically stable.

ii) μ_A is discrete-time asymptotically stable.

iii) χ_A is discrete-time asymptotically stable.

11.10 Facts on Matrix Exponential Formulas

Fact 11.10.1. Let $A \in \mathbb{R}^{n\times n}$. Then, the following statements hold:

i) If $A^2 = 0$, then $e^{tA} = I + tA$.

ii) If $A^2 = I$, then $e^{tA} = (\cosh t)I + (\sinh t)A$.

$iii)$ If $A^2 = -I$, then $e^{tA} = (\cos t)I + (\sin t)A$.

$iv)$ If $A^2 = A$, then $e^{tA} = I - A + e^t A$.

$v)$ If $A^2 = -A$, then $e^{tA} = I + A - e^{-t}A$.

$vi)$ If $\operatorname{rank} A = 1$ and $\operatorname{tr} A = 0$, then $e^{tA} = I + tA$.

$vii)$ If $\operatorname{rank} A = 1$ and $\operatorname{tr} A \neq 0$, then $e^{tA} = I + \frac{e^{(\operatorname{tr} A)t} - 1}{\operatorname{tr} A} A$.

(Remark: See [583].)

Fact 11.10.2. Let $A \triangleq \left[\begin{smallmatrix} 0 & I_n \\ I_n & 0 \end{smallmatrix} \right]$. Then,

$$e^{tA} = (\cosh t)I_{2n} + (\sinh t)A.$$

Furthermore,

$$e^{tJ_{2n}} = (\cos t)I_{2n} + (\sin t)J_{2n}.$$

Fact 11.10.3. Let $A \in \mathbb{R}^{n \times n}$, and assume that A is skew symmetric. Then, $\{e^{\theta A}: \ \theta \in \mathbb{R}\} \subseteq \mathrm{SO}(n)$ is a group. If, in addition, $n = 2$, then

$$\{e^{\theta J_2}: \ \theta \in \mathbb{R}\} = \mathrm{SO}(2).$$

(Remark: Note that $e^{\theta J_2} = \left[\begin{smallmatrix} \cos\theta & \sin\theta \\ -\sin\theta & \cos\theta \end{smallmatrix} \right]$. See Fact 3.7.13.)

Fact 11.10.4. Let $A \in \mathbb{R}^{n \times n}$, where

$$A \triangleq \begin{bmatrix} 0 & 1 & 0 & 0 & \cdots & 0 \\ 0 & 0 & 2 & 0 & \cdots & 0 \\ 0 & 0 & 0 & 3 & \cdots & 0 \\ \vdots & \vdots & \vdots & \ddots & \ddots & \vdots \\ 0 & 0 & 0 & 0 & \ddots & n-1 \\ 0 & 0 & 0 & 0 & \cdots & 0 \end{bmatrix}.$$

Then,

$$e^A = \begin{bmatrix} \binom{0}{0} & \binom{1}{0} & \binom{2}{0} & \binom{3}{0} & \cdots & \binom{n-1}{0} \\ 0 & \binom{1}{1} & \binom{2}{1} & \binom{3}{1} & \cdots & \binom{n-1}{1} \\ 0 & 0 & \binom{2}{2} & \binom{3}{2} & \cdots & \binom{n-1}{2} \\ \vdots & \vdots & \vdots & \ddots & \ddots & \vdots \\ 0 & 0 & 0 & 0 & \ddots & \binom{n-1}{n-2} \\ 0 & 0 & 0 & 0 & \cdots & \binom{n-1}{n-1} \end{bmatrix}.$$

Furthermore, if $k \geq n$, then

$$\sum_{i=1}^{k} i^{n-1} = \begin{bmatrix} 1^{n-1} & 2^{n-1} & \cdots & n^{n-1} \end{bmatrix} e^{-A} \begin{bmatrix} \binom{k}{1} \\ \vdots \\ \binom{k}{n} \end{bmatrix}.$$

(Proof: See [42].) (Remark: For related results, see [3], where A is called the *creation matrix*. See Fact 5.13.3.)

Fact 11.10.5. Let $A \in \mathbb{F}^{3\times3}$. If $\mathrm{spec}(A) = \{\lambda\}$, then

$$e^{tA} = e^{\lambda t}\left[I + t(A - \lambda I) + \tfrac{1}{2}t^2(A - \lambda I)^2\right].$$

If $\mathrm{mspec}(A) = \{\lambda, \lambda, \mu\}_{\mathrm{m}}$, where $\mu \neq \lambda$, then

$$e^{tA} = e^{\lambda t}[I + t(A - \lambda I)] + \left[\frac{e^{\mu t} - e^{\lambda t}}{(\mu - \lambda)^2} - \frac{te^{\lambda t}}{\mu - \lambda}\right](A - \lambda I)^2.$$

If $\mathrm{spec}(A) = \{\lambda, \mu, \nu\}$, then

$$e^{tA} = \frac{e^{\lambda t}}{(\lambda - \mu)(\lambda - \nu)}(A - \mu I)(A - \nu I) + \frac{e^{\mu t}}{(\mu - \lambda)(\mu - \nu)}(A - \lambda I)(A - \nu I)$$

$$+ \frac{e^{\nu t}}{(\nu - \lambda)(\nu - \mu)}(A - \lambda I)(A - \mu I).$$

(Proof: See [39].)

Fact 11.10.6. Let $x \in \mathbb{R}^3$. Then,

$$e^{K(x)} = I + \frac{\sin\theta}{\theta}K(x) + \frac{1 - \cos\theta}{\theta^2}K^2(x)$$

$$= I + \frac{\sin\theta}{\theta}K(x) + \tfrac{1}{2}\left[\frac{\sin(\theta/2)}{\theta/2}\right]^2 K^2(x)$$

$$= (\cos\theta)I + \frac{\sin\theta}{\theta}K(x) + \frac{1 - \cos\theta}{\theta^2}xx^{\mathrm{T}},$$

where $\theta \triangleq \|x\|_2$. (Remark: The cross-product matrix $K(x)$ is defined in (2.1.41). For $z \in \mathbb{R}^3$, $e^A z$ is the rotation of z about the vector x through the angle θ. See [107]. See Fact 11.10.8.) (Proof: The Cayley-Hamilton theorem or Fact 3.5.25 implies that $K^3(x) + \theta^2 K(x) = 0$. Then, every term $K^k(x)$ in the expansion of $e^{K(x)}$ can be expressed in terms of $K(x)$ or $K^2(x)$. Finally, Fact 3.5.25 implies that $\theta^2 I + K^2(x) = xx^{\mathrm{T}}$.)

Fact 11.10.7. Let $A \in \mathbb{F}^{3\times3}$, assume that A is unitary, and assume there exists $\theta \in \mathbb{R}$ such that $\mathrm{tr}\, A = 1 + 2\cos\theta$ and $|\theta| < \pi$. Then,

$$e^{[\theta/(2\sin\theta)](A - A^{\mathrm{T}})} = A.$$

(Proof: See [390, p. 364].)

Fact 11.10.8. Let $x, y \in \mathbb{R}^n$ satisfy $x^{\mathrm{T}}y = 0$, let $\theta \in [0, 2\pi]$, and define $A \in \mathbb{F}^{n\times n}$ by

$$A \triangleq I + (\sin\theta)\left(xy^{\mathrm{T}} - yx^{\mathrm{T}}\right) - (1 - \cos\theta)\left(xx^{\mathrm{T}} + yy^{\mathrm{T}}\right).$$

Then, A is orthogonal and $\det A = 1$. Now, let $n = 3$ and $z \triangleq y \times x$. Then,

$$A = (\cos\theta)I + (\sin\theta)K(z) + (1 - \cos\theta)zz^{\mathrm{T}}.$$

If, in addition, $\theta \neq \pi$, then

$$A = (I - B)(I + B)^{-1},$$

where

$$B \triangleq -\tan(\theta/2)K(z).$$

(Remark: See Fact 11.10.6.) (Problem: Represent A as a matrix exponential.)

Fact 11.10.9. Let $x, y \in \mathbb{R}^3$, and assume that x and y are nonzero. Then, there exists a skew-symmetric matrix $A \in \mathbb{R}^{3\times3}$ such that $y = e^A x$ if and only if $x^{\mathrm{T}}x = y^{\mathrm{T}}y$. If $x \neq \pm y$, then one such matrix is $A = \phi K(z)$, where

$$z \triangleq \|x \times y\|_2^{-1} x \times y,$$

and

$$\phi \triangleq \cos^{-1}(x^{\mathrm{T}}y).$$

If $x = -y$, then one such matrix is $A = \pi K(z)$, where $z \triangleq \nu \times y$ and $\nu \in \{y\}^{\perp}$ satisfies $\nu^{\mathrm{T}}\nu = 1$. (Remark: Since $\det e^A = e^{\operatorname{tr} A}$, it follows that vectors in $\mathbb{R}^3$ having the same Euclidean length are always related by a *proper rotation*. See Fact 3.7.16 and Fact 3.10.4.) (Problem: Extend this result to $\mathbb{R}^n$. See [71, 630].) (Remark: See Fact 3.7.25.) (Remark: Parameterizations of SO(3) are considered in [644, 678].)

Fact 11.10.10. Let $A \in \mathrm{SO}(3)$, let $x, y \in \mathbb{R}^3$, and assume that $x^{\mathrm{T}}x = y^{\mathrm{T}}y$. Then, $Ax = y$ if and only if, for all $t \in \mathbb{R}$,

$$Ae^{tK(x)}A^{-1} = e^{tK(y)}.$$

(Proof: See [469].)

Fact 11.10.11. Let $x, y, z \in \mathbb{R}^3$. Then, the following statements are equivalent:

    i) For every $A \in \mathrm{SO}(3)$, there exist $\alpha, \beta, \gamma \in \mathbb{R}$ such that

$$A = e^{\alpha K(x)}e^{\beta K(y)}e^{\gamma K(z)}.$$

    ii) $y^{\mathrm{T}}x = 0$ and $y^{\mathrm{T}}z = 0$.

(Proof: See [469].) (Remark: This result is due to Davenport.) (Problem: Given $A \in \mathrm{SO}(3)$, determine α, β, γ.)

Fact 11.10.12. Let $A \in \mathbb{R}^{4\times4}$, and assume that A is skew symmetric with $\operatorname{mspec}(A) = \{\jmath\omega, -\jmath\omega, \jmath\mu, -\jmath\mu\}_{\mathrm{m}}$. If $\omega \neq \mu$, then

$$e^A = a_3 A^3 + a_2 A^2 + a_1 A + a_0 I,$$

where

$$a_3 = \left(\omega^2 - \mu^2\right)^{-1}\left(\tfrac{1}{\mu}\sin\mu - \tfrac{1}{\omega}\sin\omega\right),$$

$$a_2 = \left(\omega^2 - \mu^2\right)^{-1}(\cos\mu - \cos\omega),$$

$$a_1 = \left(\omega^2 - \mu^2\right)^{-1}\left(\tfrac{\omega^2}{\mu}\sin\mu - \tfrac{\mu^2}{\omega}\sin\omega\right),$$

$$a_0 = \left(\omega^2 - \mu^2\right)^{-1}\left(\omega^2\cos\mu - \mu^2\cos\omega\right).$$

If $\omega = \mu$, then

$$e^A = (\cos\omega)I + \frac{\sin\omega}{\omega}A.$$

(Proof: See [314, p. 18] and [585].) (Remark: There are typographical errors in [314, p. 18] and [585].)

Fact 11.10.13. Let $C \in \mathbb{R}^{n \times n}$, assume that C is nonsingular, and let $k \in \mathbb{P}$. Then, there exists a matrix $B \in \mathbb{R}^{n \times n}$ such that $C^{2k} = e^B$. (Proof: Use Proposition 11.4.5 with $A = C^2$, and note that every negative eigenvalue $-\alpha < 0$ of C^2 arises as the square of complex conjugate eigenvalues $\pm\jmath\sqrt{\alpha}$ of C.)

11.11 Facts on the Matrix Exponential for One Matrix

Fact 11.11.1. Let $A \in \mathbb{F}^{n \times n}$, and assume that A is (lower triangular, upper triangular). Then, so is e^A. If, in addition, A is Toeplitz, then so is e^A. (Remark: See Fact 3.15.8.)

Fact 11.11.2. Let $A \in \mathbb{F}^{n \times n}$. Then,

$$\mathrm{sprad}\left(e^A\right) = e^{\mathrm{spabs}(A)}.$$

Fact 11.11.3. Let $A, X_0 \in \mathbb{R}^{n \times n}$. Then, the matrix differential equation

$$\dot{X}(t) = AX(t),$$
$$X(0) = X_0$$

where $t \geq 0$, has the unique solution

$$X(t) = e^{tA}X_0.$$

Fact 11.11.4. Let $A\colon [0,T] \mapsto \mathbb{R}^{n \times n}$, assume that A is continuous, and let $X_0 \in \mathbb{R}^{n \times n}$. Then, the matrix differential equation

$$\dot{X}(t) = A(t)X(t),$$
$$X(0) = X_0,$$

has a unique solution $X\colon [0,T] \mapsto \mathbb{R}^{n \times n}$. Furthermore, for all $t \in [0,T]$,

$$\det X(t) = e^{\int_0^t \mathrm{tr}\, A(\tau)\, \mathrm{d}\tau}\det X_0,$$

and thus, if X_0 is nonsingular, then $X(t)$ is nonsingular for all $t \in [0, T]$. If, in addition, for all $t_1, t_2 \in [0, T]$,

$$A(t_2) \int_{t_1}^{t_2} A(\tau) \, d\tau = \int_{t_1}^{t_2} A(\tau) \, d\tau A(t_2),$$

then, for all $t \in [0, T]$,

$$X(t) = e^{\int_0^t A(\tau) \, d\tau} X_0.$$

(Proof: See [624, pp. 64–66] and [369, pp. 507, 508].) (Remark: See Fact 11.11.4.) (Remark: The first result is *Jacobi's identity*.) (Remark: If the commutativity assumption does not hold, then the solution is given by the *Peano-Baker series*. See [624, Chapter 3]. Alternative expressions for $X(t)$ are given by the Magnus, Fer, Baker-Campbell-Hausdorff-Dynkin, Wei-Norman, Goldberg, and Zassenhaus expansions. See [130, 238, 389, 390, 504, 569, 676, 696, 766, 767, 770].)

Fact 11.11.5. Let $A\colon [0, T] \mapsto \mathbb{R}^{n \times n}$, assume that A is continuous, let $B\colon [0, T] \mapsto \mathbb{R}^{n \times m}$, assume that B is continuous, let $X\colon [0, T] \mapsto \mathbb{R}^{n \times n}$ satisfy the matrix differential equation

$$\dot{X}(t) = A(t)X(t),$$
$$X(0) = I,$$

define

$$\Phi(t, \tau) \triangleq X(t)X^{-1}(\tau),$$

let $u\colon [0, T] \mapsto \mathbb{R}^m$, and assume that u is continuous. Then, the vector differential equation

$$\dot{x}(t) = A(t)x(t) + B(t)u(t),$$
$$x(0) = x_0$$

has the unique solution

$$x(t) = X(t)x_0 + \int_0^t \Phi(t, \tau)B(\tau)u(\tau) \, d\tau.$$

(Remark: $\Phi(t, \tau)$ is the *state transition matrix*.)

Fact 11.11.6. Let $A \in \mathbb{R}^{n \times n}$, let $\lambda \in \mathrm{spec}(A)$, and let $v \in \mathbb{C}^n$ be an eigenvector of A associated with λ. Then, for all $t \geq 0$,

$$x(t) \triangleq \mathrm{Re}\left(e^{\lambda t} v\right)$$

satisfies $\dot{x}(t) = Ax(t)$.

Fact 11.11.7. Let S: $[t_0, t_1] \to \mathbb{R}^{n \times n}$ be differentiable. Then, for all $t \in [t_0, t_1]$,

$$\frac{\mathrm{d}}{\mathrm{dt}} S^2(t) = \dot{S}(t)S(t) + S(t)\dot{S}(t).$$

Let S_1: $[t_0, t_1] \to \mathbb{R}^{n \times m}$ and S_2: $[t_0, t_1] \to \mathbb{R}^{m \times l}$ be differentiable. Then, for all $t \in [t_0, t_1]$,

$$\frac{\mathrm{d}}{\mathrm{dt}} S_1(t)S_2(t) = \dot{S}_1(t)S_2(t) + S_1(t)\dot{S}_2(t).$$

Fact 11.11.8. Let $A \in \mathbb{F}^{n \times n}$, and let $A_1 = \frac{1}{2}(A + A^*)$ and $A_2 = \frac{1}{2}(A - A^*)$. Then, $A_1A_2 = A_2A_1$ if and only if A is normal. In this case, $e^{A_1}e^{A_2}$ is the polar decomposition of e^A. (Remark: See Fact 3.5.23.) (Problem: Obtain the polar decomposition of e^A when A is not normal.)

Fact 11.11.9. Let $A \in \mathbb{F}^{n \times m}$, and assume that $\operatorname{rank} A = m$. Then,

$$A^+ = \int_0^\infty e^{-tA^*A}A^* \, \mathrm{d}t.$$

Fact 11.11.10. Let $A \in \mathbb{F}^{n \times n}$, and assume that A is nonsingular. Then,

$$A^{-1} = \int_0^\infty e^{-tA^*A} \, \mathrm{d}t A^*.$$

Fact 11.11.11. Let $A \in \mathbb{F}^{n \times n}$, and let $k \triangleq \operatorname{ind} A$. Then,

$$A^{\mathrm{D}} = \int_0^\infty e^{-tA^kA^{(2k+1)*}A^{k+1}} \, \mathrm{d}t A^kA^{(2k+1)*}A^k.$$

(Proof: See [300].)

Fact 11.11.12. Let $A \in \mathbb{F}^{n \times n}$, and assume that $\operatorname{ind} A = 1$. Then,

$$A^\# = \int_0^\infty e^{-tAA^{3*}A^2} \, \mathrm{d}t AA^{3*}A.$$

(Proof: See Fact 11.11.11.)

Fact 11.11.13. Let $A \in \mathbb{F}^{n \times n}$, and let $k \triangleq \operatorname{ind} A$. Then,

$$\int_0^t e^{\tau A} \, \mathrm{d}\tau = A^{\mathrm{D}}(e^{tA} - I) + (I - AA^{\mathrm{D}})\left(tI + \tfrac{1}{2!}t^2A + \cdots + \tfrac{1}{k!}t^kA^{k-1}\right).$$

If, in particular, A is group invertible, then

$$\int\limits_0^t e^{\tau A}\, \mathrm{d}\tau = A^{\#}(e^{tA} - I) + \left(I - AA^{\#}\right)t.$$

Fact 11.11.14. Let $A \in \mathbb{F}^{n \times n}$, let $\mathrm{mspec}(A) = \{\lambda_1, \ldots, \lambda_r, 0, \ldots, 0\}_{\mathrm{m}}$, where $\lambda_1, \ldots, \lambda_r$ are nonzero, and let $t > 0$. Then,

$$\det \int\limits_0^t e^{\tau A}\, \mathrm{d}\tau = t^{n-r} \prod_{i=1}^r \lambda_i^{-1}\left(e^{\lambda_i t} - 1\right).$$

Hence, $\det \int_0^t e^{\tau A}\, \mathrm{d}\tau \neq 0$ if and only if, for all $k \in \mathbb{P}$, $2k\pi \jmath / t \notin \mathrm{spec}(A)$. Finally, $\det\left(e^{tA} - I\right) \neq 0$ if and only if $\det A \neq 0$ and $\det \int_0^t e^{\tau A}\, \mathrm{d}\tau \neq 0$.

Fact 11.11.15. Let $A \in \mathbb{F}^{n \times n}$, and assume that e^A is orthogonal. Then, either A is skew symmetric or two eigenvalues of A differ by a nonzero integer multiple of $2\pi \jmath$. (Remark: See [799].)

Fact 11.11.16. Let $A \in \mathbb{F}^{n \times n}$, and assume that $e^{A^*}e^A = e^{A^* + A}$. Then,

$$e^{A^*}e^A = e^A e^{A^*}.$$

(Proof: See [650].)

Fact 11.11.17. Let $A \in \mathbb{F}^{n \times n}$. Then, the following statements hold:

  i) The series (11.4.5) is finite, and $\log A$ exists.

 ii) If A is unipotent, then $\log A$ is nilpotent and $e^{\log A} = A$.

iii) If A is nilpotent, then e^A is unipotent and $\log e^A = A$.

(Proof: See [324, p. 60].)

11.12 Facts on the Matrix Exponential for Two or More Matrices

Fact 11.12.1. Let $A \in \mathbb{F}^{n \times n}$, $B \in \mathbb{F}^{n \times m}$, and $C \in \mathbb{F}^{m \times m}$. Then,

$$e^{t\left[\begin{smallmatrix} A & B \\ 0 & C \end{smallmatrix}\right]} = \begin{bmatrix} e^{tA} & \int_0^t e^{(t-\tau)A} B e^{\tau C}\, \mathrm{d}\tau \\ 0 & e^{tC} \end{bmatrix}.$$

Furthermore,

$$\int\limits_0^t e^{\tau A}\, \mathrm{d}\tau = \begin{bmatrix} I & 0 \end{bmatrix} e^{t\left[\begin{smallmatrix} A & I \\ 0 & 0 \end{smallmatrix}\right]} \begin{bmatrix} 0 \\ I \end{bmatrix}.$$

(Remark: The result can be extended to block-$k \times k$ matrices. See [738].

For an application to sampled-data control, see [567].)

Fact 11.12.2. Let $A, B \in \mathbb{F}^{n \times n}$, assume that $e^A e^B = e^B e^A$, and assume that either A and B are Hermitian or every entry of A and B is an algebraic number (roots of polynomials with rational coefficients). Then, $AB = BA$. (Proof. See [328, pp. 88, 89, 270–272] and [650, 771].) (Remark: The matrices $A \triangleq \begin{bmatrix} 0 & 1 \\ 0 & 2\pi j \end{bmatrix}$ and $B \triangleq \begin{bmatrix} 2\pi j & 0 \\ 0 & -2\pi j \end{bmatrix}$ do not commute but satisfy $e^A = e^B = e^{A+B} = I$.)

Fact 11.12.3. Let $A, B \in \mathbb{R}^{n \times n}$. Then,

$$\frac{d}{dt} e^{A+tB} = \int_0^1 e^{\tau(A+tB)} B e^{(1-\tau)(A+tB)} \, d\tau.$$

Hence,

$$\mathrm{Dexp}\left(e^{tA}; B\right) = \frac{d}{dt} e^{A+tB} \bigg|_{t=0} = \int_0^1 e^{\tau A} B e^{(1-\tau)A} \, d\tau.$$

Furthermore,

$$\frac{d}{dt} \mathrm{tr}\, e^{A+tB} = \mathrm{tr}\left(e^{A+tB} B\right).$$

Hence,

$$\frac{d}{dt} \mathrm{tr}\, e^{A+tB} \bigg|_{t=0} = \mathrm{tr}\left(e^A B\right).$$

(Proof: See [92, p. 175] and [463, 520, 552].)

Fact 11.12.4. Let $A, B \in \mathbb{R}^{n \times n}$. Then,

$$\frac{d}{dt} e^{A+tB} \bigg|_{t=0} = \sum_{k=0}^\infty \frac{1}{(k+1)!} \mathrm{ad}_A^k(B) e^A.$$

(Proof: See [56, p. 49].) (Remark: See Fact 2.16.5.)

Fact 11.12.5. Let $A, B \in \mathbb{F}^{n \times n}$, and assume that $e^A = e^B$. Then, the following statements hold:

i) If $|\lambda| < \pi$ for all $\lambda \in \mathrm{spec}(A) \cup \mathrm{spec}(B)$, then $A = B$.

ii) If $\lambda - \mu \neq 2k\pi j$ for all $\lambda \in \mathrm{spec}(A)$, $\mu \in \mathrm{spec}(B)$, and $k \in \mathbb{Z}$, then $[A, B] = 0$.

iii) If A is normal and $\sigma_{\max}(A) < \pi$, then $[A, B] = 0$.

iv) If A is normal and $\sigma_{\max}(A) = \pi$, then $[A^2, B] = 0$.

(Proof: See [633, 650] and [742, p. 111].) (Remark: If $[A, B] = 0$, then $[A^2, B] = 0$.)

Fact 11.12.6. Let $A, B \in \mathbb{F}^{n \times n}$, and assume that A and B are skew Hermitian. Then, $e^{tA}e^{tB}$ is unitary, and there exists a skew-Hermitian matrix $C(t)$ such that $e^{tA}e^{tB} = e^{C(t)}$. (Problem: Does (11.4.6) converge in this case? See [129, 245].)

Fact 11.12.7. Let $A, B \in \mathbb{F}^{n \times n}$, and assume that A and B are Hermitian. Then,

$$\lim_{p \to 0} \left(e^{(p/2)A} e^{pB} e^{(p/2)A} \right)^{1/p} = e^{A+B}.$$

(Proof: See [31].) (Remark: This result is the *Lie-Trotter formula*. For extensions, see [278].)

Fact 11.12.8. Let $A, B \in \mathbb{F}^{n \times n}$, and assume that A and B are Hermitian. Then,

$$\lim_{p \to \infty} \left[\tfrac{1}{2} \left(e^{pA} + e^{pB} \right) \right]^{1/p} = e^{\frac{1}{2}(A+B)}.$$

(Proof: See [108].)

Fact 11.12.9. Let $A, B \in \mathbb{F}^{n \times n}$. Then,

$$\lim_{k \to \infty} \left[e^{(1/k)A} e^{(1/k)B} e^{-(1/k)A} e^{-(1/k)B} \right]^{k^2} = e^{[A,B]}.$$

Fact 11.12.10. Let $A \in \mathbb{F}^{n \times m}$, $X \in \mathbb{F}^{m \times l}$, and $B \in \mathbb{F}^{l \times n}$. Then,

$$\frac{\mathrm{d}}{\mathrm{d}X} \operatorname{tr} e^{AXB} = Be^{AXB}A.$$

Fact 11.12.11. Let $A, B \in \mathbb{F}^{n \times n}$. Then,

$$\frac{\mathrm{d}}{\mathrm{d}t} e^{tA} e^{tB} e^{-tA} e^{-tB} \bigg|_{t=0} = 0$$

and

$$\frac{\mathrm{d}}{\mathrm{d}t} e^{\sqrt{t}A} e^{\sqrt{t}B} e^{-\sqrt{t}A} e^{-\sqrt{t}B} \bigg|_{t=0} = AB - BA.$$

Fact 11.12.12. Let $A, B, C \in \mathbb{F}^{n \times n}$, assume there exists $\beta \in \mathbb{F}$ such that $[A, B] = \beta B + C$, and assume that $[A, C] = [B, C] = 0$. Then,

$$e^{A+B} = e^A e^{\phi(\beta)B} e^{\psi(\beta)C},$$

where

$$\phi(\beta) \triangleq \begin{cases} \frac{1}{\beta}\left(1 - e^{-\beta}\right), & \beta \neq 0, \\ 1, & \beta = 0, \end{cases}$$

and

$$\psi(\beta) \triangleq \begin{cases} \frac{1}{\beta^2}\left(1 - \beta - e^{-\beta}\right), & \beta \neq 0, \\ -\frac{1}{2}, & \beta = 0. \end{cases}$$

(Proof: See [291, 688].)

Fact 11.12.13. Let $A, B \in \mathbb{F}^{n \times n}$, and assume there exist $\alpha, \beta \in \mathbb{F}$ such that $[A, B] = \alpha A + \beta B$. Then,

$$e^{t(A+B)} = e^{\phi(t)A}e^{\psi(t)B},$$

where

$$\phi(t) \triangleq \begin{cases} t, & \alpha = \beta = 0, \\ \alpha^{-1}\log(1 + \alpha t), & \alpha = \beta \neq 0, \ 1 + \alpha t > 0, \\ \int_0^t \frac{\alpha - \beta}{\alpha e^{(\alpha - \beta)\tau} - \beta} \, d\tau, & \alpha \neq \beta, \end{cases}$$

and

$$\psi(t) \triangleq \int_0^t e^{-\beta\phi(\tau)} \, d\tau.$$

(Proof: See [689].)

Fact 11.12.14. Let $A, B \in \mathbb{F}^{n \times n}$, and assume there exists nonzero $\beta \in \mathbb{F}$ such that $[A, B] = \alpha B$. Then, for all $t > 0$,

$$e^{t(A+B)} = e^{tA}e^{[(1-e^{-\alpha t})/\alpha]B}.$$

(Proof: Apply Fact 11.12.12 with $[tA, tB] = \alpha t(tB)$ and $\beta = \alpha t$.)

Fact 11.12.15. Let $A, B \in \mathbb{F}^{n \times n}$, and assume that $[[A, B], A] = 0$ and $[[A, B], B] = 0$. Then, for all $t \in \mathbb{R}$,

$$e^{tA}e^{tB} = e^{tA+tB+(t^2/2)[A,B]}.$$

In particular,

$$e^A e^B = e^{A+B+\frac{1}{2}[A,B]} = e^{A+B}e^{\frac{1}{2}[A,B]}$$

and

$$e^B e^{2A} e^B = e^{2A+2B}.$$

(Proof: See [324, pp. 64–66] and [778].)

Fact 11.12.16. Let $A, B \in \mathbb{F}^{n \times n}$, and assume that $[A, B] = B^2$. Then,

$$e^{A+B} = e^A(I + B).$$

Fact 11.12.17. Let $A, B \in \mathbb{F}^{n \times n}$. Then, for all $t \in [0, \infty)$,

$$e^{t(A+B)} = e^{tA}e^{tB} + \sum_{k=2}^{\infty} C_k t^k,$$

where, for all $k \in \mathbb{N}$,

$$C_{k+1} \triangleq \tfrac{1}{k+1}([A+B]C_k + [B, D_k]), \quad C_0 \triangleq 0,$$
$$D_{k+1} \triangleq \tfrac{1}{k+1}(AD_k + D_k B), \quad D_0 \triangleq I.$$

(Proof: See [610].)

Fact 11.12.18. Let $A \in \mathbb{R}^{2n \times 2n}$, and assume that A is symplectic and discrete-time Lyapunov stable. Then, there exists a Hamiltonian matrix $B \in \mathbb{R}^{2n \times 2n}$ such that $A = e^B$. Consequently, the map

$$\exp \colon \mathrm{sp}(n) \mapsto \{A \in \mathrm{Sp}(n) \colon A \text{ is discrete-time Lyapunov stable}\}$$

is onto. (Proof: Use Proposition 11.4.5.) (Remark: See xii) of Proposition 11.4.7.)

Fact 11.12.19. Let $A, B \in \mathbb{F}^{n \times n}$, assume that A is positive definite, and assume that B is positive semidefinite. Then,

$$A + B \le A^{1/2}e^{A^{-1/2}BA^{-1/2}}A^{1/2}.$$

Hence,

$$\frac{\det(A+B)}{\det A} \le e^{\mathrm{tr}\, A^{-1}B}.$$

Furthermore, for each inequality, equality holds if and only if $B = 0$. (Proof: For positive semidefinite A it follows that $e^A \le I + A$.)

Fact 11.12.20. Let $A, B \in \mathbb{F}^{n \times n}$, and assume that A and B are Hermitian. Then,

$$I \circ (A+B) \le \log(e^A \circ e^B).$$

(Proof: See [27, 806].) (Remark: See Fact 8.16.35.)

Fact 11.12.21. Let $A, B \in \mathbb{F}^{n \times n}$, and assume that A and B are Hermitian. Then,

$$\left(\mathrm{tr}\, e^A\right)e^{\mathrm{tr}(e^A B)/\mathrm{tr}\, e^A} \le \mathrm{tr}\, e^{A+B}.$$

(Proof: See [85].) (Remark: This inequality is equivalent to the thermodynamic inequality. See Fact 11.12.22.)

Fact 11.12.22. Let $A, B \in \mathbb{F}^{n \times n}$, and assume that A is positive definite, $\mathrm{tr}\, A = 1$, and B is Hermitian. Then,

$$\mathrm{tr}\, AB \le \mathrm{tr}(A \log A) + \log \mathrm{tr}\, e^B.$$

Furthermore, equality holds if and only if

$$A = (\operatorname{tr} e^B)^{-1} e^B.$$

(Proof: See [85].) (Remark: This result is the *thermodynamic inequality*. Equivalent forms are given by Fact 8.10.24 and Fact 11.12.21.)

Fact 11.12.23. Let $A, B \in \mathbb{F}^{n \times n}$, and assume that A and B are skew Hermitian. Then, there exist unitary matrices $S_1, S_2 \in \mathbb{F}^{n \times n}$ such that

$$e^A e^B = e^{S_1 A S_1^{-1} + S_2 B S_2^{-1}}.$$

(Proof: See [652, 695].)

Fact 11.12.24. Let $A, B \in \mathbb{F}^{n \times n}$, and assume that A and B are Hermitian. Then, there exist unitary matrices $S_1, S_2 \in \mathbb{F}^{n \times n}$ such that

$$e^{\frac{1}{2}A} e^B e^{\frac{1}{2}A} = e^{S_1 A S_1^{-1} + S_2 B S_2^{-1}}.$$

(Proof: See [651, 652, 695].) (Problem: Determine the relationship between this result and Fact 11.12.23.)

Fact 11.12.25. Let $A, B \in \mathbb{F}^{n \times n}$, assume that A and B are positive semidefinite, and assume that $B \leq A$. Furthermore, let $p, q, r, t \in \mathbb{R}$, and assume that $r \geq t \geq 0$, $p \geq 0$, $p + q \geq 0$, and $p + q + r > 0$. Then,

$$\left[e^{(r/2)A} e^{qA + pB} e^{(r/2)A} \right]^{t/(p+q+r)} \leq e^{tA}.$$

(Proof: See [731].)

Fact 11.12.26. Let $B \in \mathbb{F}^{n \times n}$, assume that B is Hermitian, let $\alpha_1, \ldots, \alpha_k \in (0, \infty)$, define $r \triangleq \sum_{i=1}^{k} \alpha_i$, assume that $r \leq 1$, let $q \in \mathbb{R}$, and define $\phi \colon \mathbf{P}^n \times \cdots \times \mathbf{P}^n \to [0, \infty)$ by

$$\phi(A_1, \ldots, A_k) \triangleq - \left[\operatorname{tr} e^{B + \sum_{i=1}^{k} \alpha_i \log A_i} \right]^q.$$

If $q \in (0, 1/r]$, then ϕ is convex. Furthermore, if $q < 0$, then $-\phi$ is convex. (Proof: See [483, 494].) (Remark: See [527].)

Fact 11.12.27. Define $\phi \colon \mathbf{P}^n \times \mathbf{P}^n \to [0, \infty)$ by

$$\phi(A, B) \triangleq -e^{[1/(2n)]\operatorname{tr}(\log A + \log B)}.$$

Then, ϕ is convex. (Proof: See [32].)

Fact 11.12.28. Let $A, B, C \in \mathbb{F}^{n \times n}$, and assume that A, B, and C are positive definite. Then,

$$\operatorname{tr} e^{\log A - \log B + \log C} \leq \operatorname{tr} \int_0^{\infty} A(B + xI)^{-1} C(B + xI)^{-1} \, dx.$$

(Proof: See [483, 494].) (Remark: $-\log B$ is correct.) (Remark: $\operatorname{tr} e^{A+B+C}$ $\leq |\operatorname{tr} e^A e^B e^C|$ is not generally true.)

Fact 11.12.29. Let $A \in \mathbb{F}^{n\times n}$ and $B \in \mathbb{F}^{m\times m}$. Then,

$$\operatorname{tr} e^{A\oplus B} = \left(\operatorname{tr} e^A\right)\left(\operatorname{tr} e^B\right).$$

Fact 11.12.30. Let $A \in \mathbb{F}^{n\times n}$, $B \in \mathbb{F}^{m\times m}$, and $C \in \mathbb{F}^{l\times l}$. Then,

$$e^{A\oplus B\oplus C} = e^A \otimes e^B \otimes e^C.$$

Fact 11.12.31. Let $A \in \mathbb{F}^{n\times n}$, $B \in \mathbb{F}^{m\times m}$, $C \in \mathbb{F}^{k\times k}$, and $D \in \mathbb{F}^{l\times l}$. Then,

$$\operatorname{tr} e^{A\otimes I\otimes B\otimes I + I\otimes C\otimes I\otimes D} = \operatorname{tr} e^{A\otimes B}\operatorname{tr} e^{C\otimes D}.$$

(Proof: By Fact 7.4.24, a similarity transformation involving the Kronecker permutation matrix can be used to reorder the inner two terms. See [657].)

11.13 Facts on the Matrix Exponential and Eigenvalues, Singular Values, and Norms for One Matrix

Fact 11.13.1. Let $A \in \mathbb{F}^{n\times n}$, and define $f\colon [0,\infty) \mapsto (0,\infty)$ by $f(t) \triangleq \sigma_{\max}(e^{At})$. Then,

$$f'(0) = \tfrac{1}{2}\lambda_{\max}(A + A^*).$$

Hence, there exists $\varepsilon > 0$ such that $f(t) \triangleq \sigma_{\max}(e^{tA})$ is decreasing on $[0,\varepsilon)$ if and only if A is dissipative. (Proof: See [759].) (Remark: The derivative is one sided.)

Fact 11.13.2. Let $A \in \mathbb{F}^{n\times n}$. Then, for all $t \geq 0$,

$$\frac{\mathrm{d}}{\mathrm{d}t}\|e^{tA}\|_{\mathrm{F}}^2 = \operatorname{tr} e^{tA}(A + A^*)e^{tA^*}.$$

Hence, if A is dissipative, then $f(t) \triangleq \|e^{tA}\|_{\mathrm{F}}$ is decreasing on $[0,\infty)$. (Proof: See [759].)

Fact 11.13.3. Let $A \in \mathbb{F}^{n\times n}$. Then,

$$\left|\operatorname{tr} e^{2A}\right| \leq \operatorname{tr} e^A e^{A^*} \leq \operatorname{tr} e^{A+A^*} \leq \left[n\operatorname{tr} e^{2(A+A^*)}\right]^{1/2} \leq \tfrac{n}{2} + \tfrac{1}{2}\operatorname{tr} e^{2(A+A^*)}.$$

In addition, $\operatorname{tr} e^A e^{A^*} = \operatorname{tr} e^{A+A^*}$ if and only if A is normal. (Proof: See [101], [369, p. 515], and [650].) (Remark: $\operatorname{tr} e^A e^{A^*} \leq \operatorname{tr} e^{A+A^*}$ is *Bernstein's inequality*. See [28].) (Remark: See Fact 3.5.8.)

Fact 11.13.4. Let $A \in \mathbb{F}^{n \times n}$. Then, for all $k = 1, \ldots, n$,

$$\prod_{i=1}^{k} \sigma_i(e^A) \leq \prod_{i=1}^{k} \lambda_i \left[e^{\frac{1}{2}(A+A^*)} \right] = \prod_{i=1}^{k} e^{\lambda_i \left[\frac{1}{2}(A+A^*) \right]} \leq \prod_{i=1}^{k} e^{\sigma_i(A)}.$$

Furthermore, for all $k = 1, \ldots, n$,

$$\sum_{i=1}^{k} \sigma_i(e^A) \leq \sum_{i=1}^{k} \lambda_i \left[e^{\frac{1}{2}(A+A^*)} \right] = \sum_{i=1}^{k} e^{\lambda_i \left[\frac{1}{2}(A+A^*) \right]} \leq \sum_{i=1}^{k} e^{\sigma_i(A)}.$$

In particular,

$$\sigma_{\max}(e^A) \leq \lambda_{\max} \left[e^{\frac{1}{2}(A+A^*)} \right] = e^{\frac{1}{2}\lambda_{\max}(A+A^*)} \leq e^{\sigma_{\max}(A)}$$

or, equivalently,

$$\lambda_{\max}(e^A e^{A^*}) \leq \lambda_{\max}(e^{A+A^*}) = e^{\lambda_{\max}(A+A^*)} \leq e^{2\sigma_{\max}(A)}.$$

Furthermore,

$$\left| \det e^A \right| = \left| e^{\operatorname{tr} A} \right| \leq e^{|\operatorname{tr} A|} \leq e^{\operatorname{tr} \langle A \rangle}$$

and

$$\operatorname{tr} \langle e^A \rangle \leq \sum_{i=1}^{n} e^{\sigma_i(A)}.$$

(Proof: See [653], Fact 8.14.4, Fact 8.14.5, and Fact 8.17.6.)

Fact 11.13.5. Let $A \in \mathbb{F}^{n \times n}$, and let $\| \cdot \|$ be a unitarily invariant norm on $\mathbb{F}^{n \times n}$. Then,

$$\left\| e^A e^{A^*} \right\| \leq \left\| e^{A+A^*} \right\|.$$

In particular,

$$\lambda_{\max}(e^A e^{A^*}) \leq \lambda_{\max}(e^{A+A^*})$$

and

$$\operatorname{tr} e^A e^{A^*} \leq \operatorname{tr} e^{A+A^*}.$$

(Proof: See [189].)

11.14 Facts on the Matrix Exponential and Eigenvalues, Singular Values, and Norms for Two or More Matrices

Fact 11.14.1. Let $A, B \in \mathbb{F}^{n \times n}$. Then,

$$\left| \operatorname{tr} e^{A+B} \right| \leq \operatorname{tr} e^{\frac{1}{2}(A+B)} e^{\frac{1}{2}(A+B)^*} \leq \operatorname{tr} e^{\frac{1}{2}(A+A^*+B+B^*)} \leq \operatorname{tr} e^{\frac{1}{2}(A+A^*)} e^{\frac{1}{2}(B+B^*)}$$

$$\leq \left(\operatorname{tr} e^{A+A^*} \right)^{1/2} \left(\operatorname{tr} e^{B+B^*} \right)^{1/2} \leq \tfrac{1}{2}\operatorname{tr}\left(e^{A+A^*} + e^{B+B^*} \right)$$

and

$$\left.\begin{array}{c} \operatorname{tr} e^A e^B \\ \frac{1}{2}\operatorname{tr}\left(e^{2A}+e^{2B}\right) \end{array}\right\} \leq \tfrac{1}{2}\operatorname{tr}\left(e^A e^{A^*}+e^B e^{B^*}\right) \leq \tfrac{1}{2}\operatorname{tr}\left(e^{A+A^*}+e^{B+B^*}\right).$$

(Proof: See [101, 190, 579] and [369, p. 514].)

Fact 11.14.2. Let $A, B \in \mathbb{F}^{n\times n}$, assume that A and B are Hermitian, and let $\|\cdot\|$ be a unitarily invariant norm on $\mathbb{F}^{n\times n}$. Then,

$$\left\|e^{A+B}\right\| \leq \left\|e^{\frac{1}{2}A}e^B e^{\frac{1}{2}A}\right\| \leq \left\|e^A e^B\right\|.$$

If, in addition, $p > 0$, then

$$\left\|e^{A+B}\right\| \leq \left\|e^{(p/2)A}e^B e^{(p/2)A}\right\|^{1/p}$$

and

$$\left\|e^{A+B}\right\| = \lim_{p\downarrow 0}\left\|e^{(p/2)A}e^B e^{(p/2)A}\right\|^{1/p}.$$

Furthermore, for all $k = 1, \ldots, n$,

$$\prod_{i=1}^k \lambda_i\left(e^{A+B}\right) \leq \prod_{i=1}^k \lambda_i\left(e^A e^B\right) \leq \prod_{i=1}^k \sigma_i\left(e^A e^B\right)$$

with equality for $k = n$, that is,

$$\prod_{i=1}^n \lambda_i\left(e^{A+B}\right) = \prod_{i=1}^n \lambda_i\left(e^A e^B\right) = \prod_{i=1}^n \sigma_i\left(e^A e^B\right) = \det\left(e^A e^B\right).$$

Furthermore, for all $k = 1, \ldots, n$,

$$\sum_{i=1}^k \lambda_i\left(e^{A+B}\right) \leq \sum_{i=1}^k \lambda_i\left(e^A e^B\right) \leq \sum_{i=1}^k \sigma_i\left(e^A e^B\right).$$

In particular,

$$\lambda_{\max}\left(e^{A+B}\right) \leq \lambda_{\max}\left(e^A e^B\right) \leq \sigma_{\max}\left(e^A e^B\right),$$

$$\operatorname{tr} e^{A+B} \leq \operatorname{tr} e^A e^B \leq \operatorname{tr}\left\langle e^A e^B\right\rangle,$$

and, for all $p > 0$,

$$\operatorname{tr} e^{A+B} \leq \operatorname{tr}\left(e^{(p/2)A}e^B e^{(p/2)A}\right).$$

(Proof: See [31], [111, p. 261], Fact 5.10.14, and Fact 8.17.6.) (Remark: $\operatorname{tr} e^{A+B} \leq \operatorname{tr} e^A e^B$ is the *Golden-Thompson inequality*.) (Remark: $\|e^{A+B}\| \leq \|e^{\frac{1}{2}A}e^B e^{\frac{1}{2}A}\|$ is *Segal's inequality*. See [28].)

Fact 11.14.3. Let $A, B \in \mathbb{F}^{n\times n}$, assume that A and B are Hermitian, let $q, p > 0$, where $q \leq p$, and let $\|\cdot\|$ be a unitarily invariant norm on $\mathbb{F}^{n\times n}$.

Then,

$$\left\|\left(e^{(q/2)A}e^{qB}e^{(q/2)A}\right)^{1/q}\right\| \leq \left\|\left(e^{(p/2)A}e^{pB}e^{(p/2)A}\right)^{1/p}\right\|.$$

(Proof: See [31].)

Fact 11.14.4. Let $A, B \in \mathbb{F}^{n \times n}$, and assume that A and B are positive semidefinite. Then,

$$e^{\sigma_{\max}^{1/2}(AB)} - 1 \leq \sigma_{\max}^{1/2}\left[(e^A - I)(e^B - I)\right]$$

and

$$e^{\sigma_{\max}^{1/3}(BAB)} - 1 \leq \sigma_{\max}^{1/3}\left[(e^B - I)(e^A - I)(e^B - I)\right].$$

(Proof: See [730].) (Remark: See Fact 8.14.23.)

Fact 11.14.5. Let $A, B \in \mathbb{F}^{n \times n}$, and let $\|\cdot\|$ be a submultiplicative norm on $\mathbb{F}^{n \times n}$. Then, for all $t \geq 0$,

$$\left\|e^{tA} - e^{tB}\right\| \leq e^{\|A\|t}\left(e^{\|A-B\|t} - 1\right).$$

Fact 11.14.6. Let $A, B \in \mathbb{R}^{n \times n}$, and assume that A is normal. Then, for all $t \geq 0$,

$$\sigma_{\max}\left(e^{tA} - e^{tB}\right) \leq \sigma_{\max}\left(e^{tA}\right)\left[e^{\sigma_{\max}(A-B)t} - 1\right].$$

(Proof: See [771].)

Fact 11.14.7. Let $A, B \in \mathbb{C}^{n \times n}$, assume that A and B are idempotent, assume that $A \neq B$, and let $\|\cdot\|$ be a norm on $\mathbb{C}^{n \times n}$. Then,

$$\left\|e^{\jmath A} - e^{\jmath B}\right\| = |e^{\jmath} - 1|\|A - B\| < \|A - B\|.$$

(Proof: See [553].) (Remark: $|e^{\jmath} - 1| \approx 0.96$.)

Fact 11.14.8. Let $A, B \in \mathbb{C}^{n \times n}$, assume that A and B are Hermitian, let $X \in \mathbb{C}^{n \times n}$, and let $\|\cdot\|$ be a unitarily invariant norm on $\mathbb{C}^{n \times n}$. Then,

$$\left\|e^{\jmath A}X - Xe^{\jmath B}\right\| \leq \|AX - XB\|.$$

(Proof: See [553].) (Remark: This result is a matrix version of xv) of Fact 1.4.23.)

Fact 11.14.9. Let $A, B \in \mathbb{F}^{n \times n}$, assume that A and B are Hermitian, and let $\alpha, \beta \in \mathbb{R}$ satisfy either $\alpha I \leq A \leq \beta I$ or $\alpha I \leq B \leq \beta I$. Then, for all $t > 0$,

$$e^{tB} \leq S\left(t, e^{\beta - \alpha}\right)e^{tA},$$

where, for all $t > 0$ and $h > 0$, $S(t, h)$ is defined by

$$S(t,h) \triangleq \begin{cases} \frac{(h^t-1)h^{t/(h^t-1)}}{et \log h}, & h \neq 1, \\ 1, & h = 1. \end{cases}$$

(Proof: See [404].) (Remark: $S(t,h)$ is *Specht's ratio*.) (Remark: See Fact 8.8.30.)

Fact 11.14.10. Let $A \in \mathbb{F}^{n \times n}$, and, for all $i = 1, \ldots, n$, define f_i: $[0, \infty) \mapsto \mathbb{R}$ by $f_i(t) \triangleq \log \sigma_i(e^{tA})$. Then, A is normal if and only if, for all $i = 1, \ldots, n$, f_i is convex. (Proof: See [52] and [241].) (Remark: The statement in [52] that convexity holds on $\mathbb{R}$ is erroneous. A counterexample is $A \triangleq \begin{bmatrix} 1 & 0 \\ 0 & -1 \end{bmatrix}$ for which $\log \sigma_1(e^{tA}) = |t|$ and $\log \sigma_2(e^{tA}) = -|t|$.)

Fact 11.14.11. Let $A \in \mathbb{F}^{n \times n}$, and, for nonzero $x \in \mathbb{F}^n$, define f_x: $\mathbb{R} \mapsto \mathbb{R}$ by $f_x(t) \triangleq \log \sigma_{\max}(e^{tA}x)$. Then, A is normal if and only if, for all nonzero $x \in \mathbb{F}^n$, f_x is convex. (Proof: See [52].) (Remark: This result is due to Friedland.)

Fact 11.14.12. Let $A, B \in \mathbb{F}^{n \times n}$, assume that A and B are positive semidefinite, and let $\| \cdot \|$ be a unitarily invariant norm on $\mathbb{F}^{n \times n}$. Then,

$$\|e^{\langle A-B \rangle} - I\| \leq \|e^A - e^B\|$$

and

$$\|e^A + e^B\| \leq \|e^{A+B} + I\|.$$

(Proof: See [34] and [111, p. 294].) (Remark: See Fact 9.9.39.)

Fact 11.14.13. Let $A, X, B \in \mathbb{F}^{n \times n}$, assume that A and B are Hermitian, and let $\| \cdot \|$ be a unitarily invariant norm on $\mathbb{F}^{n \times n}$. Then,

$$\|AX - XB\| \leq \|e^{\frac{1}{2}A}Xe^{-\frac{1}{2}B} - e^{-\frac{1}{2}B}Xe^{\frac{1}{2}A}\|.$$

(Proof: See [120].) (Remark: See Fact 9.9.40.)

11.15 Facts on Stable Polynomials

Fact 11.15.1. Let $p \in \mathbb{R}[s]$, where $p(s) = s^n + a_{n-1}s^{n-1} + \cdots + a_0$. If p is asymptotically stable, then $a_0, \ldots, a_{n-1}$ are positive. Now, assume that $a_0, \ldots, a_{n-1}$ are positive. Then, the following statements hold:

i) If $n = 1$ or $n = 2$, then p is asymptotically stable.

ii) If $n = 3$, then p is asymptotically stable if and only if

$$a_0 < a_1 a_2.$$

iii) If $n = 4$, then p is asymptotically stable if and only if

$$a_1^2 + a_0 a_3^2 < a_1 a_2 a_3.$$

iv) If $n = 5$, then p is asymptotically stable if and only if

$$a_2 < a_3 a_4,$$
$$a_2^2 + a_1 a_4^2 < a_0 a_4 + a_2 a_3 a_4,$$
$$a_0^2 + a_1 a_2^2 + a_1^2 a_4^2 + a_0 a_3^2 a_4 < a_0 a_2 a_3 + 2 a_0 a_1 a_4 + a_1 a_2 a_3 a_4.$$

(Remark: These results are special cases of the *Routh criterion*, which provides stability criteria for polynomials of arbitrary degree n. See [170].)

Fact 11.15.2. Let $\varepsilon \in [0,1]$, let $n \in \{2,3,4\}$, let $p_\varepsilon \in \mathbb{R}[s]$, where $p_\varepsilon(s) = s^n + a_{n-1}s^{n-1} + \cdots + \varepsilon a_0$, and assume that p_1 is asymptotically stable. Then, for all $\varepsilon \in (0,1]$, p_ε is asymptotically stable. Furthermore, $p_0(s)/s$ is asymptotically stable. (Remark: The result does not hold for $n = 5$. A counterexample is $p(s) = s^5 + 2s^4 + 3s^3 + 5s^2 + 2s + 2.5\varepsilon$, which is asymptotically stable if and only if $\varepsilon \in (4/5, 1]$.)

Fact 11.15.3. Let $p \in \mathbb{R}[s]$ be monic, and define $q(s) \triangleq s^n p(1/s)$, where $n \triangleq \deg p$. Then, p is asymptotically stable if and only if q is asymptotically stable. (Remark: See Fact 4.8.1 and Fact 11.15.4.)

Fact 11.15.4. Let $p \in \mathbb{R}[s]$ be monic, and assume that p is semistable. Then, $q(s) \triangleq p(s)/s$ and $\hat{q}(s) \triangleq s^n p(1/s)$ are asymptotically stable. (Remark: See Fact 4.8.1 and Fact 11.15.3.)

Fact 11.15.5. Let $p, q \in \mathbb{R}[s]$, assume that p is even, assume that q is odd, and assume that every coefficient of $p + q$ is positive. Then, $p + q$ is asymptotically stable if and only if every root of p and every root of q is imaginary, and the roots of p and the roots of q are interlaced on the imaginary axis. (Proof: See [125, 170, 365].) (Remark: This result is the *Hermite-Biehler* or *interlacing theorem*.) (Example: $s^2 + 2s + 5 = (s^2 + 5) + 2s$.)

Fact 11.15.6. Let $p \in \mathbb{R}[s]$ be asymptotically stable, and let $p(s) = \beta_n s^n + \beta_{n-1}s^{n-1} + \cdots + \beta_1 s + \beta_0$, where $\beta_n > 0$. Then, for all $i = 1, \ldots, n-2$,

$$\beta_{i-1}\beta_{i+2} < \beta_i \beta_{i+1}.$$

(Remark: This result is a necessary condition for asymptotic stability, which can be used to show that a given polynomial with positive coefficients is unstable.) (Remark: This result is due to Xie. See [801]. For alternative conditions, see [125, p. 68].)

Fact 11.15.7. Let $n \in \mathbb{P}$ be even, let $m \triangleq n/2$, let $p \in \mathbb{R}[s]$, where $p(s) = \beta_n s^n + \beta_{n-1} s^{n-1} + \cdots + \beta_1 s + \beta_0$ and $\beta_n > 0$, and assume that p is asymptotically stable. Then, for all $i = 1, \ldots, m-1$,

$$\binom{m}{i} \beta_0^{(m-i)/m} \beta_n^{i/m} \leq \beta_{2i}.$$

(Remark: This result is a necessary condition for asymptotic stability, which can be used to show that a given polynomial with positive coefficients is unstable.) (Remark: This result is due to Borobia and Dormido. See [801, 802] for extensions to polynomials of odd degree.)

Fact 11.15.8. Let $p, q \in \mathbb{R}[s]$, where $p(s) = \alpha_n s^n + \alpha_{n-1} s^{n-1} + \cdots + \alpha_1 s + \alpha_0$ and $q(s) = \beta_m s^m + \beta_{m-1} s^{m-1} + \cdots + \beta_1 s + \beta_0$. If p and q are (Lyapunov, asymptotically) stable, then $r(s) \triangleq \alpha_l \beta_l s^l + \alpha_{l-1} \beta_{l-1} s^{l-1} + \cdots + \alpha_1 \beta_1 s + \alpha_0 \beta_0$, where $l \triangleq \min\{m, n\}$, is (Lyapunov, asymptotically) stable. (Proof: See [287].) (Remark: The polynomial r is the *Schur product* of p and q. See [47, 395].)

Fact 11.15.9. Let $A \in \mathbb{R}^{n \times n}$, and assume that A is diagonalizable over $\mathbb{R}$. Then, χ_A has all positive coefficients if and only if χ_A (equivalently, A) is asymptotically stable. (Proof: Sufficiency follows from Fact 11.15.1. For necessity, note that χ_A has only real roots and that $\chi_A(\lambda) > 0$ for all $\lambda \geq 0$. Hence, $\operatorname{roots}(\chi_A) \subset (-\infty, 0)$.)

Fact 11.15.10. Let $A \in \mathbb{R}^{n \times n}$. Then, $\chi_{A \oplus A}$ has all positive coefficients if and only if $\chi_{A \oplus A}$ (equivalently, A) is asymptotically stable. (Proof: If A is not asymptotically stable, then Fact 11.17.30 implies that $A \oplus A$ has a positive eigenvalue λ. Since $\chi_{A \oplus A}(\lambda) = 0$, it follows that $\chi_{A \oplus A}$ cannot have all positive coefficients. See [277, Theorem 5].)

11.16 Facts on Lie Groups

Fact 11.16.1. The groups $\mathrm{UT}(n), \mathrm{UT}_+(n), \mathrm{UT}_{\pm 1}(n), \mathrm{SUT}(n)$, and $\{I_n\}$ are Lie groups. Furthermore, $\mathrm{ut}(n)$ is the Lie algebra of $\mathrm{UT}(n)$, $\mathrm{sut}(n)$ is the Lie algebra of $\mathrm{SUT}(n)$, and $\{0_{n \times n}\}$ is the Lie algebra of $\{I_n\}$. (Remark: See Fact 3.13.2 and Fact 3.13.3.) (Problem: Determine the Lie algebras of $\mathrm{UT}_+(n)$ and $\mathrm{UT}_{\pm 1}(n)$.)

11.17 Facts on Stable Matrices

Fact 11.17.1. Let $A \in \mathbb{F}^{n \times n}$, and assume that A is semistable. Then, A is Lyapunov stable.

Fact 11.17.2. Let $A \in \mathbb{F}^{n \times n}$, and assume that A is Lyapunov stable. Then, A is group invertible.

Fact 11.17.3. Let $A \in \mathbb{F}^{n \times n}$, and assume that A is semistable. Then, A is group invertible.

Fact 11.17.4. Let $A \in \mathbb{F}^{n \times n}$, and assume that A is semistable. Then,

$$\lim_{t \to \infty} e^{tA} = I - AA^{\#},$$

and thus

$$\lim_{t \to \infty} \frac{1}{t} \int_0^t e^{\tau A} \, \mathrm{d}\tau = I - AA^{\#}.$$

(Remark: See Fact 11.17.1, Fact 11.17.2, and Fact 10.8.3.)

Fact 11.17.5. Let $A \in \mathbb{R}^{n \times n}$, and assume that A is Lyapunov stable. Then,

$$\lim_{t \to \infty} \frac{1}{t} \int_0^t e^{\tau A} \, \mathrm{d}\tau = I - AA^{\#}.$$

(Remark: See Fact 11.17.2.)

Fact 11.17.6. Let $A, B \in \mathbb{F}^{n \times n}$. Then, $\lim_{\alpha \to \infty} e^{A + \alpha B}$ exists if and only if B is semistable. In this case,

$$\lim_{\alpha \to \infty} e^{A + \alpha B} = e^{(I - BB^{\#})A} \left(I - BB^{\#} \right) = \left(I - BB^{\#} \right) e^{A(I - BB^{\#})}.$$

(Proof: See [159].)

Fact 11.17.7. Let $A \in \mathbb{R}^{n \times n}$. Then, e^{tA} is nonnegative for all $t \geq 0$ if and only if $A_{(i,j)} \geq 0$ for all $i, j = 1, \ldots, n$ such that $i \neq j$. In this case, A is asymptotically stable if and only if, for all $i = 1, \ldots, n$, the sign of the ith leading principal subdeterminant of A is $(-1)^i$. (Proof: See [106] and [286, p. 74].) (Remark: A is *essentially nonnegative*.)

Fact 11.17.8. Let $A \in \mathbb{F}^{n \times n}$, assume that A is asymptotically stable, let $\beta > \mathrm{spabs}(A)$, and let $\| \cdot \|$ be a submultiplicative norm on $\mathbb{F}^{n \times n}$. Then, there exists $\gamma > 0$ such that, for all $t \geq 0$,

$$\left\| e^{tA} \right\| \leq \gamma e^{\beta t}.$$

(Remark: See [292, pp. 201–206] and [406].)

Fact 11.17.9. Let $A \in \mathbb{F}^{n \times n}$, assume that A is asymptotically stable, let $\beta > \mathrm{spabs}(A)$, let $\gamma \geq 1$, and let $\| \cdot \|$ be a normalized, submultiplicative norm on $\mathbb{F}^{n \times n}$. Then, for all $t \geq 0$,

$$\left\|e^{tA}\right\| \leq \gamma e^{\beta t}$$

if and only if, for all $k \in \mathbb{P}$ and $\alpha > \beta$,

$$\left\|(\alpha I - A)^{-k}\right\| \leq \frac{\gamma}{(\alpha - \beta)^k}.$$

(Remark: This result is the *Hille-Yosida theorem.*)

Fact 11.17.10. Let $A \in \mathbb{R}^{n \times n}$, assume that A is asymptotically stable, let $\beta > \mathrm{spabs}(A)$, let $P \in \mathbb{R}^{n \times n}$ satisfy

$$A^{\mathrm{T}}P + PA \leq 2\beta P,$$

and let $\| \cdot \|$ be a normalized, submultiplicative norm on $\mathbb{R}^{n \times n}$. Then, for all $t \geq 0$,

$$\left\|e^{tA}\right\| \leq \sqrt{\|P\| \|P^{-1}\|} e^{\beta t}.$$

(Remark: See [356].)

Fact 11.17.11. Let $A \in \mathbb{F}^{n \times n}$, assume that A is asymptotically stable, let $R \in \mathbb{F}^{n \times n}$, assume that R is positive definite, and let $P \in \mathbb{F}^{n \times n}$ be the positive-definite solution of $A^*P + PA + R = 0$. Then,

$$\sigma_{\max}\left(e^{tA}\right) \leq \sqrt{\frac{\sigma_{\max}(P)}{\sigma_{\min}(P)}} e^{-t\lambda_{\min}(RP^{-1})/2}$$

and

$$\left\|e^{tA}\right\|_{\mathrm{F}} \leq \sqrt{\|P\|_{\mathrm{F}} \|P^{-1}\|_{\mathrm{F}}} e^{-t\lambda_{\min}(RP^{-1})/2}.$$

If, in addition, $A + A^*$ is negative definite, then

$$\left\|e^{tA}\right\|_{\mathrm{F}} \leq e^{-t\lambda_{\min}(-A-A^*)/2}.$$

(Proof: See [506].)

Fact 11.17.12. Let $A \in \mathbb{R}^{n \times n}$, assume that A is asymptotically stable, let $R \in \mathbb{R}^{n \times n}$, assume that R is positive definite, and let $P \in \mathbb{R}^{n \times n}$ be the positive-definite solution of $A^{\mathrm{T}}P + PA + R = 0$. Furthermore, define the vector norm $\|x\|' \triangleq \sqrt{x^{\mathrm{T}}Px}$ on $\mathbb{R}^n$, let $\| \cdot \|$ denote the induced norm on $\mathbb{R}^{n \times n}$, and let $\mu(\cdot)$ denote the corresponding logarithmic derivative. Then,

$$\mu(A) = -\lambda_{\min}\left(RP^{-1}\right)/2.$$

Consequently,

$$\left\|e^{tA}\right\| \leq e^{-t\lambda_{\min}(RP^{-1})/2}.$$

(Proof: See [382] and use *xiii*) of Fact 9.10.8.) (Remark: See Fact 9.10.8 for the definition and properties of the logarithmic derivative.)

Fact 11.17.13. Let $A \in \mathbb{F}^{n \times n}$. Then, A is similar to a skew-Hermitian matrix if and only if there exists a positive-definite matrix $P \in \mathbb{F}^{n \times n}$ such that $A^*P + PA = 0$.

Fact 11.17.14. Let $A \in \mathbb{R}^{n \times n}$. Then, A and A^2 are asymptotically stable if and only if, for all $\lambda \in \mathrm{spec}(A)$, there exist $r > 0$ and $\theta \in \left(\frac{\pi}{2}, \frac{3\pi}{4}\right) \cup \left(\frac{5\pi}{4}, \frac{3\pi}{2}\right)$ such that $\lambda = re^{\jmath\theta}$.

Fact 11.17.15. Let $A \in \mathbb{R}^{n \times n}$. Then, A is group invertible and $2k\pi\jmath \notin \mathrm{spec}(A)$ for all $k \in \mathbb{P}$ if and only if

$$AA^{\#} = \left(e^A - I\right)\left(e^A - I\right)^{\#}.$$

In particular, if A is semistable, then this identity holds. (Proof: Use ii) of Fact 11.18.12 and ix) of Proposition 11.7.2.)

Fact 11.17.16. Let $A \in \mathbb{F}^{n \times n}$. Then, A is asymptotically stable if and only if A^{-1} is asymptotically stable. Hence, $e^{tA} \to 0$ as $t \to \infty$ if and only if $e^{tA^{-1}} \to 0$ as $t \to \infty$.

Fact 11.17.17. Let $A, B \in \mathbb{R}^{n \times n}$, assume that A is asymptotically stable, and assume that $\sigma_{\max}(B \oplus B) < \sigma_{\min}(A \oplus A)$. Then, $A + B$ is asymptotically stable. (Proof: Since $A \oplus A$ is nonsingular, Fact 9.12.16 implies that $A \oplus A + \alpha(B \oplus B) = (A + \alpha B) \oplus (A + \alpha B)$ is nonsingular for all $0 \leq \alpha \leq 1$. Now, suppose that $A + B$ is not asymptotically stable. Then, there exists $\alpha_0 \in (0, 1]$ such that $A + \alpha_0 B$ has an imaginary eigenvalue, and thus $(A + \alpha_0 B) \oplus (A + \alpha_0 B) = A \oplus A + \alpha_0(B \oplus B)$ is singular, which is a contradiction.) (Remark: This result provides a suboptimal solution of a nearness problem. See [352, Section 7] and Fact 9.12.16.)

Fact 11.17.18. Let $A \in \mathbb{C}^{n \times n}$, assume that A is asymptotically stable, let $\| \cdot \|$ denote either $\sigma_{\max}(\cdot)$ or $\| \cdot \|_{\mathrm{F}}$, and define

$$\beta(A) \triangleq \{\|B\|: \ B \in \mathbb{C}^{n \times n} \text{ and } A + B \text{ is not asymptotically stable}\}.$$

Then,

$$\tfrac{1}{2}\sigma_{\min}(A \otimes A) \leq \beta(A) = \min_{\gamma \in \mathbb{R}} \sigma_{\min}(A + \gamma\jmath I)$$

$$\leq \min\{\mathrm{spabs}(A), \sigma_{\min}(A), \tfrac{1}{2}\sigma_{\max}(A + A^*)\}.$$

Furthermore, let $R \in \mathbb{F}^{n \times n}$, assume that R is positive definite, and let $P \in \mathbb{F}^{n \times n}$ be the positive-definite solution of $A^*P + PA + R = 0$. Then,

$$\tfrac{1}{2}\sigma_{\min}(R)/\|P\| \leq \beta(A).$$

If, in addition, $A + A^*$ is negative definite, then

$$-\tfrac{1}{2}\lambda_{\min}(A + A^*) \leq \beta(A).$$

(Proof: See [352, 739].) (Remark: The analogous problem for real matrices and real perturbations is discussed in [598].)

Fact 11.17.19. Let $A \in \mathbb{F}^{n \times n}$, assume that A is asymptotically stable, let $V \in \mathbb{F}^{n \times n}$, assume that V is positive definite, and let $Q \in \mathbf{P}^n$ satisfy $AQ + QA^* + V = 0$. Then, for all $t \geq 0$,

$$e^{tA}e^{tA^*} \leq \kappa(Q) \operatorname{tr} e^{-tS^{-1}VS^{-*}} \leq \kappa(Q)e^{-[t/\sigma_{\max}(Q)]V},$$

where $S \in \mathbb{F}^{n \times n}$ satisfies $Q = SS^*$ and $\kappa(Q) \triangleq \sigma_{\max}(Q)/\sigma_{\min}(Q)$. (Proof: See [799].) (Remark: Fact 11.13.3 yields $e^{tA}e^{tA^*} \leq e^{t(A+A^*)}$. However, $A + A^*$ might not be asymptotically stable. See [102].)

Fact 11.17.20. Let $A \in \mathbb{R}^{n \times n}$, and assume that every entry of $A \in \mathbb{R}^{n \times n}$ is positive. Then, A is unstable. (Proof: See Fact 4.11.1.)

Fact 11.17.21. Let $A \in \mathbb{R}^{n \times n}$. Then, A is asymptotically stable if and only if there exist matrices $B, C \in \mathbb{R}^{n \times n}$ such that B is positive definite, C is dissipative, and $A = BC$. (Proof: $A = P^{-1}(-A^{\mathrm{T}}P - R)$.) (Remark: To reverse the order of factors, consider A^{T}.)

Fact 11.17.22. Let $A \in \mathbb{F}^{n \times n}$. Then, the following statements hold:

i) All of the real eigenvalues of A are positive if and only if A is the product of two dissipative matrices.

ii) A is nonsingular and $A \neq \alpha I$ for all $\alpha < 0$ if and only if A is the product of two asymptotically stable matrices.

iii) A is nonsingular if and only if A is the product of three or fewer asymptotically stable matrices.

(Proof: See [69, 797].)

Fact 11.17.23. Let $p \in \mathbb{R}[s]$, where $p(s) = s^n + \beta_{n-1}s^{n-1} + \cdots + \beta_1 s + \beta_0$ and $\beta_0, \ldots, \beta_n > 0$. Furthermore, define $A \in \mathbb{R}^{n \times n}$ by

$$A \triangleq \begin{bmatrix} \beta_{n-1} & \beta_{n-3} & \beta_{n-5} & \beta_{n-7} & \cdots & \cdots & 0 \\ 1 & \beta_{n-2} & \beta_{n-4} & \beta_{n-6} & \cdots & \cdots & 0 \\ 0 & \beta_{n-1} & \beta_{n-3} & \beta_{n-5} & \cdots & \cdots & 0 \\ 0 & 1 & \beta_{n-2} & \beta_{n-4} & \cdots & \cdots & 0 \\ \vdots & \vdots & \vdots & \vdots & \ddots & \vdots & \vdots \\ 0 & 0 & 0 & \cdots & \cdots & \beta_1 & 0 \\ 0 & 0 & 0 & \cdots & \cdots & \beta_2 & \beta_0 \end{bmatrix}.$$

If p is Lyapunov stable, then every subdeterminant of A is nonnegative. (Remark: A is *totally nonnegative*.) Furthermore, p is asymptotically stable if and only if every leading principal subdeterminant of A is positive. (Proof: See [47].) (Remark: The second statement is due to Hurwitz.) (Remark: The diagonal entries of A are $\beta_{n-1}, \ldots, \beta_0$.) (Problem: Show that this condition for stability is equivalent to the condition given in [260, p. 183] in terms of an alternative matrix $\hat{A}$.)

Fact 11.17.24. Let $A \in \mathbb{R}^{n \times n}$, assume that A is tridiagonal, and assume that $A_{(i,i)} > 0$ for all $i = 1, \ldots, n$ and $A_{(i,i+1)} A_{(i+1,i)} > 0$ for all $i = 1, \ldots, n - 1$. Then, A is asymptotically stable. (Proof: See [161].) (Remark: This result is due to Barnett and Storey.)

Fact 11.17.25. Let $A \in \mathbb{R}^{n \times n}$, and assume that A is cyclic. Then, there exists a nonsingular matrix $S \in \mathbb{R}^{n \times n}$ such that $A_S = SAS^{-1}$ is given by the tridiagonal matrix

$$
A_S = \begin{bmatrix}
0 & 1 & 0 & 0 & \cdots & 0 & 0 \\
-\alpha_n & 0 & 1 & \cdots & 0 & 0 \\
0 & -\alpha_{n-1} & 0 & \cdots & 0 & 0 \\
\vdots & \vdots & \vdots & \ddots & \vdots & \vdots \\
0 & 0 & 0 & \cdots & 0 & 1 \\
0 & 0 & 0 & \cdots & -\alpha_2 & -\alpha_1
\end{bmatrix},
$$

where $\alpha_1, \ldots, \alpha_n$ are real numbers. If $\alpha_1 \alpha_2 \cdots \alpha_n \neq 0$, then the number of eigenvalues of A in the OLHP is equal to the number of positive elements in $\{\alpha_1, \alpha_1 \alpha_2, \ldots, \alpha_1 \alpha_2 \cdots \alpha_n\}_{\mathrm{m}}$. Furthermore, $A_S^{\mathrm{T}} P + P A_S + R = 0$, where

$$
P \triangleq \mathrm{diag}(\alpha_1 \alpha_2 \cdots \alpha_n, \alpha_1 \alpha_2 \cdots \alpha_{n-1}, \ldots, \alpha_1 \alpha_2, \alpha_1)
$$

and

$$
R \triangleq \mathrm{diag}\left(0, \ldots, 0, 2\alpha_1^2\right).
$$

(Remark: A_S is in *Schwarz form*.) (Proof: See [80, pp. 52, 95].)

Fact 11.17.26. Let $\alpha_1, \alpha_2, \alpha_3 > 0$, and define $A_{\mathrm{R}}, P, R \in \mathbb{R}^{3 \times 3}$ by the tridiagonal matrix

$$
A_{\mathrm{R}} \triangleq \begin{bmatrix}
-\alpha_1 & \alpha_2^{1/2} & 0 \\
-\alpha_2^{1/2} & 0 & \alpha_3^{1/2} \\
0 & -\alpha_3^{1/2} & 0
\end{bmatrix}
$$

and the diagonal matrices

$$
P \triangleq I, \quad R \triangleq \mathrm{diag}(2\alpha_1, 0, 0).
$$

Then, $A_{\mathrm{R}}^{\mathrm{T}} P + P A_{\mathrm{R}} + R = 0$. (Remark: The matrix A_{R} is in *Routh form*. The Routh form A_{R} and the Schwarz form A_S are related by $A_{\mathrm{R}} = S_{\mathrm{RS}} A_S S_{\mathrm{RS}}^{-1}$, where

$$
S_{\mathrm{RS}} \triangleq \begin{bmatrix}
0 & 0 & \alpha_1^{1/2} \\
0 & -(\alpha_1 \alpha_2)^{1/2} & 0 \\
(\alpha_1 \alpha_2 \alpha_3)^{1/2} & 0 & 0
\end{bmatrix}.)
$$

Fact 11.17.27. Let $\alpha_1, \alpha_2, \alpha_3 > 0$, and define $A_C, P, R \in \mathbb{R}^{3 \times 3}$ by the tridiagonal matrix

$$A_C \triangleq \begin{bmatrix} 0 & 1/a_3 & 0 \\ -1/a_2 & 0 & 1/a_2 \\ 0 & -1/a_1 & -1/a_1 \end{bmatrix}$$

and the diagonal matrices

$$P \triangleq \mathrm{diag}(a_3, a_2, a_1), \quad R \triangleq \mathrm{diag}(0, 0, 2),$$

where $a_1 \triangleq 1/\alpha_1$, $a_2 \triangleq \alpha_1/\alpha_2$, and $a_3 \triangleq \alpha_2/(\alpha_1\alpha_3)$. Then, $A_C^{\mathrm{T}}P + PA_C + R = 0$. (Remark: The matrix A_C is in *Chen form*.) The Schwarz form A_S and the Chen form A_C are related by $A_S = S_{SC}A_C S_{SC}^{-1}$, where

$$S_{SC} \triangleq \begin{bmatrix} 1/(\alpha_1\alpha_3) & 0 & 0 \\ 0 & 1/\alpha_2 & 0 \\ 0 & 0 & 1/\alpha_1 \end{bmatrix}.)$$

(Proof: See [178, p. 346].) (Remark: The Schwarz, Routh, and Chen forms provide the basis for the Routh criterion. See [19, 148, 178, 577].)

Fact 11.17.28. Let $A \in \mathbb{F}^{n \times n}$. Then, the following statements are equivalent:

i) A is asymptotically stable.

ii) There exist a negative-definite matrix $B \in \mathbb{F}^{n \times n}$, a skew-Hermitian matrix $C \in \mathbb{F}^{n \times n}$, and a nonsingular matrix $S \in \mathbb{F}^{n \times n}$ such that $A = B + SCS^{-1}$.

iii) There exist a negative-definite matrix $B \in \mathbb{F}^{n \times n}$, a skew-Hermitian matrix $C \in \mathbb{F}^{n \times n}$, and a nonsingular matrix $S \in \mathbb{F}^{n \times n}$ such that $A = S(B + C)S^{-1}$.

(Proof: See [203].)

Fact 11.17.29. Let $A \in \mathbb{R}^{n \times n}$, and let $k \geq 2$. Then, there exist asymptotically stable matrices $A_1, \ldots, A_k \in \mathbb{R}^{n \times n}$ such that $A = \sum_{i=1}^{k} A_i$ if and only if $\mathrm{tr}\, A < 0$. (Proof: See [391].)

Fact 11.17.30. Let $A \in \mathbb{R}^{n \times n}$. Then, A is (Lyapunov stable, semistable, asymptotically stable) if and only if $A \oplus A$ is. (Proof: Use Fact 7.5.3 and the fact that $\mathrm{vec}(e^{tA}Ve^{tA^*}) = e^{t(A\oplus\overline{A})}\mathrm{vec}\, V$.)

Fact 11.17.31. Let $A \in \mathbb{R}^{n \times n}$ and $B \in \mathbb{R}^{m \times m}$. Then, the following statements hold:

i) If A and B are (Lyapunov stable, semistable, asymptotically stable), then so is $A \oplus B$.

ii) If $A \oplus B$ is (Lyapunov stable, semistable, asymptotically stable), then so is either A or B.

(Proof: Use Fact 7.5.3.)

Fact 11.17.32. Let $A \in \mathbb{R}^{n \times n}$, and assume that A is asymptotically stable. Then,

$$(A \oplus A)^{-1} = \int\limits_{-\infty}^{\infty} (\jmath\omega I - A)^{-1} \otimes (\jmath\omega I - A)^{-1} \, \mathrm{d}\omega$$

and

$$\int\limits_{-\infty}^{\infty} (\omega^2 I + A^2) \, \mathrm{d}\omega = -\pi A^{-1}.$$

(Proof: Use $(\jmath\omega I - A)^{-1} + (-\jmath\omega I - A)^{-1} = -2A(\omega^2 I + A^2)^{-1}$.)

Fact 11.17.33. Let $A \in \mathbb{R}^{2 \times 2}$. Then, A is asymptotically stable if and only if $\operatorname{tr} A < 0$ and $\det A > 0$.

Fact 11.17.34. Let $A \in \mathbb{C}^{n \times n}$. Then, there exists a unique asymptotically stable matrix $B \in \mathbb{C}^{n \times n}$ such that $B^2 = -A$. (Remark: This result is stated in [668]. The uniqueness of the square root for complex matrices that have no eigenvalues in $(-\infty, 0]$ is implicitly assumed in [669].) (Remark: See Fact 5.14.17.)

Fact 11.17.35. Let $A \in \mathbb{R}^{n \times n}$. Then, the following statements hold:

i) If A is semidissipative, then A is Lyapunov stable.

ii) If A is dissipative, then A is asymptotically stable.

iii) If A is Lyapunov stable and normal, then A is semidissipative.

iv) If A is asymptotically stable and normal, then A is dissipative.

v) If A is discrete-time Lyapunov stable and normal, then A is semi-contractive.

Fact 11.17.36. Let $M \in \mathbb{R}^{r \times r}$, assume that M is positive definite, let $C, K \in \mathbb{R}^{r \times r}$, assume that C and K are positive semidefinite, and consider the equation

$$M\ddot{q} + C\dot{q} + Kq = 0.$$

Furthermore, define

$$A \triangleq \begin{bmatrix} 0 & I \\ -M^{-1}K & -M^{-1}C \end{bmatrix}.$$

Then, the following statements hold:

i) A is Lyapunov stable if and only if $C + K$ is positive definite.

ii) A is Lyapunov stable if and only if rank $\left[\begin{smallmatrix} C \\ K \end{smallmatrix}\right] = r$.

iii) A is semistable if and only if $(M^{-1}K, C)$ is observable.

iv) A is asymptotically stable if and only if A is semistable and K is positive definite.

(Proof: See [103].) (Remark: See Fact 5.11.9.)

Fact 11.17.37. Let $A \in \mathbb{R}^{n \times n}$, and assume that $A_{(i,j)} \leq 0$ for all $i, j = 1, \ldots, n$ such that $i \neq j$. (Remark: A is a *Z-matrix*.) Then, the following conditions are equivalent:

i) $-A$ is asymptotically stable.

ii) There exists a matrix $B \in \mathbb{R}^{n \times n}$ such that $B \geq\geq 0$, $A = \alpha I - B$, and $\alpha > \mathrm{sprad}(B)$.

iii) If $\lambda \in \mathrm{spec}(A)$ is real, then $\lambda > 0$.

iv) $A + \alpha I$ is nonsingular for all $\alpha \geq 0$.

v) $A + B$ is nonsingular for all nonnegative, diagonal matrices $B \in \mathbb{R}^{n \times n}$.

vi) Every principal subdeterminant of A is positive.

vii) Every leading principal subdeterminant of A is positive.

viii) For all $k \in \{1, \ldots, n\}$, the sum of all $k \times k$ principal subdeterminants of A is positive.

ix) There exists a vector $x \in \mathbb{R}^n$ such that $x >> 0$ and $Ax >> 0$.

x) If $x \in \mathbb{R}^n$ and $Ax \geq\geq 0$, then $x \geq\geq 0$.

xi) A is nonsingular and $A^{-1} \geq\geq 0$.

(Proof: See [99, pp. 134–140] or [369, pp. 114–116].) (Remark: A is an *M-matrix*.)

11.18 Facts on Discrete-Time Stability

Fact 11.18.1. Let $p(s) \triangleq s^n + \beta_{n-1}s^{n-1} + \cdots + \beta_1 s + \beta_0 \in \mathbb{F}[s]$, define $\alpha \triangleq 1 + \sum_{i=0}^{n-1} |\beta_i|^2$, and let $\lambda \in \mathrm{roots}(p)$. Then,

$$\sqrt{\tfrac{1}{2}\left(\alpha - \sqrt{\alpha^2 - 4|\beta_0|^2}\right)} \leq |\lambda| \leq \sqrt{\tfrac{1}{2}\left(\alpha + \sqrt{\alpha^2 - 4|\beta_0|^2}\right)}.$$

(Proof: See [434] and Fact 9.11.6.) (Remark: See Fact 5.10.15.)

Fact 11.18.2. Let $p \in \mathbb{R}[s]$, where $p(s) = s^n + a_{n-1}s^{n-1} + \cdots + a_0$. Then, the following statements hold:

i) If $n = 1$, then p is discrete-time asymptotically stable if and only if $|a_0| < 1$.

ii) If $n = 2$, then p is discrete-time asymptotically stable if and only if $|a_0| < 1$ and $|a_1| < 1 + a_0$.

iii) If $n = 3$, then p is discrete-time asymptotically stable if and only if $|a_0| < 1$, $|a_0 + a_2| < |1 + a_1|$, and $|a_1 - a_0 a_2| < 1 - a_0^2$.

(Remark: These results are the *Schur-Cohn criterion*. See [72, p. 185]. Conditions for polynomials of arbitrary degree n follow from the *Jury test*. See [178, 405].)

Fact 11.18.3. Let $A \in \mathbb{R}^{2 \times 2}$. Then, A is discrete-time asymptotically stable if and only if $|\operatorname{tr} A| < 1 + \det A$ and $|\det A| < 1$.

Fact 11.18.4. Let $A \in \mathbb{F}^{n \times n}$. Then, A is discrete-time (Lyapunov stable, semistable, asymptotically stable) if and only if A^2 is.

Fact 11.18.5. Let $A \in \mathbb{R}^{n \times n}$, and let $\chi_A(s) = s^n + \beta_{n-1}s^{n-1} + \cdots + \beta_1 s + \beta_0$. Then, for all $k \geq 0$,

$$A^k = x_1(k)I + x_2(k)A + \cdots + x_n(k)A^{n-1},$$

where, for all $i = 1, \ldots, n$ and all $k \geq 0$, $x_i \colon \mathbb{N} \mapsto \mathbb{R}$ satisfies

$$x_i(k+n) + \beta_{n-1}x_i(k+n-1) + \cdots + \beta_1 x_i(k+1) + \beta_0 x_i(k) = 0,$$

with, for all $i, j = 1, \ldots, n$, the initial conditions

$$x_i(j-1) = \delta_{ij}.$$

(Proof: See [446].)

Fact 11.18.6. Let $A \in \mathbb{R}^{n \times n}$. Then, the following statements hold:

i) If A is semicontractive, then A is discrete-time Lyapunov stable.

ii) If A is contractive, then A is discrete-time asymptotically stable.

iii) If A is discrete-time Lyapunov stable and normal, then A is semicontractive.

iv) If A is discrete-time asymptotically stable and normal, then A is contractive.

(Problem: Prove these results by using Fact 11.13.5.)

Fact 11.18.7. Let $A \in \mathbb{F}^{n \times n}$. Then, A is discrete-time (Lyapunov stable, semistable, asymptotically stable) if and only if $A \otimes A$ is. (Proof:

Use Fact 7.4.13.)

Fact 11.18.8. Let $A \in \mathbb{R}^{n \times n}$ and $B \in \mathbb{R}^{m \times m}$. Then, the following statements hold:

i) If A and B are discrete-time (Lyapunov stable, semistable, asymptotically stable), then $A \otimes B$ is discrete-time (Lyapunov stable, semistable, asymptotically stable).

ii) If $A \otimes B$ is discrete-time (Lyapunov stable, semistable, asymptotically stable), then either A or B is discrete-time (Lyapunov stable, semistable, asymptotically stable).

(Proof: Use Fact 7.4.13.)

Fact 11.18.9. Let $A \in \mathbb{R}^{n \times n}$, and assume that A is (Lyapunov stable, semistable, asymptotically stable). Then, e^A is discrete-time (Lyapunov stable, semistable, asymptotically stable). (Problem: If $B \in \mathbb{R}^{n \times n}$ is discrete-time (Lyapunov stable, semistable, asymptotically stable), when does there exist (Lyapunov stable, semistable, asymptotically stable) $A \in \mathbb{R}^{n \times n}$ such that $B = e^A$? See Proposition 11.4.5.)

Fact 11.18.10. Let $A \in \mathbb{R}^{n \times n}$. If A is discrete-time asymptotically stable, then $B \triangleq (A + I)^{-1}(A - I)$ is asymptotically stable. Conversely, if $B \in \mathbb{R}^{n \times n}$ is asymptotically stable, then $A \triangleq (I+B)(I-B)^{-1}$ is discrete-time asymptotically stable. (Proof: See [341].) (Remark: For additional results on the Cayley transform, see Fact 3.7.22, Fact 3.7.23, Fact 3.7.24, Fact 3.12.9, and Fact 8.7.23.) (Problem: Obtain analogous results for Lyapunov-stable and semistable matrices.)

Fact 11.18.11. Let $\begin{bmatrix} P_1 & P_{12} \\ P_{12}^T & P_2 \end{bmatrix} \in \mathbb{R}^{2n \times 2n}$ be positive definite, where P_1, $P_{12}, P_2 \in \mathbb{R}^{n \times n}$. If $P_1 \geq P_2$, then $A \triangleq P_1^{-1} P_{12}^T$ is discrete-time asymptotically stable, while, if $P_2 \geq P_1$, then $A \triangleq P_2^{-1} P_{12}$ is discrete-time asymptotically stable. (Proof: If $P_1 \geq P_2$, then $P_1 - P_{12} P_1^{-1} P_1 P_1^{-1} P_{12}^T \geq P_1 - P_{12} P_2^{-2} P_{12}^T > 0$. See [184].)

Fact 11.18.12. Let $A \in \mathbb{R}^{n \times n}$, and let $\| \cdot \|$ be a norm on $\mathbb{R}^{n \times n}$. Then, the following statements hold:

i) A is discrete-time Lyapunov stable if and only if $\left\{ \|A^k\| \right\}_{k=0}^{\infty}$ is bounded.

ii) A is discrete-time semistable if and only if $A_\infty \triangleq \lim_{k \to \infty} A^k$ exists. In this case, $A_\infty = I - (A - I)(A - I)^\#$, and A_∞ is idempotent.

iii) A is discrete-time asymptotically stable if and only if $\lim_{k \to \infty} A^k = 0$.

(Remark: A proof of *ii*) is given in [535, p. 640]. See Fact 11.18.16.)

Fact 11.18.13. Let $A \in \mathbb{F}^{n \times n}$. Then, A is discrete-time Lyapunov stable if and only if

$$A_\infty \triangleq \lim_{k \to \infty} \tfrac{1}{k} \sum_{i=0}^{k-1} A^i$$

exists. In this case,

$$A_\infty = I - (A - I)(A - I)^\#.$$

(Proof: See [535, p. 633].) (Remark: A is *Cesaro summable*.) (Remark: See Fact 6.3.26.)

Fact 11.18.14. Let $A \in \mathbb{F}^{n \times n}$. Then, A is discrete-time asymptotically stable if and only if

$$\lim_{k \to \infty} A^k = 0.$$

Fact 11.18.15. Let $A \in \mathbb{F}^{n \times n}$, and assume that A is unitary. Then, A is discrete-time Lyapunov stable.

Fact 11.18.16. Let $A, B \in \mathbb{R}^{n \times n}$, assume that A is discrete-time semistable, and let $A_\infty \triangleq \lim_{k \to \infty} A^k$. Then,

$$\lim_{k \to \infty} \left(A + \tfrac{1}{k} B \right)^k = A_\infty e^{A_\infty B A_\infty}.$$

(Proof: See [133, 776].) (Remark: If A is idempotent, then $A_\infty = A$. The existence of A_∞ is guaranteed by Fact 11.18.12.)

Fact 11.18.17. Let $A \in \mathbb{R}^{n \times n}$. Then, the following statements hold:

i) A is discrete-time Lyapunov stable if and only if there exists a positive-definite matrix $P \in \mathbb{R}^{n \times n}$ such that $P - A^{\mathrm{T}} P A$ is positive semidefinite.

ii) A is discrete-time asymptotically stable if and only if there exists a positive-definite matrix $P \in \mathbb{R}^{n \times n}$ such that $P - A^{\mathrm{T}} P A$ is positive definite.

(Remark: The *discrete-time Lyapunov equation* or the *Stein equation* is $P = A^{\mathrm{T}} P A + R$.)

Fact 11.18.18. Let $\{A_k\}_{k=0}^\infty \subset \mathbb{R}^{n \times n}$ and, for $k \in \mathbb{N}$, consider the discrete-time, time-varying system

$$x_{k+1} = A_k x_k.$$

Furthermore, assume there exist real numbers $\beta \in (0, 1)$, $\gamma > 0$, and $\varepsilon > 0$ such that, for all $k \in \mathbb{N}$,

$$\mathrm{sprad}(A_k) < \beta,$$

$$\|A_k\| < \gamma,$$

$$\|A_{k+1} - A_k\| < \varepsilon,$$

where $\|\cdot\|$ is a norm on $\mathbb{R}^{n \times n}$. Then, $x_k \to 0$ as $k \to \infty$. (Proof: See [334, pp. 170–173].) (Remark: This result arises from the theory of *infinite matrix products*. See [44, 131, 132, 206, 315, 450].)

11.19 Facts on Subspace Decomposition

Fact 11.19.1. Let $A \in \mathbb{R}^{n \times n}$, and let $S \in \mathbb{R}^{n \times n}$ be a nonsingular matrix such that

$$A = S \begin{bmatrix} A_1 & A_{12} \\ 0 & A_2 \end{bmatrix} S^{-1}, \tag{11.19.1}$$

where $A_1 \in \mathbb{R}^{r \times r}$ is asymptotically stable, $A_{12} \in \mathbb{R}^{r \times (n-r)}$, and $A_2 \in \mathbb{R}^{(n-r) \times (n-r)}$. Then,

$$\mu_A^{\mathrm{s}}(A) = S \begin{bmatrix} 0 & B_{12\mathrm{s}} \\ 0 & \mu_A^{\mathrm{s}}(A_2) \end{bmatrix} S^{-1},$$

where $B_{12\mathrm{s}} \in \mathbb{R}^{r \times (n-r)}$, and

$$\mu_A^{\mathrm{u}}(A) = S \begin{bmatrix} \mu_A^{\mathrm{u}}(A_1) & B_{12\mathrm{u}} \\ 0 & \mu_A^{\mathrm{u}}(A_2) \end{bmatrix} S^{-1},$$

where $B_{12\mathrm{u}} \in \mathbb{R}^{r \times (n-r)}$ and $\mu_A^{\mathrm{u}}(A_1)$ is nonsingular. Consequently,

$$\mathcal{R}\left(S \begin{bmatrix} I_r \\ 0 \end{bmatrix} \right) \subseteq \mathcal{S}_{\mathrm{s}}(A).$$

If, in addition, $A_{12} = 0$, then

$$\mu_A^{\mathrm{s}}(A) = S \begin{bmatrix} 0 & 0 \\ 0 & \mu_A^{\mathrm{s}}(A_2) \end{bmatrix} S^{-1},$$

$$\mu_A^{\mathrm{u}}(A) = S \begin{bmatrix} \mu_A^{\mathrm{u}}(A_1) & 0 \\ 0 & \mu_A^{\mathrm{u}}(A_2) \end{bmatrix} S^{-1},$$

$$\mathcal{S}_{\mathrm{u}}(A) \subseteq \mathcal{R}\left(S \begin{bmatrix} 0 \\ I_{n-r} \end{bmatrix} \right).$$

(Proof: The result follows from Fact 4.10.9.)

Fact 11.19.2. Let $A \in \mathbb{R}^{n \times n}$, and let $S \in \mathbb{R}^{n \times n}$ be a nonsingular matrix such that

$$A = S \begin{bmatrix} A_1 & A_{12} \\ 0 & A_2 \end{bmatrix} S^{-1},$$

where $A_1 \in \mathbb{R}^{r \times r}$, $A_{12} \in \mathbb{R}^{r \times (n-r)}$, and $A_2 \in \mathbb{R}^{(n-r) \times (n-r)}$ satisfies $\mathrm{spec}(A_2)$ $\subset$ CRHP. Then,

$$\mu_A^s(A) = S \left[\begin{array}{cc} \mu_A^s(A_1) & C_{12s} \\ 0 & \mu_A^s(A_2) \end{array} \right] S^{-1},$$

where $C_{12s} \in \mathbb{R}^{r \times (n-r)}$ and $\mu_A^s(A_2)$ is nonsingular, and

$$\mu_A^u(A) = S \left[\begin{array}{cc} \mu_A^u(A_1) & C_{12u} \\ 0 & 0 \end{array} \right] S^{-1},$$

where $C_{12u} \in \mathbb{R}^{r \times (n-r)}$. Consequently,

$$\mathcal{S}_s(A) \subseteq \mathcal{R}\left(S \left[\begin{array}{c} I_r \\ 0 \end{array} \right] \right).$$

If, in addition, $A_{12} = 0$, then

$$\mu_A^s(A) = S \left[\begin{array}{cc} \mu_A^s(A_1) & 0 \\ 0 & \mu_A^s(A_2) \end{array} \right] S^{-1},$$

$$\mu_A^u(A) = S \left[\begin{array}{cc} \mu_A^u(A_1) & 0 \\ 0 & 0 \end{array} \right] S^{-1},$$

$$\mathcal{R}\left(S \left[\begin{array}{c} 0 \\ I_{n-r} \end{array} \right] \right) \subseteq \mathcal{S}_u(A).$$

Fact 11.19.3. Let $A \in \mathbb{R}^{n \times n}$, and let $S \in \mathbb{R}^{n \times n}$ be a nonsingular matrix such that

$$A = S \left[\begin{array}{cc} A_1 & A_{12} \\ 0 & A_2 \end{array} \right] S^{-1},$$

where $A_1 \in \mathbb{R}^{r \times r}$ satisfies $\mathrm{spec}(A_1) \subset$ CRHP, $A_{12} \in \mathbb{R}^{r \times (n-r)}$, and $A_2 \in \mathbb{R}^{(n-r) \times (n-r)}$. Then,

$$\mu_A^s(A) = S \left[\begin{array}{cc} \mu_A^s(A_1) & B_{12s} \\ 0 & \mu_A^s(A_2) \end{array} \right] S^{-1},$$

where $\mu_A^s(A_1)$ is nonsingular and $B_{12s} \in \mathbb{R}^{r \times (n-r)}$, and

$$\mu_A^u(A) = S \left[\begin{array}{cc} 0 & B_{12u} \\ 0 & \mu_A^u(A_2) \end{array} \right] S^{-1},$$

where $B_{12u} \in \mathbb{R}^{r \times (n-r)}$. Consequently,

$$\mathcal{R}\left(S \left[\begin{array}{c} I_r \\ 0 \end{array} \right] \right) \subseteq \mathcal{S}_u(A).$$

If, in addition, $A_{12} = 0$, then

$$\mu_A^s(A) = S \left[\begin{array}{cc} \mu_A^s(A_1) & 0 \\ 0 & \mu_A^s(A_2) \end{array} \right] S^{-1},$$

$$\mu_A^u(A) = S \begin{bmatrix} 0 & 0 \\ 0 & \mu_A^u(A_2) \end{bmatrix} S^{-1},$$

$$\mathcal{S}_s(A) \subseteq \mathcal{R}\left(S \begin{bmatrix} 0 \\ I_{n-r} \end{bmatrix}\right).$$

Fact 11.19.4. Let $A \in \mathbb{R}^{n \times n}$, and let $S \in \mathbb{R}^{n \times n}$ be a nonsingular matrix such that

$$A = S \begin{bmatrix} A_1 & A_{12} \\ 0 & A_2 \end{bmatrix} S^{-1},$$

where $A_1 \in \mathbb{R}^{r \times r}$, $A_{12} \in \mathbb{R}^{r \times (n-r)}$, and $A_2 \in \mathbb{R}^{(n-r) \times (n-r)}$ is asymptotically stable. Then,

$$\mu_A^s(A) = S \begin{bmatrix} \mu_A^s(A_1) & C_{12s} \\ 0 & 0 \end{bmatrix} S^{-1},$$

where $C_{12s} \in \mathbb{R}^{r \times (n-r)}$, and

$$\mu_A^u(A) = S \begin{bmatrix} \mu_A^u(A_1) & C_{12u} \\ 0 & \mu_A^u(A_2) \end{bmatrix} S^{-1},$$

where $\mu_A^u(A_2)$ is nonsingular and $C_{12u} \in \mathbb{R}^{r \times (n-r)}$. Consequently,

$$\mathcal{S}_u(A) \subseteq \mathcal{R}\left(S \begin{bmatrix} I_r \\ 0 \end{bmatrix}\right).$$

If, in addition, $A_{12} = 0$, then

$$\mu_A^s(A) = S \begin{bmatrix} \mu_A^s(A_1) & 0 \\ 0 & 0 \end{bmatrix} S^{-1},$$

$$\mu_A^u(A) = S \begin{bmatrix} \mu_A^u(A_1) & 0 \\ 0 & \mu_A^u(A_2) \end{bmatrix} S^{-1},$$

$$\mathcal{R}\left(S \begin{bmatrix} 0 \\ I_{n-r} \end{bmatrix}\right) \subseteq \mathcal{S}_s(A).$$

Fact 11.19.5. Let $A \in \mathbb{R}^{n \times n}$, and let $S \in \mathbb{R}^{n \times n}$ be a nonsingular matrix such that

$$A = S \begin{bmatrix} A_1 & A_{12} \\ 0 & A_2 \end{bmatrix} S^{-1},$$

where $A_1 \in \mathbb{R}^{r \times r}$ satisfies $\mathrm{spec}(A_1) \subset \mathrm{CRHP}$, $A_{12} \in \mathbb{R}^{r \times (n-r)}$, and $A_2 \in \mathbb{R}^{(n-r) \times (n-r)}$ is asymptotically stable. Then,

$$\mu_A^s(A) = S \begin{bmatrix} \mu_A^s(A_1) & C_{12s} \\ 0 & 0 \end{bmatrix} S^{-1},$$

where $C_{12s} \in \mathbb{R}^{r \times (n-r)}$ and $\mu_A^s(A_1)$ is nonsingular, and

$$\mu_A^u(A) = S \begin{bmatrix} 0 & C_{12u} \\ 0 & \mu_A^u(A_2) \end{bmatrix} S^{-1},$$

where $C_{12u} \in \mathbb{R}^{r \times (n-r)}$ and $\mu_A^u(A_2)$ is nonsingular. Consequently,

$$\mathcal{S}_u(A) = \mathcal{R}\left(S \begin{bmatrix} I_r \\ 0 \end{bmatrix} \right).$$

If, in addition, $A_{12} = 0$, then

$$\mu_A^s(A) = S \begin{bmatrix} \mu_A^s(A_1) & 0 \\ 0 & 0 \end{bmatrix} S^{-1}$$

and

$$\mu_A^u(A) = S \begin{bmatrix} 0 & 0 \\ 0 & \mu_A^u(A_2) \end{bmatrix} S^{-1},$$

Consequently,

$$\mathcal{S}_s(A) = \mathcal{R}\left(S \begin{bmatrix} 0 \\ I_{n-r} \end{bmatrix} \right).$$

Fact 11.19.6. Let $A \in \mathbb{R}^{n \times n}$, and let $S \in \mathbb{R}^{n \times n}$ be a nonsingular matrix such that

$$A = S \begin{bmatrix} A_1 & 0 \\ A_{21} & A_2 \end{bmatrix} S^{-1},$$

where $A_1 \in \mathbb{R}^{r \times r}$ is asymptotically stable, $A_{21} \in \mathbb{R}^{(n-r) \times r}$, and $A_2 \in \mathbb{R}^{(n-r) \times (n-r)}$. Then,

$$\mu_A^s(A) = S \begin{bmatrix} 0 & 0 \\ B_{21s} & \mu_A^s(A_2) \end{bmatrix} S^{-1},$$

where $B_{21s} \in \mathbb{R}^{(n-r) \times r}$, and

$$\mu_A^u(A) = S \begin{bmatrix} \mu_A^u(A_1) & 0 \\ B_{21u} & \mu_A^u(A_2) \end{bmatrix} S^{-1},$$

where $B_{21u} \in \mathbb{R}^{(n-r) \times r}$ and $\mu_A^u(A_1)$ is nonsingular. Consequently,

$$\mathcal{S}_u(A) \subseteq \mathcal{R}\left(S \begin{bmatrix} 0 \\ I_{n-r} \end{bmatrix} \right).$$

If, in addition, $A_{21} = 0$, then

$$\mu_A^s(A) = S \begin{bmatrix} 0 & 0 \\ 0 & \mu_A^s(A_2) \end{bmatrix} S^{-1},$$

$$\mu_A^u(A) = S \begin{bmatrix} \mu_A^u(A_1) & 0 \\ 0 & \mu_A^u(A_2) \end{bmatrix} S^{-1},$$

$$\mathcal{R}\left(S\begin{bmatrix} I_r \\ 0 \end{bmatrix}\right) \subseteq \mathcal{S}_{\mathrm{s}}(A).$$

Fact 11.19.7. Let $A \in \mathbb{R}^{n \times n}$, and let $S \in \mathbb{R}^{n \times n}$ be a nonsingular matrix such that

$$A = S\begin{bmatrix} A_1 & 0 \\ A_{21} & A_2 \end{bmatrix} S^{-1},$$

where $A_1 \in \mathbb{R}^{r \times r}$, $A_{21} \in \mathbb{R}^{(n-r) \times r}$, and $A_2 \in \mathbb{R}^{(n-r) \times (n-r)}$ satisfies $\mathrm{spec}(A_2) \subset \mathrm{CRHP}$. Then,

$$\mu_A^{\mathrm{s}}(A) = S\begin{bmatrix} \mu_A^{\mathrm{s}}(A_1) & 0 \\ C_{21\mathrm{s}} & \mu_A^{\mathrm{s}}(A_2) \end{bmatrix} S^{-1},$$

where $C_{21\mathrm{s}} \in \mathbb{R}^{(n-r) \times r}$ and $\mu_A^{\mathrm{s}}(A_2)$ is nonsingular, and

$$\mu_A^{\mathrm{u}}(A) = S\begin{bmatrix} \mu_A^{\mathrm{u}}(A_1) & 0 \\ C_{21\mathrm{u}} & 0 \end{bmatrix} S^{-1},$$

where $C_{21\mathrm{u}} \in \mathbb{R}^{(n-r) \times r}$. Consequently,

$$\mathcal{R}\left(S\begin{bmatrix} 0 \\ I_{n-r} \end{bmatrix}\right) \subseteq \mathcal{S}_{\mathrm{u}}(A).$$

If, in addition, $A_{21} = 0$, then

$$\mu_A^{\mathrm{s}}(A) = S\begin{bmatrix} \mu_A^{\mathrm{s}}(A_1) & 0 \\ 0 & \mu_A^{\mathrm{s}}(A_2) \end{bmatrix} S^{-1},$$

$$\mu_A^{\mathrm{u}}(A) = S\begin{bmatrix} \mu_A^{\mathrm{u}}(A_1) & 0 \\ 0 & 0 \end{bmatrix} S^{-1},$$

$$\mathcal{S}_{\mathrm{s}}(A) \subseteq \mathcal{R}\left(S\begin{bmatrix} I_r \\ 0 \end{bmatrix}\right).$$

Fact 11.19.8. Let $A \in \mathbb{R}^{n \times n}$, and let $S \in \mathbb{R}^{n \times n}$ be a nonsingular matrix such that

$$A = S\begin{bmatrix} A_1 & 0 \\ A_{21} & A_2 \end{bmatrix} S^{-1},$$

where $A_1 \in \mathbb{R}^{r \times r}$ is asymptotically stable, $A_{21} \in \mathbb{R}^{(n-r) \times r}$, and $A_2 \in \mathbb{R}^{(n-r) \times (n-r)}$ satisfies $\mathrm{spec}(A_2) \subset \mathrm{CRHP}$. Then,

$$\mu_A^{\mathrm{s}}(A) = S\begin{bmatrix} 0 & 0 \\ C_{21\mathrm{s}} & \mu_A^{\mathrm{s}}(A_2) \end{bmatrix} S^{-1},$$

where $C_{21\mathrm{s}} \in \mathbb{R}^{n-r \times r}$ and $\mu_A^{\mathrm{s}}(A_2)$ is nonsingular, and

$$\mu_A^{\mathrm{u}}(A) = S\begin{bmatrix} \mu_A^{\mathrm{u}}(A_1) & 0 \\ C_{21\mathrm{u}} & 0 \end{bmatrix} S^{-1},$$

where $C_{21u} \in \mathbb{R}^{(n-r) \times r}$ and $\mu_A^u(A_1)$ is nonsingular. Consequently,

$$S_u(A) = \mathcal{R}\left(S\begin{bmatrix} 0 \\ I_{n-r} \end{bmatrix}\right).$$

If, in addition, $A_{21} = 0$, then

$$\mu_A^s(A) = S\begin{bmatrix} 0 & 0 \\ 0 & \mu_A^s(A_2) \end{bmatrix} S^{-1}$$

and

$$\mu_A^u(A) = S\begin{bmatrix} \mu_A^u(A_1) & 0 \\ 0 & 0 \end{bmatrix} S^{-1}.$$

Consequently,

$$S_s(A) = \mathcal{R}\left(S\begin{bmatrix} I_r \\ 0 \end{bmatrix}\right).$$

Fact 11.19.9. Let $A \in \mathbb{R}^{n \times n}$, and let $S \in \mathbb{R}^{n \times n}$ be a nonsingular matrix such that

$$A = S\begin{bmatrix} A_1 & 0 \\ A_{21} & A_2 \end{bmatrix} S^{-1},$$

where $A_1 \in \mathbb{R}^{r \times r}$, $A_{21} \in \mathbb{R}^{(n-r) \times r}$, and $A_2 \in \mathbb{R}^{(n-r) \times (n-r)}$ is asymptotically stable. Then,

$$\mu_A^s(A) = S\begin{bmatrix} \mu_A^s(A_1) & 0 \\ B_{21s} & 0 \end{bmatrix} S^{-1},$$

where $B_{21s} \in \mathbb{R}^{(n-r) \times r}$, and

$$\mu_A^u(A) = S\begin{bmatrix} \mu_A^u(A_1) & 0 \\ B_{21u} & \mu_A^u(A_2) \end{bmatrix} S^{-1},$$

where $B_{21u} \in \mathbb{R}^{(n-r) \times r}$ and $\mu_A^u(A_2)$ is nonsingular. Consequently,

$$\mathcal{R}\left(S\begin{bmatrix} 0 \\ I_{n-r} \end{bmatrix}\right) \subseteq S(A).$$

If, in addition, $A_{21} = 0$, then

$$\mu_A^s(A) = S\begin{bmatrix} \mu_A^s(A_1) & 0 \\ 0 & 0 \end{bmatrix} S^{-1},$$

$$\mu_A^u(A) = S\begin{bmatrix} \mu_A^u(A_1) & 0 \\ 0 & \mu_A^u(A_2) \end{bmatrix} S^{-1},$$

$$S_u(A) \subseteq \mathcal{R}\left(S\begin{bmatrix} I_r \\ 0 \end{bmatrix}\right).$$

Fact 11.19.10. Let $A \in \mathbb{R}^{n \times n}$, and let $S \in \mathbb{R}^{n \times n}$ be a nonsingular matrix such that

$$A = S \begin{bmatrix} A_1 & 0 \\ A_{21} & A_2 \end{bmatrix} S^{-1},$$

where $A_1 \in \mathbb{R}^{r \times r}$ satisfies $\text{spec}(A_1) \subset \text{CRHP}$, $A_{21} \in \mathbb{R}^{(n-r) \times r}$, and $A_2 \in \mathbb{R}^{(n-r) \times (n-r)}$. Then,

$$\mu_A^{\text{s}}(A) = S \begin{bmatrix} \mu_A^{\text{s}}(A_1) & 0 \\ C_{12\text{s}} & \mu_A^{\text{s}}(A_2) \end{bmatrix} S^{-1},$$

where $C_{21\text{s}} \in \mathbb{R}^{(n-r) \times r}$ and $\mu_A^{\text{s}}(A_1)$ is nonsingular, and

$$\mu_A^{\text{u}}(A) = S \begin{bmatrix} 0 & 0 \\ C_{21\text{u}} & \mu_A^{\text{u}}(A_2) \end{bmatrix} S^{-1},$$

where $C_{21\text{u}} \in \mathbb{R}^{(n-r) \times r}$. Consequently,

$$\mathcal{S}_{\text{s}}(A) \subseteq \mathcal{R}\left(S \begin{bmatrix} 0 \\ I_{n-r} \end{bmatrix} \right).$$

If, in addition, $A_{21} = 0$, then

$$\mu_A^{\text{s}}(A) = S \begin{bmatrix} \mu_A^{\text{s}}(A_1) & 0 \\ 0 & \mu_A^{\text{s}}(A_2) \end{bmatrix} S^{-1},$$

$$\mu_A^{\text{u}}(A) = S \begin{bmatrix} 0 & 0 \\ 0 & \mu_A^{\text{u}}(A_2) \end{bmatrix} S^{-1},$$

$$\mathcal{R}\left(S \begin{bmatrix} I_r \\ 0 \end{bmatrix} \right) \subseteq \mathcal{S}_{\text{u}}(A).$$

Fact 11.19.11. Let $A \in \mathbb{R}^{n \times n}$, and let $S \in \mathbb{R}^{n \times n}$ be a nonsingular matrix such that

$$A = S \begin{bmatrix} A_1 & 0 \\ A_{21} & A_2 \end{bmatrix} S^{-1},$$

where $A_1 \in \mathbb{R}^{r \times r}$ satisfies $\text{spec}(A_1) \subset \text{CRHP}$, $A_{21} \in \mathbb{R}^{(n-r) \times r}$, and $A_2 \in \mathbb{R}^{(n-r) \times (n-r)}$ is asymptotically stable. Then,

$$\mu_A^{\text{s}}(A) = S \begin{bmatrix} \mu_A^{\text{s}}(A_1) & 0 \\ C_{21\text{s}} & 0 \end{bmatrix} S^{-1},$$

where $C_{21\text{s}} \in \mathbb{R}^{(n-r) \times r}$ and $\mu_A^{\text{s}}(A_1)$ is nonsingular, and

$$\mu_A^{\text{u}}(A) = S \begin{bmatrix} 0 & 0 \\ C_{21\text{u}} & \mu_A^{\text{u}}(A_2) \end{bmatrix} S^{-1},$$

where $C_{21\text{u}} \in \mathbb{R}^{(n-r) \times r}$ and $\mu_A^{\text{u}}(A_2)$ is nonsingular. Consequently,

$$\mathcal{S}_{\text{s}}(A) = \mathcal{R}\left(S \begin{bmatrix} 0 \\ I_{n-r} \end{bmatrix} \right).$$

If, in addition, $A_{21} = 0$, then

$$\mu_A^{\mathrm{s}}(A) = S \begin{bmatrix} \mu_A^{\mathrm{s}}(A_1) & 0 \\ 0 & 0 \end{bmatrix} S^{-1}$$

and

$$\mu_A^{\mathrm{u}}(A) = S \begin{bmatrix} 0 & 0 \\ 0 & \mu_A^{\mathrm{u}}(A_2) \end{bmatrix} S^{-1}.$$

Consequently,

$$\mathcal{S}_{\mathrm{u}}(A) = \mathcal{R}\left(S \begin{bmatrix} I_r \\ 0 \end{bmatrix} \right).$$

11.20 Notes

Explicit formulas for the matrix exponential are given in [1, 39, 107, 179, 332, 583, 585]. The Laplace transform (11.2.10) is given in [647, p. 34]. Computational methods are discussed in [545]. An arithmetic-mean-geometric-mean iteration for computing the matrix exponential and matrix logarithm is given in [669].

The exponential function plays a central role in the theory of Lie groups, see [90, 167, 324, 380, 386, 628, 742]. Applications to robotics and kinematics are given in [524, 551, 575]. Additional applications are discussed in [166].

The real logarithm is discussed in [197, 345, 563, 595].

An asymptotically stable polynomial is traditionally called *Hurwitz*. Semistability was first defined in [158], and is developed in [103, 109]. Stability theory is treated in [323, 467, 589] and [286, Chapter XV]. Solutions of the Lyapunov equation under weak conditions are considered in [649]. Structured solutions of the Lyapunov equation are discussed in [412].

Chapter Twelve

Linear Systems and Control Theory

This chapter considers linear state space systems with inputs and outputs. These systems are considered in both the time domain and frequency (Laplace) domain. Some basic results in control theory are also presented.

12.1 State Space and Transfer Function Models

Let $A \in \mathbb{R}^{n \times n}$ and $B \in \mathbb{R}^{n \times m}$, and, for $t \geq t_0$, consider the *state equation*

$$\dot{x}(t) = Ax(t) + Bu(t), \tag{12.1.1}$$

with the *initial condition*

$$x(t_0) = x_0. \tag{12.1.2}$$

In (12.1.1), $x(t) \in \mathbb{R}^n$ is the *state*, and $u(t) \in \mathbb{R}^m$ is the *input*.

The following result give the solution of (12.1.1) known as the *variation of constants formula*.

Proposition 12.1.1. For $t \geq t_0$ the state $x(t)$ of the dynamical equation (12.1.1) with initial condition (12.1.2) is given by

$$x(t) = e^{(t-t_0)A}x_0 + \int_{t_0}^{t} e^{(t-\tau)A}Bu(\tau)\,\mathrm{d}\tau. \tag{12.1.3}$$

Proof. Multiplying (12.1.1) by e^{-tA} yields

$$e^{-tA}[\dot{x}(t) - Ax(t)] = e^{-tA}Bu(t),$$

which is equivalent to

$$\frac{\mathrm{d}}{\mathrm{d}t}\big[e^{-tA}x(t)\big] = e^{-tA}Bu(t).$$

Integrating over $[t_0, t]$ yields

$$e^{-tA}x(t) = e^{-t_0 A}x(t_0) + \int_{t_0}^{t} e^{-\tau A}Bu(\tau)\,\mathrm{d}\tau.$$

Now, multiplying by e^{tA} yields (12.1.3).

Alternatively, let $x(t)$ be given by (12.1.3). Then, it follows from Liebniz' rule Fact 10.8.4 that

$$\dot{x}(t) = \frac{\mathrm{d}}{\mathrm{d}t}e^{(t-t_0)A}x_0 + \frac{\mathrm{d}}{\mathrm{d}t}\int_{t_0}^{t} e^{(t-\tau)A}Bu(\tau)\,\mathrm{d}\tau$$

$$= Ae^{(t-t_0)A}x_0 + \int_{t_0}^{t} Ae^{(t-\tau)A}Bu(\tau)\,\mathrm{d}\tau + Bu(t)$$

$$= Ax(t) + Bu(t). \qquad \square$$

For convenience, we can reset the clock and assume without loss of generality that $t_0 = 0$. In this case, $x(t)$ for all $t \geq 0$ is given by

$$x(t) = e^{tA}x_0 + \int_{0}^{t} e^{(t-\tau)A}Bu(\tau)\,\mathrm{d}\tau. \qquad (12.1.4)$$

If $u(t) = 0$ for all $t \geq 0$, then, for all $t \geq 0$, $x(t)$ is given by

$$x(t) = e^{tA}x_0. \qquad (12.1.5)$$

Now, let $u(t) = \delta(t)v$, where $\delta(t)$ is the *unit impulse* at $t = 0$ and $v \in \mathbb{R}^m$. Then, for all $t \geq 0$, $x(t)$ is given by

$$x(t) = e^{tA}x_0 + e^{tA}Bv. \qquad (12.1.6)$$

Let $a < b$. Then, $\delta(t)$, which has physical dimensions of 1/time, satisfies

$$\int_{a}^{b} \delta(\tau)\,\mathrm{d}\tau = \begin{cases} 0, & a > 0 \text{ or } b \leq 0, \\ 1, & a \leq 0 < b. \end{cases} \qquad (12.1.7)$$

More generally, if $g\colon \mathcal{D} \to \mathbb{R}^n$, where $[a,b] \subseteq \mathcal{D} \subseteq \mathbb{R}$, $t_0 \in \mathcal{D}$, and g is continuous at t_0, then

$$\int_{a}^{b} \delta(\tau - t_0)g(\tau)\,\mathrm{d}\tau = \begin{cases} 0, & a > t_0 \text{ or } b \leq t_0, \\ g(t_0), & a \leq t_0 < b. \end{cases} \qquad (12.1.8)$$

Alternatively, let the input $u(t)$ be constant or a *step function*, that is, $u(t) = v$ for all $t \geq 0$, where $v \in \mathbb{R}^m$. Then, by a change of variable of integration, it follows that, for all $t \geq 0$,

$$x(t) = e^{tA}x_0 + \int_0^t e^{\tau A}\, \mathrm{d}\tau Bv. \tag{12.1.9}$$

Using Fact 11.11.13, (12.1.9) can be written for all $t \geq 0$ as

$$x(t) = e^{tA}x_0 + \left[A^{\mathrm{D}}(e^{tA} - I) + (I - AA^{\mathrm{D}})\sum_{i=1}^{\mathrm{ind}\, A} (i!)^{-1} t^i A^{i-1} \right] Bv. \tag{12.1.10}$$

If A is group invertible, then, for all $t \geq 0$, (12.1.10) becomes

$$x(t) = e^{tA}x_0 + \left[A^{\#}(e^{tA} - I) + t(I - AA^{\#}) \right] Bv. \tag{12.1.11}$$

If, in addition, A is nonsingular, then, for all $t \geq 0$, (12.1.11) becomes

$$x(t) = e^{tA}x_0 + A^{-1}(e^{tA} - I)Bv. \tag{12.1.12}$$

Next, consider the *output equation*

$$y(t) = Cx(t) + Du(t), \tag{12.1.13}$$

where $t \geq 0$, $y(t) \in \mathbb{R}^l$ is the *output*, $C \in \mathbb{R}^{l \times n}$, and $D \in \mathbb{R}^{l \times m}$. Then, for all $t \geq 0$, the *total response* is

$$y(t) = Ce^{tA}x_0 + \int_0^t Ce^{(t-\tau)A}Bu(\tau)\, \mathrm{d}\tau + Du(t). \tag{12.1.14}$$

If $u(t) = 0$ for all $t \geq 0$, then the *free response* is given by

$$y(t) = Ce^{tA}x_0, \tag{12.1.15}$$

while, if $x_0 = 0$, then the *forced response* is given by

$$y(t) = \int_0^t Ce^{(t-\tau)A}Bu(\tau)\, \mathrm{d}\tau + Du(t). \tag{12.1.16}$$

Setting $u(t) = \delta(t)v$ yields, for all $t > 0$, the total response

$$y(t) = Ce^{tA}x_0 + H(t)v, \tag{12.1.17}$$

where, for all $t \geq 0$, the *impulse response function* $H(t)$ is defined by

$$H(t) \triangleq Ce^{tA}B + \delta(t)D. \tag{12.1.18}$$

The corresponding forced response is the *impulse response*

$$y(t) = H(t)v = Ce^{tA}Bv + \delta(t)Dv. \tag{12.1.19}$$

Alternatively, if $u(t) = v$ for all $t \geq 0$, then the total response is

$$y(t) = Ce^{tA}x_0 + \int_0^t Ce^{\tau A}\,\mathrm{d}\tau Bv + Dv, \qquad (12.1.20)$$

and the forced response is the *step response*

$$y(t) = \int_0^t H(\tau)\,\mathrm{d}\tau v = \int_0^t Ce^{\tau A}\,\mathrm{d}\tau Bv + Dv. \qquad (12.1.21)$$

In general, the forced response can be written as

$$y(t) = \int_0^t H(t - \tau)u(\tau)\,\mathrm{d}\tau. \qquad (12.1.22)$$

Setting $u(t) = \delta(t)v$ yields (12.1.20) by noting that

$$\int_0^t \delta(t - \tau)\delta(\tau)\mathrm{d}\tau = \delta(t). \qquad (12.1.23)$$

Proposition 12.1.2. Let $D = 0$ and $m = 1$, and assume that $x_0 = Bv$. Then, the free response and the impulse response are equal and given by

$$y(t) = Ce^{tA}x_0 = Ce^{tA}Bv. \qquad (12.1.24)$$

12.2 Laplace Transform Analysis

Now, consider the linear system

$$\dot{x}(t) = Ax(t) + Bu(t), \qquad (12.2.1)$$

$$y(t) = Cx(t) + Du(t), \qquad (12.2.2)$$

with state $x(t) \in \mathbb{R}^n$, input $u(t) \in \mathbb{R}^m$, and *output* $y(t) \in \mathbb{R}^l$, where $t \geq 0$ and $x(0) = x_0$. Taking Laplace transforms yields

$$s\hat{x}(s) - x_0 = A\hat{x}(s) + B\hat{u}(s), \qquad (12.2.3)$$

$$\hat{y}(s) = C\hat{x}(s) + D\hat{u}(s), \qquad (12.2.4)$$

where

$$\hat{x}(s) \triangleq \mathcal{L}\{x(t)\} = \int_0^\infty e^{-st}x(t)\,\mathrm{d}t, \qquad (12.2.5)$$

$$\hat{u}(s) \triangleq \mathcal{L}\{u(t)\}, \qquad (12.2.6)$$

and
$$\hat{y}(s) \triangleq \mathcal{L}\{y(t)\}. \tag{12.2.7}$$

Hence,
$$\hat{x}(s) = (sI - A)^{-1}x_0 + (sI - A)^{-1}B\hat{u}(s), \tag{12.2.8}$$

and thus
$$\hat{y}(s) = C(sI - A)^{-1}x_0 + \left[C(sI - A)^{-1}B + D\right]\hat{u}(s). \tag{12.2.9}$$

We can also obtain (12.2.9) from the time-domain expression for $y(t)$ given by (12.1.14). Using Proposition 11.2.2, it follows from (12.1.14) that

$$\hat{y}(s) = \mathcal{L}\{Ce^{tA}x_0\} + \mathcal{L}\left\{\int_0^t Ce^{(t-\tau)A}Bu(\tau)\,d\tau\right\} + D\hat{u}(s)$$

$$= C\mathcal{L}\{e^{tA}\}x_0 + C\mathcal{L}\{e^{tA}\}B\hat{u}(s) + D\hat{u}(s)$$

$$= C(sI - A)^{-1}x_0 + \left[C(sI - A)^{-1}B + D\right]\hat{u}(s), \tag{12.2.10}$$

which coincides with (12.2.9). We define

$$G(s) \triangleq C(sI - A)^{-1}B + D. \tag{12.2.11}$$

Note that $G \in \mathbb{R}^{l \times m}(s)$, that is, by Definition 4.7.2, G is a rational transfer function. Since $\mathcal{L}\{\delta(t)\} = 1$, it follows that

$$G(s) = \mathcal{L}\{H(t)\}. \tag{12.2.12}$$

Using (4.7.2), G can be written as

$$G(s) = \frac{1}{\chi_A(s)}C(sI - A)^A B + D. \tag{12.2.13}$$

It follows from (4.7.3) that G is a proper rational transfer function. Furthermore, G is a strictly proper rational transfer function if and only if $D = 0$, whereas G is an exactly proper rational transfer function if and only if $D \neq 0$. Finally, if A is nonsingular, then

$$G(0) = -CA^{-1}B + D. \tag{12.2.14}$$

Let $A \in \mathbb{R}^{n \times n}$. If $|s| > \text{sprad}(A)$, then Proposition 9.4.10 implies that

$$(sI - A)^{-1} = \frac{1}{s}\left(I - \frac{1}{s}A\right)^{-1} = \sum_{k=0}^{\infty} \frac{1}{s^{k+1}}A^k, \tag{12.2.15}$$

where the series is absolutely convergent, and thus

$$G(s) = \sum_{k=-1}^{\infty} \frac{1}{s^{k+1}}H_k, \tag{12.2.16}$$

where, for $k \geq -1$, the *Markov parameter* $H_k \in \mathbb{R}^{l \times m}$ is defined by

$$H_k \triangleq \begin{cases} D, & k = -1, \\ CA^k B, & k \geq 0. \end{cases} \tag{12.2.17}$$

It follows from (12.2.15) that $\lim_{s \to \infty} (sI - A)^{-1} = 0$, and thus

$$\lim_{s \to \infty} G(s) = D. \tag{12.2.18}$$

Finally, it follows from Definition 4.7.2 that

$$\text{reldeg}\, G = \min\{k \geq -1 \colon H_k \neq 0\}. \tag{12.2.19}$$

12.3 The Unobservable Subspace and Observability

Let $A \in \mathbb{R}^{n \times n}$ and $C \in \mathbb{R}^{l \times n}$, and, for $t \geq 0$, consider the linear system

$$\dot{x}(t) = Ax(t), \tag{12.3.1}$$
$$x(0) = x_0, \tag{12.3.2}$$
$$y(t) = Cx(t). \tag{12.3.3}$$

Definition 12.3.1. The *unobservable subspace* $\mathcal{U}_{t_f}(A, C)$ of (A, C) at time $t_f > 0$ is the subspace

$$\mathcal{U}_{t_f}(A, C) \triangleq \{x_0 \in \mathbb{R}^n \colon y(t) = 0 \text{ for all } t \in [0, t_f]\}. \tag{12.3.4}$$

Let $t_f > 0$. Then, Definition 12.3.1 states that $x_0 \in \mathcal{U}_{t_f}(A, C)$ if and only if $y(t) = 0$ for all $t \in [0, t_f]$. Since $y(t) = 0$ for all $t \in [0, t_f]$ is the free response corresponding to $x_0 = 0$, it follows that $0 \in \mathcal{U}_{t_f}(A, C)$. Now, suppose there exists a nonzero vector $x_0 \in \mathcal{U}_{t_f}(A, C)$. Then, with $x(0) = x_0$, the free response is given by $y(t) = 0$ for all $t \in [0, t_f]$, and thus x_0 cannot be determined from knowledge of $y(t)$ for all $t \in [0, t_f]$.

The following result provides explicit expressions for $\mathcal{U}_{t_f}(A, C)$.

Lemma 12.3.2. Let $t_f > 0$. Then, the following subspaces are equal:

i) $\mathcal{U}_{t_f}(A, C)$

ii) $\bigcap_{t \in [0, t_f]} \mathcal{N}(Ce^{tA})$

iii) $\bigcap_{i=0}^{n-1} \mathcal{N}(CA^i)$

iv) $\mathcal{N}\left(\begin{bmatrix} C \\ CA \\ \vdots \\ CA^{n-1} \end{bmatrix}\right)$

v) $\mathcal{N}\left(\int_0^{t_f} e^{tA^{\mathrm{T}}} C^{\mathrm{T}} C e^{tA}\, \mathrm{d}t\right)$

If, in addition, $\lim_{t_f \to \infty} \int_0^{t_f} e^{tA^T} C^T C e^{tA} dt$ exists, then the following subspace is equal to $i)$–$v)$:

$vi)$ $\mathcal{N}\left(\int_0^\infty e^{tA^T} C^T C e^{tA} dt \right)$

Proof. The proof is dual to the proof of Lemma 12.6.2. $\square$

Lemma 12.3.2 shows that $\mathcal{U}_{t_f}(A, C)$ is independent of t_f. We thus write $\mathcal{U}(A, C)$ for $\mathcal{U}_{t_f}(A, C)$, and call $\mathcal{U}(A, C)$ the *unobservable subspace* of (A, C). (A, C) is *observable* if $\mathcal{U}(A, C) = \{0\}$. For convenience, define the $nl \times n$ *observability matrix*

$$\mathcal{O}(A, C) \triangleq \begin{bmatrix} C \\ CA \\ \vdots \\ CA^{n-1} \end{bmatrix} \quad (12.3.5)$$

so that

$$\mathcal{U}(A, C) = \mathcal{N}[\mathcal{O}(A, C)]. \quad (12.3.6)$$

Define

$$p \triangleq n - \dim \mathcal{U}(A, C) = n - \operatorname{def} \mathcal{O}(A, C). \quad (12.3.7)$$

Corollary 12.3.3. For all $t_f > 0$,

$$p = \dim \mathcal{U}(A, C)^\perp = \operatorname{rank} \mathcal{O}(A, C) = \operatorname{rank} \int_0^{t_f} e^{tA^T} C^T C e^{tA} dt. \quad (12.3.8)$$

If, in addition, $\lim_{t_f \to \infty} \int_0^{t_f} e^{tA^T} C^T C e^{tA} dt$ exists, then

$$p = \operatorname{rank} \int_0^\infty e^{tA^T} C^T C e^{tA} dt. \quad (12.3.9)$$

Corollary 12.3.4. $\mathcal{U}(A, C)$ is an invariant subspace of A.

The following result shows that the unobservable subspace $\mathcal{U}(A, C)$ is unchanged by output injection $\dot{x}(t) = Ax(t) + Fy(t)$.

Proposition 12.3.5. Let $F \in \mathbb{R}^{n \times l}$. Then,

$$\mathcal{U}(A + FC, C) = \mathcal{U}(A, C). \quad (12.3.10)$$

In particular, (A, C) is observable if and only if $(A + FC, C)$ is observable.

Proof. The proof is dual to the proof of Proposition 12.6.5. $\square$

Let $\tilde{\mathcal{U}}(A, C) \subseteq \mathbb{R}^n$ be a subspace that is complementary to $\mathcal{U}(A, C)$. Then, $\tilde{\mathcal{U}}(A, C)$ is an *observable subspace* in the sense that, if $x_0 = x_0' + x_0''$, where $x_0' \in \tilde{\mathcal{U}}(A, C)$ is nonzero and $x_0'' \in \mathcal{U}(A, C)$, then it is possible to determine x_0' from knowledge of $y(t)$ for $t \in [0, t_f]$. Using Proposition 5.5.8, let $\mathcal{P} \in \mathbb{R}^{n \times n}$ be the unique idempotent matrix such that $\mathcal{R}(\mathcal{P}) = \tilde{\mathcal{U}}(A, C)$ and $\mathcal{N}(\mathcal{P}) = \mathcal{U}(A, C)$. Then, $x_0' = \mathcal{P}x_0$. The following result constructs $\mathcal{P}$ and provides an expression for x_0' in terms of $y(t)$ for $\tilde{\mathcal{U}}(A, C) \triangleq \mathcal{U}(A, C)^\perp$. In this case, $\mathcal{P}$ is a projector.

Lemma 12.3.6. Let $t_f > 0$, and define $\mathcal{P} \in \mathbb{R}^{n \times n}$ by

$$\mathcal{P} \triangleq \left(\int_0^{t_f} e^{tA^T} C^T C e^{tA} \, dt \right)^+ \int_0^{t_f} e^{tA^T} C^T C e^{tA} \, dt. \tag{12.3.11}$$

Then, $\mathcal{P}$ is the projector onto $\mathcal{U}(A, C)^\perp$, and $\mathcal{P}_\perp$ is the projector onto $\mathcal{U}(A, C)$. Hence,

$$\mathcal{R}(\mathcal{P}) = \mathcal{N}(\mathcal{P}_\perp) = \mathcal{U}(A, C)^\perp, \tag{12.3.12}$$

$$\mathcal{N}(\mathcal{P}) = \mathcal{R}(\mathcal{P}_\perp) = \mathcal{U}(A, C), \tag{12.3.13}$$

$$\operatorname{rank} \mathcal{P} = \operatorname{def} \mathcal{P}_\perp = \dim \mathcal{U}(A, C)^\perp = p, \tag{12.3.14}$$

$$\operatorname{def} \mathcal{P} = \operatorname{rank} \mathcal{P}_\perp = \dim \mathcal{U}(A, C) = n - p. \tag{12.3.15}$$

If $x_0 = x_0' + x_0''$, where $x_0' \in \mathcal{U}(A, C)^\perp$ and $x_0'' \in \mathcal{U}(A, C)$, then

$$x_0' = \mathcal{P}x_0 = \left(\int_0^{t_f} e^{tA^T} C^T C e^{tA} \, dt \right)^+ \int_0^{t_f} e^{tA^T} C^T y(t) \, dt. \tag{12.3.16}$$

Finally, (A, C) is observable if and only if $\mathcal{P} = I_n$. In this case, for all $x_0 \in \mathbb{R}^n$,

$$x_0 = \left(\int_0^{t_f} e^{tA^T} C^T C e^{tA} \, dt \right)^{-1} \int_0^{t_f} e^{tA^T} C^T y(t) \, dt. \tag{12.3.17}$$

Lemma 12.3.7. Let $\alpha \in \mathbb{R}$. Then,

$$\mathcal{U}(A + \alpha I, C) = \mathcal{U}(A, C). \tag{12.3.18}$$

The following result uses a coordinate transformation to characterize the observable dynamics of a system.

Theorem 12.3.8. There exists an orthogonal matrix $S \in \mathbb{R}^{n \times n}$ such that

$$A = S \begin{bmatrix} A_1 & 0 \\ A_{21} & A_2 \end{bmatrix} S^{-1}, \qquad C = \begin{bmatrix} C_1 & 0 \end{bmatrix} S^{-1}, \tag{12.3.19}$$

where $A_1 \in \mathbb{R}^{p \times p}$, $C_1 \in \mathbb{R}^{l \times p}$, and (A_1, C_1) is observable.

Proof. The proof is dual to the proof of Theorem 12.6.8. $\square$

Proposition 12.3.9. Let $S \in \mathbb{R}^{n \times n}$, and assume that S is orthogonal. Then, the following conditions are equivalent:

i) A and C have the form (12.3.19), where $A_1 \in \mathbb{R}^{p \times p}$, $C_1 \in \mathbb{R}^{l \times p}$, and (A_1, C_1) is observable.

ii) $\mathcal{U}(A, C) = \mathcal{R}\big(S\big[\begin{smallmatrix} 0 \\ I_{n-p} \end{smallmatrix}\big]\big)$.

iii) $\mathcal{U}(A, C)^{\perp} = \mathcal{R}\big(S\big[\begin{smallmatrix} I_p \\ 0 \end{smallmatrix}\big]\big)$.

iv) $\mathcal{P} = S\begin{bmatrix} I_p & 0 \\ 0 & 0 \end{bmatrix} S^{\mathrm{T}}$.

Proposition 12.3.10. Let $S \in \mathbb{R}^{n \times n}$, and assume that S is nonsingular. Then, the following conditions are equivalent:

i) A and C have the form (12.3.19), where $A_1 \in \mathbb{R}^{p \times p}$, $C_1 \in \mathbb{R}^{l \times p}$, and (A_1, C_1) is observable.

ii) $\mathcal{U}(A, C) = \mathcal{R}\big(S\big[\begin{smallmatrix} 0 \\ I_{n-p} \end{smallmatrix}\big]\big)$.

iii) $\mathcal{U}(A, C)^{\perp} = \mathcal{R}\big(S^{-\mathrm{T}}\big[\begin{smallmatrix} I_p \\ 0 \end{smallmatrix}\big]\big)$.

Definition 12.3.11. Let $\lambda \in \operatorname{spec}(A)$. Then, λ is an *observable eigenvalue* of (A, C) if

$$\operatorname{rank} \begin{bmatrix} \lambda I - A \\ C \end{bmatrix} = n. \tag{12.3.20}$$

Otherwise, λ is an *unobservable eigenvalue* of (A, C).

Proposition 12.3.12. Let $S \in \mathbb{R}^{n \times n}$, assume that S is nonsingular, let A and C have the form (12.3.19), where $A_1 \in \mathbb{R}^{p \times p}$, $C_1 \in \mathbb{R}^{l \times p}$, and (A_1, C_1) is observable, and let $\lambda \in \operatorname{spec}(A)$. Then, λ is an unobservable eigenvalue of (A, C) if and only if $\lambda \in \operatorname{spec}(A_2)$.

Proof. The proof is dual to the proof of Proposition 12.6.12. $\square$

Proposition 12.3.13. Let $\lambda \in \operatorname{mspec}(A)$ and $F \in \mathbb{R}^{n \times l}$. Then, λ is an unobservable eigenvalue of (A, C) if and only if λ is an unobservable eigenvalue of $(A + FC, C)$.

Lemma 12.3.14. Let $\lambda \in \operatorname{mspec}(A)$. Then,

$$\operatorname{Re} \mathcal{N}\left(\begin{bmatrix} \lambda I - A \\ C \end{bmatrix}\right) \subseteq \mathcal{U}(A, C). \tag{12.3.21}$$

Proof. Let $x_0 \in \operatorname{Re} \mathcal{N}\big(\big[\begin{smallmatrix} \lambda I - A \\ C \end{smallmatrix}\big]\big)$. Then, there exists a vector $x \in \mathcal{N}\big(\big[\begin{smallmatrix} \lambda I - A \\ C \end{smallmatrix}\big]\big)$ such that $x_0 = \operatorname{Re} x$. Since $x \in \mathcal{N}\big(\big[\begin{smallmatrix} \lambda I - A \\ C \end{smallmatrix}\big]\big)$, it follows that $Ax = \lambda x$ and $Cx = 0$. Then, for all $t \geq 0$, $y(t) = Ce^{tA}x_0 = Ce^{tA}\operatorname{Re} x = \operatorname{Re} Ce^{tA}x = \operatorname{Re} Ce^{\lambda t}x = \operatorname{Re} e^{\lambda t}Cx = 0$. Hence, $x_0 \in \mathcal{U}(A, C)$. $\square$

The next result characterizes observability in several equivalent ways.

Theorem 12.3.15. The following statements are equivalent:

i) (A, C) is observable.

ii) There exists $t > 0$ such that $\int_0^t e^{\tau A^{\mathrm{T}}} C^{\mathrm{T}} C e^{\tau A} \, \mathrm{d}\tau$ is positive definite.

iii) $\int_0^t e^{\tau A^{\mathrm{T}}} C^{\mathrm{T}} C e^{\tau A} \, \mathrm{d}\tau$ is positive definite for all $t > 0$.

iv) $\operatorname{rank} \mathcal{O}(A, C) = n$.

v) Every eigenvalue of (A, C) is observable.

If, in addition, $\lim_{t \to \infty} \int_0^t e^{\tau A^{\mathrm{T}}} C^{\mathrm{T}} C e^{\tau A} \mathrm{d}\tau$ exists, then the following condition is equivalent to *i)–v)*:

vi) $\int_0^\infty e^{t A^{\mathrm{T}}} C^{\mathrm{T}} C e^{t A} \mathrm{d}t$ is positive definite.

Proof. The proof is dual to the proof of Theorem 12.6.15.   □

12.4 Observable Asymptotic Stability

Let $A \in \mathbb{R}^{n \times n}$ and $C \in \mathbb{R}^{l \times n}$, and define $p \triangleq n - \dim \mathcal{U}(A, C)$.

Definition 12.4.1. (A, C) is *observably asymptotically stable* if

$$\mathcal{S}_{\mathrm{u}}(A) \subseteq \mathcal{U}(A, C). \tag{12.4.1}$$

Proposition 12.4.2. Let $F \in \mathbb{R}^{n \times l}$. Then, (A, C) is observably asymptotically stable if and only if $(A + FC, C)$ is observably asymptotically stable.

Proposition 12.4.3. The following statements are equivalent:

i) (A, C) is observably asymptotically stable.

ii) There exists an orthogonal matrix $S \in \mathbb{R}^{n \times n}$ such that (12.3.19) holds, where $A_1 \in \mathbb{R}^{p \times p}$ is asymptotically stable and $C_1 \in \mathbb{R}^{l \times p}$.

iii) There exists a nonsingular matrix $S \in \mathbb{R}^{n \times n}$ such that (12.3.19) holds, where $A_1 \in \mathbb{R}^{p \times p}$ is asymptotically stable and $C_1 \in \mathbb{R}^{l \times p}$.

iv) $\lim_{t \to \infty} C e^{t A} = 0$.

v) The positive-semidefinite matrix $P \in \mathbb{R}^{n \times n}$ defined by

$$P \triangleq \int_0^\infty e^{t A^{\mathrm{T}}} C^{\mathrm{T}} C e^{t A} \, \mathrm{d}t \tag{12.4.2}$$

exists.

vi) There exists a positive-semidefinite matrix $P \in \mathbb{R}^{n \times n}$ satisfying

$$A^{\mathrm{T}}P + PA + C^{\mathrm{T}}C = 0. \qquad (12.4.3)$$

In this case, the positive-semidefinite matrix $P \in \mathbb{R}^{n \times n}$ defined by (12.4.2) satisfies (12.4.3).

Proof. The proof is dual to the proof of Proposition 12.7.3. $\qquad \square$

The matrix P defined by (12.4.2) is the *observability Gramian*, and (12.4.3) is the *observation Lyapunov equation*.

Proposition 12.4.4. Assume that (A, C) is observably asymptotically stable, let $P \in \mathbb{R}^{n \times n}$ be the positive-semidefinite matrix defined by (12.4.2), and define $\mathcal{P} \in \mathbb{R}^{n \times n}$ by (12.3.11). Then, the following statements hold:

i) $PP^{+} = \mathcal{P}$.

ii) $\mathcal{R}(P) = \mathcal{R}(\mathcal{P}) = \mathcal{U}(A, C)^{\perp}$.

iii) $\mathcal{N}(P) = \mathcal{N}(\mathcal{P}) = \mathcal{U}(A, C)$.

iv) $\operatorname{rank} P = \operatorname{rank} \mathcal{P} = p$.

v) P is the only positive-semidefinite solution of (12.4.3) whose rank is p.

Proof. The proof is dual to the proof of Proposition 12.7.4. $\qquad \square$

Proposition 12.4.5. Assume that (A, C) is observably asymptotically stable, let $P \in \mathbb{R}^{n \times n}$ be the positive-semidefinite matrix defined by (12.4.2), and let $\hat{P} \in \mathbb{R}^{n \times n}$. Then, the following statements are equivalent:

i) $\hat{P}$ is positive semidefinite and satisfies (12.4.3).

ii) There exists a positive-semidefinite matrix $P_0 \in \mathbb{R}^{n \times n}$ such that $\hat{P} = P + P_0$ and $A^{\mathrm{T}}P_0 + P_0 A = 0$.

In this case,

$$\operatorname{rank} \hat{P} = p + \operatorname{rank} P_0 \qquad (12.4.4)$$

and

$$\operatorname{rank} P_0 \leq \sum_{\substack{\lambda \in \operatorname{spec}(A) \\ \lambda \in J\mathbb{R}}} \operatorname{gm}_A(\lambda). \qquad (12.4.5)$$

Proof. The proof is dual to the proof of Proposition 12.7.5. $\qquad \square$

Proposition 12.4.6. The following statements are equivalent:

i) (A, C) is observably asymptotically stable, every imaginary eigenvalue of A is semisimple, and A has no ORHP eigenvalues.

ii) (12.4.3) has a positive-definite solution $P \in \mathbb{R}^{n \times n}$.

Proof. The proof is dual to the proof of Proposition 12.7.6.   □

Proposition 12.4.7. The following statements are equivalent:

i) (A, C) is observably asymptotically stable, and A has no imaginary eigenvalues.

ii) (12.4.3) has exactly one positive-semidefinite solution $P \in \mathbb{R}^{n \times n}$.

In this case, $P \in \mathbb{R}^{n \times n}$ is given by (12.4.2) and satisfies rank $P = p$.

Proof. The proof is dual to the proof of Proposition 12.7.7.   □

Corollary 12.4.8. Assume that A is asymptotically stable. Then, the positive-semidefinite matrix $P \in \mathbb{R}^{n \times n}$ defined by (12.4.2) is the unique solution of (12.4.3) and satisfies rank $P = p$.

Proof. The proof is dual to the proof of Corollary 12.7.4.   □

Proposition 12.4.9. The following statements are equivalent:

i) (A, C) is observable, and A is asymptotically stable.

ii) (12.4.3) has exactly one positive-semidefinite solution $P \in \mathbb{R}^{n \times n}$, and P is positive definite.

In this case, $P \in \mathbb{R}^{n \times n}$ is given by (12.4.2).

Proof. The proof is dual to the proof of Proposition 12.7.9.   □

Corollary 12.4.10. Assume that A is asymptotically stable. Then, the positive-semidefinite matrix $P \in \mathbb{R}^{n \times n}$ defined by (12.4.2) exists. Furthermore, P is positive definite if and only if (A, C) is observable.

12.5 Detectability

Let $A \in \mathbb{R}^{n \times n}$ and $C \in \mathbb{R}^{l \times n}$, and define $p \triangleq n - \dim \mathcal{U}(A, C)$.

Definition 12.5.1. (A, C) is *detectable* if
$$\mathcal{U}(A, C) \subseteq \mathcal{S}_\mathrm{s}(A). \tag{12.5.1}$$

Proposition 12.5.2. Let $F \in \mathbb{R}^{n \times l}$. Then, (A, C) is detectable if and only if $(A + FC, C)$ is detectable.

Proposition 12.5.3. The following statements are equivalent:

$i)$ A is asymptotically stable.

$ii)$ (A, C) is detectable and observably asymptotically stable.

Proof. The proof is dual to the proof of Proposition 12.8.3. $\square$

Proposition 12.5.4. The following statements are equivalent:

$i)$ (A, C) is detectable.

$ii)$ There exists an orthogonal matrix $S \in \mathbb{R}^{n \times n}$ such that (12.3.19) holds, where $A_1 \in \mathbb{R}^{p \times p}$, $C_1 \in \mathbb{R}^{l \times p}$, (A_1, C_1) is observable, and $A_2 \in \mathbb{R}^{(n-p) \times (n-p)}$ is asymptotically stable.

$iii)$ There exists a nonsingular matrix $S \in \mathbb{R}^{n \times n}$ such that (12.3.19) holds, where $A_1 \in \mathbb{R}^{p \times p}$, $C_1 \in \mathbb{R}^{l \times p}$, (A_1, C_1) is observable, and $A_2 \in \mathbb{R}^{(n-p) \times (n-p)}$ is asymptotically stable.

$iv)$ Every CRHP eigenvalue of (A, C) is observable.

Proof. The proof is dual to the proof of Proposition 12.8.4. $\square$

Proposition 12.5.5. The following statements are equivalent:

$i)$ (A, C) is observably asymptotically stable and detectable.

$ii)$ A is asymptotically stable.

Proof. The proof is dual to the proof of Proposition 12.8.5. $\square$

Corollary 12.5.6. The following statements are equivalent:

$i)$ There exists a positive-semidefinite matrix $P \in \mathbb{R}^{n \times n}$ satisfying (12.4.3), and (A, C) is detectable.

$ii)$ A is asymptotically stable.

Proof. The proof is dual to the proof of Proposition 12.8.6. $\square$

12.6 The Controllable Subspace and Controllability

Let $A \in \mathbb{R}^{n \times n}$ and $B \in \mathbb{R}^{n \times m}$, and, for $t \geq 0$, consider the linear system

$$\dot{x}(t) = Ax(t) + Bu(t), \qquad (12.6.1)$$
$$x(0) = 0. \qquad (12.6.2)$$

Definition 12.6.1. The *controllable subspace* $\mathcal{C}_{t_f}(A, B)$ of (A, B) at time $t_f > 0$ is the subspace

$$\mathcal{C}_{t_f}(A, B) \triangleq \{x_f \in \mathbb{R}^n: \quad \text{there exists a continuous control } u\colon [0, t_f] \mapsto \mathbb{R}^m$$
$$\text{such that the solution } x(\cdot) \text{ of } (12.6.1), (12.6.2) \text{ satisfies } x(t_f) = x_f\}.$$
$$(12.6.3)$$

Let $t_f > 0$. Then, Definition 12.6.1 states that $x_f \in \mathcal{C}_{t_f}(A, B)$ if and only if there exists a continuous control $u\colon [0, t_f] \mapsto \mathbb{R}^m$ such that

$$x_f = \int_0^{t_f} e^{(t_f - t)A} Bu(t)\, \mathrm{d}t. \qquad (12.6.4)$$

The following result provides explicit expressions for $\mathcal{C}_{t_f}(A, B)$.

Lemma 12.6.2. Let $t_f > 0$. Then, the following subspaces are equal:

i) $\mathcal{C}_{t_f}(A, B)$

ii) $\left[\bigcap_{t \in [0, t_f]} \mathcal{N}\left(B^{\mathrm{T}} e^{tA^{\mathrm{T}}}\right) \right]^{\perp}$

iii) $\left[\bigcap_{i=0}^{n-1} \mathcal{N}(B^{\mathrm{T}} A^{i\mathrm{T}}) \right]^{\perp}$

iv) $\mathcal{R}([\, B \quad AB \quad \cdots \quad A^{n-1}B \,])$

v) $\mathcal{R}\left(\int_0^{t_f} e^{tA} BB^{\mathrm{T}} e^{tA^{\mathrm{T}}}\, \mathrm{d}t \right)$

If, in addition, $\lim_{t_f \to \infty} \int_0^{t_f} e^{tA} BB^{\mathrm{T}} e^{tA^{\mathrm{T}}}\, \mathrm{d}t$ exists, then the following subspace is equal to *i)*–*v)*:

vi) $\mathcal{R}\left(\int_0^{\infty} e^{tA} BB^{\mathrm{T}} e^{tA^{\mathrm{T}}}\, \mathrm{d}t \right)$

Proof. To prove that *i)* $\subseteq$ *ii)*, let $\eta \in \bigcap_{t \in [0, t_f]} \mathcal{N}\left(B^{\mathrm{T}} e^{tA^{\mathrm{T}}}\right)$ so that $\eta^{\mathrm{T}} e^{tA} B = 0$ for all $t \in [0, t_f]$. Now, let $u\colon [0, t_f] \mapsto \mathbb{R}^m$ be continuous. Then, $\eta^{\mathrm{T}} \int_0^{t_f} e^{(t_f - t)A} Bu(t)\, \mathrm{d}t = 0$, which implies that $\eta \in \mathcal{C}_{t_f}(A, B)^{\perp}$.

To prove that *ii)* $\subseteq$ *iii)*, let $\eta \in \bigcap_{i=0}^{n-1} \mathcal{N}(B^{\mathrm{T}} A^{i\mathrm{T}})$ so that $\eta^{\mathrm{T}} A^i B = 0$ for all $i = 0, 1, \ldots, n-1$. It follows from the Cayley-Hamilton theorem Theorem 4.4.6 that $\eta^{\mathrm{T}} A^i B = 0$ for all $i \geq 0$. Now, let $t \in [0, t_f]$. Then, $\eta^{\mathrm{T}} e^{tA} B = \sum_{i=0}^{\infty} t^i (i!)^{-1} \eta^{\mathrm{T}} A^i B = 0$, and thus $\eta \in \mathcal{N}\left(B^{\mathrm{T}} e^{tA^{\mathrm{T}}}\right)$.

To show that *iii)* $\subseteq$ *iv)*, let $\eta \in \mathcal{R}([\, B \quad AB \quad \cdots \quad A^{n-1}B \,])^{\perp}$. Then, $\eta \in \mathcal{N}\left([\, B \quad AB \quad \cdots \quad A^{n-1}B \,]^{\mathrm{T}}\right)$, which implies that $\eta^{\mathrm{T}} A^i B = 0$ for all

$i = 0, 1, \ldots, n - 1$.

To prove that $iv) \subseteq v)$, let $\eta \in \mathcal{N}\left(\int_0^{t_\mathrm{f}} e^{tA} BB^\mathrm{T} e^{tA^\mathrm{T}} \mathrm{d}t\right)$. Then,

$$\eta^\mathrm{T} \int_0^{t_\mathrm{f}} e^{tA} BB^\mathrm{T} e^{tA^\mathrm{T}} \mathrm{d}t \eta = 0,$$

which implies that $\eta^\mathrm{T} e^{tA} B = 0$ for all $t \in [0, t_\mathrm{f}]$. Differentiating with respect to t and setting $t = 0$ implies that $\eta^\mathrm{T} A^i B = 0$ for all $i = 0, 1, \ldots, n-1$. Hence, $\eta \in \mathcal{R}\left(\begin{bmatrix} B & AB & \cdots & A^{n-1}B \end{bmatrix}\right)^\perp$.

To prove that $v) \subseteq i)$, let $\eta \in \mathcal{C}_{t_\mathrm{f}}(A, B)^\perp$. Then, $\eta^\mathrm{T} \int_0^{t_\mathrm{f}} e^{(t_\mathrm{f} - t)A} Bu(t)\, \mathrm{d}t = 0$ for all continuous $u\colon [0, t_\mathrm{f}] \mapsto \mathbb{R}^m$. Letting $u(t) = B^\mathrm{T} e^{(t_\mathrm{f} - t)A^\mathrm{T}} \eta^\mathrm{T}$, implies that $\eta^\mathrm{T} \int_0^{t_\mathrm{f}} e^{tA} BB^\mathrm{T} e^{tA^\mathrm{T}} \mathrm{d}t \eta = 0$, and thus $\eta \in \mathcal{N}\left(\int_0^{t_\mathrm{f}} e^{tA} BB^\mathrm{T} e^{tA^\mathrm{T}} \mathrm{d}t\right)$. $\square$

Lemma 12.6.2 shows that $\mathcal{C}_{t_\mathrm{f}}(A, B)$ is independent of t_f. We thus write $\mathcal{C}(A, B)$ for $\mathcal{C}_{t_\mathrm{f}}(A, B)$, and call $\mathcal{C}(A, B)$ the *controllable subspace* of (A, B). (A, B) is *controllable* if $\mathcal{C}(A, B) = \mathbb{R}^n$. For convenience, define the $m \times nm$ *controllability matrix*

$$\mathcal{K}(A, B) \triangleq \begin{bmatrix} B & AB & \cdots & A^{n-1}B \end{bmatrix} \tag{12.6.5}$$

so that

$$\mathcal{C}(A, B) = \mathcal{R}[\mathcal{K}(A, B)]. \tag{12.6.6}$$

Define

$$q \triangleq \dim \mathcal{C}(A, B) = \operatorname{rank} \mathcal{K}(A, B). \tag{12.6.7}$$

Corollary 12.6.3. For all $t_\mathrm{f} > 0$,

$$q = \dim \mathcal{C}(A, B) = \operatorname{rank} \mathcal{K}(A, C) = \operatorname{rank} \int_0^{t_\mathrm{f}} e^{tA} BB^\mathrm{T} e^{tA^\mathrm{T}} \mathrm{d}t. \tag{12.6.8}$$

If, in addition, $\lim_{t_\mathrm{f} \to \infty} \int_0^{t_\mathrm{f}} e^{tA} BB^\mathrm{T} e^{tA^\mathrm{T}} \mathrm{d}t$ exists, then

$$q = \operatorname{rank} \int_0^\infty e^{tA} BB^\mathrm{T} e^{tA^\mathrm{T}} \mathrm{d}t. \tag{12.6.9}$$

Corollary 12.6.4. $\mathcal{C}(A, B)$ is an invariant subspace of A.

The following result shows that the controllable subspace $\mathcal{C}(A, B)$ is unchanged by full-state feedback $u(t) = Kx(t) + v(t)$.

Proposition 12.6.5. Let $K \in \mathbb{R}^{m \times n}$. Then,

$$\mathcal{C}(A + BK, B) = \mathcal{C}(A, B). \tag{12.6.10}$$

In particular, (A, B) is controllable if and only if $(A+BK, B)$ is controllable.

Proof. Note that

$$
\begin{aligned}
\mathcal{C}(A &+ BK, B) \\
&= \mathcal{R}[\mathcal{K}(A + BK, B)] \\
&= \mathcal{R}([\ B \quad AB + BKB \quad A^2B + ABKB + BKAB + BKBKB \quad \cdots \ ]) \\
&= \mathcal{R}[\mathcal{K}(A, B)] = \mathcal{C}(A, B). \qquad\qquad\qquad\qquad\qquad\qquad \square
\end{aligned}
$$

Let $\tilde{\mathcal{C}}(A, B) \subseteq \mathbb{R}^n$ be a subspace that is complementary to $\mathcal{C}(A, B)$. Then, $\tilde{\mathcal{C}}(A, B)$ is an *uncontrollable subspace* in the sense that, if $x_{\mathrm{f}} = x'_{\mathrm{f}} + x''_{\mathrm{f}} \in \mathbb{R}^n$, where $x'_{\mathrm{f}} \in \mathcal{C}(A, B)$ and $x''_{\mathrm{f}} \in \tilde{\mathcal{C}}(A, B)$ is nonzero, then there exists a continuous control $u\colon [0, t_{\mathrm{f}}] \to \mathbb{R}^m$ such that $x(t_{\mathrm{f}}) = x'_{\mathrm{f}}$ although there exists no continuous control such that $x(t_{\mathrm{f}}) = x_{\mathrm{f}}$. Using Proposition 5.5.8, let $\mathcal{Q} \in \mathbb{R}^{n \times n}$ be the unique idempotent matrix such that $\mathcal{R}(\mathcal{Q}) = \mathcal{C}(A, B)$ and $\mathcal{N}(\mathcal{Q}) = \tilde{\mathcal{C}}(A, B)$. Then, $x'_{\mathrm{f}} = \mathcal{Q}x_{\mathrm{f}}$. The following result constructs $\mathcal{Q}$ and a continuous control $u(\cdot)$ that yields $x(t_{\mathrm{f}}) = x'_{\mathrm{f}}$ for $\tilde{\mathcal{C}}(A, B) \triangleq \mathcal{C}(A, B)^{\perp}$. In this case, $\mathcal{Q}$ is a projector.

Lemma 12.6.6. Let $t_{\mathrm{f}} > 0$, and define $\mathcal{Q} \in \mathbb{R}^{n \times n}$ by

$$
\mathcal{Q} \triangleq \left(\int_0^{t_{\mathrm{f}}} e^{tA}BB^{\mathrm{T}}e^{tA^{\mathrm{T}}}\,\mathrm{d}t \right)^+ \int_0^{t_{\mathrm{f}}} e^{tA}BB^{\mathrm{T}}e^{tA^{\mathrm{T}}}\,\mathrm{d}t. \tag{12.6.11}
$$

Then, $\mathcal{Q}$ is the projector onto $\mathcal{C}(A, B)$, and $\mathcal{Q}_{\perp}$ is the projector onto $\mathcal{C}(A, B)^{\perp}$. Hence,

$$
\mathcal{R}(\mathcal{Q}) = \mathcal{N}(\mathcal{Q}_{\perp}) = \mathcal{C}(A, B), \tag{12.6.12}
$$

$$
\mathcal{N}(\mathcal{Q}) = \mathcal{R}(\mathcal{Q}) = \mathcal{C}(A, B)^{\perp}, \tag{12.6.13}
$$

$$
\operatorname{rank} \mathcal{Q} = \operatorname{def} \mathcal{Q}_{\perp} = \dim \mathcal{C}(A, B) = q, \tag{12.6.14}
$$

$$
\operatorname{def} \mathcal{Q} = \operatorname{rank} \mathcal{Q}_{\perp} = \dim \mathcal{C}(A, B)^{\perp} = n - q. \tag{12.6.15}
$$

Now, define $u\colon [0, t_{\mathrm{f}}] \mapsto \mathbb{R}^m$ by

$$
u(t) \triangleq B^{\mathrm{T}}e^{(t_{\mathrm{f}}-t)A^{\mathrm{T}}} \left(\int_0^{t_{\mathrm{f}}} e^{\tau A}BB^{\mathrm{T}}e^{\tau A^{\mathrm{T}}}\,\mathrm{d}\tau \right)^+ x_{\mathrm{f}}. \tag{12.6.16}
$$

If $x_{\mathrm{f}} = x'_{\mathrm{f}} + x''_{\mathrm{f}}$, where $x'_{\mathrm{f}} \in \mathcal{C}(A, B)$ and $x''_{\mathrm{f}} \in \mathcal{C}(A, B)^{\perp}$, then

$$
x'_{\mathrm{f}} = \mathcal{Q}x_{\mathrm{f}} = \int_0^{t_{\mathrm{f}}} e^{(t_{\mathrm{f}}-t)A}Bu(t)\,\mathrm{d}t. \tag{12.6.17}
$$

Finally, (A, B) is controllable if and only if $\mathcal{Q} = I_n$. In this case, for all

$x_f \in \mathbb{R}^n$,

$$x_f = \int_0^{t_f} e^{(t_f - t)A} B u(t) \, dt, \qquad (12.6.18)$$

where $u: [0, t_f] \mapsto \mathbb{R}^m$ is given by

$$u(t) = B^T e^{(t_f - t)A^T} \left(\int_0^{t_f} e^{\tau A} B B^T e^{\tau A^T} \, d\tau \right)^{-1} x_f. \qquad (12.6.19)$$

Lemma 12.6.7. Let $\alpha \in \mathbb{R}$. Then,

$$\mathcal{C}(A + \alpha I, B) = \mathcal{C}(A, B). \qquad (12.6.20)$$

The following result uses a coordinate transformation to characterize the controllable dynamics of a system.

Theorem 12.6.8. There exists an orthogonal matrix $S \in \mathbb{R}^{n \times n}$ such that

$$A = S \begin{bmatrix} A_1 & A_{12} \\ 0 & A_2 \end{bmatrix} S^{-1}, \qquad B = S \begin{bmatrix} B_1 \\ 0 \end{bmatrix}, \qquad (12.6.21)$$

where $A_1 \in \mathbb{R}^{q \times q}$, $B_1 \in \mathbb{R}^{q \times m}$, and (A_1, B_1) is controllable.

Proof. Let $\alpha < 0$ be such that $A_\alpha \triangleq A + \alpha I$ is asymptotically stable, and let $Q \in \mathbb{R}^{n \times n}$ be the positive-semidefinite solution of

$$A_\alpha Q + Q A_\alpha^T + B B^T = 0 \qquad (12.6.22)$$

given by

$$Q = \int_0^\infty e^{t A_\alpha} B B^T e^{t A_\alpha^T} \, dt.$$

It now follows from Lemma 12.6.2 and Lemma 12.6.7 that

$$\mathcal{R}(Q) = \mathcal{R}[\mathcal{C}(A_\alpha, B)] = \mathcal{R}[\mathcal{C}(A, B)].$$

Hence,

$$\operatorname{rank} Q = \dim \mathcal{C}(A_\alpha, B) = \dim \mathcal{C}(A, B) = q.$$

Next, let $S \in \mathbb{R}^{n \times n}$ be an orthogonal matrix such that $Q = S \begin{bmatrix} Q_1 & 0 \\ 0 & 0 \end{bmatrix} S^T$, where $Q_1 \in \mathbb{R}^{q \times q}$ is positive definite. Writing $A_\alpha = S \begin{bmatrix} \hat{A}_1 & \hat{A}_{12} \\ \hat{A}_{21} & \hat{A}_2 \end{bmatrix} S^{-1}$ and $B = S \begin{bmatrix} B_1 \\ B_2 \end{bmatrix}$, where $\hat{A}_1 \in \mathbb{R}^{q \times q}$ and $B_1 \in \mathbb{R}^{q \times m}$, it follows from (12.6.22) that

$$\hat{A}_1 Q_1 + Q_1 \hat{A}_1^T + B_1 B_1^T = 0,$$
$$\hat{A}_{21} Q_1 + B_2 B_1^T = 0,$$
$$B_2 B_2^T = 0.$$

Therefore, $B_2 = 0$ and $\hat{A}_{21} = 0$, and thus

$$A_\alpha = S \begin{bmatrix} \hat{A}_1 & \hat{A}_{12} \\ 0 & \hat{A}_2 \end{bmatrix} S^{-1}, \quad B = S \begin{bmatrix} B_1 \\ 0 \end{bmatrix}.$$

Furthermore,

$$A = S \begin{bmatrix} \hat{A}_1 & \hat{A}_{12} \\ 0 & \hat{A}_2 \end{bmatrix} S^{-1} - \alpha I = S \left(\begin{bmatrix} \hat{A}_1 & \hat{A}_{12} \\ 0 & \hat{A}_2 \end{bmatrix} - \alpha I \right) S^{-1}.$$

Hence,

$$A = S \begin{bmatrix} A_1 & A_{12} \\ 0 & A_2 \end{bmatrix} S^{-1},$$

where $A_1 \triangleq \hat{A}_1 - \alpha I_q$, $A_{12} \triangleq \hat{A}_{12}$, and $A_2 \triangleq \hat{A}_2 - \alpha I_{n-q}$. $\square$

Proposition 12.6.9. Let $S \in \mathbb{R}^{n \times n}$, and assume that S is orthogonal. Then, the following conditions are equivalent:

  i) A and B have the form (12.6.21), where $A_1 \in \mathbb{R}^{q \times q}$, $B_1 \in \mathbb{R}^{q \times m}$, and (A_1, B_1) is controllable.

  ii) $\mathcal{C}(A, B) = \mathcal{R}\left(S\begin{bmatrix} I_q \\ 0 \end{bmatrix}\right)$.

  iii) $\mathcal{C}(A, B)^\perp = \mathcal{R}\left(S\begin{bmatrix} 0 \\ I_{n-q} \end{bmatrix}\right)$.

  iv) $\mathcal{Q} = S \begin{bmatrix} I_q & 0 \\ 0 & 0 \end{bmatrix} S^{\mathrm{T}}$.

Proposition 12.6.10. Let $S \in \mathbb{R}^{n \times n}$, and assume that S is nonsingular. Then, the following conditions are equivalent:

  i) A and B have the form (12.6.21), where $A_1 \in \mathbb{R}^{q \times q}$, $B_1 \in \mathbb{R}^{q \times m}$, and (A_1, B_1) is controllable.

  ii) $\mathcal{C}(A, B) = \mathcal{R}\left(S\begin{bmatrix} I_q \\ 0 \end{bmatrix}\right)$.

  iii) $\mathcal{C}(A, B)^\perp = \mathcal{R}\left(S^{-\mathrm{T}}\begin{bmatrix} 0 \\ I_{n-q} \end{bmatrix}\right)$.

Definition 12.6.11. Let $\lambda \in \mathrm{spec}(A)$. Then, λ is a *controllable eigenvalue* of (A, B) if

$$\mathrm{rank} \begin{bmatrix} \lambda I - A & B \end{bmatrix} = n. \qquad (12.6.23)$$

Otherwise, λ is an *uncontrollable eigenvalue* of (A, B).

Proposition 12.6.12. Let $S \in \mathbb{R}^{n \times n}$, assume that S is nonsingular, let A and B have the form (12.6.21), where $A_1 \in \mathbb{R}^{q \times q}$, $B_1 \in \mathbb{R}^{q \times m}$, and (A_1, B_1) is controllable, and let $\lambda \in \mathrm{spec}(A)$. Then, λ is an uncontrollable eigenvalue of (A, B) if and only if $\lambda \in \mathrm{spec}(A_2)$.

Proof. Since (A_1, B_1) is controllable, it follows that

$$\operatorname{rank} \begin{bmatrix} \lambda I - A & B \end{bmatrix} = \operatorname{rank} \begin{bmatrix} \lambda I - A_1 & A_{12} & B_1 \\ 0 & \lambda I - A_2 & 0 \end{bmatrix}$$

$$= \operatorname{rank} \begin{bmatrix} \lambda I - A_1 & B_1 \end{bmatrix} + \operatorname{rank}(\lambda I - A_2)$$

$$= q + \operatorname{rank}(\lambda I - A_2).$$

Hence, $\operatorname{rank} \begin{bmatrix} \lambda I - A & B \end{bmatrix} < n$ if and only if $\operatorname{rank}(\lambda I - A_2) < n - q$, that is, if and only if $\lambda \in \operatorname{spec}(A_2)$. $\qquad\square$

Proposition 12.6.13. Let $\lambda \in \operatorname{mspec}(A)$ and $K \in \mathbb{R}^{n \times m}$. Then, λ is an uncontrollable eigenvalue of (A, B) if and only if λ is an uncontrollable eigenvalue of $(A + BK, B)$.

Proposition 12.6.14. Let $\lambda \in \operatorname{mspec}(A)$. Then,

$$\mathcal{C}(A, B) \subseteq \left[\operatorname{Re} \mathcal{R}(\begin{bmatrix} \lambda I - A & B \end{bmatrix})^\perp \right]^\perp. \qquad (12.6.24)$$

Proof. First, note that (12.6.24) is equivalent to

$$\operatorname{Re} \mathcal{N}\left(\begin{bmatrix} \bar{\lambda} I - A^{\mathrm{T}} \\ C \end{bmatrix} \right) = \operatorname{Re} \mathcal{R}(\begin{bmatrix} \lambda I - A & B \end{bmatrix})^\perp \subseteq \mathcal{C}(A, B)^\perp.$$

Let $x_{\mathrm{f}} \in \operatorname{Re} \mathcal{N}\left(\begin{bmatrix} \bar{\lambda} I - A^{\mathrm{T}} \\ B^{\mathrm{T}} \end{bmatrix} \right)$. Then, there exists a vector $x \in \mathcal{N}\left(\begin{bmatrix} \bar{\lambda} I - A^{\mathrm{T}} \\ B^{\mathrm{T}} \end{bmatrix} \right)$ such that $x_{\mathrm{f}} = \operatorname{Re} x$. Since $x \in \mathcal{N}\left(\begin{bmatrix} \bar{\lambda} I - A^{\mathrm{T}} \\ B^{\mathrm{T}} \end{bmatrix} \right)$, it follows that $A^{\mathrm{T}} x = \bar{\lambda} x$ and $B^{\mathrm{T}} x = 0$. Now, let $\mathcal{Q}$ be defined by (12.6.11). Then, since $e^{tA^{\mathrm{T}}} x = e^{\bar{\lambda} t} x$, it follows that $\mathcal{Q} x_{\mathrm{f}} = \mathcal{Q} \operatorname{Re} x = \operatorname{Re} \mathcal{Q} x = 0$. Now, (12.6.13) implies that $x_{\mathrm{f}} \in \mathcal{C}(A, B)^\perp$. $\qquad\square$

The next result characterizes controllability in several equivalent ways.

Theorem 12.6.15. The following statements are equivalent:

i) (A, B) is controllable.

ii) There exists $t > 0$ such that $\int_0^t e^{\tau A} B B^{\mathrm{T}} e^{\tau A^{\mathrm{T}}} \, d\tau$ is positive definite.

iii) $\int_0^t e^{\tau A} B B^{\mathrm{T}} e^{\tau A^{\mathrm{T}}} \, d\tau$ is positive definite for all $t > 0$.

iv) $\operatorname{rank} \mathcal{K}(A, B) = n$.

v) Every eigenvalue of (A, B) is controllable.

If, in addition, $\lim_{t \to \infty} \int_0^t e^{\tau A} B B^{\mathrm{T}} e^{\tau A^{\mathrm{T}}} \, d\tau$ exists, then the following condition is equivalent to *i)*–*v)*:

vi) $\int_0^\infty e^{tA} B B^{\mathrm{T}} e^{tA^{\mathrm{T}}} \, dt$ is positive definite.

Proof. The equivalence of i)–iv) follows from Lemma 12.6.2.

To prove iv) $\implies$ v), suppose that v) does not hold, that is, there exist $\lambda \in \mathrm{spec}(A)$ and a nonzero vector $x \in \mathbb{C}^n$ such that $x^*A = \lambda x^*$ and $x^*B = 0$. It thus follows that $x^*AB = \lambda x^*B = 0$. Similarly, $x^*A^iB = 0$ for all $i = 0, 1, \ldots, n-1$. Hence, $(\mathrm{Re}\,x)^{\mathrm{T}}\mathcal{K}(A, B) = 0$ and $(\mathrm{Im}\,x)^{\mathrm{T}}\mathcal{K}(A, B) = 0$. Since $\mathrm{Re}\,x$ and $\mathrm{Im}\,x$ are not both zero, it follows that $\dim \mathcal{C}(A, B) < n$.

Conversely, to show that v) implies iv), suppose that $\mathrm{rank}\,\mathcal{K}(A, B) < n$. Then, there exists a nonzero vector $x \in \mathbb{R}^n$ such that $x^{\mathrm{T}}A^iB = 0$ for all $i = 0, \ldots, n-1$. Now, let $p \in \mathbb{R}[s]$ be a nonzero polynomial of minimal degree such that $x^{\mathrm{T}}p(A) = 0$. Note that p is not a constant polynomial and that $x^{\mathrm{T}}\mu_A(A) = 0$. Thus, $1 \leq \deg p \leq \deg \mu_A$. Now, let $\lambda \in \mathbb{C}$ be such that $p(\lambda) = 0$, and let $q \in \mathbb{R}[s]$ be such that $p(s) = q(s)(s - \lambda)$ for all $s \in \mathbb{C}$. Since $\deg q < \deg p$, it follows that $x^{\mathrm{T}}q(A) \neq 0$. Therefore, $\eta \triangleq q(A)x$ is nonzero. Furthermore, $\eta^{\mathrm{T}}(A - \lambda I) = x^{\mathrm{T}}p(A) = 0$. Since $x^{\mathrm{T}}A^iB = 0$ for all $i = 0, \ldots, n-1$, it follows that $\eta^{\mathrm{T}}B = x^{\mathrm{T}}q(A)B = 0$. Consequently, v) does not hold. $\qquad\square$

The following result implies that arbitrary eigenvalue placement is possible for (12.6.1) when (A, B) is controllable.

Proposition 12.6.16. The pair (A, B) is controllable if and only if, for every polynomial $p \in \mathbb{R}[s]$ such that $\deg p = n$, there exists a matrix $K \in \mathbb{R}^{m \times n}$ $\mathrm{mspec}(A + BK) = \mathrm{mroots}(p)$.

Proof. For the case $m = 1$ let $A_\mathrm{c} \triangleq C(\chi_A)$ and $B_\mathrm{c} \triangleq e_n$ as in (12.9.5). Then, Proposition 12.9.3 implies that $\mathcal{K}(A_\mathrm{c}, B_\mathrm{c})$ is nonsingular, while Corollary 12.9.9 implies that $A_\mathrm{c} = S^{-1}AS$ and $B_\mathrm{c} = S^{-1}B$. Now, let $\mathrm{mroots}(p) = \{\lambda_1, \ldots, \lambda_n\}_\mathrm{m} \subset \mathbb{C}$. Letting $K \triangleq e_n^{\mathrm{T}}[C(p) - A_\mathrm{c}]S^{-1}$ it follows that

$$
\begin{aligned}
A + BK &= S(A_\mathrm{c} + B_\mathrm{c}KS)S^{-1} \\
&= S(A_\mathrm{c} + E_{n,n}[C(p) - A_\mathrm{c}])S^{-1} \\
&= SC(p)S^{-1}.
\end{aligned}
$$

The case $m > 1$ requires the multivariable controllable canonical form. See [624, p. 248]. $\qquad\square$

12.7 Controllable Asymptotic Stability

Let $A \in \mathbb{R}^{n \times n}$ and $B \in \mathbb{R}^{n \times m}$, and define $q \triangleq \dim \mathcal{C}(A, C)$.

Definition 12.7.1. (A, B) is *controllably asymptotically stable* if

$$\mathcal{C}(A, B) \subseteq \mathcal{S}_\mathrm{s}(A). \tag{12.7.1}$$

Proposition 12.7.2. Let $K \in \mathbb{R}^{m \times n}$. Then, (A, B) is controllably asymptotically stable if and only if $(A + BK, B)$ is controllably asymptotically stable.

Proposition 12.7.3. The following statements are equivalent:

i) (A, B) is controllably asymptotically stable.

ii) There exists an orthogonal matrix $S \in \mathbb{R}^{n \times n}$ such that (12.6.21) holds, where $A_1 \in \mathbb{R}^{q \times q}$ is asymptotically stable and $B_1 \in \mathbb{R}^{q \times m}$.

iii) There exists a nonsingular matrix $S \in \mathbb{R}^{n \times n}$ such that (12.6.21) holds, where $A_1 \in \mathbb{R}^{q \times q}$ is asymptotically stable and $B_1 \in \mathbb{R}^{q \times m}$.

iv) $\lim_{t \to \infty} e^{tA}B = 0$.

v) The positive-semidefinite matrix $Q \in \mathbb{R}^{n \times n}$ defined by

$$Q \triangleq \int_0^\infty e^{tA}BB^{\mathrm{T}}e^{tA^{\mathrm{T}}}\,\mathrm{d}t \tag{12.7.2}$$

exists.

vi) There exists a positive-semidefinite matrix $Q \in \mathbb{R}^{n \times n}$ satisfying

$$AQ + QA^{\mathrm{T}} + BB^{\mathrm{T}} = 0. \tag{12.7.3}$$

In this case, the positive-semidefinite matrix $Q \in \mathbb{R}^{n \times n}$ defined by (12.7.2) satisfies (12.7.3).

Proof. To prove *i)* $\Longrightarrow$ *ii)*, assume that (A, B) is controllably asymptotically stable so that $\mathcal{C}(A, B) \subseteq \mathcal{S}_{\mathrm{s}}(A) = \mathcal{N}[\mu_A^{\mathrm{s}}(A)] = \mathcal{R}[\mu_A^{\mathrm{u}}(A)]$. Using Theorem 12.6.8, it follows that there exists an orthogonal matrix $S \in \mathbb{R}^{n \times n}$ such that (12.6.21) is satisfied, where $A_1 \in \mathbb{R}^{q \times q}$ and (A_1, B_1) is controllable. Thus, $\mathcal{R}\big(S\big[{I_q \atop 0}\big]\big) = \mathcal{C}(A, B) \subseteq \mathcal{R}[\mu_A^{\mathrm{s}}(A)]$.

Next, note that

$$\mu_A^{\mathrm{s}}(A) = S\begin{bmatrix} \mu_A^{\mathrm{s}}(A_1) & B_{12\mathrm{s}} \\ 0 & \mu_A^{\mathrm{s}}(A_2) \end{bmatrix}S^{-1},$$

where $B_{12\mathrm{s}} \in \mathbb{R}^{q \times (n-q)}$, and suppose that A_1 is not asymptotically stable with CRHP eigenvalue λ. Then, $\lambda \notin \mathrm{roots}(\mu_A^{\mathrm{s}})$, and thus $\mu_A^{\mathrm{s}}(A_1) \neq 0$. Let $x_1 \in \mathbb{R}^{n-q}$ satisfy $\mu_A^{\mathrm{s}}(A_1)x_1 \neq 0$. Then, $\big[{x_1 \atop 0}\big] \in \mathcal{R}\big(S\big[{I_q \atop 0}\big]\big) = \mathcal{C}(A, B)$ and

$$\mu_A^{\mathrm{s}}(A)S\begin{bmatrix} x_1 \\ 0 \end{bmatrix} = S\begin{bmatrix} \mu_A^{\mathrm{s}}(A_1)x_1 \\ 0 \end{bmatrix},$$

and thus $\big[{x_1 \atop 0}\big] \notin \mathcal{N}[\mu_A^{\mathrm{s}}(A)] = \mathcal{S}_{\mathrm{s}}(A)$, which implies that $\mathcal{C}(A, B)$ is not contained in $\mathcal{S}_{\mathrm{s}}(A)$. Hence, A_1 is asymptotically stable.

To prove *iii*) $\Longrightarrow$ *iv*), assume there exists a nonsingular matrix $S \in \mathbb{R}^{n \times n}$ such that (12.6.21) holds, where $A_1 \in \mathbb{R}^{k \times k}$ is asymptotically stable and $B_1 \in \mathbb{R}^{k \times m}$. Thus, $e^{tA}B = \begin{bmatrix} e^{tA_1}B_1 \\ 0 \end{bmatrix} S \to 0$ as $t \to \infty$.

Next, to prove that *iv*) implies *v*), assume that $e^{tA}B \to 0$ as $t \to \infty$. Then, every entry of $e^{tA}B$ involves exponentials of t, where the coefficients of t have negative real part. Hence, so does every entry of $e^{tA}BB^{\mathrm{T}}e^{tA^{\mathrm{T}}}$, which implies that $\int_0^\infty e^{tA}BB^{\mathrm{T}}e^{tA^{\mathrm{T}}} \, \mathrm{d}t$ exists.

To prove *v*) $\Longrightarrow$ *vi*), note that, since $Q = \int_0^\infty e^{tA}BB^{\mathrm{T}}e^{tA^{\mathrm{T}}} \, \mathrm{d}t$ exists, it follows that $e^{tA}BB^{\mathrm{T}}e^{tA^{\mathrm{T}}} \to 0$ as $t \to \infty$. Thus,

$$AQ + QA^{\mathrm{T}} = \int_0^\infty \left[Ae^{tA}BB^{\mathrm{T}}e^{tA^{\mathrm{T}}} + e^{tA}BB^{\mathrm{T}}e^{tA^{\mathrm{T}}}A \right] \mathrm{d}t = \int_0^\infty \frac{\mathrm{d}}{\mathrm{d}t} e^{tA}BB^{\mathrm{T}}e^{tA^{\mathrm{T}}} \, \mathrm{d}t$$

$$= \lim_{t \to \infty} e^{tA}BB^{\mathrm{T}}e^{tA^{\mathrm{T}}} - BB^{\mathrm{T}} = -BB^{\mathrm{T}},$$

which shows that Q satisfies (12.4.3).

To prove *vi*) $\Longrightarrow$ *i*), suppose there exists a positive-semidefinite matrix $Q \in \mathbb{R}^{n \times n}$ satisfying (12.7.3). Then,

$$\int_0^t e^{tA}BB^{\mathrm{T}}e^{tA^{\mathrm{T}}} \, \mathrm{d}\tau = -\int_0^t e^{\tau A}\left(AQ + QA^{\mathrm{T}} \right)e^{tA^{\mathrm{T}}} \, \mathrm{d}\tau = -\int_0^t \frac{\mathrm{d}}{\mathrm{d}\tau} e^{\tau A}QA^{\mathrm{T}} \, \mathrm{d}\tau$$

$$= Q - e^{tA}Qe^{tA^{\mathrm{T}}} \le Q.$$

Next, it follows from Theorem 12.6.8 that there exists an orthogonal matrix $S \in \mathbb{R}^{n \times n}$ such that (12.6.21) is satisfied, where $A_1 \in \mathbb{R}^{q \times q}$, $B_1 \in \mathbb{R}^{q \times m}$, and (A_1, B_1) is controllable. Consequently, we have

$$\int_0^t e^{\tau A_1}B_1B_1^{\mathrm{T}}e^{\tau A_1^{\mathrm{T}}} \, \mathrm{d}\tau = \begin{bmatrix} I & 0 \end{bmatrix} S \int_0^t e^{\tau A}BB^{\mathrm{T}}e^{\tau A^{\mathrm{T}}} \, \mathrm{d}\tau S^{\mathrm{T}} \begin{bmatrix} I \\ 0 \end{bmatrix}$$

$$\le \begin{bmatrix} I & 0 \end{bmatrix} SQS^{\mathrm{T}} \begin{bmatrix} I \\ 0 \end{bmatrix}.$$

Thus, it follows from Proposition 8.5.3 that $Q_1 \triangleq \int_0^\infty e^{tA_1}B_1B_1^{\mathrm{T}}e^{tA_1^{\mathrm{T}}} \, \mathrm{d}t$ exists. Since (A_1, B_1) is controllable, it follows from *vii*) of Theorem 12.6.15 that Q_1 is positive definite.

Now, let λ be an eigenvalue of A_1^{T}, and let $x_1 \in \mathbb{C}^n$ be an associated eigenvector. Consequently, $\alpha \triangleq x_1^*Q_1x_1$ is positive, and

$$\alpha = x_1^* \int_0^\infty e^{\bar{\lambda}t}BB_1^{\mathrm{T}}e^{\lambda t} \, \mathrm{d}t x_1 = x_1^* B_1 B_1^{\mathrm{T}} x_1 \int_0^\infty e^{2(\mathrm{Re}\,\lambda)t} \, \mathrm{d}t.$$

Hence, $\int_0^\infty e^{2(\operatorname{Re}\lambda)t}\,dt = \alpha/x_1^* B_1 B_1^{\mathrm{T}} x_1$ exists, and thus $\operatorname{Re}\lambda < 0$. Consequently, A_1 is asymptotically stable, and thus $\mathcal{C}(A,B) \subseteq \mathcal{S}_{\mathrm{s}}(A)$, that is, (A,B) is controllably asymptotically stable. $\qquad\square$

The matrix $Q \in \mathbb{R}^{n\times n}$ defined by (12.7.2) is the *controllability Gramian*, and (12.7.3) is the *control Lyapunov equation*.

Proposition 12.7.4. Assume that (A,B) is controllably asymptotically stable, let $Q \in \mathbb{R}^{n\times n}$ be the positive-semidefinite matrix defined by (12.7.2), and define $\mathcal{Q} \in \mathbb{R}^{n\times n}$ by (12.6.11). Then, the following statements hold:

i) $QQ^+ = \mathcal{Q}$.

ii) $\mathcal{R}(Q) = \mathcal{R}(\mathcal{Q}) = \mathcal{C}(A,B)$.

iii) $\mathcal{N}(Q) = \mathcal{N}(\mathcal{Q}) = \mathcal{C}(A,B)^\perp$.

iv) $\operatorname{rank} Q = \operatorname{rank} \mathcal{Q} = q$.

v) Q is the only positive-semidefinite solution of (12.7.3) whose rank is q.

Proof. See [649] for the proof of v). $\qquad\square$

Proposition 12.7.5. Assume that (A,B) is controllably asymptotically stable, let $Q \in \mathbb{R}^{n\times n}$ be the positive-semidefinite matrix defined by (12.7.2), and let $\hat{Q} \in \mathbb{R}^{n\times n}$. Then, the following statements are equivalent:

i) $\hat{Q}$ is positive semidefinite and satisfies (12.7.3).

ii) There exists a positive-semidefinite matrix $Q_0 \in \mathbb{R}^{n\times n}$ such that $\hat{Q} = Q + Q_0$ and $AQ_0 + Q_0 A^{\mathrm{T}} = 0$.

In this case,

$$\operatorname{rank} \hat{Q} = q + \operatorname{rank} Q_0 \qquad (12.7.4)$$

and

$$\operatorname{rank} Q_0 \le \sum_{\substack{\lambda \in \operatorname{spec}(A) \\ \lambda \in \jmath\mathbb{R}}} \operatorname{gm}_A(\lambda). \qquad (12.7.5)$$

Proof. See [649]. $\qquad\square$

Proposition 12.7.6. The following statements are equivalent:

i) (A,B) is controllably asymptotically stable, every imaginary eigenvalue of A is semisimple, and A has no ORHP eigenvalues.

ii) (12.7.3) has a positive-definite solution $Q \in \mathbb{R}^{n\times n}$.

Proof. See [649]. $\qquad\square$

Proposition 12.7.7. The following statements are equivalent:

i) (A, B) is controllably asymptotically stable, and A has no imaginary eigenvalues.

ii) (12.7.3) has exactly one positive-semidefinite solution $Q \in \mathbb{R}^{n \times n}$.

In this case, $Q \in \mathbb{R}^{n \times n}$ is given by (12.7.2) and satisfies rank $Q = q$.

Proof. See [649]. □

Corollary 12.7.8. Assume that A is asymptotically stable. Then, the positive-semidefinite matrix $Q \in \mathbb{R}^{n \times n}$ defined by (12.7.2) is the unique solution of (12.7.3) and satisfies rank $Q = q$.

Proof. See [649]. □

Proposition 12.7.9. The following statements are equivalent:

i) (A, B) is controllable, and A is asymptotically stable.

ii) (12.7.3) has exactly one positive-semidefinite solution $Q \in \mathbb{R}^{n \times n}$, and Q is positive definite.

In this case, $Q \in \mathbb{R}^{n \times n}$ is given by (12.7.2).

Proof. See [649]. □

Corollary 12.7.10. Assume that A is asymptotically stable. Then, the positive-semidefinite matrix $Q \in \mathbb{R}^{n \times n}$ defined by (12.7.2) exists. Furthermore, Q is positive definite if and only if (A, B) is controllable.

12.8 Stabilizability

Let $A \in \mathbb{R}^{n \times n}$ and $B \in \mathbb{R}^{n \times m}$, and define $q \triangleq \dim \mathcal{C}(A, C)$.

Definition 12.8.1. (A, B) is *stabilizable* if

$$\mathcal{S}_{\mathrm{u}}(A) \subseteq \mathcal{C}(A, B). \tag{12.8.1}$$

Proposition 12.8.2. Let $K \in \mathbb{R}^{m \times n}$. Then, (A, B) is stabilizable if and only if $(A + BK, B)$ is stabilizable.

Proposition 12.8.3. The following statements are equivalent:

i) A is asymptotically stable.

ii) (A, B) is stabilizable and controllably asymptotically stable.

Proof. Assume that A is asymptotically stable. Then, $\mathcal{S}_{\mathrm{u}}(A) = \{0\}$, and $\mathcal{S}_{\mathrm{s}}(A) = \mathbb{R}^n$. Thus, $\mathcal{S}_{\mathrm{u}}(A) \subseteq \mathcal{C}(A, B)$, and $\mathcal{C}(A, B) \subseteq \mathcal{S}_{\mathrm{s}}(A)$. Conversely, assume that (A, B) is stabilizable and controllably asymptotically stable. Then, $\mathcal{S}_{\mathrm{u}}(A) \subseteq \mathcal{C}(A, B) \subseteq \mathcal{S}_{\mathrm{s}}(A)$, and thus $\mathcal{S}_{\mathrm{u}}(A) = \{0\}$. $\square$

Proposition 12.8.4. The following statements are equivalent:

i) (A, B) is stabilizable.

ii) There exists an orthogonal matrix $S \in \mathbb{R}^{n \times n}$ such that (12.6.21) holds, where $A_1 \in \mathbb{R}^{q \times q}$, $B_1 \in \mathbb{R}^{q \times m}$, (A_1, B_1) is controllable, and $A_2 \in \mathbb{R}^{(n-q) \times (n-q)}$ is asymptotically stable.

iii) There exists a nonsingular matrix $S \in \mathbb{R}^{n \times n}$ such that (12.6.21) holds, where $A_1 \in \mathbb{R}^{q \times q}$, $B_1 \in \mathbb{R}^{q \times m}$, (A_1, B_1) is controllable, and $A_2 \in \mathbb{R}^{(n-q) \times (n-q)}$ is asymptotically stable.

iv) Every CRHP eigenvalue of (A, B) is controllable.

Proof. To prove $i) \Longrightarrow ii)$, assume that (A, B) is stabilizable so that $\mathcal{S}_{\mathrm{u}}(A) = \mathcal{N}[\mu_A^{\mathrm{u}}(A)] = \mathcal{R}[\mu_A^{\mathrm{s}}(A)] \subseteq \mathcal{C}(A, B)$. Using Theorem 12.6.8, it follows that there exists an orthogonal matrix $S \in \mathbb{R}^{n \times n}$ such that (12.6.21) is satisfied, where $A_1 \in \mathbb{R}^{q \times q}$ and (A_1, B_1) is controllable. Thus, $\mathcal{R}[\mu_A^{\mathrm{s}}(A)] \subseteq \mathcal{C}(A, B) = \mathcal{R}\left(S\begin{bmatrix} I_q \\ 0 \end{bmatrix}\right)$.

Next, note that

$$\mu_A^{\mathrm{s}}(A) = S \begin{bmatrix} \mu_A^{\mathrm{s}}(A_1) & B_{12\mathrm{s}} \\ 0 & \mu_A^{\mathrm{s}}(A_2) \end{bmatrix} S^{-1},$$

where $B_{12\mathrm{s}} \in \mathbb{R}^{q \times (n-q)}$, and suppose that A_2 is not asymptotically stable with CRHP eigenvalue λ. Then, $\lambda \notin \mathrm{roots}(\mu_A^{\mathrm{s}})$, and thus $\mu_A^{\mathrm{s}}(A_2) \neq 0$. Let $x_2 \in \mathbb{R}^{n-q}$ satisfy $\mu_A^{\mathrm{s}}(A_2)x_2 \neq 0$. Then,

$$\mu_A^{\mathrm{s}}(A)S \begin{bmatrix} 0 \\ x_2 \end{bmatrix} = S \begin{bmatrix} B_{12\mathrm{s}}x_2 \\ \mu_A^{\mathrm{s}}(A_2)x_2 \end{bmatrix} \notin \mathcal{R}\left(S \begin{bmatrix} I_q \\ 0 \end{bmatrix}\right) = \mathcal{C}(A, B),$$

which implies that $\mathcal{S}_{\mathrm{u}}(A)$ is not contained in $\mathcal{C}(A, B)$. Hence, A_2 is asymptotically stable.

Clearly, $ii)$ implies $iii)$.

To prove $iii) \Longrightarrow iv)$, let $\lambda \in \mathrm{spec}(A)$ be a CRHP eigenvalue of A. Since A_2 is asymptotically stable, it follows that $\lambda \notin \mathrm{spec}(A_2)$. Consequently, Proposition 12.6.12 implies that λ is not an uncontrollable eigenvalue of (A, B), and thus λ is a controllable eigenvalue of (A, B).

To prove $iv) \Longrightarrow i)$, let $S \in \mathbb{R}^{n \times n}$ be nonsingular and such that A and B have the form (12.6.21), where $A_1 \in \mathbb{R}^{q \times q}$, $B_1 \in \mathbb{R}^{q \times m}$, and (A_1, B_1) is controllable. Since every CRHP eigenvalue of (A, B) is controllable, it follows from Proposition 12.6.12 that A_2 is asymptotically stable. From Fact 11.19.4 it follows that $\mathcal{S}_u(A) \subseteq \mathcal{R}(S[\begin{smallmatrix} I_q \\ 0 \end{smallmatrix}]) = \mathcal{C}(A, B)$, which implies that (A, B) is stabilizable. $\quad\square$

Proposition 12.8.5. The following statements are equivalent:

i) (A, B) is controllably asymptotically stable and stabilizable.

ii) A is asymptotically stable.

Proof. Since (A, B) is stabilizable, it follows from Proposition 12.5.4 that there exists a nonsingular matrix $S \in \mathbb{R}^{n \times n}$ such that (12.6.21) holds, where $A_1 \in \mathbb{R}^{q \times q}$, $B_1 \in \mathbb{R}^{q \times m}$, (A_1, B_1) is controllable, and $A_2 \in \mathbb{R}^{(n-q) \times (n-q)}$ is asymptotically stable. Then,

$$\int_0^\infty e^{tA} B B^\mathrm{T} e^{tA^\mathrm{T}} \, \mathrm{d}t = S \begin{bmatrix} \int_0^\infty e^{tA_1} B_1 B_1^\mathrm{T} e^{tA_1^\mathrm{T}} \, \mathrm{d}t & 0 \\ 0 & 0 \end{bmatrix} S^{-1}.$$

Since the integral on the left-hand side exists by assumption, the integral on the right-hand side also exists. Since (A_1, B_1) is controllable, it follows from *vii*) of Theorem 12.6.15 that $Q_1 \triangleq \int_0^\infty e^{tA_1} B_1 B_1^\mathrm{T} e^{tA_1^\mathrm{T}} \, \mathrm{d}t$ is positive definite.

Now, let λ be an eigenvalue of A_1^T, and let $x_1 \in \mathbb{C}^q$ be an associated eigenvector. Consequently, $\alpha \triangleq x_1^* Q_1 x_1$ is positive, and

$$\alpha = x_1^* \int_0^\infty e^{\bar\lambda t} B_1 B_1^\mathrm{T} e^{\lambda t} \, \mathrm{d}t x_1 = x_1^* B_1 B_1^\mathrm{T} x_1 \int_0^\infty e^{2(\mathrm{Re}\,\lambda)t} \, \mathrm{d}t.$$

Hence, $\int_0^\infty e^{2(\mathrm{Re}\,\lambda)t} \, \mathrm{d}t$ exists, and thus $\mathrm{Re}\,\lambda < 0$. Consequently, A_1 is asymptotically stable, and thus A is asymptotically stable. $\quad\square$

Corollary 12.8.6. The following statements are equivalent:

i) There exists a positive-semidefinite matrix $Q \in \mathbb{R}^{n \times n}$ satisfying (12.7.3), and (A, B) is stabilizable.

ii) A is asymptotically stable.

Proof. The result follows from Proposition 12.7.3 and Proposition 12.8.5. $\quad\square$

12.9 Realization Theory

Given a proper rational transfer function G we wish to determine (A, B, C, D) such that (12.2.11) holds. The following terminology is convenient.

Definition 12.9.1. Let $G \in \mathbb{R}^{l \times m}(s)$. If $l = m = 1$, then G is a *single-input/single-output (SISO)* rational transfer function; if $l = 1$ and $m > 1$, then G is a *multiple-input/single-output (MISO)* rational transfer function; if $l > 1$ and $m = 1$, then G is a *single-input/multiple-output (SIMO)* rational transfer function; and, if $l > 1$ or $m > 1$, then G is a *multiple-input/multiple output (MIMO)* rational transfer function.

Definition 12.9.2. Let $G \in \mathbb{R}^{l \times m}_{\mathrm{prop}}(s)$, and assume that $A \in \mathbb{R}^{n \times n}$, $B \in \mathbb{R}^{n \times m}$, $C \in \mathbb{R}^{l \times n}$, and $D \in \mathbb{R}^{l \times m}$ satisfy $G(s) = C(sI - A)^{-1}B + D$. Then, $\left[\begin{array}{c|c} A & B \\ \hline C & D \end{array}\right]$ is a *realization* of G, which is written as

$$G \sim \left[\begin{array}{c|c} A & B \\ \hline C & D \end{array}\right].\tag{12.9.1}$$

The *order* of the realization (12.9.1) is the order of A.

Suppose that $n = 0$. Then, A, B, and C are empty matrices, and G is given by

$$G(s) = 0_{l \times 0}(sI_{0 \times 0} - 0_{0 \times 0})^{-1}0_{0 \times m} + D = 0_{l \times m} + D = D.\tag{12.9.2}$$

Therefore, the order of the realization $\left[\begin{array}{c|c} 0_{0 \times 0} & 0_{0 \times m} \\ \hline 0_{l \times 0} & D \end{array}\right]$ is zero.

Although the realization (12.9.1) is not unique, the matrix D is unique and is given by

$$D = G(\infty).\tag{12.9.3}$$

Furthermore, note that $G \sim \left[\begin{array}{c|c} A & B \\ \hline C & D \end{array}\right]$ if and only if $G - D \sim \left[\begin{array}{c|c} A & B \\ \hline C & 0 \end{array}\right]$. Therefore, it suffices to construct realizations for strictly proper transfer functions.

The following result shows that every strictly proper, SISO rational transfer function G has a realization. In fact, two realizations are the *controllable canonical form* $G \sim \left[\begin{array}{c|c} A_c & B_c \\ \hline C_c & 0 \end{array}\right]$ and the *observable canonical form* $G \sim \left[\begin{array}{c|c} A_o & B_o \\ \hline C_o & 0 \end{array}\right]$. If G is exactly proper, then G can be replaced by $G - G(\infty)$.

Proposition 12.9.3. Let $G \in \mathbb{R}_{\mathrm{prop}}(s)$ be the SISO strictly proper rational transfer function

$$G(s) = \frac{\alpha_{n-1}s^{n-1} + \alpha_{n-2}s^{n-2} + \cdots + \alpha_1 s + \alpha_0}{s^n + \beta_{n-1}s^{n-1} + \cdots + \beta_1 s + \beta_0}.\tag{12.9.4}$$

Then, $G \sim \left[\begin{array}{c|c} A_c & B_c \\ \hline C_c & 0 \end{array}\right]$, where A_c, B_c, C_c are defined by

$$A_\mathrm{c} \triangleq \begin{bmatrix} 0 & 1 & 0 & \cdots & 0 \\ 0 & 0 & 1 & \cdots & 0 \\ \vdots & \vdots & \vdots & \ddots & \vdots \\ 0 & 0 & 0 & \cdots & 1 \\ -\beta_0 & -\beta_1 & -\beta_2 & \cdots & -\beta_{n-1} \end{bmatrix}, \quad B_\mathrm{c} \triangleq \begin{bmatrix} 0 \\ \vdots \\ 0 \\ 1 \end{bmatrix}, \qquad (12.9.5)$$

$$C_\mathrm{c} \triangleq \begin{bmatrix} \alpha_0 & \alpha_1 & \cdots & \alpha_{n-1} \end{bmatrix}, \qquad (12.9.6)$$

and $G \sim \left[\begin{array}{c|c} A_\mathrm{o} & B_\mathrm{o} \\ \hline C_\mathrm{o} & 0 \end{array} \right]$, where $A_\mathrm{o}, B_\mathrm{o}, C_\mathrm{o}$ are defined by

$$A_\mathrm{o} \triangleq \begin{bmatrix} 0 & 0 & \cdots & 0 & -\beta_0 \\ 1 & 0 & \cdots & 0 & -\beta_1 \\ 0 & 1 & \cdots & 0 & -\beta_2 \\ \vdots & \vdots & \ddots & & \vdots \\ 0 & 0 & \cdots & 1 & -\beta_{n-1} \end{bmatrix}, \quad B_\mathrm{o} \triangleq \begin{bmatrix} \alpha_0 \\ \alpha_1 \\ \vdots \\ \alpha_{n-1} \end{bmatrix}, \qquad (12.9.7)$$

$$C_\mathrm{o} \triangleq \begin{bmatrix} 0 & \cdots & 0 & 1 \end{bmatrix}. \qquad (12.9.8)$$

Furthermore, $(A_\mathrm{c}, B_\mathrm{c})$ is controllable, and $(A_\mathrm{o}, C_\mathrm{o})$ is observable. Finally, the following statements are equivalent:

i) The numerator and denominator of G given in 12.9.4 are coprime.

ii) $(A_\mathrm{c}, C_\mathrm{c})$ is observable.

iii) $(A_\mathrm{o}, B_\mathrm{o})$ is controllable.

Proof. The realizations can be verified directly. Furthermore, note that

$$\mathcal{K}(A_\mathrm{c}, B_\mathrm{c}) = \mathcal{O}(A_\mathrm{o}, C_\mathrm{o}) = \begin{bmatrix} 0 & 0 & 0 & \cdots & 0 & 1 \\ 0 & 0 & 0 & \iddots & 1 & -\beta_{n-1} \\ \vdots & \vdots & \iddots & \iddots & \iddots & \vdots \\ 0 & 0 & 1 & \iddots & -\beta_3 & -\beta_2 \\ 0 & 1 & -\beta_{n-1} & \cdots & -\beta_2 & -\beta_1 \\ 1 & -\beta_{n-1} & -\beta_{n-2} & \cdots & -\beta_1 & -\beta_0 \end{bmatrix}.$$

It follows from Fact 2.12.7 that $\det \mathcal{K}(A_\mathrm{c}, B_\mathrm{c}) = \det \mathcal{O}(A_\mathrm{o}, C_\mathrm{o}) = (-1)^{\lfloor n/2 \rfloor}$, which implies that $(A_\mathrm{c}, B_\mathrm{c})$ is controllable and $(A_\mathrm{o}, C_\mathrm{o})$ is observable.

To prove the last statement, let $p, q \in \mathbb{R}[s]$ denote the numerator and denominator, respectively, of G in (12.9.4). Then, for $n = 2$,

$$\mathcal{K}(A_\mathrm{o}, B_\mathrm{o}) = \mathcal{O}^\mathrm{T}(A_\mathrm{c}, C_\mathrm{c}) = B(p, q)\hat{I} \begin{bmatrix} 1 & -\beta_1 \\ 0 & 1 \end{bmatrix},$$

where $B(p, q)$ is the Bezout matrix of p and q. It follows from $ix)$ of Fact 4.8.6 that $B(p, q)$ is nonsingular if and only if p and q are coprime. $\square$

The following result shows that every proper rational transfer function has a realization.

Theorem 12.9.4. Let $G \in \mathbb{R}_{\text{prop}}^{l \times m}(s)$. Then, there exist matrices $A \in \mathbb{R}^{n \times n}$, $B \in \mathbb{R}^{n \times m}$, $C \in \mathbb{R}^{l \times n}$, and $D \in \mathbb{R}^{l \times m}$ such that $G \sim \left[\begin{array}{c|c} A & B \\ \hline C & D \end{array}\right]$.

Proof. By Proposition 12.9.3, every entry $G_{(i,j)}$ of G has a realization $G_{(i,j)} \sim \left[\begin{array}{c|c} A_{ij} & B_{ij} \\ \hline C_{ij} & D_{ij} \end{array}\right]$. Combining these realizations yields a realization of G. $\square$

Proposition 12.9.5. Let $G \in \mathbb{R}_{\text{prop}}^{l \times m}(s)$, let $\left[\begin{array}{c|c} A & B \\ \hline C & D \end{array}\right]$ be a realization of G, where $A \in \mathbb{R}^{n \times n}$, let $S \in \mathbb{R}^{n \times n}$, and assume that S is nonsingular. Then, $\left[\begin{array}{c|c} SAS^{-1} & SB \\ \hline CS^{-1} & D \end{array}\right]$ is also a realization of G.

Definition 12.9.6. Let $G \in \mathbb{R}_{\text{prop}}^{l \times m}(s)$, and let $\left[\begin{array}{c|c} A & B \\ \hline C & D \end{array}\right]$ and $\left[\begin{array}{c|c} \hat{A} & \hat{B} \\ \hline \hat{C} & D \end{array}\right]$ be nth-order realizations of G. Then, $\left[\begin{array}{c|c} A & B \\ \hline C & D \end{array}\right]$ and $\left[\begin{array}{c|c} \hat{A} & \hat{B} \\ \hline \hat{C} & D \end{array}\right]$ are *equivalent* if there exists a nonsingular matrix $S \in \mathbb{R}^{n \times n}$ such that $\hat{A} = SAS^{-1}$, $\hat{B} = SB$, and $\hat{C} = CS^{-1}$.

For the next result, define the *Markov block-Hankel matrix* $\mathcal{H}(A, B, C) \in \mathbb{R}^{nl \times nm}$ by

$$\mathcal{H}(A, B, C) \triangleq \mathcal{O}(A, C)\mathcal{K}(A, B), \tag{12.9.9}$$

which is the block-Hankel matrix of Markov parameters given by

$$\mathcal{H}(A, B, C) = \begin{bmatrix} CB & CAB & CA^2B & \cdots & CA^{n-1}B \\ CAB & CA^2B & \cdot^{\cdot^\cdot} & \cdot^{\cdot^\cdot} & \cdot^{\cdot^\cdot} \\ CA^2B & \cdot^{\cdot^\cdot} & \cdot^{\cdot^\cdot} & \cdot^{\cdot^\cdot} & \cdot^{\cdot^\cdot} \\ \vdots & \cdot^{\cdot^\cdot} & \cdot^{\cdot^\cdot} & \cdot^{\cdot^\cdot} & \cdot^{\cdot^\cdot} \\ CA^{n-1}B & \cdot^{\cdot^\cdot} & \cdot^{\cdot^\cdot} & \cdot^{\cdot^\cdot} & CA^{2n-2}B \end{bmatrix}. \tag{12.9.10}$$

Lemma 12.9.7. Let $G \in \mathbb{R}_{\text{prop}}^{l \times m}(s)$, and assume that G has the nth-order realization $\left[\begin{array}{c|c} A & B \\ \hline C & D \end{array}\right]$. Then, $\left[\begin{array}{c|c} A & B \\ \hline C & D \end{array}\right]$ is controllable and observable if and only if

$$\text{rank}\, \mathcal{H}(A, B, C) = n. \tag{12.9.11}$$

Proof. To prove necessity, note that, since the $n \times n$ matrices $\mathcal{O}^{\mathrm{T}}(A, C)\mathcal{O}(A, C)$ and $\mathcal{K}(A, B)\mathcal{K}^{\mathrm{T}}(A, B)$ are positive definite, it follows that

$$n = \mathrm{rank}\, \mathcal{O}^{\mathrm{T}}(A, C)\mathcal{O}(A, C)\mathcal{K}(A, B)\mathcal{K}^{\mathrm{T}}(A, B) \leq \mathrm{rank}\, \mathcal{H}(A, B, C) \leq n.$$

Conversely, $\mathrm{rank}\, \mathcal{H}(A, B, C) = n$ implies that $\mathrm{rank}\, \mathcal{O}(A, C) = \mathrm{rank}\, \mathcal{K}(A, B)$ $= n$. $\qquad\square$

Proposition 12.9.8. Let $G \in \mathbb{R}_{\mathrm{prop}}^{l \times m}(s)$, assume that G has the nth-order realizations $\left[\begin{array}{c|c} A_1 & B_1 \\ \hline C_1 & D \end{array}\right]$ and $\left[\begin{array}{c|c} A_2 & B_2 \\ \hline C_2 & D \end{array}\right]$, and assume that one of these realizations is controllable and observable. Then, both realizations are controllable and observable, they are equivalent, and there exists a unique matrix $S \in \mathbb{R}^{n \times n}$ such that

$$\left[\begin{array}{c|c} A_2 & B_2 \\ \hline C_2 & D \end{array}\right] = \left[\begin{array}{c|c} SA_1S^{-1} & SB_1 \\ \hline C_1S^{-1} & D \end{array}\right]. \tag{12.9.12}$$

In fact,

$$S = \left(\mathcal{O}_2^{\mathrm{T}}\mathcal{O}_2\right)^{-1}\mathcal{O}_2^{\mathrm{T}}\mathcal{O}_1, \qquad S^{-1} = \mathcal{K}_1\mathcal{K}_2^{\mathrm{T}}\left(\mathcal{K}_2\mathcal{K}_2^{\mathrm{T}}\right)^{-1}, \tag{12.9.13}$$

where, for $i = 1, 2$, $\mathcal{K}_i \triangleq \mathcal{K}(A_i, B_i)$ and $\mathcal{O}_i \triangleq \mathcal{O}(A_i, C_i)$.

Proof. Since the realizations $\left[\begin{array}{c|c} A_1 & B_1 \\ \hline C_1 & D \end{array}\right]$ and $\left[\begin{array}{c|c} A_2 & B_2 \\ \hline C_2 & D \end{array}\right]$ generate the same Markov parameters, it follows that $\mathcal{H}(A_1, B_1, C_1) = \mathcal{H}(A_2, B_2, C_2)$. Assume that $\left[\begin{array}{c|c} A_2 & B_2 \\ \hline C_2 & D \end{array}\right]$ is controllable and observable. Then, Lemma 12.9.7 implies that $\mathrm{rank}\, \mathcal{H}(A_2, B_2, C_2) = n$. Consequently, $\mathrm{rank}\, \mathcal{H}(A_1, B_1, C_1) = n$, and thus $\left[\begin{array}{c|c} A_1 & B_1 \\ \hline C_1 & D \end{array}\right]$ is controllable and observable.

Since $\left[\begin{array}{c|c} A_2 & B_2 \\ \hline C_2 & D \end{array}\right]$ is controllable and observable, it follows that the $n \times n$ matrices $\mathcal{K}_2\mathcal{K}_2^{\mathrm{T}}$ and $\mathcal{O}_2^{\mathrm{T}}\mathcal{O}_2$ are nonsingular. Since $\mathcal{O}_1A_1\mathcal{K}_1 = \mathcal{O}_2A_2\mathcal{K}_2$, $\mathcal{O}_1B_1 = \mathcal{O}_2B_2$, and $C_1\mathcal{K}_1 = C_2\mathcal{K}_2$, it follows that $A_2 = SA_1S^{-1}$, $B_2 = SB_1$, and $C_2 = C_1S^{-1}$.

To prove uniqueness, assume there exists a matrix $\hat{S} \in \mathbb{R}^{n \times n}$ such that $A_2 = \hat{S}A_1\hat{S}^{-1}$, $B_2 = \hat{S}B_1$, and $C_2 = C_1\hat{S}^{-1}$. Then, it follows that $\mathcal{O}_1\hat{S} = \mathcal{O}_2$. Since $\mathcal{O}_1S = \mathcal{O}_2$, it follows that $\mathcal{O}_1(S - \hat{S}) = 0$. Consequently, $S = \hat{S}$. $\qquad\square$

Corollary 12.9.9. Let $G \in \mathbb{R}_{\mathrm{prop}}(s)$ be given by (12.9.4), assume that G has the nth-order controllable and observable realization $\left[\begin{array}{c|c} A & B \\ \hline C & 0 \end{array}\right]$, and define $A_{\mathrm{c}}, B_{\mathrm{c}}, C_{\mathrm{c}}$ by (12.9.5), (12.9.6) and $A_{\mathrm{o}}, B_{\mathrm{o}}, C_{\mathrm{o}}$ by (12.9.7), (12.9.8). Furthermore, define $S_{\mathrm{c}} \triangleq [\mathcal{O}(A, B)]^{-1}\mathcal{O}(A_{\mathrm{c}}, B_{\mathrm{c}})$. Then,

$$S_{\mathrm{c}}^{-1} = \mathcal{K}(A, B)[\mathcal{K}(A_{\mathrm{c}}, B_{\mathrm{c}})]^{-1} \tag{12.9.14}$$

and

$$\left[\begin{array}{c|c} S_c A S_c^{-1} & S_c B \\ \hline C S_c^{-1} & 0 \end{array}\right] = \left[\begin{array}{c|c} A_c & B_c \\ \hline C_c & 0 \end{array}\right]. \tag{12.9.15}$$

Furthermore, define $S_o \triangleq [\mathcal{O}(A, B)]^{-1} \mathcal{O}(A_o, B_o)$. Then,

$$S_o^{-1} = \mathcal{K}(A, B)[\mathcal{K}(A_o, B_o)]^{-1} \tag{12.9.16}$$

and

$$\left[\begin{array}{c|c} S_o A S_o^{-1} & S_o B \\ \hline C S_o^{-1} & 0 \end{array}\right] = \left[\begin{array}{c|c} A_o & B_o \\ \hline C_o & 0 \end{array}\right]. \tag{12.9.17}$$

A rational transfer function $G \in \mathbb{R}^{l \times m}(s)$ can have realizations of different orders. For example, letting

$$A = 1, \qquad B = 1, \qquad C = 1, \qquad D = 0$$

and

$$\hat{A} = \left[\begin{array}{cc} 1 & 0 \\ 0 & 1 \end{array}\right], \qquad \hat{B} = \left[\begin{array}{c} 1 \\ 0 \end{array}\right], \qquad \hat{C} = \left[\begin{array}{cc} 1 & 0 \end{array}\right], \qquad \hat{D} = 0,$$

it follows that

$$G(s) = C(sI - A)^{-1}B + D = \hat{C}(sI - \hat{A})^{-1}\hat{B} + \hat{D} = \frac{1}{s-1}.$$

Generally, it is desirable to find realizations whose order is as small as possible.

Definition 12.9.10. Let $G \in \mathbb{R}^{l \times m}_{\mathrm{prop}}(s)$, and assume that $G \sim \left[\begin{array}{c|c} A & B \\ \hline C & D \end{array}\right]$. Then, $\left[\begin{array}{c|c} A & B \\ \hline C & D \end{array}\right]$ is a *minimal realization* of G if its order is less than or equal to the order of every realization of G. In this case, we write

$$G \overset{\min}{\sim} \left[\begin{array}{c|c} A & B \\ \hline C & D \end{array}\right]. \tag{12.9.18}$$

Note that the minimality of a realization is independent of D. The following result, known as the *Kalman decomposition*, is useful for constructing minimal realizations.

Proposition 12.9.11. Let $G \in \mathbb{R}^{l \times m}_{\mathrm{prop}}(s)$, where $G \sim \left[\begin{array}{c|c} A & B \\ \hline C & D \end{array}\right]$. Then, there exists a nonsingular matrix $S \in \mathbb{R}^{n \times n}$ such that

$$A = S \begin{bmatrix} A_1 & 0 & A_{13} & 0 \\ A_{21} & A_2 & A_{23} & A_{24} \\ 0 & 0 & A_3 & 0 \\ 0 & 0 & A_{43} & A_4 \end{bmatrix} S^{-1}, \quad B = S \begin{bmatrix} B_1 \\ B_2 \\ 0 \\ 0 \end{bmatrix}, \qquad (12.9.19)$$

$$C = \begin{bmatrix} C_1 & 0 & C_3 & 0 \end{bmatrix} S^{-1}, \qquad (12.9.20)$$

where $\left(\begin{bmatrix} A_1 & 0 \\ A_{21} & A_2 \end{bmatrix}, \begin{bmatrix} B_1 \\ B_2 \end{bmatrix} \right)$ is controllable and $\left(\begin{bmatrix} A_1 & A_{13} \\ 0 & A_3 \end{bmatrix}, \begin{bmatrix} C_1 & C_3 \end{bmatrix} \right)$ is observable. Furthermore, the following statements hold:

i) (A, B) is stabilizable if and only if A_3 and A_4 are asymptotically stable.

ii) (A, B) is controllable if and only if A_3 and A_4 are empty.

iii) (A, C) is detectable if and only if A_2 and A_4 are asymptotically stable.

iv) (A, C) is observable if and only if A_2 and A_4 are empty.

v) (A_1, B_1, C_1) is controllable and observable.

vi) $G \overset{\text{min}}{\sim} \left[\begin{array}{c|c} A_1 & B_1 \\ \hline C_1 & D \end{array} \right]$.

Proof. Let $\alpha \leq 0$ be such that $A + \alpha I$ is asymptotically stable, and let $Q \in \mathbb{R}^{n \times n}$ and $P \in \mathbb{R}^{n \times n}$ denote the controllability and observability Gramians of the system $(A + \alpha I, B, C)$. Then, Theorem 8.3.4 implies that there exists a nonsingular matrix $S \in \mathbb{R}^{n \times n}$ such that

$$Q = S \begin{bmatrix} Q_1 & & & 0 \\ & Q_2 & & \\ & & 0 & \\ 0 & & & 0 \end{bmatrix} S^{\mathrm{T}}, \quad P = S^{-\mathrm{T}} \begin{bmatrix} P_1 & & & 0 \\ & 0 & & \\ & & P_2 & \\ 0 & & & 0 \end{bmatrix} S^{-1},$$

where Q_1 and P_1 are the same size, and where $Q_1, Q_2, P_1,$ and P_2 are positive definite and diagonal. The form of $SAS^{-1}, SB,$ and CS^{-1} given by (12.9.20) now follows from (12.7.3) and (12.4.3) with A replaced by $A + \alpha I$, where, as in the proof of Theorem 12.6.8, $SAS^{-1} = S(A + \alpha I)S^{-1} - \alpha I$. Finally, statements i)–v) are immediate, while it can be verified directly that $\left[\begin{array}{c|c} A_1 & B_1 \\ \hline C_1 & D_1 \end{array} \right]$ is a realization of G. $\quad\square$

The following result show that the controllable and observable realization $\left[\begin{array}{c|c} A_1 & B_1 \\ \hline C_1 & D_1 \end{array} \right]$ of G in Proposition 12.9.11 is, in fact, minimal.

Corollary 12.9.12. Let $G \in \mathbb{R}^{l \times m}(s)$, and assume that $\sim \left[\begin{array}{c|c} A & B \\ \hline C & D \end{array} \right]$. Then, $\left[\begin{array}{c|c} A & B \\ \hline C & D \end{array} \right]$ is minimal if and only if it is controllable and observable.

Proof. To prove necessity, suppose that $\left[\begin{array}{c|c} A & B \\ \hline C & D \end{array}\right]$ is either not controllable or not observable. Then, Proposition 12.9.3 can be used to construct a realization of G of order less than n. Hence, $\left[\begin{array}{c|c} A & B \\ \hline C & D \end{array}\right]$ is not minimal.

To prove sufficiency, assume that $A \in \mathbb{R}^{n \times n}$, and assume that $\left[\begin{array}{c|c} A & B \\ \hline C & D \end{array}\right]$ is not minimal. Hence, G has a realization $G \sim \left[\begin{array}{c|c} \hat{A} & \hat{B} \\ \hline \hat{C} & D \end{array}\right]$ of order $\hat{n} < n$. Then, $\mathcal{H}(A, B, C) = \mathcal{H}(\hat{A}, \hat{B}, \hat{C})$, which implies that $\operatorname{rank} \mathcal{H}(A, B, C) = \operatorname{rank} \mathcal{H}(\hat{A}, \hat{B}, \hat{C}) \leq \hat{n} < n$. Therefore, $\left[\begin{array}{c|c} A & B \\ \hline C & D \end{array}\right]$ is either not controllable or not observable. $\quad\square$

Theorem 12.9.13. Let $G \in \mathbb{R}^{l \times m}_{\text{prop}}(s)$. Then, the realization $G \sim \left[\begin{array}{c|c} A & B \\ \hline C & 0 \end{array}\right]$ is minimal if and only if the McMillan degree of G is equal to the order of A.

Proof. See [624, p. 313]. $\quad\square$

Definition 12.9.14. Let $G \in \mathbb{R}^{l \times m}_{\text{prop}}(s)$, where $G \stackrel{\min}{\sim} \left[\begin{array}{c|c} A & B \\ \hline C & D \end{array}\right]$. Then, G is (asymptotically stable, semistable, Lyapunov stable) if A is.

Proposition 12.9.15. Let $G = p/q \in \mathbb{R}_{\text{prop}}(s)$, where $p, q \in \mathbb{R}[s]$, and assume that p and q are coprime. Then, G is (asymptotically stable, semistable, Lyapunov stable) if and only if q is.

Proposition 12.9.16. Let $G \in \mathbb{R}^{l \times m}_{\text{prop}}(s)$. Then, G is (asymptotically stable, semistable, Lyapunov stable) if and only if every entry of G is.

Definition 12.9.17. Let $G \in \mathbb{R}^{l \times m}_{\text{prop}}(s)$, where $G \stackrel{\min}{\sim} \left[\begin{array}{c|c} A & B \\ \hline C & D \end{array}\right]$ and A is asymptotically stable. Then, the realization $\left[\begin{array}{c|c} A & B \\ \hline C & D \end{array}\right]$ is *balanced* if the controllability and observability Gramians (12.7.2) and (12.4.2) are diagonal and equal.

Proposition 12.9.18. Let $G \in \mathbb{R}^{l \times m}_{\text{prop}}(s)$, where $G \stackrel{\min}{\sim} \left[\begin{array}{c|c} A & B \\ \hline C & D \end{array}\right]$ and A is asymptotically stable. Then, there exists a nonsingular matrix $S \in \mathbb{R}^{n \times n}$ such that the realization $G \sim \left[\begin{array}{c|c} SAS^{-1} & SB \\ \hline CS^{-1} & D \end{array}\right]$ is balanced.

Proof. It follows from Corollary 8.3.7 that there exists a nonsingular matrix $S \in \mathbb{R}^{n \times n}$ such that SQS^{T} and $S^{-\mathrm{T}}PS^{-1}$ are diagonal, where Q and P are the controllability and observability Gramians (12.7.2) and (12.4.2). Hence, the realization $\left[\begin{array}{c|c} SAS^{-1} & SB \\ \hline CS^{-1} & D \end{array}\right]$ is balanced. $\quad\square$

12.10 System Zeros

Recall Definition 4.2.4 on the rank of a matrix polynomial.

Definition 12.10.1. Let $G \in \mathbb{R}_{\text{prop}}^{l \times m}(s)$, where $G \sim \left[\begin{array}{c|c} A & B \\ \hline C & D \end{array} \right]$. Then, the *Rosenbrock system matrix* $\mathcal{Z} \in \mathbb{R}^{(n+l) \times (n+m)}[s]$ is the matrix polynomial

$$\mathcal{Z}(s) \triangleq \left[\begin{array}{cc} sI - A & B \\ C & -D \end{array} \right]. \tag{12.10.1}$$

Furthermore, $z \in \mathbb{C}$ is an *invariant zero* of the realization $\left[\begin{array}{c|c} A & B \\ \hline C & D \end{array} \right]$ if

$$\operatorname{rank} \mathcal{Z}(z) < \operatorname{rank} \mathcal{Z}. \tag{12.10.2}$$

Consider the strictly proper SISO transfer function G with minimal realization $G \overset{\text{min}}{\sim} \left[\begin{array}{c|c} A & B \\ \hline C & 0 \end{array} \right]$, where $A \in \mathbb{R}^{n \times n}$. Then,

$$G(s) = \frac{C(sI - A)^{\text{A}}B}{\det(sI - A)},$$

and thus the roots of the polynomial $C(sI - A)^{\text{A}}B$ are the classical zeros of G. In fact, it follows from Fact 2.13.2 that

$$\det \mathcal{Z}(s) = -C(sI - A)^{\text{A}}B,$$

which shows that $\operatorname{rank} \mathcal{Z} = n+1$ and the classical zeros of G are the invariant zeros of $\left[\begin{array}{c|c} A & B \\ \hline C & 0 \end{array} \right]$.

Proposition 12.10.2. Equivalent realizations have the same invariant zeros. Furthermore, invariant zeros are not changed by full-state feedback.

Proof. Let $u = Kx + v$, which leads to the rational transfer function

$$G_K \sim \left[\begin{array}{c|c} A + BK & B \\ \hline C + DK & D \end{array} \right]. \tag{12.10.3}$$

Since

$$\left[\begin{array}{cc} zI - (A + BK) & B \\ C + DK & -D \end{array} \right] = \left[\begin{array}{cc} zI - A & B \\ C & -D \end{array} \right] \left[\begin{array}{cc} I & 0 \\ -K & I \end{array} \right], \tag{12.10.4}$$

it follows that $\left[\begin{array}{c|c} A & B \\ \hline C & D \end{array} \right]$ and $\left[\begin{array}{c|c} A + BK & B \\ \hline C + DK & D \end{array} \right]$ have the same invariant zeros. $\square$

Proposition 12.10.3. Let $G \in \mathbb{R}_{\text{prop}}^{l \times m}(s)$, where $G \sim \left[\begin{array}{c|c} A & B \\ \hline C & D \end{array} \right]$, and assume that $R \triangleq D^{\text{T}}D$ is positive definite. Then, the following statements hold:

i) $\operatorname{rank} \mathcal{Z} = n + m$.

ii) $z \in \mathbb{C}$ is an invariant zero of $\left[\begin{array}{c|c} A & B \\ \hline C & D \end{array}\right]$ if and only if z is an unobservable eigenvalue of $\left(A - BR^{-1}D^{\mathrm{T}}C, \left[I - DR^{-1}D^{\mathrm{T}}\right]C\right)$.

Proof. To prove *i)*, assume that $\operatorname{rank} \mathcal{Z} < n + m$. Then, for every $s \in \mathbb{C}$, there exists a nonzero vector $\left[\begin{smallmatrix} x \\ y \end{smallmatrix}\right] \in \mathcal{N}[\mathcal{Z}(s)]$, that is,

$$\left[\begin{array}{cc} sI - A & B \\ C & -D \end{array}\right] \left[\begin{array}{c} x \\ y \end{array}\right] = 0.$$

Consequently, $Cx - Dy = 0$, which implies that $D^{\mathrm{T}}Cx - Ry = 0$, and thus $y = R^{-1}D^{\mathrm{T}}Cx$. Furthermore, since $\left(sI - A + BR^{-1}D^{\mathrm{T}}C\right)x = 0$, choosing $s \notin \operatorname{spec}\left(A - BR^{-1}D^{\mathrm{T}}C\right)$ yields $x = 0$, and thus $y = 0$, which is a contradiction.

To prove *ii)*, note that z is an invariant zero of $\left[\begin{array}{c|c} A & B \\ \hline C & D \end{array}\right]$ if and only if $\operatorname{rank} \mathcal{Z}(z) < n + m$, which holds if and only if there exists a nonzero vector $\left[\begin{smallmatrix} x \\ y \end{smallmatrix}\right] \in \mathcal{N}[\mathcal{Z}(z)]$. This condition is equivalent to

$$\left[\begin{array}{c} sI - A + BR^{-1}D^{\mathrm{T}}C \\ \left(I - DR^{-1}D^{\mathrm{T}}\right)C \end{array}\right] x = 0,$$

where $x \neq 0$. This last condition is equivalent to the fact that z is an unobservable eigenvalue of $\left(A - BR^{-1}D^{\mathrm{T}}C, \left[I - DR^{-1}D^{\mathrm{T}}\right]C\right)$. $\square$

Corollary 12.10.4. Assume that $R \triangleq D^{\mathrm{T}}D$ is positive definite, and assume that $\left(A - BR^{-1}D^{\mathrm{T}}C, \left[I - DR^{-1}D^{\mathrm{T}}\right]C\right)$ is observable. Then, $\left[\begin{array}{c|c} A & B \\ \hline C & D \end{array}\right]$ has no invariant zeros.

Recall Definition 4.7.4 on the rank of a proper rational transfer function and Definition 4.7.7 on transmission and blocking zeros.

Proposition 12.10.5. Let $G \in \mathbb{R}_{\mathrm{prop}}^{l \times m}(s)$. Then, $z \in \mathbb{C}$ is a transmission zero of G if and only if $\operatorname{rank} G(z) < \operatorname{rank} G$. Furthermore, $z \in \mathbb{C}$ is a blocking zero of G if and only if $G(z) = 0$.

Lemma 12.10.6. Let $G \in \mathbb{R}_{\mathrm{prop}}^{l \times m}(s)$, where $G \overset{\min}{\sim} \left[\begin{array}{c|c} A & B \\ \hline C & D \end{array}\right]$. If $s \in \mathbb{C}$ is an eigenvalue of A, then s is not an invariant zero of $\left[\begin{array}{c|c} A & B \\ \hline C & D \end{array}\right]$.

Proposition 12.10.7. Let $G \in \mathbb{R}_{\mathrm{prop}}^{l \times m}(s)$, where $G \sim \left[\begin{array}{c|c} A & B \\ \hline C & D \end{array}\right]$. If $s \notin \operatorname{spec}(A)$, then

$$n + \operatorname{rank} G(s) = \operatorname{rank} \mathcal{Z}(s). \tag{12.10.5}$$

Hence,

$$n + \operatorname{rank} G \le \operatorname{rank} \mathcal{Z}. \tag{12.10.6}$$

If, in addition, $\left[\begin{array}{c|c} A & B \\ \hline C & D \end{array}\right]$ is minimal, then

$$n + \operatorname{rank} G = \operatorname{rank} \mathcal{Z}. \tag{12.10.7}$$

Proof. For $s \notin \operatorname{spec}(A)$, it follows that

$$\left[\begin{array}{cc} sI - A & B \\ C & -D \end{array}\right] = \left[\begin{array}{cc} I & 0 \\ C(sI - A)^{-1} & I \end{array}\right]\left[\begin{array}{cc} sI - A & B \\ 0 & -G(s) \end{array}\right],$$

which implies (12.10.5). Therefore,

$$n + \operatorname{rank} G = n + \max_{s \in \mathbb{C} \setminus \operatorname{spec}(A)} \operatorname{rank} G(s)$$

$$= \max_{s \in \mathbb{C} \setminus \operatorname{spec}(A)} \operatorname{rank} \mathcal{Z}(s)$$

$$\leq \max_{s \in \mathbb{C}} \operatorname{rank} \mathcal{Z}(s)$$

$$= \operatorname{rank} \mathcal{Z},$$

which yields (12.10.6). The last statement follows from Lemma 12.10.6. $\quad\square$

Theorem 12.10.8. Let $G \in \mathbb{R}_{\mathrm{prop}}^{l \times m}(s)$, where $G \overset{\min}{\sim} \left[\begin{array}{c|c} A & B \\ \hline C & D \end{array}\right]$, and let $z \in \mathbb{C} \setminus \operatorname{spec}(A)$. Then, z is a transmission zero of G if and only if z is an invariant zero of $\left[\begin{array}{c|c} A & B \\ \hline C & D \end{array}\right]$.

Proof. Suppose that z is a transmission zero of G. Then, it follows from Proposition 12.10.7 that

$$\operatorname{rank} \mathcal{Z}(z) = n + \operatorname{rank} G(z) < n + \operatorname{rank} G = \operatorname{rank} \mathcal{Z},$$

which implies that z is an invariant zero of $\left[\begin{array}{c|c} A & B \\ \hline C & D \end{array}\right]$. Conversely, suppose that z is an invariant zero of $\left[\begin{array}{c|c} A & B \\ \hline C & D \end{array}\right]$. Then,

$$\operatorname{rank} G(z) = \operatorname{rank} \mathcal{Z}(z) - n < \operatorname{rank} \mathcal{Z} - n = \operatorname{rank} G,$$

which implies that z is a transmission zero of G. $\quad\square$

12.11 H$_2$ System Norm

Consider the system

$$\dot{x}(t) = Ax(t) + Bu(t), \tag{12.11.1}$$

$$y(t) = Cx(t), \tag{12.11.2}$$

where $A \in \mathbb{R}^{n \times n}$ is asymptotically stable, $B \in \mathbb{R}^{n \times m}$, and $C \in \mathbb{R}^{l \times n}$. Then, for all $t \geq 0$, the impulse response function defined by (12.1.18) is given by

$$H(t) = Ce^{tA}B. \tag{12.11.3}$$

The L_2 *norm* of $H(\cdot)$ is given by

$$\|H\|_{L_2} \triangleq \left(\int_0^\infty \|H(t)\|_F^2 \, dt \right)^{1/2}. \tag{12.11.4}$$

The following result provides expressions for $\|H(\cdot)\|_{L_2}$ in terms of the controllability and observability Gramians.

Theorem 12.11.1. Assume that A is asymptotically stable. Then, the L_2 norm of H is given by

$$\|H\|_{L_2}^2 = \operatorname{tr} CQC^T = \operatorname{tr} B^T PB, \tag{12.11.5}$$

where $Q, P \in \mathbb{R}^{n \times n}$ satisfy

$$AQ + QA^T + BB^T = 0, \tag{12.11.6}$$

$$A^T P + PA + C^T C = 0. \tag{12.11.7}$$

Proof. Note that

$$\|H\|_{L_2}^2 = \int_0^\infty \operatorname{tr} Ce^{tA}BB^T e^{tA^T}C^T dt = \operatorname{tr} CQC^T,$$

where Q satisfies (12.11.6). The dual expression (12.11.7) follows in a similar manner or by noting that

$$\operatorname{tr} CQC^T = \operatorname{tr} C^T CQ = -\operatorname{tr} \left(A^T P + PA \right) Q$$
$$= -\operatorname{tr} \left(AQ + QA^T \right) P = \operatorname{tr} BB^T P = \operatorname{tr} B^T PB. \qquad \square$$

For the following definition, note that

$$\|G(s)\|_F = \left[\operatorname{tr} G(s)G^*(s) \right]^{1/2}. \tag{12.11.8}$$

Definition 12.11.2. The H_2 *norm* of $G \in \mathbb{R}^{l \times m}(s)$ is the nonnegative number

$$\|G\|_{H_2} \triangleq \left(\frac{1}{2\pi} \int_{-\infty}^\infty \|G(\jmath\omega)\|_F^2 \, d\omega \right)^{1/2}. \tag{12.11.9}$$

The following result is *Parseval's theorem*, which relates the L_2 norm of the impulse response function to the H_2 norm of its transform.

Theorem 12.11.3. Let $G \in \mathbb{R}_{\text{prop}}^{l \times m}(s)$, where $G \sim \left[\begin{array}{c|c} A & B \\ \hline C & 0 \end{array}\right]$, and assume that $A \in \mathbb{R}^{n \times n}$ is asymptotically stable. Then,

$$\int_0^\infty H(t)H^{\mathrm{T}}(t)\,\mathrm{d}t = \tfrac{1}{2\pi}\int_{-\infty}^\infty G(\jmath\omega)G^*(\jmath\omega)\,\mathrm{d}\omega. \qquad (12.11.10)$$

Therefore,

$$\|H\|_{\mathrm{L}_2} = \|G\|_{\mathrm{H}_2}. \qquad (12.11.11)$$

Proof. First note that

$$G(s) = \mathcal{L}\{H(t)\} = \int_0^\infty H(t)e^{-st}\,\mathrm{d}t$$

and that

$$H(t) = \tfrac{1}{2\pi}\int_{-\infty}^\infty G(\jmath\omega)e^{\jmath\omega t}\,\mathrm{d}\omega.$$

Hence,

$$\int_0^\infty H(t)H^{\mathrm{T}}(t)e^{-st}\,\mathrm{d}t = \int_0^\infty \left(\tfrac{1}{2\pi}\int_{-\infty}^\infty G(\jmath\omega)e^{\jmath\omega t}\,\mathrm{d}\omega\right)H^{\mathrm{T}}(t)e^{-st}\,\mathrm{d}t$$

$$= \tfrac{1}{2\pi}\int_{-\infty}^\infty G(\jmath\omega)\left(\int_0^\infty H^{\mathrm{T}}(t)e^{-(s-\jmath\omega)t}\,\mathrm{d}t\right)\mathrm{d}\omega$$

$$= \tfrac{1}{2\pi}\int_{-\infty}^\infty G(\jmath\omega)G^{\mathrm{T}}(s-\jmath\omega)\,\mathrm{d}\omega.$$

Setting $s = 0$ yields (12.11.7), while taking the trace of (12.11.10) yields (12.11.11). $\qquad\qquad\qquad\square$

Corollary 12.11.4. Let $G \in \mathbb{R}_{\text{prop}}^{l \times m}(s)$, where $G \sim \left[\begin{array}{c|c} A & B \\ \hline C & 0 \end{array}\right]$, and assume that $A \in \mathbb{R}^{n \times n}$ is asymptotically stable. Then,

$$\|G\|_{\mathrm{H}_2}^2 = \|H\|_{\mathrm{L}_2}^2 = \operatorname{tr} CQC^{\mathrm{T}} = \operatorname{tr} B^{\mathrm{T}}PB, \qquad (12.11.12)$$

where $Q, P \in \mathbb{R}^{n \times n}$ satisfy (12.11.6) and (12.11.7), respectively.

The following corollary of Theorem 12.11.3 provides a frequency domain expression for the solution of the Lyapunov equation.

Corollary 12.11.5. Let $A \in \mathbb{R}^{n \times n}$, assume that A is asymptotically stable, let $B \in \mathbb{R}^{n \times m}$, and define $Q \in \mathbb{R}^{n \times n}$ by

$$Q = \frac{1}{2\pi} \int_{-\infty}^{\infty} (\jmath \omega I - A)^{-1} B B^{\mathrm{T}} (\jmath \omega I - A)^{-*} \, d\omega. \qquad (12.11.13)$$

Then, Q satisfies

$$AQ + QA^{\mathrm{T}} + BB^{\mathrm{T}} = 0. \qquad (12.11.14)$$

Proof. The result follows directly from Theorem 12.11.3 with $H(t) = e^{tA}B$ and $G(s) = (sI - A)^{-1}B$. Alternatively, it follows from (12.11.14) that

$$\int_{-\infty}^{\infty} (\jmath \omega I - A)^{-1} \, d\omega Q + Q \int_{-\infty}^{\infty} (\jmath \omega I - A)^{-*} \, d\omega = \int_{-\infty}^{\infty} (\jmath \omega I - A)^{-1} B B^{\mathrm{T}} (\jmath \omega I - A)^{-*} \, d\omega.$$

Assuming that A is diagonalizable with eigenvalues $\lambda_i = -\sigma_i + \jmath \omega_i$, it follows that

$$\int_{-\infty}^{\infty} \frac{d\omega}{\jmath \omega - \lambda_i} = \int_{-\infty}^{\infty} \frac{\sigma_i - \jmath \omega}{\sigma_i^2 + \omega^2} \, d\omega = \frac{\sigma_i \pi}{|\sigma_i|} - \jmath \lim_{r \to \infty} \int_{-r}^{r} \frac{\omega}{\sigma_i^2 + \omega^2} \, d\omega = \pi,$$

which implies that

$$\int_{-\infty}^{\infty} (\jmath \omega I - A)^{-1} \, d\omega = \pi I_n,$$

which yields (12.11.13). See [175] for a proof of the general case. $\qquad \square$

Proposition 12.11.6. Let $G_1, G_2 \in \mathbb{R}_{\mathrm{prop}}^{l \times m}(s)$ be asymptotically stable rational transfer functions. Then,

$$\|G_1 + G_2\|_{\mathrm{H}_2} \le \|G_1\|_{\mathrm{H}_2} + \|G_2\|_{\mathrm{H}_2}. \qquad (12.11.15)$$

Proof. Let $G_1 \overset{\min}{\sim} \left[\begin{array}{c|c} A_1 & B_1 \\ \hline C_1 & 0 \end{array} \right]$ and $G_2 \overset{\min}{\sim} \left[\begin{array}{c|c} A_2 & B_2 \\ \hline C_2 & 0 \end{array} \right]$, where $A_1 \in \mathbb{R}^{n_1 \times n_1}$ and $A_2 \in \mathbb{R}^{n_2 \times n_2}$. It follows from Proposition 12.13.2 that $G_1 + G_2 \sim \left[\begin{array}{cc|c} A_1 & 0 & B_1 \\ 0 & A_2 & B_2 \\ \hline C_1 & C_2 & 0 \end{array} \right]$. Now, Theorem 12.11.3 implies that $\|G_1\|_{\mathrm{H}_2} = \sqrt{\mathrm{tr}\, C_1 Q_1 C_1^{\mathrm{T}}}$ and $\|G_2\|_{\mathrm{H}_2} = \sqrt{\mathrm{tr}\, C_2 Q_2 C_2^{\mathrm{T}}}$, where $Q_1 \in \mathbb{R}^{n_1 \times n_1}$ and $Q_2 \in \mathbb{R}^{n_2 \times n_2}$ are the unique positive-definite matrices satisfying $A_1 Q_1 + Q_1 A_1^{\mathrm{T}} + B_1 B_1^{\mathrm{T}} = 0$ and $A_2 Q_2 + Q_2 A_2^{\mathrm{T}} + B_2 B_2^{\mathrm{T}} = 0$. Furthermore,

$$\|G_2 + G_2\|_{\mathrm{H}_2}^2 = \mathrm{tr} \left[\begin{array}{cc} C_1 & C_2 \end{array} \right] Q \left[\begin{array}{c} C_1^{\mathrm{T}} \\ C_2^{\mathrm{T}} \end{array} \right],$$

where $Q \in \mathbb{R}^{(n_1+n_2) \times (n_1+n_2)}$ is the unique, positive-semidefinite matrix satisfying

$$\begin{bmatrix} A_1 & 0 \\ 0 & A_2 \end{bmatrix} Q + Q \begin{bmatrix} A_1 & 0 \\ 0 & A_2 \end{bmatrix}^{\mathrm{T}} + \begin{bmatrix} B_1 \\ B_2 \end{bmatrix} \begin{bmatrix} B_1 \\ B_2 \end{bmatrix}^{\mathrm{T}} = 0.$$

It can be seen that $Q = \begin{bmatrix} Q_1 & Q_{12} \\ Q_{12}^{\mathrm{T}} & Q_2 \end{bmatrix}$, where Q_1 and Q_2 are as given above and where Q_{12} satisfies $A_1 Q_{12} + Q_{12} A_2^{\mathrm{T}} + B_1 B_2^{\mathrm{T}} = 0$. Now, using the Cauchy-Schwarz inequality (9.3.15) and iii) of Proposition 8.2.3, it follows that

$$\begin{aligned}
\|G_1 + G_2\|_{\mathrm{H}_2}^2 &= \mathrm{tr}\big(C_1 Q_1 C_1^{\mathrm{T}} + C_2 Q_2 C_2^{\mathrm{T}} + C_2 Q_{12}^{\mathrm{T}} C_1^{\mathrm{T}} + C_1 Q_{12} C_2^{\mathrm{T}}\big) \\
&= \|G_1\|_{\mathrm{H}_2}^2 + \|G_2\|_{\mathrm{H}_2}^2 + 2\,\mathrm{tr}\, C_1 Q_{12} Q_2^{-1/2} Q_2^{1/2} C_2^{\mathrm{T}} \\
&\le \|G_1\|_{\mathrm{H}_2}^2 + \|G_2\|_{\mathrm{H}_2}^2 + 2\,\mathrm{tr}\big(C_1 Q_{12} Q_2^{-1} Q_{12}^{\mathrm{T}} C_1^{\mathrm{T}}\big)\mathrm{tr}\big(C_2 Q_2 C_2^{\mathrm{T}}\big) \\
&\le \|G_1\|_{\mathrm{H}_2}^2 + \|G_2\|_{\mathrm{H}_2}^2 + 2\,\mathrm{tr}\big(C_1 Q_1 C_1^{\mathrm{T}}\big)\mathrm{tr}\big(C_2 Q_2 C_2^{\mathrm{T}}\big) \\
&= \big(\|G_1\|_{\mathrm{H}_2} + \|G_2\|_{\mathrm{H}_2}\big)^2. \qquad \square
\end{aligned}$$

12.12 Harmonic Steady-State Response

The following result concerns the response of a linear system to a harmonic input.

Theorem 12.12.1. For $t \ge 0$, consider the linear system

$$\dot{x}(t) = Ax(t) + Bu(t), \tag{12.12.1}$$

with harmonic input

$$u(t) = \mathrm{Re}\, u_0 e^{\jmath \omega_0 t}, \tag{12.12.2}$$

where $u_0 \in \mathbb{C}^m$ and $\omega_0 \in \mathbb{R}$ is such that $\jmath \omega_0 \notin \mathrm{spec}(A)$. Then, $x(t)$ is given by

$$x(t) = e^{tA}\big(x(0) - \mathrm{Re}\big[(\jmath \omega_0 I - A)^{-1} B u_0\big]\big) + \mathrm{Re}\big[(\jmath \omega_0 I - A)^{-1} B u_0 e^{\jmath \omega_0 t}\big]. \tag{12.12.3}$$

Proof. We have

$$x(t) = e^{tA} x(0) + \int_0^t e^{(t-\tau)A} B \mathrm{Re}(u_0 e^{\jmath \omega_0 \tau})\,\mathrm{d}\tau$$

$$= e^{tA} x(0) + e^{tA} \mathrm{Re}\left[\int_0^t e^{-\tau A} e^{\jmath \omega_0 \tau}\,\mathrm{d}\tau B u_0\right]$$

$$= e^{tA} x(0) + e^{tA} \mathrm{Re}\left[\int_0^t e^{\tau(\jmath \omega_0 I - A)}\,\mathrm{d}\tau B u_0\right]$$

$$= e^{tA}x(0) + e^{tA}\text{Re}\Big[(\jmath\omega_0 I - A)^{-1}\Big(e^{t(\jmath\omega_0 I - A)} - I\Big)Bu_0\Big]$$

$$= e^{tA}x(0) + \text{Re}\big[(\jmath\omega_0 I - A)^{-1}\big(e^{\jmath\omega_0 tI} - e^{tA}\big)Bu_0\big]$$

$$= e^{tA}x(0) + \text{Re}\big[(\jmath\omega_0 I - A)^{-1}(-e^{tA})Bu_0\big] + \text{Re}\big[(\jmath\omega_0 I - A)^{-1}e^{\jmath\omega_0 t}Bu_0\big]$$

$$= e^{tA}\big(x(0) - \text{Re}\big[(\jmath\omega_0 I - A)^{-1}Bu_0\big]\big) + \text{Re}\big[(\jmath\omega_0 I - A)^{-1}Bu_0 e^{\jmath\omega_0 t}\big]. \qquad \square$$

Theorem 12.12.1 shows that the total response $y(t)$ of the linear system $G \sim \left[\begin{array}{c|c} A & B \\ \hline C & 0 \end{array}\right]$ to a harmonic input can be written as $y(t) = y_{\text{trans}}(t) + y_{\text{hss}}(t)$, where the transient component

$$y_{\text{trans}}(t) \triangleq Ce^{tA}\big(x(0) - \text{Re}\big[(\jmath\omega_0 I - A)^{-1}Bu_0\big]\big) \qquad (12.12.4)$$

depends on the initial condition and the input, and the harmonic steady-state component

$$y_{\text{hss}}(t) = \text{Re}\big[G(\jmath\omega_0)u_0 e^{\jmath\omega_0 t}\big] \qquad (12.12.5)$$

depends only on the input.

If A is asymptotically stable, then $\lim_{t\to\infty} y_{\text{trans}}(t) = 0$, and thus $y(t)$ approaches its harmonic steady-state component $y_{\text{hss}}(t)$ for large t. Since the harmonic steady-state component is sinusoidal, it follows that $y(t)$ does not converge in the usual sense.

Finally, if A is semistable, then it follows from $vii)$ of Proposition 11.7.2 that

$$\lim_{t\to\infty} y_{\text{trans}}(t) = C(I - AA^{\#})\big(x(0) - \text{Re}\big[(\jmath\omega_0 I - A)^{-1}Bu_0\big]\big), \qquad (12.12.6)$$

which represents a constant offset to the harmonic steady-state component.

In the SISO case, let $u_0 \triangleq a_0(\sin\phi_0 + \jmath\cos\phi_0)$, and consider the input

$$u(t) = a_0\sin(\omega_0 t + \phi_0) = \text{Re}\, u_0 e^{\jmath\omega_0 t}. \qquad (12.12.7)$$

Then, writing $G(\jmath\omega_0) = \text{Re}\, Me^{\jmath\theta}$, it follows that

$$y_{\text{hss}}(t) = a_0 M\sin(\omega_0 t + \phi_0 + \theta). \qquad (12.12.8)$$

12.13 System Interconnections

Let $G \in \mathbb{R}^{l\times m}_{\text{prop}}(s)$. We define the *parahermitian conjugate* $G^{\sim}$ of G by $G^{\sim} \triangleq G^{\text{T}}(-s)$. The following result provides realizations for G^{T}, $G^{\sim}$, and G^{-1}.

Proposition 12.13.1. Let $G_{\text{prop}}^{l \times m}(s)$, and assume that $G \sim \left[\begin{array}{c|c} A & B \\ \hline C & D \end{array}\right]$. Then,

$$G^{\mathrm{T}} \sim \left[\begin{array}{c|c} A^{\mathrm{T}} & C^{\mathrm{T}} \\ \hline B^{\mathrm{T}} & D^{\mathrm{T}} \end{array}\right] \qquad (12.13.1)$$

and

$$G^{\sim} \sim \left[\begin{array}{c|c} -A^{\mathrm{T}} & -C^{\mathrm{T}} \\ \hline B^{\mathrm{T}} & D^{\mathrm{T}} \end{array}\right]. \qquad (12.13.2)$$

Furthermore, if G is square and D is nonsingular, then

$$G^{-1} \sim \left[\begin{array}{c|c} A - BD^{-1}C & -BD^{-1} \\ \hline -D^{-1}C & D^{-1} \end{array}\right]. \qquad (12.13.3)$$

Proof. Since $y = Gu$, it follows that G^{-1} satisfies $u = G^{-1}y$. Since $\dot{x} = Ax + Bu$ and $y = Cx + Du$, it follows that $u = -D^{-1}Cx + D^{-1}y$, and thus $\dot{x} = Ax + B(-D^{-1}Cx + D^{-1}y) = (A - BD^{-1}C)x + BD^{-1}y$. $\qquad \square$

Note that, if $G \in \mathbb{R}_{\text{prop}}(s)$ and $G \sim \left[\begin{array}{c|c} A & B \\ \hline C & D \end{array}\right]$, then $G \sim \left[\begin{array}{c|c} A^{\mathrm{T}} & B^{\mathrm{T}} \\ \hline C^{\mathrm{T}} & D \end{array}\right]$.

Let $G_1 \in \mathbb{R}_{\text{prop}}^{l_1 \times m_1}(s)$ and $G_2 \in \mathbb{R}_{\text{prop}}^{l_2 \times m_2}(s)$. If $m_2 = l_2$, then the *cascade interconnection* of G_1 and G_2 shown in Figure 12.13.1 is the product $G_2 G_1$, while the *parallel interconnection* shown in Figure 12.13.2 is the sum $G_1 + G_2$. Note that $G_2 G_1$ is defined only if $m_2 = l_1$, whereas $G_1 + G_2$ requires that $m_1 = m_2$ and $l_1 = l_2$.

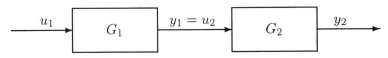

Figure 12.13.1
Cascade Interconnection of Linear Systems

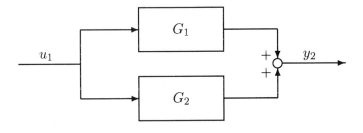

Figure 12.13.2
Parallel Interconnection of Linear Systems

Proposition 12.13.2. Let $G_1 \sim \left[\begin{array}{c|c} A_1 & B_1 \\ \hline C_1 & D_1 \end{array} \right]$ and $G_2 \sim \left[\begin{array}{c|c} A_2 & B_2 \\ \hline C_2 & D_2 \end{array} \right]$. Then,

$$G_2 G_1 \sim \left[\begin{array}{cc|c} A_1 & 0 & B_1 \\ B_2 C_1 & A_2 & B_2 D_1 \\ \hline D_2 C_1 & C_2 & D_2 D_1 \end{array} \right] \qquad (12.13.4)$$

and

$$G_1 + G_2 \sim \left[\begin{array}{cc|c} A_1 & 0 & B_1 \\ 0 & A_2 & B_2 \\ \hline C_1 & C_2 & D_1 + D_2 \end{array} \right]. \qquad (12.13.5)$$

Proof. Consider the state space equations

$$\dot{x}_1 = A_1 x_1 + B_1 u_1, \quad \dot{x}_2 = A_2 x_2 + B_2 u_2,$$
$$y_1 = C_1 x_1 + D_1 u_1, \quad y_2 = C_2 x_2 + D_2 u_2.$$

Since $u_2 = y_1$, it follows that

$$\dot{x}_2 = A_2 x_2 + B_2 C_1 x_1 + B_2 D_1 u_1,$$
$$y_2 = C_2 x_2 + D_2 C_1 x_1 + D_2 D_1 u_1,$$

and thus

$$\left[\begin{array}{c} \dot{x}_1 \\ \dot{x}_2 \end{array} \right] = \left[\begin{array}{cc} A_1 & 0 \\ B_2 C_1 & A_2 \end{array} \right] \left[\begin{array}{c} x_1 \\ x_2 \end{array} \right] + \left[\begin{array}{c} B_1 \\ B_2 D_1 \end{array} \right] u_1,$$

$$y_2 = \left[\begin{array}{cc} D_2 C_1 & C_2 \end{array} \right] \left[\begin{array}{c} x_1 \\ x_2 \end{array} \right] + D_2 D_1 u_1,$$

which yields the realization (12.13.4) of $G_2 G_1$. The realization (12.13.5) for $G_1 + G_2$ can be obtained by similar techniques. $\qquad \square$

It is sometimes useful to combine transfer functions by concatenating them into row, column, or block-diagonal transfer functions.

Proposition 12.13.3. Let $G_1 \sim \left[\begin{array}{c|c} A_1 & B_1 \\ \hline C_1 & D_1 \end{array} \right]$ and $G_2 \sim \left[\begin{array}{c|c} A_2 & B_2 \\ \hline C_2 & D_2 \end{array} \right]$. Then,

$$\left[\begin{array}{cc} G_1 & G_2 \end{array} \right] \sim \left[\begin{array}{cc|cc} A_1 & 0 & B_1 & 0 \\ 0 & A_2 & 0 & B_2 \\ \hline C_1 & C_2 & D_1 & D_2 \end{array} \right], \qquad (12.13.6)$$

$$\left[\begin{array}{c} G_1 \\ G_2 \end{array} \right] \sim \left[\begin{array}{cc|c} A_1 & 0 & B_1 \\ 0 & A_2 & B_2 \\ \hline C_1 & 0 & D_1 \\ 0 & C_2 & D_2 \end{array} \right], \qquad (12.13.7)$$

$$\begin{bmatrix} G_1 & 0 \\ 0 & G_2 \end{bmatrix} \sim \left[\begin{array}{cc|cc} A_1 & 0 & B_1 & 0 \\ 0 & A_2 & 0 & B_2 \\ \hline C_1 & 0 & D_1 & 0 \\ 0 & C_2 & 0 & D_2 \end{array} \right]. \tag{12.13.8}$$

Next, we interconnect a pair of systems G_1, G_2 by means of feedback as shown in Figure 12.13.3. It can be seen that u and y are related by

$$\hat{y} = (I + G_1 G_2)^{-1} G_1 \hat{u} \tag{12.13.9}$$

or

$$\hat{y} = G_1 (I + G_2 G_1)^{-1} \hat{u}. \tag{12.13.10}$$

The equivalence of (12.13.9) and (12.13.10) follows from the push-through identity given by Fact 2.14.15,

$$(I + G_1 G_2)^{-1} G_1 = G_1 (I + G_2 G_1)^{-1}. \tag{12.13.11}$$

A realization of this rational transfer function is given by the following result.

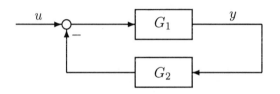

Figure 12.13.3
Feedback Interconnection of Linear Systems

Proposition 12.13.4. Let $G_1 \sim \left[\begin{array}{c|c} A_1 & B_1 \\ \hline C_1 & D_1 \end{array} \right]$ and $G_2 \sim \left[\begin{array}{c|c} A_2 & B_2 \\ \hline C_2 & D_2 \end{array} \right]$, and assume that $\det(I + D_1 D_2) \neq 0$. Then,

$$[I + G_1 G_2]^{-1} G_1$$

$$\sim \left[\begin{array}{cc|c} A_1 - B_1(I + D_2 D_1)^{-1} D_2 C_1 & -B_1(I + D_2 D_1)^{-1} C_2 & B_1(I + D_2 D_1)^{-1} \\ B_2(I + D_1 D_2)^{-1} C_1 & A_2 - B_2(I + D_1 D_2)^{-1} D_1 C_2 & B_2(I + D_1 D_2)^{-1} D_1 \\ \hline (I + D_1 D_2)^{-1} C_1 & -(I + D_1 D_2)^{-1} D_1 C_2 & (I + D_1 D_2)^{-1} D_1 \end{array} \right]. \tag{12.13.12}$$

12.14 H₂ Standard Control Problem

The standard control problem shown in Figure 12.14.1 involves four distinct signals, namely, an *exogenous input* w, a *control input* u, a *performance variable* z, and a *feedback signal* y. This system can be written as

$$\left[\begin{array}{c} \hat{z}(s) \\ \hat{y}(s) \end{array} \right] = \tilde{\mathcal{G}}(s) \left[\begin{array}{c} \hat{w}(s) \\ \hat{u}(s) \end{array} \right], \tag{12.14.1}$$

where $\mathcal{G}(s)$ is partitioned as

$$\mathcal{G} \triangleq \left[\begin{array}{cc} G_{11} & G_{12} \\ G_{21} & G_{22} \end{array} \right] \tag{12.14.2}$$

with the realization

$$\mathcal{G} \sim \left[\begin{array}{c|cc} A & D_1 & B \\ \hline E_1 & E_0 & E_2 \\ C & D_2 & D \end{array} \right]. \tag{12.14.3}$$

Consequently,

$$\mathcal{G}(s) = \left[\begin{array}{cc} E_1(sI - A)^{-1}D_1 + E_0 & E_1(sI - A)^{-1}B + E_2 \\ C(sI - A)^{-1}D_1 + D_2 & C(sI - A)^{-1}B + D \end{array} \right], \tag{12.14.4}$$

which shows that G_{11}, G_{12}, G_{21}, and G_{22} have the realizations

$$G_{11} \sim \left[\begin{array}{c|c} A & D_1 \\ \hline E_1 & E_0 \end{array} \right], \qquad G_{12} \sim \left[\begin{array}{c|c} A & B \\ \hline E_1 & E_2 \end{array} \right], \tag{12.14.5}$$

$$G_{21} \sim \left[\begin{array}{c|c} A & D_1 \\ \hline C & D_2 \end{array} \right], \qquad G_{22} \sim \left[\begin{array}{c|c} A & B \\ \hline C & D \end{array} \right]. \tag{12.14.6}$$

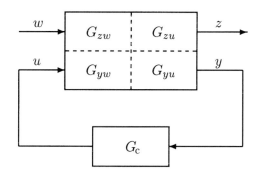

Figure 12.14.1
Standard Control Problem

Letting G_c denote a feedback controller with realization

$$G_c \sim \left[\begin{array}{c|c} A_c & B_c \\ \hline C_c & D_c \end{array} \right], \tag{12.14.7}$$

we interconnect G and G_c according to

$$\hat{u}(s) = G_c(s)\hat{y}(s). \tag{12.14.8}$$

The resulting rational transfer function $\tilde{G}$ satisfying $\hat{z}(s) = \tilde{G}(s)\hat{w}(s)$ is thus given by

$$\tilde{G} = G_{11} + G_{12}G_c(I - G_{22}G_c)^{-1}G_{21} \tag{12.14.9}$$

or

$$\tilde{G} = G_{11} + G_{12}(I - G_cG_{22})^{-1}G_cG_{21}. \tag{12.14.10}$$

A realization of $\tilde{G}$ is given by the following result.

Proposition 12.14.1. Let G and G_c have the realizations (12.14.3) and (12.14.7), and assume that $\det(I - DD_c) \neq 0$. Then,

$$\tilde{G} \sim \left[\begin{array}{cc|c} A + BD_c(I - DD_c)^{-1}C & B(I - D_cD)^{-1}C_c & D_1 + BD_c(I + DD_c)^{-1}D_2 \\ B_c(I - DD_c)^{-1}C & A_c + B_c(I - DD_c)^{-1}DC_c & B_c(I - DD_c)^{-1}D_2 \\ \hline E_1 + E_2D_c(I - DD_c)^{-1}C & E_2(I - D_cD)^{-1}C_c & E_0 + E_2D_c(I - DD_c)^{-1}D_2 \end{array} \right]. \tag{12.14.11}$$

The realization (12.14.11) can be simplified when $DD_c = 0$. For example, if $D = 0$, then

$$\tilde{G} \sim \left[\begin{array}{cc|c} A + BD_cC & BC_c & D_1 + BD_cD_2 \\ B_cC & A_c & B_cD_2 \\ \hline E_1 + E_2D_cC & E_2C_c & E_0 + E_2D_cD_2 \end{array} \right], \tag{12.14.12}$$

whereas, if $D_c = 0$, then

$$\tilde{G} \sim \left[\begin{array}{cc|c} A & BC_c & D_1 \\ B_cC & A_c + B_cDC_c & B_cD_2 \\ \hline E_1 & E_2C_c & E_0 \end{array} \right]. \tag{12.14.13}$$

Finally, if both $D = 0$ and $D_c = 0$, then

$$\tilde{G} \sim \left[\begin{array}{cc|c} A & BC_c & D_1 \\ B_cC & A_c & B_cD_2 \\ \hline E_1 & E_2C_c & E_0 \end{array} \right]. \tag{12.14.14}$$

The feedback interconnection shown in Figure 12.14.1 forms the basis for the *standard control problem* in feedback control. For this problem the signal w is an exogenous signal representing a command or a disturbance, while the signal z is the *performance variable*, that is, the variable whose behavior reflects the performance of the closed-loop system. The performance variable may or may not be physically measured. The *controlled input* or the *control* u is the output of the feedback controller G_c, while the *measurement* signal y serves as the input to the *feedback controller* G_c. The standard control problem is the following: Given knowledge of w, determine G_c to minimize a performance criterion $J(G_c)$.

12.15 Linear-Quadratic Control

Let $A \in \mathbb{R}^{n \times n}$ and $B \in \mathbb{R}^{n \times m}$, and consider the system

$$\dot{x}(t) = Ax(t) + Bu(t), \qquad (12.15.1)$$

$$x(0) = x_0, \qquad (12.15.2)$$

where $t \geq 0$. Furthermore, let $K \in \mathbb{R}^{m \times n}$, and consider the full-state-feedback control law

$$u(t) = Kx(t). \qquad (12.15.3)$$

The objective of the *linear-quadratic control problem* is to minimize the *quadratic performance measure*

$$J(K, x_0) = \int_0^\infty \left[x^{\mathrm{T}}(t) R_1 x(t) + 2x^{\mathrm{T}}(t) R_{12} u(t) + u^{\mathrm{T}}(t) R_2 u(t) \right] \mathrm{d}t, \quad (12.15.4)$$

where $R_1 \in \mathbb{R}^{n \times n}$, $R_{12} \in \mathbb{R}^{n \times m}$, and $R_2 \in \mathbb{R}^{m \times m}$. We assume that $\begin{bmatrix} R_1 & R_{12} \\ R_{12}^{\mathrm{T}} & R_2 \end{bmatrix}$ is positive semidefinite and R_2 is positive definite.

The performance measure (12.15.4) indicates the desire to maintain the state vector $x(t)$ close to the zero equilibrium without an excessive expenditure of control effort. Specifically, the term $x^{\mathrm{T}}(t) R_1 x(t)$ is a measure of the deviation of the state $x(t)$ from the zero state, where the $n \times n$ positive-semidefinite matrix R_1 determines how much weighting is associated with each component of the state. Likewise, the $m \times m$ positive-definite matrix R_2 weights the magnitude of the control input. Finally, the cross-weighting term R_{12} arises naturally when additional filters are used to shape the system response or in specialized applications.

Using (12.15.1) and (12.15.3), the closed-loop dynamic system can be written as

$$\dot{x}(t) = (A + BK)x(t) \qquad (12.15.5)$$

so that

$$x(t) = e^{t\tilde{A}} x_0, \qquad (12.15.6)$$

where $\tilde{A} \triangleq A + BK$. Thus, the performance measure (12.15.4) becomes

$$J(K, x_0) = \int_0^\infty x^{\mathrm{T}}(t) \tilde{R} x(t) \, \mathrm{d}t = \int_0^\infty x_0^{\mathrm{T}} e^{t\tilde{A}^{\mathrm{T}}} \tilde{R} e^{t\tilde{A}} x_0 \, \mathrm{d}t$$

$$= \operatorname{tr} x_0^{\mathrm{T}} \int_0^\infty e^{t\tilde{A}^{\mathrm{T}}} \tilde{R} e^{t\tilde{A}} \, \mathrm{d}t x_0 = \operatorname{tr} \int_0^\infty e^{t\tilde{A}^{\mathrm{T}}} \tilde{R} e^{t\tilde{A}} \, \mathrm{d}t x_0 x_0^{\mathrm{T}}, \qquad (12.15.7)$$

where

$$\tilde{R} \triangleq R_1 + R_{12} K + K^{\mathrm{T}} R_{12}^{\mathrm{T}} + K^{\mathrm{T}} R_2 K. \qquad (12.15.8)$$

Now, consider the standard control problem with plant

$$\mathcal{G} \sim \left[\begin{array}{c|cc} A & D_1 & B \\ \hline E_1 & 0 & E_2 \\ I_n & 0 & 0 \end{array} \right] \tag{12.15.9}$$

and full-state feedback $u = Kx$. Then, the closed-loop transfer function is given by

$$\tilde{\mathcal{G}} \sim \left[\begin{array}{c|c} A + BK & D_1 \\ \hline E_1 + E_2 K & 0 \end{array} \right]. \tag{12.15.10}$$

The following result shows that the quadratic performance measure (12.15.4) is equal to the H_2 norm of a transfer function.

Proposition 12.15.1. Assume that $D_1 = x_0$ and

$$\left[\begin{array}{cc} R_1 & R_{12} \\ R_{12}^{\mathrm{T}} & R_2 \end{array} \right] = \left[\begin{array}{c} E_1^{\mathrm{T}} \\ E_2^{\mathrm{T}} \end{array} \right] \left[\begin{array}{cc} E_1 & E_2 \end{array} \right], \tag{12.15.11}$$

and let $\tilde{\mathcal{G}}$ be given by (12.15.10). Then,

$$J(K, x_0) = \|\tilde{\mathcal{G}}\|_{\mathrm{H}_2}^2. \tag{12.15.12}$$

Proof. The result follows from Proposition 12.1.2. $\qquad\square$

For the following development, we assume that (12.15.11) holds so that R_1, R_{12}, and R_2 are given by

$$R_1 = E_1^{\mathrm{T}} E_1, \quad R_{12} = E_1^{\mathrm{T}} E_2, \quad R_2 = E_2^{\mathrm{T}} E_2. \tag{12.15.13}$$

To develop necessary conditions for the linear-quadratic control problem, we restrict K to the set of stabilizing gains

$$\mathcal{S} \triangleq \{K \in \mathbb{R}^{m \times n} : A + BK \text{ is asymptotically stable}\}. \tag{12.15.14}$$

Obviously, $\mathcal{S}$ is nonempty if and only if (A, B) is stabilizable. The following result gives necessary conditions that characterize a stabilizing solution K of the linear-quadratic control problem.

Theorem 12.15.2. Assume that (A, B) is stabilizable, assume that $K \in \mathcal{S}$ solves the linear-quadratic control problem, and assume that $(A + BK, D_1)$ is controllable. Then, there exists an $n \times n$ positive-semidefinite matrix P such that K is given by

$$K = -R_2^{-1}(B^{\mathrm{T}} P + R_{12}^{\mathrm{T}}) \tag{12.15.15}$$

and such that P satisfies

$$A_{\mathrm{R}}^{\mathrm{T}} P + P A_{\mathrm{R}} + \hat{R}_1 - P B R_2^{-1} B^{\mathrm{T}} P = 0, \tag{12.15.16}$$

where

$$A_{\mathrm{R}} \triangleq A - BR_2^{-1}R_{12}^{\mathrm{T}} \tag{12.15.17}$$

and

$$\hat{R}_1 \triangleq R_1 - R_{12}R_2^{-1}R_{12}^{\mathrm{T}}. \tag{12.15.18}$$

Furthermore, the minimal cost is given by

$$J(K) = \operatorname{tr} PV, \tag{12.15.19}$$

where $V \triangleq D_1 D_1^{\mathrm{T}}$.

Proof. Since $K \in \mathcal{S}$, it follows that $\tilde{A}$ is asymptotically stable. It then follows that $J(K)$ is given by (12.15.19), where $P \triangleq \int_0^\infty e^{t\tilde{A}^{\mathrm{T}}}\tilde{R}e^{t\tilde{A}}\, dt$ is positive semidefinite and satisfies the Lyapunov equation

$$\tilde{A}^{\mathrm{T}}P + P\tilde{A} + \tilde{R} = 0. \tag{12.15.20}$$

Note that (12.15.20) can be written as

$$(A + BK)^{\mathrm{T}}P + P(A + BK) + R_1 + R_{12}K + K^{\mathrm{T}}R_{12}^{\mathrm{T}} + K^{\mathrm{T}}R_2K = 0. \tag{12.15.21}$$

To optimize (12.15.19) subject to the constraint (12.15.20) over the open set $\mathcal{S}$, form the Lagrangian

$$\mathcal{L}(K, P, Q, \lambda_0) \triangleq \operatorname{tr}\left[\lambda_0 PV + Q\left(\tilde{A}^{\mathrm{T}}P + P\tilde{A} + \tilde{R}\right)\right], \tag{12.15.22}$$

where the Lagrange multipliers $\lambda_0 \geq 0$ and $Q \in \mathbb{R}^{n \times n}$ are not both zero. Note that the $n \times n$ Lagrange multiplier Q accounts for the $n \times n$ constraint equation (12.15.20).

The necessary condition $\partial\mathcal{L}/\partial P = 0$ implies

$$\tilde{A}Q + Q\tilde{A}^{\mathrm{T}} + \lambda_0 V = 0. \tag{12.15.23}$$

Since $\tilde{A}$ is asymptotically stable, it follows from Proposition 11.8.3 that, for all $\lambda_0 \geq 0$, (12.15.23) has a unique solution Q and, furthermore, Q is positive semidefinite. In particular, if $\lambda_0 = 0$, then $Q = 0$. Since λ_0 and Q are not both zero, we can set $\lambda_0 = 1$ so that (12.15.23) becomes

$$\tilde{A}Q + Q\tilde{A}^{\mathrm{T}} + V = 0. \tag{12.15.24}$$

Since $(\tilde{A}, D_1)$ is controllable, it follows from Corollary 12.7.10 that Q is positive definite.

Next, evaluating $\partial\mathcal{L}/\partial K = 0$ yields

$$R_2KQ + \left(B^{\mathrm{T}}P + R_{12}^{\mathrm{T}}\right)Q = 0. \tag{12.15.25}$$

Since Q is positive definite, it follows from (12.15.25) that (12.15.15) is satisfied. Furthermore, using (12.15.15), it follows that (12.15.20) is equivalent to (12.15.16). $\square$

With K given by (12.15.15) the closed-loop dynamics matrix $\tilde{A} = A + BK$ is given by

$$\tilde{A} = A - BR_2^{-1}\left(B^\mathrm{T}P + R_{12}^\mathrm{T}\right), \qquad (12.15.26)$$

where P is the solution of the *Riccati equation* (12.15.16).

12.16 Solutions of the Riccati Equation

For convenience in the following development, we assume that $R_{12} = 0$. With this assumption, the gain K given by (12.15.15) becomes

$$K = -R_2^{-1}B^\mathrm{T}P. \qquad (12.16.1)$$

Defining

$$\Sigma \triangleq BR_2^{-1}B^\mathrm{T}, \qquad (12.16.2)$$

(12.15.26) becomes

$$\tilde{A} = A - \Sigma P, \qquad (12.16.3)$$

while the Riccati equation (12.15.16) can be written as

$$A^\mathrm{T}P + PA + R_1 - P\Sigma P = 0. \qquad (12.16.4)$$

Note that (12.16.4) has the alternative representation

$$(A - \Sigma P)^\mathrm{T}P + P(A - \Sigma P) + R_1 + P\Sigma P = 0, \qquad (12.16.5)$$

which is equivalent to the Lyapunov equation

$$\tilde{A}^\mathrm{T}P + P\tilde{A} + \tilde{R} = 0, \qquad (12.16.6)$$

where

$$\tilde{R} \triangleq R_1 + P\Sigma P. \qquad (12.16.7)$$

By comparing (12.15.16) and (12.16.4), it can be seen that the linear-quadratic control problems with (A, B, R_1, R_{12}, R_2) and $(A_\mathrm{R}, B, \hat{R}_1, 0, R_2)$ are equivalent. Hence, there is no loss of generality in assuming that $R_{12} = 0$ in the following development, where A and R_1 can represent A_R and $\hat{R}_1$, respectively.

To motivate the subsequent development, the following examples demonstrate the existence of solutions under various assumptions on (A, B, E_1). In the following four examples, (A, B) is not stabilizable.

Example 12.16.1. Let $n = 1$, $A = 1$, $B = 0$, $E_1 = 0$, and $R_2 > 0$. Hence, (A, B, E_1) has an ORHP eigenvalue that is uncontrollable and unobservable. In this case, (12.16.4) has the unique solution $P = 0$. Furthermore, since $B = 0$, it follows that $\tilde{A} = A$.

Example 12.16.2. Let $n = 1$, $A = 1$, $B = 0$, $E_1 = 1$, and $R_2 > 0$. Hence, (A, B, E_1) has an ORHP eigenvalue that is uncontrollable and observable. In this case, (12.16.4) has the unique solution $P = -1/2 < 0$. Furthermore, since $B = 0$, it follows that $\tilde{A} = A$.

Example 12.16.3. Let $n = 1$, $A = 0$, $B = 0$, $E_1 = 0$, and $R_2 > 0$. Hence, (A, B, E_1) has an imaginary eigenvalue that is uncontrollable and unobservable. In this case, (12.16.4) has infinitely many solutions $P \in \mathbb{R}$. Hence, (12.16.4) has no maximal solution. Furthermore, since $B = 0$, it follows that, for every solution P, $\tilde{A} = A$.

Example 12.16.4. Let $n = 1$, $A = 0$, $B = 0$, $E_1 = 1$, and $R_2 > 0$. Hence, (A, B, E_1) has an imaginary eigenvalue that is uncontrollable and observable. In this case, (12.16.4) becomes $R_1 = 0$. Thus, (12.16.4) has no solution.

In the remaining examples, (A, B) is controllable.

Example 12.16.5. Let $n = 1$, $A = 1$, $B = 1$, $E_1 = 0$, and $R_2 > 0$. Hence, (A, B, E_1) has an ORHP eigenvalue that is controllable and unobservable. In this case, (12.16.4) has the solutions $P = 0$ and $P = 2R_2 > 0$. The corresponding closed-loop dynamics matrices are $\tilde{A} = 1 > 0$ and $\tilde{A} = -1 < 0$. Hence, the solution $P = 2R_2$ is stabilizing, and the closed-loop eigenvalue 1, which does not depend on R_2, is the reflection of the open-loop eigenvalue -1 across the imaginary axis.

Example 12.16.6. Let $n = 1$, $A = 1$, $B = 1$, $E_1 = 1$, and $R_2 > 0$. Hence, (A, B, E_1) has an ORHP eigenvalue that is controllable and observable. In this case, (12.16.4) has the solutions $P = R_2 - \sqrt{R_2^2 + R_2} < 0$ and $P = R_2 + \sqrt{R_2^2 + R_2} > 0$. The corresponding closed-loop dynamics matrices are $\tilde{A} = \sqrt{1 + 1/R_2} > 0$ and $\tilde{A} = -\sqrt{1 + 1/R_2} < 0$. Hence, the positive-definite solution $P = R_2 + \sqrt{R_2^2 + R_2}$ is stabilizing.

Example 12.16.7. Let $n = 1$, $A = 0$, $B = 1$, $E_1 = 0$, and $R_2 > 0$. Hence, (A, B, E_1) has an imaginary eigenvalue that is controllable and unobservable. In this case, (12.16.4) has the unique solution $P = 0$, which is not stabilizing.

Example 12.16.8. Let $n = 1$, $A = 0$, $B = 1$, $E_1 = 1$, and $R_2 > 0$. Hence, (A, B, E_1) has an imaginary eigenvalue that is controllable and observable. In this case, (12.16.4) has the solutions $P = -\sqrt{R_2} < 0$ and $P = \sqrt{R_2} > 0$. The corresponding closed-loop dynamics matrices are $\tilde{A} = $

$\sqrt{R_2} > 0$ and $\tilde{A} = -\sqrt{R_2} < 0$. Hence, the positive-definite solution $P = \sqrt{R_2}$ is stabilizing.

Example 12.16.9. Let $n = 2$, $A = \begin{bmatrix} 0 & 1 \\ -1 & 0 \end{bmatrix}$, $B = I_2$, $E_1 = 0$, and $R_2 = 1$. Hence, as in Example 12.16.7, both eigenvalues of (A, B, E_1) are imaginary, controllable, and unobservable. Taking the trace of (12.16.4) yields $\operatorname{tr} P^2 = 0$. Thus, the only positive-semidefinite matrix P satisfying (12.16.4) is $P = 0$, which implies that $\tilde{A} = A$. Consequently, the open-loop eigenvalues $\pm \jmath$ are unmoved by the feedback gain (12.15.15) even though (A, B) is controllable.

Example 12.16.10. Let $n = 2$, $A = 0$, $B = I_2$, $E_1 = I_2$, and $R_2 = I$. Hence, as in Example 12.16.8, both eigenvalues of (A, B, E_1) are imaginary, controllable, and observable. Furthermore, (12.16.4) becomes $P^2 = I$. Requiring that P be symmetric, it follows that P is a reflector. Hence, $P = I$ is the only positive-semidefinite solution. In fact, P is positive definite and stabilizing since $\tilde{A} = -I$.

Example 12.16.11. Let $A = \begin{bmatrix} 1 & 0 \\ 0 & 2 \end{bmatrix}$, $B = \begin{bmatrix} 1 \\ 1 \end{bmatrix}$, $E_1 = 0$, and $R_2 = 1$ so that (A, B) is controllable, although neither of the states is weighted. In this case (12.16.4) has four positive-semidefinite solutions given by

$$P_1 = \begin{bmatrix} 18 & -24 \\ -24 & 36 \end{bmatrix}, \quad P_2 = \begin{bmatrix} 2 & 0 \\ 0 & 0 \end{bmatrix}, \quad P_3 = \begin{bmatrix} 0 & 0 \\ 0 & 4 \end{bmatrix}, \quad P_4 = \begin{bmatrix} 0 & 0 \\ 0 & 0 \end{bmatrix}.$$

The corresponding feedback matrices are given by $K_1 = \begin{bmatrix} 6 & -12 \end{bmatrix}$, $K_2 = \begin{bmatrix} -2 & 0 \end{bmatrix}$, $K_3 = \begin{bmatrix} 0 & -4 \end{bmatrix}$, and $K_4 = \begin{bmatrix} 0 & 0 \end{bmatrix}$. Letting $\tilde{A}_i = A - \Sigma P_i$, it follows that $\operatorname{spec}(\tilde{A}_1) = \{-1, -2\}$, $\operatorname{spec}(\tilde{A}_2) = \{-1, 2\}$, $\operatorname{spec}(\tilde{A}_3) = \{1, -2\}$, and $\operatorname{spec}(\tilde{A}_4) = \{1, 2\}$. Thus, P_1 is the only solution that stabilizes the closed-loop system, while the solutions P_2 and P_3 partially stabilize the closed-loop system. Note also that the closed-loop poles that differ from those of the open-loop system are mirror images of the open-loop poles as reflected across the imaginary axis. Finally, note that these solutions satisfy the partial ordering $P_1 \geq P_2 \geq P_4$ and $P_1 \geq P_3 \geq P_4$, and that "larger" solutions are more stabilizing than "smaller" solutions. Moreover, letting $J(K_i) = \operatorname{tr} P_i V$, it can be seen that larger solutions incur a greater closed-loop cost, with the greatest cost incurred by the stabilizing solution P_4. However, the cost expression $J(K) = \operatorname{tr} PV$ does not follow from (12.15.4) when $A + BK$ is not asymptotically stable.

The following definition concerns solutions of the Riccati equation.

Definition 12.16.12. A matrix $P \in \mathbb{R}^{n \times n}$ is a *solution* of the Riccati equation (12.16.4) if P is symmetric and satisfies (12.16.4). Furthermore, P is the *stabilizing solution* of (12.16.4) if $\tilde{A} = A - \Sigma P$ is asymptotically stable. Finally, a solution $P_{\max}$ of (12.16.4) is the *maximal solution* to (12.16.4) if

$P \leq P_{\max}$ for every solution P to (12.16.4).

Since the ordering "$\leq$" is antisymmetric, it follows that (12.16.4) has at most one maximal solution. The uniqueness of the stabilizing solution is shown in the following section.

Next, define the $2n \times 2n$ *Hamiltonian*

$$\mathcal{H} \triangleq \begin{bmatrix} A & \Sigma \\ R_1 & -A^{\mathrm{T}} \end{bmatrix}. \tag{12.16.8}$$

Proposition 12.16.13. The following statements hold:

i) $\mathcal{H}$ is Hamiltonian.

ii) $\chi_{\mathcal{H}}$ has a spectral factorization, that is, there exists a monic polynomial $p \in \mathbb{R}[s]$ such that, for all $s \in \mathbb{C}$, $\chi_{\mathcal{H}}(s) = q(s)q(-s)$.

iii) $\chi_{\mathcal{H}}(\jmath\omega) \geq 0$ for all $\omega \in \mathbb{R}$.

iv) If either $R_1 = 0$ or $\Sigma = 0$, then $\mathrm{mspec}(\mathcal{H}) = \mathrm{mspec}(A) \cup \mathrm{mspec}(-A)$.

v) $\chi_{\mathcal{H}}$ is even.

vi) $\lambda \in \mathrm{spec}(\mathcal{H})$ if and only if $-\lambda \in \mathrm{spec}(\mathcal{H})$.

vii) If $\lambda \in \mathrm{spec}(\mathcal{H})$, then $\mathrm{am}_{\mathcal{H}}(\lambda) = \mathrm{am}_{\mathcal{H}}(-\lambda)$.

viii) Every imaginary root of $\chi_{\mathcal{H}}$ has even multiplicity.

ix) Every imaginary eigenvalue of $\mathcal{H}$ has even algebraic multiplicity.

Proof. The result follows from Proposition 4.1.1 and Fact 4.9.17. $\square$

It is important to keep in mind that spectral factorizations are not unique. For example, if $\chi_{\mathcal{H}}(s) = (s+1)(s+2)(-s+1)(-s+2)$, then $\chi_{\mathcal{H}}(s) = p(s)p(-s) = \hat{p}(s)\hat{p}(-s)$, where $p(s) = (s+1)(s+2)$ and $\hat{p}(s) = (s+1)(s-2)$. Thus, the spectral factors $p(s)$ and $p(-s)$ can "trade" roots. These roots are the eigenvalues of $\mathcal{H}$.

The following result shows that the Hamiltonian matrix $\mathcal{H}$ is closely linked to the Riccati equation (12.16.4).

Proposition 12.16.14. Let $P \in \mathbb{R}^{n \times n}$ be symmetric. Then, the following statements are equivalent:

i) P is a solution of (12.16.4).

ii) P satisfies

$$\begin{bmatrix} P & I \end{bmatrix} \mathcal{H} \begin{bmatrix} I \\ -P \end{bmatrix} = 0. \tag{12.16.9}$$

iii) P satisfies

$$\mathcal{H}\begin{bmatrix} I \\ -P \end{bmatrix} = \begin{bmatrix} I \\ -P \end{bmatrix}(A - \Sigma P). \qquad (12.16.10)$$

iv) P satisfies

$$\mathcal{H} = \begin{bmatrix} I & 0 \\ -P & I \end{bmatrix}\begin{bmatrix} A - \Sigma P & \Sigma \\ 0 & -(A - \Sigma P)^{\mathrm{T}} \end{bmatrix}\begin{bmatrix} I & 0 \\ P & I \end{bmatrix}. \qquad (12.16.11)$$

In this case, the following statements hold:

v) $\mathrm{mspec}(\mathcal{H}) = \mathrm{mspec}(A - \Sigma P) \cup \mathrm{mspec}[-(A - \Sigma P)]$.

vi) $\chi_{\mathcal{H}}(s) = (-1)^n \chi_{A-\Sigma P}(s)\chi_{A-\Sigma P}(-s)$.

vii) $\mathcal{R}\left(\begin{bmatrix} I \\ -P \end{bmatrix}\right)$ is an invariant subspace of $\mathcal{H}$.

Corollary 12.16.15. Assume that (12.16.4) has a stabilizing solution. Then, $\mathcal{H}$ has no imaginary eigenvalues.

For the next two results, P is not necessarily a solution of (12.16.4).

Lemma 12.16.16. Assume that $\lambda \in \mathrm{spec}(A)$ is an observable eigenvalue of (A, R_1), and let $P \in \mathbb{R}^{n \times n}$ be symmetric. Then, $\lambda \in \mathrm{spec}(A)$ is an observable eigenvalue of $(\tilde{A}, \tilde{R})$.

Proof. Assume that $\mathrm{rank}\begin{bmatrix} \lambda I - \tilde{A} \\ \tilde{R} \end{bmatrix} < n$. Then, there exists a nonzero vector $v \in \mathbb{C}^n$ such that $\tilde{A}v = \lambda v$ and $\tilde{R}v = 0$. Hence, $v^* R_1 v = -v^* P \Sigma P v \leq 0$, which implies that $R_1 v = 0$ and $P \Sigma P v = 0$. Hence, $\Sigma P v = 0$, and thus $Av = \lambda v$. Therefore, $\mathrm{rank}\begin{bmatrix} \lambda I - A \\ R_1 \end{bmatrix} < n$. $\square$

Lemma 12.16.17. Assume that (A, R_1) is (observable, detectable), and let $P \in \mathbb{R}^{n \times n}$ be symmetric. Then, $(\tilde{A}, \tilde{R})$ is (observable, detectable).

Lemma 12.16.18. Assume that (A, E_1) is observable, and assume that (12.16.4) has a solution P. Then, the following statements hold:

i) $\nu_-(\tilde{A}) = \nu_+(P)$.

ii) $\nu_0(\tilde{A}) = \nu_0(P) = 0$.

iii) $\nu_+(\tilde{A}) = \nu_-(P)$.

Proof. Since (A, R_1) is observable, it follows from Lemma 12.16.17 that $(\tilde{A}, \tilde{R})$ is observable. By writing (12.16.4) as the Lyapunov equation (12.16.6), the result now follows from Fact 12.21.1. $\square$

12.17 The Stabilizing Solution of the Riccati Equation

Proposition 12.17.1. The following statements hold:

i) (12.16.4) has at most one stabilizing solution.

ii) If P is the stabilizing solution of (12.16.4), then P is positive semi-definite.

iii) If P is the stabilizing solution of (12.16.4), then

$$\operatorname{rank} P = \operatorname{rank} \mathcal{O}(\tilde{A}, \tilde{R}). \qquad (12.17.1)$$

Proof. To prove *i)*, suppose that (12.16.4) has stabilizing solutions P_1 and P_2. Then,

$$A^{\mathrm{T}}P_1 + P_1 A + R_1 - P_1 \Sigma P_1 = 0,$$
$$A^{\mathrm{T}}P_2 + P_2 A + R_1 - P_2 \Sigma P_2 = 0.$$

Subtracting these equations and rearranging yields

$$(A - \Sigma P_1)^{\mathrm{T}}(P_1 - P_2) + (P_1 - P_2)(A - \Sigma P_2) = 0.$$

Since $A - \Sigma P_1$ and $A - \Sigma P_2$ are asymptotically stable, it follows from Proposition 11.8.3 and Fact 11.17.31 that $P_1 - P_2 = 0$. Hence, (12.16.4) has at most one stabilizing solution.

Next, to prove *ii)*, suppose that P is a stabilizing solution of (12.16.4). Then, it follows from (12.16.4) that

$$P = \int_0^\infty e^{t(A - \Sigma P)^{\mathrm{T}}}(R_1 + P\Sigma P)e^{t(A - \Sigma P)} \, \mathrm{d}t,$$

which shows that P is positive semidefinite.

Finally, *iii)* follows from Corollary 12.3.3. $\qquad \square$

Theorem 12.17.2. Assume that (12.16.4) has a positive-semidefinite solution P, and assume that (A, E_1) is detectable. Then, P is the stabilizing solution of (12.16.4), and thus P is the only positive-semidefinite solution of (12.16.4). If, in addition, (A, E_1) is observable, then P is positive definite.

Proof. Since (A, R_1) is detectable, it follows from Lemma 12.16.17 that $(\tilde{A}, \tilde{R})$ is detectable. Next, since (12.16.4) has a positive-semidefinite solution P, it follows from Corollary 12.8.6 that $\tilde{A}$ is asymptotically stable. Hence, P is a stabilizing solution of (12.16.4). The last statement follows from Lemma 12.16.18. $\qquad \square$

Corollary 12.17.3. Assume that (A, E_1) is detectable. Then, (12.16.4) has at most one positive-semidefinite solution.

Lemma 12.17.4. Let $\lambda \in \mathbb{C}$, and assume that λ is either an uncontrollable eigenvalue of (A, B) or an unobservable eigenvalue of (A, E_1). Then, $\lambda \in \operatorname{spec}(\mathcal{H})$.

Proof. Note that

$$\lambda I - \mathcal{H} = \begin{bmatrix} \lambda I - A & -\Sigma \\ -R_1 & \lambda I + A^{\mathrm{T}} \end{bmatrix}.$$

If λ is an uncontrollable eigenvalue of (A, B), then the first n rows of $\lambda I - \mathcal{H}$ are linearly dependent, and thus $\lambda \in \operatorname{spec}(\mathcal{H})$. On the other hand, if λ is an unobservable eigenvalue of (A, E_1), then the first n columns of $\lambda I - \mathcal{H}$ are linearly dependent, and thus $\lambda \in \operatorname{spec}(\mathcal{H})$. $\qquad\square$

The following result is a consequence of Lemma 12.17.4.

Proposition 12.17.5. Let $S \in \mathbb{R}^{n \times n}$ be a nonsingular matrix such that

$$A = S \begin{bmatrix} A_1 & 0 & A_{13} & 0 \\ A_{21} & A_2 & A_{23} & A_{24} \\ 0 & 0 & A_3 & 0 \\ 0 & 0 & A_{43} & A_4 \end{bmatrix} S^{-1}, \quad B = S \begin{bmatrix} B_1 \\ B_2 \\ 0 \\ 0 \end{bmatrix}, \tag{12.17.2}$$

$$E_1 = \begin{bmatrix} E_{11} & 0 & E_{13} & 0 \end{bmatrix} S^{-1}, \tag{12.17.3}$$

where $\left(\begin{bmatrix} A_1 & 0 \\ A_{21} & A_2 \end{bmatrix}, \begin{bmatrix} B_1 \\ B_2 \end{bmatrix} \right)$ is controllable and $\left(\begin{bmatrix} A_1 & A_{13} \\ 0 & A_3 \end{bmatrix}, \begin{bmatrix} E_{11} & E_{13} \end{bmatrix} \right)$ is observable. Then,

$$\operatorname{mspec}(A_2) \cup \operatorname{mspec}(-A_2) \subseteq \operatorname{mspec}(\mathcal{H}), \tag{12.17.4}$$

$$\operatorname{mspec}(A_3) \cup \operatorname{mspec}(-A_3) \subseteq \operatorname{mspec}(\mathcal{H}), \tag{12.17.5}$$

$$\operatorname{mspec}(A_4) \cup \operatorname{mspec}(-A_4) \subseteq \operatorname{mspec}(\mathcal{H}). \tag{12.17.6}$$

Next, we present a partial converse of Lemma 12.17.4.

Lemma 12.17.6. Let $\lambda \in \operatorname{spec}(\mathcal{H})$, and assume that $\operatorname{Re} \lambda = 0$. Then, λ is either an uncontrollable eigenvalue of (A, B) or an unobservable eigenvalue of (A, E_1).

Proof. Suppose that $\lambda = \jmath\omega$ is an eigenvalue of $\mathcal{H}$, where $\omega \in \mathbb{R}$. Then, there exist $x, y \in \mathbb{C}^n$ such that $\begin{bmatrix} x \\ y \end{bmatrix} \neq 0$ and $\mathcal{H} \begin{bmatrix} x \\ y \end{bmatrix} = \jmath\omega \begin{bmatrix} x \\ y \end{bmatrix}$. Consequently,

$$Ax + \Sigma y = \jmath\omega x, \quad R_1 x - A^{\mathrm{T}} y = \jmath\omega y.$$

Rewriting these identities as

$$(A - \jmath\omega I)x = -\Sigma y, \quad (A - \jmath\omega I)^* y = R_1 x$$

yields

$$y^*(A - \jmath\omega I)x = -y^*\Sigma y, \quad x^*(A - \jmath\omega I)^* y = x^* R_1 x.$$

Since $x^*(A - \jmath\omega I)^* y$ is real, it follows that $-y^*\Sigma y = x^* R_1 x$, and thus $y^*\Sigma y = x^* R_1 x = 0$, which implies that $B^\mathrm{T} y = 0$ and $E_1 x = 0$. Therefore,

$$(A - \jmath\omega I)x = 0, \quad (A - \jmath\omega I)^* y = 0,$$

and hence

$$\begin{bmatrix} A - \jmath\omega I \\ E_1 \end{bmatrix} x = 0, \quad y^* \begin{bmatrix} A - \jmath\omega I & B \end{bmatrix} = 0.$$

Since $\begin{bmatrix} x \\ y \end{bmatrix} \neq 0$, it follows that either $x \neq 0$ or $y \neq 0$, and thus either $\mathrm{rank} \begin{bmatrix} A - \jmath\omega I \\ E_1 \end{bmatrix} < n$ or $\mathrm{rank} \begin{bmatrix} A - \jmath\omega I & B \end{bmatrix} < n$. $\square$

The following result is a restatement of Lemma 12.17.6.

Proposition 12.17.7. Let $S \in \mathbb{R}^{n \times n}$ be a nonsingular matrix such that (12.17.2) and (12.17.3) are satisfied, where $\left(\begin{bmatrix} A_1 & 0 \\ A_{21} & A_2 \end{bmatrix}, \begin{bmatrix} B_1 \\ B_2 \end{bmatrix} \right)$ is controllable and $\left(\begin{bmatrix} A_1 & A_{13} \\ 0 & A_3 \end{bmatrix}, \begin{bmatrix} E_{11} & E_{13} \end{bmatrix} \right)$ is observable. Then,

$$\mathrm{mspec}(\mathcal{H}) \cap \jmath\mathbb{R} \subseteq \mathrm{mspec}(A_2) \cup \mathrm{mspec}(-A_2) \cup \mathrm{mspec}(A_3)$$
$$\cup \mathrm{mspec}(-A_3) \cup \mathrm{mspec}(A_4) \cup \mathrm{mspec}(-A_4). \quad (12.17.7)$$

Combining Lemma 12.17.4 and Lemma 12.17.6 yields the following result.

Proposition 12.17.8. Let $\lambda \in \mathbb{C}$, assume that $\mathrm{Re}\,\lambda = 0$, and let $S \in \mathbb{R}^{n \times n}$ be a nonsingular matrix such that (12.17.2) and (12.17.3) are satisfied, where (A_1, B_1, E_{11}) is controllable and observable, (A_2, B_2) is controllable, and (A_3, E_{13}) is observable. Then, the following statements are equivalent:

i) λ is either an uncontrollable eigenvalue of (A, B) or an unobservable eigenvalue of (A, E_1).

ii) $\lambda \in \mathrm{mspec}(A_2) \cup \mathrm{mspec}(A_3) \cup \mathrm{mspec}(A_4)$.

iii) λ is an eigenvalue of $\mathcal{H}$.

The next result gives necessary and sufficient conditions under which (12.16.4) has a stabilizing solution. This result also provides a constructive characterization of the stabilizing solution.

Theorem 12.17.9. The following statements are equivalent:

i) (A, B) is stabilizable, and every imaginary eigenvalue of (A, E_1) is observable.

ii) There exists a nonsingular matrix $S \in \mathbb{R}^{n \times n}$ such that (12.17.2) and (12.17.3) are satisfied, where $\left(\left[\begin{smallmatrix} A_1 & 0 \\ A_{21} & A_2 \end{smallmatrix} \right], \left[\begin{smallmatrix} B_1 \\ B_2 \end{smallmatrix} \right] \right)$ is controllable, $\left(\left[\begin{smallmatrix} A_1 & A_{13} \\ 0 & A_3 \end{smallmatrix} \right], \left[\begin{smallmatrix} E_{11} & E_{13} \end{smallmatrix} \right] \right)$ is observable, $\nu_0(A_2) = 0$, and A_3 and A_4 are asymptotically stable.

iii) (12.16.4) has a stabilizing solution.

In this case, let

$$M = \begin{bmatrix} M_1 & M_{12} \\ M_{21} & M_2 \end{bmatrix} \in \mathbb{R}^{2n \times 2n} \tag{12.17.8}$$

be a nonsingular matrix such that $\mathcal{H} = MZM^{-1}$, where

$$Z = \begin{bmatrix} Z_1 & Z_{12} \\ 0 & Z_2 \end{bmatrix} \in \mathbb{R}^{2n \times 2n} \tag{12.17.9}$$

and $Z_1 \in \mathbb{R}^{n \times n}$ is asymptotically stable. Then, M_1 is nonsingular, and

$$P \triangleq -M_{21}M_1^{-1} \tag{12.17.10}$$

is the stabilizing solution of (12.16.4).

Proof. The equivalence of i) and ii) is immediate.

Next, to prove i) $\Longrightarrow$ iii), first note that Lemma 12.17.6 implies that $\mathcal{H}$ has no imaginary eigenvalues. Hence, since $\mathcal{H}$ is Hamiltonian, it follows that there exists a matrix $M \in \mathbb{R}^{2n \times 2n}$ of the form (12.17.8) such that M is nonsingular and $\mathcal{H} = MZM^{-1}$, where $Z \in \mathbb{R}^{n \times n}$ is of the form (12.17.9) and $Z_1 \in \mathbb{R}^{n \times n}$ is asymptotically stable.

Next, note that $\mathcal{H}M = MZ$ implies that

$$\mathcal{H} \begin{bmatrix} M_1 \\ M_{21} \end{bmatrix} = M \begin{bmatrix} Z_1 \\ 0 \end{bmatrix} = \begin{bmatrix} M_1 \\ M_{21} \end{bmatrix} Z_1.$$

Therefore,

$$\begin{bmatrix} M_1 \\ M_{21} \end{bmatrix}^{\mathrm{T}} J_n \mathcal{H} \begin{bmatrix} M_1 \\ M_{21} \end{bmatrix} = \begin{bmatrix} M_1 \\ M_{21} \end{bmatrix}^{\mathrm{T}} J_n \begin{bmatrix} M_1 \\ M_{21} \end{bmatrix} Z_1$$

$$= \begin{bmatrix} M_1^{\mathrm{T}} & M_{21}^{\mathrm{T}} \end{bmatrix} \begin{bmatrix} M_{21} \\ -M_1 \end{bmatrix} Z_1$$

$$= LZ_1,$$

where $L \triangleq M_1^{\mathrm{T}}M_{21} - M_{21}^{\mathrm{T}}M_1$. Since $J_n\mathcal{H} = (J_n\mathcal{H})^{\mathrm{T}}$, it follows that LZ_1 is symmetric, that is, $LZ_1 = Z_1^{\mathrm{T}}L^{\mathrm{T}}$. Since, in addition, L is skew symmetric, it follows that $0 = Z_1^{\mathrm{T}}L + LZ_1$. Now, since Z_1 is asymptotically stable, it follows that $L = 0$. Hence, $M_1^\mathrm{T} M_{21} = M_{21}^\mathrm{T} M_1$, which shows that $M_{21}^\mathrm{T} M_1$ is symmetric.

To show that M_1 is nonsingular, note that it follows from the identity

$$[\ I \ \ 0\ ]\mathcal{H}\begin{bmatrix} M_1 \\ M_{21} \end{bmatrix} = [\ I \ \ 0\ ]\begin{bmatrix} M_1 \\ M_{21} \end{bmatrix} Z_1$$

that

$$AM_1 + \Sigma M_{21} = M_1 Z_1.$$

Now, let $x \in \mathbb{R}^n$ satisfy $M_1 x = 0$. We thus have

$$x^\mathrm{T} M_{21}\Sigma M_{21}x = x^\mathrm{T} M_{21}^\mathrm{T}[AM_1 + \Sigma M_{21}]x = x^\mathrm{T} M_{21}^\mathrm{T} M_1 Z_1 x$$
$$= x^\mathrm{T} M_1^\mathrm{T} M_{21} Z_1 x = 0,$$

which implies that $B^\mathrm{T} M_{21}x = 0$. Hence, $M_1 Z_1 x = (AM_1 + \Sigma M_{21})x = 0$. Thus, $Z_1 \mathcal{N}(M_1) \subseteq \mathcal{N}(M_1)$.

Now, suppose that M_1 is singular. Since $Z_1 \mathcal{N}(M_1) \subseteq \mathcal{N}(M_1)$, it follows that there exists $\lambda \in \mathrm{spec}(Z_1)$ and $x \in \mathbb{C}^n$ such that $Z_1 x = \lambda x$ and $M_1 x = 0$. Forming

$$[\ 0 \ \ I\ ]\mathcal{H}\begin{bmatrix} M_1 \\ M_{21} \end{bmatrix} x = [\ 0 \ \ I\ ]\begin{bmatrix} M_1 \\ M_{21} \end{bmatrix} Z_1 x$$

yields $-A^\mathrm{T} M_{21}x = M_{21}\lambda Z$, and thus $(\lambda I + A^\mathrm{T})M_{21}x = 0$. Since, in addition, as shown above, $B^\mathrm{T} M_{21}x = 0$, it follows that $x^* M_{21}^\mathrm{T}[\ -\bar{\lambda}I - A \ \ B\ ] = 0$. Since $\lambda \in \mathrm{spec}(Z_1)$, it follows that $\mathrm{Re}(-\bar{\lambda}) > 0$. Furthermore, since, by assumption (A, B) is stabilizable, it follows that $\mathrm{rank}[\ \bar{\lambda}I - A \ \ B\ ] = n$. Therefore, $M_{21}x = 0$. Combining this fact with $M_1 x = 0$ yields $\begin{bmatrix} M_1 \\ M_{21} \end{bmatrix}x = 0$. Since x is nonzero, it follows that M is singular, which is a contradiction. Consequently, M_1 is nonsingular. Next, define $P \triangleq -M_{21}M_1^{-1}$ and note that, since $M_1^\mathrm{T} M_{21}$ is symmetric, it follows that $P = -M_1^{-\mathrm{T}}(M_1^\mathrm{T} M_{21})M_1^{-1}$ is also symmetric.

Since $\mathcal{H}\begin{bmatrix} M_1 \\ M_{21} \end{bmatrix} = \begin{bmatrix} M_1 \\ M_{21} \end{bmatrix}Z_1$, it follows that

$$\mathcal{H}\begin{bmatrix} I \\ M_{21}M_1^{-1} \end{bmatrix} = \begin{bmatrix} I \\ M_{21}M_1^{-1} \end{bmatrix}M_1 Z_1 M_1^{-1},$$

and thus

$$\mathcal{H}\begin{bmatrix} I \\ -P \end{bmatrix} = \begin{bmatrix} I \\ -P \end{bmatrix}M_1 Z_1 M_1^{-1}.$$

Multiplying on the left by $[\ P \ \ I\ ]$ yields

$$0 = [\ P \ \ I\ ]\mathcal{H}\begin{bmatrix} I \\ -P \end{bmatrix} = A^\mathrm{T} P + PA + R_1 - P\Sigma P,$$

which shows that P is a solution of (12.16.4). Similarly, multiplying on the left by $\begin{bmatrix} I & 0 \end{bmatrix}$ yields $A - \Sigma P = M_1 Z_1 M_1^{-1}$. Since Z_1 is asymptotically stable, it follows that $A - \Sigma P$ is also asymptotically stable.

To prove $iii) \implies i)$, note that the existence of a stabilizing solution P implies that (A, B) is stabilizable, and that (12.16.11) implies that $\mathcal{H}$ has no imaginary eigenvalues. $\qquad \square$

Corollary 12.17.10. Assume that (A, B) is stabilizable and (A, E_1) is detectable. Then, (12.16.4) has a stabilizing solution.

12.18 The Maximal Solution of the Riccati Equation

In this section we consider the existence of the maximal solution of (12.16.4). Example 12.16.3 shows that the assumptions of Proposition 12.19.1 are not sufficient to guarantee that (12.16.4) has a maximal solution.

Theorem 12.18.1. The following statements are equivalent:

$i)$ (A, B) is stabilizable.

$ii)$ (12.16.4) has a solution $P_{\max}$ that is positive semidefinite, maximal, and satisfies $\operatorname{spec}(A - \Sigma P_{\max}) \subset \text{CLHP}$.

Proof. The result $i) \implies ii)$ is given by Theorem 2.1 and Theorem 2.2 of [295]. See also (i) of Theorem 13.11 of [818]). The converse result follows from Corollary 3 of [631]. $\qquad \square$

Proposition 12.18.2. Assume that (12.16.4) has a maximal solution $P_{\max}$, let P be a solution of (12.16.4), and assume that $\operatorname{spec}(A - \Sigma P_{\max}) \subset$ CLHP and $\operatorname{spec}(A - \Sigma P) \subset$ CLHP. Then, $P = P_{\max}$.

Proof. It follows from $i)$ of Proposition 12.16.14 that $\operatorname{spec}(A - \Sigma P) = \operatorname{spec}(A - \Sigma P_{\max})$. Since $P_{\max}$ is the maximal solution of (12.16.4), it follows that $P \leq P_{\max}$. Consequently, it follows from the contrapositive form of the second statement of Theorem 8.4.9 that $P = P_{\max}$. $\qquad \square$

Proposition 12.18.3. Assume that (12.16.4) has a solution P such that $\operatorname{spec}(A - \Sigma P) \subset$ CLHP. Then, P is stabilizing if and only if $\mathcal{H}$ has no imaginary eigenvalues

It follows from Proposition 12.18.2 that (12.16.4) has at most one positive-semidefinite solution P such that $\operatorname{spec}(A - \Sigma P) \subset$ CLHP. Consequently, (12.16.4) has at most one positive-semidefinite stabilizing solution.

Theorem 12.18.4. The following statements hold:

i) (12.16.4) has at most one stabilizing solution.

ii) If P is the stabilizing solution of (12.16.4), then P is positive semi-definite.

iii) If P is the stabilizing solution of (12.16.4), then P is maximal.

Proof. To prove *i)*, assume that (12.16.4) has stabilizing solutions P_1 and P_2. Then, (A, B) is stabilizable, and Theorem 12.18.1 implies that (12.16.4) has a maximal solution $P_{\max}$ such that $\mathrm{spec}(A - \Sigma P_{\max}) \subset \mathrm{CLHP}$. Now, Proposition 12.18.2 implies that $P_1 = P_{\max}$ and $P_2 = P_{\max}$. Hence, $P_1 = P_2$.

Alternatively, suppose that (12.16.4) has the stabilizing solutions P_1 and P_2. Then,

$$A^{\mathrm{T}}P_1 + P_1 A + R_1 - P_1 \Sigma P_1 = 0,$$
$$A^{\mathrm{T}}P_2 + P_2 A + R_1 - P_2 \Sigma P_2 = 0.$$

Subtracting these equations and rearranging yields

$$(A - \Sigma P_1)^{\mathrm{T}}(P_1 - P_2) + (P_1 - P_2)(A - \Sigma P_2) = 0.$$

Since $A - \Sigma P_1$ and $A - \Sigma P_2$ are asymptotically stable, it follows from Proposition 11.8.3 and Fact 11.17.31 that $P_1 - P_2 = 0$. Hence, (12.16.4) has at most one stabilizing solution.

Next, to prove *ii)*, suppose that P is a stabilizing solution of (12.16.4). Then, it follows from (12.16.4) that

$$P = \int_0^\infty e^{t(A - \Sigma P)^{\mathrm{T}}}(R_1 + P \Sigma P) e^{t(A - \Sigma P)} \, \mathrm{d}t,$$

which shows that P is positive semidefinite.

To prove *iii)*, let P' be a solution of (12.16.4). Then, it follows that

$$(A - \Sigma P)^{\mathrm{T}}(P - P') + (P - P')(A - \Sigma P) + (P - P')\Sigma(P - P') = 0,$$

which implies that $P' \le P$. Thus, P is also the maximal solution of (12.16.4). $\qquad\square$

The following results concerns the monotonicity of solutions of the Riccati equation (12.16.4).

Proposition 12.18.5. Assume that (A, B) is stabilizable, and let $P_{\max}$ denote the maximal solution of (12.16.4). Furthermore, let $\hat{R}_1 \in \mathbb{R}^{n \times n}$ be positive semidefinite, let $\hat{R}_2 \in \mathbb{R}^{m \times m}$ be positive definite, let $\hat{A} \in \mathbb{R}^{n \times n}$, let

$\hat{B} \in \mathbb{R}^{n \times m}$, define $\hat{\Sigma} \triangleq \hat{B}\hat{R}_2^{-1}B^{\mathrm{T}}$, assume that

$$\begin{bmatrix} \hat{R}_1 & \hat{A}^{\mathrm{T}} \\ \hat{A} & -\hat{\Sigma} \end{bmatrix} \le \begin{bmatrix} R_1 & A^{\mathrm{T}} \\ A & -\Sigma \end{bmatrix},$$

and let $\hat{P}$ be a solution of

$$\hat{A}^{\mathrm{T}}\hat{P} + \hat{P}\hat{A} + \hat{R}_1 - \hat{P}\hat{\Sigma}\hat{P} = 0. \tag{12.18.1}$$

Then,

$$\hat{P} \le P_{\max}. \tag{12.18.2}$$

Proof. The result is given by Theorem 1 of [785]. $\square$

Corollary 12.18.6. Assume that (A, B) is stabilizable, let $\hat{R}_1 \in \mathbb{R}^{n \times n}$ be positive semidefinite, assume that $\hat{R}_1 \le R_1$, and let $P_{\max}$ and $\hat{P}_{\max}$ denote, respectively, the maximal solutions of (12.16.4) and

$$A^{\mathrm{T}}P + PA + \hat{R}_1 - P\Sigma P = 0. \tag{12.18.3}$$

Then,

$$\hat{P}_{\max} \le P_{\max}. \tag{12.18.4}$$

Proof. The result follows from Proposition 12.18.5 or Theorem 2.3 of [295]. $\square$

The following result shows that, if $R_1 = 0$, then the closed-loop eigenvalues of the closed-loop dynamics obtained from the maximal solution consist of the CLHP open-loop eigenvalues and reflections of the ORHP open-loop eigenvalues.

Proposition 12.18.7. Assume that (A, B) is stabilizable, assume that $R_1 = 0$, and let $P \in \mathbb{R}^{n \times n}$ be a positive-semidefinite solution of (12.16.4). Then, P is the maximal solution of (12.16.4) if and only if

$$\mathrm{mspec}(A - \Sigma P) = [\mathrm{mspec}(A) \cap \mathrm{CLHP}] \cup [\mathrm{mspec}(-A) \cap \mathrm{OLHP}]. \tag{12.18.5}$$

Proof. Sufficiency follows from Proposition 12.18.2. To prove necessity, note that it follows from the definition (12.16.8) of $\mathcal{H}$ with $R_1 = 0$ and from $iv)$ of Proposition 12.16.14 that

$$\mathrm{mspec}(A) \cup \mathrm{mspec}(-A) = \mathrm{mspec}(A - \Sigma P) \cup \mathrm{mspec}[-(A - \Sigma P)].$$

Now, Theorem 12.18.1 implies that $\mathrm{mspec}(A - \Sigma P) \subseteq \mathrm{CLHP}$, which implies that (12.18.5) is satisfied. $\square$

Corollary 12.18.8. Let $R_1 = 0$, and assume that $\mathrm{spec}(A) \subset \mathrm{CLHP}$. Then, $P = 0$ is the only positive-semidefinite solution of (12.16.4).

12.19 Positive-Semidefinite and Positive-Definite Solutions of the Riccati Equation

The following result gives sufficient conditions under which (12.16.4) has a positive-semidefinite solution.

Proposition 12.19.1. Suppose that there exists a nonsingular matrix $S \in \mathbb{R}^{n \times n}$ such that (12.17.2) and (12.17.3) are satisfied, where $\left(\left[\begin{smallmatrix} A_1 & 0 \\ A_{21} & A_2 \end{smallmatrix} \right] , \left[\begin{smallmatrix} B_1 \\ B_2 \end{smallmatrix} \right] \right)$ is controllable, $\left(\left[\begin{smallmatrix} A_1 & A_{13} \\ 0 & A_3 \end{smallmatrix} \right] , \left[E_{11} \; E_{13} \right] \right)$ is observable, and A_3 is asymptotically stable. Then, (12.16.4) has a positive-semidefinite solution.

Proof. First, rewrite (12.17.2) and (12.17.3) as

$$A = S \begin{bmatrix} A_1 & A_{13} & 0 & 0 \\ 0 & A_3 & 0 & 0 \\ A_{21} & A_{23} & A_2 & A_{24} \\ 0 & A_{43} & 0 & A_4 \end{bmatrix} S^{-1}, \qquad B = S \begin{bmatrix} B_1 \\ 0 \\ B_2 \\ 0 \end{bmatrix},$$

$$E_1 = \begin{bmatrix} E_{11} & E_{13} & 0 & 0 \end{bmatrix} S^{-1},$$

where $\left(\left[\begin{smallmatrix} A_1 & 0 \\ A_{21} & A_2 \end{smallmatrix} \right] , \left[\begin{smallmatrix} B_1 \\ B_2 \end{smallmatrix} \right] \right)$ is controllable, $\left(\left[\begin{smallmatrix} A_1 & A_{13} \\ 0 & A_3 \end{smallmatrix} \right] , \left[E_{11} \; E_{13} \right] \right)$ is observable, and A_3 is asymptotically stable. Since $\left(\left[\begin{smallmatrix} A_1 & A_{13} \\ 0 & A_3 \end{smallmatrix} \right] , \left[\begin{smallmatrix} B_1 \\ 0 \end{smallmatrix} \right] \right)$ is stabilizable, it follows from Theorem 12.18.1 that there exists a positive-semidefinite matrix $\hat{P}_1$ that satisfies

$$\begin{bmatrix} A_1 & A_{13} \\ 0 & A_3 \end{bmatrix}^{\mathrm{T}} \hat{P}_1 + \hat{P}_1 \begin{bmatrix} A_1 & A_{13} \\ 0 & A_3 \end{bmatrix} + \begin{bmatrix} E_{11}^{\mathrm{T}} E_{11} & E_{11}^{\mathrm{T}} E_{13} \\ E_{13}^{\mathrm{T}} E_{11} & E_{13}^{\mathrm{T}} E_{13} \end{bmatrix} - \hat{P}_1 \begin{bmatrix} B_1 R_2^{-1} B_1^{\mathrm{T}} & 0 \\ 0 & 0 \end{bmatrix} \hat{P}_1 = 0.$$

Consequently, $P \triangleq S^{\mathrm{T}} \mathrm{diag}(\hat{P}_1, 0, 0) S$ is a positive-semidefinite solution of (12.16.4). $\square$

Corollary 12.19.2. Suppose that (A, B) is stabilizable. Then, (12.16.4) has a positive-semidefinite solution P. If, in addition, (A, E_1) is detectable, then P is the stabilizing solution of (12.16.4), and thus P is the only positive-semidefinite solution of (12.16.4). Finally, if (A, E_1) is observable, then P is positive definite.

Proof. The first statement is given by Theorem 12.18.1. Next, assume that (A, E_1) is detectable. Then, Theorem 12.17.2 implies that P is a stabilizing solution of (12.16.4), which is the only positive-semidefinite solution of (12.16.4). Finally, Theorem 12.17.2 implies that, if (A, E_1) is observable, then P is positive definite. $\square$

The next result gives necessary and sufficient conditions under which (12.16.4) has a positive-definite solution.

Proposition 12.19.3. The following statements are equivalent:

i) (12.16.4) has a positive-definite solution.

ii) There exists a nonsingular matrix $S \in \mathbb{R}^{n \times n}$ such that (12.17.2) and (12.17.3) are satisfied, where $\left(\begin{bmatrix} A_1 & 0 \\ A_{21} & A_2 \end{bmatrix}, \begin{bmatrix} B_1 \\ B_2 \end{bmatrix} \right)$ is controllable, $\left(\begin{bmatrix} A_1 & A_{13} \\ 0 & A_3 \end{bmatrix}, \begin{bmatrix} E_{11} & E_{13} \end{bmatrix} \right)$ is observable, A_3 is asymptotically stable, $-A_2$ is asymptotically stable, $\mathrm{spec}(A_4) \subset \jmath\mathbb{R}$, and A_4 is semisimple.

In this case, (12.16.4) has exactly one positive-definite solution if and only if A_4 is empty, and infinitely many positive-definite solutions if and only if A_4 is not empty.

Proof. See [609]. $\qquad\qquad\qquad\qquad\qquad\qquad\qquad\qquad\qquad$ $\square$

Proposition 12.19.4. Assume that (12.16.4) has a stabilizing solution P, and let $S \in \mathbb{R}^{n \times n}$ be a nonsingular matrix such that (12.17.2) and (12.17.3) are satisfied, where (A_1, B_1, E_{11}) is controllable and observable, (A_2, B_2) is controllable, (A_3, E_{13}) is observable, $\nu_0(A_2) = 0$, and A_3 and A_4 are asymptotically stable. Then,

$$\mathrm{def}\, P = \nu_-(A_2). \tag{12.19.1}$$

Hence, P is positive definite if and only if $\mathrm{spec}(A_2) \subset \mathrm{ORHP}$.

12.20 Facts on Stability, Observability, and Controllability

Fact 12.20.1. For all $v \in \mathbb{R}^m$, the step response

$$y(t) = \int_0^t C e^{tA} \, \mathrm{d}\tau B v + D v$$

is bounded on $[0, \infty)$ if and only if A is Lyapunov stable and nonsingular.

Fact 12.20.2. Let $A \in \mathbb{R}^{n \times n}$, $C \in \mathbb{R}^{p \times n}$, assume that (A, C) is detectable, and assume that $y(t) \to 0$ as $t \to \infty$, where $\dot{x}(t) = Ax(t)$ and $y(t) = Cx(t)$. Then, $x(t) \to 0$ as $t \to \infty$.

Fact 12.20.3. Let $x(0) = x_0$, and let $x_\mathrm{f} - e^{t_\mathrm{f} A} x_0 \in \mathcal{C}(A, B)$. Then, for all $t \in [0, t_\mathrm{f}]$, the control $u \colon [0, t_\mathrm{f}] \mapsto \mathbb{R}^m$ defined by

$$u(t) \triangleq B^\mathrm{T} e^{(t_\mathrm{f} - t) A^\mathrm{T}} \left(\int_0^{t_\mathrm{f}} e^{\tau A} B B^\mathrm{T} e^{\tau A^\mathrm{T}} \, \mathrm{d}\tau \right)^+ \left(x_\mathrm{f} - e^{t_\mathrm{f} A} x_0 \right)$$

yields $x(t_\mathrm{f}) = x_\mathrm{f}$.

Fact 12.20.4. Let $x(0) = x_0$, let $x_f \in \mathbb{R}^n$, and assume that (A, B) is controllable. Then, for all $t \in [0, t_f]$, the control u: $[0, t_f] \mapsto \mathbb{R}^m$ defined by

$$u(t) \triangleq B^T e^{(t_f - t)A^T} \left(\int_0^{t_f} e^{\tau A} B B^T e^{\tau A^T} \, d\tau \right)^{-1} \left(x_f - e^{t_f A} x_0 \right)$$

yields $x(t_f) = x_f$.

Fact 12.20.5. Let $A \in \mathbb{R}^{n \times n}$, let $B \in \mathbb{R}^{n \times m}$, assume that A is skew symmetric, and assume that (A, B) is controllable. Then, for all $\alpha > 0$, $A - \alpha BB^T$ is asymptotically stable.

Fact 12.20.6. Let $A \in \mathbb{R}^{n \times n}$ and $B \in \mathbb{R}^{n \times m}$. Then, (A, B) is (controllable, stabilizable) if and only if (A, BB^T) is (controllable, stabilizable). Now, assume that B is positive semidefinite. Then, (A, B) is (controllable, stabilizable) if and only if $(A, B^{1/2})$ is (controllable, stabilizable).

Fact 12.20.7. Let $A \in \mathbb{R}^{n \times n}, B \in \mathbb{R}^{n \times m}$, and $\hat{B} \in \mathbb{R}^{n \times \hat{m}}$, and assume that (A, B) is (controllable, stabilizable) and $\mathcal{R}(B) \subseteq \mathcal{R}(\hat{B})$. Then, $(A, \hat{B})$ is also (controllable, stabilizable).

Fact 12.20.8. Let $A \in \mathbb{R}^{n \times n}$ and $B \in \mathbb{R}^{n \times m}$. Then, the following statements are equivalent:

i) (A, B) is controllable.

ii) There exists $\alpha \in \mathbb{R}$ such that $(A + \alpha I, B)$ is controllable.

iii) $(A + \alpha I, B)$ is controllable for all $\alpha \in \mathbb{R}$.

Fact 12.20.9. Let $A \in \mathbb{R}^{n \times n}$ and $B \in \mathbb{R}^{n \times m}$. Then, the following statements are equivalent:

i) (A, B) is stabilizable.

ii) There exists $\alpha \leq \max\{0, -\text{spabs}(A)\}$ such that $(A + \alpha I, B)$ is stabilizable.

iii) $(A + \alpha I, B)$ is stabilizable for all $\alpha \leq \max\{0, -\text{spabs}(A)\}$.

Fact 12.20.10. Let $A \in \mathbb{R}^{n \times n}$, assume that A is diagonal, and let $B \in \mathbb{R}^{n \times 1}$. Then, (A, B) is controllable if and only if the diagonal entries of A are distinct and every entry of B is nonzero. (Proof: Note that

$$\det \mathcal{K}(A, B) = \det \begin{bmatrix} b_1 & & 0 \\ & \ddots & \\ 0 & & b_n \end{bmatrix} \begin{bmatrix} 1 & a_1 & \cdots & a_1^{n-1} \\ \vdots & \vdots & \ddots & \vdots \\ 1 & a_n & \cdots & a_n^{n-1} \end{bmatrix}$$

$$= \left(\prod_{i=1}^{n} b_i\right) \prod_{i<j}(a_i - a_j).)$$

Fact 12.20.11. Let $A \in \mathbb{R}^{n \times n}$ and $B \in \mathbb{R}^{n \times 1}$, and assume that (A, B) is controllable. Then, A is cyclic. (Proof: See Fact 5.12.6.)

Fact 12.20.12. Let $A \in \mathbb{R}^{n \times n}$ and $B \in \mathbb{R}^{n \times m}$. Then, the following conditions are equivalent:

i) (A, B) is (controllable, stabilizable) and A is nonsingular.

ii) (A, AB) is (controllable, stabilizable).

Fact 12.20.13. Let $A \in \mathbb{R}^{n \times n}$ and $B \in \mathbb{R}^{n \times m}$, and assume that (A, B) is controllable. Then, $(A, B^{\mathrm{T}}S^{-\mathrm{T}})$ is observable, where $S \in \mathbb{R}^{n \times n}$ is a nonsingular matrix satisfying $A^{\mathrm{T}} = S^{-1}AS$.

Fact 12.20.14. Let (A, B) be controllable, let $t_1 > 0$, and define

$$P = \left(\int_0^{t_1} e^{-tA}BB^{\mathrm{T}}e^{-tA^{\mathrm{T}}} \mathrm{d}t\right)^{-1}.$$

Then, $A - BB^{\mathrm{T}}P$ is asymptotically stable. (Proof: P satisfies

$$(A - BB^{\mathrm{T}}P)^{\mathrm{T}}P + P(A - BB^{\mathrm{T}}P) + P\left(BB^{\mathrm{T}} + e^{t_1 A}BB^{\mathrm{T}}e^{t_1 A^{\mathrm{T}}}\right)P = 0.$$

Since $\left(A - BB^{\mathrm{T}}P, BB^{\mathrm{T}} + e^{t_1 A}BB^{\mathrm{T}}e^{t_1 A^{\mathrm{T}}}\right)$ is observable and P is positive definite, Proposition 11.8.6 implies that $A - BB^{\mathrm{T}}P$ is asymptotically stable.) (Remark: This result is due to Lukes and Kleinman. See [625, p. 113–114].)

12.21 Facts on the Lyapunov Equation and Inertia

Fact 12.21.1. Let $A \in \mathbb{F}^{n \times n}$ and $C \in \mathbb{R}^{l \times n}$, assume that (A, C) is observable, and assume there exists a Hermitian matrix $P \in \mathbb{F}^{n \times n}$ such that $A^*P + PA + C^*C = 0$. Then, the following statements hold:

i) $\nu_-(A) = \nu_+(P)$.

ii) $\nu_0(A) = \nu_0(P) = 0$.

iii) $\nu_+(A) = \nu_-(P)$.

If, in addition, P is positive definite, then A is asymptotically stable. (Proof: See [37] and [455, p. 448].)

Fact 12.21.2. Let $A \in \mathbb{F}^{n \times n}$, and assume there exists a Hermitian matrix $P \in \mathbb{F}^{n \times n}$ such that $A^*P + PA$ is negative definite. Then, the following statements hold:

  i) $\nu_-(A) = \nu_+(P)$.

  ii) $\nu_0(A) = \nu_0(P) = 0$.

  iii) $\nu_+(A) = \nu_-(P)$.

(Proof: See [455, p. 445] and [568]. This result follows from Fact 12.21.1.) (Remark: This result is due to Krein, Ostrowski, and Schneider.)

Fact 12.21.3. Let $A \in \mathbb{F}^{n \times n}$, and assume there exists a nonsingular Hermitian matrix $P \in \mathbb{F}^{n \times n}$ such that $A^*P + PA$ is negative semidefinite. Then, the following statements hold:

  i) $\nu_-(A) \leq \nu_+(P)$.

  ii) $\nu_+(A) \leq \nu_-(P)$.

(Proof: See [455, p. 447].)

Fact 12.21.4. Let $A \in \mathbb{F}^{n \times n}$. Then, the following statements are equivalent:

  i) There exists a nonsingular Hermitian matrix $P \in \mathbb{F}^{n \times n}$ such that $A^*P + PA$ is negative definite.

  ii) $\nu_0(A) = 0$.

In this case, $\text{In}(-A) = \text{In } P$. (Proof: The result *i)* $\Longrightarrow$ *ii)* follows from Fact 12.21.2. For the result *ii)* $\Longrightarrow$ *i)*, see [455, p. 445]. See [30].)

Fact 12.21.5. Let $A \in \mathbb{F}^{n \times n}$, assume that $\nu_0(A) = 0$, and assume there exists a Hermitian matrix $P \in \mathbb{F}^{n \times n}$ such that $A^*P + PA$ is negative semidefinite. Then, the following statements hold:

  i) $\nu_-(P) \leq \nu_+(A)$.

  ii) $\nu_+(P) \leq \nu_-(A)$.

(Proof: See [455, p. 447].)

Fact 12.21.6. Let $A \in \mathbb{F}^{n \times n}$, assume that $\nu_0(A) = 0$, and assume there exists a nonsingular Hermitian matrix $P \in \mathbb{F}^{n \times n}$ such that $A^*P + PA$ is negative semidefinite. Then, the following statements hold:

  i) $\nu_-(A) = \nu_+(P)$.

$ii)$ $\nu_+(A) = \nu_-(P)$.

(Proof: Combine Fact 12.21.3 and Fact 12.21.5. See [455, p. 448].) (Remark: This result is due to Carlson and Schneider.)

Fact 12.21.7. Let $A \in \mathbb{F}^{n \times n}$, assume that $\nu_0(A) = 0$, and assume there exists a Hermitian matrix $P \in \mathbb{F}^{n \times n}$ such that $A^*P + PA$ is negative definite. Then, the following statements hold:

$i)$ $\nu_-(A) = \nu_+(P)$.

$ii)$ P is nonsingular.

$iii)$ $\nu_+(A) = \nu_-(P)$.

(Proof: The result follows from $iv)$ of Fact 12.21.8.)

Fact 12.21.8. Let $A, P, R \in \mathbb{F}^{n \times n}$, assume that P is Hermitian, let $C \in \mathbb{F}^{l \times n}$, and assume that $A^*P + PA + C^*C = 0$. Then, the following statements hold:

$i)$ $|\nu_-(A) - \nu_+(P)| \leq n - \operatorname{rank} \mathcal{O}(A, C)$.

$ii)$ $|\nu_+(A) - \nu_-(P)| \leq n - \operatorname{rank} \mathcal{O}(A, C)$.

$iii)$ If $\nu_0(A) = 0$, then

$$|\nu_-(A) - \nu_+(P)| + |\nu_+(A) - \nu_-(P)| \leq n - \operatorname{rank} \mathcal{O}(A, C).$$

$iv)$ If (A, C) is observable, then $\operatorname{In}(-A) = \operatorname{In} P$ and $\nu_0(A) = \nu_0(P) = 0$.

(Proof: See [493].) (Remark: Statement $iv)$ is given by Fact 12.21.1. The result that $\nu_0(P) = 0$ does not follow from $i)$ and $ii)$.) (Remark: For related results, see [568] and references given in [493]. See also [205].)

Fact 12.21.9. Let $A, P \in \mathbb{R}^{n \times n}$, assume that $\nu_0(A) = n$, and assume that P is positive semidefinite. Then, $A^{\mathrm{T}}P + PA$ is either zero or has at least one positive eigenvalue and one negative eigenvalue. (Proof: See [729].)

Fact 12.21.10. Let $A, B \in \mathbb{F}^{n \times n}$, assume that A is nonsingular and Hermitian, and assume that $B + B^*$ is positive definite. Then, $\operatorname{In} A = \operatorname{In} AB$. (Remark: This result is the *classical Lyapunov theorem*.)

Fact 12.21.11. Let $A, B \in \mathbb{F}^{n \times n}$, assume that A is nonsingular and Hermitian, and assume that B has no eigenvalues in $[0, -\infty)$. Then, $\operatorname{In} A = \operatorname{In} AB$. (Remark: This result is the *generalized Reid theorem*.)

Fact 12.21.12. Let $A_1 \in \mathbb{R}^{n_1 \times n_1}$, $A_2 \in \mathbb{R}^{n_2 \times n_2}$, $B \in \mathbb{R}^{n_1 \times m}$, and $C \in \mathbb{R}^{m \times n_2}$, assume that $A_1 \oplus A_2$ is nonsingular, and assume that $\operatorname{rank} B = \operatorname{rank} C = m$. Furthermore, let $X \in \mathbb{R}^{n_1 \times n_2}$ be the unique solution of

$$A_1 X + X A_2 + BC = 0.$$

Then,
$$\operatorname{rank} X \leq \min\{\operatorname{rank} \mathcal{K}(A_1, B), \operatorname{rank} \mathcal{O}(A_2, C)\}.$$

Furthermore, equality holds if $m = 1$. (Proof: See [213].) (Remark: Related results are given in [782, 786].)

Fact 12.21.13. Let $A_1, A_2 \in \mathbb{R}^{n \times n}$, $B \in \mathbb{R}^n$, $C \in \mathbb{R}^{1 \times n}$, assume that $A_1 \oplus A_2$ is nonsingular, let $X \in \mathbb{R}^{n \times n}$ satisfy
$$A_1 X + X A_2 + B C = 0,$$
and assume that (A_1, B) is controllable and (A_2, C) is observable. Then, X is nonsingular. (Proof: See Fact 12.21.12 and [786].)

Fact 12.21.14. Let $A, P, R \in \mathbb{R}^{n \times n}$, and assume that P and R are positive semidefinite, $A^{\mathrm{T}} P + PA + R = 0$, and $\mathcal{N}[\mathcal{O}(A, R)] = \mathcal{N}(A)$. Then, A is semistable. (Proof: See [109].)

Fact 12.21.15. Let $A, V \in \mathbb{R}^{n \times n}$, assume that A is asymptotically stable, assume that V is positive semidefinite, and let $Q \in \mathbb{R}^{n \times n}$ be the unique, positive-definite solution to $AQ + QA^{\mathrm{T}} + V = 0$. Furthermore, let $C \in \mathbb{R}^{l \times n}$, and assume that CVC^{T} is positive definite. Then, CQC^{T} is positive definite.

Fact 12.21.16. Let $A, R \in \mathbb{R}^{n \times n}$, assume that A is asymptotically stable, assume that $R \in \mathbb{R}^{n \times n}$ is positive semidefinite, and let $P \in \mathbb{R}^{n \times n}$ satisfy $A^{\mathrm{T}} P + PA + R = 0$. Then, for all $i, j = 1, \ldots, n$, there exist $\alpha_{ij} \in \mathbb{R}$ such that
$$P = \sum_{i,j=1}^{n} \alpha_{ij} A^{(i-1)\mathrm{T}} R A^{j-1}.$$

In particular, for all $i, j = 1, \ldots, n$, $\alpha_{ij} = \hat{P}_{(i,j)}$, where $\hat{P} \in \mathbb{R}^{n \times n}$ satisfies $\hat{A}^{\mathrm{T}} \hat{P} + \hat{P} \hat{A} + \hat{R} = 0$, where $\hat{A} = C(\chi_A)$ and $\hat{R} = E_{1,1}$. (Proof: See [648].) (Remark: This identity is *Smith's method*. See [226, 498] for finite-sum solutions of linear matrix equations.)

Fact 12.21.17. Let $A, R \in \mathbb{R}^{n \times n}$, assume that $R \in \mathbb{R}^{n \times n}$ is positive semidefinite, let $q, r \in \mathbb{R}$, where $r > 0$, and assume that there exists a positive-definite matrix $P \in \mathbb{R}^{n \times n}$ satisfying
$$[A - (q+r)I]^{\mathrm{T}} P + P[A - (q+r)I] + \tfrac{1}{r} A^{\mathrm{T}} P A + R = 0,$$

Then, the spectrum of A is contained in a disk centered at $q + \jmath 0$ with radius r. (Remark: The disk is an *eigenvalue inclusion region*. See [75, 320, 758] for related results concerning elliptical, parabolic, hyperbolic, sector, and vertical strip regions.)

12.22 Facts on Realizations and the H₂ System Norm

Fact 12.22.1. Let $G \sim \left[\begin{array}{c|c} A & B \\ \hline C & D \end{array}\right]$, let $a, b \in \mathbb{R}$, where $a \neq 0$, and define $H(s) \triangleq G(as + b)$. Then,

$$H \sim \left[\begin{array}{c|c} a^{-1}(A - bI) & B \\ \hline a^{-1}C & D \end{array}\right].$$

Fact 12.22.2. Let $G \sim \left[\begin{array}{c|c} A & B \\ \hline C & D \end{array}\right]$, where A is nonsingular, and define $H(s) \triangleq G(1/s)$. Then,

$$H \sim \left[\begin{array}{c|c} A^{-1} & -A^{-1}B \\ \hline CA^{-1} & D - CA^{-1}B \end{array}\right].$$

Fact 12.22.3. Let $G(s) = C(sI - A)^{-1}B$. Then,

$$G(\jmath\omega) = -CA(\omega^2 I + A^2)^{-1}B - \jmath\omega C(\omega^2 I + A^2)^{-1}B.$$

Fact 12.22.4. Let $G \sim \left[\begin{array}{c|c} A & B \\ \hline C & 0 \end{array}\right]$ and $H(s) = sG(s)$. Then,

$$H \sim \left[\begin{array}{c|c} A & B \\ \hline CA & CB \end{array}\right].$$

Consequently,

$$sC(sI - A)^{-1}B = CA(sI - A)^{-1}B + CB.$$

Fact 12.22.5. Let $G = \left[\begin{array}{cc} G_{11} & G_{12} \\ G_{21} & G_{22} \end{array}\right]$, where $G_{ij} \sim \left[\begin{array}{c|c} A_{ij} & B_{ij} \\ \hline C_{ij} & D_{ij} \end{array}\right]$ for all $i, j = 1, 2$. Then,

$$\left[\begin{array}{cc} G_{11} & G_{12} \\ G_{21} & G_{22} \end{array}\right] \sim \left[\begin{array}{cccc|cc} A_{11} & 0 & 0 & 0 & B_{11} & 0 \\ 0 & A_{12} & 0 & 0 & 0 & B_{12} \\ 0 & 0 & A_{21} & 0 & B_{21} & 0 \\ 0 & 0 & 0 & A_{22} & 0 & B_{22} \\ \hline C_{11} & C_{12} & 0 & 0 & D_{11} & D_{12} \\ 0 & 0 & C_{21} & C_{22} & D_{21} & D_{22} \end{array}\right].$$

Fact 12.22.6. Let $G \sim \left[\begin{array}{c|c} A & B \\ \hline C & 0 \end{array}\right]$, where $G \in \mathbb{R}^{l \times m}(s)$, and let $M \in \mathbb{R}^{m \times l}$. Then,

$$[I + GM]^{-1} \sim \left[\begin{array}{c|c} A - BMC & B \\ \hline -C & I \end{array}\right]$$

and

$$[I + GM]^{-1}G \sim \left[\begin{array}{c|c} A - BMC & B \\ \hline C & 0 \end{array}\right].$$

Fact 12.22.7. Let $G \sim \left[\begin{array}{c|c} A & B \\ \hline C & D \end{array}\right]$, where $G \in \mathbb{R}^{l \times m}(s)$. If D has a left inverse $D^{\mathrm{L}} \in \mathbb{R}^{m \times l}$, then

$$G^{\mathrm{L}} \sim \left[\begin{array}{c|c} A - BD^{\mathrm{L}}C & BD^{\mathrm{L}} \\ \hline -D^{\mathrm{L}}C & D^{\mathrm{L}} \end{array}\right]$$

satisfies $G^{\mathrm{L}}G = I$. If D has a right inverse $D^{\mathrm{R}} \in \mathbb{R}^{m \times l}$, then

$$G^{\mathrm{R}} \sim \left[\begin{array}{c|c} A - BD^{\mathrm{R}}C & BD^{\mathrm{R}} \\ \hline -D^{\mathrm{R}}C & D^{\mathrm{R}} \end{array}\right]$$

satisfies $GG^{\mathrm{R}} = I$.

Fact 12.22.8. Let $G \sim \left[\begin{array}{c|c} A & B \\ \hline C & 0 \end{array}\right]$ be a SISO rational transfer function, and let $\lambda \in \mathbb{C}$. Then, there exists a rational function H such that

$$G(s) = \frac{1}{(s + \lambda)^r} H(s)$$

and such that λ is neither a pole nor a zero of H if and only if the Jordan form of A has exactly one block associated with λ, which is of size $r \times r$.

Fact 12.22.9. Let $A \in \mathbb{R}^{n \times n}$, $B \in \mathbb{R}^{n \times m}$, and $C \in \mathbb{R}^{m \times n}$. Then,

$$\det[sI - (A + BC)] = \det\left[I - C(sI - A)^{-1}B\right]\det(sI - A).$$

(Proof: Note that

$$\det\left[I - C(sI - A)^{-1}B\right]\det(sI - A) = \det\left[\begin{array}{cc} sI - A & B \\ C & I \end{array}\right]$$

$$= \det\left[\begin{array}{cc} sI - A & B \\ C & I \end{array}\right]\left[\begin{array}{cc} I & 0 \\ -C & I \end{array}\right]$$

$$= \det\left[\begin{array}{cc} sI - A - BC & B \\ 0 & I \end{array}\right]$$

$$= \det(sI - A - BC).)$$

Fact 12.22.10. Let $A \in \mathbb{R}^{n \times n}$, $B \in \mathbb{R}^{n \times m}$, $C \in \mathbb{R}^{m \times n}$, and $K \in \mathbb{R}^{m \times n}$, and assume that $A + BK$ is nonsingular. Then,

$$\det\left[\begin{array}{cc} A & B \\ C & 0 \end{array}\right] = (-1)^m \det(A + BK)\det\left[C(A + BK)^{-1}B\right].$$

Hence, $\left[\begin{smallmatrix} A & B \\ C & 0 \end{smallmatrix}\right]$ is nonsingular if and only if $C(A + BK)^{-1}B$ is nonsingular. (Proof: Note that

$$\det \begin{bmatrix} A & B \\ C & 0 \end{bmatrix} = \det \begin{bmatrix} A & B \\ C & 0 \end{bmatrix} \begin{bmatrix} I & 0 \\ K & I \end{bmatrix}$$

$$= \det \begin{bmatrix} A + BK & B \\ C & 0 \end{bmatrix}$$

$$= \det(A + BK) \det \left[-C(A + BK)^{-1}B \right].)$$

Fact 12.22.11. Let $G_1 \in \mathbb{R}^{l_1 \times m}(s)$ and $G_2 \in \mathbb{R}^{l_2 \times m}(s)$ be strictly proper. Then,

$$\left\| \begin{bmatrix} G_1 \\ G_2 \end{bmatrix} \right\|_{H_2}^2 = \|G_1\|_{H_2}^2 + \|G_2\|_{H_2}^2.$$

Fact 12.22.12. Let $G_1, G_2 \in \mathbb{R}^{m \times m}(s)$ be strictly proper. Then,

$$\left\| \begin{bmatrix} G_1 \\ G_2 \end{bmatrix} \right\|_{H_2} = \left\| \begin{bmatrix} G_1 & G_2 \end{bmatrix} \right\|_{H_2}.$$

Fact 12.22.13. Let $G(s) \triangleq \frac{\alpha}{s+\beta}$, where $\beta > 0$. Then,

$$\|G\|_{H_2} = \frac{|\alpha|}{\sqrt{2\beta}}.$$

Fact 12.22.14. Let $G(s) \triangleq \frac{\alpha_1 s + \alpha_0}{s^2 + \beta_1 s + \beta_0}$, where $\beta_0, \beta_1 > 0$. Then,

$$\|G\|_{H_2} = \sqrt{\frac{\alpha_0^2}{2\beta_0\beta_1} + \frac{\alpha_1^2}{2\beta_1}}.$$

Fact 12.22.15. Let $G_1(s) = \frac{\alpha_1}{s+\beta_1}$ and $G_2(s) = \frac{\alpha_2}{s+\beta_2}$, where $\beta_1, \beta_2 > 0$. Then,

$$\|G_1 G_2\|_{H_2} \le \|G_1\|_{H_2} \|G_2\|_{H_2}$$

if and only if $\beta_1 + \beta_2 \ge 2$. (Remark: The H_2 norm is not submultiplicative.)

12.23 Facts on the Riccati Equation

Fact 12.23.1. Assume that (A, B) is stabilizable, and assume that $\mathcal{H}$ defined by (12.16.8) has an imaginary eigenvalue λ. Then, every Jordan block of $\mathcal{H}$ associated with λ has even size. (Proof: Let P be a solution of (12.16.4), and let $\mathcal{J}$ denote the Jordan form of $A - \Sigma P$. Then, there exists a nonsingular $2n \times 2n$ block-diagonal matrix S such that $\hat{\mathcal{H}} \triangleq S^{-1}\mathcal{H}S = \begin{bmatrix} \mathcal{J} & \hat{\Sigma} \\ 0 & -\mathcal{J}^T \end{bmatrix}$, where $\hat{\Sigma}$ is positive semidefinite. Next, let $\mathcal{J}_\lambda \triangleq \lambda I_r + N_r$ be a Jordan block of $\mathcal{J}$ associated with λ, and consider the submatrix of $\lambda I - \hat{\mathcal{H}}$ consisting of the rows and columns of $\lambda I - \mathcal{J}_\lambda$ and $\lambda I + \mathcal{J}_\lambda^T$. Since (A, B) is stabilizable, it follows that the rank of this submatrix is $2r - 1$. Hence, every Jordan block of $\mathcal{H}$ associated with λ has even size.) (Remark: Canonical forms for symplectic and Hamiltonian matrices are discussed in [457].)

Fact 12.23.2. Let $A, B \in \mathbb{C}^{n \times n}$, and assume that the $2n \times 2n$ matrix

$$\begin{bmatrix} A & -2I \\ 2B - \frac{1}{2}A^2 & A \end{bmatrix}$$

is simple. Then, there exists a matrix $X \in \mathbb{C}^{n \times n}$ satisfying

$$X^2 + AX + B = 0.$$

(Proof: See [725].)

Fact 12.23.3. Let $P_0 \in \mathbb{R}^{n \times n}$, assume that P_0 is positive definite, and, for all $t \geq 0$, let $P(t) \in \mathbb{R}^{n \times n}$ satisfy

$$\dot{P}(t) = A^{\mathrm{T}}P(t) + P(t)A + P(t)VP(t),$$
$$P(0) = P_0.$$

Then, for all $t \geq 0$,

$$P(t) = e^{tA^{\mathrm{T}}}\left[P_0^{-1} - \int_0^t e^{\tau A}Ve^{\tau A^{\mathrm{T}}}\mathrm{d}\tau\right]^{-1}e^{tA}.$$

(Remark: $P(t)$ satisfies a Riccati differential equation.)

Fact 12.23.4. Let $G_{\mathrm{c}} \sim \left[\begin{array}{c|c} A_{\mathrm{c}} & B_{\mathrm{c}} \\ \hline C_{\mathrm{c}} & 0 \end{array}\right]$ denote an nth-order dynamic controller for the standard control problem. If G_{c} minimizes $\|\tilde{\mathcal{G}}\|_2$, then G_{c} is given by

$$A_{\mathrm{c}} \triangleq A + BC_{\mathrm{c}} - B_{\mathrm{c}}C - B_{\mathrm{c}}DC_{\mathrm{c}},$$
$$B_{\mathrm{c}} \triangleq (QC^{\mathrm{T}} + V_{12})V_2^{-1},$$
$$C_{\mathrm{c}} \triangleq -R_2^{-1}(B^{\mathrm{T}}P + R_{12}^{\mathrm{T}}),$$

where P and Q are positive-semidefinite solutions to the algebraic Riccati equations

$$A_{\mathrm{R}}^{\mathrm{T}}P + PA_{\mathrm{R}} - PBR_2^{-1}B^{\mathrm{T}}P + \hat{R}_1 = 0,$$
$$A_{\mathrm{E}}Q + QA_{\mathrm{E}}^{\mathrm{T}} - QC^{\mathrm{T}}V_2^{-1}CQ + \hat{V}_1 = 0,$$

where A_{R} and $\hat{R}_1$ are defined by

$$A_{\mathrm{R}} \triangleq A - BR_2^{-1}R_{12}^{\mathrm{T}}, \quad \hat{R}_1 \triangleq R_1 - R_{12}R_2^{-1}R_{12}^{\mathrm{T}},$$

and A_{E} and $\hat{V}_1$ are defined by

$$A_{\mathrm{E}} \triangleq A - V_{12}V_2^{-1}C, \quad \hat{V}_1 \triangleq V_1 - V_{12}V_2^{-1}V_{12}^{\mathrm{T}}.$$

Furthermore, the eigenvalues of the closed-loop system are given by

$$\text{mspec}\left(\begin{bmatrix} A & BC_\text{c} \\ B_\text{c}C & A_\text{c} + B_\text{c}DC_\text{c} \end{bmatrix}\right) = \text{mspec}(A + BC_\text{c}) \cup \text{mspec}(A - B_\text{c}C).$$

Fact 12.23.5. Let $G_\text{c} \sim \left[\begin{array}{c|c} A_\text{c} & B_\text{c} \\ \hline C_\text{c} & D_\text{c} \end{array}\right]$ denote an nth-order dynamic controller for the discrete-time standard control problem. If G_c minimizes $\|\tilde{\mathcal{G}}\|_2$, then G_c is given by the following expressions. If $D \neq 0$, then

$$A_\text{c} \triangleq A + BC_\text{c} - B_\text{c}C - B_\text{c}DC_\text{c},$$
$$B_\text{c} \triangleq (AQC^\text{T} + V_{12})(V_2 + CQC^\text{T})^{-1},$$
$$C_\text{c} \triangleq -(R_2 + B^\text{T}PB)^{-1}(R_{12}^\text{T} + B^\text{T}PA),$$
$$D_\text{c} \triangleq 0,$$

and the eigenvalues of the closed-loop system are given by

$$\text{mspec}\left(\begin{bmatrix} A & BC_\text{c} \\ B_\text{c}C & A_\text{c} + B_\text{c}DC_\text{c} \end{bmatrix}\right) = \text{mspec}(A + BC_\text{c}) \cup \text{mspec}(A - B_\text{c}C).$$

Alternatively, if $D = 0$, then

$$A_\text{c} \triangleq A + BC_\text{c} - B_\text{c}C - BD_\text{c}C,$$
$$B_\text{c} \triangleq (AQC^\text{T} + V_{12})(V_2 + CQC^\text{T})^{-1} + BD_\text{c},$$
$$C_\text{c} \triangleq -(R_2 + B^\text{T}PB)^{-1}(R_{12}^\text{T} + B^\text{T}PA) - D_\text{c}C,$$
$$D_\text{c} \triangleq (R_2 + B^\text{T}PB)^{-1}[B^\text{T}PAQC^\text{T} + R_{12}^\text{T}QC^\text{T} + B^\text{T}PV_{12}](V_2 + CQC^\text{T})^{-1},$$

and the eigenvalues of the closed-loop system are given by

$$\text{mspec}\left(\begin{bmatrix} A + BD_\text{c}C & BC_\text{c} \\ B_\text{c}C & A_\text{c} \end{bmatrix}\right) = \text{mspec}(A + BC_\text{c}) \cup \text{mspec}(A - B_\text{c}C).$$

In both cases, P and Q are positive-semidefinite solutions to the discrete-time algebraic Riccati equations

$$P = A_\text{R}^\text{T}PA_\text{R} - A_\text{R}^\text{T}PB(R_2 + B^\text{T}PB)^{-1}B^\text{T}PA_\text{R} + \hat{R}_1,$$
$$Q = A_\text{E}QA_\text{E}^\text{T} - A_\text{E}QC^\text{T}(V_2 + CQC^\text{T})^{-1}CQA_\text{E}^\text{T} + \hat{V}_1,$$

where A_R and $\hat{R}_1$ are defined by

$$A_\text{R} \triangleq A - BR_2^{-1}R_{12}^\text{T}, \quad \hat{R}_1 \triangleq R_1 - R_{12}R_2^{-1}R_{12}^\text{T},$$

and A_E and $\hat{V}_1$ are defined by

$$A_\text{E} \triangleq A - V_{12}V_2^{-1}C, \quad \hat{V}_1 \triangleq V_1 - V_{12}V_2^{-1}V_{12}^\text{T}.$$

(Proof: See [321].)

12.24 Notes

Linear system theory is treated in [145, 624, 724, 790]. Time-varying linear systems are considered in [201, 624]. The equivalence of *iv*) and *v*) of Theorem 12.6.15 is the *PBH test*, due to [340]. Spectral factorization results are given in [185]. Stabilization aspects are discussed in [235]. Observable asymptotic stability and controllable asymptotic stability were introduced and used to analyze Lyapunov equations in [649]. Zeros are treated in [12, 257, 407, 410, 500, 578, 627, 635]. Matrix-based methods for linear system identification are developed in [571], while stochastic theory is considered in [330].

Solutions of the LQR problem under weak conditions are given in [288]. Solutions of the Riccati equation are considered in [441, 443, 453, 454, 518, 609, 780, 785, 788]. Proposition 12.16.16 is based on Theorem 3.6 of [794, p. 79]. A variation of Theorem 12.18.1 is given without proof by Theorem 7.2.1 of [392, p. 125].

There are numerous extensions to the results given in this chapter to various generalizations of (12.16.4). These generalizations include the case in which R_1 is indefinite [295, 783, 784] as well as the case in which Σ is indefinite [631]. The latter case is relevant to H_∞ optimal control theory [104]. Additional extensions include the Riccati inequality $A^{\mathrm{T}}P + PA + R_1 - P\Sigma P \geq 0$ [604] as well as the discrete-time Riccati equation [388] and extensions to fixed-order controllers [384]. Riccati equations for discrete-time systems are discussed in [4, 388, 453, 604].

Bibliography

[1] R. Ablamowicz, "Matrix Exponential via Clifford Algebras," *J. Nonlin. Math. Phys.*, Vol. 5, pp. 294–313, 1998.

[2] H. Abou-Kandil, G. Freiling, V. Ionescu, and G. Jank, *Matrix Riccati Equations in Control and Systems Theory.* Basel: Birkhauser, 2003.

[3] L. Aceto and D. Trigiante, "The Matrices of Pascal and Other Greats," *Amer. Math. Monthly*, Vol. 108, pp. 232–245, 2001.

[4] C. D. Ahlbrandt and A. C. Peterson, *Discrete Hamiltonian Systems: Difference Equations, Continued Fractions, and Riccati Equations.* Dordrecht: Kluwer, 1996.

[5] A. C. Aitken, *Determinants and Matrices*, 9th ed. Edinburgh: Oliver and Boyd, 1956.

[6] M. Al-Ahmar, "An Identity of Jacobi," *Amer. Math. Monthly*, Vol. 103, pp. 78–79, 1996.

[7] A. A. Albert and B. Muckenhoupt, "On Matrices of Trace Zero," *Michigan Math. J.*, Vol. 4, pp. 1–3, 1957.

[8] A. E. Albert, "Conditions for Positive and Nonnegative Definiteness in Terms of Pseudoinverses," *SIAM J. Appl. Math.*, Vol. 17, pp. 434–440, 1969.

[9] A. E. Albert, *Regression and the Moore-Penrose Pseudoinverse.* New York: Academic, 1972.

[10] R. Aldrovandi, *Special Matrices of Mathematical Physics: Stochastic, Circulant and Bell Matrices.* Singapore: World Scientific, 2001.

[11] M. Alic, B. Mond, J. E. Pecaric, and V. Volenec, "Bounds for the Differences of Matrix Means," *SIAM J. Matrix Anal. Appl.*, Vol. 18, pp. 119–123, 1997.

[12] H. Aling and J. M. Schumacher, "A Nine-fold Decomposition for Linear Systems," *Int. J. Contr.*, Vol. 39, pp. 779–805, 1984.

[13] G. Alpargu and G. P. H. Styan, "Some Remarks and a Bibliography on the Kantorovich Inequality," in *Multidimensional Statistical Analysis and Theory of Random Matrices.* Utrecht: VSP, 1996, pp. 1–13.

[14] R. C. Alperin, "The Matrix of a Rotation," *College Math. J.*, Vol. 20, p. 230, 1989.

[15] S. L. Altmann, *Rotations, Quaternions, and Double Groups.* New York: Oxford University Press, 1986.

[16] H. Alzer, "A Lower Bound for the Difference between the Arithmetic and Geometric Means," *Nieuw. Arch. Wisk.*, Vol. 8, pp. 195–197, 1990.

[17] B. D. O. Anderson, "Orthogonal Decomposition Defined by a Pair of Skew-Symmetric Forms," *Lin. Alg. Appl.*, Vol. 8, pp. 91–93, 1974.

[18] B. D. O. Anderson, "Weighted Hankel-Norm Approximation: Calculation of Bounds," *Sys. Contr. Lett.*, Vol. 7, pp. 247–255, 1986.

[19] B. D. O. Anderson, E. I. Jury, and M. Mansour, "Schwarz Matrix Properties for Continuous and Discrete Time Systems," *Int. J. Contr.*, Vol. 23, pp. 1–16, 1976.

[20] B. D. O. Anderson and J. B. Moore, "A Matrix Kronecker Lemma," *Lin. Alg. Appl.*, Vol. 15, pp. 227–234, 1976.

[21] W. N. Anderson, "Shorted Operators," *SIAM J. Appl. Math.*, Vol. 20, pp. 520–525, 1971.

[22] W. N. Anderson and R. J. Duffin, "Series and Parallel Addition of Matrices," *J. Math. Anal. Appl.*, Vol. 26, pp. 576–594, 1969.

[23] W. N. Anderson, E. J. Harner, and G. E. Trapp, "Eigenvalues of the Difference and Product of Projections," *Lin. Multilin. Alg.*, Vol. 17, pp. 295–299, 1985.

[24] W. N. Anderson and M. Schreiber, "On the Infimum of Two Projections," *Acta Sci. Math.*, Vol. 33, pp. 165–168, 1972.

[25] W. N. Anderson and G. E. Trapp, "Shorted Operators II," *SIAM J. Appl. Math.*, Vol. 28, pp. 60–71, 1975.

[26] W. N. Anderson and G. E. Trapp, "Symmetric Positive Definite Matrices," *Amer. Math. Monthly*, Vol. 95, pp. 261–262, 1988.

[27] T. Ando, "Concavity of Certain Maps on Positive Definite Matrices and Applications to Hadamard Products," *Lin. Alg. Appl.*, Vol. 26, pp. 203–241, 1979.

[28] T. Ando, "Majorizations and Inequalities in Matrix Theory," *Lin. Alg. Appl.*, Vol. 199, pp. 17–67, 1994.

[29] T. Ando, "Majorization Relations for Hadamard Products," *Lin. Alg. Appl.*, Vol. 223–224, pp. 57–64, 1995.

[30] T. Ando, "Lowner Inequality of Indefinite Type," *Lin. Alg. Appl.*, Vol. 385, pp. 73–80, 2004.

[31] T. Ando and F. Hiai, "Log-Majorization and Complementary Golden-Thompson Type Inequalities," *Lin. Alg. Appl.*, Vol. 197/198, pp. 113–131, 1994.

[32] T. Ando and F. Hiai, "Holder Type Inequalities for Matrices," *Math. Ineq. Appl.*, Vol. 1, pp. 1–30, 1998.

[33] T. Ando, C.-K. Li, and R. Mathias, "Geometric Means," *Lin. Alg. Appl.*, Vol. 385, pp. 305–334, 2004.

[34] T. Ando and X. Zhan, "Norm Inequalities Related to Operator Monotone Functions," *Math. Ann.*, Vol. 315, pp. 771–780, 1999.

[35] E. Andruchow, G. Corach, and D. Stojanoff, "Geometric Operator Inequalities," *Lin. Alg. Appl.*, Vol. 258, pp. 295–310, 1997.

[36] E. Angel, *Interactive Computer Graphics*, 3rd ed. Reading: Addison-Wesley, 2002.

[37] A. C. Antoulas and D. C. Sorensen, "Lyapunov, Lanczos, and Inertia," *Lin. Alg. Appl.*, Vol. 326, pp. 137–150, 2001.

[38] J. D. Aplevich, *Implicit Linear Systems*. Berlin: Springer, 1991.

[39] T. M. Apostol, "Some Explicit Formulas for the Exponential Matrix," *Amer. Math. Monthly*, Vol. 76, pp. 289–292, 1969.

[40] H. Araki, "On an Inequality of Lieb and Thirring," *Lett. Math. Phys.*, Vol. 19, pp. 167–170, 1990.

[41] A. Arimoto, "A Simple Proof of the Classification of Normal Toeplitz Matrices," *Elec. J. Lin. Alg.*, Vol. 9, pp. 108–111, 2002.

[42] T. Arponen, "A Matrix Approach to Polynomials," *Lin. Alg. Appl.*, Vol. 359, pp. 181–196, 2003.

[43] M. Artin, *Algebra*. Englewood Cliffs: Prentice-Hall, 1991.

[44] M. Artzrouni, "A Theorem on Products of Matrices," *Lin. Alg. Appl.*, Vol. 49, pp. 153–159, 1983.

[45] H. Aslaksen, "Laws of Trigonometry on SU(3)," *Trans. Amer. Math. Soc.*, Vol. 317, pp. 127–142, 1990.

[46] H. Aslaksen, "Quaternionic Determinants," *Math. Intell.*, Vol. 18, no. 3, pp. 57–65, 1996.

[47] B. A. Asner, "On the Total Nonnegativity of the Hurwitz Matrix," *SIAM J. Appl. Math.*, Vol. 18, pp. 407–414, 1970.

[48] Y.-H. Au-Yeung, "A note on Some Theorems on Simultaneous Diagonalization of Two Hermitian Matrices," *Proc. Cambridge Phil. Soc.*, Vol. 70, pp. 383–386, 1971.

[49] Y.-H. Au-Yeung, "Some Inequalities for the Rational Power of a Nonnegative Definite Matrix," *Lin. Alg. Appl.*, Vol. 7, pp. 347–350, 1973.

[50] J. S. Aujla, "Some Norm Inequalities for Completely Monotone Functions," *SIAM J. Matrix Anal. Appl.*, Vol. 22, pp. 569–573, 2000.

[51] J. S. Aujla and F. C. Silva, "Weak Majorization Inequalities and Convex Functions," *Lin. Alg. Appl.*, Vol. 369, pp. 217–233, 2003.

[52] B. Aupetit and J. Zemanek, "A Characterization of Normal Matrices by Their Exponentials," *Lin. Alg. Appl.*, Vol. 132, pp. 119–121, 1990.

[53] M. Avriel, *Nonlinear Programming: Analysis and Methods.* Englewood Cliffs: Prentice-Hall, 1976, reprinted by Dover, Mineola, 2003.

[54] O. Axelsson, *Iterative Solution Methods.* Cambridge: Cambridge University Press, 1994.

[55] J. C. Baez, "The Octonions," *Bull. Amer. Math. Soc.*, Vol. 39, pp. 145–205, 2001.

[56] A. Baker, *Matrix Groups: An Introduction to Lie Group Theory.* New York: Springer, 2001.

[57] J. K. Baksalary and O. M. Baksalary, "Nonsingularity of Linear Combinations of Idempotent Matrices," *Lin. Alg. Appl.*, Vol. 388, pp. 25–29, 2004.

[58] J. K. Baksalary, F. Pukelsheim, and G. P. H. Styan, "Some Properties of Matrix Partial Orderings," *Lin. Alg. Appl.*, Vol. 119, pp. 57–85, 1989.

[59] J. K. Baksalary and G. P. H. Styan, "Generalized Inverses of Bordered Matrices," *Lin. Alg. Appl.*, Vol. 354, pp. 41–47, 2002.

[60] K. Ball, E. Carlen, and E. Lieb, "Sharp Uniform Convexity and Smoothness Inequalities for Trace Norms," *Invent. Math.*, Vol. 115, pp. 463–482, 1994.

[61] C. S. Ballantine, "Products of Positive Semidefinite Matrices," *Pac. J. Math.*, Vol. 23, pp. 427–433, 1967.

[62] C. S. Ballantine, "Products of Positive Definite Matrices II," *Pac. J. Math.*, Vol. 24, pp. 7–17, 1968.

[63] C. S. Ballantine, "Products of Positive Definite Matrices III," *J. Algebra*, Vol. 10, pp. 174–182, 1968.

[64] C. S. Ballantine, "Products of Positive Definite Matrices IV," *Lin. Alg. Appl.*, Vol. 3, pp. 79–114, 1970.

[65] C. S. Ballantine, "Products of EP Matrices," *Lin. Alg. Appl.*, Vol. 12, pp. 257–267, 1975.

[66] C. S. Ballantine, "Some Involutory Similarities," *Lin. Multilin. Alg.*, Vol. 3, pp. 19–23, 1975.

[67] C. S. Ballantine, "Products of Involutory Matrices I," *Lin. Multilin. Alg.*, Vol. 5, pp. 53–62, 1977.

[68] C. S. Ballantine, "Products of Idempotent Matrices," *Lin. Alg. Appl.*, Vol. 19, pp. 81–86, 1978.

[69] C. S. Ballantine and C. R. Johnson, "Accretive Matrix Products," *Lin. Multilin. Alg.*, Vol. 3, pp. 169–185, 1975.

[70] R. B. Bapat and B. Zheng, "Generalized Inverses of Bordered Matrices," *Elec. J. Lin. Alg.*, Vol. 10, pp. 16–30, 2003.

[71] I. Y. Bar-Itzhack, D. Hershkowitz, and L. Rodman, "Pointing in Real Euclidean Space," *J. Guid. Contr. Dyn.*, Vol. 20, pp. 916–922, 1997.

[72] E. J. Barbeau, *Polynomials.* New York: Springer, 1989.

[73] S. Barnett, "A Note on the Bezoutian Matrix," *SIAM J. Appl. Math.*, Vol. 22, pp. 84–86, 1972.

[74] S. Barnett, "Inversion of Partitioned Matrices with Patterned Blocks," *Int. J. Sys. Sci.*, Vol. 14, pp. 235–237, 1983.

[75] S. Barnett, *Polynomials and Linear Control Systems.* New York: Marcel Dekker, 1983.

[76] S. Barnett, *Matrices in Control Theory*, revised ed. Malabar: Krieger, 1984.

[77] S. Barnett, "Leverrier's Algorithm: A New Proof and Extensions," *SIAM J. Matrix Anal. Appl.*, Vol. 10, pp. 551–556, 1989.

[78] S. Barnett, *Matrices: Methods and Applications.* Oxford: Clarendon, 1990.

[79] S. Barnett and P. Lancaster, "Some Properties of the Bezoutian for Polynomial Matrices," *Lin. Multilin. Alg.*, Vol. 9, pp. 99–110, 1980.

[80] S. Barnett and C. Storey, *Matrix Methods in Stability Theory.* New York: Barnes and Noble, 1970.

[81] J. Barria and P. R. Halmos, "Vector Bases for Two Commuting Matrices," *Lin. Multilin. Alg.*, Vol. 27, pp. 147–157, 1990.

[82] F. L. Bauer, J. Stoer, and C. Witzgall, "Absolute and Monotonic Norms," *Numer. Math.*, Vol. 3, pp. 257–264, 1961.

[83] D. S. Bayard, "An Optimization Result with Application to Optimal Spacecraft Reaction Wheel Orientation Design," in *Proc. Amer. Contr. Conf.*, Arlington, VA, June 2001, pp. 1473–1478.

[84] M. S. Bazaraa, H. D. Sherali, and C. M. Shetty, *Nonlinear Programming*, 2nd ed. Wiley, 1993.

[85] N. Bebiano, J. da Providencia, and R. Lemos, "Matrix Inequalities in Statistical Mechanics," *Lin. Alg. Appl.*, Vol. 376, pp. 265–273, 2003.

[86] N. Bebiano, R. Lemos, and J. da Providencia, "Inequalities for Quantum Relative Entropy," *Lin. Alg. Appl.*, preprint.

[87] E. F. Beckenbach and R. Bellman, *Inequalities*.   Berlin: Springer, 1965.

[88] T. N. Bekjan, "On Joint Convexity of Trace Functions," *Lin. Alg. Appl.*, Vol. 390, pp. 321–327, 2004.

[89] P. A. Bekker, "The Positive Semidefiniteness of Partitioned Matrices," *Lin. Alg. Appl.*, Vol. 111, pp. 261–278, 1988.

[90] J. G. Belinfante, B. Kolman, and H. A. Smith, "An Introduction to Lie Groups and Lie Algebras with Applications," *SIAM Rev.*, Vol. 8, pp. 11–46, 1966.

[91] G. R. Belitskii and Y. I. Lyubich, *Matrix Norms and Their Applications*.   Basel: Birkhauser, 1988.

[92] R. Bellman, *Introduction to Matrix Analysis*, 2nd ed.   New York: McGraw-Hill, 1960, reprinted by SIAM, Philadelphia, 1995.

[93] R. Bellman, "Some Inequalities for the Square Root of a Positive Definite Matrix," *Lin. Alg. Appl.*, Vol. 1, pp. 321–324, 1968.

[94] A. Ben-Israel, "A Note on Partitioned Matrices and Equations," *SIAM Rev.*, Vol. 11, pp. 247–250, 1969.

[95] A. Ben-Israel, "The Moore of the Moore-Penrose Inverse," *Elect. J. Lin. Alg.*, Vol. 9, pp. 150–157, 2002.

[96] A. Ben-Israel and T. N. E. Greville, *Generalized Inverses: Theory and Applications*, 2nd ed.   New York: Springer, 2003.

[97] A. Ben-Tal and A. Nemirovski, *Lectures on Modern Convex Optimization*.   Philadelphia: SIAM, 2001.

[98] L. D. Berkovitz, *Convexity and Optimization in $\mathbb{R}^n$*.   New York: Wiley, 2002.

[99] A. Berman, M. Neumann, and R. J. Stern, *Nonnegative Matrices in Dynamic Systems*.   New York: Wiley, 1989.

[100] A. Berman and R. J. Plemmons, *Nonnegative Matrices in the Mathematical Sciences*.   New York: Academic, 1979, reprinted by SIAM, Philadelphia, 1994.

[101] D. S. Bernstein, "Inequalities for the Trace of Matrix Exponentials," *SIAM J. Matrix Anal. Appl.*, Vol. 9, pp. 156–158, 1988.

[102] D. S. Bernstein, "Some Open Problems in Matrix Theory Arising in Linear Systems and Control," *Lin. Alg. Appl.*, Vol. 162–164, pp. 409–432, 1992.

[103] D. S. Bernstein and S. P. Bhat, "Lyapunov Stability, Semistability, and Asymptotic Stability of Matrix Second-Order Systems," *ASME Trans. J. Vibr. Acoustics*, Vol. 117, pp. 145–153, 1995.

[104] D. S. Bernstein and W. M. Haddad, "LQG Control with an H_∞ Performance Bound: A Riccati Equation Approach," *IEEE Trans. Autom. Contr.*, Vol. 34, pp. 293–305, 1989.

[105] D. S. Bernstein, W. M. Haddad, D. C. Hyland, and F. Tyan, "Maximum Entropy-Type Lyapunov Functions for Robust Stability and Performance Analysis," *Sys. Contr. Lett.*, Vol. 21, pp. 73–87, 1993.

[106] D. S. Bernstein and D. C. Hyland, "Compartmental Modeling and Second-Moment Analysis of State Space Systems," *SIAM J. Matrix Anal. Appl.*, Vol. 14, pp. 880–901, 1993.

[107] D. S. Bernstein and W. So, "Some Explicit Formulas for the Matrix Exponential," *IEEE Trans. Autom. Contr.*, Vol. 38, pp. 1228–1232, 1993.

[108] K. V. Bhagwat and R. Subramanian, "Inequalities between Means of Positive Operators," *Math. Proc. Camb. Phil. Soc.*, Vol. 83, pp. 393–401, 1978.

[109] S. P. Bhat and D. S. Bernstein, "Nontangency-Based Lyapunov Tests for Convergence and Stability in Systems Having a Continuum of Equilibria," *SIAM J. Contr. Optim.*, Vol. 42, pp. 1745–1775, 2003.

[110] R. Bhatia, *Perturbation Bounds for Matrix Eigenvalues*. Essex: Longman Scientific and Technical, 1987.

[111] R. Bhatia, *Matrix Analysis*. New York: Springer, 1997.

[112] R. Bhatia, "Linear Algebra to Quantum Cohomology: The Story of Alfred Horn's Inequalities," *Amer. Math. Monthly*, Vol. 108, pp. 289–318, 2001.

[113] R. Bhatia and C. Davis, "More Matrix Forms of the Arithmetic-Geometric Mean Inequality," *SIAM J. Matrix Anal. Appl.*, Vol. 14, pp. 132–136, 1993.

[114] R. Bhatia and F. Kittaneh, "Norm Inequalities for Partitioned Operators and an Application," *Math. Anal.*, Vol. 287, pp. 719–726, 1990.

[115] R. Bhatia and F. Kittanch, "On the Singular Values of a Product of Operators," *SIAM J. Matrix Anal. Appl.*, Vol. 11, pp. 272–277, 1990.

[116] R. Bhatia and F. Kittaneh, "Norm Inequalities for Positive Operators," *Lett. Math. Phys.*, Vol. 43, pp. 225–231, 1998.

[117] R. Bhatia and F. Kittaneh, "Cartesian Decompositions and Schatten Norms," *Lin. Alg. Appl.*, Vol. 318, pp. 109–116, 2000.

[118] R. Bhatia and F. Kittaneh, "Notes on Matrix Arithmetic-Geometric Mean Inequalities," *Lin. Alg. Appl.*, Vol. 308, pp. 203–211, 2000.

[119] R. Bhatia and F. Kittaneh, "Clarkson Inequalities with Several Operators," *Bull. London Math. Soc.*, Vol. 36, pp. 820–832, 2004.

[120] R. Bhatia and K. R. Parthasarathy, "Positive Definite Functions and Operator Inequalities," *Bull. London Math. Soc.*, Vol. 32, pp. 214–228, 2000.

[121] R. Bhatia and P. Rosenthal, "How and Why to Solve the Operator Equation $AX - XB = Y$," *Bull. London Math. Soc.*, Vol. 29, pp. 1–21, 1997.

[122] R. Bhatia and P. Semrl, "Orthogonality of Matrices and Some Distance Problems," *Lin. Alg. Appl.*, Vol. 287, pp. 77–85, 1999.

[123] R. Bhatia and X. Zhan, "Norm Inequalities for Operators with Positive Real Part," *J. Operator Theory*, Vol. 50, pp. 67–76, 2003.

[124] R. Bhattacharya and K. Mukherjea, "On Unitary Similarity of Matrices," *Lin. Alg. Appl.*, Vol. 126, pp. 95–105, 1989.

[125] S. P. Bhattacharyya, H. Chapellat, and L. Keel, *Robust Control: The Parametric Approach.* Englewood Cliffs: Prentice-Hall, 1995.

[126] M. R. Bicknell, "The Lambda Number of a Matrix: The Sum of Its n^2 Cofactors," *Amer. Math. Monthly*, Vol. 72, pp. 260–264, 1965.

[127] K. Binmore, *Fun and Games: A Text on Game Theory.* Lexington: D. C. Heath and Co., 1992.

[128] A. Bjorck, *Numerical Methods for Least Squares Problems.* Philadelphia: SIAM, 1996.

[129] S. Blanes and F. Casas, "On the Convergence and Optimization of the Baker-Campbell-Hausdorff Formula," *Lin. Alg. Appl.*, Vol. 378, pp. 135–158, 2004.

[130] S. Blanes, F. Casas, J. A. Oteo, and J. Ros, "Magnus and Fer Expansions for Matrix Differential Equations: The Convergence Problem," *J. Phys. A: Math. Gen.*, Vol. 31, pp. 259–268, 1998.

[131] V. Blondel and J. N. Tsitsiklis, "When Is a Pair of Matrices Mortal?" *Inform. Proc. Lett.*, Vol. 63, pp. 283–286, 1997.

[132] V. Blondel and J. N. Tsitsiklis, "The Boundedness of All Products of a Pair of Matrices Is Undecidable," *Sys. Contr. Lett.*, Vol. 41, pp. 135–140, 2000.

[133] W. Boehm, "An Operator Limit," *SIAM Rev.*, Vol. 36, p. 659, 1994.

[134] A. Borck, *Numerical Methods for Least Squares Problems.* Philadelphia: SIAM, 1996.

[135] J. M. Borwein and A. S. Lewis, *Convex Analysis and Nonlinear Optimization.* New York: Springer, 2000.

[136] A. J. Bosch, "The Factorization of a Square Matrix into Two Symmetric Matrices," *Amer. Math. Monthly*, Vol. 93, pp. 462–464, 1986.

[137] A. J. Bosch, "Note on the Factorization of a Square Matrix into Two Hermitian or Symmetric Matrices," *SIAM Rev.*, Vol. 29, pp. 463–468, 1987.

[138] T. L. Boullion and P. L. Odell, *Generalized Inverse Matrices*. New York: Wiley, 1971.

[139] J.-C. Bourin, "Some Inequalities for Norms on Matrices and Operators," *Lin. Alg. Appl.*, Vol. 292, pp. 139–154, 1999.

[140] S. Boyd, "Entropy and Random Feedback," in *Open Problems in Mathematical Systems and Control Theory*, V. D. Blondel, E. D. Sontag, M. Vidyasagar, and J. C. Willems, Eds. New York: Springer, 1998, pp. 71–74.

[141] S. Boyd and L. Vandenberghe, *Convex Optimization*. Cambridge: Cambridge University Press, 2004.

[142] J. L. Brenner, "Expanded Matrices from Matrices with Complex Elements," *SIAM Rev.*, Vol. 3, pp. 165–166, 1961.

[143] J. L. Brenner and J. S. Lim, "The Matrix Equations $A = XYZ$ and $B = ZYX$ and Related Ones," *Canad. Math. Bull.*, Vol. 17, pp. 179–183, 1974.

[144] J. W. Brewer, "Kronecker Products and Matrix Calculus in System Theory," *IEEE Trans. Circ. Sys.*, Vol. CAS–25, pp. 772–781, 1978, Correction: CAS–26:360, 1979.

[145] R. Brockett, *Finite Dimensional Linear Systems*. New York: Wiley, 1970.

[146] E. T. Browne, *Introduction to the Theory of Determinants and Matrices*. Chapel Hill: University of North Carolina Press, 1958.

[147] R. A. Brualdi and J. Q. Massey, "Some Applications of Elementary Linear Algebra in Combinatorics," *College Math. J.*, Vol. 24, pp. 10–19, 1993.

[148] R. A. Brualdi and S. Mellendorf, "Regions in the Complex Plane Containing the Eigenvalues of a Matrix," *Amer. Math. Monthly*, Vol. 101, pp. 975–985, 1994.

[149] R. A. Brualdi and H. J. Ryser, *Combinatorial Matrix Theory*. Cambridge: Cambridge University Press, 1991.

[150] R. A. Brualdi and H. Schneider, "Determinantal Identities: Gauss, Schur, Cauchy, Sylvester, Kronecker, Jacobi, Binet, Laplace, Muir, and Cayley," *Lin. Alg. Appl.*, Vol. 52/53, pp. 769–791, 1983.

[151] P. S. Bullen, *A Dictionary of Inequalities*. Essex: Longman, 1998.

[152] P. S. Bullen, *Handbook of Means and Their Inequalities*. Dordrecht: Kluwer, 2003.

[153] P. S. Bullen, D. S. Mitrinovic, and P. M. Vasic, *Means and Their Inequalities*. Dordrecht: Reidel, 1988.

[154] A. Bultheel and M. Van Barel, *Linear Algebra, Rational Approximation and Orthogonal Polynomials*. Amsterdam: Elsevier, 1997.

[155] F. Burns, D. Carlson, E. V. Haynsworth, and T. L. Markham, "Generalized Inverse Formulas Using the Schur-Complement," *SIAM J. Appl. Math*, Vol. 26, pp. 254–259, 1974.

[156] P. J. Bushell and G. B. Trustrum, "Trace Inequalities for Positive Definite Matrix Power Products," *Lin. Alg. Appl.*, Vol. 132, pp. 173–178, 1990.

[157] S. L. Campbell, *Singular Systems*. London: Pitman, 1980.

[158] S. L. Campbell and C. D. Meyer, *Generalized Inverses of Linear Transformations*. London: Pitman, 1979, reprinted by Dover, Mineola, 1991.

[159] S. L. Campbell and N. J. Rose, "Singular Perturbation of Autonomous Linear Systems," *SIAM J. Math. Anal.*, Vol. 10, pp. 542–551, 1979.

[160] E. A. Carlen and E. H. Lieb, "A Minkowski Type Trace Inequality and Strong Subadditivity of Quantum Entropy," *Amer. Math. Soc. Transl.*, Vol. 189, pp. 59–62, 1999.

[161] D. Carlson, "Controllability, Inertia, and Stability for Tridiagonal Matrices," *Lin. Alg. Appl.*, Vol. 56, pp. 207–220, 1984.

[162] D. Carlson, "What Are Schur Complements Anyway?" *Lin. Alg. Appl.*, Vol. 74, pp. 257–275, 1986.

[163] D. Carlson, E. V. Haynsworth, and T. L. Markham, "A Generalization of the Schur Complement by Means of the Moore-Penrose Inverse," *SIAM J. Appl. Math.*, Vol. 26, pp. 169–175, 1974.

[164] D. Carlson, C. R. Johnson, D. C. Lay, and A. D. Porter, Eds., *Linear Algebra Gems: Assets for Undergraduate Mathematics*. Washington, DC: Mathematical Association of America, 2002.

[165] D. Carlson, C. R. Johnson, D. C. Lay, A. D. Porter, A. E. Watkins, and W. Watkins, Eds., *Resources for Teaching Linear Algebra*. Washington, DC: Mathematical Association of America, 1997.

[166] P. Cartier, "Mathemagics, A Tribute to L. Euler and R. Feynman," in *Noise, Oscillators and Algebraic Randomness*, M. Planat, Ed. New York: Springer, 2000, pp. 6–67.

[167] D. I. Cartwright and M. J. Field, "A Refinement of the Arithmetic Mean-Geometric Mean Inequality," *Proc. Amer. Math. Soc.*, Vol. 71, pp. 36–38, 1978.

[168] F. S. Cater, "Products of Central Collineations," *Lin. Alg. Appl.*, Vol. 19, pp. 251–274, 1978.

[169] N. N. Chan and M. K. Kwong, "Hermitan Matrix Inequalities and a Conjecture," *Amer. Math. Monthly*, Vol. 92, pp. 533–541, 1985.

[170] H. Chapellat, M. Mansour, and S. P. Bhattacharyya, "Elementary Proofs of Some Classical Stability Criteria," *IEEE Trans. Educ.*, Vol. 33, pp. 232 239, 1990.

[171] G. Chartrand, *Introductory Graph Theory*. New York: Dover, 1984.

[172] F. Chatelin, *Eigenvalues of Matrices*. New York: Wiley, 1993.

[173] J.-J. Chattot, *Computational Aerodynamics and Fluid Dynamics*. Berlin: Springer, 2002.

[174] V.-S. Chellaboina and W. M. Haddad, "Is the Frobenius Matrix Norm Induced?" *IEEE Trans. Autom. Contr.*, Vol. 40, pp. 2137–2139, 1995.

[175] V.-S. Chellaboina and W. M. Haddad, "Solution to 'Some Matrix Integral Identities'," *SIAM Rev.*, Vol. 39, pp. 763–765, 1997.

[176] V.-S. Chellaboina, W. M. Haddad, D. S. Bernstein, and D. A. Wilson, "Induced Convolution Operator Norms of Linear Dynamical Systems," *Math. Contr. Sig. Sys.*, Vol. 13, pp. 216–239, 2000.

[177] B. M. Chen, Z. Lin, and Y. Shamash, *Linear Systems Theory: A Structural Decomposition Approach*. Boston: Birkhauser, 2004.

[178] C.-T. Chen, *Linear System Theory and Design*. New York: Holt, Rhinehart, Winston, 1984.

[179] H.-W. Cheng and S. S.-T. Yau, "More Explicit Formulas for the Matrix Exponential," *Lin. Alg. Appl.*, Vol. 262, pp. 131–163, 1997.

[180] S. Cheng and Y. Tian, "Two Sets of New Characterizations for Normal and EP Matrices," *Lin. Alg. Appl.*, Vol. 375, pp. 181–195, 2003.

[181] J. Chollet, "Some Inequalities for Principal Submatrices," *Amer. Math. Monthly*, Vol. 104, pp. 609–617, 1997.

[182] M. T. Chu, R. E. Funderlic, and G. H. Golub, "A Rank-One Reduction Formula and Its Application to Matrix Factorizations," *SIAM Rev.*, Vol. 37, pp. 512–530, 1995.

[183] J. Chuai and Y. Tian, "Rank Equalities and Inequalities for Kronecker Products of Matrices with Applications," *Appl. Math. Comp.*, Vol. 150, pp. 129–137, 2004.

[184] N. L. C. Chui and J. M. Maciejowski, "Realization of Stable Models with Subspace Methods," *Automatica*, Vol. 32, pp. 1587–1595, 1996.

[185] D. J. Clements, B. D. O. Anderson, A. J. Laub, and J. B. Matson, "Spectral Factorization with Imaginary-Axis Zeros," *Lin. Alg. Appl.*, Vol. 250, pp. 225–252, 1997.

[186] R. E. Cline, "Representations for the Generalized Inverse of a Partitioned Matrix," *SIAM J. Appl. Math.*, Vol. 12, pp. 588–600, 1964.

[187] R. E. Cline and R. E. Funderlic, "The Rank of a Difference of Matrices and Associated Generalized Inverses," *Lin. Alg. Appl.*, Vol. 24, pp. 185–215, 1979.

[188] M. J. Cloud and B. C. Drachman, *Inequalities with Applications to Engineering.* New York: Springer, 1998.

[189] J. E. Cohen, "Spectral Inequalities for Matrix Exponentials," *Lin. Alg. Appl.*, Vol. 111, pp. 25–28, 1988.

[190] J. E. Cohen, S. Friedland, T. Kato, and F. P. Kelly, "Eigenvalue Inequalities for Products of Matrix Exponentials," *Lin. Alg. Appl.*, Vol. 45, pp. 55–95, 1982.

[191] D. K. Cohoon, "Sufficient Conditions for the Zero Matrix," *Amer. Math. Monthly*, Vol. 96, pp. 448–449, 1989.

[192] J. C. Conway and D. A. Smith, *On Quaternions and Octonions: Their Geometry, Arithmetic, and Symmetry.* Natick: A. K. Peters, 2003.

[193] P. J. Costa and S. Rabinowitz, "Matrix Differentiation Identities," *SIAM Rev.*, Vol. 36, pp. 657–659, 1994.

[194] C. R. Crawford and Y. S. Moon, "Finding a Positive Definite Linear Combination of Two Hermitian Matrices," *Lin. Alg. Appl.*, Vol. 51, pp. 37–48, 1983.

[195] C. G. Cullen, "A Note on Normal Matrices," *Amer. Math. Monthly*, Vol. 72, pp. 643–644, 1965.

[196] C. G. Cullen, *Matrices and Linear Transformations*, 2nd ed. Reading: Addison-Wesley, 1972, reprinted by Dover, Mineola, 1990.

[197] W. J. Culver, "On the Existence and Uniqueness of the Real Logarithm of a Matrix," *Proc. Amer. Math. Soc.*, Vol. 17, pp. 1146–1151, 1966.

[198] M. L. Curtis, *Matrix Groups*, 2nd ed. New York: Springer, 1984.

[199] P. J. Daboul and R. Delbourgo, "Matrix Representations of Octonions and Generalizations," *J. Math. Phys.*, Vol. 40, pp. 4134–4150, 1999.

[200] R. D'Andrea, "Extension of Parrott's Theorem to Nondefinite Scalings," *IEEE Trans. Autom. Contr.*, Vol. 45, pp. 937–940, 2000.

[201] H. D'Angelo, *Linear Time-Varying Systems: Analysis and Synthesis.* Boston: Allyn and Bacon, 1970.

[202] F. M. Dannan, "Matrix and Operator Inequalities," *J. Inequal. Pure. Appl. Math.*, Vol. 2, no. 3/34, pp. 1–7, 2001.

[203] R. Datko and V. Seshadri, "A Characterization and a Canonical Decomposition of Hurwitzian Matrices," *Amer. Math. Monthly*, Vol. 77, pp. 732–733, 1970.

[204] B. N. Datta, *Numerical Linear Algebra and Applications*. Pacific Grove: Brooks/Cole, 1995.

[205] B. N. Datta, "Stability and Inertia," *Lin. Alg. Appl.*, Vol. 302–303, pp. 563–600, 1999.

[206] I. Daubechies and J. C. Lagarias, "Sets of Matrices all Infinite Products of Which Converge," *Lin. Alg. Appl.*, Vol. 162, pp. 227–263, 1992.

[207] P. J. Davis, *Circulant Matrices*, 2nd ed. New York: Chelsea, 1994.

[208] J. Day, W. So, and R. C. Thompson, "Some Properties of the Campbell-Baker-Hausdorff Formula," *Lin. Multilin. Alg.*, Vol. 29, pp. 207–224, 1991.

[209] C. de Boor, "An Empty Exercise," *SIGNUM*, Vol. 25, pp. 2–6, 1990.

[210] P. P. N. de Groen, "A Counterexample on Vector Norms and the Subordinate Matrix Norms," *Amer. Math. Monthly*, Vol. 97, pp. 406–407, 1990.

[211] J. de Pillis, "Transformations on Partitioned Matrices," *Duke Math. J.*, Vol. 36, pp. 511–515, 1969.

[212] J. de Pillis, "Inequalities for Partitioned Semidefinite Matrices," *Lin. Alg. Appl.*, Vol. 4, pp. 79–94, 1971.

[213] E. de Souza and S. P. Bhattacharyya, "Controllability, Observability and the Solution of $AX - XB = C$," *Lin. Alg. Appl.*, Vol. 39, pp. 167–188, 1981.

[214] H. P. Decell, "An Application of the Cayley-Hamilton Theorem to Generalized Matrix Inversion," *SIAM Rev.*, Vol. 7, pp. 526–528, 1965.

[215] J. W. Demmel, *Applied Numerical Linear Algebra*. Philadelphia: SIAM, 1997.

[216] E. D. Denman and A. N. Beavers, "The Matrix Sign Function and Computations in Systems," *Appl. Math. Computation*, Vol. 2, pp. 63–94, 1976.

[217] C. R. DePrima and C. R. Johnson, "The Range of $A^{-1}A^*$ in GL(n, C)," *Lin. Alg. Appl.*, Vol. 9, pp. 209–222, 1974.

[218] C. A. Desoer and H. Haneda, "The Measure of a Matrix as a Tool to Analyze Computer Algorithms for Circuit Analysis," *IEEE Trans. Circ. Thy.*, Vol. 19, pp. 480–486, 1972.

[219] E. Deutsch and M. Mlynarski, "Matricial Logarithmic Derivatives," *Lin. Alg. Appl.*, Vol. 19, pp. 17–31, 1978.

[220] L. Dieci, "Real Hamiltonian Logarithm of a Symplectic Matrix," *Lin. Alg. Appl.*, Vol. 281, pp. 227–246, 1998.

[221] J. Ding, "Perturbation of Systems in Linear Algebraic Equations," *Lin. Multilin. Alg.*, Vol. 47, pp. 119–127, 2000.

[222] J. Ding, "Lower and Upper Bounds for the Perturbation of General Linear Algebraic Equations," *Appl. Math. Lett.*, Vol. 14, pp. 49–52, 2001.

[223] J. Ding and W. C. Pye, "On the Spectrum and Pseudoinverse of a Special Bordered Matrix," *Lin. Alg. Appl.*, Vol. 331, pp. 11–20, 2001.

[224] A. Dittmer, "Cross Product Identities in Arbitrary Dimension," *Amer. Math. Monthly*, Vol. 101, pp. 887–891, 1994.

[225] G. M. Dixon, *Division Algebras: Octonions, Quaternions, Complex Numbers and the Algebraic Design of Physics*. Dordrecht: Kluwer, 1994.

[226] T. E. Djaferis and S. K. Mitter, "Algebraic Methods for the Study of Some Linear Matrix Equations," *Lin. Alg. Appl.*, Vol. 44, pp. 125–142, 1982.

[227] D. Z. Djokovic, "Product of Two Involutions," *Arch. Math.*, Vol. 18, pp. 582–584, 1967.

[228] D. Z. Djokovic, "On Some Representations of Matrices," *Lin. Multilin. Alg.*, Vol. 4, pp. 33–40, 1976.

[229] D. Z. Djokovic and O. P. Lossers, "A Determinant Inequality," *Amer. Math. Monthly*, Vol. 83, pp. 483–484, 1976.

[230] D. Z. Djokovic and J. Malzan, "Products of Reflections in the Unitary Group," *Proc. Amer. Math. Soc.*, Vol. 73, pp. 157–160, 1979.

[231] D. Z. Dokovic, "On the Product of Two Alternating Matrices," *Amer. Math. Monthly*, Vol. 98, pp. 935–936, 1991.

[232] D. Z. Dokovic, "Unitary Similarity of Projectors," *Aequationes Mathematicae*, Vol. 42, pp. 220–224, 1991.

[233] D. Z. Dokovic, F. Szechtman, and K. Zhao, "An Algorithm that Carries a Square Matrix into Its Transpose by an Involutory Congruence Transformation," *Elec. J. Lin. Alg.*, Vol. 10, pp. 320–340, 2003.

[234] W. F. Donoghue, *Monotone Matrix Functions and Analytic Continuation*. New York: Springer, 1974.

[235] V. Dragan and A. Halanay, *Stabilization of Linear Systems*. Boston: Birkhauser, 1999.

[236] M. P. Drazin, "A Note on Skew-Symmetric Matrices," *Math. Gaz.*, Vol. 36, pp. 253–255, 1952.

[237] S. W. Drury and G. P. H. Styan, "Normal Matrix and a Commutator," *IMAGE*, Vol. 31, p. 24, 2003.

[238] I. Duleba, "On a Computationally Simple Form of the Generalized Campbell-Baker-Hausdorff-Dynkin Formula," *Sys. Contr. Lett.*, Vol. 34, pp. 191–202, 1998.

[239] G. E. Dullerud and F. Paganini, *A Course in Robust Control Theory: A Convex Approach*, 2nd ed. New York: Springer, 1999.

[240] H. G. Eggleston, *Convexity*. Cambridge: Cambridge University Press, 1958.

[241] L. Elsner and K. D. Ikramov, "Normal Matrices: An Update," *Lin. Alg. Appl.*, Vol. 285, pp. 291–303, 1998.

[242] L. Elsner, C. R. Johnson, and J. A. D. DaSilva, "The Perron Root of a Weighted Geometric Mean of Nonnegative Matrices," *Lin. Multilin. Alg.*, Vol. 24, pp. 1–13, 1988.

[243] L. Elsner and M. H. C. Paardekooper, "On Measures of Nonnormality of Matrices," *Lin. Alg. Appl.*, Vol. 92, pp. 107–124, 1987.

[244] L. Elsner and T. Szulc, "Convex Sets of Schur Stable and Stable Matrices," *Lin. Multilin. Alg.*, Vol. 48, pp. 1–19, 2000.

[245] K. Engo, "On the BCH formula in so(3)," *Numer. Math. BIT*, Vol. 41, pp. 629–632, 2001.

[246] S. Fallat and M. J. Tsatsomeros, "On the Cayley Transform of Positivity Classes of Matrices," *Elec. J. Lin. Alg.*, Vol. 9, pp. 190–196, 2002.

[247] K. Fan, "Generalized Cayley Transforms and Strictly Dissipative Matrices," *Lin. Alg. Appl.*, Vol. 5, pp. 155–172, 1972.

[248] K. Fan, "On Real Matrices with Positive Definite Symmetric Component," *Lin. Multilin. Alg.*, Vol. 1, pp. 1–4, 1973.

[249] K. Fan, "On Strictly Dissipative Matrices," *Lin. Alg. Appl.*, Vol. 9, pp. 223–241, 1974.

[250] Y. Fang, K. A. Loparo, and X. Feng, "Inequalities for the Trace of Matrix Product," *IEEE Trans. Autom. Contr.*, Vol. 39, pp. 2489–2490, 1994.

[251] R. W. Farebrother, "A Class of Square Roots of Involutory Matrices," *IMAGE*, Vol. 28, pp. 26–28, 2002.

[252] R. W. Farebrother, J. Gross, and S.-O. Troschke, "Matrix Representation of Quaternions," *Lin. Alg. Appl.*, Vol. 362, pp. 251–255, 2003.

[253] R. W. Farebrother and I. Wrobel, "Regular and Reflected Rotation Matrices," *IMAGE*, Vol. 29, pp. 24–25, 2002.

[254] A. Fassler and E. Stiefel, *Group Theoretical Methods and Their Applications.* Boston: Birkhauser, 1992.

[255] A. E. Fekete, *Real Linear Algebra.* New York: Marcel Dekker, 1985.

[256] B. Q. Feng, "Equivalence Constants for Certain Matrix Norms," *Lin. Alg. Appl.*, Vol. 374, pp. 247–254, 2003.

[257] P. G. Ferreira and S. P. Bhattacharyya, "On Blocking Zeros," *IEEE Trans. Autom. Contr.*, Vol. AC-22, pp. 258–259, 1977.

[258] J. H. Ferziger and M. Peric, *Computational Methods for Fluid Dynamics*, 3rd ed. Berlin: Springer, 2002.

[259] M. Fiedler, "A Note on the Hadamard Product of Matrices," *Lin. Alg. Appl.*, Vol. 49, pp. 233–235, 1983.

[260] M. Fiedler, *Special Matrices and Their Applications in Numerical Mathematics.* Dordrecht: Martinus Nijhoff, 1986.

[261] M. Fiedler and T. L. Markham, "A Characterization of the Moore-Penrose Inverse," *Lin. Alg. Appl.*, Vol. 179, pp. 129–133, 1993.

[262] M. Fiedler and T. L. Markham, "An Observation on the Hadamard Product of Hermitian Matrices," *Lin. Alg. Appl.*, Vol. 215, pp. 179–182, 1995.

[263] M. Fiedler and V. Ptak, "A New Positive Definite Geometric Mean of Two Positive Definite Matrices," *Lin. Alg. Appl.*, Vol. 251, pp. 1–20, 1997.

[264] J. A. Fill and D. E. Fishkind, "The Moore-Penrose Generalized Inverse for Sums of Matrices," *SIAM J. Matrix Anal. Appl.*, Vol. 21, pp. 629–635, 1999.

[265] P. A. Fillmore, "On Similarity and the Diagonal of a Matrix," *Amer. Math. Monthly*, Vol. 76, pp. 167–169, 1969.

[266] H. Flanders, "Methods of Proof in Linear Algebra," *Amer. Math. Monthly*, Vol. 63, pp. 1–15, 1956.

[267] H. Flanders, "On the Norm and Spectral Radius," *Lin. Multilin. Alg.*, Vol. 2, pp. 239–240, 1974.

[268] T. M. Flett, *Differential Analysis.* Cambridge: Cambridge University Press, 1980.

[269] J. Foley, A. van Dam, S. Feiner, and J. Hughes, *Computer Graphics Principles and Practice*, 2nd ed. Reading: Addison-Wesley, 1990.

[270] E. Formanek, "Polynomial Identities and the Cayley-Hamilton Theorem," *Mathematical Intelligencer*, Vol. 11, pp. 37–39, 1989.

[271] E. Formanek, *The Polynomial Identities and Invariants of $n \times n$ Matrices.* Providence: American Mathematical Society, 1991.

[272] B. A. Francis, *A Course in H_∞ Control Theory*. New York: Springer, 1987.

[273] J. Franklin, *Matrix Theory*. Englewood Cliffs: Prentice-Hall, 1968.

[274] M. Frazier, *An Introduction to Wavelets through Linear Algebra*. New York: Springer, 1999.

[275] P. A. Fuhrmann, *A Polynomial Approach to Linear Algebra*. New York: Springer, 1996.

[276] J. I. Fujii, M. Fujii, T. Furuta, and R. Nakamoto, "Norm Inequalities Equivalent to Heinz Inequality," *Proc. Amer. Math. Soc.*, Vol. 118, pp. 827–830, 1993.

[277] A. T. Fuller, "Conditions for a Matrix to Have Only Characteristic Roots with Negative Real Parts," *J. Math. Anal. Appl.*, Vol. 23, pp. 71–98, 1968.

[278] T. Furuta, "Convergence of Logarithmic Trace Inequalities via Generalized Lie-Trotter Formulae," *Lin. Alg. Appl.*, preprint.

[279] T. Furuta, "$A \geq B \geq 0$ Assures $(B^r A^p B^r)^{1/q} \geq B^{(p+2r)/q}$ for $r \geq 0$, $p \geq 0$, $q \geq 1$ with $(1 + 2r)q \geq p + 2r$," *Proc. Amer. Math. Soc.*, Vol. 101, pp. 85–88, 1987.

[280] T. Furuta, "Norm Inequalities Equivalent to Loewner-Heinz Theorem," *Rev. Math. Phys.*, Vol. 1, pp. 135–137, 1989.

[281] T. Furuta, "Simple Proof of the Concavity of Operator Entropy $f(A) = -A \log A$," *Math. Ineq. Appl.*, Vol. 3, pp. 305–306, 2000.

[282] T. Furuta, *Invitation to Linear Operators: From Matrices to Bounded Linear Operators on a Hilbert Space*. London: Taylor and Francis, 2001.

[283] T. Furuta, "Spectral Order $A \succ B$ if and only if $A^{2p-r} \geq \left(A^{-r/2} B^p A^{-r/2}\right)^{(2p-r)/(p-r)}$ for all $p > r \geq 0$ and Its Application," *Math. Ineq. Appl.*, Vol. 4, pp. 619–624, 2001.

[284] F. Gaines, "A Note on Matrices with Zero Trace," *Amer. Math. Monthly*, Vol. 73, pp. 630–631, 1966.

[285] F. R. Gantmacher, *The Theory of Matrices*. New York: Chelsea, 1959, Vol. I.

[286] F. R. Gantmacher, *The Theory of Matrices*. New York: Chelsea, 1959, Vol. II.

[287] J. Garloff and D. G. Wagner, "Hadamard Products of Stable Polynomials Are Stable," *J. Math. Anal. Appl.*, Vol. 202, pp. 797–809, 1996.

[288] T. Geerts, "A Necessary and Sufficient Condition for Solvability of the Linear-Quadratic Control Problem without Stability," *Sys. Contr. Lett.*, Vol. 11, pp. 47–51, 1988.

[289] A. Gerrard and J. M. Burch, *Introduction to Matrix Methods in Optics.* New York: Wiley, 1975.

[290] R. Gilmore, *Lie Groups, Lie Algebras, and Some of Their Applications.* New York: Wiley, 1974.

[291] M. L. Glasser, "Exponentials of Certain Hilbert Space Operators," *SIAM Rev.*, Vol. 34, pp. 498–500, 1992.

[292] S. K. Godunov, *Modern Aspects of Linear Algebra.* Providence: American Mathematical Society, 1998.

[293] I. Gohberg, P. Lancaster, and L. Rodman, *Matrix Polynomials.* New York: Academic, 1982.

[294] I. Gohberg, P. Lancaster, and L. Rodman, *Invariant Subspaces of Matrices with Applications.* New York: Wiley, 1986.

[295] I. Gohberg, P. Lancaster, and L. Rodman, "On Hermitian Solutions of the Symmetric Algebraic Riccati Equation," *SIAM J. Contr. Optim.*, Vol. 24, pp. 1323–1334, 1986.

[296] M. Goldberg, "Mixed Multiplicativity and l_p Norms for Matrices," *Lin. Alg. Appl.*, Vol. 73, pp. 123–131, 1986.

[297] M. Goldberg, "Equivalence Constants for l_p Norms of Matrices," *Lin. Multilin. Alg.*, Vol. 21, pp. 173–179, 1987.

[298] M. Goldberg, "Multiplicativity Factors and Mixed Multiplicativity," *Lin. Alg. Appl.*, Vol. 97, pp. 45–56, 1987.

[299] G. H. Golub and C. F. Van Loan, *Matrix Computations*, 3rd ed. Baltimore: Johns Hopkins University Press, 1996.

[300] N. C. Gonzalez, J. J. Koliha, and Y. Wei, "Integral Representation of the Drazin Inverse," *Elect. J. Lin. Alg.*, Vol. 9, pp. 129–131, 2002.

[301] N. Gordon and D. Salmond, "Bayesian Pattern Matching Technique for Target Acquisition," *J. Guid. Contr. Dyn.*, Vol. 22, pp. 68–77, 1999.

[302] W. Govaerts and B. Sijnave, "Matrix Manifolds and the Jordan Structure of the Bialternate Matrix Product," *Lin. Alg. Appl.*, Vol. 292, pp. 245–266, 1999.

[303] R. Gow, "The Equivalence of an Invertible Matrix to Its Transpose," *Lin. Alg. Appl.*, Vol. 8, pp. 329–336, 1980.

[304] R. Gow and T. J. Laffey, "Pairs of Alternating Forms and Products of Two Skew-Symmetric Matrices," *Lin. Alg. Appl.*, Vol. 63, pp. 119–132, 1984.

[305] A. Graham, *Kronecker Products and Matrix Calculus with Applications.* Chichester: Ellis Horwood, 1981.

[306] J. F. Grcar, "A Matrix Lower Bound," Lawrence Berkeley National Laboratory, Report LBNL–50635, 2002.

[307] W. Greub, *Linear Algebra*. New York: Springer, 1981.

[308] T. N. E. Greville, "Solutions of the Matrix Equation $XAX = X$ and Relations between Oblique and Orthogonal projectors," *SIAM J. Appl. Math*, Vol. 26, pp. 828–832, 1974.

[309] R. Grone, C. R. Johnson, E. M. Sa, and H. Wolkowicz, "Normal Matrices," *Lin. Alg. Appl.*, Vol. 87, pp. 213–225, 1987.

[310] J. Gross, "On the Product of Orthogonal Projectors," *Lin. Alg. Appl.*, Vol. 289, pp. 141–150, 1999.

[311] J. Gross and G. Trenkler, "Nonsingularity of the Difference of Two Oblique Projectors," *SIAM J. Matrix Anal. Appl.*, Vol. 21, pp. 390–395, 1999.

[312] J. Gross, G. Trenkler, and S.-O. Troschke, "Quaternions: Further Contributions to a Matrix Oriented Approach," *Lin. Alg. Appl.*, Vol. 326, pp. 205–213, 2001.

[313] A. K. Gupta and D. K. Nagar, *Matrix Variate Distributions*. Boca Raton: CRC, 1999.

[314] K. Gurlebeck and W. Sprossig, *Quaternionic and Clifford Calculus for Physicists and Engineers*. New York: Chichester, 1997.

[315] L. Gurvits, "Stability of Discrete Linear Inclusion," *Lin. Alg. Appl.*, Vol. 231, pp. 47–85, 1995.

[316] K. E. Gustafson, "Matrix Trigonometry," *Lin. Alg. Appl.*, Vol. 217, pp. 117–140, 1995.

[317] K. E. Gustafson and D. K. M. Rao, *Numerical Range*. New York: Springer, 1997.

[318] W. H. Gustafson, P. R. Halmos, and H. Radjavi, "Products of Involutions," *Lin. Alg. Appl.*, Vol. 13, pp. 157–162, 1976.

[319] W. M. Haddad and D. S. Bernstein, "Robust Stabilization with Positive Real Uncertainty: Beyond the Small Gain Theorem," *Sys. Contr. Lett.*, Vol. 17, pp. 191–208, 1991.

[320] W. M. Haddad and D. S. Bernstein, "Controller Design with Regional Pole Constraints," *IEEE Trans. Autom. Contr.*, Vol. 37, pp. 54–69, 1992.

[321] W. M. Haddad, V. Kapila, and E. G. Collins, "Optimality Conditions for Reduced-Order Modeling, Estimation, and Control for Discrete-Time Linear Periodic Plants," *J. Math. Sys. Est. Contr.*, Vol. 6, pp. 437–460, 1996.

[322] W. W. Hager, "Updating the Inverse of a Matrix," *SIAM Rev.*, Vol. 31, pp. 221–239, 1989.

[323] W. Hahn, *Stability of Motion.* Berlin: Springer, 1967.

[324] B. C. Hall, *Lie Groups, Lie Algebras, and Representations: An Elementary Introduction.* New York: Springer, 2003.

[325] P. R. Halmos, *Finite-Dimensional Vector Spaces.* Princeton: Van Nostrand, 1958, reprinted by Springer, New York, 1974.

[326] P. R. Halmos, *A Hilbert Space Problem Book*, 2nd ed. New York: Springer, 1982.

[327] P. R. Halmos, "Bad Products of Good Matrices," *Lin. Alg. Appl.*, Vol. 29, pp. 1–20, 1991.

[328] P. R. Halmos, *Problems for Mathematicians Young and Old.* Washington, DC: Mathematical Association of America, 1991.

[329] P. R. Halmos, *Linear Algebra Problem Book.* Washington, DC: Mathematical Association of America, 1995.

[330] E. J. Hannan and M. Deistler, *The Statistical Theory of Linear Systems.* New York: Wiley, 1988.

[331] L. A. Harris, "The Inverse of a Block Matrix," *Amer. Math. Monthly*, Vol. 102, pp. 656–657, 1995.

[332] W. A. Harris, J. P. Fillmore, and D. R. Smith, "Matrix Exponentials–Another Approach," *SIAM Rev.*, Vol. 43, pp. 694–706, 2001.

[333] G. W. Hart, *Multidimensional Analysis.* New York: Springer, 1995.

[334] D. J. Hartfiel, *Nonhomogeneous Matrix Products.* Singapore: World Scientific, 2002.

[335] R. E. Hartwig, "Block Generalized Inverses," *Arch. Rat. Mech. Anal.*, Vol. 61, pp. 197–251, 1976.

[336] R. E. Hartwig, "A Note on the Partial Ordering of Positive Semi-Definite Matrices," *Lin. Multilin. Alg.*, Vol. 6, pp. 223–226, 1978.

[337] R. E. Hartwig and I. J. Katz, "On Products of EP Matrices," *Lin. Alg. Appl.*, Vol. 252, pp. 339–345, 1997.

[338] R. E. Hartwig and K. Spindelbock, "Matrices for which A^* and A^+ Commute," *Lin. Multilin. Alg.*, Vol. 14, pp. 241–256, 1984.

[339] D. A. Harville, *Matrix Algebra from a Statistician's Perspective.* New York: Springer, 1997.

[340] M. L. J. Hautus, "Controllability and Observability Conditions of Linear Autonomous Systems," *Proc. Koniklijke Akademic Van Wetenshappen*, Vol. 72, pp. 443–448, 1969.

[341] T. Haynes, "Stable Matrices, the Cayley Transform, and Convergent Matrices," *Int. J. Math. Math. Sci.*, Vol. 14, pp. 77–81, 1991.

[342] E. V. Haynsworth, "Applications of an Inequality for the Schur Complement," *Proc. Amer. Math. Soc.*, Vol. 24, pp. 512–516, 1970.

[343] E. Hecht, *Optics*, 4th ed. Reading: Addison Wesley, 2002.

[344] U. Helmke and P. A. Fuhrmann, "Bezoutians," *Lin. Alg. Appl.*, Vol. 122–124, pp. 1039–1097, 1989.

[345] B. W. Helton, "Logarithms of Matrices," *Proc. Amer. Math. Soc.*, Vol. 19, pp. 733–738, 1968.

[346] H. V. Henderson, F. Pukelsheim, and S. R. Searle, "On the History of the Kronecker Product," *Lin. Multilin. Alg.*, Vol. 14, pp. 113–120, 1983.

[347] H. V. Henderson and S. R. Searle, "The Vec-Permutation Matrix, The Vec Operator and Kronecker Products: A Review," *Lin. Multilin. Alg.*, Vol. 9, pp. 271–288, 1981.

[348] D. Hestenes, *Space-Time Algebra*. New York: Gordon and Breach, 1966.

[349] F. Hiai and D. Petz, "The Golden-Thompson Trace Inequality Is Complemented," *Lin. Alg. Appl.*, Vol. 181, pp. 153–185, 1993.

[350] F. Hiai and X. Zhan, "Submultiplicativity vs Subadditivity for Unitarily Invariant Norms," *Lin. Alg. Appl.*, Vol. 377, pp. 155–164, 2004.

[351] N. J. Higham, "Newton's Method for the Matrix Square Root," *Math. Computation*, Vol. 46, pp. 537–549, 1986.

[352] N. J. Higham, "Matrix Nearness Problems and Applications," in *Applications of Matrix Theory*, M. J. C. Gover and S. Barnett, Eds. Oxford: Oxford University Press, 1989, pp. 1–27.

[353] N. J. Higham, "Estimating the Matrix p-Norm," *Numer. Math.*, Vol. 62, pp. 539–555, 1992.

[354] N. J. Higham, *Accuracy and Stability of Numerical Algorithms*, 2nd ed. Philadelphia: SIAM, 2002.

[355] G. N. Hile and P. Lounesto, "Matrix Representations of Clifford Algebras," *Lin. Alg. Appl.*, Vol. 128, pp. 51–63, 1990.

[356] D. Hinrichsen, E. Plischke, and F. Wirth, "Robustness of Transient Behavior," in *Unsolved Problems in Mathematical Systems and Control Theory*, V. D. Blondel and A. Megretski, Eds. Princeton: Princeton University Press, 2004.

[357] M. W. Hirsch and S. Smale, *Differential Equations, Dynamical Systems, and Linear Algebra*. San Diego: Academic Press, 1974.

[358] M. W. Hirsch, S. Smale, and R. L. Devaney, *Differential Equations, Dynamical Systems and an Introduction to Chaos*, 2nd ed. New York: Elsevier, 2003.

[359] O. Hirzallah and F. Kittaneh, "Matrix Young Inequalities for the Hilbert-Schmidt Norm," *Lin. Alg. Appl.*, Vol. 308, pp. 77–84, 2000.

[360] O. Hirzallah and F. Kittaneh, "Commutator Inequalities for Hilbert-Schmidt Norm," *J. Math. Anal. Appl.*, Vol. 268, pp. 67–73, 2002.

[361] O. Hirzallah and F. Kittaneh, "Non-Commutative Clarkson Inequalities for Unitarily Invariant Norms," *Pac. J. Math.*, Vol. 202, pp. 363–369, 2002.

[362] O. Hirzallah and F. Kittaneh, "Norm Inequalities for Weighted Power Means of Operators," *Lin. Alg. Appl.*, Vol. 341, pp. 181–193, 2002.

[363] A. Hmamed, "A Matrix Inequality," *Int. J. Contr.*, Vol. 49, pp. 363–365, 1989.

[364] K. Hoffman and R. Kunze, *Linear Algebra*, 2nd ed. Englewood Cliffs: Prentice-Hall, 1971.

[365] O. Holtz, "Hermite-Biehler, Routh-Hurwitz, and Total Positivity," *Lin. Alg. Appl.*, Vol. 372, pp. 105–110, 2003.

[366] Y. Hong and R. A. Horn, "The Jordan Canonical Form of a Product of a Hermitian and a Positive Semidefinite Matrix," *Lin. Alg. Appl.*, Vol. 147, pp. 373–386, 1991.

[367] R. A. Horn and C. R. Johnson, *Matrix Analysis*. Cambridge: Cambridge University Press, 1985.

[368] R. A. Horn and C. R. Johnson, "Hadamard and Conventional Submultiplicativity for Unitarily Invariant Norms on Matrices," *Lin. Multilin. Alg.*, Vol. 20, pp. 91–106, 1987.

[369] R. A. Horn and C. R. Johnson, *Topics in Matrix Analysis*. Cambridge: Cambridge University Press, 1991.

[370] R. A. Horn and R. Mathias, "An Analog of the Cauchy-Schwarz Inequality for Hadamard Products and Unitarily Invariant Norms," *SIAM J. Matrix Anal. Appl.*, Vol. 11, pp. 481–498, 1990.

[371] R. A. Horn and R. Mathias, "Cauchy-Schwarz Inequalities Associated with Positive Semidefinite Matrices," *Lin. Alg. Appl.*, Vol. 142, pp. 63–82, 1990.

[372] R. A. Horn and R. Mathias, "Block-Matrix Generalizations of Schur's Basic Theorems on Hadamard Products," *Lin. Alg. Appl.*, Vol. 172, pp. 337–346, 1992.

[373] R. A. Horn and I. Olkin, "When Does $A^*A = B^*B$ and Why Does One Want to Know?" *Amer. Math. Monthly*, Vol. 103, pp. 470–482, 1996.

[374] R. A. Horn and G. G. Piepmeyer, "Two Applications of the Theory of Primary Matrix Functions," *Lin. Alg. Appl.*, Vol. 361, pp. 99–106, 2003.

[375] R. A. Horn and F. Zhang, "Basic Properties of the Schur Complement," in *The Schur Complement and Its Applications*, F. Zhang, Ed. New York: Springer, 2004, pp. 17–46.

[376] B. G. Horne, "Lower Bounds for the Spectral Radius of a Matrix," *Lin. Alg. Appl.*, Vol. 263, pp. 261–273, 1997.

[377] S.-H. Hou, "A Simple Proof of the Leverrier-Faddeev Characteristic Polynomial Algorithm," *SIAM Rev.*, Vol. 40, pp. 706–709, 1998.

[378] A. S. Householder, *The Theory of Matrices in Numerical Analysis*. New York: Blaisdell, 1964, reprinted by Dover, New York, 1975.

[379] A. S. Householder, "Bezoutiants, Elimination and Localization," *SIAM Rev.*, Vol. 12, pp. 73–78, 1970.

[380] R. Howe, "Very Basic Lie Theory," *Amer. Math. Monthly*, Vol. 90, pp. 600–623, 1983.

[381] P.-F. Hsieh and Y. Sibuya, *Basic Theory of Ordinary Differential Equations*. New York: Springer, 1999.

[382] G.-D. Hu and G.-H. Hu, "A Relation between the Weighted Logarithmic Norm of a Matrix and the Lyapunov Equation," *Numer. Math. BIT*, Vol. 40, pp. 606–610, 2000.

[383] C. H. Hung and T. L. Markham, "The Moore-Penrose Inverse of a Partitioned Matrix," *Lin. Alg. Appl.*, Vol. 11, pp. 73–86, 1975.

[384] D. C. Hyland and D. S. Bernstein, "The Optimal Projection Equations for Fixed-Order Dynamic Compensation," *IEEE Trans. Autom. Contr.*, Vol. AC-29, pp. 1034–1037, 1984.

[385] D. C. Hyland and E. G. Collins, "Block Kronecker Products and Block Norm Matrices in Large-Scale Systems Analysis," *SIAM J. Matrix Anal. Appl.*, Vol. 10, pp. 18–29, 1989.

[386] N. H. Ibragimov, *Elementary Lie Group Analysis and Ordinary Differential Equations*. Chichester: Wiley, 1999.

[387] Y. Ikebe and T. Inagaki, "An Elementary Approach to the Functional Calculus for Matrices," *Amer. Math. Monthly*, Vol. 93, pp. 390–392, 1986.

[388] V. Ionescu, C. Oar, and M. Weiss, *Generalized Riccati Theory and Robust Control*. Chichester: Wiley, 1999.

[389] A. Iserles, "Solving Linear Ordinary Differential Equations by Exponentials of Iterated Commutators," *Numer. Math.*, Vol. 45, pp. 183–199, 1984.

[390] A. Iserles, H. Z. Munthe-Kaas, S. P. Norsett, and A. Zanna, "Lie-Group Methods," *Acta Numerica*, Vol. 9, pp. 215–365, 2000.

[391] Y. Ito, S. Hattori, and H. Maeda, "On the Decomposition of a Matrix into the Sum of Stable Matrices," *Lin. Alg. Appl.*, Vol. 297, pp. 177–182, 1999.

[392] D. H. Jacobson, D. H. Martin, M. Pachter, and T. Geveci, *Extensions of Linear Quadratic Control Theory*. Berlin: Springer, 1980.

[393] A. Jennings and J. J. McKeown, *Matrix Computation*, 2nd ed. New York: Wiley, 1992.

[394] C. R. Johnson, "An Inequality for Matrices Whose Symmetric Part Is Positive Definite," *Lin. Alg. Appl.*, Vol. 6, pp. 13–18, 1973.

[395] C. R. Johnson, "Closure Properties of Certain Positivity Classes of Matrices under Various Algebraic Operations," *Lin. Alg. Appl.*, Vol. 97, pp. 243–247, 1987.

[396] C. R. Johnson, M. Neumann, and M. J. Tsatsomeros, "Conditions for the Positivity of Determinants," *Lin. Multilin. Alg.*, Vol. 40, pp. 241–248, 1996.

[397] C. R. Johnson and M. Newman, "A Surprising Determinantal Inequality for Real Matrices," *Math. Ann.*, Vol. 247, pp. 179–186, 1980.

[398] C. R. Johnson and P. Nylen, "Monotonicity Properties of Norms," *Lin. Alg. Appl.*, Vol. 148, pp. 43–58, 1991.

[399] C. R. Johnson, K. Okubo, and R. Beams, "Uniqueness of Matrix Square Roots," *Lin. Alg. Appl.*, Vol. 323, pp. 51–60, 2001.

[400] C. R. Johnson and R. Schreiner, "The Relationship between AB and BA," *Amer. Math. Monthly*, Vol. 103, pp. 578–582, 1996.

[401] C. R. Johnson and H. Shapiro, "The Relative Gain Array $A \circ A^{-T}$," *SIAM J. Alg. Disc. Meth.*, Vol. 7, pp. 627–644, 1986.

[402] M. Jolly, "On the Calculus of Complex Matrices," *Int. J. Contr.*, Vol. 61, pp. 749–755, 1995.

[403] A. Joseph, A. Melnikov, and R. Rentschler, Eds., *Studies in Memory of Issai Schur*. Cambridge: Birkhauser, 2002.

[404] M. Jujii, Y. Seo, and M. Tominaga, "Golden-Thompson Type Inequalities Related to a Geometric Mean via Specht's Ratio," *Math. Ineq. Appl.*, Vol. 5, pp. 573–582, 2002.

[405] E. I. Jury, *Inners and Stability of Dynamic Systems*, 2nd ed. Malabar: Krieger, 1982.

[406] J. B. Kagstrom, "Bounds and Perturbation Bounds for the Matrix Exponential," *Numer. Math. BIT*, Vol. 17, pp. 39–57, 1977.

[407] T. Kailath, *Linear Systems.* Englewood Cliffs: Prentice-Hall, 1980.

[408] D. Kalman and J. E. White, "Polynomial Equations and Circulant Matrices," *Amer. Math. Monthly*, Vol. 108, pp. 821–840, 2001.

[409] I. Kaplansky, *Linear Algebra and G eometry: A Second Course.* New York: Chelsea, 1974, reprinted by Dover, Mineola, 2003.

[410] N. Karcanias, "Matrix Pencil Approach to Geometric System Theory," *Proc. IEE*, Vol. 126, pp. 585–590, 1979.

[411] S. Karlin and F. Ost, "Some Monotonicity Properties of Schur Powers of Matrices and Related Inequalities," *Lin. Alg. Appl.*, Vol. 68, pp. 47–65, 1985.

[412] E. Kaszkurewicz and A. Bhaya, *Matrix Diagonal Stability in Systems and Computation.* Boston: Birkhauser, 2000.

[413] T. Kato, "Spectral Order and a Matrix Limit Theorem," *Lin. Multilin. Alg.*, Vol. 8, pp. 15–19, 1979.

[414] T. Kato, *Perturbation Theory for Linear Operators.* Berlin: Springer, 1980.

[415] J. Y. Kazakia, "Orthogonal Transformation of a Trace Free Symmetric Matrix into One with Zero Diagonal Elements," *Int. J. Eng. Sci.*, Vol. 26, pp. 903–906, 1988.

[416] N. D. Kazarinoff, *Analytic Inequalities.* New York: Holt, Rinehart and Winston, 1961, reprinted by Dover, Mineola, 2003.

[417] M. G. Kendall, *A Course in the Geometry of n Dimensions.* London: Griffin, 1961, reprinted by Dover, Mineola, 2004.

[418] C. Kenney and A. J. Laub, "Controllability and Stability Radii for Companion Form Systems," *Math. Contr. Sig. Sys.*, Vol. 1, pp. 239–256, 1988.

[419] C. Kenney and A. J. Laub, "Rational Iteration Methods for the Matrix Sign Function," *SIAM J. Matrix Anal. Appl.*, Vol. 12, pp. 273–291, 1991.

[420] H. Kestelman, "Eigenvectors of a Cross-Diagonal Matrix," *Amer. Math. Monthly*, Vol. 93, p. 566, 1986.

[421] N. Keyfitz, *Introduction to the Mathematics of Population.* Reading: Addison-Wesley, 1968.

[422] W. Khalil and E. Dombre, *Modeling, Identification, and Control of Robots.* New York: Taylor and Francis, 2002.

[423] C. G. Khatri and S. K. Mitra, "Hermitian and Nonnegative Definite Solutions of Linear Matrix Equations," *SIAM J. Appl. Math.*, Vol. 31, pp. 579–585, 1976.

[424] C. King and M. Nathanson, "New Trace Norm Inequalities for 2×2 Blocks of Diagonal Matrices," *Lin. Alg. Appl.*, preprint.

[425] F. Kittaneh, "Inequalities for the Schatten p-Norm III," *Commun. Math. Phys.*, Vol. 104, pp. 307–310, 1986.

[426] F. Kittaneh, "Inequalities for the Schatten p-Norm. IV," *Commun. Math. Phys.*, Vol. 106, pp. 581–585, 1986.

[427] F. Kittaneh, "On Zero-Trace Matrices," *Lin. Alg. Appl.*, Vol. 151, pp. 119–124, 1991.

[428] F. Kittaneh, "Norm Inequalities for Fractional Powers of Positive Operators," *Lett. Math. Phys.*, Vol. 27, pp. 279–285, 1993.

[429] F. Kittaneh, "Singular Values of Companion Matrices and Bounds on Zeros of Polynomials," *SIAM J. Matrix Anal. Appl.*, Vol. 16, pp. 333–340, 1995.

[430] F. Kittaneh, "Norm Inequalities for Certain Operator Sums," *J. Funct. Anal.*, Vol. 143, pp. 337–348, 1997.

[431] F. Kittaneh, "Some Norm Inequalities for Operators," *Canad. Math. Bull.*, Vol. 42, pp. 87–96, 1999.

[432] F. Kittaneh, "Commutator Inequalities Associated with the Polar Decomposition," *Proc. Amer. Math. Soc.*, Vol. 130, pp. 1279–1283, 2001.

[433] F. Kittaneh, "Norm Inequalities for Sums of Positive Operators," *J. Operator Theory*, Vol. 48, pp. 95–103, 2002.

[434] F. Kittaneh, "A Numerical Radius Inequality and an Estimate for the Numerical Radius of the Frobenius Companion Matrix," *Studia Mathematica*, Vol. 158, pp. 11–17, 2003.

[435] F. Kittaneh, "Norm Inequalities for Sums and Differences of Positive Operators," *Lin. Alg. Appl.*, Vol. 383, pp. 85–91, 2004.

[436] F. Kittaneh and H. Kosaki, "Inequalities for the Schatten p-Norm. V," *Publ. RIMS Kyoto Univ.*, Vol. 23, pp. 433–443, 1987.

[437] J. J. Koliha, V. Rakocevic, and I. Straskraba, "The Difference and Sum of Projectors," *Lin. Alg. Appl.*, Vol. 388, pp. 279–288, 2004.

[438] R. H. Koning, H. Neudecker, and T. Wansbeek, "Block Kronecker Products and the vecb Operator," *Lin. Alg. Appl.*, Vol. 149, pp. 165–184, 1991.

[439] T. Koshy, *Fibonacci and Lucas Numbers with Applications.* New York: Wiley, 2001.

[440] O. Krafft, "An Arithmetic-Harmonic-Mean Inequality for Nonnegative Definite Matrices," *Lin. Alg. Appl.*, Vol. 268, pp. 243–246, 1998.

[441] W. Kratz, *Quadratic Functionals in Variational Analysis and Control Theory*. New York: Wiley, 1995.

[442] E. Kreindler and A. Jameson, "Conditions for Nonnegativeness of Partitioned Matrices," *IEEE Trans. Autom. Contr.*, Vol. AC-17, pp. 147–148, 1972.

[443] V. Kucera, "On Nonnegative Definite Solutions to Matrix Quadratic Equations," *Automatica*, Vol. 8, pp. 413–423, 1972.

[444] J. B. Kuipers, *Quaternions and Rotation Sequences: A Primer with Applications to Orbits, Aerospace, and Virtual Reality*. Princeton: Princeton University Press, 1999.

[445] K. Kwakernaak and R. Sivan, *Linear Optimal Control Systems*. New York: Wiley, 1972.

[446] M. Kwapisz, "The Power of a Matrix," *SIAM Rev.*, Vol. 40, pp. 703–705, 1998.

[447] K. R. Laberteaux, "Hermitian Matrices," *Amer. Math. Monthly*, Vol. 104, p. 277, 1997.

[448] T. J. Laffey, "Products of Skew-Symmetric Matrices," *Lin. Alg. Appl.*, Vol. 68, pp. 249–251, 1985.

[449] T. J. Laffey and S. Lazarus, "Two-Generated Commutative Matrix Subalgebras," *Lin. Alg. Appl.*, Vol. 147, pp. 249–273, 1991.

[450] J. C. Lagarias and Y. Wang, "The Finiteness Conjecture for the Generalized Spectral Radius of a Set of Matrices," *Lin. Alg. Appl.*, Vol. 214, pp. 17–42, 1995.

[451] S. Lakshminarayanan, S. L. Shah, and K. Nandakumar, "Cramer's Rule for Non-Square Matrices," *Amer. Math. Monthly*, Vol. 106, p. 865, 1999.

[452] P. Lancaster, *Lambda-matrices and Vibrating Systems*. Oxford: Pergamon, 1966, reprinted by Dover, Mineola, 2002.

[453] P. Lancaster and L. Rodman, "Solutions of the Continuous and Discrete Time Algebraic Riccati Equations: A Review," in *The Riccati Equation*, S. Bittanti, J. C. Willems, and A. Laub, Eds. New York: Springer, 1991, pp. 11–51.

[454] P. Lancaster and L. Rodman, *Algebraic Riccati Equations*. Oxford: Clarendon, 1995.

[455] P. Lancaster and M. Tismenetsky, *The Theory of Matrices*, 2nd ed. Orlando: Academic, 1985.

[456] A. J. Laub, *Matrix Analysis for Scientists and Engineers*. Philadelphia: SIAM, 2004.

[457] A. J. Laub and K. Meyer, "Canonical Forms for Symplectic and Hamiltonian Matrices," *Celestial Mechanics*, Vol. 9, pp. 213–238, 1974.

[458] C. Laurie, B. Mathes, and H. Radjavi, "Sums of Three Idempotents," *Lin. Alg. Appl.*, Vol. 208/209, pp. 175–197, 2004.

[459] C. L. Lawson, *Solving Least Squares Problems*. Englewood Cliffs: Prentice-Hall, 1974, reprinted by SIAM, Philadelphia, 1995.

[460] J. D. Lawson and Y. Lim, "The Geometric Mean, Matrices, Metrics, and More," *Amer. Math. Monthly*, Vol. 108, pp. 797–812, 2001.

[461] P. D. Lax, *Linear Algebra*. New York: Wiley, 1997.

[462] S. R. Lay, *Convex Sets and Their Applications*. New York: Wiley, 1982.

[463] K. J. LeCouteur, "Representation of the Function $\mathrm{Tr}(\exp(A\text{-}\lambda B))$ as a Laplace Transform with Positive Weight and Some Matrix Inequalities," *J. Phys. A Math. Gen.*, Vol. 13, pp. 3147–3159, 1980.

[464] A. Lee, "On S-Symmetric, S-Skewsymmetric, and S-Orthogonal Matrices," *Periodica Math. Hungar.*, Vol. 7, pp. 71–76, 1976.

[465] A. Lee, "Centrohermitian and Skew-Centrohermitian Matrices," *Lin. Alg. Appl.*, Vol. 29, pp. 205–210, 1980.

[466] J. M. Lee and D. A. Weinberg, "A Note on Canonical Forms for Matrix Congruence," *Lin. Alg. Appl.*, Vol. 249, pp. 207–215, 1996.

[467] S. H. Lehnigk, *Stability Theorems for Linear Motions*. Englewood Cliffs: Prentice-Hall, 1966.

[468] T.-G. Lei, C.-W. Woo, and F. Zhang, "Matrix Inequalities by Means of Embedding," *Elect. J. Lin. Alg.*, Vol. 11, pp. 66–77, 2004.

[469] F. S. Leite, "Bounds on the Order of Generation of $so(n, r)$ by One-Parameter Subgroups," *Rocky Mountain J. Math.*, Vol. 21, pp. 879–911, 1183–1188, 1991.

[470] E. Leonard, "The Matrix Exponential," *SIAM Rev.*, Vol. 38, pp. 507–512, 1996.

[471] G. Letac, "A Matrix and Its Matrix of Reciprocals Both Positive Semidefinite," *Amer. Math. Monthly*, Vol. 82, pp. 80–81, 1975.

[472] J. S. Lew, "The Cayley Hamilton Theorem in n Dimensions," *Z. Angew. Math. Phys.*, Vol. 17, pp. 650–653, 1966.

[473] A. S. Lewis, "Convex Analysis on the Hermitian Matrices," *SIAM J. Optim.*, Vol. 6, pp. 164–177, 1996.

[474] D. C. Lewis, "A Qualitative Analysis of S-Systems: Hopf Bifurcations," in *Canonical Nonlinear Modeling*, E. O. Voit, Ed. New York: Van Nostrand Reinhold, 1991, pp. 304–344.

[475] A.-L. Li and C.-K. Li, "Isometries for the Vector (p, q) Norm and the Induced (p, q) Norm," *Lin. Multilin. Alg.*, Vol. 21, pp. 315–332, 1995.

[476] C.-K. Li and R.-C. Li, "A Note on Eigenvalues of Perturbed Hermitian Matrices," *Lin. Alg. Appl.*, preprint.

[477] C.-K. Li and R. Mathias, "The Determinant of the Sum of Two Matrices," *Bull. Austral. Math. Soc.*, Vol. 52, pp. 425–429, 1995.

[478] C.-K. Li and R. Mathias, "The Lidskii-Mirsky-Wielandt Theorem– Additive and Multiplicative Versions," *Numer. Math.*, Vol. 81, pp. 377–413, 1999.

[479] C.-K. Li and R. Mathias, "Extremal Characterizations of the Schur Complement and Resulting Inequalities," *SIAM Rev.*, Vol. 42, pp. 233–246, 2000.

[480] C.-K. Li and R. Mathias, "Inequalities on Singular Values of Block Triangular Matrices," *SIAM J. Matrix Anal. Appl.*, Vol. 24, pp. 126–131, 2002.

[481] C.-K. Li and S. Nataraj, "Some Matrix Techniques in Game Theory," *Math. Ineq. Appl.*, Vol. 3, pp. 133–141, 2000.

[482] C.-K. Li and H. Schneider, "Orthogonality of Matrices," *Lin. Alg. Appl.*, Vol. 347, pp. 115–122, 2002.

[483] E. H. Lieb, "Convex Trace Functions and the Wigner-Yanase-Dyson Conjecture," *Adv. Math.*, Vol. 11, pp. 267–288, 1973.

[484] E. H. Lieb and M. B. Ruskai, "Some Operator Inequalities of the Schwarz Type," *Adv. Math.*, Vol. 12, pp. 269–273, 1974.

[485] E. H. Lieb and W. E. Thirring, "Inequalities for the Moments of the Eigenvalues of the Schrodinger Hamiltonian and Their Relation to Sobolev Inequalities," in *Studies in Mathematical Physics*, E. Lieb, B. Simon, and A. Wightman, Eds. Princeton: Princeton University Press, 1976, pp. 269–303.

[486] T.-P. Lin, "The Power Mean and the Logarithmic Mean," *Amer. Math. Monthly*, Vol. 81, pp. 879–883, 1974.

[487] B. Liu and H.-J. Lai, *Matrices in Combinatorics and Graph Theory*. Dordrecht: Kluwer, 2000.

[488] J. Liu and J. Wang, "Some Inequalities for Schur Complements," *Lin. Alg. Appl.*, Vol. 293, pp. 233–241, 1999.

[489] R.-W. Liu and R. J. Leake, "Exhaustive Equivalence Classes of Optimal Systems with Separable Controls," *SIAM Rev.*, Vol. 4, pp. 678–685, 1966.

[490] S. Liu, "Several Inequalities Involving Khatri-Rao Products of Positive Semidefinite Matrices," *Lin. Alg. Appl.*, Vol. 354, pp. 175–186, 2002.

[491] S. Liu and H. Neudecker, "Several Matrix Kantorovich-Type Inequalities," *Math. Anal. Appl.*, Vol. 197, pp. 23–26, 1996.

[492] E. Liz, "A Note on the Matrix Exponential," *SIAM Rev.*, Vol. 40, pp. 700–702, 1998.

[493] R. Loewy, "An Inertia Theorem for Lyapunov's Equation and the Dimension of a Controllability Space," *Lin. Alg. Appl.*, Vol. 260, pp. 1–7, 1997.

[494] M. Loss and M. B. Ruskai, Eds., *Inequalities: Selecta of Elliott H. Lieb.* New York: Springer, 2002.

[495] D. G. Luenberger, *Optimization by Vector Space Methods.* New York: Wiley, 1969.

[496] D. G. Luenberger, *Introduction to Linear and Nonlinear Programming*, 2nd ed. Reading: Addison-Wesley, 1984.

[497] H. Lutkepohl, *Handbook of Matrices.* Chichester: Wiley, 1996.

[498] E.-C. Ma, "A Finite Series Solution of the Matrix Equation $AX - XB = C$," *SIAM J. Appl. Math.*, Vol. 14, pp. 490–495, 1966.

[499] C. C. MacDuffee, *The Theory of Matrices.* New York: Chelsea, 1956.

[500] A. G. J. Macfarlane and N. Karcanias, "Poles and Zeros of Linear Multivariable Systems: A Survey of the Algebraic, Geometric, and Complex-Variable Theory," *Int. J. Contr.*, Vol. 24, pp. 33–74, 1976.

[501] D. S. Mackey, N. Mackey, and F. Tisseur, "Structured Tools for Structured Matrices," *Elec. J. Lin. Alg.*, Vol. 10, pp. 106–145, 2003.

[502] J. R. Magnus, *Linear Structures.* London: Griffin, 1988.

[503] J. R. Magnus and H. Neudecker, *Matrix Differential Calculus with Applications in Statistics and Econometrics.* Chichester: Wiley, 1988.

[504] W. Magnus, "On the Exponential Solution of Differential Equations for a Linear Operator," *Commun. Pure Appl. Math.*, Vol. VII, pp. 649–673, 1954.

[505] K. N. Majindar, "On Simultaneous Hermitian Congruence Transformations of Matrices," *Amer. Math. Monthly*, Vol. 70, pp. 842–844, 1963.

[506] A. N. Malyshev and M. Sadkane, "On the Stability of Large Matrices," *J. Comp. Appl. Math.*, Vol. 102, pp. 303–313, 1999.

[507] L. E. Mansfield, *Linear Algebra with Geometric Application.* New York: Marcel-Dekker, 1976.

[508] M. Marcus, "An Eigenvalue Inequality for the Product of Normal Matrices," *Amer. Math. Monthly*, Vol. 63, pp. 173–174, 1956.

[509] M. Marcus, "Two Determinant Condensation Formulas," *Lin. Multilin. Alg.*, Vol. 22, pp. 95–102, 1987.

[510] M. Marcus and N. A. Khan, "A Note on the Hadamard Product," *Canad. Math. J.*, Vol. 2, pp. 81–83, 1959.

[511] M. Marcus and H. Minc, *A Survey of Matrix Theory and Matrix Inequalities.* Boston: Prindle, Weber, and Schmidt, 1964, reprinted by Dover, New York, 1992.

[512] T. L. Markham, "An Application of Theorems of Schur and Albert," *Proc. Amer. Math. Soc.*, Vol. 59, pp. 205–210, 1976.

[513] T. L. Markham, "Oppenheim's Inequality for Positive Definite Matrices," *Amer. Math. Monthly*, Vol. 93, pp. 642–644, 1986.

[514] G. Marsaglia and G. P. H. Styan, "Equalities and Inequalities for Ranks of Matrices," *Lin. Multilin. Alg.*, Vol. 2, pp. 269–292, 1974.

[515] J. E. Marsden and T. S. Ratiu, *Introduction to Mechanics and Symmetry.* New York: Springer, 1994.

[516] A. W. Marshall and I. Olkin, *Inequalities: Theory of Majorization and Its Applications.* New York: Academic, 1979.

[517] A. W. Marshall and I. Olkin, "Matrix Versions of the Cauchy and Kantorovich Inequalities," *Aequationes Math.*, Vol. 40, pp. 89–93, 1990.

[518] K. Martensson, "On the Matrix Riccati Equation," *Inform. Sci.*, Vol. 3, pp. 17–49, 1971.

[519] A. M. Mathai, *Jacobians of Matrix Transformations and Functions of Matrix Argument.* Singapore: World Scientific, 1997.

[520] R. Mathias, "Evaluating the Frechet Derivative of the Matrix Exponential," *Numer. Math.*, Vol. 63, pp. 213–226, 1992.

[521] R. Mathias, "An Arithmetic-Geometric-Harmonic Mean Inequality Involving Hadamard Products," *Lin. Alg. Appl.*, Vol. 184, pp. 71–78, 1993.

[522] R. Mathias, "A Chain Rule for Matrix Functions and Applications," *SIAM J. Matrix Anal. Appl.*, Vol. 17, pp. 610–620, 1996.

[523] J. E. McCarthy, "Pick's Theorem–What's the Big Deal?" *Amer. Math. Monthly*, Vol. 110, pp. 36–45, 2003.

[524] J. M. McCarthy, *Geometric Design of Linkages.* New York: Springer, 2000.

[525] J. P. McCloskey, "Characterizations of r-Potent Matrices," *Math. Proc. Camb. Phil. Soc.*, Vol. 96, pp. 213–222, 1984.

[526] A. R. Meenakshi and C. Rajian, "On a Product of Positive Semidefinite Matrices," *Lin. Alg. Appl.*, Vol. 295, pp. 3–6, 1999.

[527] C. L. Mehta, "Some Inequalities Involving Traces of Operators," *J. Math. Phys.*, Vol. 9, pp. 693–697, 1968.

[528] Y. A. Melnikov, *Influence Functions and Matrices.* New York: Marcel Dekker, 1998.

[529] J. K. Merikoski, H. Sarria, and P. Tarazaga, "Bounds for Singular Values Using Traces," *Lin. Alg. Appl.*, Vol. 210, pp. 227–254, 1994.

[530] R. Merris, "Inequalities Involving the Inverses of Positive Definite Matrices," *Proc. Edinburgh Math. Soc.*, Vol. 22, pp. 11–15, 1979.

[531] R. Merris, *Multilinear Algebra.* Amsterdam: Gordon and Breach, 1997.

[532] R. Merris and S. Pierce, "Monotonicity of Positive Semidefinite Hermitian Matrices," *Proc. Amer. Math. Soc.*, Vol. 31, pp. 437–440, 1972.

[533] C. D. Meyer, "The Moore-Penrose Inverse of a Bordered Matrix," *Lin. Alg. Appl.*, Vol. 5, pp. 375–382, 1972.

[534] C. D. Meyer, "Generalized Inverses and Ranks of Block Matrices," *SIAM J. Appl. Math*, Vol. 25, pp. 597–602, 1973.

[535] C. D. Meyer, *Matrix Analysis and Applied Linear Algebra.* Philadelphia: SIAM, 2000.

[536] J.-M. Miao, "General Expressions for the Moore-Penrose Inverse of a 2×2 Block Matrix," *Lin. Alg. Appl.*, Vol. 151, pp. 1–15, 1991.

[537] L. Mihalyffy, "An Alternative Representation of the Generalized Inverse of Partitioned Matrices," *Lin. Alg. Appl.*, Vol. 4, pp. 95–100, 1971.

[538] K. S. Miller, *Some Eclectic Matrix Theory.* Malabar: Krieger, 1987.

[539] G. A. Milliken and F. Akdeniz, "A Theorem on the Difference of the Generalized Inverses of Two Nonnegative Marices," *Commun. Statist. Theory Methods*, Vol. 6, pp. 73–79, 1977.

[540] N. Minamide, "An Extension of the Matrix Inversion Lemma," *SIAM J. Alg. Disc. Meth.*, Vol. 6, pp. 371–377, 1985.

[541] H. Miranda and R. C. Thompson, "A Trace Inequality with a Subtracted Term," *Lin. Alg. Appl.*, Vol. 185, pp. 165–172, 1993.

[542] L. Mirsky, *An Introduction to Linear Algebra.* Oxford: Clarendon, 1972, reprinted by Dover, Mineola, 1990.

[543] D. S. Mitrinovic, J. E. Pecaric, and A. M. Fink, *Classical and New Inequalities in Analysis.* Dordrecht: Kluwer, 1993.

[544] B. Mityagin, "An Inequality in Linear Algebra," *SIAM Rev.*, Vol. 33, pp. 125–127, 1991.

[545] C. Moler and C. F. Van Loan, "Nineteen Dubious Ways to Compute the Exponential of a Matrix, Twenty-Five Years Later," *SIAM Rev.*, Vol. 45, pp. 3–49, 2000.

[546] B. Mond and J. E. Pecaric, "Inequalities for the Hadamard Product of Matrices," *SIAM J. Matrix Anal. Appl.*, Vol. 19, pp. 66–70, 1998.

[547] V. V. Monov, "On the Spectrum of Convex Sets of Matrices," *IEEE Trans. Autom. Contr.*, Vol. 44, pp. 1009–1012, 1992.

[548] T. Mori, "Comments on 'A Matrix Inequality Associated with Bounds on Solutions of Algebraic Riccati and Lyapunov Equation'," *IEEE Trans. Autom. Contr.*, Vol. 33, p. 1088, 1988.

[549] T. Muir, *The Theory of Determinants in the Historical Order of Development.* New York: Dover, 1966.

[550] W. W. Muir, "Inequalities Concerning the Inverses of Positive Definite Matrices," *Proc. Edinburgh Math. Soc.*, Vol. 19, pp. 109–113, 1974–75.

[551] R. M. Murray, Z. Li, and S. S. Sastry, *A Mathematical Introduction to Robotic Manipulation.* Boca Raton: CRC, 1994.

[552] I. Najfeld and T. F. Havel, "Derivatives of the Matrix Exponential and Their Computation," *Adv. Appl. Math.*, Vol. 16, pp. 321–375, 1995.

[553] R. Nakamoto, "A Norm Inequality for Hermitian Operators," *Amer. Math. Monthly*, Vol. 110, pp. 238–239, 2003.

[554] Y. Nakamura, "Any Hermitian Matrix is a Linear Combination of Four Projections," *Lin. Alg. Appl.*, Vol. 61, pp. 133–139, 1984.

[555] A. W. Naylor and G. R. Sell, *Linear Operator Theory in Engineering and Science.* New York: Springer, 1986.

[556] T. Needham, *Visual Complex Analysis.* Oxford: Oxford University Press, 1997.

[557] C. N. Nett and W. M. Haddad, "A System-Theoretic Appropriate Realization of the Empty Matrix Concept," *IEEE Trans. Autom. Contr.*, Vol. 38, pp. 771–775, 1993.

[558] M. F. Neuts, *Matrix-Geometric Solutions in Stochastic Models.* Baltimore: Johns Hopkins University Press, 1981, reprinted by Dover, Mineola, 1994.

[559] R. W. Newcomb, "On the Simultaneous Diagonalization of Two Semidefinite Matrices," *Quart. Appl. Math.*, Vol. 19, pp. 144–146, 1961.

[560] M. Newman, W. So, and R. C. Thompson, "Convergence Domains for the Campbell-Baker-Hausdorff Formula," *Lin. Multilin. Alg.*, Vol. 24, pp. 301–310, 1989.

[561] K. Nishio, "The Structure of a Real Linear Combination of Two Projections," *Lin. Alg. Appl.*, Vol. 66, pp. 169–176, 1985.

[562] B. Noble and J. W. Daniel, *Applied Linear Algebra*, 3rd ed. Engle-wood Cliffs: Prentice-Hall, 1988.

[563] J. Nunemacher, "Which Real Matrices Have Real Logarithms?" *Math. Mag.*, Vol. 62, pp. 132–135, 1989.

[564] H. Ogawa, "An Operator Pseudo-Inversion Lemma," *SIAM J. Appl. Math.*, Vol. 48, pp. 1527–1531, 1988.

[565] I. Olkin, "An Inequality for a Sum of Forms," *Lin. Alg. Appl.*, Vol. 52–53, pp. 529–532, 1983.

[566] J. M. Ortega, *Matrix Theory, A Second Course*. New York: Plenum, 1987.

[567] S. L. Osburn and D. S. Bernstein, "An Exact Treatment of the Achiev-able Closed-Loop H_2 Performance of Sampled-Data Controllers: From Continuous-Time to Open-Loop," *Automatica*, Vol. 31, pp. 617–620, 1995.

[568] A. Ostrowski and H. Schneider, "Some Theorems on the Inertia of General Matrices," *J. Math. Anal. Appl.*, Vol. 4, pp. 72–84, 1962.

[569] J. A. Oteo, "The Baker-Campbell-Hausdorff Formula and Nested Commutator Identities," *J. Math. Phys.*, Vol. 32, pp. 419–424, 1991.

[570] D. A. Overdijk, "Skew-symmetric Matrices in Classical Mechanics," Eindhoven University, Memorandum COSOR 89–23, 1989.

[571] P. V. Overschee and B. De Moor, *Subspace Identification for Linear Systems: Theory, Implementation, Applications*. Dordrecht: Kluwer, 1996.

[572] B. P. Palka, *An Introduction to Complex Function Theory*. New York: Springer, 1991.

[573] C. V. Pao, "Logarithmic Derivatives of a Square Matrix," *Lin. Alg. Appl.*, Vol. 6, pp. 159–164, 1973.

[574] J. G. Papastravridis, *Tensor Calculus and Analytical Dynamics*. Boca Raton: CRC, 1998.

[575] F. C. Park, "Computational Aspects of the Product-of-Exponentials Formula for Robot Kinematics," *IEEE Trans. Autom. Contr.*, Vol. 39, pp. 643–647, 1994.

[576] P. Park, "On the Trace Bound of a Matrix Product," *IEEE Trans. Autom. Contr.*, Vol. 41, pp. 1799–1802, 1996.

[577] P. C. Parks, "A New Proof of the Routh-Hurwitz Stability Crite-rion Using the Second Method of Liapunov," *Proc. Camb. Phil. Soc.*, Vol. 58, pp. 694–702, 1962.

[578] R. V. Patel, "On Blocking Zeros in Linear Multivariable Systems," *IEEE Trans. Autom. Contr.*, Vol. AC-31, pp. 239–241, 1986.

[579] R. V. Patel and M. Toda, "Trace Inequalities Involving Hermitian Matrices," *Lin. Alg. Appl.*, Vol. 23, pp. 13–20, 1979.

[580] M. C. Pease, *Methods of Matrix Algebra.* New York: Academic, 1965.

[581] S. Perlis, *Theory of Matrices.* Reading: Addison-Wesley, 1952, reprinted by Dover, New York, 1991.

[582] I. R. Petersen and C. V. Hollot, "A Riccati Equation Approach to the Stabilization of Uncertain Systems," *Automatica*, Vol. 22, pp. 397–411, 1986.

[583] L. A. Pipes, "Applications of Laplace Transforms of Matrix Functions," *J. Franklin Inst.*, Vol. 285, pp. 436–451, 1968.

[584] A. O. Pittenger and M. H. Rubin, "Convexity and Separation Problem of Quantum Mechanical Density Matrices," *Lin. Alg. Appl.*, Vol. 346, pp. 47–71, 2002.

[585] T. Politi, "A Formula for the Exponential of a Real Skew-Symmetric Matrix of Order 4," *Numer. Math. BIT*, Vol. 41, pp. 842–845, 2001.

[586] D. S. G. Pollock, "Tensor Products and Matrix Differential Calculus," *Lin. Alg. Appl.*, Vol. 67, pp. 169–193, 1985.

[587] B. Poonen, "A Unique $(2k + 1)$-th Root of a Matrix," *Amer. Math. Monthly*, Vol. 98, p. 763, 1991.

[588] B. Poonen, "Positive Deformations of the Cauchy Matrix," *Amer. Math. Monthly*, Vol. 102, pp. 842–843, 1995.

[589] V. M. Popov, *Hyperstability of Control Systems.* Berlin: Springer, 1973.

[590] G. J. Porter, "Linear Algebra and Affine Planar Transformations," *College Math. J.*, Vol. 24, pp. 47–51, 1993.

[591] B. H. Pourciau, "Modern Multiplier Rules," *Amer. Math. Monthly*, Vol. 87, pp. 433–452, 1980.

[592] V. V. Prasolov, *Problems and Theorems in Linear Algebra.* Providence: American Mathematical Society, 1994.

[593] J. S. Przemieniecki, *Theory of Matrix Structural Analysis.* New York: McGraw-Hill, 1968.

[594] P. J. Psarrakos, "On the *m*th Roots of a Complex Matrix," *Elec. J. Lin. Alg.*, Vol. 9, pp. 32–41, 2002.

[595] N. J. Pullman, *Matrix Theory and Its Applications: Selected Topics.* New York: Marcel Dekker, 1976.

[596] S. Puntanen and G. P. H. Styan, "Historical Introduction: Issai Schur and the Early Development of the Schur Complement," in *The Schur Complement and Its Applications*, F. Zhang, Ed. New York: Springer, 2004, pp. 1–16.

[597] R. X. Qian and C. L. DeMarco, "An Approach to Robust Stability of Matrix Polytopes through Copositive Homogeneous Polynomials," *IEEE Trans. Autom. Contr.*, Vol. 37, pp. 848–852, 1992.

[598] L. Qiu, B. Bernhardsson, A. Rantzer, E. J. Davison, P. M. Young, and J. C. Doyle, "A Formula for Computation of the Real Stability Radius," *Automatica*, Vol. 31, pp. 879–890, 1995.

[599] V. Rabanovich, "Every Matrix Is a Linear Combination of Three Idempotents," *Lin. Alg. Appl.*, Vol. 390, pp. 137–143, 2004.

[600] H. Radjavi, "Decomposition of Matrices into Simple Involutions," *Lin. Alg. Appl.*, Vol. 12, pp. 247–255, 1975.

[601] H. Radjavi and P. Rosenthal, *Simultaneous Triangularization.* New York: Springer, 2000.

[602] H. Radjavi and J. P. Williams, "Products of Self-Adjoint Operators," *Michigan Math. J.*, Vol. 16, pp. 177–185, 1969.

[603] V. Rakocevic, "On the Norm of Idempotent Operators in a Hilbert Space," *Amer. Math. Monthly*, Vol. 107, pp. 748–750, 2000.

[604] A. C. M. Ran and R. Vreugdenhil, "Existence and Comparison Theorems for Algebraic Riccati Equations for Continuous- and Discrete-Time Systems," *Lin. Alg. Appl.*, Vol. 99, pp. 63–83, 1988.

[605] A. Rantzer, "On the Kalman-Yakubovich-Popov Lemma," *Sys. Contr. Lett.*, Vol. 28, pp. 7–10, 1996.

[606] C. R. Rao and S. K. Mitra, *Generalized Inverse of Matrices and Its Applications.* New York: Wiley, 1971.

[607] J. V. Rao, "Some More Representations for the Generalized Inverse of a Partitioned Matrix," *SIAM J. Appl. Math.*, Vol. 24, pp. 272–276, 1973.

[608] P. A. Regalia and S. K. Mitra, "Kronecker Products, Unitary Matrices and Signal Processing Applications," *SIAM Rev.*, Vol. 31, pp. 586–613, 1989.

[609] T. J. Richardson and R. H. Kwong, "On Positive Definite Solutions to the Algebraic Riccati Equation," *Sys. Contr. Lett.*, Vol. 7, pp. 99–104, 1986.

[610] A. N. Richmond, "Expansions for the Exponential of a Sum of Matrices," in *Applications of Matrix Theory*, M. J. C. Gover and S. Barnett, Eds. Oxford: Oxford University Press, 1989, pp. 283–289.

[611] J. R. Ringrose, *Compact Non-Self-Adjoint Operators.* New York: Van Nostrand Reinhold, 1971.

[612] R. S. Rivlin, "Further Remarks on the Stress Deformation Relations for Isotropic Materials," *J. Rational Mech. Anal.*, Vol. 4, pp. 681–702, 1955.

[613] J. W. Robbin, *Matrix Algebra Using MINImal MATlab*. Wellesley: A. K. Peters, 1995.

[614] R. T. Rockafellar, *Convex Analysis*. Princeton: Princeton University Press, 1990.

[615] R. T. Rockafellar and R. J. B. Wets, *Variational Analysis*. Berlin: Springer, 1998.

[616] L. Rodman, "Products of Symmetric and Skew Symmetric Matrices," *Lin. Multilin. Alg.*, Vol. 43, pp. 19–34, 1997.

[617] G. S. Rogers, *Matrix Derivatives*. New York: Marcel Dekker, 1980.

[618] C. A. Rohde, "Generalized Inverses of Partitioned Matrices," *SIAM J. Appl. Math.*, Vol. 13, pp. 1033–1035, 1965.

[619] J. Rohn, "Computing the Norm $||A||_{\infty,1}$ Is NP-Hard," *Lin. Multilin. Alg.*, Vol. 47, pp. 195–204, 2000.

[620] O. Rojo, "Further Bounds for the Smallest Singular Value and the Spectral Condition Number," *Computers Math. Appl.*, Vol. 38, pp. 215–228, 1999.

[621] K. H. Rosen, Ed., *Handbook of Discrete and Combinatorial Mathematics*. Boca Raton: CRC, 2000.

[622] M. Rosenfeld, "A Sufficient Condition for Nilpotence," *Amer. Math. Monthly*, Vol. 103, pp. 907–909, 1996.

[623] W. Rossmann, *Lie Groups: An Introduction Through Linear Groups*. Oxford: Oxford University Press, 2002.

[624] W. J. Rugh, *Linear System Theory*, 2nd ed. Upper Saddle River: Prentice Hall, 1996.

[625] D. L. Russell, *Mathematics of Finite-Dimensional Control Systems*. New York: Marcel Dekker, 1979.

[626] A. Saberi, P. Sannuit, and B. M. Chen, H_2 *Optimal Control*. New York: Prentice Hall, 1995.

[627] M. K. Sain and C. B. Schrader, "The Role of Zeros in the Performance of Multiinput, Multioutput Feedback Systems," *IEEE Trans. Educ.*, Vol. 33, pp. 244–257, 1990.

[628] D. H. Sattinger and O. L. Weaver, *Lie Groups and Algebras with Applications to Physics, Geometry, and Mechanics*. New York: Springer, 1986.

[629] A. H. Sayed, *Fundamentals of Adaptive Filtering*. New York: Wiley, 2003.

[630] H. Schaub, P. Tsiotras, and J. L. Junkins, "Principal Rotation Representations of Proper $N \times N$ Orthogonal Matrices," *Int. J. Eng. Sci.*, Vol. 33, pp. 2277–2295, 1995.

[631] C. W. Scherer, "The Algebraic Riccati Equation and Inequality for Systems with Uncontrollable Modes on the Imaginary Axis," *SIAM J. Matrix Anal. Appl.*, Vol. 16, pp. 1308–1327, 1995.

[632] P. Scherk, "On the Decomposition of Orthogonalities into Symmetries," *Proc. Amer. Math. Soc.*, Vol. 1, pp. 481–491, 1950.

[633] C. Schmoeger, "On the Operator Equation $e^A = e^B$," *Lin. Alg. Appl.*, Vol. 359, pp. 169–179, 2003.

[634] H. Schneider, "Olga Taussky-Todd's Influence on Matrix Theory and Matrix Theorists," *Lin. Multilin. Alg.*, Vol. 5, pp. 197–224, 1977.

[635] C. B. Schrader and M. K. Sain, "Research on System Zeros: A Survey," *Int. J. Contr.*, Vol. 50, pp. 1407–1433, 1989.

[636] A. J. Schwenk, "Tight Bounds on the Spectral Radius of Asymmetric Nonnegative Matrices," *Lin. Alg. Appl.*, Vol. 75, pp. 257–265, 1986.

[637] S. R. Searle, *Matrix Algebra Useful for Statistics*. New York: Wiley, 1982.

[638] P. Sebastian, "On the Derivatives of Matrix Powers," *SIAM J. Matrix Anal. Appl.*, Vol. 17, pp. 640–648, 1996.

[639] D. Serre, *Matrices: Theory and Applications*. New York: Springer, 2002.

[640] C. Shafroth, "A Generalization of the Formula for Computing the Inverse of a Matrix," *Amer. Math. Monthly*, Vol. 88, pp. 614–616, 1981.

[641] H. Shapiro, "Notes from Math 223: Olga Taussky Todd's Matrix Theory Course, 1976–1977," *Mathematical Intelligencer*, Vol. 19, no. 1, pp. 21–27, 1997.

[642] R. Shaw and F. I. Yeadon, "On $(a \times b) \times c$," *Amer. Math. Monthly*, Vol. 96, pp. 623–629, 1989.

[643] G. E. Shilov, *Linear Algebra*. Englewood Cliffs: Prentice-Hall, 1971, reprinted by Dover, New York, 1977.

[644] M. D. Shuster, "A Survery of Attitude Representations," *J. Astron. Sci.*, Vol. 41, pp. 439–517, 1993.

[645] D. D. Siljak, *Large-Scale Dynamic Systems: Stability and Structure*. New York: North-Holland, 1978.

[646] S. F. Singer, *Symmetry in Mechanics: A Gentle, Modern Introduction*. Boston: Birkhauser, 2001.

[647] R. E. Skelton, T. Iwasaki, and K. Grigoriadis, *A Unified Algebraic Approach to Linear Control Design*. London: Taylor and Francis, 1998.

[648] R. A. Smith, "Matrix Calculations for Lyapunov Quadratic Forms," *J. Diff. Eqns.*, Vol. 2, pp. 208–217, 1966.

[649] J. Snyders and M. Zakai, "On Nonnegative Solutions of the Equation $AD + DA' = C$," *SIAM J. Appl. Math.*, Vol. 18, pp. 704–714, 1970.

[650] W. So, "Equality Cases in Matrix Exponential Inequalities," *SIAM J. Matrix Anal. Appl.*, Vol. 13, pp. 1154–1158, 1992.

[651] W. So, "The High Road to an Exponential Formula," *Lin. Alg. Appl.*, Vol. 379, pp. 69–75, 2004.

[652] W. So and R. C. Thompson, "Product of Exponentials of Hermitian and Complex Symmetric Matrices," *Lin. Multilin. Alg.*, Vol. 29, pp. 225–233, 1991.

[653] W. So and R. C. Thompson, "Singular Values of Matrix Exponentials," *Lin. Multilin. Alg.*, Vol. 47, pp. 249–258, 2000.

[654] E. D. Sontag, *Mathematical Control Theory: Deterministic Finite-Dimensional Systems*, 2nd ed. New York: Springer, 1998.

[655] A. R. Sourour, "A Factorization Theorem for Matrices," *Lin. Multilin. Alg.*, Vol. 19, pp. 141–147, 1986.

[656] W.-H. Steeb, *Matrix Calculus and Kronecker Product with Applications and C++ Programs.* Singapore: World Scientific, 2001.

[657] W.-H. Steeb and F. Wilhelm, "Exponential Functions of Kronecker Products and Trace Calculation," *Lin. Multilin. Alg.*, Vol. 9, pp. 345–346, 1981.

[658] J. M. Steele, *The Cauchy-Schwarz Master Class.* Washington, DC: Mathematical Association of America, 2004.

[659] R. F. Stengel, *Flight Dynamics.* Princeton: Princeton University Press, 2004.

[660] C. Stepniak, "Ordering of Nonnegative Definite Matrices with Application to Comparison of Linear Models," *Lin. Alg. Appl.*, Vol. 70, pp. 67–71, 1985.

[661] H. J. Stetter, *Numerical Polynomial Algebra.* Philadelphia: SIAM, 2004.

[662] G. W. Stewart, *Introduction to Matrix Computations.* New York: Academic, 1973.

[663] G. W. Stewart, "On the Perturbation of Pseudo-Inverses, Projections and Linear Least Squares Problems," *SIAM Rev.*, Vol. 19, pp. 634–662, 1977.

[664] G. W. Stewart, *Matrix Algorithms Volume I: Basic Decompositions.* Philadelphia: SIAM, 1998.

[665] G. W. Stewart, "On the Adjugate Matrix," *Lin. Alg. Appl.*, Vol. 283, pp. 151–164, 1998.

[666] G. W. Stewart, *Matrix Algorithms Volume II: Eigensystems.* Philadelphia: SIAM, 2001.

[667] G. W. Stewart and J. Sun, *Matrix Perturbation Theory.* Boston: Academic, 1990.

[668] E. U. Stickel, "Fast Computation of Matrix Exponential and Logarithm," *Analysis*, Vol. 5, pp. 163–173, 1985.

[669] E. U. Stickel, "An Algorithm for Fast High Precision Computation of Matrix Exponential and Logarithm," *Analysis*, Vol. 10, pp. 85–95, 1990.

[670] L. Stiller, "Multilinear Algebra and Chess Endgames," in *Games of No Chance*, R. Nowakowski, Ed. Berkeley: Mathematical Sciences Research Institute, 1996, pp. 151–192.

[671] J. Stoer, "On the Characterization of Least Upper Bound Norms in Matrix Space," *Numer. Math*, Vol. 6, pp. 302–314, 1964.

[672] J. Stoer and C. Witzgall, *Convexity and Optimization in Finite Dimensions I.* Berlin: Springer, 1970.

[673] M. G. Stone, "A Mnemonic for Areas of Polygons," *Amer. Math. Monthly*, Vol. 93, pp. 479–480, 1986.

[674] G. Strang, *Linear Algebra and Its Applications*, 3rd ed. San Diego: Harcourt, Brace, Jovanovich, 1988.

[675] G. Strang, "The Fundamental Theorem of Linear Algebra," *Amer. Math. Monthly*, Vol. 100, pp. 848–855, 1993.

[676] R. S. Strichartz, "The Campbell-Baker-Hausdorff-Dynkin Formula and Solutions of Differential Equations," *J. Funct. Anal.*, Vol. 72, pp. 320–345, 1987.

[677] T. Strom, "On Logarithmic Norms," *SIAM J. Numer. Anal.*, Vol. 12, pp. 741–753, 1975.

[678] J. Stuelpnagel, "On the Parametrization of the Three-Dimensional Rotation Group," *SIAM Rev.*, Vol. 6, pp. 422–430, 1964.

[679] K. N. Swamy, "On Sylvester's Criterion for Positive-Semidefinite Matrices," *IEEE Trans. Autom. Contr.*, Vol. AC-18, p. 306, 1973.

[680] G. Szep, "Simultaneous Triangularization of Projector Matrices," *Acta Math. Hung.*, Vol. 48, pp. 285–288, 1986.

[681] T. Szirtes, *Applied Dimensional Analysis and Modeling.* New York: McGraw-Hill, 1998.

[682] O. Taussky, "Positive-Definite Matrices and Their Role in the Study of the Characteristic Roots of General Matrices," *Adv. Math.*, Vol. 2, pp. 175–186, 1968.

[683] O. Taussky, "The Role of Symmetric Matrices in the Study of General Matrices," *Lin. Alg. Appl.*, Vol. 5, pp. 147–154, 1972.

[684] O. Taussky, "How I Became a Torchbearer for Matrix Theory," *Amer. Math. Monthly*, Vol. 95, pp. 801–812, 1988.

[685] O. Taussky and J. Todd, "Another Look at a Matrix of Mark Kac," *Lin. Alg. Appl.*, Vol. 150, pp. 341–360, 1991.

[686] O. Taussky and H. Zassenhaus, "On the Similarity Transformation between a Matrix and Its Transpose," *Pac. J. Math.*, Vol. 9, pp. 893–896, 1959.

[687] W. Tempelman, "The Linear Algebra of Cross Product Operations," *J. Astron. Sci.*, Vol. 36, pp. 447–461, 1988.

[688] R. E. Terrell, "Solution to 'Exponentials of Certain Hilbert Space Operators'," *SIAM Rev.*, Vol. 34, pp. 498–500, 1992.

[689] R. E. Terrell, "Matrix Exponentials," *SIAM Rev.*, Vol. 38, pp. 313–314, 1996.

[690] R. C. Thompson, "On Matrix Commutators," *J. Washington Academy Sci.*, Vol. 48, pp. 306–307, 1958.

[691] R. C. Thompson, "A Determinantal Inequality for Positive Definite Matrices," *Canad. Math. Bull.*, Vol. 4, pp. 57–62, 1961.

[692] R. C. Thompson, "Some Matrix Factorization Theorems," *Pac. J. Math.*, Vol. 33, pp. 763–810, 1970.

[693] R. C. Thompson, "A Matrix Inequality," *Comment. Math. Univ. Carolinae*, Vol. 17, pp. 393–397, 1976.

[694] R. C. Thompson, "Matrix Type Metric Inequalities," *Lin. Multilin. Alg.*, Vol. 5, pp. 303–319, 1978.

[695] R. C. Thompson, "Proof of a Conjectured Exponential Formula," *Lin. Multilin. Alg.*, Vol. 19, pp. 187–197, 1986.

[696] R. C. Thompson, "Convergence Proof for Goldberg's Exponential Series," *Lin. Alg. Appl.*, Vol. 121, pp. 3–7, 1989.

[697] R. C. Thompson, "Pencils of Complex and Real Symmetric and Skew Matrices," *Lin. Alg. Appl.*, Vol. 147, pp. 323–371, 1991.

[698] R. C. Thompson, "High, Low, and Quantitative Roads in Linear Algebra," *Lin. Alg. Appl.*, Vol. 162–164, pp. 23–64, 1992.

[699] Y. Tian, "EP Matrices Revisited," preprint.

[700] Y. Tian, "On Mixed-Type Reverse Order Laws for the Moore-Penrose Inverse of a Matrix Product," preprint.

[701] Y. Tian, "The Moore-Penrose Inverse of $m \times n$ Block Matrices and Their Applications," *Lin. Alg. Appl.*, Vol. 283, pp. 35–60, 1998.

[702] Y. Tian, "Matrix Representations of Octonions and Their Applications," *Adv. Appl. Clifford Algebras*, Vol. 10, pp. 61–90, 2000.

[703] Y. Tian, "Commutativity of EP Matrices," *IMAGE*, Vol. 27, pp. 25–27, 2001.

[704] Y. Tian, "How to Characterize Equalities for the Moore-Penrose Inverse of a Matrix," *Kyungpook Math. J.*, Vol. 41, pp. 1–15, 2001.

[705] Y. Tian, "How to Express a Parallel Sum of k Matrices," *J. Math. Anal. Appl.*, Vol. 266, pp. 333–341, 2002.

[706] Y. Tian, "Upper and Lower Bounds for Ranks of Matrix Expression Using Generalized Inverses," *Lin. Alg. Appl.*, Vol. 355, pp. 187–214, 2002.

[707] Y. Tian, "A Range Equality for Idempotent Matrix," *IMAGE*, Vol. 30, pp. 26–27, 2003.

[708] Y. Tian, "A Range Equality for the Difference of Orthogonal Projectors," *IMAGE*, Vol. 30, p. 36, 2003.

[709] Y. Tian, "Rank Equalities for Block Matrices and Their Moore-Penrose Inverses," *Houston J. Math.*, Vol. 30, pp. 483–510, 2004.

[710] Y. Tian, "Using Rank Formulas to Characterize Equalities for Moore-Penrose Inverses of Matrix Products," *Appl. Math. Comp.*, Vol. 147, pp. 581–600, 2004.

[711] Y. Tian and G. P. H. Styan, "How to Establish Universal Block-Matrix Factorizations," *Electr. J. Lin. Alg.*, Vol. 8, pp. 115–127, 2001.

[712] Y. Tian and G. P. H. Styan, "Rank Equalities for Idempotent and Involutory Matrices," *Lin. Alg. Appl.*, Vol. 335, pp. 101–117, 2001.

[713] Y. Tian and G. P. H. Styan, "A New Rank Formula for Idempotent Matrices with Applications," *Comment. Math. Univ. Carolinae*, Vol. 43, pp. 370–384, 2002.

[714] Y. Tian and G. P. H. Styan, "When Does rank(ABC) = rank(AB) + rank(BC) − rank(B) Hold?" *Int. J. Math. Educ. Sci. Tech.*, Vol. 33, pp. 127–137, 2002.

[715] A. Tonge, "Equivalence Constants for Matrix Norms: A Problem of Goldberg," *Lin. Alg. Appl.*, Vol. 306, pp. 1–13, 2000.

[716] G. E. Trapp, "Hermitian Semidefinite Matrix Means and Related Matrix Inequalities–An Introduction," *Lin. Multilin. Alg.*, Vol. 16, pp. 113–123, 1984.

[717] L. N. Trefethen and D. Bau, *Numerical Linear Algebra.* Philadelphia: SIAM, 1997.

[718] D. Trenkler, G. Trenkler, C.-K. Li, and H. J. Werner, "Square Roots and Additivity," *IMAGE*, Vol. 29, p. 30, 2002.

[719] G. Trenkler, "A Trace Inequality," *Amer. Math. Monthly*, Vol. 102, pp. 362–363, 1995.

[720] G. Trenkler, "A Condition for an Idempotent Matrix to be Hermitian," *IMAGE*, Vol. 31, pp. 41–43, 2003.

[721] G. Trenkler, "A Matrix Related to an Idempotent Matrix," *IMAGE*, Vol. 31, pp. 39–40, 2003.

[722] G. Trenkler, "On the Product of Orthogonal Projectors," *IMAGE*, Vol. 31, p. 43, 2003.

[723] G. Trenkler and H. J. Werner, "Partial Isometry and Idempotent Matrices," *IMAGE*, Vol. 29, pp. 30–32, 2002.

[724] H. L. Trentelman, A. A. Stoorvogel, and M. L. J. Hautus, *Control Theory for Linear Systems.* New York: Springer, 2001.

[725] P. Treuenfels, "The Matrix Equation $X^2 - 2AX + B = 0$," *Amer. Math. Monthly*, Vol. 66, pp. 145–146, 1959.

[726] S. H. Tung, "On Lower and Upper Bounds of the Difference between the Arithmetic and the Geometric Mean," *Math. Comput.*, Vol. 29, pp. 834–836, 1975.

[727] D. A. Turkington, *Matrix Calculus and Zero-One Matrices.* Cambridge: Cambridge University Press, 2002.

[728] H. W. Turnbull, *The Theory of Determinants, Matrices and Invariants.* London: Blackie, 1950.

[729] F. Tyan and D. S. Bernstein, "Global Stabilization of Systems Containing a Double Integrator Using a Saturated Linaer Controller," *Int. J. Robust Nonlin. Contr.*, Vol. 9, pp. 1143–1156, 1999.

[730] M. Uchiyama, "Norms and Determinants of Products of Logarithmic Functions of Positive Semi-Definite Operators," *Math. Ineq. Appl.*, Vol. 1, pp. 279–284, 1998.

[731] M. Uchiyama, "Some Exponential Operator Inequalities," *Math. Ineq. Appl.*, Vol. 2, pp. 469–471, 1999.

[732] F. E. Udwadia and R. E. Kalaba, *Analytical Dynamics: A New Approach.* Cambridge: Cambridge University Press, 1996.

[733] F. Uhlig, "A Recurring Theorem About Pairs of Quadratic Forms and Extensions: A Survey," *Lin. Alg. Appl.*, Vol. 25, pp. 219–237, 1979.

[734] F. Uhlig, "Constructive Ways for Generating (Generalized) Real Orthogonal Matrices as Products of (Generalized) Symmetries," *Lin. Alg. Appl.*, Vol. 332–334, pp. 459–467, 2001.

[735] F. A. Valentine, *Convex Sets*. New York: McGraw-Hill, 1964.

[736] M. Van Barel, V. Ptak, and Z. Vavrin, "Bezout and Hankel Matrices Associated with Row Reduced Matrix Polynomials, Barnett-Type Formulas," *Lin. Alg. Appl.*, Vol. 332–334, pp. 583–606, 2001.

[737] P. Van Dooren, "The Computation of Kronecker's Canonical Form of a Singular Pencil," *Lin. Alg. Appl.*, Vol. 27, pp. 103–140, 1979.

[738] C. F. Van Loan, "Computing Integrals Involving the Matrix Exponential," *IEEE Trans. Autom. Contr.*, Vol. AC-23, pp. 395–404, 1978.

[739] C. F. Van Loan, "How Near Is a Stable Matrix to an Unstable Matrix," *Contemporary Math.*, Vol. 47, pp. 465–478, 1985.

[740] C. F. Van Loan, *Computational Frameworks for the Fast Fourier Transform*. Philadelphia: SIAM, 1992.

[741] C. F. Van Loan, "The Ubiquitous Kronecker Product," *J. Comp. Appl. Math.*, Vol. 123, pp. 85–100, 2000.

[742] V. S. Varadarajan, *Lie Groups, Lie Algebras, and Their Representations*. New York: Springer, 1984.

[743] A. I. G. Vardulakis, *Linear Multivariable Control: Algebraic Analysis and Synthesis Methods*. Chichester: Wiley, 1991.

[744] R. S. Varga, *Matrix Iterative Analysis*. Englewood Cliffs: Prentice-Hall, 1962.

[745] R. Vein and P. Dale, *Determinants and Their Applications in Mathematical Physics*. New York: Springer, 1999.

[746] W. J. Vetter, "Matrix Calculus Operations and Taylor Expansions," *SIAM Rev.*, Vol. 15, pp. 352–369, 1973.

[747] M. Vidyasagar, "On Matrix Measures and Convex Liapunov Functions," *J. Math. Anal. Appl.*, Vol. 62, pp. 90–103, 1978.

[748] G. Visick, "Majorizations of Hadamard Products of Matrix Powers," *Lin. Alg. Appl.*, Vol. 269, pp. 233–240, 1998.

[749] G. Visick, "A Quantitative Version of the Observation that the Hadamard Product Is a Principal Submatrix of the Kronecker Product," *Lin. Alg. Appl.*, Vol. 304, pp. 45–68, 2000.

[750] G. Visick, "Another Inequality for Hadamard Products," *IMAGE*, Vol. 29, pp. 32–33, 2002.

[751] B.-Y. Wang and M.-P. Gong, "Some Eigenvalue Inequalities for Positive Semidefinite Matrix Power Products," *Lin. Alg. Appl.*, Vol. 184, pp. 249–260, 1993.

[752] B.-Y. Wang, B.-Y. Xi, and F. Zhang, "Some Inequalities for Sum and Product of Positive Semidefinite Matrices," *Lin. Alg. Appl.*, Vol. 293, pp. 39–49, 1999.

[753] B.-Y. Wang and F. Zhang, "A Trace Inequality for Unitary Matrices," *Amer. Math. Monthly*, Vol. 101, pp. 453–455, 1994.

[754] B.-Y. Wang and F. Zhang, "Trace and Eigenvalue Inequalities for Ordinary and Hadamard Products of Positive Semidefinite Hermitian Matrices," *SIAM J. Matrix Anal. Appl.*, Vol. 16, pp. 1173–1183, 1995.

[755] B.-Y. Wang and F. Zhang, "Schur Complements and Matrix Inequalities of Hadamard Products," *Lin. Multilin. Alg.*, Vol. 43, pp. 315–326, 1997.

[756] D. Wang, "The Polar Decomposition and a Matrix Inequality," *Amer. Math. Monthly*, Vol. 96, pp. 517–519, 1989.

[757] Q.-G. Wang, "Necessary and Sufficient Conditions for Stability of a Matrix Polytope with Normal Vertex Matrices," *Automatica*, Vol. 27, pp. 887–888, 1991.

[758] Y. W. Wang and D. S. Bernstein, "Controller Design with Regional Pole Constraints: Hyperbolic and Horizontal Strip Regions," *AIAA J. Guid. Contr. Dyn.*, Vol. 16, pp. 784–787, 1993.

[759] Y. W. Wang and D. S. Bernstein, "L_2 Controller Synthesis with L_∞-Bounded Closed-Loop Impulse Response," *Int. J. Contr.*, Vol. 60, pp. 1295–1306, 1994.

[760] J. Warga, *Optimal Control of Differential and Functional Equations*. New York: Academic, 1972.

[761] W. E. Waterhouse, "A Determinant Identity with Matrix Entries," *Amer. Math. Monthly*, Vol. 97, pp. 249–250, 1990.

[762] W. Watkins, "Convex Matrix Functions," *Proc. Amer. Math. Soc.*, Vol. 44, pp. 31–34, 1974.

[763] W. Watkins, "A Determinantal Inequality for Correlation Matrices," *Lin. Alg. Appl.*, Vol. 104, pp. 59–63, 1988.

[764] J. R. Weaver, "Centrosymmetric (Cross-Symmetric) Matrices, Their Basic Properties, Eigenvalues, and Eigenvectors," *Amer. Math. Monthly*, Vol. 92, pp. 711–717, 1985.

[765] R. Webster, *Convexity*. Oxford: Oxford University Press, 1994.

[766] J. Wei and E. Norman, "Lie Algebraic Solution of Linear Differential Equations," *J. Math. Phys.*, Vol. 4, pp. 575–581, 1963.

[767] J. Wei and E. Norman, "On Global Representations of the Solutions of Linear Differential Equations as a Product of Exponentials," *Proc. Amer. Math. Soc.*, Vol. 15, pp. 327–334, 1964.

[768] M. Wei, "Reverse Order Laws for Generalized Inverses of Multiple Matrix Products," *Lin. Alg. Appl.*, Vol. 293, pp. 273–288, 1999.

[769] Y. Wei, "Expressions for the Drazin Inverse of a 2×2 Block Matrix," *Lin. Multilin. Alg.*, Vol. 45, pp. 131–146, 1998.

[770] G. H. Weiss and A. A. Maradudin, "The Baker-Hausdorff Formula and a Problem in Crystal Physics," *J. Math. Phys.*, Vol. 3, pp. 771–777, 1962.

[771] E. M. E. Wermuth, "Two Remarks on Matrix Exponentials," *Lin. Alg. Appl.*, Vol. 117, pp. 127–132, 1989.

[772] H. J. Werner, "On the Product of Orthogonal Projectors," *IMAGE*, Vol. 32, pp. 35–36, 2004.

[773] P. Wesseling, *Principles of Computational Fluid Dynamics.* Berlin: Springer, 2001.

[774] J. R. Westlake, *A Handbook of Numerical Matrix Inversion and Solution of Linear Equations.* New York: Wiley, 1968.

[775] N. A. Wiegmann, "Normal Products of Matrices," *Duke Math. J.*, Vol. 15, pp. 633–638, 1948.

[776] Z. Wiener, "An Interesting Matrix Exponent Formula," *Lin. Alg. Appl.*, Vol. 257, pp. 307–310, 1997.

[777] E. P. Wigner and M. M. Yanase, "On the Positive Semidefinite Nature of a Certain Matrix Expression," *Canad. J. Math.*, Vol. 16, pp. 397–406, 1964.

[778] R. M. Wilcox, "Exponential Operators and Parameter Differentiation in Quantum Physics," *J. Math. Phys.*, Vol. 8, pp. 962–982, 1967.

[779] J. H. Wilkinson, *The Algebraic Eigenvalue Problem.* London: Oxford University Press, 1965.

[780] J. C. Willems, "Least Squares Stationary Optimal Control and the Algebraic Riccati Equation," *IEEE Trans. Autom. Contr.*, Vol. AC-16, pp. 621–634, 1971.

[781] D. A. Wilson, "Convolution and Hankel Operator Norms for Linear Systems," *IEEE Trans. Autom. Contr.*, Vol. AC-34, pp. 94–97, 1989.

[782] H. K. Wimmer, "Inertia Theorems for Matrices, Controllability and Linear Vibrations," *Lin. Alg. Appl.*, Vol. 8, pp. 337–343, 1974.

[783] H. K. Wimmer, "The Algebraic Riccati Equation without Complete Controllability," *SIAM J. Alg. Disc. Math.*, Vol. 3, pp. 1–12, 1982.

[784] H. K. Wimmer, "The Algebraic Riccati Equation: Conditions for the Existence and Uniqueness of Solutions," *Lin. Alg. Appl.*, Vol. 58, pp. 441–452, 1984.

[785] H. K. Wimmer, "Monotonicity of Maximal Solutions of Algebraic Riccati Equations," *Sys. Contr. Lett.*, Vol. 5, pp. 317–319, 1985.

[786] H. K. Wimmer, "Linear Matrix Equations, Controllability and Observability, and the Rank of Solutions," *SIAM J. Matrix Anal. Appl.*, Vol. 9, pp. 570–578, 1988.

[787] H. K. Wimmer, "On the History of the Bezoutian and the Resultant Matrix," *Lin. Alg. Appl.*, Vol. 128, pp. 27–34, 1990.

[788] H. K. Wimmer, "Lattice Properties of Sets of Semidefinite Solutions of Continuous-time Algebraic Riccati Equations," *Automatica*, Vol. 31, pp. 173–182, 1995.

[789] H. Wolkowicz and G. P. H. Styan, "Bounds for Eigenvalues Using Traces," *Lin. Alg. Appl.*, Vol. 29, pp. 471–506, 1980.

[790] W. A. Wolovich, *Linear Multivariable Systems.* New York: Springer, 1974.

[791] M. J. Wonenburger, "A Decomposition of Orthogonal Transformations," *Canad. Math. Bull.*, Vol. 7, pp. 379–383, 1964.

[792] M. J. Wonenburger, "Transformations Which are Products of Two Involutions," *J. Math. Mech.*, Vol. 16, pp. 327–338, 1966.

[793] C. S. Wong, "Characterizations of Products of Symmetric Matrices," *Lin. Alg. Appl.*, Vol. 42, pp. 243–251, 1982.

[794] W. M. Wonham, *Linear Multivariable Control: A Geometric Approach*, 2nd ed. New York: Springer, 1979.

[795] P. Y. Wu, "Products of Nilpotent Matrices," *Lin. Alg. Appl.*, Vol. 96, pp. 227–232, 1987.

[796] P. Y. Wu, "Products of Positive Semidefinite Matrices," *Lin. Alg. Appl.*, Vol. 111, pp. 53–61, 1988.

[797] P. Y. Wu, "The Operator Factorization Problems," *Lin. Alg. Appl.*, Vol. 117, pp. 35–63, 1989.

[798] Z.-G. Xiao and Z.-H. Zhang, "The Inequalities $G \leq L \leq I \leq A$ in n Variables," *J. Inequal. Pure. Appl. Math.*, Vol. 4, no. 2/39, pp. 1–6, 2003.

[799] H. Xu, "Two Results About the Matrix Exponential," *Lin. Alg. Appl.*, Vol. 262, pp. 99–109, 1997.

[800] T. Yamazaki, "Further Characterizations of Chaotic Order via Specht's Ratio," *Math. Ineq. Appl.*, Vol. 3, pp. 259–268, 2000.

[801] X. Yang, "Necessary Conditions of Hurwitz Polynomials," *Lin. Alg. Appl.*, Vol. 359, pp. 21–27, 2003.

[802] X. Yang, "Some Necessary Conditions for Hurwitz Stability," *Automatica*, Vol. 40, pp. 527–529, 2004.

[803] Z. P. Yang and X. X. Feng, "A Note on the Trace Inequality for Products of Hermitian Matrix Power," *J. Inequal. Pure. Appl. Math.*, Vol. 3, no. 5/78, pp. 1–12, 2002.

[804] D. M. Young, *Iterative Solution of Large Linear Systems*. New York: Academic, 1971, reprinted by Dover, Mineola, 2003.

[805] X. Zhan, "Inequalities for Unitarily Invariant Norms," *SIAM J. Matrix Anal. Appl.*, Vol. 20, pp. 466–470, 1998.

[806] X. Zhan, *Matrix Inequalities*. New York: Springer, 2002.

[807] X. Zhan, "On Some Matrix Inequalities," *Lin. Alg. Appl.*, Vol. 376, pp. 299–303, 2004.

[808] F. Zhang, *Linear Algebra: Challenging Problems for Students*. Baltimore: Johns Hopkins University Press, 1996.

[809] F. Zhang, "Quaternions and Matrices of Quaternions," *Lin. Alg. Appl.*, Vol. 251, pp. 21–57, 1997.

[810] F. Zhang, "A Compound Matrix with Positive Determinant," *Amer. Math. Monthly*, Vol. 105, p. 958, 1998.

[811] F. Zhang, *Matrix Theory: Basic Results and Techniques*. New York: Springer, 1999.

[812] F. Zhang, "Schur Complements and Matrix Inequalities in the Lowner Ordering," *Lin. Alg. Appl.*, Vol. 321, pp. 399–410, 2000.

[813] F. Zhang, "Matrix Inequalities by Means of Block Matrices," *Math. Ineq. Appl.*, Vol. 4, pp. 481–490, 2001.

[814] F. Zhang, "Inequalities Involving Square Roots," *IMAGE*, Vol. 29, pp. 33–34, 2002.

[815] F. Zhang, "Block Matrix Techniques," in *The Schur Complement and Its Applications*, F. Zhang, Ed. New York: Springer, 2004, pp. 83–110.

[816] F. Zhang, "A Matrix Identity on the Schur Complement," *Lin. Multilin. Alg.*, Vol. 52, pp. 367–373, 2004.

[817] L. Zhang, "A Characterization of the Drazin Inverse," *Lin. Alg. Appl.*, Vol. 335, pp. 183–188, 2001.

[818] K. Zhou, *Robust and Optimal Control*. Upper Saddle River: Prentice-Hall, 1996.

[819] S. Zlobec, "An Explicit Form of the Moore-Penrose Inverse of an Arbitrary Complex Matrix," *SIAM Rev.*, Vol. 12, pp. 132–134, 1970.

[820] D. Zwillinger, *Standard Mathematical Tables and Formulae*, 31st ed. Boca Raton: Chapman and Hall/CRC, 2003.

Author Index

Index

symmetric matrix
factorization
Fact 5.14.22, 218

Frobenius canonical form
definition, 222

Frobenius matrix
definition, 222

Frobenius norm
adjugate
Fact 9.8.11, 368
definition, 348
determinant
Fact 9.8.30, 372
dissipative matrix
Fact 11.13.2, 460
eigenvalue bound
Fact 9.10.7, 386
eigenvalue
perturbation
Fact 9.10.9, 387
Fact 9.10.10, 388
Fact 9.10.11, 388
Hermitian matrix
Fact 9.9.26, 379
inequality
Fact 9.9.18, 377
Kronecker product
Fact 9.12.22, 398
matrix difference
Fact 9.9.18, 377
matrix exponential
Fact 11.13.2, 460
maximum singular
value bound
Fact 9.11.16, 391
normal matrix
Fact 9.10.10, 388
outer-product
matrix
Fact 9.7.13, 366
polar decomposition
Fact 9.9.27, 380
positive-semidefinite
matrix
Fact 9.8.30, 372
Fact 9.9.9, 376

Fact 9.9.10, 376
Fact 9.9.14, 377
Schatten norm, **349**
spectral radius
Fact 5.10.27, 200
trace
Fact 9.10.2, 385
Fact 9.10.5, 386
trace norm
Fact 9.9.8, 375
unitary matrix
Fact 9.9.27, 380

Frobenius' inequality
rank of partitioned
matrix
Fact 2.10.28, 54
Fact 6.4.25, 238

Fujii
Furuta inequality
Fact 8.8.25, 299

full column rank
definition, **32**
equivalent properties
Theorem 2.6.1, 34
nonsingular
equivalence
Corollary 2.6.4, 37

full rank
definition, **32**

full row rank
definition, **32**
equivalent properties
Theorem 2.6.1, 34
nonsingular
equivalence
Corollary 2.6.4, 37

full-state feedback
controllable subspace
Proposition 12.6.5, 501
controllably
asymptotically
stable
Proposition 12.7.2, 507
determinant
Fact 12.22.10, 557
invariant zero

Proposition 12.10.2,
520
Fact 12.22.10, 557
stabilizability
Proposition 12.8.2, 510
uncontrollable
eigenvalue
Proposition 12.6.13,
505
unobservable
subspace
Proposition 12.3.5, 493

function
definition, **4**

function composition
matrix
multiplication
Theorem 2.1.2, 17

fundamental theorem of algebra
definition, **122**

fundamental theorem of linear algebra
rank and defect
Corollary 2.5.5, 33

Furuta inequality
positive-definite
matrix
Fact 8.8.25, 299
spectral ordering
Fact 8.8.27, 299

Furuta's inequality
positive-semidefinite
matrix inequality
Proposition 8.5.6, 279

G

Gastinel
distance to
singularity of a
nonsingular matrix
Fact 9.12.4, 393

generalized eigenvalues